ECOLOGY
Individuals, Populations and Communities

ECOLOGY

INDIVIDUALS, POPULATIONS, AND COMMUNITIES

Michael Begon
University of Liverpool

John L. Harper
University College of North Wales, Bangor

Colin R. Townsend
University of East Anglia

Sinauer Associates, Inc. • Publishers

Sunderland, Massachusetts

THE COVER
Black mangrove (*Avicennia germinans* L.) pneumatophores
with young seedling in Lignumvitae Key, Florida.
Photograph by Thomas Eisner. Cover design by Joseph Vesely.

ECOLOGY: INDIVIDUALS, POPULATIONS AND COMMUNITIES

Published in the United States and Canada by
Sinauer Associates Inc., Publishers
Sunderland, Massachusetts 01375

Library of Congress Cataloging-in-Publication Data

Begon, Michael.
  Ecology: individuals, populations and communities.

  Bibliography: p.
  Includes index.
  1. Ecology.   I. Harper, John L.   II. Townsend,
Colin R.   III. Title.
QH541.B415 1986   574.5   85-22168
ISBN 0-87893-051-5

Printed in U.S.A.

5   4   3   2   1

Contents

Preface

Ecology is probably the oldest science. Soon after our primitive ancestors had their first conscious thoughts, they must have realized the value of knowing where they might find palatable plants and catchable animals, and where they might be safe from the enemies that sought to attack them. Today, however, we can go much further than this. We know that if we are to make use of the products of the natural world without destroying them, if we are to produce food for ourselves rather than for our pests, and if we are to predict 'what happens next' in our environments, then we must understand the natural world and the organisms that comprise it. This, then, is a book designed to document and promote such understanding. It is a book about the distribution and abundance of different types of organism over the face of the earth, and about the physical, chemical but especially the biological features and interactions that determine these distributions and abundances.

Unlike some other sciences, the subject matter of ecology is apparent to everybody; to the extent that most people have observed and pondered nature, most people are ecologists of sorts. But ecology is not an easy science, and it has particular subtlety and complexity. It must deal explicitly with three levels of the biological hierarchy—the organisms, the populations of organisms, and the communities of populations—and, as we shall see, it ignores at its peril the details of the biology of individuals, or the pervading influences of historical, evolutionary and geological events. It feeds, in a peripheral way, on advances in our knowledge of biochemistry, behaviour, climatology, plate tectonics and so on, but it feeds back to our understanding of vast areas of biology too. If, as T. H. Dobzhansky said, 'Nothing in biology makes sense, except in the light of evolution', then, equally, very little in evolution makes sense, except in the light of ecology—that is, in terms of the interactions between organisms and their physical, chemical and biological environments.

Ecology also has the distinction of being peculiarly confronted with uniqueness: millions of different species, countless billions of genetically distinct individuals, all living and interacting in a varied and ever-changing world. The beauty of ecology is that it challenges us to develop an understanding of very basic and apparent problems, in a way that recognizes the uniqueness and complexity of all aspects of nature but seeks patterns and predictions within this complexity rather than being swamped by it. As L. C. Birch has pointed out, the physicist Whitehead's recipe for science is never more apposite than when applied to ecology: seek simplicity, but distrust it.

In this book, we have aimed to build an understanding by moving from organisms to populations to communities. But sometimes, and especially in Chapters 14 and 15, we have looked back to see the light that the more complex interactions throw on the simpler levels of the hierarchy. Ecology is not a science with a simple linear structure: everything affects everything else. Different chapters contain different proportions of descriptive natural history, physiology, behaviour, rigorous laboratory and field experimentation, careful field monitoring and censusing, and mathematical modelling (a form of simplicity that it is essential to seek but equally essential to distrust). These varying proportions to some extent reflect the progress made in different areas. They also reflect intrinsic differences in various aspects of ecology. Whatever progress is made, ecology will remain a meeting-ground for the naturalist, the experimentalist, the field biologist and the mathematical modeller. We believe that all ecologists should to some extent try to combine all these facets.

The book is aimed at all those whose degree programme includes ecology. Certain aspects of the subject, particularly the mathematical ones, will prove difficult for some, but our coverage is designed to ensure that wherever our readers' strengths lie—in field or laboratory, in theory or in practice—a balanced and up-to-date view should emerge.

One technical feature of the book is the incorporation of marginal notes as signposts throughout the text. These, we hope, will serve a number of purposes. In the first place, they constitute a series of sub-headings highlighting the detailed structure of the text. However, because they are numerous and often informative in their own right, they can also be read in sequence along with the conventional sub-headings, as an outline of each chapter. In this same form, they should act too as a revision aid for students—indeed, they are similar to the annotations that students themselves often add to their textbooks. Finally, because the marginal notes generally summarize the take-home message of the paragraph or paragraphs that they accompany, they can act as a continuous assessment of comprehension: if you can see that the signpost *is* the take-home message of what you have just read, then you have understood.

Having completed this book, it is a pleasure to record our gratitude to the people who, in a variety of ways, helped us with its conception, its development and its birth. Many colleagues read and made valuable comments on various drafts of the text. John Lawton and Bob May both took on the task of examining all the chapters. Jo Anderson, Roy Anderson, David Atkinson, Leo Buss, Arthur Cain, James Chubb, Andrew Cossins, Robert Cowie, Michael Crawley, John Farrar, Michael Hassell, William Heal, Alan Hildrew, Michael Hutchings, Robert James, Richard Law, Richard Palmer, Mary Price, Derek Ranwell, Persio de Souza Santos Jr., Dave Thompson, Des Thompson, James Thomson, Jeff Waage, Richard Wall, Hilary Wallace, Andrew Watkinson and John Whittaker all helped us with one or more chapters. At Blackwell Scientific Publications we were fortunate to have the help and encouragement of Simon Rallison, who took our text and turned it into a book, and Bob Campbell, who coaxed and cajoled us throughout. Andy Sinauer of Sinauer Associates provided feedback on our original prospectus from a number of North American ecologists and coordinated the chapter reviews in North America. The text itself emerged from our scrawlings because of the cheerful and patient typing of Anita Callaghan, Jane Farrell, Gweneth Kell, Susan Scott and Maria Woolley. Lastly, we are glad to thank our wives and families for supporting us, listening to us and ignoring us precisely as was required.

Introduction: Ecology and its Domain

The word 'ecology' was first used by Ernest Haeckel in 1869. Paraphrasing Haeckel we can define ecology as the scientific study of the interactions between organisms and their environment. The word is derived from the Greek *oikos,* meaning 'home'. Ecology might therefore be thought of as the study of the 'home life' of living organisms. A more informative, much less vague definition has been suggested by Krebs (1972): 'Ecology is the scientific study of the interactions that determine the distribution and abundance of organisms '. This definition has the merit of pinpointing the ultimate subject matter of ecology: the distribution and abundance of organisms—where they occur, how many there are and what they do. However, Krebs's definition does not use the word 'environment', and to see why, it is necessary to define the word. The environment of an organism consists of all those factors and phenomena outside the organism that influence it, whether those factors be physical and chemical (abiotic) or other organisms (biotic). But the interactions in Krebs's definition are, of course, interactions with these very factors. The environment has therefore retained the central position that Haeckel gave it in the definition of ecology.

As far as the subject matter of ecology is concerned, 'the distribution and abundance of organisms' is pleasantly succinct. But we need to expand it. Ecology deals with three levels of concern: the individual *organism,* the *population* (consisting of individuals of the same species) and the *community* (consisting of a greater or lesser number of populations).

At the level of the organism, ecology deals with how individuals are affected by (and how they affect) their biotic and abiotic environment. At the level of the population, ecology deals with the presence or absence of particular species, with their abundance or rarity, and with the trends and fluctuations in their numbers. There are, however, two approaches at this level. One deals first with the attributes of individual organisms, and then considers the way in which these combine to determine the characteristics of the population. The alternative deals directly with the characteristics of populations, and tries to relate these to aspects of the environment. Both approaches have their uses, and both will be used. Indeed, these two approaches are also both useful at the level of the community. Community ecology deals with the composition or *structure* of communities, and with the pathways followed by energy, nutrients and other chemicals as they pass through them (the *functioning* of communities). We can pursue an understanding of these patterns and processes from a consideration of the component populations; but we can also look directly at properties of the communities themselves, like species diversity, the rate of biomass production and so on. Again, both approaches will be used. Ecology is a central discipline in biology, and therefore, not surprisingly, it overlaps with many other disciplines, especially genetics, evolution, behaviour and physiology. The particular concern of ecology is, however, with those characteristics that most affect distribution and abundance—the processes of birth, death and migration.

It is worth stressing at this early stage that ecologists are concerned not only with communities, populations and organisms *in nature,* but also with man-made or man-influenced environments (orchards, wheat fields, grain stores, nature reserves and so on). What is more, ecologists often become interested in laboratory systems and mathematical models. These have played a crucial role in the development of ecology, and they are bound to continue to do so. They will be examined frequently in this book. Ultimately, however, it is the real world that we are interested in, and the worth of models and simple laboratory experiments must always be judged in terms of the light they throw on the working of more natural systems. They are a means to an end—never an end in themselves. A major aim of science is to simplify, and thereby make it easier to understand the complexity of the real world. Ecology is a science that underpins natural history and seeks to explain it.

Explanation, description, prediction and control

At all levels of the ecological hierarchy we can try to do a number of different things. In the first place we can try to *explain* or *understand*. This is a search for knowledge in the pure scientific tradition. In order to do this, however, it is necessary first to describe. This, too, adds to our knowledge of the living world. Obviously, in order to understand something, we must first have a description of whatever it is that we wish to understand. Equally, but less obviously, the most valuable descriptions are those carried out with a particular problem or 'need for understanding' in mind. All descriptions are selective, but undirected description, carried out for its own sake, is often found afterwards to have selected the wrong things.

Ecologists also often try to *predict* what will happen to an organism, a population or a community under a particular set of circumstances; and on the basis of these predictions we try to control or exploit them. We try to minimize the effects of locust plagues by predicting when they are likely to occur and taking appropriate action. We try to protect crops by predicting when conditions will be favourable to the crop and unfavourable to its enemies. We try to preserve rare species by predicting the conservation policy that will enable us to do so. Some prediction and control can be carried out without explanation or understanding. But confident predictions, precise predictions and predictions of what will happen in unusual circumstances can be made only when we can also explain what is going on.

It is important to realize that there are two different classes of explanation in biology: proximal and ultimate explanations. For example, the present distribution and abundance of a particular species of bird may be 'explained' in terms of the physical environment that the bird tolerates, the food that it eats and the parasites and predators that attack it. This is a *proximal* explanation. However, we may also ask how this species of bird has come to have these properties that now appear to govern its life. This question has to be answered by an explanation in evolutionary terms. The *ultimate* explanation of the present distribution and abundance of this bird lies in the ecological experiences of its ancestors.

There are many problems in ecology that demand evolutionary, ultimate explanations. 'How does it come about that coexisting species are often similar but rarely the same?' (Chapter 7), 'What causes predators to adopt particular patterns of foraging behaviour?' (Chapter 9), and 'How have organisms come to possess particular combinations of size, developmental rate, reproductive output and so on?' (Chapter 14). These problems are as much a part of modern ecology as are the prevention of plagues, the protection of crops and the preservation of rarities. Our ability to control and exploit cannot fail to be improved by an ability to explain and understand. In the search for understanding, we must combine both proximal and ultimate explanations.

PART 1

Organisms

Introduction

We have chosen to start this book with chapters about organisms, then to consider the ways in which they interact with each other, and lastly to consider the properties of the communities that they form. We could, quite sensibly, have treated the subject the other way round—starting with a discussion of the complex communities of natural and man-made habitats, proceeding to analyse them at ever finer scales, and ending with chapters on the characteristics of the individual organisms. The problem is a little like writing a book about clocks and watches. We could start with a survey and classification of the different types and end with chapters on the springs, cogwheels, batteries and crystals that make them work. We chose to start with organisms because it is these that are acted upon by evolutionary forces (they live or die, leave descendants or fail to do so); and the nature of communities must ultimately be explained (rather than simply described) in terms of their parts—individual populations composed of individual organisms.

We consider first the sorts of correspondences that we can detect between organisms and the environments in which they live. It would be trite and superficial to start with the view that every organism is in some way ideally fitted to live where it does. We may reach this conclusion after exhaustive study, but logically it cannot be a premise. Nonetheless, we emphasize that organisms are as they are, and live where they do, because of their evolutionary history (Chapter 1). We then consider the ways in which environmental conditions vary from place to place and from time to time, and the ways in which organisms differ in the conditions that they tolerate or prefer (Chapter 2). Next we look at the resources that different types of organisms consume, and the nature of their interaction with these resources (Chapter 3).

The sorts of species present in a community, their numbers and/or their mass, give that community much of its ecological interest. Abundance is determined by the balance between birth rates and death rates, and between immigration and emigration. In Chapter 4 we consider some of the variety in the patterns of birth and death that we will later see having profound effects on the behaviour of populations.

Some ability to disperse characterizes every species of plant and animal. It determines the rate at which organisms escape from environments that are or become unfavourable, and the rate of discovery of environments that are ripe for colonization and exploitation. Dispersal of organisms is considered in Chapter 5.

1 The Match between Organisms and their Environments

1.1 Introduction

Ecology deals with organisms and their environments, and it is important that we understand the relationship between them. Probably the most fundamental statement that we can make about this relationship is that different kinds of organism are not distributed at random amongst different kinds of environment: there is a correspondence between the two. This correspondence is part of our sense of the order of things. But what exactly is the nature of the match between organisms and their environment?

1.1.1 Natural selection—adaptation or abaptation?

The phrase that, in everyday speech, is most commonly used to describe this match is: 'organism X is adapted to' followed by a description of the conditions where the organism is found. Thus we often hear that 'fish are adapted to life in aquatic environments', or 'cactuses are adapted to the conditions of drought found in deserts', and so on. Paradoxically, the surest way to understand the *true* nature of the match between organisms and their environment is to establish why 'adapted to' is a wholly inappropriate phrase in this context.

Charles Darwin's (1859) theory of evolution by natural selection may be thought of, simply, as resting on a series of propositions.

evolution by natural selection

(i) The individuals that make up a population of a species are *not identical*: they vary, though sometimes only slightly, in size, rate of development, response to temperature, and so on.

(ii) Some, at least, of this variation is *heritable*. In other words, the characteristics of an individual are determined to some extent by its genetic make-up. Offspring receive their genes from their parents, and offspring therefore have a tendency to share characteristics with their parents.

(iii) All populations have the *potential* to populate the whole earth, and they would do so if each individual survived and each individual produced its maximum number of offspring. But they do not: many individuals die prior to reproduction, and most (if not all) reproduce at a less than maximal rate.

(iv) Different individuals leave *different numbers of descendants*. This means more than saying that different individuals have different numbers of offspring. It includes also the chances of survival of individuals to reproductive age, the number of offspring they produce, the survival and reproduction of these offspring, the survival and reproduction of *their* offspring in turn, and so on.

(v) Finally, the number of descendants that an individual leaves depends, not entirely but crucially, on *the interaction between the characteristics of the individual and the environment of the individual.*

An individual will survive, reproduce and leave descendants in some environments but not in others. This is the sense in which nature may be loosely thought of as *selecting*. It is in this sense that some environments may be described as favourable or unfavourable, and it is in this same sense that some organisms can be considered to be fit or not. If as a consequence of some individuals leaving more descendants than others the heritable characteristics of a population change from generation to generation, then evolution by natural selection is said to have occurred.

It would be reasonable to say that the organisms of a particular generation are *ab*apted by the environments of previous generations. Past environments act as a filter through which combinations of characters have passed on their way to the present. But organisms appear to be *ad*apted (fitted) to their present environment only because present environments tend to be similar to past environments. The word 'adaptation' gives an erroneous impression of prediction, forethought or, at the very least, design. Organisms are not designed for, or adapted to, the present or the future—they are consequences of, and therefore abapted by, their past.

1.1.2 Fitness

The fittest individuals in a population are by definition those that leave the greatest number of descendants. In practice, the term is often applied not to a single individual, but to a typical individual or a type: for example, we may say that in sand-dunes, yellow-shelled snails are fitter than brown-shelled snails (i.e. they are more likely to survive and leave larger numbers of descendants).

fitness is the *proportionate* contribution of individuals to future generations

Fitness, however, is a relative not an absolute term. The numbers of seeds produced by a plant, or eggs produced by an insect, are not direct measures of their fitness; nor indeed are the *numbers* of descendants that they leave. Rather, it is the proportionate contribution that an individual makes to future generations that determines its fitness: the fittest individuals in a population are those that leave the greatest number of descendants *relative to* the number of descendants left by other, less fit individuals in the population. And those individuals that leave the greatest proportion of descendants in a population have the greatest influence on the heritable characteristics of that population.

the theory of natural selection does not predict perfection

No population of organisms can contain all the possible genetic variants that might exist and might influence fitness. It follows from this that natural selection is unlikely to lead to the evolution of perfect, 'maximally fit' individuals. It favours those that are fittest *from amongst those available,* and this may be a very restricted choice. The powers of evolutionary forces are far from limitless. Darwin's theory does not predict perfection, not even in an environment that remains unchanged from generation to generation. It predicts only that some individuals will leave more descendants than others, and will therefore have most influence in determining the characteristics of future generations.

This is the most fundamental limitation on the extent to which organisms match their environment. The very essence of natural selection is that organisms come to match their environments by being 'the fittest available' or 'the fittest yet': they are not 'the best imaginable'. The reasons why we do not expect to find the evolution of perfection in nature are developed further in papers by Jacob (1977), Gould & Lewontin (1979) and Harper (1982).

There are, moreover, other limitations and qualifications to the manner and extent to which organisms match their environment. Some of these will be dealt with in the remainder of this chapter.

1.2 Historical factors

It is particularly important to realize that past events on the earth can have profound repercussions for the present. Our world has not been constructed by someone taking each organism in turn, and testing it against each environment so that every organism finds its perfect place. It is a world in which organisms live where they do for reasons that are often, at least in part, accidents of history.

1.2.1 Movements of land masses

The curious distributions of organisms between continents, seemingly inexplicable in terms of dispersal over vast distances, led biologists, especially Wegener (1915), to suggest that the continents themselves must have moved. This was vigorously denied by geologists until geomagnetic measurements required the same, apparently wildly improbable, explanation. The movement of the tectonic plates of the earth's crust, with consequent migration of continents, reconciles geologist and biologist (Figure 1.1). Thus, while major evolutionary developments were occurring in the plant and animal kingdoms, populations were being split and separated, and land areas were moving across climatic zones.

the distributions of large flightless birds

Figure 1.2 shows just one example of a major group of organisms (the large flightless birds) whose distributions begin to make sense only in the light of the movement of land masses. It would be unwarranted to say that the emus and cassowaries are where they are because they represent the best match to Australian environments, whereas the rheas and tinamous are where *they* are because *they* represent the best match to South American environments. Each has evolved in its own continent and has been abapted by past environments there. But their very disparate distributions are essentially determined by the prehistoric movements of the continents, and the subsequent impossibility of each reaching the others' environment.

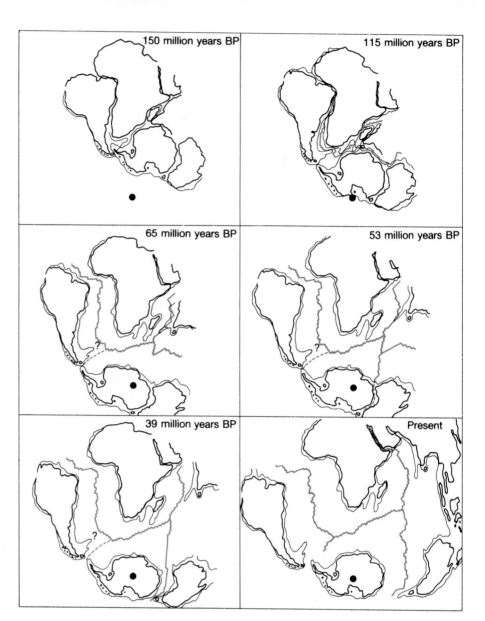

Figure 1.1. Reconstructions of the successive stages in the break-up of the ancient super-continent of Gondwana-land, showing the separation of the southern continents and the formation of the South Atlantic and Indian Oceans. (After Norton and Sclater, 1979.)

1.2.2 Climatic changes

Changes in climate have occurred on shorter time-scales than the movements of land masses, and much of what we see in the present distribution of species represents phases in a recovery from past climatic shifts. The Pleistocene ice-ages in particular, are responsible for distributions affected by historical changes in climate as much as by precise fits of organisms to their present environment.

A classic example of species' distributions that may need to be explained in terms of history is the presence of isolated pockets of often very specialized cold-tolerant species of flowering plant in the arctic–alpine floras of northern America and northern Europe. Frequently such species are found in only one locality. Other species occur in two or more curiously isolated regions (i.e.

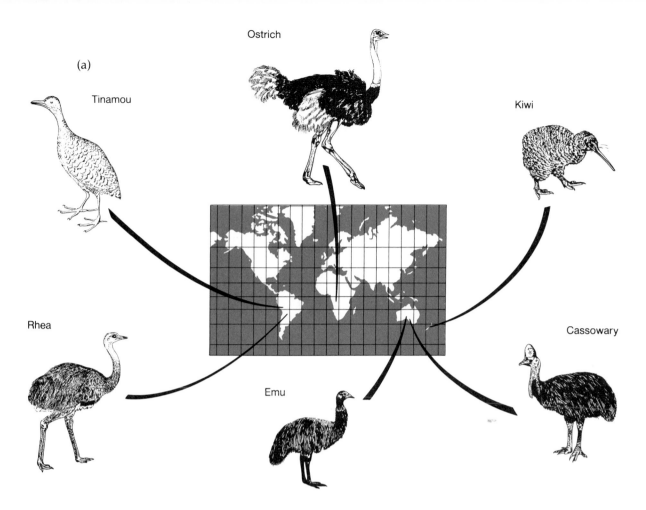

(a)

Ostrich

Tinamou

Kiwi

Rhea

Emu

Cassowary

Figure 1.2. (a) The distributions and degrees of relatedness of a group of large flightless birds can be at least
partly explained by continental drift (see Figure 1.1).
(b) The degrees of relatedness have been measured by a
DNA hybridization technique. The double-stranded
DNA is separated into single strands by heating. The
strands from different species can then be combined,
and again separated by heating. The more similar they
are, the higher is the temperature required to separate
them, $\Delta T_{50}H$. The temperature for separation then gives
a measure of the relatedness of the species and an
estimate of the time at which they diverged.
Myr = millions of years. The earliest divergence was
that of tinamous from the remainder (the ratites). The
subsequent divergences agree well with the timing of
the break-up of Gondwanaland and the subsequent
continental drift (Figure 1.1): (i) the rift between
Australia and the other southern continents; (ii) the
opening of the Atlantic between Africa and South
America; (iii) the opening of the Tasman Sea about 80
Myr ago, probably followed by island hopping by
ancestors of the kiwi across this divide to New Zealand
40 Myr ago. The divergence of the various species of
kiwi appears to be very recent. (After Diamond, 1983,
from data of Sibley and Ahlquist.)

(b)

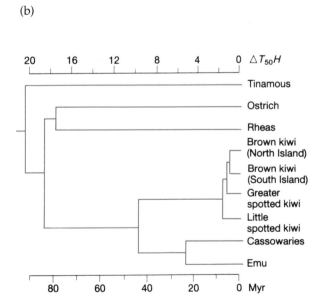

they have bicentric or polycentric disjunct distributions). The bicentric distribution of *Campanula uniflora* in Norway (Figure 1.3) is a typical example.

isolated distributions and nunataks

One view of these separate centres of distribution is that they represent locally suitable habitats, providing the unique specialized conditions required for the growth of plants of the species. The plant species' propagules are supposed to have dispersed to these sites across intervening areas where the habitats are unsuitable. There may (on this argument) be other areas in which the species might perpetuate itself, but these have not yet been reached. Certainly, many of these oddly distributed plant species are found in patches of calcareous soils or associated with particular temperature conditions (see Figure 1.3), but this is not always the case.

A different interpretation is that the present distributions are relicts of populations once distributed more widely. As the ice sheet moved down from the north, some high-altitude areas remained free of ice (though still intensely cold). Populations of some plant species persisted in these unglaciated areas or 'nunataks', and when the ice retreated they remained in their isolated fastnesses. Geological evidence can be found that supports this idea (though it is not conclusive). There is also some intriguing evidence from the distri-

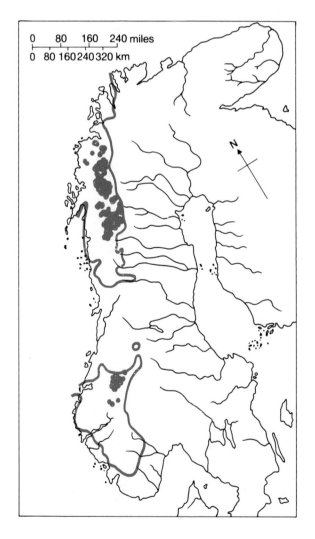

Figure 1.3. The distribution of *Campanula uniflora* in Norway: a characteristic bicentric distribution. The thick lines are 'isotherms': lines joining places where the mean maximum summer temperature is 22°C. (After Ives, 1974.)

bution of flightless beetles in supposed nunataks that adds support to the 'relict' argument. (These arguments are reviewed by Ives, 1974.)

In fact, the argument about nunataks is only a tiny part of a broader argument between those who interpret the present distribution of organisms as showing their match to present conditions, and those who interpret much of the present as a hangover from the past. The extent of climatic and biotic change in the past 2 million years (the Pleistocene) is only beginning to be unravelled as the technology for discovering, analysing and dating biological remains becomes more sophisticated (particularly by the analysis of buried pollen samples). These methods increasingly allow us to determine just how much of the present distribution of organisms represents a precise local specialization to present conditions, and how much is a fingerprint left by the hand of history.

Techniques for the measurement of oxygen isotopes in ocean cores indicate that there may have been as many as 16 glacial cycles in the Pleistocene, each lasting for about 125 000 years (Figure 1.4). Contrary to popular belief, it seems that each glacial phase may have lasted for as long as 50–100 000 years with brief intervals of 10–20 000 years when the temperatures rose to, or above, those we experience today. If these time scales are correct, it is present floras and faunas that are unusual, because they have developed towards the end of one of a series of unusual catastrophic warm events!

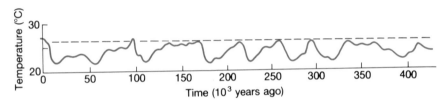

Figure 1.4. An estimate of the temperature variations with time during glacial cycles over the last 400 000 years. The estimates were obtained by comparing oxygen isotope ratios in fossils taken from ocean cores in the Caribbean. The dashed line corresponds to the ratio 10 000 years ago, at the start of the present warm period. This suggests that interglacial periods as warm as this have been rare events, and most of the past 400 000 years has experienced glacial climates. (After Emiliani, 1966, and Davis, 1976.)

the distributions of trees have changed gradually since the last glaciation

The rate at which vegetation has changed since the last glaciation at Rogers Lake, Connecticut (as evidenced by pollen records) is illustrated in Figure 1.5, showing the changes over the 14 000 years before the present. The woody species which dominate such pollen profiles have arrived in turn, spruce first and chestnut most recently. Each new arrival has added to the number of the species present, which has increased continually over the 14 000 year period. The same picture is repeated in European profiles.

As the number of pollen records has increased, it has become possible not only to plot the changes in vegetation at a point in space, but to begin to map the movements of the various species as they have spread across the continents. It appears that different species have spread at different speeds, and not always in the same direction. The composition of the vegetation types has, at the same time, been shifting in balance, and is almost certainly continuing to do so.

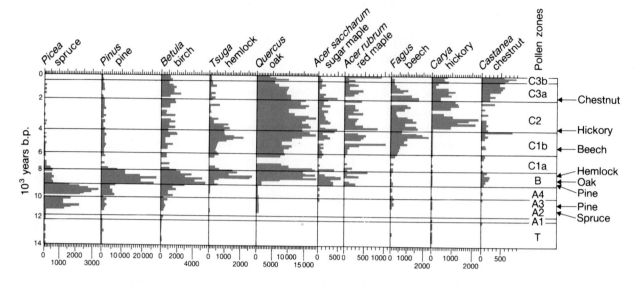

Figure 1.5. The profiles of pollen accumulated from late glacial times to the present in the sediments of Rogers Lake, Connecticut. The estimated date of arrival of each species in Connecticut is shown at the right of the figure. (After Davis *et al.*, 1973.)

In the invasions that followed the retreat of the ice in North America, spruce invaded first, followed by jack pine or red pine, which spread northwards at a rate of 350–500 metres per year for several thousands of years. White pine started its migration about 1000 years later, at the same time as oak. Hemlock was also one of the rapid invaders (200–300 metres per year), and arrived at most sites about 1000 years after white pine. Chestnut moved slowly (100 metres per year) but became a dominant species once it had arrived; it appeared on the Allegheny Mountains some 3000 years before it reached Connecticut. (Even 100 metres a year seems a remarkably rapid rate of advance for a plant species that bears heavy seeds and has a long period of juvenile growth before it starts to produce them.) As the ice retreated, the rate of seed dispersal may have limited the speed with which some plant species colonized the environments that were created.

We do not have such good records for the post-glacial spread of the animals associated with the changing forests, but it is at least certain that many species could not have spread faster than the trees on which they feed. Some of the animals may still be catching up.

The composition of northern temperate forests has continually changed over the past 10 000 years and continues to alter. The rate of change appears to be limited by the rates of spread of the main species. As Davis (1976) has remarked, the fact that '. . . forest trees are still migrating into deglaciated areas, even at the end of the Holocene (i.e. the present) interglacial interval, implies that the time span of an average interglacial is too short for the attainment of floristic equilibrium'.

Very little is known about the floras of previous interglacial periods, but presumably these repeated again and again some of what has been discovered from more recent records. Species populations that did not retreat fast enough when the ice advanced (and it advanced faster than it withdrew) presumably became extinct, with a few perhaps surviving in nunataks. Those that survived, those that we now know, were simply the ones that could keep pace with (or survive through) the successive cycles of the ice's invasion and retreat. Moreover, there is no reason to suppose that the genetic composition

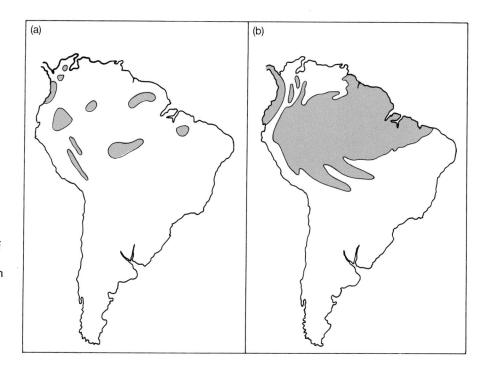

Figure 1.6. (a) The estimated distribution of tropical forest in South America at the time when the last glaciation was at its highest. (b) The present distribution. (Redrawn from Simberloff, 1983 and Pielou, 1979.)

of those species that did survive the migrations remained the same. For example, we know that quite rapid changes can be produced in laboratory populations of both animals and plants under artificial selection for cold tolerance. In particular, we might expect that after a an episode of migration, each species would have been selected for increased rates of colonization. The communities of temperate countries are ephemeral, risk-ridden and perpetually readjusting to changing environments. They have never had time to be otherwise.

the present interglacial recovery is the most recent of many

The records of climatic change in the tropics are far less complete than those for temperate regions. There is therefore the temptation to imagine that while dramatic climatic shifts and ice invasions were dominating temperate regions, the tropics persisted in the state we know today. This is almost certainly wrong. Instead, a picture is emerging of vegetational changes parallel those occurring in temperate regions, as the extent of tropical forest increased in warmer, wetter periods, and contracted as savanna dominated the vegetation in the cooler and drier periods. The present distributions of both plant and animal species provide evidence of the positions once occupied by these 'tropical islands in a sea of savanna' (Figure 1.6).

climatic change in the tropics

1.2.3 Island patterns

The fauna and flora of islands (whether surrounded by ocean or by a 'sea' of different vegetation) have several features that distinguish them from the fauna and flora of continents. These are discussed lucidly and in some detail in Williamson's (1981) book, and are touched on in several chapters here, especially Chapter 20. In the present context though, island biotas illustrate three important, related points: (i) the historical element in the match between organisms and environments, (ii) the fact that there is not just one perfect organism for each type of environment, and (iii) the power of natural selection in acting on disparate types of organism and matching them to the same type of environment.

Put very simply, the fauna and flora of islands have two distinguishing features. There are fewer species on islands than in comparable areas of mainland of the same size; and many of the species on islands are either subtly or profoundly different from those on the nearest comparable area of mainland. Again put simply, there are two main reasons for this. First, the fauna and flora of an island are limited to those types having an ancestor that managed to disperse to the island, though the extent of this limitation depends on the distance of the island from the mainland (or other islands) and varies from group to group of organism depending on their intrinsic dispersal ability. Secondly, because of this isolation, the rate of evolutionary change on an island may often be fast enough to outweigh the effects of the exchange of genetic material between the island population and the parent population on the mainland.

Let us explain this second point more fully. Biologists recognize organisms as members of the same species if the individuals are capable of breeding with each other, freely exchanging genetic material and producing fertile offspring. Such genetic exchange tends to result in homogeny in the genetic character of a population (as does the reassortment of genes resulting from genetic recombination). Reproductive isolation on the other hand (as, for instance, between island and mainland populations) will allow the more localized evolution of matches between organisms and their environments. Indeed, it seems that reproductive isolation is an essential step in the splitting of one ancestral species of animal into two. This undoubtedly goes a long way towards explaining why islands contain many species unique to themselves, as well as many differentiated 'races' or 'sub-species' that are distinguishable from mainland forms but not to a sufficient extent to be called separate species.

The *Drosophila* fruit-flies of the Hawaiian islands provide one of the most spectacular examples of species formation on islands, and it is certainly the example best studied genetically. The Hawaiian chain of islands (Figure 1.7) are volcanic in origin, and have been formed gradually over the last 40 million years as the centre of the Pacific tectonic plate has moved steadily over a 'hot-spot' in a south-easterly direction (Niihau is therefore the most ancient of the islands, Hawaii itself the most recent). The richness of the Hawaiian *Drosophila* is amazing: there are probably about 1500 *Drosophila* species worldwide, but at least 500 of these are found only in the Hawaiian islands. This is probably at least partly due to the fact that there are islands within islands in Hawaii, because lava-flows have frequently isolated areas of vegetation known as 'kipukas'.

Of particular interest are the 100 or so species of 'picture-winged' *Drosophila*. The lineages through which these species have evolved can be traced by analysing the banding patterns on the giant chromosomes found in the salivary glands of their larvae. The evolutionary tree that emerges is shown in Figure 1.7, with each species lined up above the island on which it is found (there are only two species found on more than one island). The

the isolation of islands favours and promotes the formation of new species

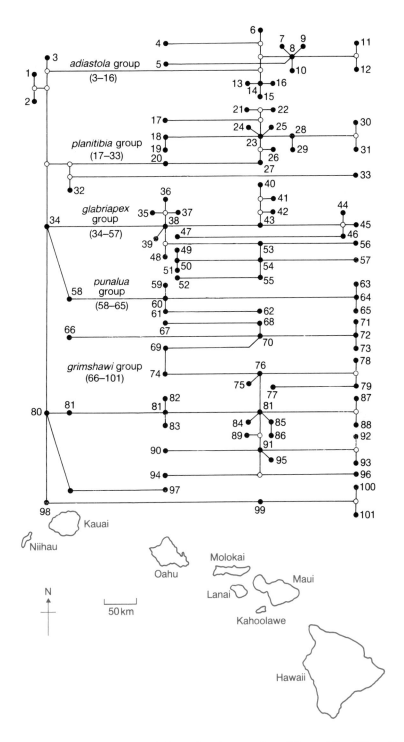

Figure 1.7. An evolutionary tree linking the picture-winged *Drosophila* of Hawaii, traced by the analysis of chromosomal banding patterns. The most ancient species are *D. primaeva* (species 1) and *D. attigua* (species 2), found only on the most ancient island (Kauai). Other species are represented by solid circles; hypothetical species, needed to link the present-day ones, are represented by open circles. Each species has been placed above the island or islands on which it is found (though Molokai, Lanai and Maui are grouped together). Niihau and Kahoolawe support no *Drosophila*. (After Carson & Kaneshiro, 1976 and Williamson, 1981.) The first Drosophila colonist probably reached the Hawaiian archipelago 40 million years ago, before any of the present islands existed (Beverly & Wilson, 1985).

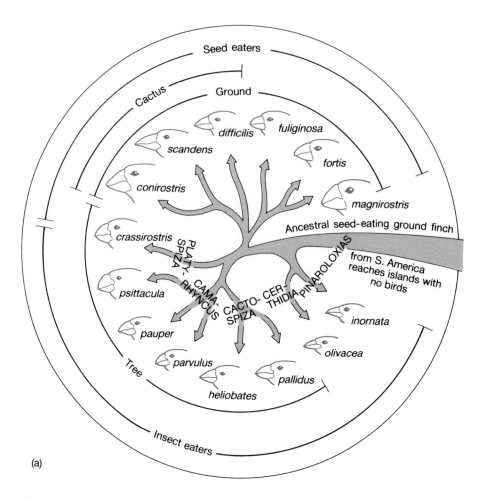

Figure 1.8. (a) Darwin's finches exploit a diverse range of food-types and habitats, and exhibit a diverse range of forms of beak, despite being closely related. (b, facing page) The distribution of the different species on the Galapagos (and Cocos) islands. The number of species on each island is shown on the map. The distribution of species between the islands is shown in the table and within each species a different letter is used for each subspecies.

the picture-winged *Drosophila* **of Hawaii**

historical element in 'what lives where' is plainly apparent: the more ancient species live on the more ancient islands, and as new islands have been formed, rare dispersers have reached them and eventually evolved to new species. At least some of these species appear to match the same environment as others on different islands. Of the closely related species *D. adiastola* (species 8) and *D. setosimentum* (species 11), for example, the former is only found on Maui and the latter only on Hawaii, but the environments that they live in are apparently indistinguishable (Heed, 1968). What is most noteworthy, of course, is the power and importance of isolation (coupled with natural selection) in generating new species—and thus new matches to the environment.

The most celebrated and familiar example of evolution and speciation on islands is the case of Darwin's finches in the Galapagos (Lack, 1947; William-

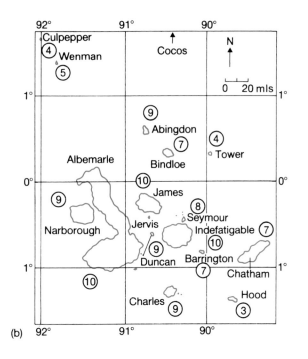

(b)

	Large central				Small central				Medium outlying				Small outlying			
	James	Santa Cruz	Isabella	Fernandina	Jervis	Seymour	Duncan	Barrington	Pinta	Marehena	San Cristobal	Foreana	Culpepper	Wenman	Tower	Hood
Geospiza																
magnirostris	A	A	A	A	A	A	A	A	A	A	—	B	—	A	A	—
fortis	A	A	A	A	A	A	A	A	A	A	A	A	—	—	—	—
fuliginosa	A	A	A	A	A	A	A	A	A	A	A	A	—	—	—	A
difficilis	A	A	A	A	—	—	—	—	B	—	—	C	D	D	B	—
scandens	A	B	B	—	A	B	B	B	C	D	E	B	—	—	—	—
conirostris	—	—	—	—	—	—	—	—	—	—	—	—	A	B	B	C
Platyspiza																
crassirostris	A	A	A	A	A	—	A	—	A	A	A	A	—	—	—	—
Camarhynchus																
psittacula	A	A	B	B	A	A	A–B	A	C	C	—	A	—	—	—	—
pauper	—	—	—	—	—	—	—	—	—	—	—	A	—	—	—	—
parculus	A	A	A	A	A	A	A	A	A	—	B	A	—	—	—	—
(Cactospiza)																
pallidus	A	A	B	B	A	A	A	—	—	—	C	—	—	—	—	—
heliobates	—	—	A	A	—	—	—	—	—	—	—	—	—	—	—	—
Certhidea																
oliracea	A	A	A	A	A	A	A	B	C	C	D	E	F	F	G	H
No. of resident species	10	10	11	10	9	8	9	6	9	7	7	10	3	3–4	4	3

In each species, a different letter is used for each subspecies.
After Lack, 1969a. Island names from Harris, 1973a.

son, 1981). These 13 species comprise approximately 40% of the bird species on the Galapagos, and between them they exploit a diverse range of food-types and habitats (Figure 1.8); but they are all closely related, and probably evolved from something similar to the present-day Cocos finch. Natural selection has acted on the material that happened to reach the isolated Galapagos islands; it has produced organisms matching environments that are matched by quite different species elsewhere.

1.3 Convergents and parallels

A match between the nature of organisms and their environment can often be seen as a similarity in form and behaviour between organisms living in a similar environment but belonging to different phyletic lines (i.e. different branches of the evolutionary tree). Such similarities also undermine further the idea that for every environment there is one, and only one, perfect organism. The evidence is particularly persuasive when the phyletic lines are far removed from each other, and when similar roles are played by structures that have quite different evolutionary origins, i.e. when the structures are *analogous* (similar in superficial form or function) but not *homologous* (derived from an equivalent structure in a common ancestry). When this is seen to occur, we speak of *convergent evolution*.

Large swimming carnivores have evolved in four quite distinct groups: among fish, reptiles, birds and mammals. The convergence of form (Figure 1.9) is remarkable, because it conceals profound differences in internal structures and metabolism, indicating that the organisms are far removed from each other in their evolutionary history.

For many groups of higher plants, the agents of seed dispersal are birds. These are attracted by the provision of fleshy, sugary tissue, which is eaten and digested, and which surrounds seeds that are protected by thick coats and that pass through the digestive tract unharmed or even promoted to germinate. The flesh of such fruits has been derived from quite different structures in different parts of the plant kingdom. A number of examples are shown in Figure 1.10. Many of these come from widely separated phylogenetic lines; for example, the yew is a gymnosperm and the remainder are angiosperms. However, even within a single family, the Rosaceae, a considerable variety of structures have been brought into service to form a fleshy, attractive cover. The strawberry, the apple and the peach each develop attractive flesh and/or protect their seeds in quite different ways.

In these examples (there are very many more) we can argue that similar selective forces have acted so that the same property has been acquired from quite different evolutionary starting points. A comparable series of examples can be used to show the parallels in the evolutionary pathways of phylo-genetically *related* groups that have radiated after they were isolated from each other.

The classic example of such *parallel evolution* is the radiation among the placental and marsupial mammals. Marsupials arrived on the Australian continent in the Cretaceous period (around 90 million years ago), when the only other mammals present were the curious egg-laying monotremes (now

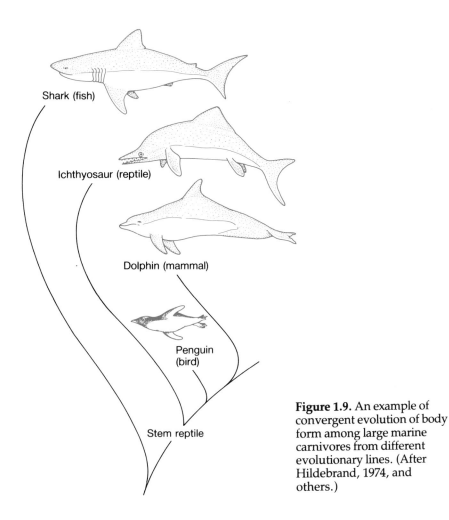

Shark (fish)

Ichthyosaur (reptile)

Dolphin (mammal)

Penguin
(bird)

Stem reptile

Figure 1.9. An example of convergent evolution of body form among large marine carnivores from different evolutionary lines. (After Hildebrand, 1974, and others.)

marsupial and placental mammals show parallel evolution

represented only by the spiny ant-eaters and the duck-billed platypus). An evolutionary process of radiation then occurred that in many ways accurately paralleled that occurring in the mammals on other continents. Some ecological parallels between marsupials and placentals are illustrated in Figure 1.11. The subtlety of the parallels in both the form of the organisms and their life-style is so striking that it is hard to escape the view that the environments of placentals and marsupials contained 'ecological niches' into which the evolutionary process has neatly 'fitted' ecological equivalents. (It is important to remember that, in contrast with convergent evolution, the marsupials and placentals started their radiative evolution with an essentially common set of constraints and potentials because they sprang from a common ancestral line.)

It is a fascinating exercise to collect examples to illustrate convergent or parallel evolution in plant and animal groups, but there are serious risks. One such risk is illustrated by the 'rosette treelets' or megaphytes (Plate I, p. 84). These plants, from many different groups, usually bear only one main stem which continues growth for a long period, often with no branching at all. They are characteristically topped with a mop of very large leaves, and when they flower they produce gigantic inflorescences. In many species the act of

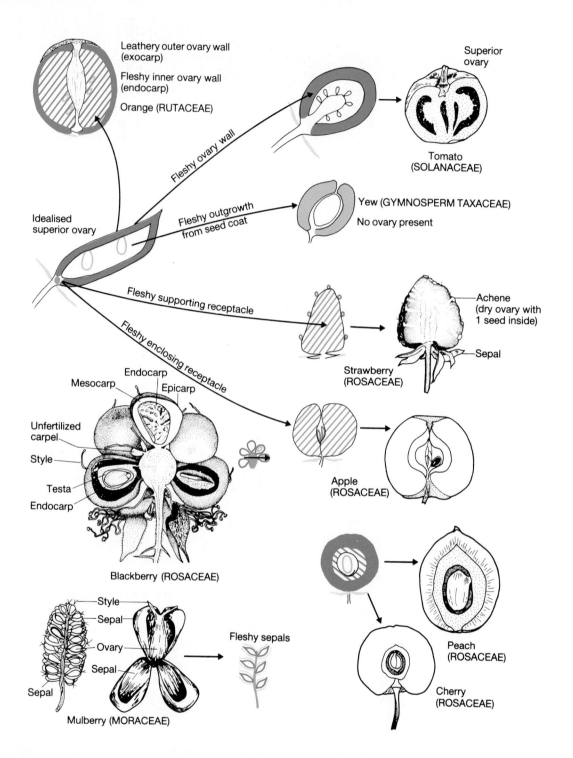

Figure 1.10. Convergent evolution of fleshy fruits from different groups of higher plants and from different morphological structures. Tissues that are homologous are shaded in the same way.

flowering is lethal, or there is a long delay before the next flowering episode. In addition, they often carry a mass of dead leaves or leaf bases, so that their trunks often falsely appear to be enormously thick (this is particularly apparent in many palms, espeletias and giant senecios).

We might expect that such striking similarity in form among a group of often quite unrelated plants implies an equally striking similarity in their

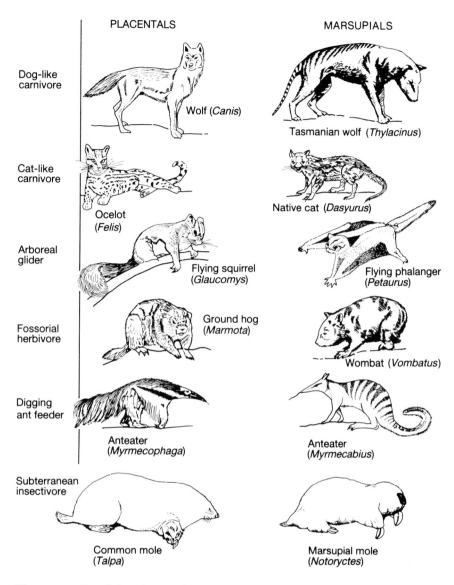

PLACENTALS | MARSUPIALS

Dog-like carnivore — Wolf (*Canis*) | Tasmanian wolf (*Thylacinus*)

Cat-like carnivore — Ocelot (*Felis*) | Native cat (*Dasyurus*)

Arboreal glider — Flying squirrel (*Glaucomys*) | Flying phalanger (*Petaurus*)

Fossorial herbivore — Ground hog (*Marmota*) | Wombat (*Vombatus*)

Digging ant feeder — Anteater (*Myrmecophaga*) | Anteater (*Myrmecabius*)

Subterranean insectivore — Common mole (*Talpa*) | Marsupial mole (*Notoryctes*)

Figure 1.11. Parallel evolution of marsupial and placental mammals. The pairs of species are similar in both appearance and habit.

do the megaphytes show convergent evolution?

environments. In fact they 'fit' into a very odd diversity of vegetation types (see Table 1.1). The high-mountain and often island habitats of the dicotyledonous megaphytes seem to have little in common with those of the understorey palms and tree ferns on the one hand, and with the arid habitats of many aloes and agaves on the other.

A partial explanation for the resemblance among the plants may be that they each represent a correlated set of characters that come together as a package, but that selection has acted on only one, and not necessarily the same one, of these characters in each case. It may be that the less a given plant body branches, the larger are the parts that develop on these branches. Perhaps selection operating to increase the size of inflorescences necessarily has the consequence of increasing leaf size, leading to an increase in the

Table 1.1. Megaphytes or rosette treelets and their habitats (see also Plate I).

Tree ferns	Tropical and subtropical forest—usually understorey
Cycads	Tropical and subtropical forest—usually understorey
Angiosperms	
Monocotyledons	
Palms	Tropical and subtropical forest—often understorey
Liliaceae and Amaryllidaceae e.g. *Yucca, Agave, Aloe, Cordyline, Dracaena*	Tropical and subtropical—many in arid zones and deserts but others forming understorey in tropical and subtropical forest
Xanthorrhoeaceae (grass trees)	Arid regions—exposed or understorey
Dicotyledons	
Boraginaceae, e.g. *Echium*	Canary Islands
Campanulaceae, e.g. tree *Lobelias*	Hawaii and African mountains
Compositae, e.g. *Espeletia, Senecio*	High Andes and African mountains

support given to the stems by the leaf bases. In other habitats it may be that selection for large leaves brings with it a necessary increase in the size of inflorescences.

Certainly, there are examples from plant breeding which tend to support this view. The cultivated sunflower has been bred from wild ancestors that are many-branched and bear small leaves, many small flowers and small seeds. The cultivated form has been selected to produce large seeds which are easily harvested. It also has a single, huge inflorescence, an unbranched stem and very large leaves. In many ways it has become a typical megaphyte. Plant breeders have also produced a variety of profound changes within the genus *Brassica*. From wild forms that are branched with relatively small leaves have been bred the cauliflower and broccolli—typical megaphytes producing a stout, unbranched stem with very large leaves and a hugh inflorescence. Another *Brassica* has its unbranched stem so strongly developed (6–8 feet high), with a mop of large leaves at the top, that it is used commercially to make walking sticks!

Whatever the cause of the megaphytes' convergence, however, the selective forces are less easily envisaged than those that acted on the large swimming carnivores or the fruits.

Another problem arises because most convergences and parallels are recognized by striking visual resemblances. Yet although the marsupial bandicoot resembles a placental rabbit (the long ears make the resemblance particularly striking), and both species can excavate burrows, the bandicoot is largely carnivorous, feeding on insect larvae, while the rabbit is a herbivore. By contrast, the kangaroo and the sheep show virtually no resemblance in form, although both are large grazing herbivores which, when they live side by side, have very similar diets (Griffiths & Barker, 1966).

convergences within guilds need not involve visual convergences

Root (1967) introduced the word 'guild' to describe a group of species that exploit the same class of environmental resources in a similar way. The kangaroo and the sheep are apparently members of similar guilds, without any very obvious likeness in appearance. The rabbit and the bandicoot have evolved a likeness in appearance but belong to quite different guilds. There is, in other words, an important sense in which the match of organisms to their environment may sometimes be better illustrated by parallels in feeding habits (what they feed on and what feeds on them) than by the more obviously striking visual resemblances.

1.4 Convergences between and divergences within communities

A consideration of convergent and parallel evolution stresses that a list of the names of species, genera or even families present in two areas may tell us virtually nothing about the ecological similarities between these areas. Moreover, although a particular type of organism is often characteristic of a particular ecological situation, it will almost inevitably be only part of a diverse community of forms that vary dramatically from each other. Only in the rarest situations do we find natural communities occupied by just one unique species. (This is when there is one overwhelmingly important physical factor in the environment, so that, for example, a curious species of square bacterium is often the sole occupant of concentrated brine (Walsby, 1980).) In the vast majority of cases, natural communities contain a spectrum of life forms and biologies. A fully satisfactory account of the match between organisms and their environment, therefore, must do more than pick out organisms from different communities that have strong resemblances in form and behaviour. It should also be able to relate whole communities of organisms to features of the environment, and should be able to explain the differences between organisms in the same environment as well the similarities.

1.4.1 Convergences between communities

A classic vegetation type such as the 'Mediterranean' maquis can be recognized around the Mediterranean itself, in California, Chile, South Africa and South Australia. These are all areas with very similar climates. The similarity in vegetation is apparent in aerial photographs or during a fast car-ride through these areas. Yet the taxonomist's list of the species present (even the families they represent) gives no indication of the similarity. These similarities have proved extraordinarily difficult to describe and measure. Often they are features of the 'architectures' of the various plants present, and these are not readily quantified. Hence loose, qualitative terms such as shrub, hummock, tussock, sward and scrub are often used. However, serious attempts have been made to develop and refine the ways in which the life and form of higher plants can be described, irrespective of their taxonomy and systematics.

Still the simplest, and in many ways the most satisfying, classification of plant forms disregarding their systematics is that of the Danish botanist Raunkiaer (1934). The growth of the shoots of higher plants depends on the initiation of tissues at apices (meristems), and Raunkiaer classified plants according to the way in which their meristems were held and protected (Table

Raunkiaer's classification of plant communities suggests community convergence

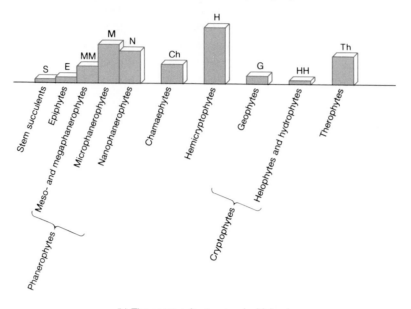

(a) The 'normal' spectrum obtained by Raunkiaer by sampling from Index Kewensis

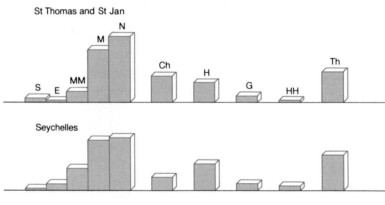

(b) The spectra for two tropical islands

St Thomas and St Jan

Seychelles

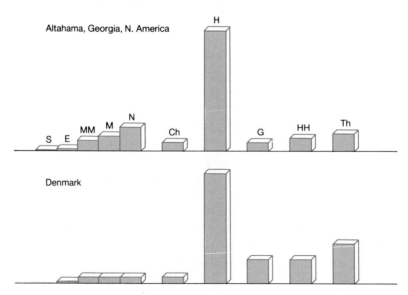

(c) The spectra for two temperate environments

Altahama, Georgia, N. America

Denmark

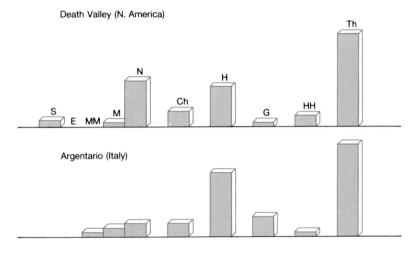

(d) The spectra for two arid regions

Death Valley (N. America)

Argentario (Italy)

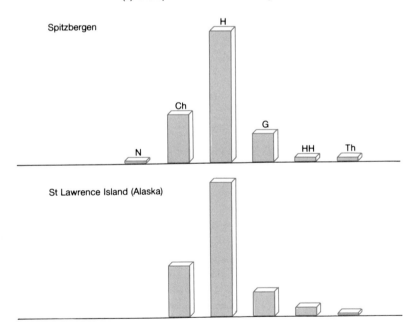

(e) The spectra for two arctic regions

Spitzbergen

St Lawrence Island (Alaska)

Figure 1.12. Whole plant communities can be compared by the method of Raunkiaer. Facing page. (a) The relative numbers of species of different life-forms present in a sample of the world flora (Index Kewensis). (b) Life-form spectra for the vegetation of two tropical islands. (c) Life-form spectra for two temperate environments. Above. (d) Life-form spectra for two arid environments. (e) Life-form spectra for two arctic regions.

Table 1.2. The life-form classification of Raunkiaer.

Phanerophytes	The surviving buds or shoot apices are borne on shoots which project into the air	
	(a) Evergreens without bud covering (b) Evergreens with bud covering (c) Deciduous with bud covering	} More than 2 m high
	(d) Less than 2 m high	
Chamaephytes	The surviving buds or shoot apices are borne on shoots very close to the ground	
	(a) Suffruticose chamaephytes, i.e. those bearing erect shoots which die back to the portion that bears the surviving buds (b) Passive chamaephytes with persistent weak shoots that trail on or near the ground (c) Active chamaephytes that trail on or near the ground because they are persistent and have horizontally directed growth (d) Cushion plants	
Hemicryptophytes	The surviving buds or shoot apices are situated in the soil surface	
	(a) Protohemicryptophytes with aerial shoots that bear normal foliage leaves but of which the lower ones are less perfectly developed (b) Partial rosette plants bearing most of their leaves (and the largest) on short internodes near ground level (c) Rosette plants bearing all their foliage leaves in a basal rosette	
Cryptophytes	The surviving buds or shoot apices are buried in the ground (or under water)	
	(a) Geocryptophytes or geophytes which include forms with (i) rhizomes, (ii) bulbs, (iii) stem tubers, (iv) root tubers (b) Marsh plants (helophytes) (c) Aquatic plants	
Therophytes	Plants that complete their life cycle from seed to seed and die within a season (this group also includes species that germinate in autumn and flower and die in the spring of the following year)	

1.2). Some examples of Raunkiaer spectra are shown in Figure 1.12 for the floras of different regions, compared with the spectrum that Raunkiaer obtained by sampling from the Index Kewensis (the compendium of all species known and described in his time, biased by the fact that the tropics were, and still are, relatively unexplored). Striking similarities between vegetation types are apparent, demonstrating matches not just of organisms to environments but of whole community complexes.

Raunkiaer's methods are clearly a step on the way to a definitive ecological classification of communities, and more recently attempts have been made to improve on Raunkiaer's methods (see, for example, Box, 1981). These all share a number of important problems: the difficulties of pigeon-holing species, and the subjective nature of the criteria by which they and their environments are classified. Although these methods remain only steps in the long process of determining the nature and magnitude of the match between whole communities and their environments, they do represent the beginning of a real test of whether there is indeed such a match. Comparable studies of the match between animal communities and their environment have seldom been attempted. A rare example is a study of animal communities in Mediterranean environments by Cody and Mooney (1978).

1.4.2 Diversity within communities

There are many explanations for the diversity that exists *within* communities.

(a) There are no homogeneous environments in nature. Even the seemingly homogeneous laboratory test-tube is heterogeneous in that it has to have a boundary, the walls of the tube. Microbiologists are often made aware of this when forms of cultured organisms subdivide into two forms: one that sticks to the walls and the other that remains free in the medium. The heterogeneity of the environment is something that is dependent on the scale of the organism that senses it. To a mustard seed, a grain of soil is a mountain; and to a caterpillar a single leaf may represent a lifetime's diet. A seed lying in the shadow of a leaf may be inhibited in its germination, while a seed lying outside that shadow may germinate freely. What appears to the observing ecologist as a homogeneous environment may, to an organism within it, be a mosaic of the intolerable and the adequate.

the heterogeneity of environments

(b) Related to this, most (perhaps all) environments contain within them gradients of conditions or of available resources. These may be gradients in space or gradients in time, and these latter, in their turn, may be rhythmic (like daily and seasonal cycles), directional (like the accumulation of a pollutant in a lake) or erratic (like fires, hailstorms and typhoons).

(c) The existence of one type of organism in an area immediately diversifies it for others. The very presence of an organism implies that its dead body will eventually be available as food, and during its life it may add dung, urine or dead leaves to the environment. Its own living body may serve as a place to live for other species, or as a resource for predators, pathogens and parasites that had no place in the community until it was there.

The diversity within natural communities may be due to all three of these types of heterogeneity. To some extent, the 'explanation' of this diversity is a trivial exercise. It comes as no surprise to anyone that a plant utilizing sunlight, a fungus living on the plant, a herbivore eating the plant, and a parasitic worm living in the herbivore should all coexist in the same community. On the other hand, most communities contain a variety of different species that are all constructed in a fairly similar way and all living (at least superficially) a fairly similar life. The explanation or understanding of this type of diversity is a far from trivial exercise.

The antarctic seals illustrate some of the differences that exist between

the antarctic seals and the coexistence of similar species

species that occur together in the same community. It is thought that the ancestral seals evolved in the Northern Hemisphere, where they are present as Miocene fossils, but one group of seals moved south into warmer waters and probably colonized the Antarctic in the late Miocene or early Pliocene (about 5 million years ago). When they entered the Antarctic, the Southern Ocean was probably rich in food and free from major predators, as it is again today. It was within this environment that the group appears to have undergone radiative evolution (Figure 1.13). The Weddell seal, for example, feeds primarily on fish and has unspecialized dentition; the crabeater seal feeds almost exclusively on krill and its teeth are suited to filtering these from the sea water; the Ross seal has small, sharp teeth and feeds mainly on pelagic squid; and the leopard seal has large, cusped, grasping teeth and feeds on a wide variety of foods, including other seals and, in some seasons, penguins. Indeed, the differences between these seal species could possibly determine the extent to which they coexist in mixed communities. Where two or more species feed on similar diets at similar depths (e.g. elephant and Ross) they are separated geographically with only slight overlap. However, when mixed communities of several seal species occur, they comprise species that exploit different food organisms (crabeater, Ross and leopard, or leopard and elephant, or Weddell and leopard—Figure 1.13).

The interpretation of patterns such as this (or at least as full an interpretation as we can currently achieve) must be postponed until

Figure 1.13. The food, feeding depth and jaw structure of a group of antarctic seals (after Laws, 1984).

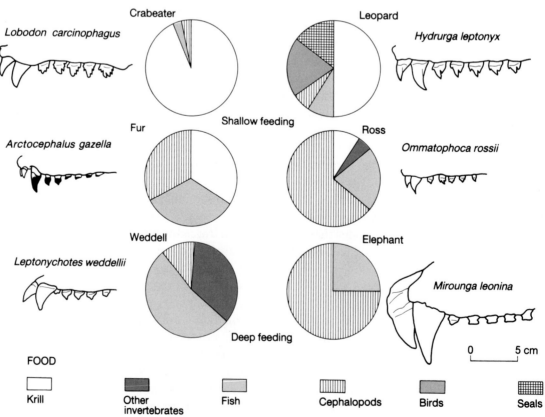

coexistence and competition are discussed in later chapters. Even at this early stage, though, a number of important points can be made.

(i) Coexisting species commonly differ in subtle ways, but each matches its environment and exists because the environment itself is heterogeneous.

(ii) This differentiation is most easily recognized in closely related species (as with the seals), because the differences can be seen most clearly against a background of many common features. But it may be at least as important to understand the coexistence of members of the same guild (p. 19) that are not necessarily closely related. Krill, for instance, is an abundant resource in antarctic waters, and the crabeater seal and to a lesser extent the leopard seal specialize on this diet. But krill is also eaten by squid, fish, birds and baleen whales, and a complete analysis of the fits in the jigsaw puzzle of such a community must therefore take account not just of a particular taxonomic group but of whole guilds.

(iii) The seal example provides many more questions than answers. Are coexisting species *necessarily* different? If so, *how* different do they need to be: is there some *limit* to their similarity? Do species like the seals *interact* with one another at the present time, or has radiative evolution in the past led to the absence of such interactions in contemporary guilds? None of these questions has a straightforward answer, but each of them will be addressed in later chapters.

1.5 Specialization within species

Up to this point we have been considering species (or larger taxa) as the units that match the environment, but it is *within* species that some of the most specialized fits to the environment can be traced. When islands were discussed, it was stressed that the homogenizing effects of genetic exchange (and recombination) tend to mean that most of the more obvious heterogeneities in populations have arisen when parts of the population become isolated geographically and cease to interbreed. However, if the local forces of natural selection are very powerful, they may override the homogenizing forces of sexual reproduction and recombination. Then, even where there is some interbreeding, locally favoured genotypes may be at such an advantage that ill-favoured combinations are continually eliminated. Gene-flow continues to occur, populations remain parts of the same species, but local specialized races appear within it.

The exchange of genetic material through a population depends on the mobility of whole organisms or, in the case of sessile organisms, on the mobility of their gametes, pollen or seeds. If the members of a population wander freely in search of mates (or their gametes, spores or seeds are spread widely), then local, special forces of natural selection are less likely to be effective in producing local, special matches between the organisms and their local environments.

sessile organisms must match their environment; mobile organisms can match their environment to themselves

Local, specialized populations become differentiated most conspicuously among organisms that are sessile for most of their lives. This is because motile organisms have a large measure of control over the environment in which they live; they can recoil or retreat from a lethal or unfavourable environment and actively seek another. However, the sessile higher plant, seaweed and coral have no such freedom. After dispersal they must live, or die, in the conditions where they settle. As Bradshaw (1972) has remarked '. . . the plant cannot run away to a new environment or hide in a protected corner'. The most that a higher plant can do is to search out resources or escape from an unfavourable site by growing from one place to another; it can never uproot itself and choose to transplant itself elsewhere. Its descendants (seeds or pollen or gametes) are hazarded to the vagaries of passive dispersal on the wind or water, or in or on the bodies of animals. Populations of non-mobile organisms are therefore exposed to forces of natural selection in a peculiarly intense form.

The contrast between the ways in which mobile and fixed organisms 'match' their environment is seen at its most dramatic on the sea-shore, where the intertidal environment continually oscillates between being terrestrial and being aquatic. The fixed algae, hydroids, sponges, bryozoans, mussels and barnacles all meet and tolerate life in the two extremes. But the mobile members of the community, the shrimps, crabs and fish travel with and track their aquatic habitat as it moves, while the shore-feeding birds move back and forth, following the advance and retreat of their terrestrial habitat. The fixed organisms have to tolerate the whole daily cycle of change in their environment, but those that are mobile have no need for such tolerance—they move with the tides. There is a sense in which the match of such mobile organisms to their environment enables them to escape many of the forces of natural selection. Mobility enables the organism to match its environment to itself. The immobile organism must match itself to its environment.

1.5.1 Ecotypes

It was for certain plants that the word 'ecotype' was first coined, to describe genetically determined local matches between organisms and their environment *within species*. Major differences within species but between habitats have been demonstrated by transplanting plants from a variety of natural habitats into a common habitat, and allowing them to develop through one or more growing seasons. It is essential in such studies that the plants from the different places of origin should be grown and compared in the same environment, since some of the differences between field populations may be environmentally determined phenotypic responses that are not due to differences in genotype. For example, the creeping stolons of clover are much branched in a heavily grazed pasture and less so in a hay meadow, but such differences can be elicited from different parts of the same individual by placing them in the different environments (Figure 1.14).

The kinds of differences between races of a species that have been recognized as ecotypes are illustrated in Table 1.3.

The extremely fine scale over which local specialization can occur among plants has been most clearly shown in studies of tolerance to toxic heavy metals (lead, zinc, copper and so on) (Antonovics & Bradshaw, 1970). At the edge of areas that have been contaminated after mining activities, the inten-

heavy metal tolerant ecotypes

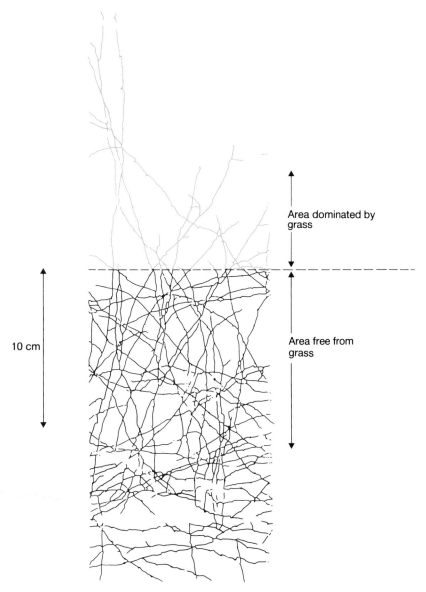

10 cm

Area dominated by grass

Area free from grass

Figure 1.14. As an individual plant of white clover (*Trifolium repens*) grows from an area with no grass into an area dominated by grass (*Lolium perenne*), its growth form changes. The figure shows a map of the creeping stolons of white clover on the ground surface after the clover has begun to invade the grass area. Note that the clover stolons are sparser, have longer internodes and rarely branch when they are in the area dominated by grass. (Data of Solangarachi.)

sity of selection against susceptible genotypes changes abruptly, and populations on contaminated areas may differ sharply in their tolerance of heavy metals over distances of less than 100 m (in sweet vernal grass, populations are differentiated over less than 1.5 m). This occurs in spite of the fact that pollen is freely borne across such boundaries by wind or insects. In some cases it has been possible to date the time since which such selection has occurred. Heavy metal tolerance has apparently evolved since mining started 300 years ago in Swansea (South Wales); zinc tolerance has evolved beneath a

Table 1.3. Examples of differences between ecotypes in higher plants.

Aspects of form or life cycle	Difference between ecotypes	Investigator(s) associated with the studies
Growth form	Contrast between upright and prostrate *Plantago maritima*	Gregor (1930)
Water requirements Earliness of growth	Contrasts between alpine, sub-alpine and lowland forms of *Poa alpina*	Turesson (1922)
Annual growth cycle	Contrasts between forms along an altitudinal gradient with different flowering and fruiting times, e.g. *Hemizonia, Achillea*	Clausen, Keck and Hiesey (1941)
Longevity and vegetative vigour	Contrasts between grasses of various species in man-managed grasslands	Stapledon (1928)
Flowering time	Geographic differences in photoperiodic requirement for flowering	McMillan (1957)
Nutritional responses	Variation within species of grass and white clover in response to nitrogen, phosphorus, calcium, etc.	Snaydon and Bradshaw (1969)
Tolerance of toxic metals	Contrasts between forms on heavy metal mine spoils and elsewhere	Antonovics and Bradshaw (1970)

galvanized iron fence within 25 years; a single generation of selection has been sufficient to establish a metal-tolerant population when a high density of seed was experimentally applied to a heavily contaminated soil. Both the speed of selection and the sharpness of the boundaries show how relatively powerless is the mixing power of gene flow against very strong forces of selection.

1.5.2 Genetic polymorphism

On a finer scale even than ecotypes, biologists are increasingly able to detect levels of selectively relevant variation *within* small local populations. Such variation is known as *polymorphism*. Specifically, genetic polymorphism is 'the occurrence together in the same habitat of two or more discontinuous forms of a species in such proportions that the rarest of them cannot merely be maintained by recurrent mutation or immigration' (Ford, 1940). Not all such variation represents a match between organism and environment; indeed, some of it may represent a mis-match. Mis-matches may arise through the dispersal of propagules from one specialized form into the habitat of another. They may also arise if conditions in a habitat have changed so that one form is in the process of being replaced by another that is fitter in the changed conditions. Such polymorphisms are called transient. As all communities are always changing, a lot of the variation that we observe in nature may be of this transient character, representing the extent to which populations will always be out of step with their environment and unable to anticipate change.

Many polymorphisms, however, are actively maintained in a population

transient polymorphisms and mismatches

by natural selection, and there are a number of ways in which this may occur. (a) Heterozygotes may be of superior fitness, but because of the mechanics of Mendelian genetics they continually generate less fit homozygotes within the population. Such 'heterosis' is seen in human sickle-cell anaemia where malaria is prevalent. Sickle-cell heterozygotes suffer only slightly from anaemia and are little affected by malaria; but they continually generate homozygotes that are either dangerously anaemic or susceptible to malaria. (b) There may be gradients of selective forces selecting for one form (morph) at one end of the gradient, and another form at the other. This can produce polymorphic populations at intermediate positions in the gradient. (c) There may be frequency-dependent selection in which each of the morphs of a species is fittest when it is rarest. This is believed to be the case when rare colour forms of a prey are fit because they go unrecognized and are therefore ignored by their predators. (d) Selective forces may operate in different directions within different patches of a fine mosaic in the population. In this case the maintenance of a fit between organisms and environment has to depend on large numbers of propagules being dispersed, so that each type has a reasonable chance of landing in the part of the patchwork in which it is fittest. Alternatively, the organism may be long-lived and able to wander between and exploit the appropriate parts of the patchwork.

A striking example of polymorphism in a natural population is provided

the active maintenance of genetic polymorphisms

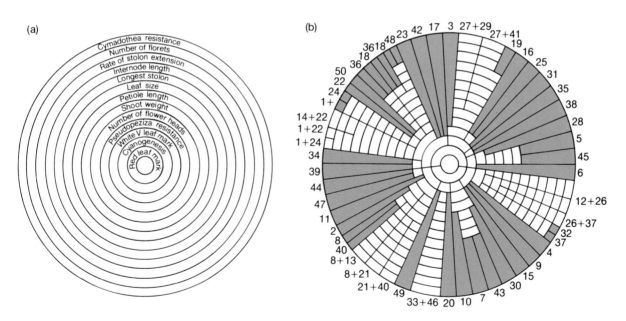

Figure 1.15. The diversity of genotypes of 50 plants (1–50) of white clover (*Trifolium repens*) sampled from a one hectare field of permanent pasture. The variety of factors examined and under known genetic control is shown in the left-hand diagram. On the right-hand figure we start at the centre with 50 plants and each genetic character is used in turn to subdivide this population. Thus the initial population is divided into approximately equal numbers with and without a red leaf mark. Each of these populations is then further subdivided into four classes according to the genotype controlling hydrogen cyanide production. The process continues to the outermost circle, which divides resistance from and susceptibility to the disease *Cymadothea*. Each unique genotype identified in this way is represented by a shaded sector. (After Burdon, 1980.)

by a series of studies of white clover in a field of permanent grassland (at least 80 years old) in North Wales. In one study, 50 plants of this species were sampled from the field. The plants were grown in a glasshouse, and each plant was described with respect to the genes it carried for a number of characters, almost all of which were of known selective importance (Figure 1.15). Among the 50 clones, any one differed from any other in a combination of qualities that might be expected to affect its fitness in the field.

Turkington and Harper (1979) attempted to test how far the different variants in the clover population represent local matches with local features of the field. They removed plants from the field, multiplied them in the glasshouse, and then transplanted them back into the field. They returned a sample of each clone to the place from which it had been taken, and also to the places from which the others had come. Samples had been taken from patches dominated by different species of pasture grass. Each clover plant was transplanted back into patches dominated by its own grass species and into patches of the other grass species. Each type of clover grew best when it was returned to the species of grass neighbour from which it had been taken.

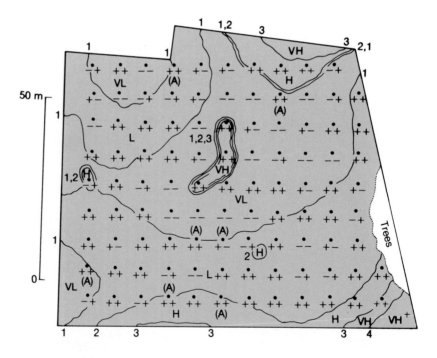

Figure 1.16. A map of the distribution of the slug *Dendroceras* in a small field of permanent grassland. The contours represent the density of slugs and distinguish zones with very high (VH), high (H), low (L) and very low (VL) densities. On the same map are shown the genotypes of white clover sampled at intersections of a grid across the field. Plants are distinguished as ++ (*AcLi*), +− (*Acli*), −+ (*acLi*) and −− (*acli*). Only ++ plants release HCN when they are damaged; they carry the alleles *Ac* and *Li*, which determine the production of the cyanogenic glycoside and the enzyme that releases HCN from it. X^2 test of the null hypothesis that clover genotypes and the density of slugs are unrelated firmly rejects the hypothesis. (From Dirzo & Harper, 1982.)

This was strong and direct evidence that clover genotypes in the pasture were distributed in such a way that they matched their local biotic environment.

In the same field, Dirzo and Harper (1982) studied the distribution of a specific polymorphism in white clover which controls whether the plant releases hydrogen cyanide when it is damaged (e.g. nibbled). Molluscs are known to avoid eating cyanogenic clover, and a highly significant association was found within the field between the distribution of cyanogenic forms and areas of high mollusc density (Figure 1.16).

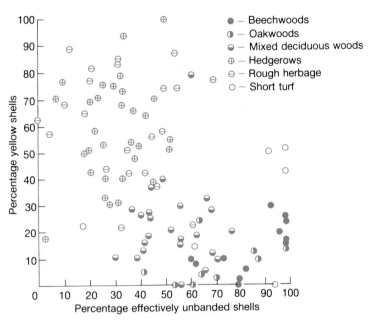

Figure 1.17. The distribution of yellow (as opposed to pink and brown) and unbanded (as opposed to banded) shells of the snail *Cepaea nemoralis* in different types of habitat. There are good matches between the commonest morph and its background: for example pink and brown shells in beech woodland; yellow-banded (effectively green) shells in hedgerows. (After Cain & Sheppard, 1954.)

Clearly, some at least of the variation in white clover within this small field is genetic, and represents a match between organisms and environment.

polymorphism in a snail

An example of polymorphism in an animal is provided by the snail *Cepaea nemoralis*. Snails have low mobility, and their populations may extend over a variety of habitats where the pressures from predators and the physical features of the environment differ sharply. This has led to the local differentiation of populations with different shell markings (Figure 1.17). Each population is polymorphic with respect to one at least of two characters, shell colour and shell banding. The percentage of different morphs in different habitats varies, but the commonest morph in each habitat shows a good match to its own background. They appear to be camouflaged from marauding thrushes against the backgrounds of the different soils and vegetations.

1.6 The match of organisms to varying environments

No environment is constant over time, but some are more constant than others. No form or behaviour of an organism can fit a changing environment unless it too changes. Three major categories of environmental change can be recognized.

(a) Cyclic changes—rhythmically repetitive, like the cycles of the seasons, the movements of the tides and the light and dark periods within a day.

(b) Directional changes—in which the direction of a change is maintained over a period that may be long in relation to the life-span of the organisms that experience it. Examples are the progressive erosion of a coastline, the progressive deposition of silt in an estuary and the cycles of glaciation.

(c) Erratic change—this includes all those environmental changes that have no rhythm and no consistent direction: for example, the variation in the time of arrival of monsoon rains, the erratic course and timing of hurricanes and cyclones, flash storms, and fires caused by lightning.

organisms may respond directly to change or utilize a cue

The optimal fit of organisms to varying environments must involve some compromise between matching the variation and tolerating it. The repeated experience of cyclic change by successive generations of an organism has selected many patterns of behaviour that are in themselves cyclic: diapause in insects, the annual shedding of leaves from deciduous trees, the diurnal movements of leaves, the rhythm of tidal movement in crabs, the annual cycle of breeding systems and the seasonal cycle of fur thickness in mammals.

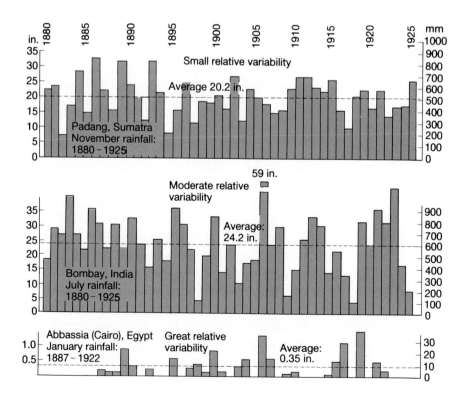

Figure 1.18. Variability and unpredictability in rainfall. The amount of rain received each year at three locations, during the month which, on average, is the rainiest at that location. (From Stahler, 1960. Data from H. H. Clayton, Smithsonian Institute.)

There are two main ways in which organisms time their responses to variations in their environment: (a) by changing in response to the environmental change or (b) by using a cue that anticipates the change. If the cycle in conditions is weak and contains much variation, the organisms may best match the changing conditions by responding to them directly. Figure 1.18 illustrates this unpredictability of rainfall, even at locations where there is a marked seasonal rhythm. This variation can be seen dramatically in the bottom diagram, which shows rainfall figures for Cairo in January. Under such circumstances a plant that is cued to germinate at the start of the 'rainy season' would rarely survive. Instead, most desert plants germinate in direct response to the arrival of rain, rather than responding to some cue that indicates that the rainy season is coming.

There is, however, a disadvantage, a price to be paid by organisms that respond directly to environmental change. A mammal that changes the thickness of its coat as a reaction to the weather becoming cold will have to shiver until the process of replacing its coat has taken place. But if it reacts not to the onset of cold, but to an environmental cue that is correlated with, and therefore predicts, the onset of cold, such as the shortening of day-length, it may start to develop the thicker coat in advance of the event. The use of such cues is common among both plants and animals that live in environments in which there is a strong and repeated cycle of environmental change, and where the variation in the cycle is relatively weak.

There is a temptation to write about organisms as if they make predictions (or about environments having predictability) and about cues that enable the organism to plan ahead. Patterns of cyclic behaviour give the appearance of being predictive—but of course they are not. They are characteristics selected by the repeated experiences of the past. The form and behaviour of an organism is programmed by its ancestors' experience of life and death in the past, rather than by its own anticipation of the future.

Events like the darkness of night and the cold of winter are tied to the passage of time, and they can be measured and timed by using the sun itself as a clock (the height of the sun during the day, or the length of the photoperiod during a year). Some organisms have added to this external clock their own internal physiological clock (an 'endogenous rhythm'). Such clocks require to be repeatedly set right by reference to the solar clock, but they may continue for long intervals without such a reference.

the way change is experienced depends on the length of the life cycle

The way in which any organism experiences the rhythmic changes of the environment depends on the length of its life cycle. One cell in a multiplying population of a unicellular alga like *Chlorella* may live over only one cycle of day and night before dividing to form two other cells. An oak tree, an elephant, a whale or a man includes within one life cycle the repeated experiences of spring, summer, autumn and winter, and the associated famines and gluts in the seasonal rarity and abundance of food. If the life cycle is short, the organism may spend the whole of its active life concentrated in a small segment of the year, and may be constrained within a sharply defined programme of development and behaviour restricted to one short season. The gall wasps have just this type of life, emerging at a time that is tightly synchronized with the availability of the leaf or flower of their particular host species. If, on the other hand, the life cycle is long relative to the annual cycle of the seasons, the organism cannot so easily become specialized. It may endure the periods of drought or cold by aestivating or hibernating, sleeping if it is a dormouse or dropping its leaves if it is a deciduous tree; or it may have an all-purpose physiology that merely continues faster or slower through the seasons (as does the polar bear or the evergreen tree). There must be a conflict in the evolution of most organisms between being active for a very short part of the year, tightly matching the environment of that part, and being a generalist, a jack of all trades but perhaps master of none.

birds and mammals often change or move with the seasons

The life cycles of most birds and mammals are longer than a calendar year, and many of the smaller mammals in seasonal climates pass through winter by hibernating: lowering their rate of metabolism and resting in protected places. However, there are many others that meet the variation in conditions by varying the extent to which their bodies are insulated. The most dramatic changes of fur thickness are found in the larger mammals (Hart, 1956). The winter fur of the black bear has an insulating power 92% greater than that of its summer coat. This sort of cyclic change in thickness (and colour) of insulation is characteristic of the seasonal responses of animals that cannot escape from the adverse seasons; but by far the most conspicuous seasonal rhythms in the behaviour of mobile animals involve movement from place to place. This may be movement to a place of shelter, or migration to another climate (the annual migrations of reindeer, bison and many birds). But, of course, such movement is denied to the rooted higher plant or sessile animal, and it is in these organisms that some of the most profound seasonal rhythms can be seen, affecting most particularly their form.

variation and somatic polymorphism in aquatic plants

For many aquatic plants, changes in the depth of the water represent a major seasonal cycle in the environment; but these seasonal rhythms may be affected by erratic change during serious drought or in flash floods. In the zoned habitat of the water's edge there is a variety of characteristic plant forms. At the margins, most plants bear aerial foliage that is photosynthetically inefficient or even damaged when it is submerged. In somewhat deeper water, species commonly bear floating foliage. In deeper water still, species have entirely submerged foliage with long, flexible, band-like leaves that are not damaged in a fast water-flow. And in the deep water zone, there are species that have submerged, finely dissected foliage. There are many species that bear one or other of these specialized leaf types. In addition, there are many aquatic plants capable of bearing more than one sort of leaf on a single plant. Such *somatic polymorphism* allows these species a potentially wider range of water levels at which they can function effectively.

Even within the single genus *Ranunculus*, subgenus *Batrachium* (the water crowfoots), various types of match to a varying environment can be found. There are species that are *monomorphic*, and match the water level that they most commonly experience. Two species, *Ranunculus hederaceus* and *R. omiophyllus*, bear only floating foliage and are usually found growing in mud or shallow water. Three species, *R. fluitans*, *R. circinatus* and *R. trichophyllus*, produce only submerged, finely dissected leaves. These species are usually found in deeper or fast-flowing water.

Within a further group of *Ranunculus* species there are differences in the ways by which the transition from one leaf form to another is initiated. In *R. aquatilis*, the type of leaf is determined by the photoperiod and the temperature at the time of leaf initiation, *not* by the level of the water but by a seasonal cue (see, for example, Bradshaw, 1965). The plant begins growing in spring, and grows for a time producing submerged leaves. It reaches the surface, but may continue extending its submerged leaves for some time. Then, in response to the stimulus of temperature and photoperiod, it suddenly produces floating leaves on the surface. Yet if such plants are kept under a regime of short days, they continue to produce finely divided leaves, irrespective of whether they are submerged or exposed. In other species, for instance *R. flabellaris*, it is the prevailing conditions themselves that influence the type of leaf that is developed. The form of leaf (either floating or submerged and band-like) is determined by the conditions affecting the plant when the leaves are initiated, and if the environment suddenly changes, the plant finds itself with the 'wrong' leaf form when the water level falls.

somatic polymorphism in desert plants

For an organism that cannot run away from adverse conditions, seasonal changes in its form may be the most effective solution to problems of survival in a changing environment. In arid environments, somatic polymorphism may be even more extreme than in the case of aquatic plants. Some species produce three crops of leaves within a year, each of different morphology. In *Teucrium pollium*, relatively large leaves are formed in the wetter season. These fall off and are replaced by small leaves or scales during the drier season. And this second foliage may itself be lost, so that the plant spends the driest period with only green stems and thorns (Orshan, 1963).

1.7 Pairs of species

Some of the most strongly developed matches between organisms and their environment are those in which one species has developed a dependence upon another. This is the case in many relationships between consumers and their foods, such as the dependence of koala bears on *Eucalyptus* foliage or giant pandas on bamboo shoots. Whole syndromes of form, behaviour and metabolism constrain the animal within its narrow food niche, and deny it access to what might otherwise appear suitable alternative foods. Similar tight matches are characteristic of the relationships between some parasites and their hosts (Chapter 11). The host-specific rust fungi, for example, fit narrow and precisely defined environments: their unique hosts.

Where two species have evolved a mutual dependence, the fit may be even tighter. The mutualistic association of nitrogen-fixing bacteria with the roots of leguminous plants, and the often extremely precise relationship between insect pollinators and their flowers are two good examples (Chapter 12). The closest matches between organisms and their environments have evolved where the most critical factor in the life of one species is the presence of another: the whole environment of one organism may then be another organism.

When a population has been exposed to variations in the physical factors of the environment, for example a short growing season, a high risk of frost or drought, or the repeated application of a herbicide, a once-and-for-all tolerance may ultimately evolve. The physical factor cannot itself change or evolve as a result of the evolution of the organisms. By contrast, when members of two species interact, the change in each produces alterations in the life of the other, and each may generate selective forces that direct the evolution of the other. In such a *coevolutionary* process the interaction between two species may continually escalate. What we then see in nature may be pairs of species that have driven each other into ever-narrowing ruts of specialization.

When we look at the diversity of nature it is difficult not to be overwhelmed by feelings of wonder, admiration and enchantment at what can so easily be interpreted as perfection. We need to remember that the abilities to wonder and admire are special features of our own biology! As a scientist the ecologist searches for causes and effects and must not be satisfied by 'explanations' that seek only to show how, at this moment, all is 'for the best in the best of all possible worlds'.

2 Conditions

2.1 Introduction

In order to understand the distribution and abundance of a species we need to know many things: the history of that species (Chapter 1), the resources that it requires (Chapter 3), the individuals' rates of birth, death and migration (Chapters 4 and 5), their interactions with their own and other species (Chapters 6 to 12) and the effects of environmental conditions. This chapter deals with the limitation of organisms by environmental conditions.

We define a condition as an abiotic environmental factor which varies in space and time, and to which organisms are differentially responsive. Examples include temperature, relative humidity, pH, salinity, stream flow velocity and the concentration of pollutants. A condition may be modified by the presence of other organisms: soil pH may be changed by the presence of plants; temperature and humidity may be altered under a forest canopy. But unlike *resources* (Chapter 3), conditions are not consumed or used up by an organism or made unavailable or less available to others.

Ideally, we could define for a given species an optimum concentration or level of a condition at which it performs best, with its biological activity tailing off at both lower and higher levels (Figure 2.1). The problem comes in defining 'performs best'. The optimal conditions should be those under which the individuals of the species leave most descendants (are fittest), but

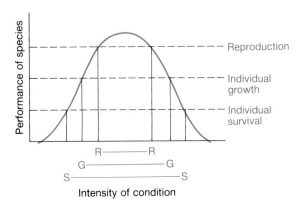

Figure 2.1. A generalized graphical representation of the manner in which the performance of a species is related to the intensity of an environmental condition. The narrow range over which reproduction can occur (R–R) usually dictates where continued existence of the species is possible (though some plants can apparently persist indefinitely by vegetative growth alone).

these are usually extremely difficult to determine in practice. Instead we have to take short cuts. We measure the effect of conditions on some special, chosen properties like growth rate, reproduction, respiration rate or survival. However, the effect of a range of conditions on these various properties will often not be the same; for example, organisms can usually survive over a wider range of conditions than permit them to grow or reproduce (Figure 2.1). Moreover, the precise shape of the curve of response to a condition—whether it is symmetrical or skewed, broad or narrow—will vary from condition to condition, from species to species, and will depend on which of the organism's responses we have chosen to measure.

In this chapter we consider one condition—temperature—in some detail. This is perhaps the single most important condition affecting the lives of organisms. We derive some general principles and then consider other conditions in the light of these principles.

2.2 Temperature and individuals

2.2.1 Classifying relationships

When examining the relationships between organisms and environmental temperature, it is usual to divide organisms into two types. One possible division is between the 'warm-blooded' and the 'cold-blooded', but these terms are subjective and inaccurate; they will not be used here. A more satisfactory classification divides organisms into *homeotherms* and *poikilotherms*: as environmental temperature varies, homeotherms maintain an approximately constant body temperature, while poikilotherms have one that varies. One major problem with this classification, however, is that even classic homeotherms like birds and mammals experience reduced temperatures during periods of hibernation or torpor, while some poikilotherms (e.g. antarctic fish) experience temperature variation of only tenths of a degree because their environmental temperature hardly varies. Moreover, many supposed poikilotherms have at least some regulatory powers, even if

organisms can be divided into homeotherms and poikilotherms

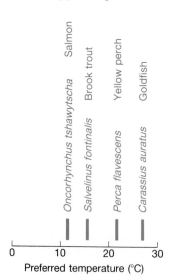

Figure 2.2. Characteristic preferred temperatures at which individuals congregate in a laboratory temperature gradient for four species of fish previously acclimated at 15°C (from Ferguson, 1958).

these only involve the behavioural response of moving in the appropriate direction along a temperature gradient. For example, individuals of four species of fish, when placed in a laboratory temperature gradient, congregate at characteristic preferred temperatures (Figure 2.2).

organisms can be divided into endotherms and ectotherms

An even more satisfactory distinction, therefore, is between *endotherms* and *ectotherms*. Endotherms regulate their temperature by the production of heat within their own bodies; ectotherms rely on external sources of heat. Very broadly, this is a distinction between the birds and mammals in the first case, and the other animals, plants, fungi and protists in the second case, but even here the distinction is not clear cut. There are a number of reptiles, fish and insects (e.g. certain bees, moths and dragonflies) that utilize heat generated in their own bodies in order to regulate body temperature for limited periods; in some plants, metabolic heat maintains a relatively constant temperature in the flowers (e.g. *Philodendron* and the skunk cabbage); and there are some birds and mammals that relax or suspend their endothermic abilities at the most extreme temperatures (Bartholomew, 1982).

Despite all these reservations, the distinctions between endotherms and ectotherms and between homeotherms and poikilotherms can be useful, and between them they represent a basis from which to proceed. Here we begin by discussing ectotherms in some detail before moving on to consider endotherms in section 2.2.11.

2.2.2 Heat exchange in ectotherms

All organisms gain heat from and lose heat to their environment as well as producing heat themselves (if only as a by-product of metabolism). Typical pathways of heat exchange for an ectotherm are shown diagrammatically in Figure 2.3, but almost all ectotherms modify or moderate the heat that is actually exchanged along one or more of these pathways (see Bartholomew, 1982). Amongst the mechanisms they use, some are fixed properties of particular species (like the reflective, shiny or silvery leaves of many desert plants) and some are simple behavioural responses (such as the shade-seeking at high temperatures of many reptiles); some are more sophisticated patterns of behaviour (like the various sunbasking postures adopted by locusts) and some are complex aspects of their physiology (as when bumble-bees shiver their flight muscles).

ectotherms modify their gain and loss of heat— but only to a limited extent

Despite these properties that prevent the organism from changing its temperature like a lifeless, immobile black box, the body temperature of an ectotherm varies significantly with that of its environment, for three reasons. First, the regulatory powers of many ectotherms (especially plants) are severely limited. Secondly, ectotherms are generally dependent (by definition) on external sources of heat: an animal can only move to a warmer (or cooler) spot if there is a warmer spot to move to, and it can only raise its temperature by basking if the sun is actually shining. Thirdly, there are costs associated with temperature regulation. Energy must be expended on properties that modify the heat budget: for instance in providing a reflective cuticle or finding a suitable location. And an animal that exposes itself to the sun's rays may also be exposing itself to predators. If the costs of a regulated

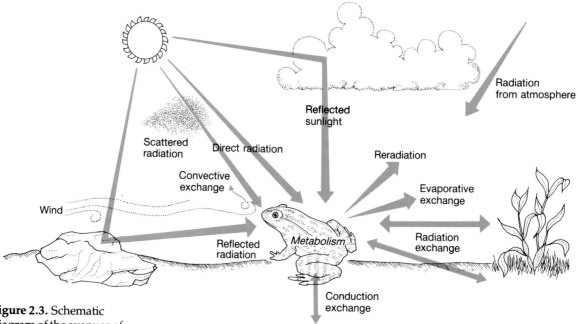

Figure 2.3. Schematic diagram of the avenues of heat exchange between an ectotherm and a variety of physical aspects of its environment (from Hainsworth, 1981, after Gates & Porter, 1970).

temperature, reaction rates, and the number of body processes that take place

temperature in a particular environment exceed the benefits, regulation will be selected against. The extent to which an organism regulates its temperature will therefore be a compromise between costs and benefits.

2.2.3 Temperature and metabolism

The effects of different temperatures on ectotherms follow a typical pattern, though there is variation from species to species. There are, in essence, three temperature ranges of interest: the dangerously low, the dangerously high, and the temperatures in between. Over the middle range, the rates of metabolic reactions increase with temperature (Figure 2.4). There is, characteristically, a fairly good approximation to an exponential relationship which is usually described by a 'temperature coefficient' or Q_{10}. The Q_{10} of 2.5 in Figure 2.4, a typical example, means that for every 10°C rise in temperature the reaction rate increases 2.5 times. An obvious consequence is that ectothermic organisms take in and metabolize resources only slowly at low temperatures, but much more rapidly at higher temperatures. Another common consequence is that certain processes (like respiration) occur throughout the temperature range (albeit at varying rates), but other processes, which perhaps demand a more rapid and consistent flow of energy and matter, only occur at relatively high temperatures. Typically, reproduction in ectotherms occurs over a more restricted range of temperatures than growth, which occurs over a more restricted range of temperatures than mere survival.

2.2.4 Physiological time: the day–degree concept

Within the non-lethal range, arguably the most important effect of temperature on ectotherms from an ecological point of view is its effect on rates of

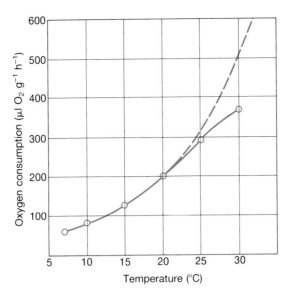

Figure 2.4. The rate of oxygen consumption of the Colorado beetle increases with temperature (solid line). Over most of the range the rate increases 2.5 times for a 10°C rise in temperature (Q_{10} = 2.5). The broken line shows the expected line had Q_{10} stayed constant at 2.5 rather than decreasing at the highest temperatures. (From Schmidt-Nielsen, 1983, after Marzusch, 1952.)

development and growth. Two typical examples are illustrated in Figure 2.5, and their most important feature is readily apparent: when rate of development is plotted against body temperature there is an extended range of temperatures over which the relationship is linear. Moreover, the deviation from linearity at the lowest temperatures is only minor; and organisms typically spend almost all of their time below the high-temperature non-linear region. It is therefore often assumed simply that the rate of development rises linearly with temperature above a developmental threshold (Figure 2.5b).

Figure 2.5 also illustrates the most significant consequence of the temperature–development relationship. In Figure 2.5a, for instance, development to hatching takes 17.5 days at 20°C (4°C above threshold), but only 5 days at 30°C (14°C above threshold). At both temperatures, therefore, development requires 70 *day–degrees* (or, more properly, 'day–degrees above threshold'), i.e. 17.5×4 = 70 and 5×14 = 70. This is also the requirement for development in the grasshopper at other temperatures within the non-lethal range. In a similar way, the butterflies in Figure 2.5b require 174 day–degrees above their threshold to complete their development. Thus, unlike ourselves for instance, and endotherms in general, ectotherms cannot be said to require a certain length of *time* for development. What they require is a combination of time and temperature, and this combination is often referred to as *physiological time*. To put it another way: time is temperature-dependent for ectotherms, and it can indeed 'stand still' if the temperature drops below the threshold.

The importance of the day–degree (or, generally, time–temperature) concept lies in the ability it gives us to understand the timing of events, and

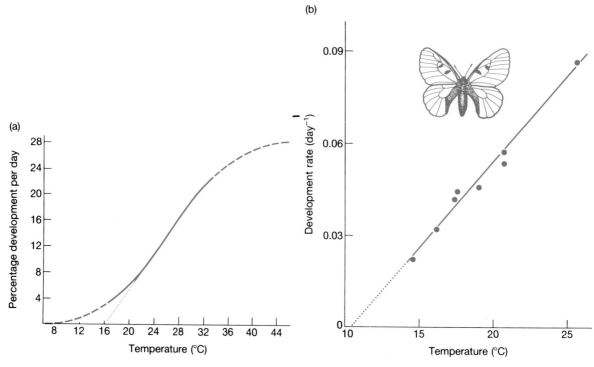

Figure 2.5. Development requires a given amount of physiological time. (a) Development of post-diapause eggs of the grasshopper *Austroicetes cruciata* requires 70 day–degrees above a developmental 'threshold' of 16°C; the relationship departs from linearity only at the lowest and highest temperature (from Davidson, 1944). (b) Development of the cabbage white butterfly, *Pieris rapae*, from egg hatch to pupa requires 174 day–degrees above a threshold temperature of 10.5°C (from Gilbert, 1984).

thus the dynamics of ectotherm populations. In the field, the grasshoppers and the butterflies from Figure 2.5 will emerge and appear in large numbers on different dates each year. But they will always appear when approximately the same number of day–degrees above threshold have accumulated since the inception of development during the winter. Their appearance can therefore only be understood, and can certainly only be predicted, on the basis of a physiological time-scale—a scale that measures time from the organism's not our own (homeothermic) point of view.

In practice there can be considerable difficulties in determining a particular organism's physiological time-scale, especially when the effects of fluctuating temperatures are taken into account; the linear relationship itself is never more than an approximation; and there are always problems in monitoring the effective body temperature of an organism in the field. Overall though, it is important to at least bear the concept of physiological time in mind. All ecologists must learn to see the world through the eyes of the organism being studied.

2.2.5 Temperature as a stimulus

We have seen that temperature as a condition affects the rate at which ectotherms develop. It may also act as a stimulus, determining whether or not the organism starts its development at all. For instance, for many species of temperate, arctic and alpine herbs, a period of chilling or freezing (or even of alternating high and low temperatures) is necessary before germination will occur. A cold experience (physiological evidence that winter has passed) is required before the plant can start on its cycle of growth and development. Temperature may also interact with other stimuli (e.g. photoperiod) to break dormancy and so time the onset of growth. The seeds of the birch (*Betula pubescens*) require a photoperiodic stimulus before they will germinate, but if the seed has been chilled it starts into growth without a light stimulus. However, the temperatures that break dormancy are often quite different from those that control the subsequent rates of growth and development. All of these comments apply in a similar fashion to the effects of temperature on diapause (essentially dormancy) in insects (see Chapter 5).

2.2.6 Acclimatization

The responses of an individual ectotherm to temperature are not fixed: they are influenced by the temperatures it has experienced in the past. Exposure of an individual to a relatively high temperature for several days (or even less) can shift its whole temperature response upwards along the temperature scale (Figure 2.6); and several days' exposure to a relatively low temperature can shift its response downwards. The process is usually referred to as *acclimation* if the changes are induced by laboratory conditions (as in Figure 2.6) and as *acclimatization* if they occur naturally. Note though that the temperature difference between the response curves is less than the difference between the previous regimes: there is compensation but it is only a partial compensation. In nature, the changes may improve the match between an ectotherm's temperature response and the changing temperature of its environment, but acclimatization takes time. Moreover, too rapid an acclimatization may be disastrous. An organism that has altered its tolerances after a period of warm weather may suffer as a consequence if there is a sudden cold spell.

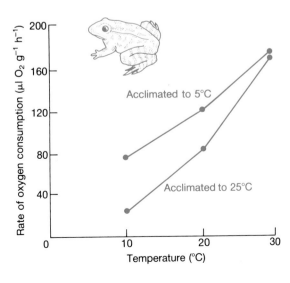

Figure 2.6. The rate of oxygen consumption of frogs (*Rana pipiens*) at a given temperature depends on the temperature of acclimation (from Rieck *et al.*, 1960).

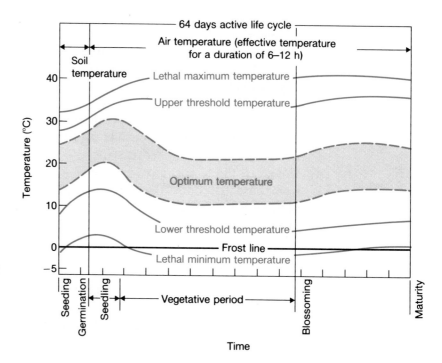

Figure 2.7. The temperature responses of canning peas (*Pisum sativum*) are different at different stages of their development (after Wang, 1960).

Acclimatization aside, individuals commonly vary in their temperature response depending on which stage in their development they have reached (e.g. Figure 2.7). Probably the most extreme form of this is when an organism has a dormant stage in its life cycle which typically differs from other stages in being more tolerant of extremes of temperature and in its metabolic response to temperature generally.

2.2.7 High temperatures

Perhaps the most important thing about dangerously high temperatures is that they usually lie only a few degrees above the metabolic optimum. This is largely an unavoidable consequence of the physico-chemical properties of enzymes (see Heinrich, 1977). High temperatures may thus be dangerous in that they lead to the inactivation or even the denaturation of enzymes. They can also be dangerous if they unbalance the various components of metabolism; plants at high temperatures, for instance, respire faster than they photosynthesize, and they therefore 'starve' because they consume metabolites faster than they produce them (Sutcliffe, 1977). In practice though, high temperatures often exert their ill effects indirectly by leading to dehydration. All terrestrial organisms need to conserve water, and at high temperatures the rate of water loss can be lethal. But counteracting this by protecting evaporative surfaces (e.g. closing stomata in plants or spiracles in insects) is also dangerous, since evaporation is an important means of reducing body temperature. Thus the significance of high temperatures in cases like this is crucially dependent on the relative humidity of the organism's environment: the lower this is, the greater the dangers of dehydration. In this sense, temperature and humidity cannot really be treated as separate conditions.

death at high temperatures can result from enzyme inactivation, metabolic imbalance, or dehydration

Death at high temperatures also results from an interaction between temperature and exposure time. For instance, many ectothermic animals have a range of high temperatures where a 'heat coma' is induced from which they recover (apparently unaffected) if the temperature is subsequently reduced. It is only at even higher temperatures that death occurs rapidly.

As with temperature responses generally, the reactions to high temperatures are subject to the effects of acclimatization. There may also be stages in an organism's life-history that are particularly resistant to the effects of high temperature. This is especially true of dormant structures such as the resting spores of fungi, cysts of nematodes, and seeds of plants, probably as a result of their naturally dehydrated state; dry wheat grains can withstand 90°C for up to 10 minutes, but after being soaked for 24 hours, 60°C kills them in about 1 minute (Sutcliffe, 1977).

2.2.8 Low temperatures

There is considerable variation amongst ectotherms in what constitutes a dangerously high temperature, but generally less variation at the other end of the scale. Many species are killed by temperatures below approximately −1°C because of the damaging effects of the formation of ice crystals, especially within cells. Those species that survive at temperatures lower than this (and often much lower) do so because they have mechanisms that prevent the formation of ice crystals within their cells. The crystals themselves may actually damage cell walls or other structures, but more important is the fact that the crystals effectively absorb water, leaving behind a solution which may be concentrated to a dangerously high extent. Some ectothermic animals that repeatedly or habitually experience low temperatures accumulate solutes which act as an anti-freeze, preventing crystal formation. Sutcliffe (1977) discusses the mechanisms conferring low-temperature tolerance in plants.

Low-temperature tolerance in plants is, in almost every case, associated with a period of acclimatization or what horticulturalists call 'hardening'. A plant may be unable to withstand any frost in the summer, but by, say, mid-October in the northern hemisphere it will have acquired resistance; it will have hardened (Figure 2.8). Moreover, resistance to freezing injury alters markedly with the plant's stage of development. Most seeds, even of frost-sensitive species, are resistant to temperatures that would kill a young seedling. Similar comments apply to ectothermic animals: those that live through freezing winters do so as a specific resistant, dormant stage of their life cycle.

Temperatures though can be lethal without dropping as low as freezing. Generally, at sufficiently low temperatures, metabolic reactions slow down and virtually stop. Ectothermic animals become moribund, and ectotherms of all types effectively cease to carry out the normal functions of maintenance and repair (to say nothing of growth and reproduction). This is often not harmful in the short term, but in the longer term it may so weaken an organism that it dies or is at least made more susceptible to all other sources of mortality.

More specifically, many plants are liable to injury by *chilling* at tem-

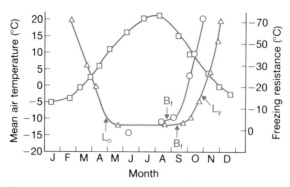

Figure 2.8. Hardening in plants: seasonal changes in the freezing resistance of *Salix pauciflora* (O———O) and *Salix sachalinensis* (△———△) on Mount Kurodake, and mean air temperature (□———□) throughout the year. L_o = leaves opening; L_y = leaves yellowing; B_f = buds forming. (From Sutcliffe, 1977, after Sakai & Otsuka, 1970.)

peratures around 10°C. The probable cause is a disruption of membrane structure, though this often expresses itself as a malfunction in the plant's uptake or retention of water. Unlike freezing injury, exposure to chilling temperatures often needs to be prolonged before injury occurs; there is an interaction between temperature and exposure time similar to that occurring at dangerously high temperatures. There are also many animals that are susceptible to chilling injury, especially those that are not normally exposed to low temperatures.

2.2.9 Interspecific and inter-racial variation

The variations in tolerance that occur within individuals are relatively slight compared with the differences between species that habitually occupy different types of habitat. For instance, some species of fish are as active at 5°C as others are at 30°C, and some live in polar waters at between −1 and −2°C while others live in shallow, tropical waters near 40°C. There are flies that recover repeatedly from exposure to temperatures of −6°C, and water beetles that live in water as hot as 50°C. And there are algae that live and reproduce at temperatures above 70°C, and mosses and lichens that can withstand temperatures of −70°C. The potentialities of living matter in the hands of evolution far exceed those of any one species.

However, even within species there are often differences in temperature response between populations from different locations, and these differences have frequently been found to be the result of genetic differences rather than being attributable solely to acclimatization (Figure 2.9). It can therefore be misleading on several counts to think of a species as having *one* temperature response. In practice though, the effects of geographic differentiation are usually limited (Wallace (1960) discusses why this may be so), and most species have a temperature response by which they can be characterized. This limits the temperature range within which they can live, and therefore limits the range of environments that they can occupy.

There are, however, striking cases where the geographic range of a crop has been extended by deliberate selection for cold tolerance. For instance, the breeding of shorter-season and more cold-tolerant varieties of corn has greatly extended the area of the United States over which the crop can be profitably grown. Over the period 1920–30 to 1940–49, for example, the production of corn in Iowa and Illinois increased by 21.6 and 27.3%, whereas that in the colder state of Wisconsin increased by 54.3%. Indeed, in 1953 the average grain yield in Wisconsin exceeded that of any other state for the first time in history, and a dramatic extension of range has continued. It is dangerous to assume that the tolerances of a species are immutably fixed. The natural selection operating during glacial or inter-glacial episodes was probably as potent as any activity of a modern plant breeder.

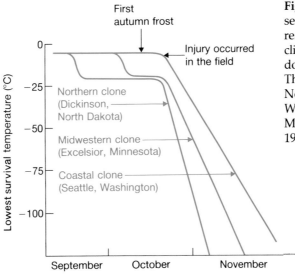

Figure 2.9. Variation in the seasonal patterns of cold resistance shown by three climatic races of red-osier dogwood, *Cornus stolonifera.* The curves are for clones from North Dakota, Minnesota and Washington grown in Minnesota. (After Weiser, 1970.)

2.2.10 A resumé for ectotherms

An overall picture for ectotherms can now be drawn. They are liable to be killed by even short exposures to very low temperatures and by more prolonged exposure to moderately low temperatures (although the exact temperature depends on the stage in development and the species concerned). Temperatures only a few degrees higher than the metabolic optimum are also liable to be lethal. However, just as important as these 'extreme' responses is what may happen at intermediate temperatures. At a lower than optimum temperature, an organism may be slow to obtain its resources and inept at escaping its predators. Most significant of all, it may be capable of growing, developing or reproducing only slowly. The effects of temperature on these different processes are integrated into effects on the organism's whole life cycle and its ability to leave descendants. An ectotherm therefore has an optimal environmental temperature, and it has upper and lower lethal limits of temperature which may be especially important at a particular stage of development. But on either side of its optimum, the environmental temperature becomes increasingly less conducive to the prolonged existence of the organism. These non-lethal but sub-optimal temperatures are of crucial importance.

2.2.11 Endotherms

The picture for endotherms is in fact not fundamentally different from that for ectotherms. They differ in the extent to which they are able to maintain a constant body temperature, but underlying the endotherm's homeothermy is a relationship like the one in Figure 2.10. Over a certain temperature range (the thermal neutral zone) an endotherm consumes energy at a basal rate. But at environmental temperatures further and further above or below that zone, the endotherm consumes more and more energy in maintaining a constant body temperature. Moreover, even in the thermal neutral zone an endotherm typically consumes energy many times more rapidly than an ectotherm of comparable body size.

endotherms regulate temperature effectively—but they can expend large amounts of energy in doing so

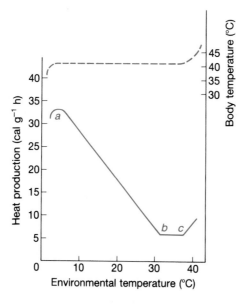

Figure 2.10. Thermostatic heat production by an endotherm increases between the lower critical temperature (*b*) and the maximum possible rate of heat production (*a*). At environmental temperatures above the lower critical temperature, metabolic rate is constant in the thermoneutral zone (to *c*). Above the thermoneutral zone, metabolic rate and body temperature both increase. (After Hainsworth, 1981.)

Endotherms produce heat at a rate controlled by a thermostat in the brain. They usually maintain a constant body temperature between 35 and 40°C, and they therefore tend to lose heat to their environment; but this loss is moderated by insulation in the form of fur, feathers and fat, by controlling blood flow near the skin surface, and so on. And when it is necessary to increase the rate of heat loss, this too can be achieved by the control of surface blood flow, and by a number of other mechanisms shared with ectotherms like panting and the simple choice of an appropriate habitat (Bartholomew, 1982). Together, all these mechanisms and properties give endotherms a powerful (though not a perfect) capability for regulating their body temperature (Figure 2.10), and the benefits they obtain from this are a constancy of performance and a much greater 'peak' or 'burst' performance. But the price they pay for this capability is a large expenditure of energy, and thus a concomitantly large requirement for food to provide that energy.

It would be misleading to say that ectotherms are 'primitive' and endotherms 'advanced'. Rather, endotherms have a high-benefit strategy which demands a high cost, while ectotherms have a low-cost strategy which may sometimes bring only low benefits. Endotherms, like ectotherms, therefore have an optimal environmental temperature (where costs are lowest), and upper and lower lethal limits beyond which their thermoregulatory powers cannot cope. But on either side of their optimal temperature, the environment becomes increasingly less conducive to their prolonged existence, as it demands an increasingly high cost in return for the standard homeotherm's benefit.

2.3 Environmental temperatures

Having described the effects that different temperatures have on organisms, it is appropriate to examine the variations in temperature that are found in nature. These variations and their effects together define the potential role of temperature in determining the distribution and abundance of organisms. Variations can be described under seven main headings: latitudinal, altitudinal, continental, seasonal, diurnal, microclimatic, and variations

latitudinal and seasonal variation

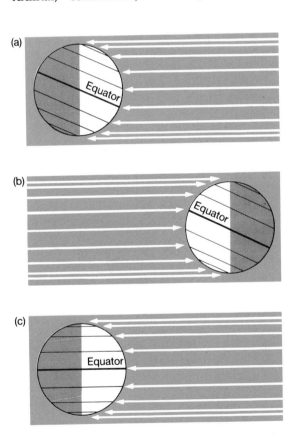

Figure 2.11. (a) Position of earth on 22 June: summer begins in the northern hemisphere; winter begins in the southern hemisphere. Long days towards the north pole, short days towards the south pole. The sun's rays are most concentrated north of the equator. (b) Position of earth on 22 December: the reverse of (a), above. (c) Position of earth on 21 March and 23 September: spring and autumn begin. The length of the day is 12 hours at all latitudes. The sun's rays are most concentrated on the equator itself.

associated with depth. Many of the essential points about these variations are, of course, part of common knowledge.

Latitudinal and seasonal variations cannot really be separated. As Figure 2.11 shows, the angle at which the earth is tilted relative to the sun changes with the seasons. This leads to the very generalized temperature zones shown in Figure 2.12, although it should be realized that the hottest temperatures occur in the middle latitudes rather than at the equator: almost everywhere in the United States has at some time reached well over 38°C, but neither Colon in Panama nor Belem on the equator in Brazil has ever exceeded 35°C (MacArthur, 1972).

Superimposed on these broad geographical trends are the influences of altitude and 'continentality'. There is a drop of 1°C for every 100 m increase in altitude in dry air, and a drop of 0.6°C in moist air. This is the result of the 'adiabatic' expansion of air as atmospheric pressure falls with increasing altitude. The effects of continentality are largely attributable to different rates of heating and cooling of the land and the sea. The land surface reflects less heat than the water, so the surface warms more quickly, but it also loses heat more quickly. The sea therefore has a moderating, 'maritime' effect on the temperatures of coastal regions and especially islands; both daily and seasonal variations in temperature are far less marked than at more inland, continental locations at the same latitude (Figure 2.13). Moreover, there are comparable effects within land masses: dry, bare areas like deserts suffer

altitudinal variation

continentality

Figure 2.12. A simple classification of climates into five major types and their distribution on the surface of the earth. (The Canary Islands are marked with an X—see text.)

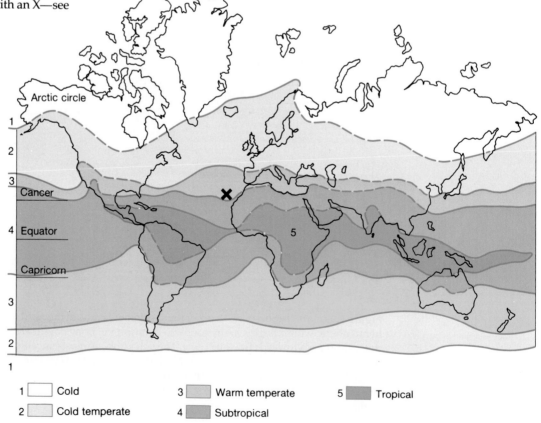

1 ☐ Cold	3 ▨ Warm temperate	5 ▨ Tropical
2 ▨ Cold temperate	4 ▨ Subtropical	

greater daily and seasonal extremes of temperature than do wetter areas like forests.

microclimate

Thus a global map of temperature zones like Figure 2.12 hides a great deal of local variation. It is much less widely appreciated, however, that on a smaller scale still there can be a great deal of *microclimatic* variation. To give a few examples (Geiger, 1955): the sinking of dense, cold air into the bottom of a valley at night can make it 31°C colder than the side of the valley only 100 m higher; the winter sun, shining on a cold day, can heat the south-facing side of a tree (and the habitable cracks and crevices within it) to as high as 30°C; and the air temperature in a patch of vegetation can vary by 10°C over a vertical distance of 2.6 m from the soil surface to the top of the canopy. Hence we need not confine our attention to global or geographical patterns when seeking evidence for the influence of temperature on the distribution and abundance of organisms.

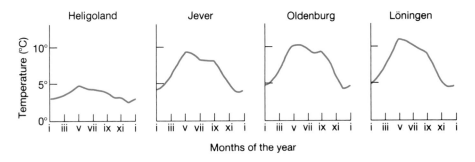

Figure 2.13. Average daily temperature change on the coast and inland. The range increases with distance from the coast as the moderating influence of the sea is lost. Heligoland is an island. Jever, Oldenburg and Löningen are respectively 11, 30 and 80 km from the North Sea coast of West Germany. (From Roth, 1981.)

depth

This is true, too, when we look at the effects of depth on temperature. Specifically, depth (in soil or in water) has two effects on the fluctuations in temperature that occur on the surface. First, the fluctuations are diminished or 'damped', and secondly, related to this, they lag behind the fluctuations at the surface; the strength of these effects increases with depth itself and decreases with the thermal conductivity of the medium (low in the soil but higher in water). A metre or so below the soil surface, daily air temperature fluctuations of many tens of degrees are essentially damped out altogether, and even annual fluctuations are damped out at a depth of a few metres.

2.4 Temperature, distribution and abundance

general evidence for the effects of temperature on distribution

On one level it would be foolish to even question the importance of temperature (and climate generally) in determining the distribution and abundance of organisms. The distributions of the major *biomes* over the earth are obviously a reflection of the major temperature zones: tundra near the poles, coniferous forests slightly nearer the equator, tropical rainforests around the equator and so on (see Chapter 15). Similarly, a climb up a tropical mountain would reveal species of animals and plants appearing and disappearing as an obvious reflection of the gradually declining temperature and the changing climate generally (Figure 2.14). And it has been understood for some time (Hooker, 1866, quoted in Turrill, 1964) that the animals and plants on islands are typically characteristic of the mainland at higher (more polar) latitudes (as with the 'Mediterranean' flora of the Canaries, situated off the Atlantic coast of Africa (Figure 2.12)).

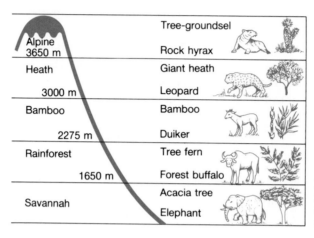

Figure 2.14. Altitudinal zonation of plants and animals on Mount Kenya (from Cox *et al.*, 1976).

However, it is more difficult to attribute a precise role to temperature when single species are examined. In some cases we can relate the distributional limits of a species to a lethal temperature which precludes the species' existence. Injury by frost, for instance, is probably the single most important factor limiting plant distribution (and it is also a major cause of damage to crops). To take one example: the saguaro cactus is liable to be killed when temperatures remain below freezing for 36 hours, but if there is a daily thaw it is under no threat. In Arizona, the northern and eastern edges of the cactus's distribution correspond to a line joining places where on occasional days it fails to thaw (Figure 2.15). Thus the saguaro is absent where there are lethal conditions *occasionally*—an individual need only be killed once.

freezing temperatures and the saguaro cactus

A more complicated distribution of lethal conditions explains why peaches grow well in the Niagara Peninsular of Ontario, Canada. They do so not because it is so warm, but because by the time it *is* warm (and the peaches have acclimatized and lost their hardiness) there is very little chance of a late frost (MacArthur, 1972). Further south, away from the moderating influence of Lake Ontario, the climate is warmer on average, but there are more late frosts and no peaches. This shows that even when lethal conditions control distributions, they may do so in a subtle way.

the subtle effects of frost on peach distribution

A more widespread type of relationship between temperature and distribution is illustrated by two examples in Figure 2.16, both of which show a close (but not perfect) correspondence between the distributional limits of a species and an 'isotherm'. An isotherm is a line on a map that joins places having the same mean temperature at a particular time of year, and it is therefore apparent in Figure 2.16 that distribution is correlated with tempera-

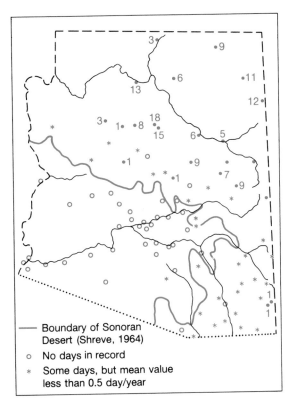

Figure 2.15. There is a close correspondence between the northern and eastern edges of the range of the saguaro cactus in the Sonoran Desert (———) and the line beyond which it occasionally fails to thaw (. . . . .). The numbers are mean numbers of days per year with no rise above freezing. (From MacArthur, 1972, after Hastings & Turner, 1965.)

Boundary of Sonoran Desert (Shreve, 1964)
○ No days in record
* Some days, but mean value less than 0.5 day/year

the wild madder and its isotherm

ture. On the other hand, there is only a limited significance that attaches to the isotherms themselves. The wild madder in Figure 2.16a, for instance, produces new shoots in January for the following spring, and shoot production is undoubtedly inhibited by low temperatures. Yet 4.5°C (the temperature of the isotherm) is not a threshold below which production ceases. Rather, a January average of less than 4.5°C is simply an indication of a relatively cold location—a location, that is, where temperatures are frequently too low for shoot production and a viable population of the wild madder cannot be maintained.

Balanus—and what its isotherm does and does not explain

Another, more complicated example is provided by Figure 2.16b, which shows the relationship between the southern limit of the barnacle *Balanus balanoides* and the winter sea-surface 8°C isotherm (Barnes, 1957; also Lewis, 1976). This arctic species is killed by high air temperatures, approaching 25°C, and it has an optimal range for feeding and respiration that lies between water temperatures of 2°C and 18°C. On the other hand, breeding (which occurs in winter) requires a temperature which is sufficiently low: field observations in North America suggested that gonad maturation and fertilization only take place when air temperatures drop below 10°C and are maintained at that level for some time (perhaps 20 days). Overall, Figure 2.16b suggests that it is this breeding requirement which is of paramount importance, since the species cannot extend further south than the winter sea-surface 8°C isotherm. Yet the connection between the distribution and the isotherm is not direct. It derives only from the fact that winter sea temperatures of this type will have been preceded by suitably low air temperatures in the autumn.

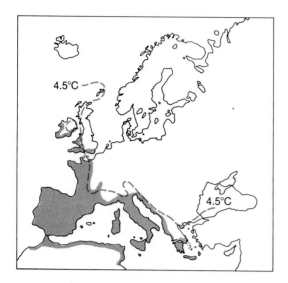

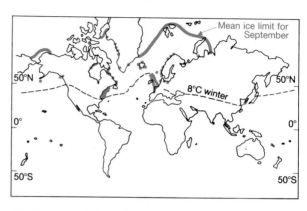

Figure 2.16. (a) The northern limit of the distribution of the wild madder (*Rubia peregrina*) is closely correlated with the position of the January 4.5°C isotherm (from Cox *et al.*, 1976). (b) The southern limit of the distribution of the barnacle *Balanus balanoides* is closely correlated with the winter minimum water-surface 8°C isotherm (after Barnes, 1957).

In fact, in Europe it appears that both temperature barriers are important. South of the southern limit, the summer air temperatures are too high for adult survival and the winter air temperatures are too high for gonad maturation. On the Atlantic coast of North America, however, the seasonal variation in temperature is far greater, and this alters the situation. In South Carolina, for instance, 480 km south of the species' southern limit, the winter conditions are still suitable for breeding. Here, therefore, it is only the summers which are too hot, and the southernmost records are all from north-facing, shaded locations. Moreover, on the Pacific coast of North America, temperature conditions are such that *B. balanoides* would seem able to extend 1600 km south of its present southern limit. The fact that it does not do so is attributed to the adverse effects of competition from indigenous barnacles (*B. balanoides* has migrated to the Pacific via the Bering Strait relatively recently). In other words, the temperature conditions in which *B. balanoides* fails to establish a population because it is outcompeted are far less extreme than those in which it is simply unable to survive, and it is therefore not limited by temperature alone but by the effect of temperature on a biological interaction (i.e. competition).

MacArthur (1972) pointed out that most of us are confronted every day by examples of the same phenomenon. Gardens and arboretums are full of plants that are found naturally in some other, typically much warmer location. They are able to survive and sometimes to flourish, but only if their beds are weeded; that is, only if natural competitors are removed by the gardener. In other words, the natural limits to the distributions of some of these plants are determined not by temperatures which kill, but by conditions that make them inept competitors. Note, however, that individuals of many other exotic species can survive but do not leave descendants.

Competition is not the only biological interaction that combines with temperature to limit distributions. Many animals, for instance, have a distribution which is correlated with temperature, while they are in fact limited by the occurrence or quality of their food. For example, the small moth *Coleophora alticolella* feeds over the Cumbrian moors in England on the seeds of the heath rush, *Juncus squarrosus*, but is absent above an altitude of approximately 600 m. *Juncus* grows above this height and produces flowers, and the adult female moths lay their eggs on the flowers as they do elsewhere; but the temperatures here are generally too low for successful seed production, and the newly hatched moth larvae therefore starve (Randall, 1982). Once again, temperature can be seen to act not directly, but by altering the nature of a biological interaction.

In many cases, variations in temperature are intimately related to variations in some other environmental condition or resource such that the two are not really separable. Probably the most widely understood example of this is the relationship between temperature and relative humidity, which is considered in the next section. Another example, amongst swimming aquatic animals, is the relationship between temperature and oxygen concentration in rivers. Strictly speaking, we must consider oxygen to be a resource rather than a condition because its consumption by one individual reduces, at least potentially, its availability to other individuals. However, in rivers the level of oxygen availability is set mainly by physical forces, being high in upstream regions where turbulence is great and air is effectively mixed with water, and low in downstream regions where higher temperatures mean lower gaseous concentrations. The latter point is most important in the present context; solubility of oxygen in water decreases with a rise in temperature. Table 2.1 lists some fish typical of different regions in British rivers, and also their requirements in terms of temperature and oxygen concentration. There is clearly a close correlation between these environmental factors and the

Table 2.1. There is a close correspondence in British rivers between the temperature and oxygen requirements of different species and the conditions where they are found. Temperatures are lowest and oxygen concentrations are highest in upstream waters; temperatures are highest and oxygen concentrations lowest in downstream waters. (From Varley, 1964.)

Exemplar	Distribution	Oxygen requirement for survival (ml l^{-1})	Upper lethal temperature limit (°C)	Optimum temperature for growth (°C)
Brown trout	Upstream	5–11	< 28	7–17
Pike	Midstream	4	28–34	14–23
Carp	Downstream	0.5	> 34	20–28

species' distributions. An upstream species like the brown trout is limited in its distribution, at least in part, because as temperature increases downstream its requirement for oxygen goes up while the oxygen concentration of the water goes down. Similar comments apply to the downstream limits of midstream species like the pike. In these cases it is impossible to isolate temperature from oxygen concentration: their physiological effects are interrelated and they affect distribution in concert. On the other hand, the upstream limits of these species must be related more to sub-optimally low temperatures, to high velocity flow, or possibly to competition or predation from other species.

Allen's rule, Bergmann's rule, and geographical variations generally

As has already been explained, the effects of temperature on individuals may be moderated by behaviour, by acclimatization, and by evolved differences either between different species or between different races of the same species. Inevitably, these factors also influence distribution and abundance. As examples of the influence of evolved responses, endothermic animals from cold climates usually have shorter extremities (e.g. ears and limbs) than animals with otherwise similar characteristics from warmer climates ('Allen's rule'); and foxes, deer and other mammals with a wide distribution are often larger in colder areas ('Bergmann's rule'). The apparent explanation in both cases is that the endotherms in colder climates should, relative to their volume, have a smaller surface area across which they lose heat, and animals of the same shape necessarily experience a decrease in the ratio of surface area to volume as they get larger. However, while Allen's rule, involving only a change in shape, appears to be very widely applicable, Bergmann's rule, involving body size which must be subject to very many other selective forces, is much less universally true. In a comparable but more specific vein, Figure 2.9 shows the patterns of cold resistance in three climatic races of the red-osier dogwood, when grown under the same conditions in the natural habitat of one of the races (Minnesota, U.S.A.). Most temperate-zone plants are remarkably resistant to midwinter cold, but their cold hardiness is as much a matter of timing as of absolute resistance. Appropriately, resistance was apparent earlier in the year in the race which had evolved in the colder conditions.

close to their geographical limits, species may be restricted to particular microhabitats

As temperature varies geographically, its effects may also be moderated by the restriction of a species to particular microhabitats. This was the case with the barnacle *B. balanoides*. Another example is provided by the rufous grasshopper (*Gomphocerripus rufus*), which is distributed widely in Europe but in Britain reaches its northern limit only 150 km from the south coast. Significantly, the British populations are restricted to steep, south-facing and therefore relatively sun-drenched and warm grassy slopes. Yet in the more southerly and generally warmer parts of its distribution there is no such restriction. Similarly, the prostrate shrub *Dryas octopetala* around its southern limit in north Wales is restricted to altitudes exceeding 650 m, whereas in Sutherland in Scotland it is found right down to sea-level. Thus the geographical limits of a species may be paralleled by microenvironmental limits with the same underlying cause.

a summary of the effects of temperature and of conditions in general

This discussion of temperature can now be summarized in a number of major points. The summary is important, because the principles involved can be applied to any condition (others are discussed briefly in the following pages) simply by replacing 'temperature' with 'humidity', 'salinity' or whatever the case may be.

1. Lethal conditions may limit distributions, but when they do so they need only occur occasionally.

2. Distributions are limited more often by conditions that are regularly sub-optimal (rather than lethal) leading to a reduction in growth or reproduction or an increased chance of mortality.

3. Sub-optimal conditions often act by altering the outcome of a biological interaction between the species concerned and one or more other species.

4. Sub-optimal conditions often interact with other factors, so that it is often impossible to isolate a single condition as a factor in its own right.

5. The ill effects of sub-optimal conditions are often moderated by the evolutionary, physiological and behavioural responses of the organisms.

6. Towards the edge of a species' range, it occupies patches in which conditions are closest to those found in the centre of its range.

2.5 Moisture in terrestrial environments: relative humidity

Living matter is entirely dependent on water. Life is thought to have originated in the sea, and the biochemical and physiological functioning of all organisms ultimately take place in water within their organs, tissues and cells. Terrestrial animals live in air which has a lower concentration of water than they have themselves. They all therefore tend to lose water by evaporation, and also by excreting waste products of metabolism with water. These losses can be reduced, however, by limiting the extent and the exposure of their evaporative surfaces, and by producing dry excretory products; and they are counteracted by water gained from metabolism, from food and/or from drinking. In this context, the 'condition' important for terrestrial animal life is the *relative humidity* of the aerial environment. The greater this is, the lower is the differential between the animal and its environment; and the lower the differential, the less need an animal has to reduce or counteract its water losses. The essential point is that animals differ in their abilities to reduce and counteract these losses, and they therefore differ in the relative humidities that allow or favour their existence.

terrestrial organisms need to conserve water—but differ in their ability to do so

Terrestrial plants differ from terrestrial animals in two important respects. First, although their aerial parts suffer the same losses of water as animals, their underground parts (i.e. their roots) are in direct contact with a medium from which they can obtain water directly with more or less ease depending on the soil water content. Secondly, water is as much a resource for plants as it is a condition, since it is combined with carbon dioxide in photosynthesis as the basis of plant nutrition. (Water is therefore discussed in some detail in the following chapter, in which we consider resources.) Nevertheless, relative humidity, as a determinant of the rate of evaporative water loss, is an environmental condition of importance to terrestrial plants.

water provides a condition and a resource for plants

In practice, the effects of relative humidity are often difficult to disentangle from those of temperature. This is simply because a rise in temperature leads to an increased rate of evaporation. A relative humidity that is acceptable to an organism at a low temperature may become unacceptable at a higher temperature. Moreover, relative humidity and temperature may themselves interact with wind speed; rapid movement of air across an evaporative surface maintains the moisture gradient and increases the rate of evaporation. And finally, it is often impossible to separate the relative humidity of an environment from the general availability of water within it. It would be meaningless, for instance, to discuss whether the important thing about a desert is its low relative humidity or its lack of available water. The two things share a common underlying cause, and they combine to charac-

relative humidity interacts with temperature and wind speed—and is difficult to disentangle from water availability generally terize an environment that demands specializations in morphology, physiology, behaviour and life-history. In a similar vein, it is worth noting that the global distribution of the major biomes (tundra, temperate forest, etc.) can be explained either by the combined effects of temperature and mean annual precipitation, or by the combined effects of temperature and relative humidity (Whittaker, 1975).

In addition to these global differences, the microclimatic variations in relative humidity can be even more marked than those involving temperature. It is not unusual, for instance, for the relative humidity to be almost 100% (i.e. almost saturated) at ground level amongst dense vegetation, while the air immediately above the vegetation, perhaps 40 cm away, has a relative humidity of only 50%.

many 'terrestrial' animals are confined to very damp habitats

The organisms most obviously affected by humidity in their distribution are those 'terrestrial' animals that are actually, in terms of the way they control their water balance, 'aquatic'. Amphibians, terrestrial isopods, nematodes, earthworms and molluscs are all, at least in their active stages, confined to microenvironments where the relative humidity is at or very close to 100%. The major group of animals to escape such confinement are the terrestrial arthropods, and in particular the insects. Even here though, the evaporative loss of water is often of crucial importance. This can be seen, for instance, in the case of the fruit-fly *Drosophila subobscura*. Like other insects, *D. subobscura* has an impermeable exoskeleton; but it loses water across its respiratory surfaces when its spiracles are open, which they need to be during periods of activity. *D. subobscura* is largely confined to woodlands (high relative humidity) making only occasional forays into open areas (low relative humidity). More particularly, *D. subobscura* only flies (its most energetic activity) at particular times of the day: just after dawn and just before dusk. These are the periods of daylight with the lowest temperature and the highest relative humidity. During the middle of the day a flying *Drosophila* would rapidly dehydrate. Moreover, its periods of activity are even more narrowly circumscribed in arid, open areas than they are in damper woodlands (Inglesfield & Begon, 1981).

relative humidity and the daily activity of a fruit-fly

relative humidity, temperature and the distribution of the moss Tetraphis

A classic example of the effect of relative humidity, combined with temperature, on the distribution of a species is provided by Forman's (1964) study of the distribution of the moss *Tetraphis pellucida* in North America. Forman examined the effects on growth in the laboratory of various combinations of temperature, relative humidity, pH and light intensity, but he found that in nature these last two factors were never limiting. To relate his experimental growth data to climatic conditions in the field, he utilized information collected at meteorological stations throughout North America. Mean monthly maximum and minimum temperatures and mean monthly night-time relative humidity were used to draw up a theoretical distribution of the moss (expected on the basis of the laboratory growth data), and this distribution was then compared with the actual known distribution. As Figure 2.17a shows, the correspondence was close but by no means perfect. It is highly significant, however, that the correspondence was far closer than when Forman used either temperature alone or relative humidity alone, or combinations of simpler measures like mean annual temperature and humidity (e.g. Figure 2.17b–d).

The lack of a perfect correspondence in Figure 2.17a probably has three main causes. First, some other unknown but important condition may have been ignored. Secondly, some other more detailed characterization of the temperature and humidity regimes may have been more appropriate. And thirdly, the climatic data (from meteorological stations) take no account of microclimatic variations which are certain to be crucial to the moss. However, in spite of these imperfections, the data do provide a good example of the correlation between humidity and distribution and abundance; and they provide a good indication, too, of the sort of information that is necessary to forge a firm link between the responses of individuals and their distribution in nature.

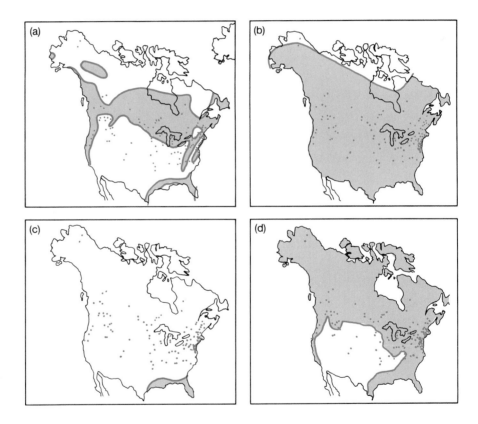

Figure 2.17. (a) The actual distribution of the moss *Tetraphis pellucida* (dots) is fairly closely correlated with theoretical distributional limits drawn up using mean monthly maximum and minimum temperatures and mean monthly night-time relative humidities. The correlation is far better than when limits are predicted on the basis of (b) mean monthly maximum and minimum temperatures alone, (c) mean annual temperature and mean annual night-time relative humidity, or (d) mean monthly night-time relative humidity alone. (From Forman, 1964.)

2.6 pH of soil and water

The pH of soil in terrestrial environments or of water in aquatic ones is a condition which can exert a powerful influence on the distribution and abundance of organisms.

The protoplasm of the root cells of most vascular plants is damaged as a direct result of toxic concentrations of H^+ or OH^- ions in soils below pH 3 or above pH 9, respectively. Further, indirect effects occur because soil pH influences the availability of nutrients and/or the concentration of toxins (Figure 2.18). Below pH 4.0–4.5, mineral soils contain such a high concentration of aluminium ions (Al^{3+}) as to be severely toxic to most plants. In addition, both manganese (Mn^{2+}) and iron (Fe^{3+}) can be present at toxic concentrations at low pH, even though they are essential plant nutrients. At the other end of the pH scale, scarcity of the same nutrients can be a significant problem. In alkaline soils, iron, manganese, phosphate (PO_4^{3-}) and certain trace elements are fixed in relatively insoluble compounds so that plants are poorly supplied with them. Tolerance limits for pH vary among plant species, but only a minority are able to grow and reproduce at a pH below about 4.5.

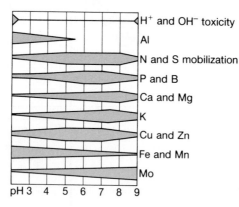

Figure 2.18. Soil pH influences the availability to plants of minerals (indicated by the widths of the bands) (from Larcher, 1975).

The situation is similar for animals inhabiting streams, ponds and lakes; typically, species diversity decreases in acid waters. Increased acidity may act in three ways: (i) directly, by upsetting osmoregulation, enzyme activity or gaseous exchange across respiratory surfaces; (ii) indirectly, by increasing toxic heavy metal concentrations, particularly of aluminium, through cation exchange with the sediment; and (iii) indirectly, by reducing the quality and range of food sources available to animals (for example, fungal growth is reduced at low pH (Hildrew *et al.*, 1984) and the aquatic flora is often absent or less diverse).

There are some important parallels in the role of pH in soil and aquatic systems. In both cases the effects may be direct (toxic concentrations of H^+ or OH^-) or indirect. Indirect effects involve either an interaction between two different conditions (pH and concentration of a toxic chemical), or an interaction between a *condition* (pH) and the availability of a *resource* (less phosphate at high pH for a terrestrial plant, or reduced fungal biomass as food for an invertebrate in a stream at low pH).

pH has both direct and indirect effects in both terrestrial and aquatic environments

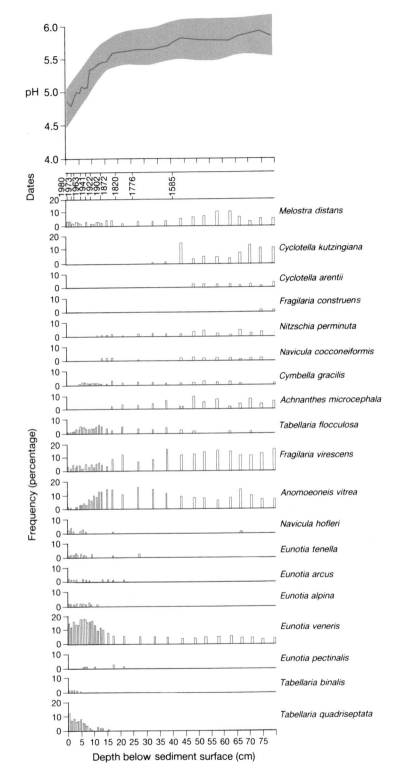

Figure 2.19. Percentage frequency of selected diatom species at various depths in a sediment core from Round Loch of Glenhead in Scotland. Dates corresponding to particular depths have been obtained by an isotope radioactive decay technique (^{210}Pb). The pH history has been reconstructed on the basis of knowledge of pH tolerance ranges of the species concerned (on the assumption that these have not changed over time). (From Flower & Battarbee, 1983.)

Our discussion in this chapter of the relationship between species and conditions has so far stressed the spatial aspect of distribution. In contrast, recent studies of diatoms (microscopic algae) within single lakes have emphasized the way in which changing conditions can induce temporal changes in the relative abundance of various species. Figure 2.19 shows how species composition has changed over the past 400 years or more in Round Loch of Glenhead in Scotland. A sediment core containing the accumulated identifiable remains of diatoms has been analysed section by section, and there are clear indications of a rapid and dramatic decline in species known to be found only rarely below pH 5.5 (e.g. *Anomoeoneis vitrea* and *Tabellaria flocculosa*), along with a concomitant increase in species typical of acid conditions (e.g. *Eunotia veneris* and *Tabellaria quadriseptata*). The main reason for this change, which has been described for a variety of European and North American lakes, seems to be *acid 'rain'* (Battarbee, 1984). Since the Industrial Revolution, the burning of fossil fuels and the consequent emission to the atmosphere of various pollutants, notably sulphur dioxide, has produced a deposition of dry acidic particles and rain which is essentially dilute sulphuric acid. Our knowledge of the pH tolerances of diatom species enables an approximate pH history of a lake to be constructed, and the one for Round Loch of Glenhead is shown in Figure 2.19. Since about 1850 the pH has declined from about 5.5 to about 4.6.

the effects of acid rain on diatoms

2.7 Salinity

In aquatic environments, the availability of water as such is no problem. In freshwater or brackish habitats, however, there is a tendency for water to move in from the environment to the organism by osmosis, and the organism must prevent or counteract this tendency. In marine habitats, the majority of organisms are isotonic to their environment so that there is no net flow of water: but there are many that are hypotonic so that water flows out from the organism to the environment, putting them in a similar position to terrestrial organisms. Thus for many aquatic organisms the regulation of body fluid concentration is a vital and sometimes an energetically expensive process; and the salinity of an aquatic environment can have an important influence on distribution and abundance, especially in places like estuaries where there is a particularly sharp gradient between truly marine and freshwater habitats.

Figure 2.20, for instance, shows the typical distribution along British rivers of three closely related species of amphipod crustacean (Spooner, 1947). *Gammarus locusta* is an estuarine species and is found where the salt concentration never falls below 25 parts per thousand. *Gammarus zaddachi* is a moderately salt-tolerant species and is found in the region where the salt concentration exhibits considerable variation as a result of the tidal cycle, averaging 10–20 parts per thousand. And *Gammarus pulex* is a truly freshwater species found only where the river shows no signs of tidal influence.

estuarine species

Salinity can also have an important effect on distribution in terrestrial habitats bordering the sea, especially in the case of plants (see Ranwell, 1972). Salt-marshes in particular encompass a range of salt concentrations running from full strength sea-water salinity down to totally non-saline. The distributions of plants in salt-marshes, therefore, reflect their differing tolerances to these salinities, i.e. their differing abilities to prevent, counteract or tolerate the influx of unwanted ions, especially sodium ions (Ranwell, 1972). The salinity of a particular location, however, varies considerably as the tides

salt-marsh plants

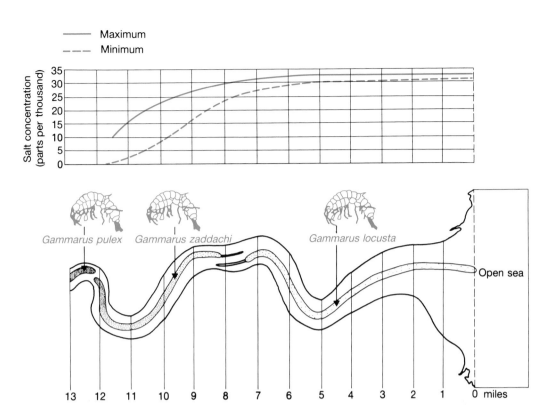

Figure 2.20. The distribution along British rivers of three closely related species of amphipod crustacean, relative to the concentration of salt in the water (from Cox *et al.*, 1976, after Spooner, 1947).

follow their twice-daily and seasonal cycles. Here, therefore, as with other environmental conditions, a species may be limited not by average values but by occasional peaks (or troughs).

2.8 Current flow

In streams and rivers, the current influences the distribution of both animals and plants. The average velocity of flow generally increases in a downstream direction, but the greatest influence on the benthic (bottom-dwelling) community is in upstream regions, because the water here is turbulent and shallow, and the animals and plants are exposed to their greatest danger of being washed away. The only plants to be found in the most extreme flows are literally 'low-profile' species like encrusting and filamentous algae, mosses and liverworts. Where the flow is slightly less extreme there are plants like the water crowfoot, which is streamlined, offering little resistance to flow, and which anchors itself around an immovable object by means of a dense development of adventitious roots. Plants such as the free-floating duckweed are usually only found where the flow is negligible or slow.

A diversity of morphological specializations is to be found among the invertebrates of turbulent, headwater streams. Extreme flattening of the body permits some species to live in the relatively still 'boundary layer' just above the substratum of the stream bed, or to live in crevices under stones and thus avoid the current altogether. Some species are able to maintain their position by means of hooks or suckers, while others, such as the mayfly nymph *Baetis*, have long tail cerci which serve to keep their body facing into the current where the streamlined body shape reduces drag to a minimum (Townsend, 1980).

2.9 Soil structure and substrata

Another factor which influences both animals and plants in aquatic environments is the physical nature of the substratum. Amongst invertebrates in streams, for example, species that inhabit the crevices beneath and between stones can only live where the substratum is stony; while others that need a firm base for a sedentary way of life, like blackflies and net-spinning caddis larvae, are not found where the substratum is unstable or fine grained. Conversely, the burrowing nymph of the mayfly *Ephemera simulans* requires a substratum of fine particles into which its modified front legs can dig effectively (Townsend, 1980).

soil structure and seed germination

The nature of the substratum (i.e. soil) is also important to terrestrial plants and soil-dwelling animals. When seeds germinate, for instance, the chances of their doing so depend on the microtopography (essentially the roughness) of the soil in which they are found. Thus when the seeds of two grasses, *Bromus rigidus* and *Bromus madritensis,* were sown under experimental conditions in soils at a range of surface microtopographies, the seedlings that germinated were predominantly *B. rigidus* on the smoothest soil and predominantly *B. madritensis* on the roughest (Harper *et al.,* 1965). This occurred because the seeds of *B. madritensis* are small and light with a curved awn, while those of *B. rigidus* are heavier with a straight awn. Thus on smooth soil, *B. madritensis* seeds tend to roll onto their convex surface with the embryo pointing impotently into the air; but *B. rigidus* seeds lie flat on the ground, making good contact with the soil and its water. On rough surfaces, however, *B. rigidus* seeds tend to lie across soil clods with their embryos suspended in the air; while *B. madritensis* seeds tend to curl around the clods, making firm contact with the soil surface.

2.10 Sea-shore zonation

The plants and animals that live on rocky sea-shores are influenced by environmental conditions in a very profound and often particularly obvious way. The shore community is overwhelmingly marine in character, and the single most important influence on distribution is the varying extent to which different species can withstand exposure to the aerial environment. This expresses itself in the *zonation* of the organisms, with different species at different heights up the shore; yet as Figure 2.21 makes clear, the precise nature of this zonation is crucially dependent on the physical characteristics of the particular shore. On a very calm and sheltered shore (Figure 2.21a) variations in exposure are almost entirely the result of the twice-daily, tidal changes in sea-level. But on a very exposed shore, waves and sea-spray change the pattern of the exposure. They extend the shore environment upwards and they alter the detailed nature of zonation (though not its underlying pattern; Figure 2.21b).

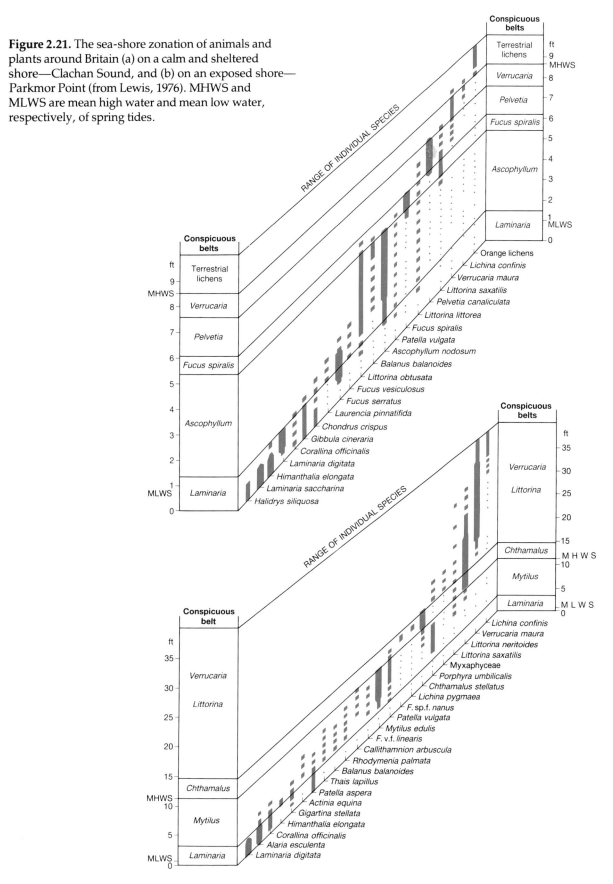

Figure 2.21. The sea-shore zonation of animals and plants around Britain (a) on a calm and sheltered shore—Clachan Sound, and (b) on an exposed shore—Parkmor Point (from Lewis, 1976). MHWS and MLWS are mean high water and mean low water, respectively, of spring tides.

To talk of 'zonation as a result of exposure', however, is to over-simplify the matter greatly (see, for example, Lewis (1976) for a more extended discussion). In the first place, 'exposure' can mean a variety or a combination of many different things: desiccation, extremes of temperature, changes in salinity, excessive illumination and so on. Secondly, as ever, exposure may only set the scene for a biological interaction rather than being limiting in its own right. Amongst the seaweeds around British coasts, for example, the red algae of the lower shore appear in unusual luxuriance in the mid-shore wherever the heavier, blanketing layers of competing *Fucus serratus* and *Ascophyllum* have been experimentally removed (Lewis, 1976). The third point is that 'exposure' only really explains the *upper* limits of these essentially marine species, and yet zonation depends on them having lower limits too. For some species there can be too little exposure in the lower zones. Green algae, for instance, would be starved of blue and especially red light if they were submerged for long periods too low down the shore. For many other species though, a lower limit to distribution is set by competition and predation. The seaweed *Fucus spiralis* will readily extend lower down the shore than usual in Britain whenever the competing midshore fucoids are scarce.

2.11 Pollutants

One environmental condition which is, regrettably, becoming increasingly important is the concentration in the environment of the toxic by-products of man's activities. Sulphur dioxide emitted from power stations, and metals like copper, zinc and lead, dumped around mines or deposited around refineries, are just some of the pollutants that limit distributions, especially of plants. These particular pollutants, and many others too, are present naturally at low concentrations, and some are indeed essential nutrients for plants. But in polluted areas their concentrations can rise to lethal levels. Yet it is rare to find even the most inhospitable areas completely devoid of vegetation; there are usually at least a few individuals of a few species that can tolerate the conditions.

pollution allows us to observe evolution in action

This is the case because even natural populations from unpolluted areas often contain a low frequency of tolerant individuals (Figure 2.22); this is part of the genetic variability present in sexually reproducing populations. These individuals may be the only ones to survive as pollutant levels rise, or they may be the only seeds to germinate and become established in the bare, polluted soil; but either way they can be the founders of a tolerant population to which they have passed on their 'tolerance' genes. Pollution therefore provides us with perhaps our best opportunity of observing evolution in action. However, sufficient genetic variability is by no means present in all populations; some species repeatedly have tolerant individuals giving rise to tolerant populations, while others rarely if ever do (Table 2.2).

Thus, in very simple terms, a pollutant has a twofold effect. When it is newly arisen or is at extremely high concentrations, there will be few individuals of any species present (the exceptions being naturally tolerant

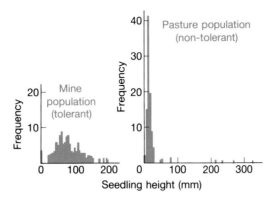

Figure 2.22. On copper-contaminated soil, only copper-tolerant plants grow successfully; but even when a generally non-tolerant pasture population of bent grass (*Agrostis tenuis*) is grown on copper-contaminated soil (right-hand figure), there is a low frequency of individuals with a tolerance to copper comparable with that found in populations growing near a copper mine (left-hand figure). (Walley *et al.*, 1974).

variants or their immediate descendants). Subsequently, however, the polluted area is likely to support a much higher density of individuals, but these will be representatives of a much smaller range of species than would be present in the absence of the pollutant. Such newly evolved, species-poor communities are now an established part of man's environment (see, for example, Bradshaw & McNeilly, 1981).

Table 2.2. Some grass species are much more likely to exhibit copper tolerance than others: variability in copper tolerance available for selection determined by screening normal populations by growth on mine soils. (From Bradshaw & McNeilly, 1981 after Gartside & McNeilly, 1974.)

Species	Survivors per 5000 plants		Copper tolerance (%)		
	On mine soil	On ameliorated mine soil	Five best survivors	Five best offspring	Unselected material
Found on copper waste					
Agrostis tenuis	2	4	55	35	6
Not found on copper waste					
Arrhenatherum elatius	0	4	8	6	0
Poa trivialis	0	2	5		0
Cynosurus cristatus	0	2	6		0
Lolium perenne	0	5	11	9	0
*Dactylis glomerata**	0	6	48	43	5

*Now known to grow on copper waste

2.12   The ecological niche

The list of environmental conditions that have now been considered is not exhaustive, but general principles have emerged, supported by a range of different examples. We are now in a position to consider a concept which is central to much of ecological thinking, and the definition of which depends crucially on the responses of organisms to environmental conditions. This concept is the ecological niche.

The term ecological niche has been part of the ecologist's vocabulary for more than half a century, but for the first 30 years its meaning was rather vague (see Vandermeer, 1972, for a historical review). The current, generally accepted definition (Hutchinson, 1957) is best illustrated by example.

Organisms of any given species can survive, grow, reproduce and maintain a viable population only within certain temperature limits. This range of temperature is the species' ecological niche in one dimension (i.e. the dimension 'temperature': Figure 2.23a). Examples of optimum temperature ranges for a variety of plant species are illustrated in Figure 2.24a. (It should be noted that the niche of a species in any given dimension cannot always be conceived of simply in terms of a maximum and a minimum value between which survival and reproduction are possible. Sometimes we need to consider more complex 'regimes' which include, for example, allowable daily or seasonal fluctuations.)

conditions act as dimensions of a niche

Of course, an organism is not affected by temperature in isolation, nor by any other single condition. Thus, organisms of the species in question may also be able to survive and reproduce only within certain limits of relative humidity. Taking temperature and humidity together, the niche becomes *two*-dimensional and can be visualized as an *area* (Figure 2.23b). A real example is given in Figure 2.24b, showing that optimum conditions for the Mediterranean fruit-fly lie between 16 and 32°C, and 75 and 85% relative humidity. Populations thrive under these conditions, and can reach agricultural pest proportions where the conditions persist, as they often do, for instance around Tel Aviv in Israel. A second example shows the niche of a sand shrimp in two dimensions—temperature and salinity.

If further conditions are brought into consideration (as they should be) then the next step is a *three*-dimensional description of the niche, a *volume* (Figure 2.23c); but the incorporation of more than three dimensions is impossible to visualize. Nevertheless, the process can be continued in an abstract way, by analogy, and a species' true ecological niche can be thought

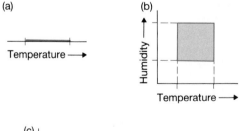

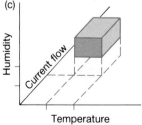

Figure 2.23. Ecological niches. (a) In one dimension (temperature); (b) in two dimensions (temperature and humidity); (c) in three dimensions (temperature, humidity and, say, current flow).

72   CHAPTER 2

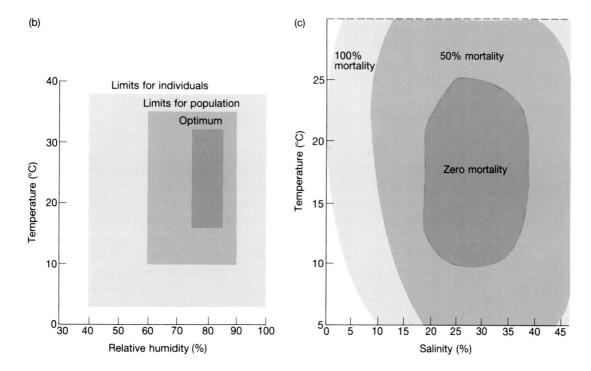

Figure 2.24. (a) Niches in one dimension: optimum temperature ranges for net photosynthesis (at low light intensity, 70 W m⁻²) in species originating from warm temperate lowland (80–250 m in altitude), mountain valley (530–900 m), treeline (1900 m) and high mountain regions (2500–2600 m) of the Alps (from Pisek *et al.*, 1973). (b) A niche in two dimensions for the Mediterranean fruit-fly. The inner rectangle includes conditions optimal for growth, the middle rectangle conditions suitable for development, while the outer rectangle shows the extreme tolerance range (from Allee *et al.*, 1949). (c) A niche in two dimensions for the sand shrimp (*Crangon septemspinosa*). Contours represent 0, 50 and 100% mortality in experiments in aerated water with egg-bearing females at a range of temperatures and salinities (after Haefner, 1970).

of as an *n-dimensional hypervolume* within which it can maintain a viable population.

This, in essence, is the niche concept developed by Hutchinson (1957), except that in addition Hutchinson proposed a separate niche dimension not only for each important environmental condition, but also for each of the resources that the organism requires (e.g. solar radiation, water, mineral nutrients for plants; food, nest sites, etc., for animals—see the next chapter). This view is now one of the cornerstones of ecological thought.

Provided that a location is characterized by conditions within acceptable limits for a given species, and provided also that it contains all necessary resources, then the species can, potentially, occur and persist there. Whether or not it does so depends on two further factors.

First, it must be able to reach the location, and this depends in turn on its powers of colonization and the remoteness of the site.

Secondly, its occurrence may be precluded by the action of individuals of other species which compete with or prey upon it. Throughout this chapter it has been apparent that a species typically has a larger ecological niche in the absence of competitors and predators than it has in their presence. In other words, there are certain combinations of conditions and resources which *can* allow a species to maintain a viable population, but only if it is not being adversely affected by enemies. This led Hutchinson to distinguish between the *fundamental* and the *realized* niche. The former describes the overall potentialities of a species; the latter describes the more limited spectrum of conditions and resources which allows a species to maintain a viable population even in the presence of competitors and predators. Fundamental and realized niches will receive more attention in Chapter 7, when we look at interspecific competition.

It is important to realize that a niche is not something that can be seen. Nor is it necessary to make measurements along each and every niche dimension for it to be a useful idea. It is an abstract concept that brings together, in a single descriptive term, all of an organism's requirements, i.e. all of the environmental conditions that are necessary for an organism to maintain a viable population, and the amounts of each of the resources that it requires to do so. An ecological niche is therefore a characteristic of an organism or, by extension, a characteristic of a species. Habitats, by contrast, are actual places, and as such they provide numerous niches. A woodland habitat, for example, may provide niches for warblers, oak trees, spiders and myriads of other species. The niches of the species occurring in any one habitat are likely to differ, sometimes markedly so. This is a point to which we return in Chapter 7.

3 Resources

3.1 Introduction

Science adopts many words from common speech, and they can carry with them a cloud of nuances that bear little relation to the science. This is true of words like 'conditions' and 'resources', which have become assimilated into the science of ecology. Yet the very act of searching for what is meant by a word can deepen understanding.

According to Tilman (1982), *all things consumed by an organism* are resources for it. 'Just as nitrate, phosphate and light may be resources for a plant, so may nectar, pollen and a hole in a log be resources for a bee, and so may acorns, walnuts, other seeds, and a larger hole in a log be resources for a squirrel.' But even Tilman's broad definition poses further questions. For example, what is meant by 'consumed'? It cannot simply mean *eaten* (although nectar and acorns are eaten) nor yet *incorporated into biomass* (although this happens to nitrate, nectar and acorns): bees and squirrels do not eat holes or incorporate them into biomass. Nevertheless, a hole that has been occupied is no longer available to another bee or squirrel, just as the atom of nitrogen, the sip of nectar or the mouthful of acorn is no longer available to other consumers. Similarly, females that have already mated may be unavailable to other mates. All these things have been consumed in the sense that the stock or supply has been reduced. Thus they are resources, not conditions, because they represent *quantities* that can be reduced by the activity of the organism.

what is a resource?

The resources of living organisms are mainly the stuffs of which their bodies are made, the energy that is involved in their activities and the places or spaces in which they act out the cycles of their lives. The body of a green plant is assembled from inorganic ions and molecules which represent its *food* resources, while solar radiation, trapped in photosynthesis, provides the *energy* resource. Green plants themselves represent packages of food resources for herbivores which, in turn, form the packages of resources for carnivores. The bodies of organisms also represent food resources for parasites and, when dead, for microflora and detritivores. A large part of ecology is about the assembly of inorganic resources by green plants and the reassembly of these packages at each successive stage in a web of consumer interactions. Thinking about resources becomes particularly important when we come to consider how what one organism consumes affects what is available to others (of the same or of different species). We will develop this theme in Chapters 6 and 7.

3.2 Radiation as a resource

Solar radiation is the only source of energy that can be used in metabolic activities by green plants. It differs in many ways from all other resources.

Radiant energy comes to the plant as a flux of radiation from the sun, either directly or after it has been diffused by the atmosphere or reflected or transmitted by other objects. The relative amounts of direct and diffused radiation arriving at an exposed leaf depend on the amount of dust in the air and, in particular, the thickness of the scattering air layer between the sun and the plant. The direct fraction is highest at low latitudes (Figure 3.1).

When a leaf intercepts radiant energy it may be reflected (with its wavelength unchanged), transmitted (after some wavebands have been filtered out) or absorbed. Part of the fraction that is absorbed may reach the chloroplasts and drive the photosynthetic process (Figure 3.2).

Radiant energy is converted during photosynthesis into energy-rich chemical compounds of carbon which will subsequently be broken down in respiration (either by the plant itself or by those that eat or decompose it). But unless the radiation is captured and fixed at the instant it falls on the leaf, it is irretrievably lost. Radiant energy that has been fixed in photosynthesis passes just once through the world. This is in complete contrast to an atom of nitrogen or carbon or a molecule of water that may cycle repeatedly through endless generations of organisms.

radiant energy must be captured or is lost forever

Figure 3.1. Global map of the solar radiation absorbed annually in the earth–atmosphere system, from data obtained with a radiometer on the Nimbus 3 meteorological satellite. The units are cal cm^{-2} min^{-1} (1 cal = 4.2 joules). (From Raushke *et al.*, 1973.)

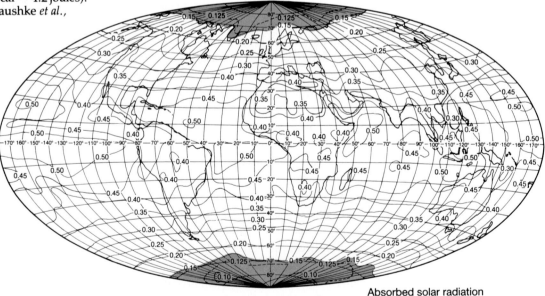

Absorbed solar radiation

radiation is a spectrum of which photosynthesis uses only a part

Solar radiation is a resource continuum—a spectrum of different wavelengths—but the photosynthetic apparatus is able to gain access to energy in only a restricted band of this spectrum. All green plants depend on chlorophyll pigments for the photosynthetic fixation of carbon, and these pigments

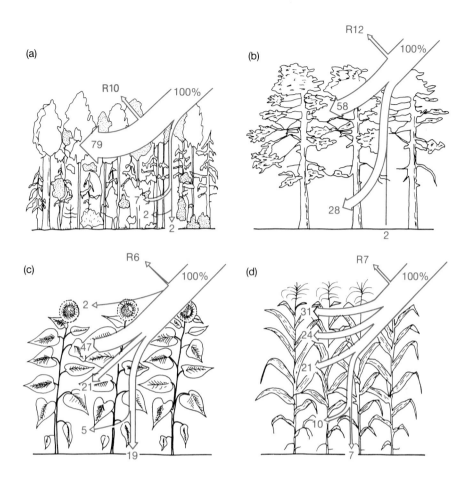

Figure 3.2. The attenuation of solar radiation falling on various plant communities. The arrows show the percentage of incident radiation reaching various levels in the vegetation. (a) A boreal forest of mixed birch and spruce, (b) a pine forest, (c) a field of sunflowers and (d) a field of corn (maize). These figures represent data obtained in particular communities and great variation will occur depending on the stage of growth of the forest or crop canopy and on the time of day and season at which the measurements are taken. (Redrawn from Larcher, 1980, and a variety of sources.)

fix radiation in a waveband between 380 and 710 nm (or, in broad terms 400–700 nm). This is the band of 'photosynthetically active radiation' (PAR). Only about 44% of the total solar radiation incident on the earth's surface at sea level lies in this band; the remainder is unavailable to the green plant as an energy resource. Wavelengths outside the PAR range may serve as stimuli and determine conditions, but they are not resources. The nature of the chlorophyll system therefore puts a fundamental constraint on the activity of the green plant. This, in turn, limits the energy that flows from the green plant to the ecosystem as a whole. There are, however, photosynthetic pigments in prokaryotes that operate in regions outside the green plant's PAR. Bacteriochlorophyll, for example, has its peak absorption at 800, 850 and 870–890 nm.

3.2.1 The rate of photosynthesis is only partly dependent on the light intensity

The response of a leaf to variations in the level of incident radiation can be measured as an increase or decrease in dry matter (photosynthesis minus respiration). This is known as *net assimilation*. Net assimilation is negative in darkness, when respiration exceeds photosynthesis, and increases with light intensity. There is a point at which photosynthesis exactly compensates for respiration—the compensation point. At higher light intensities, assimilation increases to a plateau in so-called C_3 plants, or continues to increase, though according to a law of diminishing returns, in C_4 plants (see p. 87 and legend to Figure 3.3). In either case the higher the light intensity, the lower is the proportion that is used in assimilation.

C_3 and C_4 plants

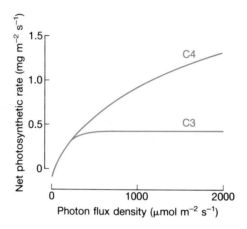

Figure 3.3. Typical light response curves for the photosynthesis of C_3 and C_4 plants. See also Table 3.1. Note that the leaves of the C_3 plants may become saturated with light at intensities commonly encountered in the field, whereas leaves of C_4 plants reach higher absolute rates of photosynthesis and do not become light saturated within any normally experienced range of light intensities (see also p. 87 for characteristics of C_3 and C_4 species).

For part of a bright, sunny day, an exposed leaf may be incapable of taking full advantage of much of the incident radiation. There are two ways in which the form of the plant can influence this. First, leaves may be exposed at acute angles to incident radiation. This has the effect of spreading an incident beam of radiation over a larger leaf area, which effectively reduces its intensity. A light intensity that is super-optimal for photosynthesis when it strikes a leaf at 90° may therefore be optimal for a leaf inclined at an acute angle. Secondly, leaves may be superimposed into a multi-layered canopy. In bright sunshine even the shaded leaves in lower layers may have positive assimilation rates and make a contribution to the assimilation of the plant to which they are attached.

leaves at angles and leaves in multi-layered canopies

The rate at which a leaf photosynthesizes also depends on the demands that are made on it by other parts of the plant. In the absence of any vigorously growing parts, which serve as a 'sink' for photosynthetic products, photosynthesis may be reduced even though conditions are otherwise ideal.

3.2.2 Variation in resource supply

Most leaves in the real world live in a light regime that varies throughout the day and the year, and in an environment of other leaves which modify the quality and the quantity of the light received. This illustrates two vital properties of all resources: their supply can vary both systematically and unsystematically. The ways in which an organism or an organ reacts to a systematic (predictable) or an unsystematic (unpredictable) supply of resource reflects both its present physiology and its past evolution.

systematic variation in the supply of radiation

The systematic elements in the variations of light intensity are the diurnal and annual rhythms of solar radiation (Figure 3.4). The green plant passes through periods of famine and glut of its light resource every 24 hours (except near the poles) and seasons of famine and glut every year (except in the tropics). The seasonal shedding of leaves by deciduous trees in temperate regions *in part* reflects the annual rhythm of light intensity—they are shed when they are least useful. In consequence, an evergreen leaf of an understorey species may experience a further systematic change, because the seasonal cycle of leaf production of overstorey species determines what is left over to penetrate to the understorey. The daily movement of leaves in many species also reflects the changing intensity and direction of light.

unsystematic variation in the supply of radiation

Less systematic variations in the light received by a leaf are caused by the nature and position of its neighbouring leaves and those that overtop it. Each canopy, each plant and each leaf, by intercepting light, creates a resource depletion zone (RDZ)—a moving band of shadow in which other leaves of the same plant, or of other plants, may lie. Thus, for example, a leaf of an understorey shrub will meet unsystematic variation over the course of its life (as overstorey trees grow up and die, creating and filling gaps in the canopy) and even during the course of a single day (as the angle of the sun's rays change and as leaves of the canopy move and create a continually changing pattern of flecks of light (Holmes, 1983)).

The shadows (light depletion zones) become less well defined deep in a canopy because much of the light loses its original direction by diffusion and reflection. The composition of radiation that has passed through leaves in a canopy is also altered—it is photosynthetically less useful because PAR has been reduced. The character of the light as a resource (and as a condition) has been changed.

strategic and tactical responses to variation in supply

Among animals and plants there are classic responses to varying environments, and the responses of green plants to variations in light are typical. Where the environmental variation is systematic and repeated, there is commonly a determinate pattern of response—an evolved, genotypically fixed programme that allows little flexibility or plasticity. The analogy in military terms is with a rigid strategy leaving little room for tactical manoeuvre. By contrast, the typical response to non-systematic, unpredictable variation in the environment is plasticity in individual response, and the ability to make tactical changes in behaviour.

Figure 3.4. (a) The daily totals of solar radiation received throughout the year at Wageningen (Holland) and Kabanyolo (Equatorial Africa) (redrawn from de Wit, 1965). (b) Monthly average of daily radiation recorded at Coimbra (Spain), Poona (India) and Bergen (Norway) (from Van Wijk, 1963).

(a) Annual cycles

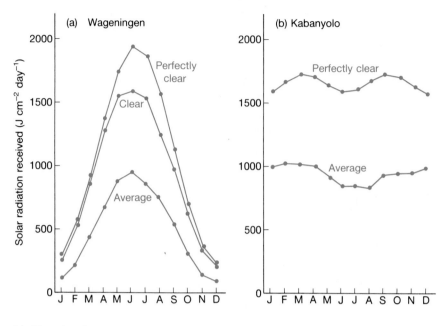

(b) Diurnal cycles

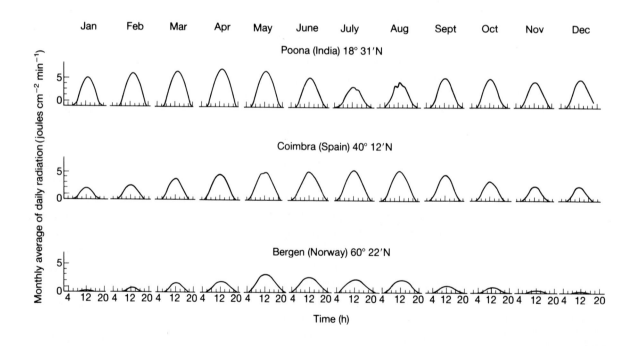

sun and shade species

The major 'strategic' differences between species in their reaction to light intensity are the evolved differences between 'sun species' and 'shade species'. In general, plant species that are characteristic of shaded habitats use light at low intensities more efficiently than sun species, but reach a plateau of photosynthetic rate at lower intensities (Figure 3.5). In addition, the leaves of shade species tend to respire at lower rates. Thus the *net* assimilation rates

80   CHAPTER 3

of shade species are higher than those of sun species when both are growing in the shade. Moreover, C_4 plants (which include a number of grasses and dicotyledons, particularly of the tropics and arid zones, among these some important crops such as sorghum and maize) are capable of increasing their photosynthetic rate in response to increasing light far beyond any intensities likely to be met in the field (Figures 3.3 and 3.5). Given such variation among species in their response to different intensities of radiation, it is not surprising that naturally established vegetation tends to be formed of layers of plants whose ability to use radiation corresponds with their positions in the canopy.

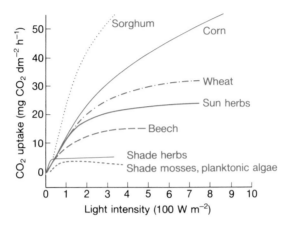

Figure 3.5. The response of photosynthesis to light intensity in various plants at optimal temperatures and with the natural supply of CO_2. Note that corn and sorghum are C_4 plants and the remainder are C_3 (compare with Figure 3.3.). (Redrawn from Larcher, 1980, after various authors.)

sun and shade leaves

In contrast to these 'strategic' differences, it may also happen that as a plant grows, its leaves develop differently as a direct response to the light environment in which the leaf (or its parent bud) developed. This often leads to the formation of 'sun leaves' and 'shade leaves' within the canopy of a single plant. Sun leaves are typically smaller in area, thicker, and have more cells per unit area, denser veins, more densely packed chloroplasts and greater dry weight per unit area of leaf. The shade leaves have correspondingly a large area relative to their dry weight and are often much more translucent. The lower shade leaves of a tree may not make a very great contribution to the energy budget of the plant of which they are a part, but, with their lower compensation points, they may at least pay for their own respiration.

These 'tactical manoeuvres', then, tend to occur not at the level of the whole plant, but at the level of the individual leaf or even its parts. Nevertheless they take time. To form sun and shade leaves as a response to the position in which they are growing, the plant, its bud or the developing leaf must sense the leaf's environment and respond by growing a leaf with appropriate structure. There is a time lag involved in making a new leaf so that it is impossible for the plant to change its form fast enough to track the changes in light intensity between a cloudy and a clear day, for example. It can, however, change its rate of photosynthesis extremely rapidly, reacting even to the passing of a fleck of sunlight.

3.2.3 Light as a population resource

The farmer, forester or nature conservationist is usually less interested in explaining the behaviour of an individual leaf or plant than in explaining how a whole population (or community or area of land) behaves. The canopy of a crop or natural forest is a population of leaves. This can be described holistically by a measure such as the 'leaf area index' (LAI) which defines the area of leaves borne above an area of ground. The light depletion zones produced by individual leaves in a canopy together produce a gradient of light intensity within it. The shape of this light extinction curve depends in large measure on the angles at which the leaves are borne. A canopy of leaves which are borne nearly horizontally, such as clover, produces an abrupt fall-off in light intensity when the sun is overhead. In contrast, the leaves of a dense sward of grasses allow much light to penetrate and to be internally reflected deep into the canopy (Figure 3.6).

the leaf area index

In most vegetation the leaves are crowded together (even on sparse desert shrubs) with some leaves in full sun and others in shade. Most of the photosynthesis will occur in the fully exposed part of the canopy, but the higher the light intensity the greater is the contribution made by lower layers. All living things respire, however, and if leaves occur too low in the canopy, respiration may exceed photosynthesis. Such leaves would have a negative net assimilation rate and be a drain on the energy-fixing rate of the canopy as a whole. There will be, for a population of any given species, an optimal leaf area index: one that gives the highest rate of energy fixation per unit of land surface. At high values of LAI the most shaded leaves or plants may be a drag on the assimilatory potential of the community as a whole: a population of plants may have too many leaves! This was demonstrated experimentally in a classic experiment by Watson (1958), who removed the lower leaves from a crowded crop of kale and showed that the rate of dry matter production of the crop increased as a result.

optimal LAIs ...

An intelligent manager of a crop, a forest or even a natural ecosystem might wish to maintain it at its optimal foliar density—that at which it fixes most solar radiation. This is easier said than done. The optimal LAI for a population of plants (assuming that water and nutrients are not limiting) depends on the shape and disposition of leaves in the canopy, on the angle of the sun and on the intensity of its radiation. When the sun is low in the sky, radiation will not penetrate in the same way as it does at high noon. More important, as light intensity increases, the point at which photosynthesis exceeds respiration will fall deeper in the canopy. Thus the optimal LAI for a stand of vegetation will change from season to season, from day to day and even during the passage of a single day (Figure 3.7). The consequence is that most (probably all) vegetation spends nearly half of its life with a sub-optimal

... which are forever changing

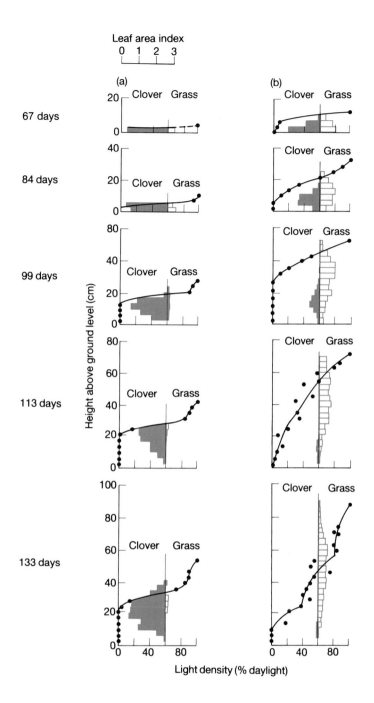

Figure 3.6. The vertical distribution of leaf area and of light intensity within a mixed canopy of grass and clover (*Trifolium repens*). The figures show the leaf area index (area of leaf per area of ground) within successive intervals above ground, with values for clover plotted to the left and for grass to the right. (a) A clover dominated sward and (b) a grass-dominated sward. Note the sharp reduction in light intensity at the top of the clover canopy, in which the leaves tend to be horizontal, and the smooth attenuation of light down through the grass canopy, where the leaves are held at steeper angles. (Redrawn from Stern & Donald, 1962.)

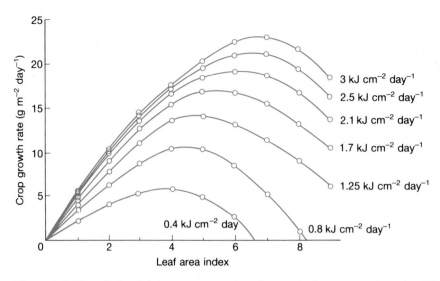

Figure 3.7. The relationship between crop growth rate of subterranean clover (*Trifolium subterraneum*) and leaf area index (LAI) at various intensities of radiation. Note that the leaf area index at which crop growth rate is maximal depends on the light intensity. (Redrawn from Black, 1963.)

LAI and nearly half its life at a super-optimal LAI. Only momentarily will it be optimal.

3.2.4 A resumé for radiation

The limitations of green plants as users of light as a resource can be summarized as follows.

1. For much of the year in temperate climates, and for the whole of the year in arid climates, the leaf canopy does not cover the land surface, so that most of the incident radiation falls on bare branches or bare ground.

2. Only about 44% of solar radiation lies in the photosynthetically active range.

3. Leaves seem to achieve their maximal photosynthetic rate only when the products are being actively withdrawn (to developing buds, tubers, etc.). Even in bright sunlight a leaf may be photosynthesizing below its full capacity.

4. Even when a canopy is fully developed and covers the ground surface, the leaf area index will rarely be optimal, because of the difficulty of adjusting the number and angle of leaves fast enough to compensate for changes in the direction and intensity of light.

5. The highest efficiency of light utilization that has been found in plants is 3–4.5%, and this value has been obtained from cultured marine microalgae at rather low light intensities. In tropical forests values fall within the range 1–3%, and in temperate forests 0.6–1.2%. The approximate efficiency of temperate crops is *c.* 0.6%. It is on such levels of utilization of light as a resource that the energetics of all ecosystems depend.

The following references give a more detailed account of light as a resource in plant growth and contain references that provide a lead into a

Plate I. Rosette treelets—a characteristic plant form that has arisen from a variety of phylogenetic lines and in a variety of environments. (see p.19 *et seq.*). (a) *Macrozemia macdonellii* (Cycadaceae), Alice Springs, central Australia. (b) *Xanthorrhoea* sp. (grass tree, Xanthorrhoeaceae), Australia. (c) *Yucca* (Liliaceae), North end, Wassau Island.

(d) *Sorenoa repens* (Palmaceae), Mt. St. Catherine, Grenada. (e) *Senecio keniodendron* (tree groundsel, Compositae), Mt. Kenya. (Photographs: a and b, J.N.A. Lott, University of Guelph/BPS; c and d, B.J. Miller, Fairfax VA/BPS; e, Peter Cummings, Tom Stack & Associates/BPS.)

(a)

(b)

(c)

Plate II. Aposematism, mimicry and crypsis in two butterflies, the monarch (*Danaus plexippus*) and viceroy (*Limenitis archippus*) (see p.115). The palatable adult viceroy (d) mimics the aposematic colouring of the distasteful adult monarch (c). However, while the monarch caterpillar (b) is also distasteful and aposematic, the viceroy caterpillar (a) is camouflaged as a bird dropping. (Photographs: a, L.E. Gilbert, University of Texas at Austin/BPS; b, c and d, R. Humbert, Stanford University/BPS.)

(d)

Plate III. Stages in the development of potato blight (*Phytophthora infestans*) in a population of a single variety of potato (see p.451). Photographs taken at five successive dates show the spread of disease from a primary focus. (a) 24 August, (b) 30 August, (c) 5 September, (d) 19 September, (e) 1 October. (Photographs by courtesy of D. Skidmore.)

(a)

(b)

(c)

(d)

(e)

Plate IV. Flowerhead specialization in the Ranunculaceae. Complexity of the nectaries restricts the variety of species that can act as pollinators. (see p.469). (a) Creeping buttercup, *Ranunculus repens*, (b) lesser celandine, *R. ficaria*, (c) monk's hood, *Aconitum anglicum*, (d) *Delphinium*, (e) yellow columbine, *Aquilegia chrysantha*. (Photographs: Heather Angel.)

large and extensive literature: Whatley & Whatley (1980), Grace (1983), Fitter & Hay (1981) and Larcher (1980). The last describes in English a lot of research originally published in French and German.

3.3 Inorganic molecules as resources

Three resources, light, carbon dioxide and water, are involved directly in the process of photosynthesis, and they interact in a complex way. Energy intercepted as radiant energy by chlorophyll is used to split water molecules, carbon dioxide is reduced and oxygen is released.

3.3.1 Carbon dioxide

The carbon dioxide used in photosynthesis is obtained almost entirely from the atmosphere, where its concentration varies little around 300 parts per million, except in the immediate region of actively respiring or photosynthesizing organisms. In a terrestrial ecosystem the flux of carbon dioxide at night is upwards, from the soil and vegetation to the atmosphere; on sunny days above a photosynthesizing canopy there is a downward flux. In one study, the concentration difference between points 48 and 138 cm above a crop of corn was found to be of the order of only 2–12 ppm during the night and 2–4 ppm in the daytime (Wright & Lemon, 1966). When photosynthesis was proceeding at its fastest, the concentration of CO_2 within the crop did not fall below c. 264 ppm, yet under laboratory conditions a maize leaf is capable of reducing CO_2 concentration to as low as 5–10 ppm in a closed container. This suggests that leaves of a crop in the field are usually nowhere near the level at which CO_2 could be a limiting resource. The observed reduction in the CO_2 concentration within the corn crop might have reduced the peak rate of photosynthesis by 12% at the most.

CO_2 varies little in its availability

Carbon dioxide diffuses freely in air and its depletion is likely to affect members of a plant population evenly—there is no obvious way in which neighbouring plants can easily be seen to gain an advantage, one over another, by capturing what is a rather freely available resource.

3.3.2 Water

The amount of water that is used in photosynthesis is infinitesimal compared with the volume that passes through a plant during the course of photosynthesis. No organisms have evolved membranes that allow the passage of CO_2 without allowing the passage of water vapour (because H_2O is a smaller molecule). Hence any terrestrial organism that obtains CO_2 from (or loses it to) the atmosphere will lose water at the same time. Even in an atmosphere saturated with water vapour, a plant will lose water while it absorbs CO_2 if the temperature of the leaf is higher than that of the atmosphere.

Hydration is a necessary condition for metabolic reactions to proceed within the organism; water is the medium in which the reactions occur. Because no organism is completely watertight its water content needs continual replenishment. Most terrestrial animals drink free water and/or obtain it from the food they eat. Some water is generated by the metabolism of food and body materials, and there are extreme cases in which animals of arid zones may derive all their water in this way, especially from the metabolism of fats. It may be that the metabolic water produced from the respiration of fats is also important in the water economy of some seeds in arid regions.

For the terrestrial animal that depends on a source of drinking water, the availability and proximity of water holes may place absolute limits on its distribution and abundance. The intensity with which areas of grassland are grazed may depend on their distance from a water hole, leading to heavy overgrazing near to, and undergrazing far from, the water supply. Elegant models of this effect are provided by Goodall (1967).

For the terrestrial plant, the compromises imposed by problems of water economy are complicated, not only because water can move out whenever CO_2 moves in, but also because the plants are rooted to their site and, unlike most animals, cannot wander in search of water.

If the green plant is to absorb CO_2 it will lose water, but there are five main ways in which water loss may be reduced with minimal hindrance to carbon assimilation.

1. Rhythms of stomatal (leaf pore) opening and closure may ensure that the aerial parts of the plant remain more or less watertight except during controlled periods of active photosynthesis. These rhythms may be diurnal; alternatively, the movements may be quickly responsive to the plant's internal water status. Stomatal movement may even be triggered directly by conditions at the leaf surface itself—the plant then responds to desiccating conditions at the very site that, and at the same time as, the conditions are first sensed.

It is usual to find that the number of stomata per unit area of leaf is highest in plants that are likely to suffer drought, and that these same species have more heavily cuticularized (waterproofed) leaf surfaces. This has the effect of allowing ready exchanges of gas and water during photosynthesis but making the plant more watertight at other times. Thus the exposed sun leaves of trees usually have a higher stomatal density and thicker cuticle than shade leaves on the same plant.

2. Structural features of leaves such as hairs, sunken stomata and the restriction of stomata to specialized areas on the lower surface of a leaf, all make the diffusion pathway for water vapour from the wet cell surfaces of the mesophyll to the outside atmosphere less steep and so slow down water loss. But at the same time these morphological features make the diffusion gradient for CO_2 into the plant less steep and reduce its rate of entry. These features are not responsive from day to day or minute to minute like stomatal closure, and they simply slow down both water loss and photosynthesis. Waxy and hairy leaf surfaces may, however, reflect a greater proportion of radiation that is not in the PAR range and so keep the leaf temperature down and reduce water loss.

3. Quite different leaf forms may be developed by some plants at different seasons, each being shed in turn. For example, some shrubs of the Israeli desert (e.g. *Teucrium polium*) bear finely divided, thin-cuticled leaves during a season when soil water is freely available. These are replaced by

undivided, small, thick-cuticled leaves in more drought-prone seasons, and these in turn fall and may leave only green spines or thorns (Orshan, 1963). In these species, there is a sequential polymorphism of the leaf canopy through the season, each leaf morph being replaced by a less photosynthetically active but more watertight structure.

CAM plants

4. The process of carbon dioxide uptake by a plant may be dissociated from the process of photosynthesis itself. This is accomplished by maintaining stomata open (and absorbing CO_2) at night and closed in the daytime. Plants with this behaviour are called CAM (crassulacean acid metabolism) plants, because it was in the family Crassulaceae that the process was first recognized, and because the absorbed CO_2 is accumulated in organic acids. The dark-fixed CO_2 is incorporated into the photosynthetic pathway during the daytime, while the stomata are closed. The process seems a remarkably effective way of conserving water but there are presumably some hidden costs. CAM species have not come to inherit the earth—they remain generally restricted to arid and usually open environments. The system is, however, now known in a wide variety of families, not just the Crassulaceae.

C_4 plants

5. The gradient of CO_2 into the leaf may be made steeper than the gradient of water vapour out of it. This is what happens in the group of plants called C_4 plants (so called because CO_2 is first assimilated into four-carbon carboxylic acids). Table 3.1 lists some characteristic C_3, C_4 and CAM plants. In C_4 plant the enzymic process of CO_2 fixation reduces the concentration of CO_2 within the leaf to much lower levels than in the more common C_3 species. As a consequence, the rate of loss of water vapour is unaffected but the speed of

Table 3.1. Examples of plants with C_3, C_4 and CAM photosynthetic systems. A very full list of species with the various photosynthetic systems is given by Evans (1971).

Typical C_3 plants

Triticum vulgare (wheat), *Secale cereale* (rye), *Lolium perenne* (rye-grass), *Dactylis glomerata* (orchard grass), *Vicia faba* (field bean), *Phaseolus multiflorus* (French bean), *Trifolium repens* (white clover), *Medicago sativa* (alfalfa)
All species of *Quercus* (oaks), *Fagus* (beech), *Betula* (birch), *Pinus* (pines)

Typical C_4 plants

Zea mays (corn), *Saccharum officinale* (sugar cane), *Panicum miliaceum* (millet), *Sorghum vulgare* (sorghum), *Echinochloa crus-galli* (cockspur), *Setaria italica* (Italian millet).
Many species of the family Amaranthaceae (amaranths), Portulacaceae and Chenopodiaceae

Typical CAM (crassulacean acid metabolism) plants

Echinocactus fendleri, *Ferocactus acanthodes* and *Opuntia polyacantha* (all desert succulents), *Tillandsia recurvata* (an epiphyte)

Mixed genera. Some genera contain both C_3 and C_4 species

Euphorbia maculata (C_4), *E. corollata* (C_3), *Cyperus rotundus* (C_4), *C. papyrus* (C_3), *Atriplex rosea* (C_4) and *A. hastata* (C_3)

CO₂ diffusion is increased. The water-use efficiency of C_4 plants (the amount of carbon fixed per unit of water transpired) may be double that of C_3 plants.

Again, one wonders how C_4 plants, with such high water-use efficiency, have failed to dominate the vegetation of the world; but in this case there are clear costs to set against the gains. The C_4 system has a high light compensation point and is inefficient at low light intensities; C_4 species are therefore ineffective as shade plants. Moreover, the temperature optima for growth of C_4 plants are higher than those of C_3 species, and most C_4 plants are found in arid regions or in the tropics. The few C_4 species that extend into temperate regions (e.g. *Spartina* spp.) are found in marine or other saline environments where osmotic conditions may especially favour species with efficient water use. Perhaps the most remarkable feature of the C_4 plants is that they do not seem to use their high water-use efficiency in faster shoot growth, but instead devote a greater fraction of the plant body to a well developed root system. This is perhaps a hint that the rate of carbon assimilation is not in itself a major limit to the growth of these plants, but that the shortage of water and/or nutrients matters more.

the uptake of water

Above ground, the water and CO₂ economies of a plant are tightly coupled, but below ground CO₂ has no significance as a resource. The terrestrial plant has access to water by directly intercepting rainfall or condensing dew, and absorbing it through the leaf surface; but this is probably of only minor importance. The main water resource for terrestrial plants is in the soil, which serves as a reservoir. Water enters this reservoir as rain or melting snow and passes into the soil pores. Just because water has entered the soil does not mean it is available to plants. What happens to it then depends on the size of the pores, which may hold it by capillary forces against gravity. If the pores are wide, as in a sandy soil, much of the water will drain away, passing down through the soil profile until it reaches some impediment and either accumulates as a rising water table or drains away ultimately into streams or rivers. The water held by soil pores against the force of gravity is called the 'field capacity' of the soil. This is the upper limit of the water-holding capacity of a freely drained soil. There is also a lower limit on the capacity of the soil to hold water as a useable resource for plant growth (Figure 3.8). This lower limit is determined by the ability of plants to exert sufficient suction force to extract water from the narrower soil pores, and is known as the 'permanent wilting point'—the soil water content at which plants that are allowed to transpire slowly will enter a state of permanent wilting, from which they cannot recover. The soil water content at the wilting point does not differ significantly among plant species. Solutes in the soil solution add osmotic forces to the capillary forces that the plant must match when it absorbs water from the soil. These osmotic forces become particularly important in saline soils in arid areas. Here much water movement is upwards from the soil to the atmosphere and salts rise to the surface, creating osmotically lethal salt pans.

the soil as a reservoir: the 'field capacity' and the 'permanent wilting point'

the capture of water by roots

A predator may acquire prey by running after it or, as with a spider and its web, waiting for the prey to come to it. There are analogies with both processes in the way a root captures water and nutrients. Water may move

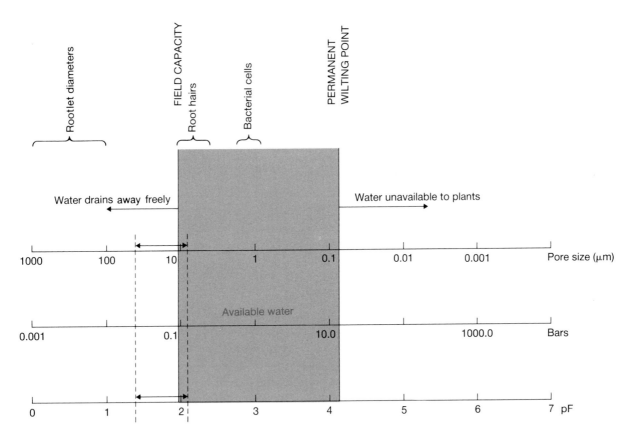

Figure 3.8. The status of water in the soil, showing the relationship between three measures of water status: (i) pF = the force with which water is held expressed as the logarithm of the height of the column of water that the soil would support; (ii) this force expressed as atmospheres or bars; (iii) the diameter of soil pores that remain water-filled. The size of water-filled pores may be compared in the figure with the sizes of rootlets, root hairs and bacterial cells.

through the soil towards a root *and* the root may grow through the soil towards water. As a root withdraws water from the soil pores at its surface, it creates water depletion zones around it. These determine gradients of water potential between the interconnected soil pores. Water flows in the capillary pores, along the gradient into the depleted zones, supplying further water to the root.

This simple process is made much more complex because the more the soil around the roots is depleted of water, the more resistance there is to water flow. As the root starts to withdraw water from the soil, the first water that it obtains is from the wider pores because they hold the water with weaker capillary forces. This leaves only the narrower, more tortuous water-filled paths through which flow can occur, and so increases the resistance to water flow. Thus when the root draws water from the soil very rapidly, the resource depletion zone becomes sharply defined and water can move across it only slowly. For this reason, rapidly transpiring plants may wilt in a soil which contains abundant water. The fineness and degree of ramification of the root system through the soil then become important in determining the access of the plant to the water in the soil reservoir. The more intimately the soil

89   RESOURCES

volume is interwoven with the weft of roots, the shorter are the paths along which water must flow.

the distribution of water in the soil

Water that arrives on a soil surface as rain or as melting snow does not distribute itself evenly through the soil mass. Instead it tends to bring the surface layer to field capacity, and further rain extends this layer further and further down into the soil profile (Figure 3.9). This means that different parts of the same plant root system may encounter water held with quite different forces. In arid areas, where rainfall is in rare, short showers, the surface layers may be brought to field capacity while the rest of the soil stays at or below wilting point. This is a potential hazard in the life of a seedling which may, after rain, germinate in the wet surface layers lying above a soil mass that cannot provide the water resource to support its further growth. A variety of specialized dormancy-breaking mechanisms are found in species living in such habitats, which protect against too quick a response to an insufficient rain.

exploration of the soil and the capture of water

Most roots have features that ensure that they are explorers. Roots elongate before they produce laterals (shoots initiate leaves before they elongate) and this ensures that exploration precedes exploitation. Branch roots usually develop so that they emerge on radii of the parent root, secondary roots radiate from these primaries, and tertiaries from the secondaries. These rules of growth guide the exploration of a volume of soil, and they reduce the chance that two branches of the same root will forage in the same soil particle and enter each other's depletion zones.

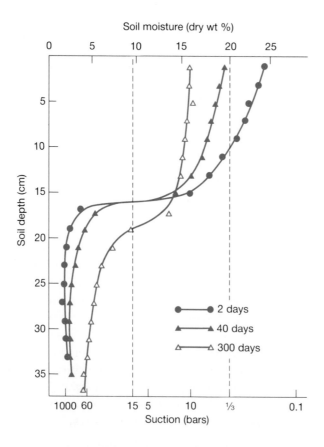

Figure 3.9. The rate of movement of a limited volume of water into a dry loam soil (from Russell, 1973).

The root system that a plant establishes early in its life can determine its responsiveness to future events. Many plants that are waterlogged early in their life develop a superficial root system which is inhibited from growth into anaerobic, water-filled parts of the soil. Later in the season, when water is in short supply, these same plants may suffer from drought because their root system has not developed into the deeper layers. In an environment in which most water is received as occasional showers on a dry substrate, a seedling

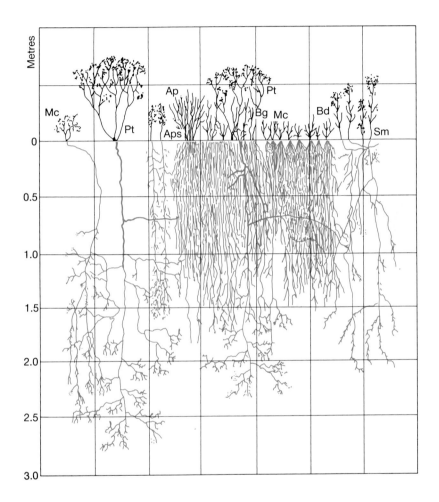

Figure 3.10. The root systems of plants in a typical short-grass prairie after a run of years with average rainfall (Hays, Kansas). Al, *Allionia linearis*; Ap, *Aristida purpurea*; Aps, *Ambrosia psilostachya*; Bd, *Buchloe dactyloides*; Bg, *Bouteloua gracilis*; Kg, *Kuhnia glutinosa*; Lj, *Lygodesmia juncea*; Mc, *Malvastrum coccineum*; Pt, *Psoralia tenuiflora*; Sm, *Solidago mollis*; Ss, *Sideranthus spinulosus*. (Redrawn from Albertson, 1937 and Weaver & Albertson, 1943.)

with a developmental programme that puts its early energy into a deep taproot will gain little from subsequent showers. By contrast, a programme that determines that the taproot will be formed early in life may guarantee continual access to water in an environment in which heavy rains fill a soil reservoir to depth in the spring but are followed by a long period of drought.

Although coarse differences in developmental programmes can be recognized between the roots of different species (Figure 3.10) and some of these are of major importance in the 'match between the organism and its environment', it is the ability of root systems to override strict programmes and be opportunistic that makes them effective exploiters of the soil.

roots as opportunists

Roots pass through a medium in which they meet obstacles and encounter heterogeneity—patches of nutrient and water-supply that vary on the same scale as the diameter of a root itself. In a centimetre of growth, a root

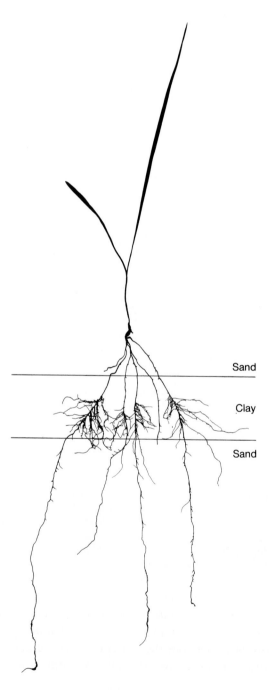

Sand

Clay

Sand

Figure 3.11. The root system developed by a plant of wheat grown through a sandy soil containing a layer of clay. Note the responsiveness of root development to the localized environment that it encounters. (Courtesy of J.V. Lake.)

may encounter a boulder, pebbles and sand grains, a dead root, and the decomposing body of a worm. It may pass through layers of clay soil with fine pores and layers of sand or loam with much coarser pores (Figure 3.11). As a root passes through a heterogeneous soil (and all soils are heterogeneous seen from a 'root's-eye view') it responds by branching freely in zones that supply resources, and scarcely branching in less rewarding patches. That it can do so depends on the individual rootlet's ability to react on an extremely local scale to the conditions that it meets. A gross example of this has been seen when tree roots have entered a decaying human corpse in a peat bog. The root has ramified within the nutrient-rich corpse and the root system has taken the form of the human body.

The military analogy of strategy and tactics can be extended to the comparison of root and shoot. The development of a shoot is under strong strategic (genotypic) control, with allowance for only minor tactical modification (of internode length, branch angle, and leaf size and shape). A root system is under loose strategic control, allowing very great freedom of tactical (phenotypic) response to life in a heterogeneous, remarkably unpredictable and unsystematic environment.

A detailed review of the role of water in the soil–plant–atmosphere continuum is given by Passioura (1983).

3.3.3 Mineral nutrients

It takes more than light, CO_2 and water to make a plant. Mineral resources are also needed. The mineral resources that the plant must obtain from the soil (or, in the case of aquatic plants, from the surrounding water) include macronutrients (i.e. those needed in relatively large amounts)—N, P, S, K, Ca, Mg and Fe—and a series of trace elements—e.g. Mn, Zn, Cu and Bo (Figure 3.12). (Many of these chemicals are also essential to animals, although it is more common for animals to obtain them in organic form in their food than as inorganic chemicals.) Some plant groups have special requirements. For example, aluminium is a necessary nutrient for some ferns, silicon for diatoms, and selenium for certain planktonic algae. Cobalt is required in the mutualistic association between leguminous plants and the nitrogen-fixing bacteria of their root nodules.

Green plants do not obtain their mineral resources as a single package. Each element enters the plant independently as an ion or a molecule, and each has its own characteristic properties of absorption in the soil and of diffusion, affecting its accessibility to the plant even before any selective processes of uptake occur at the root membranes.

all green plants need the same 'essential' elements but may differ in the proportions they require

All green plants require all of the 'essential' elements listed in Figure 3.12. so there is no opportunity (unlike animals) for different species to specialize on different resources. However, plants of different species do not use mineral resources in the same proportion, and there are some quite striking differences between the mineral compositions of plant tissues of different species and between the different parts of a single plant (Figure 3.13). These differences may play an important role in limiting particular plants to particular soil types.

ESSENTIAL FOR MOST ORGANISMS

1. Essential to all living organisms
2. Essential to animals

ESSENTIAL TO RESTRICTED GROUPS OF ORGANISMS

1. Boron – Some vascular plants and algae
2. Chromium – Probably essential in higher animals
3. Cobalt – Essential in ruminants and N-fixing legumes
4. Fluorine – Beneficial to bone and tooth formation
5. Iodine – Higher animals
6. Selenium — Some higher animals?
7. Silicon – Diatoms
8. Vanadium – Tunicates, echinoderms and some algae

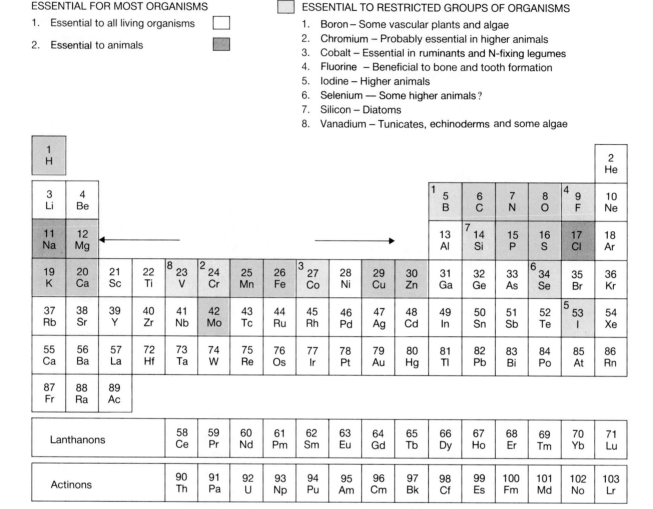

Figure 3.12. The Periodic Table of the elements showing those that are essential resources in the life of various organisms.

water and minerals share characteristics as resources . . .

. . . and they may interact as resources

Many of the points made about water as a resource, and about roots as extractors of this resource, apply equally to mineral nutrients. Again, the resources may be distributed unevenly in the soil; and whether or not a plant gains effective access to the resources will depend in part on the programmed strategies (genotypic instructions) of root development that determine the basic exploration pattern, and in part on its responsiveness to patches of local riches. Again too, the capture of resources may involve *both* the growth of the root to the resource *and* the movement of the resource to the root.

There are not only major similarities between water and soil minerals as plant resources—there are also strong interactions between them. Roots will not grow freely into soil zones that lack available water, and so nutrients in these zones will not be exploited. Plants deprived of essential minerals make less growth and may then fail to reach volumes of soil that contain available water. There are similar interactions between mineral resources. A plant starved of nitrogen makes poor root growth and so may fail to 'forage' in areas that contain available phosphate or indeed contain more nitrogen.

Although there are strong *interactions* between the uptake of water and minerals from the soil, the *correlation* between the two is only really strong in

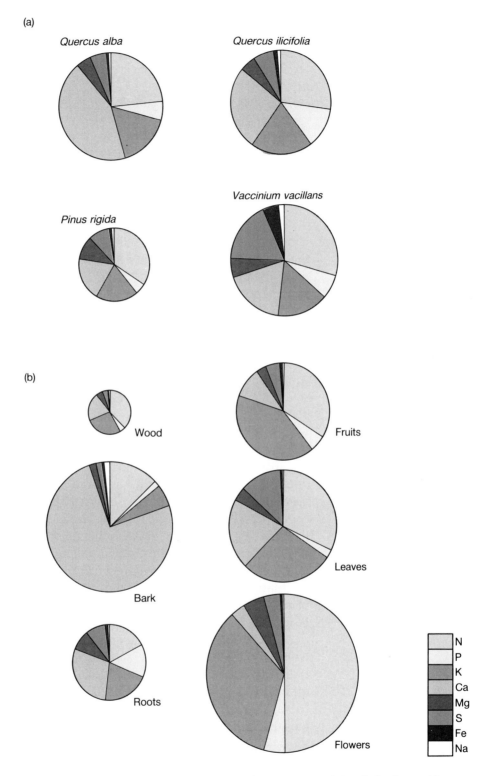

Figure 3.13. (a) The relative concentration of various minerals in whole plants of four species in the Brookhaven Forest, New York. (b) The relative concentration of various minerals in different tissues of the white oak (*Quercus alba*) in the Brookhaven Forest. (Data from Woodwell *et al.*, 1975.)

the case of nitrate ions. Of all the major plant nutrients, nitrate ions (NO_3^-) move most freely in the soil solution and are carried to the root surface in the mass flow of water through the capillaries. Nitrate ions will be carried from as far away from the root surface as water is carried. The movement of water is most rapid in soils at or near field capacity, and in soils with wide pores. Hence, it is under these conditions that nitrate will be most mobile. The resource depletion zones (RDZs) for nitrate will then be wide, and concentration gradients around the roots will be shallow. When RDZs are wide, those produced around individual roots will be more likely to overlap. Competition can then occur (even between the roots of a single plant) because the depletion of a resource by one organ only begins to affect another when they begin to draw on resources accessible to both—i.e. when their RDZs overlap. The lower the amount of available water in the soil, the more slowly it moves and the more slowly nitrate ions will be carried to the root surfaces of the plant. RDZs then become smaller with less overlap between them. Thus roots are less likely to compete with each other for nitrate when water is in short supply.

the architecture of a root determines access to mineral resources

The concept of resource depletion zones is important not only in visualizing how one organism influences the resources available to another, but also in understanding how the architecture of the root system affects the capture of these resources. For a plant growing in an environment in which water moves freely to the root surface, those nutrients that are freely in solution will move with the water. They will then be most effectively captured by wide ranging, but not intimately branched, root systems. The less freely water moves in the soil, the narrower will be the depletion zones, and the more will it 'pay' to explore the soil intensively rather than extensively.

the minerals 'available' are a biased sample of the minerals present

Just because a root system has access to the nutrient solution in the soil does not mean that it has equal access to all nutrients. The soil solution that flows through soil pores to the root surface has a biased mineral composition compared with what is potentially available. This is because different mineral ions are held by different forces in the soil. Ions such as nitrate, calcium and sodium may, in a fertile agricultural soil, be carried to the root surface faster than they are accumulated in the body of the plant. By contrast, the phosphate and potassium content of the soil solution will often fall far short of the plant's requirements. Phosphate is bound on soil colloids by surfaces that bear calcium, aluminium and ferric ions. The rate at which phosphate ions are released into the soil solution then depends on (a) the rate at which the root withdraws them from the soil solution and (b) the rate at which the concentration is replenished by release from the colloids and by diffusion. The diffusion rate is the main factor that determines the width of an RDZ for ions that are bound on soil surfaces. In dilute solutions the diffusion coefficients of ions that are not absorbed, such as nitrate, are of the order of 10^{-5} cm^2 s^{-1} and for cations such as calcium, magnesium, ammonium and potassium they are 10^{-7} cm^2 s^{-1}. For strongly absorbed anions such as phosphate, the coefficients are as low as 10^{-9} cm^2 s^{-1}.

For resources like phosphate that are not carried in bulk in the mass flow of water through the soil, and that have low diffusion coefficients, the RDZs will be narrow (Figure 3.14); roots or root hairs will only tap common pools of resource (i.e. will compete) if they are very close together. It has been estimated that more than 90% of the phosphate absorbed by a root hair will have come from the soil within 0.1 mm of its surface (in a four day period). Two roots will therefore only draw on the same phosphate resource in this period if they are less than 0.2 mm apart. If phosphate is in short supply, the intimate branching of the root, and particularly its extension by root hairs and associated mycorrhizal fungi, will dramatically affect its potential for absorption of phosphate, compared to that for more mobile nutrients. A widely spaced, extensive root system tends to maximize access to nitrates, while a narrowly spaced, intensively branched root system tends to maximize access to phosphates (Nye & Tinker, 1977). Plants with different shapes of root system may therefore tolerate different levels of soil mineral resources, and different species may deplete different mineral resources to different extents. This may be of the greatest importance in allowing a variety of plant species to cohabit in the same area (Chapters 7 and 18).

mycorrhizae

The parts of this chapter that concern water and nutrient resources in the soil have been written on the assumption that plants have roots. In fact most plants do not have roots—they have mycorrhizae. These are associations of fungal and root tissue which have resource-gathering properties differing greatly from those observed in laboratory studies of sterile roots. The properties of mycorrhizae are discussed in the chapter on mutualism (Chapter 13).

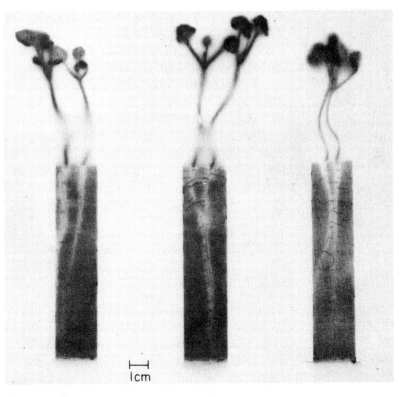

Figure 3.14. Radioautograph of soil in which seedlings of mustard have been grown. The soil was supplied with radioactively labelled phosphate (^{32}P) and the zones that have been depleted by the activity of the roots show up clearly as white. (From Nye & Tinker, 1977.)

3.3.4 Oxygen as a resource

Oxygen is a resource for both animals and plants. Only a few prokaryotes can do without it. Its diffusibility and solubility in water are very low and so it becomes limiting most quickly in aquatic and waterlogged environments. Its solubility in water decreases rapidly with increasing temperature. When organic matter decomposes in an aquatic environment, microbial respiration makes a demand for oxygen and this 'biological oxygen demand' may constrain the types of higher animal that can persist. High biological oxygen demands are particularly characteristic of still waters into which leaf litter or organic pollutants are deposited and become most acute during periods of high temperature.

Because oxygen diffuses so slowly in water, aquatic animals must either maintain a continual flow of water over their respiratory surfaces (e.g. the gills of fish), or have very large surface areas relative to body volume (many of the aquatic crustacea have large feathery appendages—see Figure 5.11), or have specialized respiratory pigments or a slow respiration rate (e.g. the midge larvae that live in still and nutrient-rich waters), or continually return to the surface to breath (e.g. whales, dolphins, turtles, newts).

The roots of many higher plants fail to grow into waterlogged soil, or die if the water table rises after they have penetrated deeply. These reactions may in part be direct responses to oxygen deficiency and in part responses to the accumulation of gases such as hydrogen sulphide, methane and ethylene that are produced by microorganisms engaged in anaerobic decomposition. Even if roots do not die when starved of oxygen, they may cease to absorb mineral nutrients so that the plants suffer from mineral deficiency.

Not all of the oxygen needs of a plant's roots are supplied from the soil. Some oxygen diffuses down through the roots from the shoot (Luxmoore *et al.*, 1970). Indeed in some circumstances the soil may become locally aerated in this way, and the roots then support a population of aerobic organisms on their surfaces (Greenwood, 1969). The diffusion of oxygen down through root systems of plants is probably quite a common phenomenon, particularly in species where the roots are thick or are filled with air spaces (in rice for example).

3.4 Organisms as food resources

3.4.1 Introduction

autotrophs and
heterotrophs

decomposers, parasites,
predators and grazers

Autotrophic organisms (green plants and certain bacteria) assimilate inorganic resources into 'packages' of organic molecules (proteins, carbohydrates, etc.). These become the resources for heterotrophs (organisms that require resources in organic, energy-rich form) and take part in a chain of events in which each consumer of a resource becomes, in turn, a resource for another consumer. At each link in this *food chain* we can usually recognize three pathways to the next trophic level.

(i) *Decomposition,* in which the bodies (or parts of bodies) of organisms die and, together with waste and secretory products, become a food resource for 'decomposers' (bacteria, fungi and detritivorous animals)—a group which cannot use the organisms while these are alive.

(ii) *Parasitism,* in which the living organism is used as a resource while it is still alive. We define a parasite as a consumer that usually does not kill its food organism and feeds from only one or a very few host organisms in its lifetime. Examples are tapeworms, aphids extracting phloem sap from the leaves of trees, and obligate fungal parasites of plants, such as the rusts, which do not kill the host cells that they penetrate.

(iii) *Predation,* the final category, applies to those cases in which the food organism, or part of it, is eaten and killed. Examples of predator–prey interactions are a seedling attacked by the fungus *Pythium,* a water flea consuming phytoplankton cells, an acorn eaten by a beetle or squirrel (consumers of seeds bring an end to the life of the embryos within them), a mountain lion consuming a rabbit, a whale eating krill, and perhaps even a pitcher-plant drowning a mosquito. *Grazing* can be regarded as a type of predation but the food (prey) organism is not killed; only part of the prey is taken, leaving the remainder with the potential to regenerate. Many defoliating caterpillars and grazing mammals take leaves but rarely eat the buds from which a new crop of leaves may develop. Grazers, in contrast to parasites but like most other predators, are likely to feed from many prey during their lifetime.

The three categories are not neat and tidy and can be endlessly subdivided by the pedantic. For example, there are decomposers that may kill host tissues in advance of penetration (some fungi, e.g. *Botrytis* spp.); and there are 'parasites' that inevitably kill their hosts, for instance parasitoids such as larvae of the ichneumon fly (actually a wasp) which develop within the living body of their insect host but leave it dead when they mature. Nevertheless, this basic framework is useful for categorizing a diversity of types of resource–consumer interactions and, in particular, for highlighting the effect of the consumer on the continued life, or otherwise, of the food organism. Of course, there is a nearly infinite number of ways in which consumers could conceivably be classified, and two more deserve mention here: the degree to which a consumer's diet is specialized or generalized in nature; and the relationship between the life-spans of the consumer and its resource.

generalists and specialists

Consumers may be generalists (polyphagous), taking a wide variety of prey species—though often with a rank order of acceptability among the foods that they take. Alternatively, a consumer may specialize on specific parts of its prey but range over a number of species. This is most common among herbivores because, as we shall see, different parts of plants are quite different in their composition. Thus a variety of birds specialize on seeds though they are seldom restricted to a particular species; many grazing animals specialize on leaves and do not usually take roots; and certain species of eelworm (nematodes) and the larvae of some beetle species are specialized root feeders, taking roots from various species but not leaves. Among consumers of animals, parasites (including liver flukes, for example) often attack particular tissues or organs but are not necessarily restricted to a single host species. Cichlid fish of the species *Genychromis meuto* are unusual in their

Honeybee *Apis mellifera*
(Hymenoptera)

Imbibes nectar through proboscis

Mosquito *Culex* (Diptera)
Piercing – sucking type

Chiasognathus grantii
(Coleoptera)

Mandibles show extreme
development in male, exceeding
the body length

Tabanus atratus (Diptera)
Blood-sucking. Mandibles
and maxillae flattened

Manduca quinquemaculata
(Lepidoptera)

Flower-visiting moth.
Proboscis coiled when
not in use

Shoebill (*Balaenicops rex*)
Fish-eater, but also uses large,
hooked bill to dig lungfish out of mud

Black skimmer (*Rhynchops nigra*)
Knife-like lower mandible cuts the water
in flight. Bill snaps shut on striking a fish

Greater flamingo
(*Phoenicapterus ruber*)

Filtering mechanism in bill, for
feeding on minute algae

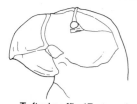

Tufted puffin (*Fratercula
corniculata*)

Holds several fish in bill while
capturing more

Falcon (*Falco*)
Toothed beak for plucking
feathers and shearing

Figure 3.15. The specializations of organisms with respect to the nature of their food resources are illustrated by the variety of their mouthparts. Insects and birds, illustrated here, show such specialization to a marked degree. (Redrawn from Figures in Daly *et al.*, 1978; Richards & Davies, 1977; Snodgrass, 1944; and the *Encyclopaedia Britannica*.)

specialized feeding on scales from other fish. Finally, a consumer may specialize on a single species or a narrow range of closely related species (when it is said to be monophagous). Examples are caterpillars of the cinnabar moth (which eat leaves, flower buds and very young stems of species of *Senecio*), the raspberry fruit fly (p. 102) and the oak gall wasps (p. 106).

The more specialized the food resource required by an organism, the more the organism is constrained to live in patches of that resource *or* to spend time and energy in searching for it among a mixture. Such specialization may be fixed by peculiar structures, particularly mouthparts, which make it possible to deal efficiently with the resource. The stylet of the aphid can be interpreted as an exquisite product of the evolutionary process that has given the aphid access to a uniquely valuable food resource *or* as an example of the ever-deepening rut of specialization that has evolved, narrowed and constrained what aphids can feed on. Figure 3.15 further illustrates mouthpart specializations amongst insects and birds.

the seasonal availability of resources

Many food resources are seasonal in their availability, and this is well illustrated by reference to a population of wild raspberries in a temperate woodland. In winter, the plant is a mass of tiny twigs but in spring it develops an abundance of young, protein-rich buds and juvenile leaves. Flowering brings a short period of nectar production offering a quite new type of resource, but for the flowering period only. In turn, as fruit is set, a new flush of resources is displayed in ripening and ripe fruit (Figure 3.16). Such seasonal structures may be reliable resources *either* to generalist herbivores that can turn to other foods when raspberry is out of season *or* to specialists which have an active life concentrated in the appropriate season and which spend the rest of the year making no demands on food (dormant or in diapause). The bird species feeding on raspberry fruits take these only as a seasonal component of a diet that varies continuously through the year—and some species may be present only as migrant summer visitors. In complete contrast, the raspberry beetle (*Byturus tomentosus*) lays eggs in the flower and the larva completes its life cycle within the developing fruit. It then remains in diapause as a pupa until the next raspberry flowering season in 10–11 months' time. The larva of the raspberry moth (*Lampronia rubiella*) has a longer life on a more consistently present resource—the pith within the woody stems.

This example illustrates how a single resource (the raspberry plant) can be used by a variety of types of consumer; it also shows how many seemingly unrelated consumers may nonetheless interact with each other via a shared resource (see Chapter 7).

3.4.2 The nutritional content of plants and animals as food

As a 'package' of resources, the body of a green plant is quite different from the body of an animal. This has a tremendous effect on the value of these resources as potential food. The most important contrast is that plant cells are bounded by walls of cellulose, lignin and/or other structural materials. It is these cell walls which give plant material its high fibre content. The presence of cell walls is also largely responsible for the high fixed carbon content of

C:N ratios in animals and plants

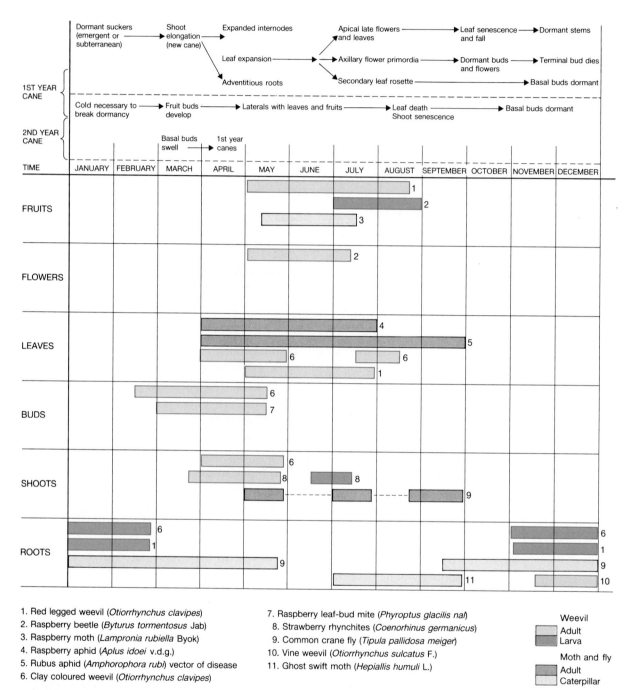

Figure 3.16. The life cycle and phenology of the raspberry plant (*Rubus idaeus*) and of some of the animals that use it as a food resource (data from a variety of sources).

1. Red legged weevil (*Otiorrhynchus clavipes*)
2. Raspberry beetle (*Byturus tormentosus* Jab)
3. Raspberry moth (*Lampronia rubiella* Byok)
4. Raspberry aphid (*Aplus idoei* v.d.g.)
5. Rubus aphid (*Amphorophora rubi*) vector of disease
6. Clay coloured weevil (*Otiorrhynchus clavipes*)

7. Raspberry leaf-bud mite (*Phyroptus glacilis* nal)
8. Strawberry rhynchites (*Coenorhinus germanicus*)
9. Common crane fly (*Tipula pallidosa meiger*)
10. Vine weevil (*Otiorrhynchus sulcatus* F.)
11. Ghost swift moth (*Hepiallis humuli* L.)

Weevil
Adult
Larva

Moth and fly
Adult
Caterpillar

plant tissues and the high ratio of carbon to other important elements. For example, the C:N ratio of plant tissues varies from 40:1 to 20:1, in contrast to the ratios of approximately 8:1 or 10:1 in bacteria and fungi, detritivores, herbivores and carnivores. Unlike plants, animal tissues contain no structural carbohydrate or fibre component but are rich in fat and, in particular, protein. The dramatic differences between the body compositions of plants and their consumers are illustrated in Figure 3.17.

Herbivores which consume living plant material—and bacteria, fungi and detritivores which consume dead plant material—all utilize a food resource that is rich in carbon and poor in protein. The transition from plant to consumer involves a massive burning off of carbon as the C:N ratio is lowered. The main waste products of organisms that consume plants are therefore carbon-rich compounds (carbon dioxide and fibre). In contrast, herbivores and their consumers are remarkably similar in composition. The greater part of the energy requirements of carnivores is obtained from the protein and fats of their prey, and their main excretory products are in consequence nitrogenous.

cellulose—which very few organisms can digest

The large amounts of fixed carbon in plant materials means that they are potentially rich sources of energy. Yet most of that energy is not directly available to consumers. To exploit the full energy resource of plants it is necessary to have enzymes capable of mobilizing cellulose and lignins. A limited number of species of bacteria and many fungi possess cellulases, and a few protozoa (e.g. *Vampyrella*) can dissolve holes in the cellulose walls of algae and gain access to the cell contents. The salivary glands of slugs and snails are a rich source of cellulases and a few other animal species are believed to possess them too. But the overwhelming majority of species in both the plant and animal kingdoms lack the necessary enzymes. Hence, the major energy-rich component of most plant tissue is quite unavailable as a direct energy resource to plants or to plant-eaters. Of all the many constraints that put limits on what living organisms can do, the failure of the majority of living organisms to evolve cellulolytic enzymes is one of the most remarkable evolutionary puzzles.

even excluding cell walls, plants have a high C:N ratio

Even if the cell wall fraction is excluded from a consideration of plants as articles of diet, the C:N ratio is high in green plants compared with other organisms. This is illustrated by the feeding behaviour of aphids, which gain direct access to cell contents by driving their stylets into the phloem transport system of the plant, and extracting phloem sap which is rich in soluble sugars (Figure 3.18). The aphid uses only a fraction of this energy resource, and excretes the rest as the carbohydrate mellibiose in the honeydew that may drip as a rain from an aphid-infested tree. It appears that for most herbivores and decomposers, the body of a plant is a super-abundant energy and carbon resource; it is other components of the diet (e.g. nitrogen) that are more likely to be limiting.

Because most animals lack cellulases, the cell wall material of plants hinders the access of digestive enzymes to the contents of plant cells. The acts of chewing by the grazing mammal and grinding in the gizzard of birds (geese for example) are necessary precursors to digestion because they break down the cellular structure of the plant diet. The carnivore, by contrast, can more safely gulp its food.

organisms that possess cellulases

Those organisms that do possess cellulases have access to a food resource for which they compete only among themselves. Their activity makes two striking contributions to the availability of resources to other organisms.
(i) The alimentary canal of herbivores may provide a miniature ecosystem in which cellulolytic bacteria gain especially easy access to cell wall material. The rumen or the caecum in warm-blooded animals is a temperature-regulated

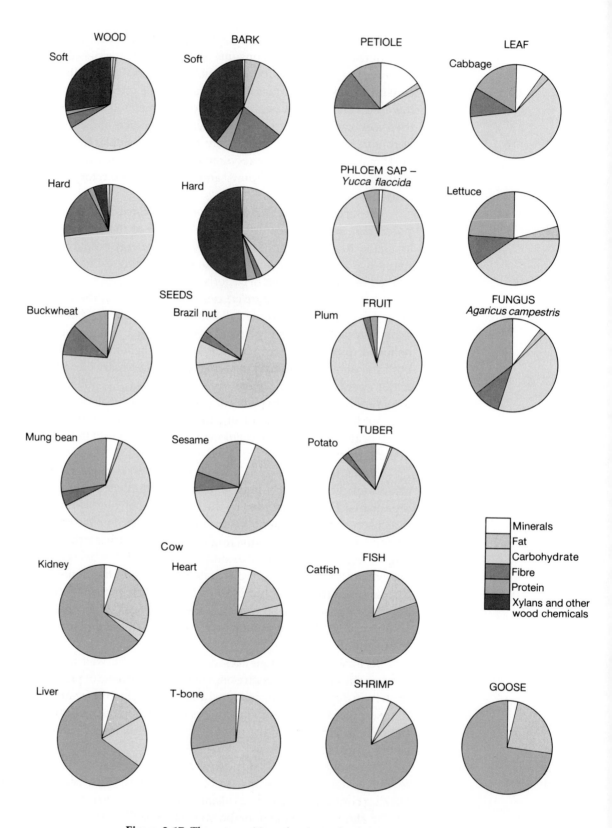

Figure 3.17. The composition of various plant parts and of the bodies of animals that serve as food resources for other organisms (data from a variety of sources).

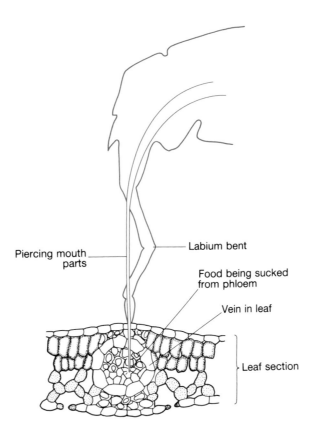

Figure 3.18. The head of a feeding aphid showing the stylet penetrating into the phloem tissue of a leaf.

Piercing mouth parts

Labium bent

Food being sucked from phloem

Vein in leaf

Leaf section

culture chamber into which already partially fragmented cell walls flow continually, like the chemostat of an industrial fermentation. Microbial cellulases are predominantly surface enzymes and the intimate contact of bacteria with the chewed food mass speeds up the rate of cell wall breakdown. In the ruminants, the by-products of this bacterial fermentation are in part absorbed by the host (see Chapter 13).

(ii) When plant parts are decomposed, material with a high carbon content is converted to microbial bodies with a relatively low carbon content—the limitations on microbial growth and multiplication are resources other than carbon. Thus when microbes multiply on a decaying plant part, they withdraw nitrogen and other mineral resources from their surroundings and build them into their own microbial bodies. For this reason, and because microbial tissue is more readily digested and assimilated, plant detritus which has been richly colonized by microorganisms is generally preferred by detritivorous animals. Conversely, from the living plant's point of view, local microbial activity can have adverse consequences. The incorporation into microbial tissue of mineral resources has the effect of lowering their levels of availability, and higher plants growing in soil close by may then suffer from mineral deficiencies. The effects can be seen after straw has been ploughed into the soil, when soil nitrogen becomes unavailable to crops and they may show signs of nitrogen deficiency.

a plant is a population of parts that differ greatly in composition and food value

The cellular packages of plants are aggregated into tissues of similar cells and into organs composed of heterogeneous assemblages of cell types. The highest concentrations of nitrogen and mineral elements are found in the young growing tips, in axillary buds and in seeds, and the highest concentrations of carbohydrate in phloem sieve tubes and in storage organs such as tubers and some seeds. The highest concentrations of cellulose and lignin are in old and dead tissues such as wood and bark. The dietary value of different tissues and organs is so different that it is no surprise to find that most small herbivores are specialists—not only on particular species or plant groups, but on particular plant parts: meristems, leaves, roots, stems, etc. The smaller the herbivore, the finer is the scale of heterogeneity of the plant on which it may specialize. Extreme examples can be found in the larvae of the oak gall wasps, of which some species may specialize on young leaves, some on old, some on vegetative buds, some on male flowers and others on root tissues (Figure 3.19). But even the most catholic feeders usually display some

Figure 3.19. Specialization of the larvae of gall wasps (Cynipidae: Hymenoptera). (a) Galled acorns on *Quercus carris* caused by *Callirhytis erythrocephalum*. (b) Gall on lateral bud of oak caused by *Biorhiza pallida*. (c) Oak leaf galls caused by *Neuroterus numismalis* and *N. lenticulatus*. (d) Gall on male catkin caused by *Neuroterus quercus-baccarum*. (All photographs kindly provided by R.R. Askew.)

preferences, avoiding if possible woody stems and selecting the more nutritious parts.

Seeds are the most nutritionally complete part of plants. They are particularly concentrated resources of carbohydrate, fat, protein and minerals, and thus provide the food resource for many different sorts of herbivore. A single seed may provide a lifetime's food supply for a bruchid beetle. The egg is laid in or on the seed and the larva may complete its life cycle to pupation within that single seed. The same seed might, however, have formed just part of one day's diet for a bird or have been cached in the winter stores of a rodent. Similarly, a leaf of clover in a pasture may form part of one bite for a sheep, a whole day's diet for a slug or snail, or a lifetime's diet for a weevil, a leaf-mining caterpillar or a fungal leaf pathogen (Figure 3.20).

Although plants and their parts may differ widely in the resources they offer to potential consumers, the composition of the bodies of different herbivores is remarkably similar. Moreover, there is little difference between the body composition (in nutritional terms) of a carnivore and a herbivore. In terms of the content of protein, carbohydrate, fat, water and minerals per gram there is very little to choose between a diet of caterpillars or cod, or of earthworms, shrimps or venison. The packages may be differently parcelled (and the taste may be different) but the contents are essentially the same. Carnivores, then, are not faced with problems of digestion (and they vary rather little in their digestive apparatus) but rather with difficulties in finding, catching and handling their prey (see Chapter 8).

3.4.3 Food resources are often defended against their consumers

So far we have discussed food resources only in terms of their nutritional quality and, of course, this is of fundamental significance to a potential consumer. However, this is not the only property that influences whether particular consumers and resources interact—the extent to which resources are defended is also a crucial consideration. Not only does the commitment of resources to defences in potential prey reduce the resources that they may have available for reproduction, but such defences can have a dramatic effect on the potential food value of a food item to a consumer (Edmunds, 1974).

All organisms are potentially food resources for other organisms, and it is in the interests of any organism to stay alive (since it is then likely to leave more descendants—it will have greater fitness in the evolutionary sense). Not surprisingly, therefore, organisms have evolved physical, chemical, morphological and/or behavioural defences against being attacked or eaten. These defences serve to reduce the chance of an encounter with a consumer and/or increase the chance of surviving such an encounter. But the interaction does not necessarily stop there. A better defended food resource itself exerts a selection pressure on consumers. Those best able to deal with the defence will leave more descendants, and their traits are likely to spread through the consumer population. A continuing evolutionary interaction can be envisaged (akin to a sequence in the development of weapons and defences by opposing military forces). In such situations, where reciprocal evolutionary pressures operate to make the evolution of each taxon partially dependent

Slug damage

Sheep damage

Weevil damage

Pigeon damage

Figure 3.20. Typical bite marks made by animals that eat the leaves of white clover, *Trifolium repens* (redrawn from Peters, 1980).

upon the evolution of the other, the interaction is referred to as coevolution. Although an enticing idea, it should be noted that the role of coevolution as a widespread and important evolutionary force is by no means fully established.

At this point it should be noted that the resources of green plants (and of autotrophs in general) do not grow, do not reproduce and do not evolve. With the exception of radioactive decay, atoms are immutable. Plants do not have to be concerned with overcoming evolved 'defences' in their resources, and coevolution between consumer and resource is in this case impossible. Co-evolution is also not possible between decomposer organisms and their dead food resources. However, bacteria, fungi and detritivorous animals will often have to contend with residual effects of physical and, in particular, chemical defences in their food. Strong interactions may nonetheless occur among autotrophs or among decomposers in competition for these resources (Chapters 6 and 7).

physical defences: spines

The spiny leaves of holly are not eaten by larvae of the oak eggar moth, but if the spines are removed the leaves are eaten quite readily. No doubt a similar result would be achieved in equivalent experiments with perch or foxes as predators and de-spined sticklebacks or hedgehogs as prey! Spines can be an effective deterrent.

Many small zooplanktonic invertebrates that live in the water column of lakes, including rotifers and cladoceran water fleas, exhibit a remarkable degree of variation in the occurrence and size of spines and crests and other appendages which have been shown to reduce their vulnerability to predation (Figure 3.21). It may seem surprising that a predator, simply by its presence, can induce the development of such physical defences but this is just what is observed among many of these species. Thus, for example, spine development in the offspring of an egg-bearing rotifer, *Keratella cochlearis*, is promoted if their mother was cultured in a medium conditioned by the predatory rotifer *Asplachna priodonta* (Stemberger & Gilbert, 1984). Similarly, the development of the crest in the water flea *Daphnia pulex* is promoted in the presence of predatory phantom midge larvae (*Chaoborus*) (Krueger & Dodson, 1981).

Many plant surfaces are clothed in epidermal hairs (trichomes) and in some species these develop thick secondary walls to form strong hooks or points which may trap or impale insects. The bean *Phaseolus vulgaris* has on the undersurface of its leaves hooked trichomes, 0.06–0.11 mm long, which impale both nymphs and adult leafhoppers (*Empoasca fabae*). The trapped insects stop feeding and many dehydrate or starve (Figure 3.22). Likewise, the trichomes on *Passiflora* leaves are a very effective defence against most caterpillars (but at least one species escapes the problem—*Mechanitis isthmia* caterpillars cooperate to build a web suspended below the leaf and feed from the undefended edges (Rathcke & Poole, 1975)).

protective coatings— perhaps utilizing what is most readily available

Any feature of an organism's life-style that increases the energy that a consumer spends in discovering or handling it is a defence if, as a consequence, the consumer eats less of it. The thick shell of a nut or the fibrous cone on a pine increases the time spent by an animal in extracting a unit of

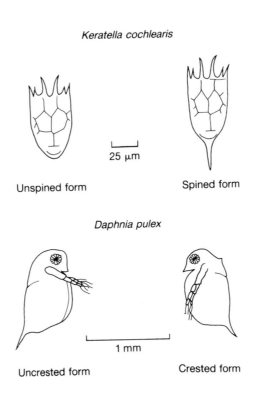

Figure 3.21. Zooplanktonic invertebrates with and without physical defences that can be induced by the presence of predators. The rotifer *Keratella cochlearis* with and without a basal spine and immature females of the cladoceran *Daphnia pulex* with and without a head crest.

effective food, and this may reduce the number of nuts or seeds in a cone that are eaten. The green plant uses none of its energetic resources in running away and so may have relatively more available to invest in energy-rich defence structures. Moreover, it has already been suggested that most green plants are relatively over-provided with energy resources in the form of cellulose and lignin. It may therefore be cheap to build husks and shells around seeds (and woody spines on stems) if these defence tissues contain rather little protein or other limiting nutrients, and if what is protected are the real riches—the scarce resources of nitrogen, phosphorus, potassium, etc., in embryos and meristems.

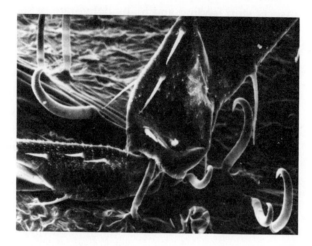

Figure 3.22. A leafhopper nymph (*Empoasca fabae*) trapped on trichomes of the bean *Phaseolus vulgaris* (from Pillemer & Tingey, 1976).

A predator of mussels faces much the same problem as a predator of walnuts. Adult mussels do not move about, but they have shells which at least demand a substantial handling time and may protect the animal completely against some predators. Even a slow-moving snail has some of the same properties. The mussel may be carried up into the air by a crow and dropped onto rocks to break the shell; the snail may be broken on a stone by a thrush; the nut may be cracked by the teeth of a squirrel—in each case the shell wastes the time and energy of the predator.

Janzen (1981) has pointed out that our ancestors have probably known for the past five million years that truly hard biotic containers often contain otherwise undefended contents. Generally the only seeds that can be eaten with confidence are those encased in thick, hard walls. In short, a tough castle can substitute for other defences such as running away or the presence of toxic chemicals.

different plant parts are defended to different extents:

As we might expect, the more valuable a part of an organism is to its fitness, the more strongly is it usually defended. The most concentrated food resources offered by plants are the growing points (root and shoot meristems) and the seeds. The meristems represent the potential for new growth; the seeds represent the potential descendants. These are the regions in which most scarce resources are concentrated, and where protection from herbivores is most likely to increase the number of descendants that a plant leaves.

... buds ...

Shoot meristems are commonly protected against small predators (those of the same order of size as the bud themselves) by the presence of cuticularized and often lignified scales. Such protection, however, will usually be no defence against larger predators like birds; the bullfinch, for instance, concentrates on the developing flower buds of fruit trees in the spring. Buds may escape predation from larger herbivores too by being an integral part of a larger woody stem or twig, and may be further protected by spines intimately associated with the bud itself.

... and seeds—
dissipation or
protection?

Seeds are most at risk to predators when they have just ripened and are still attached, in a cone or ovary, to the parent plant. Seeds massed together in a capsule during the ripening process offer a potentially marvellous food package to a predator, but its value is literally dissipated as soon as the capsule opens and the seeds are shed. The poppy weeds that live in cornfields and the cultivated oil seed poppy illustrate this point (Figure 3.23). The seeds of the wild poppies are shed through a series of pores at the apex of the capsule as it waves in the wind. Two of the species, *Papaver rhoeas* and *P. dubium*, open these pores as soon as the seed is ripe and the capsules are often empty by the following day. Two other species, *Papaver argemone* and *P. hybridum*, have seeds that are large relative to the size of the capsule pores and dispersal is a slow process over the autumn and winter months. The capsules of these species carry spines. Occasionally, spineless capsuled forms of *Papaver argemone* have been found and these are the only ones which are attacked by birds. This suggests that the evolution of spines may have been selected because it protects the capsule against bird attack.

The cultivated poppy (*Papaver somniferum*), by contrast, has been selected by man not to disperse its seeds—the capsule pores do not open. Birds can therefore be a serious pest on crops of the cultivated poppy; they tear open the capsules within which an oil- and protein-rich reward is concentrated. Man has, of course, selected most of his crops to retain and not disperse their seeds. Harvesting the seeds of a wheat or a rice crop after they have dispersed onto the ground would scarcely be a rewarding operation! One consequence is that the seeds of most agricultural crops represent a

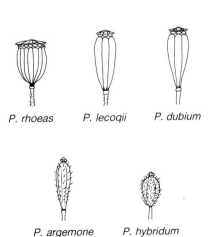

P. rhoeas   P. lecoqii   P. dubium

P. argemone   P. hybridum

Figure 3.23. The capsules of various species of poppy (*Papaver* spp.). The smooth-capsuled wild species have large pores from which the seeds are rapidly dispersed. The spiny, protected capsules are of species with small pores and the seeds disperse slowly.

sitting target for seed-eating birds. The ancestors of the crop plants, which dispersed their seeds on ripening, were much less at risk.

In addition to the bewildering array of physical defences exhibited by different organisms there is a battery of chemical defences. The plant kingdom is very rich in chemicals that apparently play no role in the normal pathways of plant biochemistry. These 'secondary' chemicals range from simple molecules like oxalic acid and cyanide to the more complex glycosides, alkaloids, terpenoids, saponins, flavonoids and tannins (Futuyma, 1983).

A defensive function is generally ascribed to these chemicals (Schild-knecht, 1971). Whilst caution is necessary before jumping to such a conclusion, in some cases a defensive function has been demonstrated un-equivocally. Populations of white clover, *Trifolium repens,* are commonly polymorphic for the ability to release hydrogen cyanide when the tissues are attacked. The cyanogenic property is dependent on two pairs of alleles, one determining the presence or absence of linamarin (a glycoside) and the other of a β-glucosidase enzyme that releases hydrogen cyanide from the linamarin when the tissues are damaged, for instance by chewing. Plants that lack the glycoside or the enzyme, or both, are eaten by slugs and snails. The cyano-genic forms, however, are nibbled but then rejected (Table 3.2).

Table 3.2. The grazing activity of slugs (*Agriolimax reticulatus*) on cyanogenic (*Ac Li*) and acyanogenic (*ac li*) plants of white clover (*Trifolium repens*). Two plants, one of each genotype, were grown in plastic containers and slugs were allowed to graze for seven successive nights. The figures show the numbers of leaves that suffered various categories of grazing damage. (+) and (−) indicate more or less than expected in this category by chance. The difference from expectation is significant at $P < 0.001$. (From Dirzo & Harper, 1982.)

	Conditions of leaves after grazing			
	Not damaged	Nibbled	Up to 50% of leaf removed	More than 50% of leaf removed
Cyanogenic plants (*Ac Li*)	160 (+)	22 (+)	38 (−)	9 (−)
Acyanogenic plants (*ac li*)	87 (−)	7 (−)	50 (+)	65 (+)

Noxious plant chemicals have been broadly classified into two types: toxic (or qualitative) chemicals which are poisonous even in small quantities, and digestion-reducing (or quantitative) chemicals which act in proportion to their concentration. Tannins are an example of the second type. They act by binding proteins and render tissues that contain them, such as mature oak leaves, relatively undigestible. The growth of caterpillars of the winter moth (*Operophtera brumata*) decreases with an increase in the concentration of tannin included in the diet (Feeny, 1968).

Short-lived, ephemeral (unapparent) plants gain a measure of protection from consumers because of the unpredictability of their appearance in space and time. It has been argued that they need to invest less in defence than predictable, long-lived species (Rhoades & Cates, 1976). In addition, the former are predicted to contain 'qualitative' defences that protect only against generalist predators which come across them, whereas the predictable, long-lived (apparent) species need to possess quantitative defences, relatively effective against all kinds of consumer and less susceptible to coevolution of a specialist predator (Feeny, 1976). Many plant species produce more than one secondary chemical and it is likely that some vary their investment in qualitative and quantitative defences as the season progresses. In the bracken fern (*Pteridium aquilinum*), for example, the young leaves which push up through the soil in spring are less apparent to potential herbivores than the luxuriant foliage in late summer. Intriguingly, the young leaves are rich in cyanogenic glucosides while the tannin content steadily increases in concentration to its maximum in mature leaves (Figure 3.24).

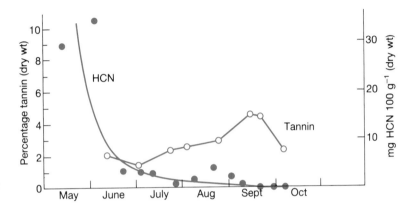

Figure 3.24. Seasonal variation in the concentration of cyanide and tannin in bracken, *Pteridium aquilinum* (from Rhoades & Cates, 1976).

chemical defences in animals . . .

Animals have more options than plants when it comes to defending themselves, but some still make use of chemicals. For example, defensive secretions of sulphuric acid of pH 1 or 2 occur in some marine gastropod groups, including the cowries. They are produced by what are probably modified mucus glands, which may have originally had a lubricative function. The bombadier beetle (*Bradinus crepitans*) possesses a reservoir in its abdomen filled with hydroquinone and hydrogen peroxide (Figure 3.25). When threatened, these chemicals are ejected into an explosion chamber, and mixed with a peroxidase enzyme which allows the hydrogen peroxide to oxidize hydroquinone to noxious quinone. The accompanying release of oxygen gas causes the fluid to be ejected explosively as a spray.

. . . often obtained from their poisonous food plants

Other animals, which can tolerate the chemical defences of their plant food, may actually be able to store the plant toxins and use them in their own defence. A classic example is that of the monarch butterfly, whose caterpillars feed on milkweeds (*Asclepias* spp.). Milkweeds contain secondary chemicals, cardiac glycosides, which affect the vertebrate heartbeat and are poisonous to mammals and birds. Caterpillars of the monarch butterfly can store the poison, and it is still present in the adult which in consequence is completely unacceptable to bird predators. For example, a naïve blue jay (i.e. one that has not tried a monarch butterfly before) will vomit violently after eating one, and, once it recovers, will reject all others on sight. In contrast, monarchs reared on cabbage, or on one of the few milkweed species that lack the glycoside, are edible (Brower & Corvino, 1967).

'one man's meat is another man's poison'

Chemical defences are not equally effective against all consumers. Indeed, what is unacceptable to some animals may be the chosen, even unique diet of others: many herbivores, particularly insects, specialize on one

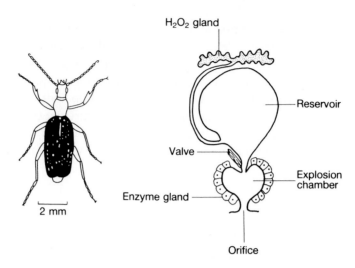

Figure 3.25. The bombadier beetle (*Brachinus crepitans*) and its defensive apparatus which opens close to the anus. The mixture of chemicals ejected into the explosion chamber causes a spray of quinone to be ejected. (Gland based on a figure in Eisner & Meinwald, 1966.)

or a few plant species whose particular defence they have overcome. An important step in the evolutionary interaction of plants and herbivores is the evolution of tolerance of, and then even attraction to and utilization of, the chemical defence and the plant producing it. For example, gravid females of the cabbage root fly (*Delia brassicae*) home in on a brassica crop from distances as far as 15 m downwind of the plants (Figure 3.26). It is probably hydrolysed glucosinolates (toxic to many other species) that provide the attractive odour.

morphology and colour as defence: crypsis

An animal may reduce the likelihood of a predator spotting it by matching its background, by possessing a pattern which disrupts its outline or by resembling an inedible feature of its environment. Straightforward examples of such *crypsis* are the green coloration of many grasshoppers and caterpillars and the transparency of many planktonic animals that inhabit the surface layers of oceans and lakes. More dramatic cases are the sargassum fish, whose body outline mimics the sargassum weed in which it is found, the many small invertebrates that closely resemble twigs, leaves and flower parts (Figure 3.27), and the caterpillar of the viceroy butterfly which is camouflaged as a bird dropping (Plate II, p. 84). Cryptic animals may be highly palatable,

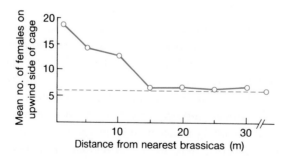

Figure 3.26. The mean numbers of female cabbage root flies (*Delia brassicae*) orientating upwind towards a brassica crop, in relation to their distance from the plants (Hawkes, 1974).

but their morphology and colour (and behavioural selection of the appropriate background) reduce the likelihood that they will be used as a resource.

warning coloration

While crypsis may be a defence strategy of a palatable organism, noxious or dangerous animals often seem to advertise the fact by bright, conspicuous colours and patterns. These are said to be aposematically coloured and the phenomenon is referred to as *aposematism.* The monarch butterfly, discussed above in the context of its chemical defence and distastefulness to birds, is aposematically coloured, as is the caterpillar which actually sequesters the defensive cardiac glycosides from its food (Plate II). One attempt by a bird to eat such a butterfly is so memorable that others are subsequently avoided for some time. But clearly the advantage will only accrue to other monarch butterflies if they *look like* those which have actually been tasted (and killed). Thus there is a strong selection pressure for monarch butterflies to look like their conspecifics and for their form and colour to be easily memorized by a predator.

Batesian mimicry

The adoption of memorable body patterns by distasteful prey immediately opens the door for deceit by other species because there will be a clear selection advantage to a palatable prey (the mimic) if it looks like an unpalatable species (the model). Such *Batesian mimicry* is a widespread phenomenon. In one sense it is a special case of camouflage (crypsis), because the mimic resembles something else in the environment, but it differs because the mimic, rather than remaining inconspicuous, produces clear signals likely to be detected by the predator. Developing the story of the monarch butterfly a little further, we can note that the adult of the palatable viceroy butterfly mimics the distasteful monarch, and a blue jay that has learned to avoid monarchs will also avoid viceroys. Intriguingly, while the distasteful caterpillar of the monarch, like the adult, is aposematically coloured, the caterpillar of the palatable viceroy is cryptically coloured and resembles a bird dropping (Plate II).

Mimicry of a distasteful organism appears not to be confined to animals. The mistletoes, for instance, often show a remarkable resemblance to their distasteful host plants.

behavioural defences

By living in holes, animals (millipedes, moles) may avoid stimulating the sensory receptors of predators; by 'playing dead' (opossum, African ground squirrel, many beetles and grasshoppers) animals may fail to stimulate a killing response. Animals which withdraw to a prepared retreat (rabbits and prairie dogs to their burrows, snails to their shells) or which roll up and protect their vulnerable parts by a tough exterior (armadillos, hedgehogs, pill millipedes) reduce their chance of capture. However, once the rabbit or armadillo has withdrawn into its 'castle' it stakes its life on the chance that the attacker will not be able to breach its walls (a rabbit burrow is no defence against a weasel), and it has also sacrificed its ability to tell what is going on outside. Other animals seem to try to bluff themselves out of trouble by threat displays. The startle response of moths and butterflies which suddenly expose eye spots on their wings is one example. Others include the rattling of quills by the African porcupine and tail erection and foot-stamping by the skunk.

No doubt the most common behavioural response of an animal in danger

Figure 3.27. Cryptic animals. Top left: Buff-tip moth (*Phalera bucephala*). Right: Bush cricket (*Lamprophyllum sp.*), Barro Colorado Island. Bottom left: Marbled gecko (*Phyllodactylus porphreus*). (Photographs: Heather Angel.)

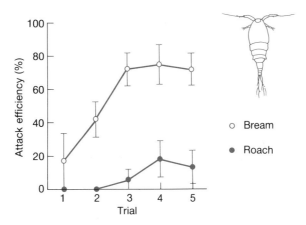

Figure 3.28. With increasing experience, young bream (*Abramis brama*) learn to capture the elusive copepod prey (*Cyclops vicinus*—shown in vignette) but roach (*Rutilus rutilus*) never achieve more than a 20% success rate. (From Winfield *et al.*, 1983.)

of being preyed upon is to flee. The small lake-dwelling copepod *Cyclops vicinus*, for example, when a young fish is near, exhibits a rapid, darting motion that brings it safely through an encounter with all but the most effective of predators. Figure 3.28 shows how a young bream, with its protrusible mouth and ability to develop strong suction pressure, can learn to deal successfully with *Cyclops*, while the less well equipped roach cannot cope efficiently with such elusive prey.

This example illustrates three important points. First, like all defensive behaviours, 'flight' reduces the likelihood that the animal will end up as a food resource. Secondly, and perhaps inevitably, one or more predators exist which can overcome the defence. Thirdly, the defence costs energy, and the rewards (escape) must be seen as having to be paid for by the sacrifice of energy or other resources that might have been used in other activities.

3.5 Space as a resource

space as a portmanteau term

All living organisms occupy space, and in a crude sense they can often be said to compete for it—for example, a plant may be said to compete with another for a space in a canopy. But it is usually more accurate to say that the plants are competing for the light, or some other resource, that may be captured in that space. Similarly, plants may be said to require root space, but this seems inevitably to be saying that they require the resources of water and nutrients that are in the space. The word 'space' may then be used as a 'portmanteau' word to describe the resources that may be captured within it rather than regarding space as a resource itself. There are, however, a few cases in which one can stretch a point and recognize that the corms of a crocus may become so crowded that they force each other out of the ground. However, a forest community will contain abundant space for more trees, and the surface of a pond may contain enough space for many more layers of floating duck-weeds—though in both cases the supply of light or mineral resources in that space will ultimately limit growth, not a shortage of the space itself.

Space becomes a potentially limiting resource only when the physical packing of the organism limits what they can do, even with an abundance of food. Barnacles and mussels may pack on a rock surface at densities at which there is no physical room for others, and the territorial behaviour of a bird may define a space that is defended. The space contains resources but these are obtained by virtue of the occupation of the space. There is a sense in which the behaviour of a territorial animal has itself made space into a resource. This becomes important when we consider the ways in which organisms compete with each other for resources. We shall see in Chapter 7 (when competition between organisms is considered) that it is possible to distinguish on the one hand situations in which individual A reduces the level of a resource and individual B reacts to this reduced level. The two organisms respond not to each other's presence but to the level of resource depletion that each produces (exploitation competition). On the other hand, particularly in higher animals and birds, the level of conflict may be shifted so that A and B interact directly by capturing space (territory)—they then react directly to each other (interference competition), rather than to the level to which they have depleted the resource.

exploitation and interference

In other cases, it is abundantly clear that space itself, and not the resources it contains, comprises the resource. Thus we can envisage competition between lizards for warm basking sites on rocks. Lizards can hardly be said to be consuming temperature (a condition) but they certainly use up favourable microsites, making them unavailable for others. By definition, such microsites are resources. In addition, we can identify nest sites and hiding places as potential resources for many kinds of animal.

sites as resources

3.6 A classification of resources

Having discussed a variety of resources, we will now describe a system, developed by Tilman (1982), by which resources can be classified. This should serve as a useful summary and synthesis of a number of ideas from the present chapter, but it will also, in later chapters, help to clarify our thinking about the ways in which organisms compete with each other.

We have seen that every plant requires 20–30 distinct resources to complete its life cycle, and most plants require the same set of resources though in subtly different proportions. Each of the resources has to be obtained independently of the others, and often by quite different uptake mechanisms—some as ions (K^+), some as molecules (CO_2), some in solution, some as gases. Carbon cannot be substituted by nitrogen, nor phosphorus by potassium. Only a very few of the resources needed by higher plants may be substituted in whole or in part. Nitrogen can be taken up by most plants either as nitrate or as ammonium ions, but there is no substitute for nitrogen itself. In complete contrast, for many carnivores, most prey of about the same size are wholly substitutable one for another as articles of diet. This contrast between resources that are individually *essential* for an organism, and those that are *substitutable*, can be extended into a classification of resources taken in pairs (Figure 3.29).

In this classification, the concentration or quantity of one resource is plotted on the *x*-axis, and that of the other resource on the *y*-axis. We know that different combinations of the two resources will support different growth rates for the organism in question (this can be individual growth or population growth, i.e. survival and reproduction). Thus we can join together points (i.e. combinations of resources) with the same growth rates, and these are therefore contours or 'isoclines' of equal growth. In Figure 3.29, for example, line B in each case is an isocline of zero net growth. In other words, each of the resource combinations on these lines allows the organism just to maintain itself, neither increasing or decreasing. The A isoclines, therefore, with less resources than B, join combinations giving the same *negative* growth rate; while the C isoclines, with more resources than B, join combinations giving the same *positive* growth rate. As we shall see, it is the shapes of the isoclines that vary with the nature of the resources.

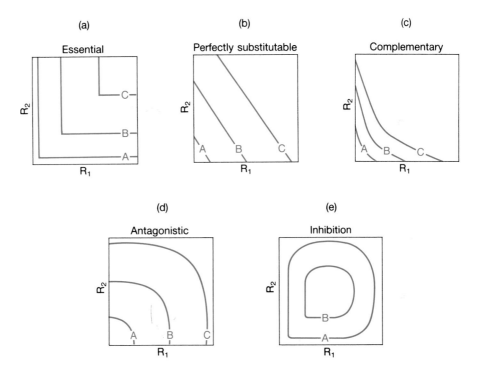

Figure 3.29. Resource-dependent growth isoclines. Each of the growth isoclines represents the amounts of two resources R_1 and R_2 that would have to exist in a habitat for a population to have a given growth rate. Because this rate increases with resource availability, isoclines further from the origin represent higher population growth rates—isocline A has the lower rate, isocline B an intermediate rate, and isocline C the higher rate. (a) Essential resources; (b) perfectly substitutable; (c) complementary; (d) antagonistic; and (e) inhibition. (From Tilman, 1982.)

3.6.1 Essential resources

Two resources are said to be *essential* when one is unable to substitute for another. Thus, the growth that can be supported on resource 1 is absolutely dependent on the amount available of resource 2. This is denoted in Figure 3.29a by the isoclines running parallel to both axes. They do so because the amount available of one resource defines a maximum possible growth rate irrespective of the amount of the other resource. This growth rate is achieved unless the amount available of the other resource defines an even lower growth rate. It will be true for nitrogen and potassium as resources in the growth of green plants, and for the two obligate hosts in the life of a parasite or pathogen which is required to alternate in its life cycle (Chapter 12). In the life of *Heliconius* butterflies, resource 1 might be the leaves of a particular species of *Passiflora* vine for the larval food, and resource 2 would be the pollen of the cucurbit plant *Gurania* for the adult butterfly.

3.6.2 Substitutable resources

Two resources are said to be *perfectly substitutable* when either can wholly replace the other. For most green plants this will be true of nitrate and ammonium ions as sources of nitrogen nutrition. It will also be true for seeds of wheat or barley in the diet of a farmyard chicken, or for zebra and gazelle in the diet of a lion. Note that we do not imply that the two resources are as good as each other. In fact nitrate ions will be more costly to the plant as a nitrogen source than ammonium, because the plant must expend metabolic energy in reducing nitrate to ammonium before it can be assimilated into protein. This feature (perfectly substitutable but not necessarily as good as each other) is included in Figure 3.29b by the isoclines having slopes that do not cut both axes at the same distance from the origin. Thus, in Figure 3.29b, in the absence of resource 2, the organism needs relatively little of resource 1, but in the absence of resource 1 it needs a relatively large amount of resource 2.

complementary resources

Substitutable resources are defined as *complementary* if the isoclines bow inwards towards the origin (Figure 3.29c). This shape means that a species requires less of two resources when taken together than when consumed separately. Humans eating certain kinds of beans together with rice can increase the usable protein content of their food by 40% (Lappe, 1971). The beans are rich in lysine, an essential amino acid poorly represented in rice, while rice is rich in sulphur-containing amino acids that are present only in low abundance in beans.

antagonistic resources

A pair of substitutable resources with isoclines that bow away from the origin are defined as *antagonistic* (Figure 3.29d). The shape indicates that a species requires proportionately more resource to maintain a given rate of increase when two resources are consumed together than when consumed separately. This could arise, for example, if the resources contain different toxic compounds which act synergistically (more than just additively) on their consumer. For example, D,L-pipecolic acid and djenkolic acid (two secondary chemicals believed to have a defensive function in certain seeds) had no significant effect on the growth of the seed-eating larva of a bruchid beetle if consumed separately, but they had a pronounced effect if taken together (Janzen *et al.*, 1977). If the seed of one species contained one of these compounds, and another seed the second compound, then a mixed diet would produce an adverse effect on growth.

Finally, Figure 3.29e illustrates the phenomenon of *inhibition* at high resource levels (for a pair of essential resources). It is not difficult to find examples of resources that are essential but become toxic or damaging when in excess. Carbon dioxide, water and mineral nutrients such as iron are all required for photosynthesis, but each is lethal in excess. Similarly, light leads to increased growth rates in plants through a broad range of intensities, but can inhibit growth at very high intensities. In such cases, the isoclines form closed curves because growth decreases with an increase in resources at very high levels. This is a situation in which what are *resources* at one level become limiting *conditions* at another.

3.7 Resource dimensions of the ecological niche

In Chapter 2 we developed the concept of the ecological niche as an *n*-dimensional hypervolume. This defines the limits within which a given species can survive and reproduce, for a large number (*n*) of environmental factors, including both conditions (Chapter 2) and resources (this chapter). Note, therefore, that the zero growth isoclines in Figure 3.29 (the B lines) define niche boundaries in two dimensions. Resource combinations to one side of the line allow the organisms to thrive—but to the other side of the line the organisms decline.

The resource dimensions of a species' niche can sometimes be represented in a manner similar to that adopted for conditions, with lower and upper limits within which a species can thrive. Thus, for a size-selective predator, limits to its ability to detect and handle prey mean that it is able to exploit only a limited range within a continuum from tiny to very large potential prey. For other resources, such as mineral nutrients for plants, a lower limit to nutrient concentration may be defined below which individuals cannot grow and reproduce—but an upper limit may not exist (Figure 3.29a–d). Given a range of nitrate concentrations in a patchy soil environment, individuals of a particular plant species may be able to thrive in *all* locations above a minimum concentration. However, as we noted in section 3.6.2, for some plant resources too much of a good thing may be deleterious (e.g. water, Fe^{2+} ions, solar radiation), and in these cases an upper limit to the resource dimension may also be defined (Figure 3.29e). Finally, many resources cannot be described as continuous variables; rather, they must be viewed as discrete entities. Larvae of butterflies in the genus *Heliconius* require *Passiflora* leaves to eat; those of the monarch butterfly specialize on plants in the milkweed family; and various species of animals require nest sites with particular specifications. These resource requirements cannot be arranged along a continuous graph axis labelled, for example, 'food plant species'. Instead, the food plant or nest site dimension of their niches needs to be defined simply by a restricted list of the appropriate resources.

Resource dimensions, however they are defined, are crucial components of the *n*-dimensional niche. We will see that the niche concept is at its most powerful when considering problems related to the use of limiting, or potentially limiting, resources. It will figure prominently in later chapters which consider the potential role of interspecific competition for resources in determining community composition and diversity (Chapters 7, 18 and 22).

There are some striking differences in the ways in which plant and animal ecologists have studied environmental resources and their consumption. Much of the history of plant ecology has been dominated by 'ecophysiology' and particularly by the ways in which *individual* plants obtain their resources of light, water and nutrients. There has been a strong emphasis on laboratory studies made on individual plants. In contrast, the animal ecologist has more often concentrated on resources as the object of competition between organisms, or as the material of predator–prey interactions. Some of this historical difference in approach is apparent in the structure of this chapter. As ecology becomes an increasingly integrated science, these differences in approach may largely disappear.

4! Life and Death in Unitary and Modular Organisms

4.1 Introduction: an ecological fact of life

In this chapter we change the emphasis of our approach. We will not be concerned so much with the interaction between individuals and their environment, but rather with the numbers of individuals and the processes leading to changes in the numbers of individuals.

In this regard, there is a fundamental and unalterable ecological fact of life—and this is it:

$$N_{now} = N_{then} + B - D + I - E.$$

In other words, the numbers of a particular organism presently occupying a particular site of interest (N_{now}) is equal to the numbers previously there (N_{then}), plus the number of births between then and now (B), minus the number of deaths (D), plus the number of immigrants (I), minus the number of emigrants (E). Or, to put it in a slightly different way:

$$N_{future} = N_{now} + B - D + I - E.$$

where B, in this case, is the number of births between now and some time in the future, and so on.

These facts of life define the main aim of ecology: to describe, explain and understand the distribution and abundance of organisms. Thus, if ecologists study the effects of temperature, or light, or a pollutant such as mercury, on a particular organism, then they will probably concentrate on just one phase or aspect of the organism's life; but the study has ecological relevance only insofar as the particular phase or aspect affects the birth, death or migration of the organism. Ultimately the aim is to improve our understanding of N_{now} or to predict N_{future}. If other ecologists study the effects of an insect pest on a crop, then they will probably be trying to find ways of increasing N_{future} for the crop by reducing N_{future} for the pest; and if they study the distribution of a rare plant in a protected site, then they will monitor the variation in N_{now} from place to place and probably try to find out whether particular micro-sites favour colonization and birth or hasten death. Finally, ecologists studying the animal and plant community of a polluted stream will probably catalogue N_{now} for a wide range of species, and compare these with data from a similar but unpolluted stream. In all cases, therefore, ecologists are interested in the number of individuals, the distributions of individuals, the *demographic processes* (birth, death and migration) which influence these, and the ways in which these demographic processes are themselves influenced by environmental factors.

ecologists are interested in demographic processes, their consequences, and the influences on them

It is for this reason that the present chapter lays foundations of ecology by examining patterns of birth and death, and to a lesser extent migration. (Migration will be considered in much greater detail in Chapter 5.) Particular attention will also be paid to the ways in which these patterns are quantified, and to discovering if there are generalizations linking apparently dissimilar types of organisms.

4.2 What is an individual? Unitary and modular organisms

First, however, an important assumption, implicit in what has been said so far, must be recognized and rejected. The 'ecological fact of life' was framed in terms of 'individuals'. This implies that any one individual is just like any other, which is patently false on a number of counts.

In the first place, almost all species pass through a number of *stages* in their life cycle. Insects metamorphose from eggs to larvae to adults, and some have a pupal stage as well; plants pass from seeds to seedlings to photo-synthesizing adults; and so on. In all such cases, the different stages are likely to be influenced by different factors and to have different rates of migration, death and of course reproduction. These stages need to be recognized and treated separately.

individuals differ in their life cycle stage and their quality

Secondly, even within a stage, or where there are no separate stages, individuals can differ in 'quality'. The most obvious aspect of this is size, but it is also common for individuals to differ in the amount of stored reserves they possess, as Figure 4.1 shows for the quantity of stored fat in brown trout (Elliott, 1976).

The most important area in which the simplistic view of an individual breaks down, however, is whenever an organism is *modular* rather than *unitary*. Thus, before we go on to examine and quantify patterns of birth and death (section 4.3 onwards), we must discuss the differences between unitary and modular organisms.

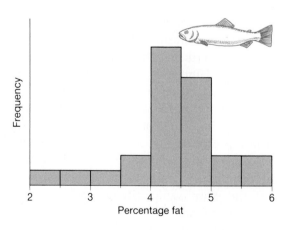

Figure 4.1. Variation in individual quality: fat content of a sample of small trout all weighing close to 11 g (after Elliott, 1976).

4.2.1 Unitary and modular organisms

unitary organisms

In unitary organisms, form is highly determinate. Barring aberrations, all dogs have four legs, all locusts have six, all fish have one mouth and all squid have two eyes. Humans are perfect examples of unitary organisms. A life begins when a sperm fertilizes an egg to form a zygote. This implants in the wall of the uterus, and the complex processes of embryonic development commence. By six weeks the foetus has a recognizable nose, eyes, ears and limbs with digits, and accidents apart, it will remain in this form until it dies. The foetus continues to grow until birth, and then the infant grows until perhaps the eighteenth year of life; but the only changes in form (as opposed to size) are the relatively minor ones associated with sexual maturity. The reproductive phase lasts for perhaps thirty years in females and rather longer in males. This is followed, by or may merge into, a phase of senescence. Death can intervene at any time, but for surviving individuals the succession of phases is, like form, entirely predictable.

modular organisms

In modular organisms, on the other hand, the zygote develops into a unit of construction (a module) which then produces further modules like the first. The product is almost always branched and, except for a juvenile phase, immobile. In contrast to unitary organisms, individual modular organisms are composed of a highly variable number of basic elements, and their programme of development is unpredictable and strongly dependent on their interaction with their environment. Most plants are modular, and plants are undoubtedly for many people the most obvious group of modular organisms. There are, however, many important groups of modular animals (indeed, some 19 phyla, including sponges, hydroids, corals, bryozoans and colonial ascidians) and many modular protists and fungi.

Many ecological and evolutionary generalizations have been made in the past as if the unitary animal (such as a man or a mosquito) in some way typifies the living world. This is dangerous. Modular organisms such as seaweeds, corals, forest trees and grasses dominate large parts of the terrestrial and aquatic environments.

growth in higher plants

In the growth of a higher plant, the fundamental module of construction above ground is the leaf with its axillary bud and the attendant internode of the stem. As a bud develops and grows, it in turn produces leaves, each bearing buds in their axils. The plant grows by accumulating these modules. At some stage in development a new sort of modified module appears, associated with reproduction—these are the flowers of a higher plant (or the gonophores of a hydroid, etc.)—ultimately giving rise to new zygotes which develop to an embryonic stage. Modules that are specialized for reproduction usually cease to give rise to new modules (though this is not true of all modular animals). The programme of development in modular organisms is typically determined by the proportion of modules that are allocated to different roles (e.g. to reproduction or to continued growth).

modules may be physiologically separate

Among plants (and modular organisms generally) we can recognize two broad categories of form: those concentrating on vertical growth (in which placing leaves higher than those of neighbours may have dominated their evolution), and those spreading themselves laterally, expanding their modules on or in a substrate. Many plants that spread laterally produce new root systems associated with the laterally extending stem: these are the rhizomatous and stoloniferous plants. The connections between the parts of such plants may die and rot away, so that the product of the original zygote becomes represented by physiologically separated parts (the modules with the potential for separate existence are known as 'ramets'). The products of a single zygote are called a clone, and amongst clonal plants that 'fall to pieces' as they grow, the most extreme examples are the many species of floating aquatics like duckweeds (*Lemna*), the water hyacinth (*Eichornia*) and the water lettuce (*Pistia*), in which whole ponds, lakes or rivers may be filled with the separate and independent parts produced by a single zygote.

The peculiar feature distinguishing trees and shrubs from most herbs is the connecting system linking modules together and connecting them to the root system. This does not rot away, but thickens with wood, conferring perenniality. Most of the structure of such a woody tree is dead, with a thin layer of living material lying immediately below the bark. The living layer, however, continually regenerates new tissue, and adds further layers of dead material to the trunk of the tree. Most of a tree is a sort of cemetery in which dead stem tissues of the past are interred. Trees represent, *par excellence*, plants whose growth is concentrated vertically. Most of the characteristic form of a tree such as cypress, spruce, hemlock or oak, is determined by the ways in which its modules are displayed.

We can often recognize two or more levels of modular construction. The fundamental units of higher plants—the leaves with their axillary buds—are assembled into clusters with a form that is itself continually repeated. The strawberry is a good example of this. Typical strawberry leaves are repeatedly developed from a bud, but these leaves are arranged into rosettes. The strawberry plant grows (a) by adding new leaves to a rosette, and (b) by producing new rosettes on stolons grown from axils of its rosette leaves. In trees too we can recognize modularity at several levels: the leaf with its axillary bud, the whole shoot on which the leaves are arranged, and whole branch systems that repeat a characteristic pattern that is reiterated after damage. The form or architecture of modular organisms is determined, essentially, by the angles between successive modules and the length of the stems or internodes that link them.

A variety of growth forms and architectures produced by modular growth are illustrated in Figure 4.2. The growth of modular animals can be illustrated by a hydrozoan like *Obelia*. Development begins when a short-lived, free-swimming planula larva attaches itself to a solid object. It gives rise to a horizontal root-like structure which bears a number of branched stalks. The basic *Obelia* modules, the polyps (which are both feeding and defensive structures), are borne on these stalks. The terminal polyp of each branch is

Figure 4.2. A range of modular organisms: plants to the left and animals to the right. (a) Modular organisms that fall to pieces as they grow: duckweed (*Lemna* sp.) and *Hydra* sp. (a non-colonial cnidarian). Freely branching organisms that are relatively short-lasting: the annual hare's foot clover (*Trifolium arvense*) and *Pennaria* sp. (Cnidaria: Hydrozoa). (c) Rhizomatous and stoloniferous organisms in which clones spread laterally: the common polypody (*Polypodium vulgare*) and *Campanularia* sp. (Cnidaria: Hydrozoa). (d) Tussock-formers comprising tightly packed modules: six-weeks fescue grass (*Festuca octoflora*) and *Cryptosula* sp. (an encrusting bryozoan). (e) Persistent, multiple-branched organisms: an oak tree (*Quercus* sp.) and *Gorgonia* sp., the common sea fan (a coral—Cnidaria: Anthozoa).

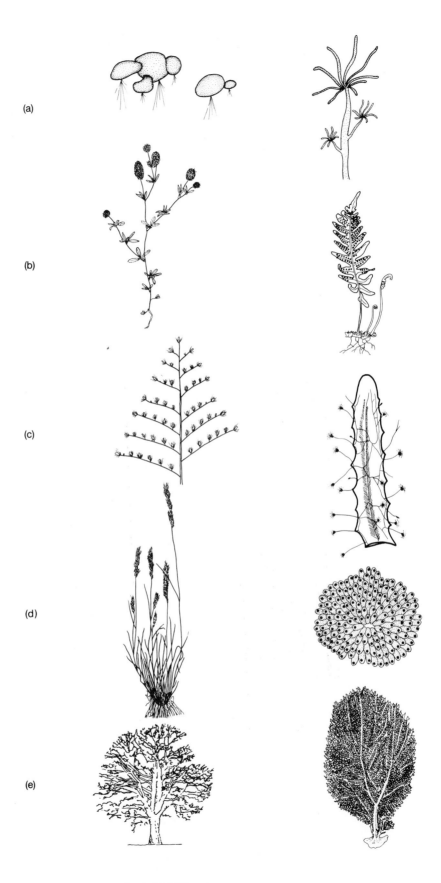

temporarily the youngest, but is overgrown by the next one to develop, which arises as a bud at its base. The branched stalks remain as an interconnecting network between all the polyps in a colony. The basic similarity between this and the modularity of plants is clear. Reproduction in *Obelia* begins when tiny, free-swimming jellyfish are budded off from modified polyps called gonophores; these jellyfish then reproduce sexually to produce the planula larvae. Other 'colonial' animals differ in their precise method of growth and reproduction. They are nonetheless modular, and must be considered as such when examining their ecological characteristics.

individual modular organisms: genets

Returning to the question of what is an individual, one can determine the number of individual rabbits in a field by counting their ears and dividing by two, or their legs and dividing by four—the result is the number of zygotes that have survived. But there is no divisor that makes such a calculation possible from the number of leaves of a higher plant, the fronds of a fern, or the zooids of an ascidian or bryozoan. In other words, N_{now} may represent the current number of surviving zygotes, but it can give only a partial and misleading impression of the 'size' of the population if the organism is modular. This led Kays and Harper (1974) to coin the word 'genet' to describe the 'genetic individual': the product of a zygote. The genet may then be contrasted with the module that forms its parts—be the module a ramet, a shoot, a tiller, a zooid, a polyp, or whatever.

module abundance is often more important than genet abundance

In modular organisms, therefore, while the distribution and abundance of genets (individuals) is important, it is also necessary, and often more useful, to study the distribution and abundance of modules. The amount of grass in a field available to a herd of cattle, for instance, is not determined by the number of grass genets but by the number of leaves, the number of modules; and if the changing abundance of food is to be monitored, then it is the births and deaths of modules which are important, not the births and deaths of genets. Modular organisms should therefore be studied at two interacting levels. For them, the 'ecological fact of life' comprises at least two equations: the one already established, plus another:

$$\text{modules}_{now} = \text{modules}_{then} + \text{modular birth} - \text{modular death}.$$

modularity can lead to extreme individual variability

As has already been noted, there are individual differences amongst unitary organisms. But amongst modular organisms, the potentialities for individual difference are far greater. An individual of the annual plant *Chenopodium album* may, if grown in poor or crowded conditions, flower and set seed when only 50 mm high. Yet, given more ideal conditions, it may reach 1 m in height, and produce 50000 times as many seeds as its depauperate counterpart. It is modularity, and the differing birth and death rates of plant *parts*, which gives rise to this plasticity. The vital processes of birth, senescence and death, then, occur not only at the level of the whole organism, but also at the level of the module. In fact there is often no programmed senescence of the whole organism, and indeed clonal plants appear to have perpetual somatic youth, continually losing their old tissues. Even in trees which accumulate their dead stem tissues, or gorgonian corals which accumulate old calcified branches, death often results from becoming too big or succumbing to disease rather than from programmed senescence.

The most dramatic example of modular senescence is the annual death of the leaves on a deciduous tree—but roots, buds and flowers all pass through phases of youth, middle age, senescence and death, and modular animals such as a coral or *Obelia* may be composed of parts in all of these stages at the same time. The growth of the individual genet is the resultant of these processes.

modular individuals have an age-structure

Thus the body of an individual modular organism has an age-structure; it is composed of young and developing, actively functioning, and senescent parts. This means that we can describe the age-structure of a population of modular organisms in two ways: the ages of the genets, or the ages of the parts of which they are composed. There is no such problem in describing the age-structure of a population of rabbits. The issue is important, moreover, because a module such as a leaf or root changes its activity as it ages, and may also change its dietary value and attraction to a herbivore. It usually becomes tougher, has a higher fibre content and lower protein content, and becomes less digestible. For the ecologist to treat all leaves in a pasture or forest canopy as equal is to ignore the fact that other organisms will discriminate between them. Figure 4.3 shows the changes in age-structure of shoots of the sedge *Carex arenaria* brought about by the application of NPK fertilizer. The total number of shoots present was scarcely affected by the treatment, but the age-structure was dramatically changed. The fertilizer plots became dominated by young shoots, as the older shoots which were common on control plots were forced into early death.

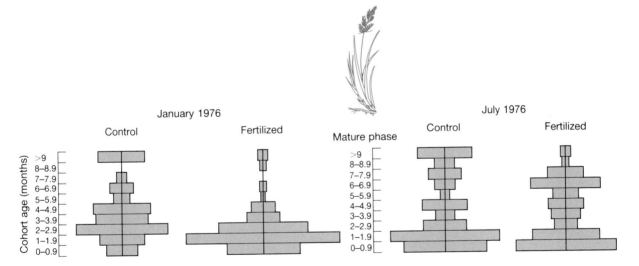

Figure 4.3. The age-structure of shoots in clones of the sand sedge *Carex arenaria* growing on sand dunes in North Wales. Clones are composed of shoots of different ages. The effect of applying fertilizer is to change this age-structure. The clones become dominated by young shoots and the older shoots die. (After Noble *et al.*, 1979.)

Finally, there are two further important differences between unitary and modular organisms.

(i) The taxonomic features by which we distinguish species of modular organism are mainly features of the module, not of the whole organism (we use the form of polyps, flowers and leaves, rather than the form of the whole). (ii) The way in which modular organisms interact with their environment is determined by their architecture—how their modules are placed in relation to the modules of other organisms. There is little or no mobility, search or escape other than can be achieved by growing from one place to another or by releasing specialized dispersal units (or, in some aquatic plants, falling to pieces and floating away).

Thus any consideration of ecology which is to encompass a broad spectrum of life-forms must take as its currency not only birth and death but *modular growth* (i.e. modular birth and death) as well.

4.3 Counting individuals

If we are going to study birth, death and modular growth seriously, we must quantify them. This means counting individuals and (where appropriate) modules. Often in modular organisms we cannot recognize individual genets—their parts can become separated, and clones can intermingle. We can then count only modules.

numerical change is often monitored without monitoring demographic processes

The remainder of this chapter will examine and quantify patterns of birth, death and modular growth. However, many studies concern themselves not with birth and death but with the consequences of them, i.e. the total number of individuals present, and the way these numbers vary with time. Such studies are obviously limited in scope, but they can nevertheless often be useful. Figure 4.4a, for instance, shows that the number of sheep in Tasmania rose steadily from 1820 to 1850 and remained at a fluctuating but fairly stable plateau thereafter; Figure 4.4b shows that the number of water fleas in an experiment fluctuated persistently with a regular periodicity of approximately 40 days; while Figure 4.4c shows the decline in abundance in recent years experienced by three species of whale. Of course, the best studies provide data on the numbers of individuals *and* the processes affecting numbers.

the meaning of 'population'

It is usual to use the term *population* to describe a group of individuals of one species under investigation. However, what actually constitutes a population will vary from species to species and from study to study. In some cases, the boundaries of a population are readily apparent: the sticklebacks occupying a small, homogeneous, isolated lake are obviously 'the stickleback population of that lake'. In other cases, boundaries are determined by an investigator's purpose or convenience: it is possible to study the population of lime aphids inhabiting one leaf, one tree, one stand of trees, or a whole woodland. In yet other cases—and there are many of these—individuals are distributed continuously over a wide area, and an investigator must define the limits of his population arbitrarily. In such cases especially, it is often more

density

convenient to consider the *density* of a population. This is usually defined as 'numbers per unit area', but in certain circumstances 'numbers per leaf', 'numbers per host' or some other measure may be appropriate.

The most straightforward approach to determining the number of individuals in a population is simply to count every individual present. This is a

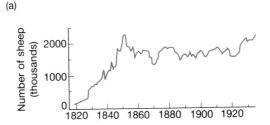

(a)

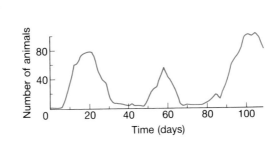

(b)

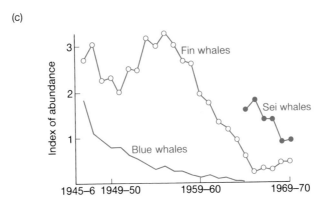

(c)

Figure 4.4. Numerical change in populations. (a) The rise to a fluctuating plateau in Tasmanian sheep (after Davidson, 1938). (b) Regular fluctuations of the water flea *Daphnia magna* (after Pratt, 1943). (c) The decline in abundance of three whale species in the Antarctic, as estimated by an index of abundance, the number caught per 1000 catcher–ton–days (after Gulland, 1971).

complete enumerations

possibility with many plants, and with animals that are either sessile or very slow, or large and conspicuous enough for a whole population to be surveyed without missing individuals or counting others twice. Even when this is possible, however, such 'complete enumerations' frequently require large investments of time and trained manpower.

the sampling of populations

The most common alternative is to *sample* the population. Samples are taken from one or (much more sensibly) several small portions of the population. These portions are of known size and are usually chosen at random so as to be representative of the population as a whole. For plants and animals living on the ground surface, the sample unit is generally a small area known as a *quadrat* (which is also the name given to the square or rectangular device used to demarcate the boundaries of the area on the ground). For soil-dwelling organisms the unit is usually a volume of soil, and for lake-dwellers

131   LIFE AND DEATH

a volume of water; for many herbivorous insects the unit is one typical plant or leaf, and so on. In each case, every individual or module within the unit is counted, and it is therefore assumed that individuals remain within the unit while this is going on. The data from sampling lead immediately to estimates of population density. Density can then be converted to population size if, for instance, in the case of ground surface, the ratio of sampled area to total population area is known. Further details of sampling methods, and of methods for counting individuals generally, can be found in one of many texts devoted to ecological methodology (e.g. Southwood, 1978a; Kershaw, 1973).

capture–recapture For animals especially, there are two further methods of estimating population size or density. The first is known as capture–recapture. Very simply, this involves catching a random sample of a population, marking individuals so that they can be recognized subsequently, releasing them so that they re-mix with the rest of the population, and then catching a further random sample. Population size can be estimated from the proportion of this second sample which bear a mark. Roughly speaking, the proportion of marked animals in the second sample will be high when the population is relatively small, and low when the population is relatively large. Further details are provided by Begon (1979).

indices of abundance The final method is to use an *index* of abundance. This can provide information on the *relative* size of a population, but can usually, by itself, give little indication of absolute size. The relative abundance of fruit-flies, for instance, can be estimated by monitoring the numbers attracted each day to a standard bait. As another example, Figure 4.4c is based on an index of abundance for whales: the numbers caught per unit fishing effort (where effort is measured in terms of 'catcher–ton–days', combining the number of ships, their size, and the length of time they fish). Despite their shortcomings, even indices of abundance can provide valuable information.

4.4 Life cycles and the quantification of death and birth

Having recognized the complexities of individuality and the necessities of counting, it is possible now to examine patterns of birth, death and growth. To a large extent, these are a reflection of the organism's life cycle, of which there are five main types (though there are many life cycles which defy this simple classification). Birth, death and growth in these various life cycles will be discussed in some detail in the remainder of this chapter. The life cycles are summarized diagrammatically in Figure 4.5.

The medium we will be using most often to tabulate and examine patterns of mortality (and its opposite, survivorship) is known as a *life-table*. Frequently this will allow us to construct a *survivorship curve*. A survivorship curve traces the decline in numbers, over time, of a group of newly born or newly emerged individuals; or it can be thought of as a plot of the probability, for a representative newly born individual, of surviving to various ages. Patterns of birth amongst individuals of different ages are often monitored at the same time as life-tables are constructed. These patterns are displayed in what are known as *fecundity schedules*.

life-tables, survivorship curves, and fecundity schedules

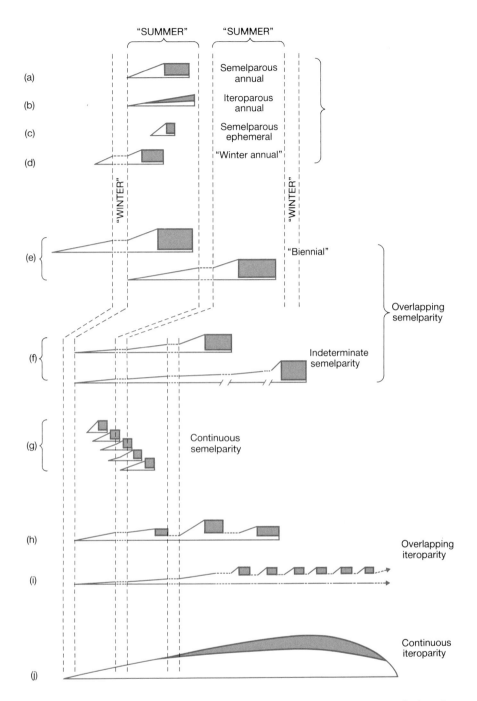

Figure 4.5. A range of life cycles. In each case, the length of a bar represents the length of an individual's life, and the height of a bar represents individual size. The hatched region indicates the proportion of available resources allocated to reproduction. Semelparous species, when they reproduce, allocate a large proportion of resources to reproduction and die soon thereafter. Iteroparous species tend to allocate less to current reproduction but may survive for further reproductive bouts. Most life cycles should be self-explanatory. In overlapping semelparity, species may have a strict two-year (or three-year, etc.) cycle (e.g. part (e)), or they may have a cycle of indeterminate length. In overlapping iteroparity, species may be relatively short-lived (h) or long-lived (i).

Life-tables, survivorship curves and fecundity schedules are of the utmost importance because they contain the raw material of our 'ecological fact of life'. Without them we have little hope of understanding the N_{now}'s of the species that interest us, and still less hope of predicting the N_{future}'s.

4.5 Annual species

Annual life cycles (Figure 4.5a–d) take approximately 12 months or rather less to complete. Usually, every individual in a population breeds during one particular season of the year, but then dies before the same season in the next year. Generations are therefore said to be *discrete*, in that each generation is distinguishable from every other; the only overlap of generations is between breeding adults and their offspring during and immediately after the breeding season. Species with discrete generations need not be annual, since generation lengths other than one year are conceivable. In practice, however, most are: the regular annual cycle of climates provides the major pressure in favour of synchrony. Annual species may be either *semelparous*, in which case each individual has only a single reproductive event in its life, after which it dies, or *iteroparous*, in which case there are many reproductive events continuing throughout a season or the whole year.

semelparity and iteroparity defined

4.5.1 Simple annuals: cohort life-tables

One annual species, living throughout much of Europe, is the common field grasshopper, *Chorthippus brunneus*. The first-instar nymphs emerge from their eggs in late spring, and then feed, grow, and eventually metamorphose into larger but otherwise fairly similar second-instar nymphs. Two other nymphal instars follow, until in mid-summer the fourth instars moult into winged adults. The adult females lay eggs in the soil (in egg pods containing about 11 eggs each), but by mid-November all the adults will have died, so ending that particular generation. Meanwhile, however, the next generation is starting: the eggs begin development, which proceeds until the following spring, when first-instar nymphs emerge once again.

A life-table and fecundity schedule for an isolated population of *C. brunneus* from near Ascot, England is presented in Table 4.1 (Richards & Waloff, 1954). The life-table is known as a 'cohort' (or 'dynamic' or 'horizontal') life-table, because a single cohort of individuals (i.e. a group of individuals born within the same short interval of time) was followed from birth to the death of the last survivor. With an annual species like *C. brunneus*, there is no other way of constructing a life-table.

the columns of a life-table

The first column of Table 4.1 sets out (and numbers) the various stages of the life cycle that have been distinguished. The second column, a_x, then lists the major part of Richards and Waloff's raw data: it gives the total number of individuals observed in the population at each stage. The problem with any a_x column is that the information in it is specific to one population in one year, making comparisons with other populations and other years very difficult. The data have therefore been standardized in a column of l_x-values. This is headed by an l_0-value of 1.000, and all succeeding figures have been brought

Table 4.1. A cohort life-table for the common field grasshopper, *Chorthippus brunneus*. The columns are explained in the text. (After Richards & Waloff, 1954.)

Stage (x)	Number observed at start of each stage a_x	Proportion of original cohort surviving to start of each stage l_x	Proportion of original cohort dying during each stage d_x	Mortality rate q_x	$\log_{10} a_x$	$\log_{10} l_x$	$\log_{10} a_x - \log_{10} a_{x+1}$ $= k_x$	Eggs produced in each stage F_x	Eggs produced per surviving individual in each stage m_x	Eggs produced per original individual in each stage $l_x m_x$
Eggs (0)	44000	1.000	0.920	0.92	4.64	0.00	1.09	—	—	—
Instar I (1)	3513	0.080	0.022	0.28	3.55	-1.09	0.15	—	—	—
Instar II (2)	2529	0.058	0.014	0.24	3.40	-1.24	0.12	—	—	—
Instar III (3)	1922	0.044	0.011	0.25	3.28	-1.36	0.12	—	—	—
Instar IV (4)	1461	0.033	0.003	0.11	3.16	-1.48	0.05	—	—	—
Adults (5)	1300	0.030	—	—	3.11	-1.53	—	22617	17	0.51

$$R_0 = \Sigma \, l_x \, m_x = \frac{\Sigma F_x}{a_0} = 0.51$$

into line accordingly (e.g. $l_2 = 2529 \times 1.000/44000 = 0.575$). Thus, while the a_0-value of 44000 is peculiar to this set of data, *all* studies have an l_0-value of 1.000, making all studies comparable. The l_x-values are best thought of as the proportion of the original cohort surviving to the start of each stage.

To consider mortality more explicitly, the proportion of the original cohort dying during each stage (d_x) has been computed, being simply the difference between l_x and l_{x+1}; for example $d_1 = 0.080–0.058 = 0.022$. The stage-specific mortality *rate*, q_x, has also been computed. This considers d_x as a fraction of l_x, so that, for instance, q_3 (the fraction dying during the third nymphal instar) is $0.011/0.044 = 0.25$. Note too that q_x may be thought of as the average 'chance' or 'probability' of an individual dying. It is therefore equivalent to $(1-p_x)$ where p refers to the probability of survival, i.e. the fraction dying and the fraction surviving must always add up to 1.

The advantage of the d_x-values is that they can be summed: the proportion of the cohort dying as nymphs was $d_1+d_2+d_3+d_4$ (= 0.050). The disadvantage is that the individual values give no real idea of the intensity or importance of mortality during a particular stage. This is because the more individuals there are, the more are available to die, so the d_x-values are larger. The q_x-values, on the other hand, are an excellent measure of the intensity of mortality. In the present example, for instance, it is clear from the q_x column that the mortality rate remained almost constant throughout the early nymphal stages; this is not clear from the d_x column. The q_x-values, however, have the disadvantage of not being liable to summation: $q_1+q_2+q_3+q_4$ does *not* give the overall mortality rate for the nymphs. The advantages are combined, though, in the last column of the life-table, which contains k_x-values (Haldane, 1949; Varley & Gradwell, 1970). The value k_x is defined simply as $\log_{10} a_x - \log_{10} a_{x+1}$ (or, equivalently, $\log_{10} a_x/a_{x+1}$), and is sometimes referred to as a 'killing-power'. Like the q_x's, the k_x's reflect the intensity or rate of mortality (as Table 4.1 shows); but unlike the q_x's, summing k_x's is a legitimate procedure. Thus the killing-power or k-value for the nymphal period is $0.15+0.12+0.12+0.05 = 0.44$, which is also the value of $\log_{10} a_1 - \log_{10} a_5$. Note too that the k_x-values can be computed from the l_x-values as well as from the a_x-values (Table 4.1); and that, like the l_x's the k_x's are standardized, and are therefore appropriate for comparing quite separate studies. In this and later chapters, k_x-values will be used repeatedly.

k-values

a fecundity schedule

The fecundity schedule in Table 4.1 (the final three columns) begins with a column of raw data, F_x: the total number of eggs deposited during each stage. This is followed in the next column by m_x: the individual fecundity or birth rate, i.e. the mean number of eggs produced per surviving individual. (Note that the number of eggs produced per *female* would be roughly twice this value.) Because only the adults reproduced, there are, in the present case, entries in the fecundity schedule only for this final stage.

the basic reproductive rate R_0

Perhaps the most important summary term that can be extracted from a life-table and fecundity schedule is the *basic reproductive rate*, denoted by R_0. This is the mean number of offspring produced per original individual by the end of the cohort, and it therefore indicates, in annual species, the overall extent by which the population has increased or decreased over that time. (As we shall see in section 4.7, the situation becomes more complicated when generations overlap or species breed continuously.)

There are two ways in which R_0 can be computed. The first is from the formula:

$$R_0 = \frac{\Sigma F_x}{a_0}$$

i.e. the total number of eggs produced during one generation divided by the original number of individuals ('ΣF_x' means the sum of the values in the F_x column). The more usual way of calculating R_0, however, is from the formula:

$$R_0 = \Sigma l_x m_x$$

i.e. the sum of the number of eggs produced per original individual during each of the stages (the final column of the fecundity schedule). As Table 4.1 shows, the basic reproductive rate is the same, whichever formula is used.

Another cohort life-table and fecundity schedule is set out in Table 4.2, this time for the annual plant *Phlox drummondii* in Nixon, Texas (Leverich & Levin, 1979). The most obvious difference between this and Table 4.1 is apparent from the first column. Leverich and Levin divided the life cycle of *Phlox* not into a number of stages but into a number of *age-classes*. They took a census of the seed population on various occasions before germination, and then took further censuses at regular intervals until all individuals had flowered and died. The advantage of using age-classes is that it allows an observer to look in detail at the patterns of birth and mortality *within* a stage. The disadvantage is that the age of an individual is not necessarily the best, nor even a satisfactory, measure of the individual's biological 'status'. The grasshoppers, for instance, develop at a rate dependent on their temperature,

a plant life-table based on age-classes not stages

Table 4.2. A cohort life-table for *Phlox drummondii*. The columns are explained in the text. (After Leverich & Levin, 1979.)

Age interval (days) $x-x'$	Number surviving to day x a_x	Proportion of original cohort surviving to day x l_x	Proportion of original cohort dying during interval d_x	Mortality rate per day q_x	$\log_{10} l_x$	Daily killing power k_x	F_x	m_x	$l_x m_x$
0–63	996	1.000	0.329	0.005		0.003	—	—	—
63–124	668	0.671	0.375	0.009	−0.17	0.006	—	—	—
124–184	295	0.296	0.105	0.006	−0.53	0.003	—	—	—
184–215	190	0.191	0.014	0.002	−0.72	0.001	—	—	—
215–264	176	0.177	0.004	0.001	−0.75	< 0.001	—	—	—
264–278	172	0.173	0.005	0.002	−0.76	0.001	—	—	—
278–292	167	0.168	0.008	0.003	−0.78	0.002	—	—	—
292–306	159	0.160	0.005	0.002	−0.80	0.001	53.0	0.33	0.05
306–320	154	0.155	0.007	0.003	−0.81	0.001	485.0	3.13	0.49
320–334	147	0.148	0.043	0.021	−0.83	0.011	802.7	5.42	0.80
334–348	105	0.105	0.083	0.057	−0.98	0.049	972.7	9.26	0.97
348–362	22	0.022	0.022	1.000	−1.66	—	94.8	4.31	0.10
362–	0	0		—		—	—	—	—
							2408.2		2.41

$$R_0 = \Sigma l_x m_x = \frac{\Sigma F_x}{a_0} = 2.41$$

which they regulate by basking in direct sunlight. Thus a 20-day-old grasshopper may be a second instar in a cloudy year or a fourth instar in a sunny year. Similarly, in many long-lived plants (see below), individuals of the same age may be reproducing actively, growing vegetatively but not reproducing, or doing neither. In such cases, a classification based on developmental stages (as opposed to ages) is clearly appropriate. In other cases, the balance of advantage will favour the use of age-classes. Leverich and Levin's decision to use age-classes in *Phlox* was based on the small number of stages, the variation within each, and the synchronous development of the whole population.

It is apparent from Table 4.2 that the use of age-classes has exposed the detailed patterns of fecundity and mortality. (Note that the variable lengths of the age-classes has necessitated the conversion of the q_x and k_x columns to 'daily' rates.) The pattern of fecundity has been plotted in Figure 4.6. The age-specific fecundity, m_x (the fecundity per *surviving* individual), demonstrates the existence of a pre-productive period, a gradual rise to a peak, and then a rapid decline. The reproductive output of the whole population, F_x, parallels this pattern to a large extent, but also takes into account the fact that while the age-specific fecundity was changing, the size of the population was gradually declining. This combination of fecundity and survivorship is an important property of F_x-values, shared by the basic reproductive rate (R_0). It reinforces the point that actual reproduction depends both on reproductive potential *and* on survivorship.

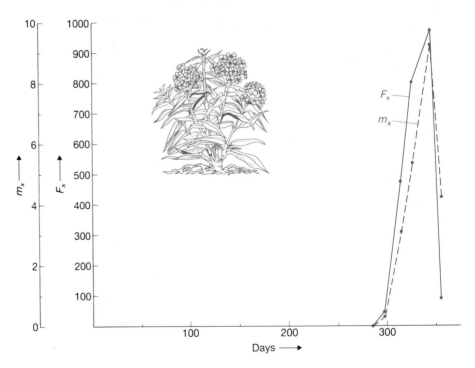

Figure 4.6. The pattern of reproduction in the life cycle of *Phlox drummondii*, as indicated by the seeds produced per individual (m_x) and the output of the whole population (F_x). The two differ because the number of surviving individuals declines with time. (After Leverich & Levin, 1979.)

a realistic picture
requires data for several
years

In the case of the *Phlox*, R_0 was 2.4. This means that there was a 2.4-fold increase in the size of the population over one generation. For the grasshopper the value was 0.51 (the population declined to 0.51 of its former size). If such values were maintained from generation to generation, the *Phlox* population would grow ever larger and soon cover the globe, while the grasshopper population would quickly decline to extinction. Neither of these events has happened. The reasons why will be considered in later chapters. The important point to realise now is that, as judged by these R_0 values, Table 4.1 and Table 4.2 must be atypical. In other words, life-tables and fecundity schedules drawn up in different years would exhibit rather different patterns. Therefore, a balanced and realistic picture of life and death in the grasshopper, *Phlox* or any other species can only emerge from several or many years data.

The pattern of mortality in the *Phlox* population is illustrated in Figure 4.7a, using both q_x and k_x values. Mortality rate was fairly high at the beginning of the seed stage but became very low towards the end. Then, amongst the adults, there was a period where mortality rate fluctuated about a moderate level, followed finally by a sharp increase to high levels during the last weeks of the generation. The same pattern is shown in a different form in Fig. 4.7b. This is a *survivorship curve*, and follows the decline of $\log_{10} l_x$ with age. When the mortality rate is roughly constant, the survivorship curve is straight; when the rate increases the curve is convex; and when the rate decreases, the curve is concave. Thus the curve is concave towards the end of the seed stage, and convex towards the end of the generation. Survivorship curves are the most widely used way of depicting patterns of mortality.

Pearl (1928), in a much-quoted attempt at classification (Figure 4.8), described convex, straight and concave survivorship curves as 'type I', 'type

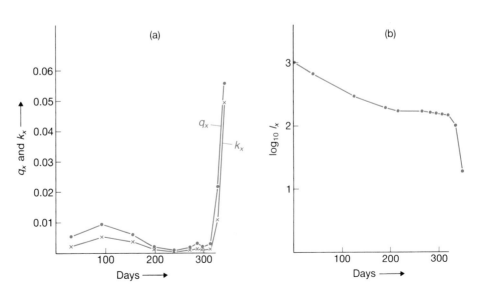

Figure 4.7. Mortality and survivorship in the life cycle of *Phlox drummondii*. (a) The age-specific daily mortality rate (q_x) and daily killing-power (k_x). (b) The survivorship curve: $\log_{10} l_x$ plotted against age. (After Leverich & Levin, 1979.)

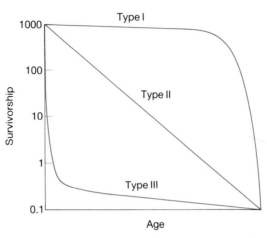

Figure 4.8. A classification of survivorship curves (after Pearl, 1928, and Deevey, 1947). Type I (convex)—epitomized perhaps by humans in rich countries or cosseted animals in a zoo—describes the situation in which mortality is concentrated at the end of the maximum lifespan. Type II (straight) indicates that the probability of death remains constant with age, and may well apply to the buried seed banks of many plant populations. Type III (concave) indicates extensive early mortality, with those that remain having a high rate of survival subsequently. This is true, for instance, of many marine fish which produce millions of eggs of which very few survive to become adults.

II' and 'type III', respectively (also discussed in some depth by Deevey (1947)). In practice, however, most species, when observed throughout their whole life, do not display a single type of curve. Instead, like *Phlox*, they display a succession of shapes as stage follows stage.

'death before birth'

Most survivorship curves show changes in numbers of individuals from the time they are born until they die. This completely misses a stage at which massive and ecologically important mortality may occur. The life of an individual starts with the formation of the zygote, and in mammals and seed plants the first part of an individual's life is spent within the mother's body. Competition between the individuals may start at this stage, as developing embryos compete for limited maternal resources. Those failing to obtain sufficient resources may be resorbed by the mother or may simply cease growth and abort. There is extraordinarily little information about this part of the survivorship curve for most animals or plants, but Wiens (1984) gives data which suggest that abortion rates of ovules are of the order of 15% for annuals and 50% for perennial plants. In humans only about 31% of zygotes survive to birth (Biggers, 1981).

the use of logarithms

The reader may have noticed that the *y*-axis in Figure 4.7b is logarithmic, and the use of logarithms in survivorship curves is important. This can be seen most clearly by comparing two imaginary investigations of the same population. In the first, the whole population is censused. In the second, samples are taken which are equivalent to one-tenth of the population, though this proportion is not known to the investigator. Suppose that in the first case, the population declines in one time interval from 1000 to 500 individuals. Then, in the second, the index of density would decline from 100 to 50. The slope of an *arithmetic* survivorship curve would be -500 in the first case but -50 in the second, despite the fact that the two cases are *biologically* identical, i.e. the rate or probability of death per individual (the 'per capita' rate) is the same. This identity would be reflected, however, in logarithmic survivorship curves: in both cases the slope would be -0.301. The use of logarithms therefore gives to survivorship curves a desirable quality which 'rates' like q_x, k_x and m_x have by definition. (Notice, in passing, that plotting numbers on a logarithmic scale will also indicate when per capita rates of *increase* are identical.) 'Log numbers' will therefore often be used in preference to 'numbers' when numerical change is being plotted.

semelparity and iteroparity illustrated

Finally, there is another difference between the grasshopper and *Phlox* which is not apparent from Tables 4.1 and 4.2. Although the reproductive season for the *Phlox* population lasts for 56 days, each individual plant is semelparous. It has a single reproductive phase during which all of its seed develop synchronously (or nearly so). The extended reproductive season occurs because different individuals enter this phase at different times. The grasshopper on the other hand is iteroparous. Each adult female, assuming she survives for long enough, produces a number of egg pods, with a period of perhaps a week or more between each. During these intervals between pods, active maintenance of body tissues persists, and stored reserves may be laid down.

In fact, semelparity is the more common pattern for annual plants; while for animals, iteroparity is more usual. There are exceptions, however. In the freshwater isopod *Asellus aquaticus,* for instance, females generally lay a single clutch of eggs which they hold, externally, beneath their abdomen, 'brooding' them there until the independent juveniles can fend for themselves. Indeed, if for some reason these eggs are not fertilized, the semelparous female will die without ever laying a replacement clutch. Conversely, a number of iteroparous annual plants, unlike *Phlox*, start to flower when small, and continue to grow, flower and set seed until they die from some extrinsic cause, such as frost or drought. This is the case with many species of *Veronica* (speedwell), with *Poa annua* (annual meadow grass) and with *Senecio vulgaris* (groundsel).

4.5.2 Seed banks

Using *Phlox* as an example of an annual plant has, to a certain extent, been misleading, because the group of seedlings developing in one year *is* a true cohort: it derives entirely from seed set by adults in the previous year, and seed which does not germinate one year will not survive till the next year. In most 'annual' plants this is not the case. Instead, seed accumulates in the soil in a buried *seed bank*. At any one time, therefore, seeds of a variety of ages are likely to occur together in the seed bank; and when they germinate, the seedlings will also be of varying ages (age being the length of time since the seed was first produced).

The formation of something comparable to a 'seed bank' is rather rarer amongst animals, though there are examples to be seen among the eggs of nematodes, mosquitos and fairy shrimps, the gemmules of sponges, the statocysts of bryozoans, and so on. In the case of the fairy shrimp, *Strepto-cephalus vitreus*, in Kenya, resistant eggs are present in the dry sediment after a pool has dried out. Only a proportion hatch at the onset of the next rains, and this happens within three days of the pool re-forming. The other eggs present require at least one more drying and wetting sequence before they hatch. Even eggs produced by the same female in the same batch show variation in hatching—some hatch after one drying–wetting sequence, others only after two or more such episodes (Hildrew, 1985). This presumably serves to spread the risk of eggs hatching and failing to reproduce before the pool dries.

Figure 4.9 illustrates the results of an experiment designed to follow the fates of some annual weed seeds in horticultural soil (Roberts, 1964). No new seed was allowed to enter the soil after the start of the experiment (year 0) but, as Figure 4.9 shows, many of the seeds remained viable in the soil for periods greater than a year, and seedlings therefore developed from seeds of various ages.

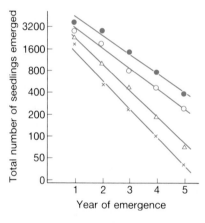

Figure 4.9. The consequence of a seed bank. The number of seedlings emerging in successive years from a horticultural soil into which there was no immigration of seed after year zero. ● = *Thlaspi arvense*, △ = *Stellaria media*, ○ = *Capsella bursa-pastoris*, × = *Senecio vulgaris*. The data represent the combined results of experiments started in 1953, 1954 and 1955. (After Roberts, 1964.)

Figure 4.9 also shows that the rate at which numbers in the seed bank declined varies from species to species. In fact, it is likely that the seed banks of all these species declined more rapidly than would have been the case in undisturbed soils. To take an extreme example, Ødum (1978) took soil from an eleventh century grave in Denmark, and germinated from it an individual of *Verbascum thapsiforme* which had lain dormant in the seed bank for 850 years. At the other end of the spectrum are plants like *Phlox*, in which no seeds persist in a seed bank for more than one year.

different species make very different contributions to the seed bank

As a general rule, dormant seeds, which enter and make a significant contribution to seed banks, are more common in annuals and other short-lived plant species than they are in longer-lived species. A notable con-

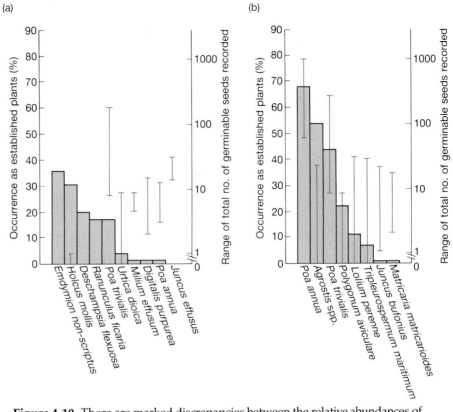

Figure 4.10. There are marked discrepancies between the relative abundances of mature plants (histograms) and of seed in the soil (horizontal bars) for the major species of (a) a deciduous woodland and (b) an arable field. (After Thompson & Grime, 1979.)

sequence is that seeds of short-lived species tend to predominate in buried seed banks, even when most of the established plants above them belong to much longer-lived species (Figure 4.10).

An important point in the present context is that no organism with a 'seed bank' can truly be considered an annual, even if it progresses from germination to reproduction within a year. This is because some of the seeds destined to germinate each year will already be more than 12 months old. Nevertheless, there are many plants conforming to such a description that are generally referred to as 'annuals'. All we can do is bear this fact in mind, and note that it is just one example of real organisms spoiling our attempts to fit them neatly into clear-cut categories.

4.5.3 Ephemeral and facultative annuals

In fact, annual species with seed banks are not the only ones for which the term 'annual' is, strictly speaking, inappropriate. There are, for example, many 'annual' plant species living in deserts that are far from seasonal in their appearance. They have a substantial buried seed bank, with germination occurring on the rare, unpredictable occasions when conditions are favourable. Subsequent development is usually rapid, so that the period from

143   LIFE AND DEATH

germination to seed production is short. Such plants are best described as semelparous *ephemerals* (Figure 4.5c).

Data concerning the appearance of one such species, *Eriophyllum wallacei*, is illustrated in Figure 4.11 (Juhren *et al.*, 1956). Some degree of consistency is apparent: *E. wallacei* germinated usually when temperatures were relatively low, and when there had been either heavy rainfall or a period of moderate rainfalls. Under these conditions, *E. wallacei*, like many desert annuals, developed rapidly with high survivorship and rather low individual fecundity. More apparent than the consistency, however, is the unpredictability. Over a period of four years, this opportunistic ephemeral germinated at various times between early September and late May. It was therefore an annual only in the limited sense of having short generations, which were generally discrete. In this particular case, the generations did occur on average approximately once per year; other desert annuals germinate less frequently.

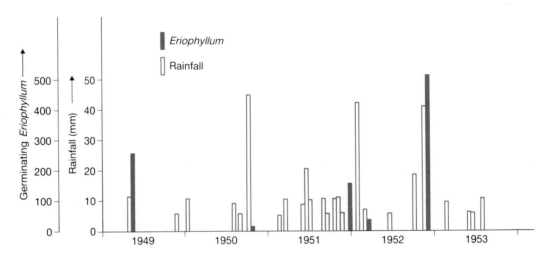

Figure 4.11. A desert annual. *Eriophyllum wallacei* living in Californian deserts is an annual in the sense that it has discrete generations lasting less than one year. But it germinates at various times of the year linked loosely to the rainfall in the immediately preceding period. (After Juhren *et al.*, 1956.)

Another group of species to which a simple 'annual' label does not apply are those in which the majority of individuals in each generation are annual, but in which a small number postpone reproduction until their second summer. This is true, for example, of the terrestrial isopod *Philoscia muscorum* living in north-east England (Sunderland *et al.*, 1976). Approximately 90% of females bred only in the first summer after they were born; the other 10% bred only in their second summer. In some other species, the difference in numbers between those that reproduce in their first or second years is so slight that the description 'annual–biennial' is most appropriate.

In short, it is clear that annual life cycles merge into more complex ones without any sharp discontinuity.

4.6 Overlapping iteroparity

Species with this type of life cycle not surprisingly have overlapping generations and iteroparous reproduction (Figure 4.5h–i). They breed repeatedly (assuming they survive long enough), but they nevertheless have a specific breeding season. Among the more obvious examples are long-lived, seasonally breeding vertebrates, some corals, temperate-region trees and other iteroparous perennial plants; in all of these, individuals of a range of ages breed side by side. (Note, however, that some species in this category, some grasses for example, live for relatively short periods—Figure 4.5h.)

4.6.1 Cohort life-tables

Constructing a cohort life-table for such species is even more difficult than constructing one for an annual species. A cohort must be recognized and followed (often for many years), even though the organisms within it are coexisting and intermingling with organisms from many other cohorts, older and younger. This was possible, though, as part of an extensive study of red deer (*Cervus elaphus*) on the small island of Rhum, Scotland (Lowe, 1969).

following a cohort of red deer

The deer live for up to 16 years, and the females (hinds) are capable of breeding each year from their fourth summer onwards. In 1957, Lowe and his co-workers made a very careful count of the total number of deer on the island, and the total number of calves (less than a year old); Lowe's cohort consisted of the deer that were calves in 1957. Then, each year from 1957 to 1966, every one of the deer that was shot under the rigorously controlled conditions of this Nature Conservancy Council reserve was examined. Extensive searches for the carcasses of deer that died from natural causes were also carried out. All of these deer were aged reliably by examining tooth replacement, eruption and wear. It was therefore possible to identify those dead deer that had been calves in 1957; and by 1966, 92% of this cohort had been observed dead, and their age at death noted. The cohort life-table for hinds is presented in Table 4.3 and the survivorship curve is shown in Figure 4.12. There appears to be a fairly consistent increase in the risk of mortality with age (the curve is convex).

Table 4.3. Cohort life-table for red deer hinds on the island of Rhum that were calves in 1957 (after Lowe, 1969).

Age (years) x	Proportion of original cohort surviving to the beginning of age-class x l_x	Proportion of original cohort dying during age-class x d_x	Mortality rate q_x
1	1.000	0	0
2	1.000	0.061	0.061
3	0.939	0.185	0.197
4	0.754	0.249	0.330
5	0.505	0.200	0.396
6	0.305	0.119	0.390
7	0.186	0.054	0.290
8	0.132	0.107	0.810
9	0.025	0.025	1.0

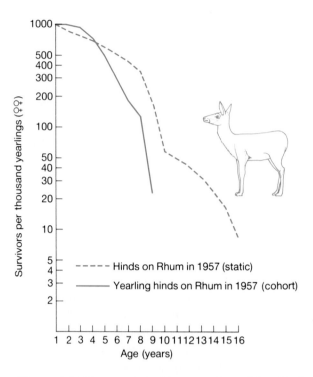

Figure 4.12. Two survivorship curves for red deer hinds on the island of Rhum. As explained in the text, one is based on the cohort life-table for the 1957 calves and therefore applies to the post-1957 period; the other is based on the static life-table of the 1957 population and therefore applies to the pre-1957 period. (After Lowe, 1969.)

cohort life-tables are often more easily constructed for sessile organisms

The difficulties of constructing a cohort life-table for an organism with overlapping generations are eased somewhat when the organism is sessile. In such a case, newly arrived or newly emerged individuals can be mapped, photographed or even marked in some way, so that they (or their exact location) can be recognized whenever the site is revisited subsequently. Thus, as one of many studies involving sessile animals, Connell (1970) constructed a cohort life-table for the barnacle *Balanus glandula* (see Table 4.6, below). A number of workers have followed cohorts of grasses. Over a rather longer period, workers in a vegetation reserve in South Australia were able to follow

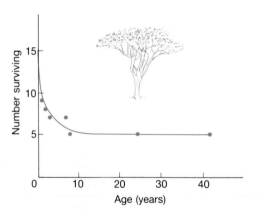

Figure 4.13. Survival of a cohort of *Acacia burkittii* individuals recorded regularly in permanent plots in South Australia since their germination in 1929 (after Crisp & Lange, 1976).

a cohort of the shrub *Acacia burkittii* in permanently established plots, from germination in 1928 until 1970 (Crisp & Lange, 1976). The survivorship curve is shown in Figure 4.13.

4.6.2. Static life-tables

Taken overall, however, the practical problems have tended to deter ecologists from constructing cohort life-tables for long-lived iteroparous organisms with overlapping generations, even when the individuals are sessile. But there is an alternative: the construction of a *static* life-table (also called 'time-specific' or 'vertical'). As will become clear, this alternative is noticeably imperfect. The major argument in its defence is that it is often better than nothing at all.

An interesting example of a static life-table emerges from Lowe's (1969) study of red deer on Rhum. As has already been explained, Lowe and his colleagues had access to a large proportion of the deer that died from 1957 to 1966, and these dead deer could be aged reliably. If, for example, a deer was examined in 1961 and found to be six years old, it was known that in 1957 the deer was alive and two years old. Lowe was therefore eventually able to reconstruct the age-structure of the 1957 population, and *age-structures are the basis for static life-tables*. Of course, the age-structure of the 1957 population could have been ascertained by shooting and examining large deer in 1957; but since the ultimate aim of this project was the enlightened conservation of the deer, this method would have been somewhat inappropriate. (Note that Lowe's results did not represent the total numbers alive in 1957, because a few carcasses must have decomposed before they could be discovered and examined.) Lowe's raw data for red deer hinds are presented in column two of Table 4.4.

Table 4.4. A static life-table for red deer hinds on the island of Rhum, based on the reconstructed age-structure of the population in 1957. (After Lowe, 1969.)

Age (years) x	Number of individuals observed of age x a_x	l_x	d_x	q_x	Smoothed l_x	d_x	q_x
1	129	1.000	0.116	0.116	1.000	0.137	0.137
2	114	0.884	0.008	0.009	0.863	0.085	0.097
3	113	0.876	0.048	0.055	0.778	0.084	0.108
4	81	0.625	0.023	0.037	0.694	0.084	0.121
5	78	0.605	0.148	0.245	0.610	0.084	0.137
6	59	0.457	−0.047	—	0.526	0.084	0.159
7	65	0.504	0.078	0.155	0.442	0.085	0.190
8	55	0.426	0.232	0.545	0.357	0.176	0.502
9	25	0.194	0.124	0.639	0.181	0.122	0.672
10	9	0.070	0.008	0.114	0.059	0.008	0.141
11	8	0.062	0.008	0.129	0.051	0.009	0.165
12	7	0.054	0.038	0.704	0.042	0.008	0.198
13	2	0.016	0.008	0.500	0.034	0.009	0.247
14	1	0.08	−0.023	—	0.025	0.008	0.329
15	4	0.031	0.015	0.484	0.017	0.008	0.492
16	2	0.016	—	—	0.009	0.009	1.000

Remember that the data in Table 4.4 refer to ages in 1957. They can be used as the basis for a life-table, but only if it is assumed that there had been *no* year-to-year variation prior to 1957 in either the total number of births or the age-specific survival rates. In other words, it must be assumed that the 59 six-year-old deer alive in 1957 were the survivors of 78 five-year-old deer alive in 1956, which were themselves the survivors of 81 four-year-olds in 1955, and so on; or, in short, that the data in Table 4.4 are the same as *would* have been obtained if a single cohort *had* been followed.

Having made these assumptions, l_x, d_x and q_x columns have been constructed. It is clear, however, that the assumptions are false. There were actually more animals in their seventh year than in their sixth year, and more in their fifteenth year than in their fourteenth year. There were therefore 'negative' deaths and meaningless mortality rates; and the pitfalls of constructing such static life-tables (and equating age-structures with survivorship curves) are amply illustrated.

static life-tables can be useful . . .

Nevertheless the data are by no means valueless. Lowe's aim was to provide a *general* idea of the population's age-specific survival rate prior to 1957 (when culling of the population began). He could then compare this with the situation after 1957, as illustrated by the cohort life-table previously discussed. He was more concerned with general trends than with the particular changes occurring from one year to the next. He therefore 'smoothed out' the variations in population size between ages 2–8 and 10–16 in order to ensure a steady decline during both of these periods. The results of this process are shown in the final three columns of Table 4.4, and the survivorship curve is plotted in Figure 4.12. A general picture does indeed emerge: the introduction of culling on the island appears to have had a significant, depressant effect on overall survivorship, overcoming any possible compensatory decreases in natural mortality.

. . . within very strict limits

Notwithstanding this success in the case of the red deer, the interpretation of static life-tables generally, and the age-structures from which they stem, is fraught with difficulty: the belief that age-structures offer an easy short cut to understanding population dynamics is a snare and a delusion.

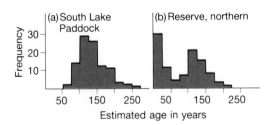

Figure 4.14. Age-structures of *Acacia burkittii* populations at two sites in South Australia (after Crisp & Lange, 1976).

There are occasions, however, where background knowledge greatly facilitates the interpretation of population age-structures, making their construction worthwhile. Crisp and Lange (1976), for instance, apart from following a cohort of *Acacia burkittii*, were also able to examine the age-structure of a number of populations (Figure 4.14). (The age of an individual could be estimated from a number of size measurements, since, in a sample of trees of known age, there were several highly significant age–size relationships.) The two populations in Figure 4.14 are obviously different in age-structure. But it was also known that the South Lake Paddock (Figure 4.14a) had been grazed by sheep from *c.* 1865 to 1970 (when the data were collected) and by rabbits from *c.* 1885 to 1970, while the Reserve (Figure 4.14b) had been fenced in 1925 to exclude sheep. It was therefore possible for Crisp and Lange to draw a number of conclusions: (a) grazing from 1865 onwards led to drastic reductions in the number of successful new recruits to the populations; (b) fencing in 1925 allowed more successful recruitment into the Reserve population; and (c) even after fencing, rabbit grazing led to levels of recruitment lower than those prior to 1865 (the 1920 to 1945 age-class, for instance, is much smaller than the 1845 to 1860 class, even though the latter has survived for an additional 75 years).

4.6.3 Fecundity schedules

'Static fecundity schedules' for iteroparous species with overlapping generations can provide useful information, especially if fecundity schedules from successive breeding seasons are obtained. This is illustrated in Table 4.5 by data from a population of great tits (*Parus major*) in Wytham Wood, near Oxford (Perrins, 1965). The table shows (a) that mean fecundity rose to a peak in two-year-old birds and declined gradually thereafter, and (b) that mean fecundity nevertheless varied considerably from year to year. These data were only obtainable, of course, because individual birds could be aged (in this case, because they had been marked with individually recognizable leg-rings soon after hatching). Note too that the great tit is an example of a species which is iteroparous and has overlapping generations, but cannot really be described as long-lived; few great tits live much longer than two years.

Table 4.5. Mean clutch size and age of great tits in Wytham Wood, near Oxford. (After Perrins, 1965.)

Age	1961		1962		1963	
	Number of birds	Mean clutch size	Number of birds	Mean clutch size	Number of birds	Mean clutch size
Yearlings	128	7.7	54	8.5	54	9.4
2	18	8.5	43	9.0	33	10.0
3	14	8.3	12	8.8	29	9.7
4			5	8.2	9	9.7
5			1	8.0	2	9.5
6					1	9.0

mast years in trees

Some variation in individual fecundity from breeding season to breeding season is found in all iteroparous species. The phenomenon is most marked, however, in a number of forest trees which exhibit 'masting'. A mast year is a year in which a tree produces a massive crop of seeds, fruit or cones, and the effect is often amplified by the synchronous masting of many or most of the trees of that species in a forest. Mast years are generally interspersed amongst years in which trees are sterile or nearly so, as Figure 4.15 shows for two examples: Scots pine and Norway spruce in northern Sweden (Hagner,

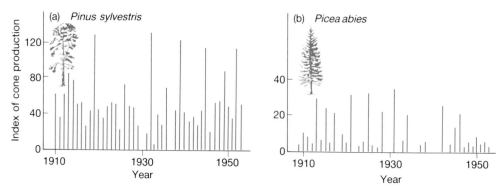

Figure 4.15. Masting: fluctuations in the cone crop of Scots pine (*Pinus sylvestris*) and Norway spruce (*Picea abies*) in northern Sweden. Based on counts made from reports by rangers (After Hagner, 1965.)

1965). The possible advantages and consequences of masting will be discussed in Chapter 9.

In spite of these year-to-year variations, most iteroparous species (like the great tit) show an age- or stage-related pattern of fecundity. Figure 4.16 shows the stage-dependent (in this case size-dependent) fecundity of three species of oak (Downs & McQuilkin, 1944).

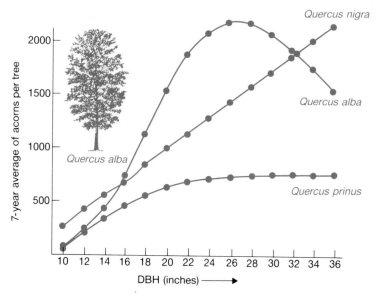

Figure 4.16. Size-specific variation in fecundity for three species of oak tree; DBH = diameter at breast height (after Downs & McQuilkin, 1944).

4.6.4 Modular iteroparous perennials

The difficulties of constructing any sort of life-table for organisms that are not only iteroparous with overlapping generations but also modular are emphasized by Figure 4.17. This illustrates data for the sedge *Carex bigelowii* growing in a lichen heath in Norway (Callaghan, 1976). *C. bigelowii* has an extensive underground rhizome system which produces tillers (aerial shoots) at intervals along its length as it grows. It initiates modular growth by producing a lateral meristem in the axil of a leaf belonging to a 'parent' tiller. The meristem is at first completely dependent on the parent tiller, and consists of one or a few non-photosynthetic leaves; but each lateral is potentially capable of developing into a vegetative parent tiller itself, and is also potentially capable of flowering, which it does when it has produced a total of 16 or more leaves. Flowering, however, is always followed by tiller death, i.e. the tillers are semelparous though the genets are iteroparous.

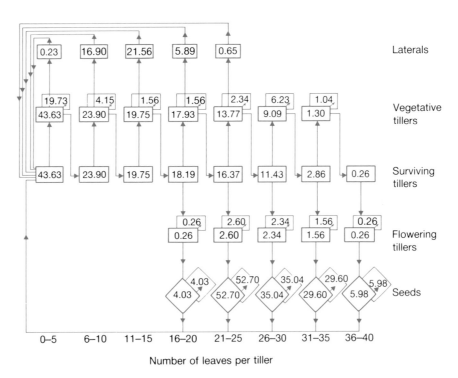

Figure 4.17. A reconstructed static life-table for the modules (tillers) of a *Carex bigelowii* population. Rows represent tiller types while columns depict size-classes of tiller. Thin-walled boxes represent dead tiller compartments, and arrows denote pathways between size-classes, death or reproduction. Within each box, the density per m² is presented. (After Callaghan, 1976.)

modular stage-structure and a modular static life-table

Callaghan took a number of well separated young tillers, and excavated their rhizome systems through progressively older generations of parent tillers. This was made possible by the persistence of tillers after death. He excavated 23 such systems containing a total of 360 tillers, and was able, as Figure 4.17 shows, to construct a type of static life-table (and fecundity schedule) based on stages rather than age-classes. There were for example 1.04 dead vegetative tillers (per square metre) with 31–35 leaves. Thus, since there were also 0.26 tillers in the next (36–40) stage, it can be assumed that a total of 1.30 (i.e. 1.04+0.26) living vegetative tillers entered the 31–35 leaf stage. And because there were 1.30 vegetative tillers and 1.56 flowering tillers in the 31–35 leaf stage, 2.86 tillers must have survived from the 26–30 stage. It is in this way that the life-table was constructed.

Note, however, (a) that the life-table applies not to individual genets but to tillers (i.e. modules); (b) that there appeared to be no successful seed germination in this particular population (no new genets), so that tiller numbers were being maintained by modular growth alone; and (c) that a 'modular growth schedule', analogous to a fecundity schedule, can sensibly be constructed for this population. Note most of all though, how difficult it is to collect data and construct a life-table which fully conveys the modularity of modular organisms.

the importance of stage not age in modular organisms

The use of stages rather than age-classes is something which is almost always desirable and necessary when dealing with modular iteroparous organisms. This is because variability stemming from modular growth accumulates year upon year, making age a particularly poor measure of biological status. Figure 4.18, for instance, shows the relationship between age (determined by the number of growth rings) and shoot volume in the creosote bush, *Larrea tridentata* (Chew & Chew, 1965): shrubs of the same volume may differ in age by up to 40 years, while shrubs of the same age may differ 60-fold in volume.

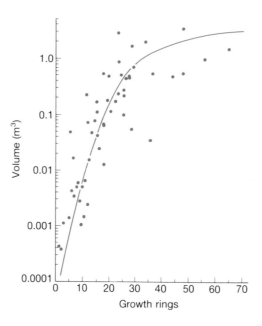

Figure 4.18. The relationship between shoot volume and age (measured by number of growth rings) in *Larrea tridentata* (creosote bush) in San Simon Valley, Arizona. Age is a poor indicator of shoot volume in this modular species. (After Chew & Chew, 1965.)

In addition, many tree populations contain individuals which linger for a number of years in a highly stunted condition way below the tree canopy. These are individuals that have germinated at a time when the canopy is effectively closed above them, so that there is insufficient incident light for them to grow and invade it. An example is the striped maple, *Acer pensylvanicum*, in the eastern U.S.A., where such stunted individuals can persist in the population for up to 20 years suffering little mortality. If an opening in the canopy appears, these individuals grow quickly to fill it, after which they soon flower and reproduce (Hibbs & Fischer, 1979). Such long-lived stunted individuals in a forest have much the same role in population dynamics as the buried seed populations of shorter-lived plants—they remain stunted until the unpredictable happens and an opening appears in the canopy (few trees have dormant seeds). Their behaviour also emphasizes that size rather than age is the appropriate indicator of a tree's chances of dying, growing or reproducing.

4.7 Reproductive rates, generation lengths, and rates of increase

In the previous section we saw that the life-tables and fecundity schedules drawn up for species with overlapping generations are at least superficially similar to those constructed for species with discrete generations. With discrete generations, we were able to compute the basic reproductive rate (R_0) as a summary term describing the overall outcome of the patterns of survivorship and fecundity (section 4.5.1). We turn now to the question of whether a comparable summary term can be computed when generations overlap.

In order to address this question, we must realise first that for species with discrete generations, R_0 describes two separate population parameters. It is the number of offspring produced on average by an individual over the course of its life; but it is also the multiplication factor which converts an orginal population size into a new population size, one generation hence. With overlapping generations, when a *cohort* life-table is available, the basic reproductive rate can be calculated using the same formula:

$$R_0 = \Sigma l_x m_x,$$

but it now refers *only* to the average number of offspring produced by an individual. Further manipulations of the data are necessary before we can talk about the rate at which a population increases or decreases in size—or, for that matter, about the length of a generation. The difficulties are much greater still when only a static life-table (i.e. an age-structure) is available (see below).

We will proceed by deriving a general relationship that links population size, the rate of increase, and time, but which is not limited to measuring time in terms of generations. Imagine a population which starts with 10 individuals, and which, after successive intervals of time, rises to 20, 40, 80, 160 individuals and so on. We refer to the initial population size as N_0 (meaning the population size when no time has elapsed). The population size after one time interval is N_1, after two time intervals it is N_2, and in general after t time intervals it is N_t. In the present case:

$$N_0 = 10$$

$$\text{and } N_1 = 20,$$

and we can say that

$$N_1 = N_0 R,$$

where R, which is 2 in the present case, is known as 'the fundamental net reproductive rate' or 'the fundamental net per capita rate of increase'. Clearly, populations will increase when $R > 1$, and decrease when $R < 1$.

R combines the birth of new individuals with the survival of existing individuals. Thus, in the present example, each individual could give rise to two offspring but die itself, or give rise to only one offspring and remain alive: in either case R (birth plus survival) would be 2. Note too that in the present case R remains the same over the successive intervals of time, i.e.

$$N_2 = 40 = N_1 R,$$

$$N_3 = 80 = N_2 R \text{ and so on.}$$

Thus: $N_3 = N_1 R \times R = N_0 R \times R \times R = N_0 R^3$

and in general terms: $N_{t+1} = N_t R$ $\hfill$ (4.1)

and $\qquad\qquad N_t = N_0 R^t.$ $\hfill$ (4.2)

These equations (4.1 and 4.2) link together population size, rate of increase, and time; and we can now link these in turn with R_0, the basic reproductive rate, and with the generation length (defined as lasting T intervals of time). Looking back to section 4.5.1, we can see that R_0 is the multiplication factor that converts one population size to another population size, one generation later—i.e. T time intervals later.

Thus: $N_T = N_0 R_0.$

But we can see from equation 4.2 that

$$N_T = N_0 R^T.$$

Therefore: $R_0 = R^T,$

or, if we take natural logarithms of both sides:

$$\ln R_0 = T \ln R.$$

The term $\ln R$ is usually denoted by r, and is called 'the intrinsic rate of natural increase'. It is the rate at which the population increases in size, i.e. the change in population size per individual per unit time. Clearly, populations will increase in size for $r > 0$, and decrease for $r < 0$; and we can note from the preceding equation that

$$r = \frac{\ln R_0}{T}.$$ $\hfill$ (4.3)

Summarizing so far, we have a relationship between the average number of offspring produced by an individual in its lifetime, R_0, the increase in population size per unit time, r ($= \ln R$), and the generation time, T. Previously, with discrete generations (section 4.5.1) the unit of time *was* a generation. It was for this reason that R_0 was the same as R.

In populations with overlapping generations (or continuous breeding), r is the intrinsic rate of natural increase that the population is capable of achieving; but it will only be able to achieve this rate of increase if the survivorship and fecundity schedules remain steady over a long period of time. If they do, r will be approached gradually (and thereafter maintained), and over the same period the population will gradually approach a stable

when fecundity and survivorship schedules remain steady, a population achieves a rate of increase of r, and a stable age-structure

age-structure (i.e. one in which the proportion of the population in each age-class remains constant over time; see below). If, on the other hand, the fecundity and survivorship schedules alter over time, then the rate of increase will continually change, and it will be impossible to characterize in a single figure. This may seem to set very severe limits to the usefulness of calculating r from a life-table with overlapping generations, since the schedules will only rarely remain steady for long periods of time. However, it can often be useful to characterize a population in terms of its potential, especially when the aim is to make a comparison (e.g. comparing various populations of the same species in different environments, to see which environment appears to be the most favourable for the species).

The most precise way to calculate r is from the equation:

an exact but biologically obscure equation for r

$$\Sigma\, e^{-rx} l_x m_x = 1 \qquad\qquad \text{(4.4) (Lotka, 1907)}$$

where the l_x and m_x values are taken from a cohort life-table, and e is the base of natural logarithms. However, this is a so-called 'implicit' equation, which cannot be solved directly (only by iteration, nowadays usually on a computer), and it is an equation without any clear biological meaning. It is therefore customary to use instead an approximation to equation 4.3, namely:

an approximate but biologically transparent equation for r based on T_c, the cohort generation time

$$r \simeq \ln R_0 / T_c, \qquad\qquad (4.5)$$

where T_c is the 'cohort generation time' (computed presently). This equation shares with equation 4.3 the advantage of making explicit the dependence of r on the reproductive output of individuals (R_0) and the length of a generation (T). Equation 4.5 is a good approximation when $R_0 \simeq 1$ (i.e. population size stays approximately constant), or when there is little variation in generation length, or for some combination of these two things (May, 1976).

We can estimate r from equation 4.5, if we know the value of the cohort generation time T_c. This can be computed if we recognize that it is the average length of time between the birth of an individual and the birth of one of its own offspring. This, like any average, is the sum of all those lengths of time from all offspring, divided by the total number of offspring, i.e.

$$T_c = \frac{\Sigma\, x l_x m_x}{\Sigma\, l_x m_x}$$

an equation for T_c

$$\text{or} \qquad T_c = \frac{\Sigma\, x l_x m_x}{R_0}. \qquad\qquad (4.6)$$

This is only approximately equal to the true generation time T, because it takes no account of the fact that some offspring may themselves develop and give birth during the reproductive life of the parent.

an example of the calculation of summary terms for a species with overlapping generations

Thus, equation 4.6 and 4.5 allow us to calculate T_c, and to approximate to r, from a cohort life-table of a population with overlapping generations or continuous breeding; that is, they allow us to produce the summary terms we require. A worked example is set out in Table 4.6, using Connell's (1970) data for the barnacle *Balanus glandula*. It is of interest that the precise value of r, from equation 4.4, is 0.085, compared to the approximation 0.080; while T, calculated from equation 4.3, is 2.9 compared to $T_c = 3.1$. The simpler and

Table 4.6. A cohort life-table and a fecundity schedule for the barnacle *Balanus glandula* at Pile Point, San Juan Island, Washington (Connell, 1970). The computations for R_0, T_c and the approximate value of r are explained in the text. Numbers marked with an asterisk were interpolated from the survivorship curve.

Age, years x	a_x	l_x	m_x	$l_x m_x$	$x l_x m_x$
0	1 000 000	1.000	0	0	
1	62	0.0000620	4600	0.285	0.285
2	34	0.0000340	8700	0.296	0.592
3	20	0.0000200	11600	0.232	0.696
4	15.5*	0.0000155	12700	0.197	0.788
5	11	0.0000110	12700	0.140	0.700
6	6.5*	0.0000065	12700	0.082	0.492
7	2	0.0000020	12700	0.025	0.175
8	2	0.0000020	12700	0.025	0.200
				1.282	3.928

$$R_0 = 1.282$$

$$T_c = \frac{3.928}{1.282} = 3.1$$

$$r \approx \frac{\ln R_0}{T_c} = 0.08014$$

biologically transparent approximations are clearly satisfactory in this case. They show that if the schedules had remained steady, the population would have increased in size slowly (i.e. r is somewhat greater than zero); and that this increase would have been the combined result of (a) individuals producing 1.282 offspring on average, and (b) a generation length of approximately 3 years.

a model population achieving a rate of increase of r and a stable age-structure

Finally in this section, we return to the age-structures of populations where survivorship and fecundity schedules remain steady over long periods. Table 4.7 show what happens in a model population where this is the

Table 4.7. A hypothetical cohort life-table and fecundity schedule, and a model population growing under the influence of these. Values for R_0, T_c, approximate and exact r, and R have been calculated from the cohort data. Values for R and r have also been estimated from the rate of increase of the population between time intervals 8 and 9.

Cohort life-table data x	l_x	m_x	Inital population	After 1 time interval	After 2 time intervals	After 8 time intervals	After 9 time intervals
0	1.000	0	1600 (1.00)	1880 (1.00)	2560 (1.00)	12305 (1.000)	15990 (1.000)
1	0.100	5	100 (0.063)	160 (0.085)	188 (0.073)	948 (0.077)	1231 (0.077)
2	0.060	15	60 (0.038)	40 (0.032)	96 (0.038)	437 (0.036)	569 (0.036)
3	0.018	10	20 (0.013)	18 (0.010)	18 (0.007)	101 (0.008)	130 (0.008)
4	0	—	0	0	0	0	0
						13791	17920

Age-structure (with static life-table entries in brackets)

$R_0 = 1.58$
$T_c = 1.80$

$$r \approx \frac{\ln R_0}{T_c} = 0.254$$

$r = 0.262$
$e^r = R = 1.30$

$$\frac{17920}{13791} = 1.30 = \text{estimated } R$$

estimated $r = \ln R = 0.262$

case. Initially, there are 100, 60 and 20 individuals in age-classes 1, 2 and 3 respectively; and at the start of the first time interval these produce the 1600 individuals in age-class 0, as a result of the age-specific birth rates in the m_x column ($1600 = 100 \times 5 + 60 \times 15 + 20 \times 10$). During the first time interval, the numbers in each age-class are subjected to the mortality rates implied by the entries in the l_x column of the imaginary cohort life-table, giving the numbers in age-classes 1, 2 and 3 'after one time interval'. The numbers in age class 0 (i.e. 1880) are then arrived at in the same way as before. The process has been repeated to show the nature of the population after eight and nine time intervals.

Values for R_0, T_c, approximate and exact r, and R have all been computed from the cohort life-table using equations 4.4, 4.5 and 4.6; and it is apparent that after eight time intervals the population did indeed increase at a rate equal to r. It is also apparent that the age-structure of the population after this time was stable (as shown by the entries of the static life-tables). However, the static life-tables clearly give a misleading impression of the true survivorship schedule shown in the cohort life-table. In fact, each entry in the stable age-structure static life-table is given by the following formula: $l_x e^{-rx}$, where l_x is the entry from the cohort life-table. This shows that the cohort and static life-tables are the same only when the population remains of a constant size ($r = 0$, $e^{-rx} = 1$). In a population that increases steadily in size ($r > 0$, $e^{-rx} < 1$), the older age-classes will be increasingly poorly represented in static life-tables and age-structures; while in a population declining in size ($r < 0$, $e^{-rx} > 1$) the older age-classes will be increasingly well represented. Thus, the age-specific survivorship and fecundity schedules of a population, its potential rate of increase, and its stable age-structure are all inter-linked. We return to this point when we look at human populations in section 4.10.

why cohort and static life-tables are usually different

4.8 Overlapping semelparity

Returning to the various types of life cycle, in overlapping semelparity (Figure 4.5 e–f) there is a distinct breeding season during which breeding individuals occur together with developing but non-breeding individuals; but each individual reproduces only once and then dies. It is seen in its simplest form in organisms that are strictly biennial, i.e. where each individual takes two summers and the intervening winter to develop, but has only a single reproductive phase, in its second summer. An example of this is the sweet white clover, *Melilotus alba*, growing in New York State (Klemow & Raynal, 1981). Two cohort survivorship curves for one clover population are shown in Figure 4.19. These were constructed by carefully mapping the position of every plant in a number of randomly selected permanent plots, and then up-dating these maps every two weeks from 1976 to 1978 (winter mapping was prevented by snow). The survivorship curves show that, as usual, the two years were different in detail (the summer of 1977 was hotter and drier than that of 1976). Nevertheless, the two cohorts displayed a similar pattern of relatively high mortality during the first growing season, followed by much lower mortality until the end of the second summer when survivorship decreased rapidly. This pattern is tied closely to the strictly biennial life cycle

sweet white clover: a strictly biennial plant

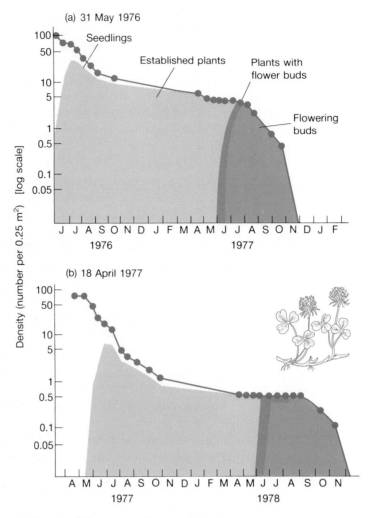

Figure 4.19. Survivorship curves for two cohorts of *Melilotus alba* (sweet white clover), a strict biennial. Survivorship was moderate in seedlings, good in established plants and very poor after flowering. (After Klemow & Raynal, 1981.)

of *M. alba*. There was high mortality while seedlings were developing into established plants (first summer). Subsequently very few plants died until after they had flowered and reproduced, which *all* survivors did in their second summer. No plants survived to a third summer.

Thus in *M. alba* there is a overlap of two generations at most; in the summer and autumn there are newly emerged, growing plants, and plants which are reproducing or about to do so. A more typical example of a semelparous plant with overlapping generations is provided by the composite *Grindelia lanceolata*, studied by Baskin and Baskin (1979) in Tennessee (Figure 4.20). This figure, like Figure 4.19, contains two cohort survivorship curves. In the case of *G. lanceolata*, these were obtained by placing wire rings around all newly germinated seedlings at the ends of the germinating seasons in 1972 and 1973, and then following the marked cohorts carefully until all individuals had died. (Despite the overlapping generations,

a semelparous plant with a less determinate life cycle

cohort life-tables are obviously easier to construct in these relatively short-lived species than they are in the iteroparous organisms considered in section 4.6.)

Figure 4.20 is, in certain respects, similar to Figure 4.19. The major difference is that *G. lanceolata*, rather than flowering in its second year, flowered in its third, fourth or fifth years. But whenever an individual did flower, it died soon after. This has led to near-vertical drops down the survivorship curve associated with each flowering season. However, apart from these drops, the curves (like Figure 4.19) indicate low survivorship during establishment, but high survivorship of established plants until they reproduce.

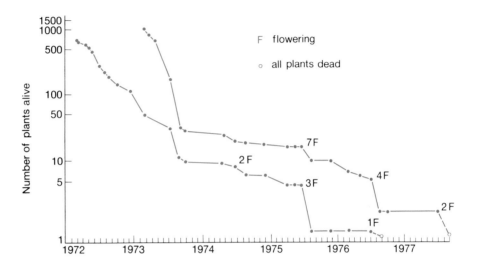

Figure 4.20. Survivorship curves of two cohorts of *Grindelia lanceolata.* Survivorship was moderate in young plants, good in established plants and very poor after flowering. (After Baskin & Baskin, 1979.)

once again: stage not age

The behaviour of these indeterminate semelparous plants is, to a large extent, a reflection of their modularity and the fact that size (i.e. the extent of modular growth) is more important than age in determining an individual's fate. Werner (1975) has demonstrated this in populations of the semelparous teazel, *Dipsacus fullonum* (Figure 4.21). Individuals were sown in sites where they had previously been absent, and followed for a period of five years from emergence. During this time, rosette size and the vegetative or flowering condition of individuals were noted. The risks of pre-reproductive death and the chances of flowering are both clearly related to rosette size. This relationship to size was essentially the same whether the rosettes were 2, 3 or 4 years old.

Semelparous species with overlapping generations are not as common amongst animals as they are amongst plants. This is largely due to the paucity of semelparous animals generally. Nevertheless, they do occur. The common octopus, *Octopus vulgaris*, for instance, has a life-span in the Mediterranean

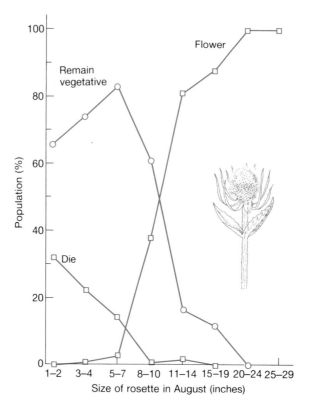

Figure 4.21. The fate of individual teazel plants (*Dipsacus fullonum*) in the following year depends on rosette size in August (rather than age) (after Werner, 1975).

of 15–24 months (Nixon, 1969). When reproduction is initiated in the females, protein synthesis in the muscles is suppressed, and there is a rapid growth of the ovaries, with a concomitant loss of weight elsewhere (O'Dor & Wells, 1978). The final release of the single brood of offspring is followed rapidly by the death of the adult. Another, well-known example is provided by the many species of salmon which grow without reproducing for several years before making a single determined effort to migrate up-river to spawn.

4.9 Continuous semelparity

The distinction between this type of life cycle (Figure 4.5g) and the previous type, is the lack of a specific breeding season. Each individual has only a single reproductive event in its life, but life may begin and reproduction may occur at any time during a large portion of the year. Of the five main types of life cycle, this is probably the least important, mainly because there are relatively few semelparous animals and relatively few plants which breed continuously. One major group of exceptions, are those semelparous animals that live in a continuously favourable environment. The octopus *Octopus cyanea* for instance is, like *O. vulgaris,* semelparous. *O. cyanea,* however, lives in the Indo-Pacific region, where conditions fluctuate little and breeding is possible at all times of the year (van Heukelem, 1973).

4.10 Continuous iteroparity: human demography

The meaning of this final type of life cycle (Figure 4.5j) should, by now, be clear: individuals reproduce repeatedly and can do so at any time throughout the year. This life cycle occurs, for instance, in environments which are not seasonal. This is the case in some tropical species, in many parasites, and in many inhabitants of artificial environments kept constant by man (e.g. grain stores). Such a life cycle can also occur in species which are essentially oblivious to seasonal changes in the environment, because of their sophisticated physiologies. The most obvious example of this is *Homo sapiens*.

The quantitative study of the size, distribution, structure and dynamics of human populations is a mature discipline in its own right. Data of various sorts are essential for national and international planning and policy-making, and life-table analysis was in fact originally developed for devising pension and insurance schemes on the basis of individuals' life-expectancies. Thus the

Figure 4.22. A population pyramid for France on January 1, 1967 (after Shryock *et al.*, 1976).

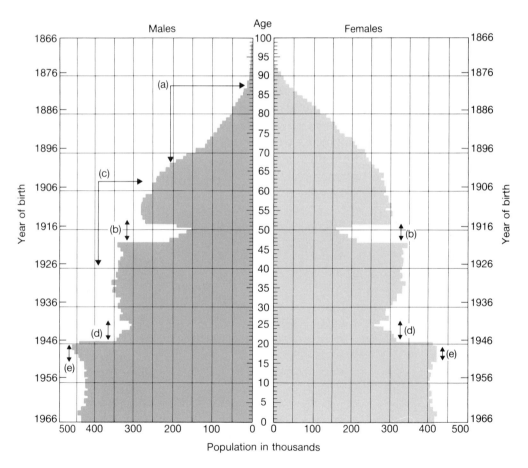

(a) Military losses in World War I
(b) Deficit of births during World War I
(c) Military losses in World War II
(d) Deficit of births during World War II
(c) Rise of births due to demobilization after World War II

data that are available for human populations have a scope and an accuracy which cannot be matched by any other species.

the population pyramid

Figures 4.22 and 4.23 contain age-structure data in a form which is commonly used with human populations: the population pyramid. Each age-class is represented by a horizontal bar, the length of which represents the size or relative size of the age-class. The pyramid is constructed by arranging the age-classes vertically with the youngest at the base and oldest at the apex. Figure 4.22 shows how, with humans, such data can be combined with historical and sociological information to provide an intelligible picture of a country's population (in this case France). Figure 4.23 contrasts the age distributions of the total populations of 'developing' (i.e. poor) countries and of developed countries in 1980, and also shows the corresponding United Nations projections for the year 2000. The age distribution for the developing nations tapers sharply from its base to its apex, while that for the developed nations has near-vertical or even overhanging sides until the older age-classes. This contrast is partly due to the higher birth rates and lower survivorship rates in the poorer countries; but as explained in section 4.7, it also reflects the fact that the populations of the developing nations are expanding rapidly, while those of the developed nations are increasing slowly if at all.

prospects for the year 2000

The projection for the year 2000 in the developing nations is bound to cause concern when we think of the problems of feeding these multitudes. Indeed, even though the birth rates and survivorships are likely to alter and promote slower rates of increase over the next 20 years, the populations will still increase rapidly because of the large number of people still to enter their reproductive years in developing countries. As May (1980) has discussed, this phenomenon—known as the 'momentum of population growth'—can be

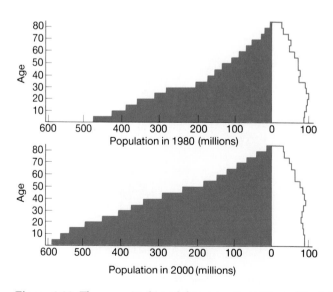

Figure 4.23. The magnitude and the age composition of the total population in developing countries (shaded, on the left) and in developed countries (unshaded, on the right). The top figure is for 1980, and the bottom is the U.N. projection for the year 2000. (After May, 1980.)

expressed most dramatically by noting that even if an intrinsic *r*-value of zero could be achieved overnight, the developing countries would still be committed to a rough doubling of total population before coming to equilibrium (we saw in section 4.7 that intrinsic rates of increase take some time to achieve). Thus, while we can derive some comfort (or complacency?) from the recent declines in birth rates in developing countries, we must still face up to a problem of quite frightening proportions.

4.11 Finalé: looking forward

Having examined a wide variety of life cycles, it is natural to wonder how it comes about that some species exhibit one type of life cyle, and other species another type. It would be foolish to pretend that ecologists know the detailed answer to this question; but we can be confident that the answer will be based on natural selection, and that, in a very important sense, the life cycles match the environments of the species exhibiting them. Trying to understand the correspondence between life cycles and environments is one of the exciting challenges of modern ecology; but before we take up this challenge here, we must look much more fully at the nature of the environment. In particular, we must recognize that for the majority of organisms the most important aspects of the environment are the other organisms living alongside them. It is therefore necessary to postpone our attempts to understand life cycles until after we have examined the interactions that occur within and between species (Chapters 6–12). We will return to life cycles in Chapter 14.

Migration and Dispersal in Space and Time

5.1 Introduction

All organisms in nature are where we find them because they have moved there. This is true for even the most apparently sedentary of organisms, such as oysters and redwood trees. Their movements range from the passive transport that affects many plant seeds to the active movement of many mobile animals and their larvae. The effects of such movements are also varied. In some cases they aggregate members of a population into clumps; in others they continually redistribute them and shuffle them amongst each other; and in others they spread the individuals out and 'dilute' their density.

the meaning of 'migration' and 'dispersal'

The terms 'dispersal' and 'migration' are used to describe certain aspects of the movement of organisms. *Migration* is most often taken to mean the mass directional movements of large numbers of a species from one location to another. The term therefore applies to classic migrations (the movements of locust swarms, the intercontinental journeys of birds, the transatlantic movements of eels), but also to less obvious examples like the to-and-fro movements of shore animals following the tidal cycle (see section 5.3). *Dispersal* is most often taken to mean a spreading of individuals away from others (e.g. their parents or siblings), and it may involve active (walking, swimming, flying) or passive movements (carriage in water or wind). Dispersal is therefore an appropriate description for several kinds of movements: (a) of plant seeds or starfish larvae away from each other and their parents; (b) of voles from one area of grassland to another, usually leaving residents behind and being counterbalanced by the dispersal of other voles in the other direction; (c) of land birds amongst an archipelago of islands (or aphids amongst a mixed stand of plants) in the search for a suitable habitat.

In this usage of the terms dispersal and migration, both are defined for a *group* of organisms. However, it is the individual that actually moves (migration is mass movement, and an individual can only disperse, literally, by exploding, though see section 5.8 for dispersal of parts of a genet). Many dispersing organisms (especially plant seeds and many marine larvae) are simply hazarded into the world at large, and they must do the best they can wherever they land (though features that influence the spatial scale and direction of dispersal may have evolved to increase the chances of the organisms finding a suitable habitat). And dispersers, just as much as migrators, may move towards, or until they find, a habitat suitable for their continued existence. At the level of the individual, therefore, there is no sharp distinction between migration and dispersal. Nevertheless, their effects are

no sharp distinction between dispersal and migration

sufficiently different for them to be discussed separately in the course of this chapter.

5.2 Distributional patterns

The movements of organisms affect the distributions of organisms. Before we look at movement, therefore, we must consider the general nature of distributions. There are three main types (Figure 5.1), though they form part of a continuum.

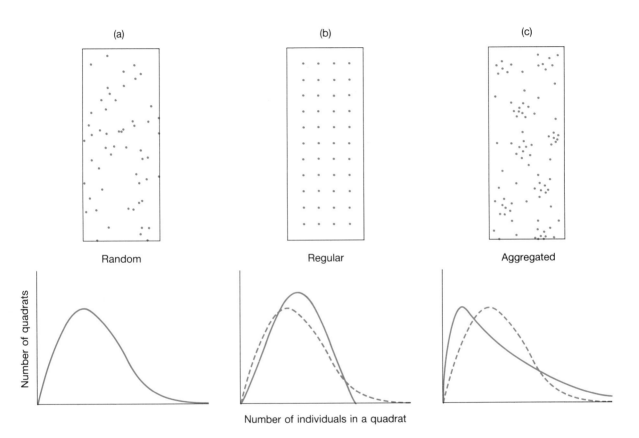

Figure 5.1. The three main types of distribution of organisms in space (above) and the consequent frequency with which quadrat samples containing different numbers of individuals would be obtained from such distributions (below). (a) random; (b) regular; (c) aggregated. In (b) and (c) the frequency distribution from (a) is included for comparison (dashed lines). In a regular distribution most quadrats have approximately the same number of individuals in them. In an aggregated distribution there are relatively large numbers of quadrats with very few individuals and also a relatively large number with very many.

random, regular and aggregated distributions

A *random* distribution (Figure 5.1a) occurs when there is an equal probability of an organism occupying any point in space, and when the presence of one individual does not influence the presence of another. If a series of small quadrat samples (p. 131) is taken from a randomly distributed population then the distribution of individuals per quadrat is of the type shown in the lower part of Figure 5.1a.

A *regular* distribution (also called 'uniform', 'even' or 'over-dispersed': Figure 5.1b) occurs either when each individual has a tendency to avoid all other individuals, or when individuals that are too close to other individuals die or leave the population altogether. In a series of quadrat samples, the most common frequency is shifted to the right relative to a random distribution (Figure 5.1b).

An *aggregated* distribution (also called 'contagious', 'clumped' or 'under-dispersed': Figure 5.1c) occurs either when individuals all tend to be attracted to (or are more likely to survive in) particular parts of the environment, or when the presence of one individual at a location attracts or gives rise to other individuals at the same location. In a series of quadrat samples, the most common frequency is shifted to the left relative to a random distribution (Figure 5.1c).

The distribution exhibited by a group of organisms depends on the spatial scale on which the organisms are studied. This can be illustrated by a hypothetical but realistic example of an aphid living on a particular species of tree. With any fairly large frame of reference, the aphids are aggregated: concentrated in the terrestrial environment, concentrated in particular parts of the world, and concentrated in woodland as opposed to other types of habitat. Even within a woodland the aphids will be aggregated: on their host tree species rather than unsuitable trees. However, within a tree, quadrats of 25 cm² (about the size of leaf) might reveal that the aphid is randomly distributed over the tree as a whole; while quadrats of 1 cm² might indicate a regular distribution because aphids on a single leaf avoid one another. The problems of measuring and interpreting distributions are discussed by Southwood (1978a), Kershaw (1975) and Greig-Smith (1984), and in Chapter 12 we show how the binomial distribution can be used as a further means of characterizing patterned and aggregated distributions.

5.3 Patterns of migration

5.3.1 Diurnal and tidal movements

The populations of many species move from one habitat to another and back again repeatedly during their life. The time-scale involved may be hours, days, months or years. In some cases, these movements have the effect of maintaining the organism in the same type of environment; this is the case in the movement of crabs on a shoreline, moving with the advance and retreat of the tide.

By contrast, many migrations ensure that during its life an individual passes backwards and forwards from one type of environment to another. For example, planktonic algae both in the sea and in freshwater lakes descend to the depths at night but move to the surface during the day. It appears that they accumulate phosphorus and perhaps other nutrients in the deeper water (the hypolimnion) before returning to near the surface (the epilimnion) and photosynthesizing (Salonen *et al.*, 1984). Birds, bats, slugs, snails and a wide variety of other mobile animals move from one habitat to another regularly within the course of a 24-hour cycle of rest and activity. As a rule, their movements aggregate them into tight populations during the resting period and separate them from each other when feeding. Most snails, for example, rest in confined humid microhabitats by day, but forage actively for their food by night. Thus these are all species with fundamental niches that can only be satisfied by alternating life in two distinct habitats within each day of their lives.

5.3.2 Seasonal movement between habitats

Many motile organisms make seasonal moves between habitats. The patches of the environment in which resources are available change with the changing seasons, and populations move from one type of patch in the environment to

Figure 5.2. Migration patterns in relation to life cycles.

A. THE MULTIPLE RETURN TICKET

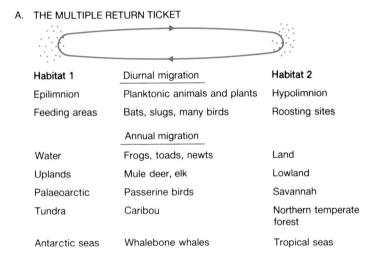

Habitat 1	Diurnal migration	Habitat 2
Epilimnion	Planktonic animals and plants	Hypolimnion
Feeding areas	Bats, slugs, many birds	Roosting sites
	Annual migration	
Water	Frogs, toads, newts	Land
Uplands	Mule deer, elk	Lowland
Palaeoarctic	Passerine birds	Savannah
Tundra	Caribou	Northern temperate forest
Antarctic seas	Whalebone whales	Tropical seas

B. THE ONE RETURN ONLY TICKET

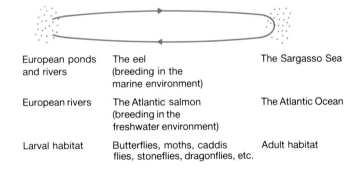

European ponds and rivers	The eel (breeding in the marine environment)	The Sargasso Sea
European rivers	The Atlantic salmon (breeding in the freshwater environment)	The Atlantic Ocean
Larval habitat	Butterflies, moths, caddis flies, stoneflies, dragonflies, etc.	Adult habitat

C. THE ONE WAY ONLY TICKET

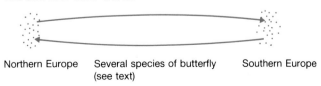

Northern Europe	Several species of butterfly (see text)	Southern Europe

another. The altitudinal migration of grazing animals in mountainous regions is one example, where, for instance, the American elk and mule deer move up into high mountain areas in the summer and down to the valleys in the winter. (Such annual altitudinal migration has been conspicuous in the way man has managed his livestock in mountainous areas. Cattle, sheep, goats and even pigs were moved to high pastures during the summer—often with the women and younger children as shepherds—while the men harvested hay from valley pastures. The livestock were returned to the lowlands during the winter to be fed on the stored hay.) In these situations, the migrations tend to ensure that the animals always forage where foraging conditions are best; the animals move seasonally and escape the major changes in food supply and climate that they would meet if they stayed always in one area.

This pattern can be contrasted with the 'migration' of amphibians (frogs, toads, newts) between an aquatic breeding habitat in spring and a terrestrial environment for the remainder of the year. The young develop (as tadpoles) in water with a different food resource from that which they will later eat on land. They will return to the same aquatic habitat for mating, aggregate into dense populations for a time, and then separate to lead more isolated lives on land. During its lifetime an individual may make several return journeys (Figure 5.2).

5.3.3 Long-distance migration

The most remarkable habitat shifts are those that involve travelling very long distances. For terrestrial animals in the northern hemisphere, most of these involve a spring movement north, where food supplies may be abundant only during the warm summer period, and an autumn movement south, to savannas where food is abundant only after the rainy season. The migrations seem, in virtually every case, to involve transit between areas both of which supply abundant food but only for a limited period. They are areas in which seasons of comparative glut and famine alternate, and which cannot support large resident populations all the year round. For example, swallows which migrate seasonally to South Africa vastly outnumber a related resident species. The all-year-round food supply can support only a small population of residents, but a seasonal glut exceeds what the residents can consume. Of those species that breed in the Palaearctic (temperate Europe and Asia) and leave for the winter, 98% travel south to Africa. They spend the winter in areas of thornbush and savanna (deciduous vegetation), and their arrival usually coincides with the ripening of an immense crop of seeds of the dominant grasses.

Such migrations can greatly increase the local diversity of a fauna. Moreau (1952) calculated that of the 589 species of birds (excluding seabirds) which breed in the Palaearctic, 40% spend the winter elsewhere.

the costs and benefits of long-distance migration

There is clearly a metabolic cost to travelling between such geographically separated areas, but we must presume that the benefits of increased food availability outweigh these metabolic costs. Alaskan wheatears make a twice-yearly journey of 7000 miles to Africa. The arctic tern travels from its arctic breeding ground to the antarctic pack ice and back each

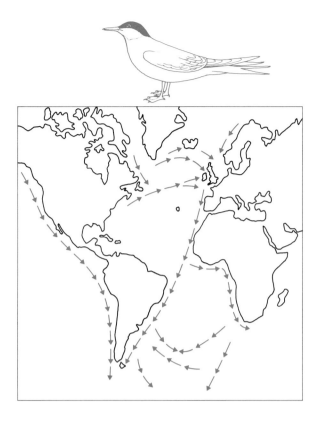

Figure 5.3. Migration routes of the arctic tern (*Sterna paradisaea*) (redrawn from Kullenberg, 1946).

year—about 10 000 miles each way (though unlike many other migrants it can feed on its way). The terns that breed in eastern Canada cross the Atlantic twice on their way south (Figure 5.3).

The same species may behave in different ways in different places. All robins (*Erithacus rubecula*) leave Finland and Sweden in winter, but on the Canary Islands the same species is resident the whole year round. In most of the intervening countries, part of the population migrates and part remains resident (Lack, 1954).

On the whole, migrating species are not characteristic of tropical rain-forest or evergreen montane forest, where productivity is less seasonal and all available resources are more likely to be consumed by the population of resident species.

We know most about the migratory habits of birds, because marking by leg rings is so easy and there are so many ornithologist observers. However, the phenomenon is widespread in other groups too. Baleen whales in the southern hemisphere move south in summer to feed in the food-rich waters of the Antarctic. In winter they move north to breed (but scarcely to feed) in tropical and subtropical waters. Caribou travel several hundred miles a year from northern forests to the tundra and back; while tunny fish breed in the Mediterranean in April and May, and then migrate to the northern part of the North Sea. In all of these examples (Figure 5.2) an individual of the migrating species may make the return journey several times. It is therefore at least possible that it learns a route, and uses landmarks, in combination with magnetic information about position, and/or information from the sun and stars (see Baker, 1982).

5.3.4 'One return journey' migration

Many migrant species make only one return journey during their lifetime. They are born in one habitat, make their major growth in another habitat, but then return to breed and die in the home of their infancy. The eel and the migratory salmon are classic examples. The European eel is thought to travel from European rivers and ponds across the Atlantic to the Sargasso Sea, where it reproduces and dies (though spawning adults and eggs have never actually been caught there). The American eel makes a comparable journey from areas ranging between the Guianas in the south to south-west Greenland in the north. The marine juveniles then travel back and grow to maturity as freshwater adults before making their return journey to the Sargasso.

The salmon makes a comparable transition, but from a freshwater egg and juvenile phase to mature as a marine adult. It then returns to freshwater sites to lay eggs. After spawning, all Pacific salmon die without ever returning to the sea. Many Atlantic salmon also die after spawning, but some survive to return to the sea and then migrate back upstream to spawn again.

It is not easy to 'explain away' this type of migration as a movement between two habitats, each of which offers a seasonal abundance of resource. Nor is it easy in these situations to envisage any learned process of direction-finding. Nevertheless, a very accurate return of individuals to the river of their birth is well known among salmon.

The monarch butterfly (*Danaus plexippus*) has a comparable migration pattern (Figure 5.4). Populations travel north in the United States and Canada to breed in summer, and south to Florida and California to overwinter but not to breed. The same individuals may return to the north in the spring but it appears that each individual makes no more than one return journey. Such 'one return ticket' migrations can also be seen on a smaller scale in the life cycles of many insects: butterflies, moths, caddis-flies and stoneflies. The individual spends the early part of its life in one habitat (the larval food plant

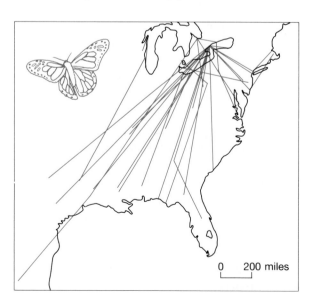

Figure 5.4. Lines joining the points of release and recapture of marked monarch butterflies (*Danaus plexippus*), and examples of distances travelled on the southward migration in the eastern part of North America (from Urquhart, 1960).

0   200 miles

for the butterfly, the aquatic stage for the caddis), migrates to another habitat as a sexually mature, reproductive adult, and returns to the juvenile habitat only to lay eggs.

5.3.5 'One way only' migration

In some migratory species, the journey for an individual is on a strictly one-way ticket. In Europe, the butterflies clouded yellow (*Colias croceus*), red admiral (*Vanessa atalanta*) and painted lady (*Vanessa caddui*) breed at both ends of their migrations. The individuals which reach Britain in the summer breed there, and their offspring fly south in autumn and breed in the Mediterranean region—the offspring of these in turn come north in the following summer.

5.3.6 Forces favouring aggregation

The examples of migration considered so far all tend to lead to the aggregation of individuals. Life cycles tend also to be *synchronized* in time, so that the time of migration is concentrated at a precise season. Moreover, the times of seedling emergence, breaking of insect dormancy and bud-break of trees in spring, and the time of birth and recruitment of new mammals and birds into a population, all tend to occur in a narrow band of season within the year (see section 5.7).

aggregation can result from a shared habitat selection

The most pervasive reason for the synchronous behaviour of populations (aggregation in time), and for their aggregation in space, is the simultaneous but independent association of large numbers of individuals at a suitable time of the year or in a suitable location. In other words, each organism is individually associated with a suitable habitat, and a population aggregated in this habitat is the consequence. There are, however, specific ways in which organisms may gain from being close to neighbours in space and time. One advantage that has sometimes been suggested is the benefit accruing to the population or species as a whole. For example, altruistic warning cries from one individual would serve to protect other aggregated individuals but not isolated individuals. However, the conditions under which evolution by such group selection might operate are stringent and limited. We are therefore forced to look for an advantage accruing to individuals if we wish to identify important benefits of aggregation (see the discussion in Dawkins's readable book *The Selfish Gene* (1976)).

aggregation can result from the attraction of individuals: the selfish herd

An elegant theory identifying such selective advantages to the individual was put forward by Hamilton (1971) in his paper 'Geometry for the selfish herd'. In essence, Hamilton argued that the risk to an individual from a predator may be lessened if it places another potential prey individual between itself and the predator. The consequence of many individuals doing this is bound to be an aggregation. The 'domain of danger' for individuals in a herd is at the edge, so that an individual capable of exerting its social status by joining a herd and assimilating into the centre of it would gain in fitness by doing so. Where a tendency to form flocks has arisen in the way Hamilton envisaged, we might expect subordinate individuals to be forced into the domain of danger on the outside of the flock. This seems to be the case, for example, in reindeer and in woodpigeons, where a newcomer may have to join the herd or flock at its risky perimeter and can only establish itself in a more protected position within the flock after social interaction (Murton *et al.*, 1966). Individuals may also gain in fitness from group-living because the proximity of other individuals helps in detection of predators, because individuals cooperate to fight off the predator, or for other reasons (see Pulliam & Curaco, 1984).

predator satiation in time

The principle of the selfish herd as described for the aggregation of organisms in space is just as appropriate for the synchronous appearance of organisms in time. The individual that is precocious or delayed in its appearance, outside the norm for its population, may be at greater risk from predators than those conformist individuals that take part in 'flooding the market' and thereby dilute their own risk. Among the most remarkable examples of synchrony are the periodic cicadas, the adults of which emerge at simultaneously after 13 or 17 years of life underground as nymphs. In many species of bamboo too, the individuals flower and set seed synchronously, in some species after intervals of as much as 100 years. As a consequence, much of the bumper seed crop is not consumed by the predators, whose populations cannot change fast enough to consume this glut (Janzen, 1976).

distributions as compromises

There are many other situations in which the individual may gain by assembling with others. Food supplies may be more efficiently located (particularly if they themselves are aggregated). There are fluid-dynamic effects that give greater lift to birds flying in appropriately shaped flocks and possible advantages to fish that swim in shoals. On the other hand, there are also selective pressures that can act *against* the formation of aggregations in space or time. The foremost of these is certain to be the increase in crowding, depletion of resources, and competition (Chapter 6) that is likely to result from many individuals being in the same place or appearing at the same time. In addition, a group of individuals may actually attract a predator's attention. Overall, all distributions are bound to be the resultants of opposing forces: the attractant forces of shared habitat requirements and benefits of group-living, and the repellent forces of overcrowding and cost of group-living.

5.4 Dispersal

5.4.1 Dispersal as escape and discovery

Dispersal is the term applied to the process by which individuals escape from the immediate environment of their parents and neighbours, and become less aggregated; dispersal may therefore relieve local congestion. But dispersal is not always just 'escape'; it can also often involve a large element of discovery. A useful distinction can be made between two types of such 'discovery dispersal' (Baker, 1978). First, there is dispersal in which individuals visit and 'explore' a large number of sites before finally returning and settling in one of those sites (presumably a suitable one). Secondly, there is dispersal in which individuals visit a succession of locations, but then just cease to move (with no element of 'return' to a site previously explored). In fact, this latter category can be split further into cases where the cessation of movement is under the dispersing organism's control and cases where it is not.

The dispersal of plant seeds is non-exploratory and beyond the control of the seed itself. The discovery aspect of seed dispersal is therefore a matter of chance (though the chances of reaching a suitable site may be increased by the specializations for dispersal that the seeds possess). Animal dispersal, on the other hand, can fall into any of the three categories. Some animals have essentially the same type of dispersal as plant seeds (e.g. the simple freshwater invertebrates described on p. 182 below). Many other animals cannot be said to explore, but they certainly control their settlement and cease movement only when an acceptable site has been found. Most aphids, for example, even in their winged form, have powers of flight which are too weak

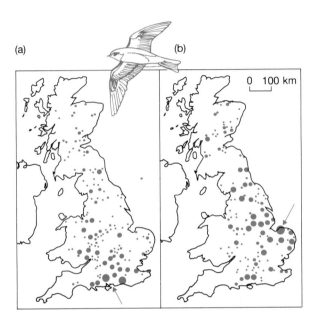

Figure 5.5. Migrations of the sand martin (*Riparia riparia*) in the post fledgling stage. The arrows show the positions of two roosts in England: (a) Chichester and (b) Fenland. The size of a dot represents the number of individuals associated with the roost in question that were caught while visiting one of a large number of colonies in Great Britain marked by the dots. The data included records of birds that were ringed at the roost and then captured as juveniles at the roost, and those that had been ringed elsewhere and were captured at the roost later in the same year. (Redrawn from Mead & Harrison, 1979.)

to counteract the forces of prevailing winds. But they control their take-off from their site of origin, they control when they drop out of the windstream, and they make additional, often small-scale flights if their original site of settlement is unsatisfactory. Their dispersal, therefore, involves 'discovery', over which they have control.

The number of animals exhibiting truly exploratory dispersal is difficult to judge, mainly because the appropriate data are lacking. It is not sufficient to know where an animal started from and where it finished; it is also necessary to know everywhere the animal went in between. However, such data as are available suggest that exploratory dispersal might be surprisingly widespread (Baker, 1982). Figure 5.5, for example, shows the sites visited by juvenile sand martins in Britain in the late-summer weeks after they fledged (Mead & Harrison, 1979). After this dispersal, they migrated south in the autumn and then returned the following spring. Significantly, 55% of the surviving juveniles bred in the immediate vicinity of their natal colony; 87% settled within 10 km; and only 2% settled more than 100 km away. The post-fledgling dispersal certainly appears to have been an exploratory assessment of potentially suitable sites.

some species disperse more than others . . .

Whatever the precise nature of a species' dispersal, all species disperse—but some are more dispersive than others. Insects living in habitats that are, by nature, temporary have a more pronounced dispersive phase than insects living in more permanent habitats (Table 5.1). In a similar way, 'super-tramp' bird species (Diamond, 1973; Diamond & May, 1976) are good dispersers and good early colonizers of islands, but they do not persist for long when the habitat becomes colonized by other species. The pigeon *Macropygia mackinlayi* (Figure 5.6), for example, is a classic super-tramp, occurring only on small and species-poor islands of the Bismarck archipelago (east of New Guinea). It contrasts with the cuckoo *Centropus violaceus,* which is a late colonizer and then only of islands already well stocked with other species.

. . . but all must disperse to some extent

In general, dispersal is essential for a good chance of survival in individuals that exploit temporary stages in a changing community (a community 'succession'—Chapter 16). The descendants of individuals of all successional species are doomed in their local habitats. Yet even the species of so-called 'climax' communities (the relatively stable end-points of successions) are doomed in the long run unless they colonize new areas. The movements of forests following the advance and retreat of ice sheets (Chapter 1), or of tropical forests following arid periods, are on a different time-scale to that usually associated with the dispersal of organisms. But they make the point that in the life of all terrestrial organisms, home is sooner or later a dangerous place to stay in.

Table 5.1. Examples of species of insects able, or unable, to migrate or to fly, in relation to the permanence of their habitat. (Permanent: lakes, rivers, streams, canals, trees, bushes, salt-marshes, woods. Temporary: ponds, ditches, pools, annual plants, perennial plants but not climax vegetation, arable lands.) (After Johnson 1969.)

	Permanent habitats	Temporary habitats
British Anisoptera		
Total number of species	20	23
Pronounced dispersive phase	6	13
British Macrolepidoptera		
Total number of species	594	181
Pronounced dispersive phase	14	42
British water beetles		
Total number of species	52	127
Able to fly	13	81
Unable to fly	20	9
Flight variable	19	37

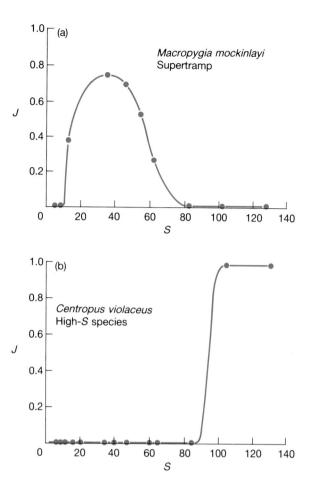

Figure 5.6. The incidence of the 'super-tramp' pigeon *Macropygia mackinlayi* the cuckoo *Centropus violaceus* on islands in the Bismarck Archipelago. *S* is the number of bird species on an island and *J* is the fraction of the islands on which the pigeon (a) or the cuckoo (b) are present. The pigeon tends to be an early colonist of islands but not to persist in species-rich communities. The cuckoo, in contrast, is found only on islands that have been colonized by a rich bird fauna. (From Diamond & May, 1976.)

5.4.2 The argument that all organisms should have some dispersive properties

Dispersal itself tends to be risky, and there will always be a balance of risk between living longer in an already occupied habitat and hazarding resources in an act of colonization. As Gadgil (1971) has argued:

'In a very general way, the factor favouring evolution of dispersal would be the chance of colonising a site more favourable than the one that is presently inhabited. . . . An organism should disperse if the chance of reaching a better site exceeds the loss from the risk of death during dispersal or the chance of reaching a poorer habitat.'

There is, however, a very simple argument that leads to the conclusion that even in very stable habitats, all organisms in their evolutionary history will have been under some selective pressure to disperse (Hamilton & May, 1977). Imagine a population in which the majority of organisms have a stay-at-home, non-dispersive genotype O, but in which a rare mutant genotype, X, keeps some offspring at home but commits others to dispersal. The disperser X is certain to replace itself in its own habitat. This is because the other, O-type individuals make no attempt to displace it. However, X also has at least some chance of displacing some of the non-dispersers. The disperser X will therefore increase in frequency in the population. On the other hand, if the majority of the population are type X while O is the rare mutant, O will still do worse than X. This is because O can never displace any of the X's, while O itself has to contend with several or many dispersers. Dispersal is therefore said to be an evolutionarily stable strategy or ESS (Maynard Smith, 1972; Parker, 1984). A population of non-dispersers will evolve towards the ubiquitous possession of a dispersive strategy; a population of dispersers will show no tendency to evolve towards the loss of that strategy.

dispersal as an ESS

Hamilton and May's models were incapable (except in extreme, unrealistic cases) of predicting precisely what proportion of a parent's offspring should be committed to dispersal. This will depend on, amongst other things, the pattern of appearance of sites for colonization, both at home and abroad, and on the relative chances of successful colonization of dispersers and residents. The important point, though, is that some dispersal is to be expected even in uniform and predictable environments.

5.4.3 The demographic significance of dispersal

The ecological fact of life propounded on p. 123 emphasized that dispersal can have a potentially profound effect on the dynamics of populations. In practice, however, many studies have paid little attention to dispersal. The reason often given is that emigration and immigration are approximately equal, and they therefore cancel one another out. One suspects, though, that the real reason is that dispersal is usually extremely difficult to quantify. The studies that *have* looked carefully at dispersal have tended to bear out what the ecological fact of life suggests. In a long-term and intensive study of a population of great tits in Wytham Wood, Oxford, England (Greenwood *et al.*, 1978; see p. 189, below) it was observed that 57% of breeding birds were immigrants rather than residents (i.e. born in the population). The bird itself was described as having a 'low level of dispersal'.

dispersal—when studied—is typically found to be important demographically

In a review of dispersal in mice and voles (Gaines & McClenaghan, 1980), there were a number of cases in which either more than half of a study population emigrated each week, or more than half of a population were individuals that had immigrated during the previous week. Significantly, when populations of the voles *Microtus pennsylvanicus* and *M. ochrogaster* were enclosed to prevent emigration, densities rose to abnormally high levels and the vegetation was overgrazed in a way not normally observed (Krebs *et al.*, 1969).

In a study of the spruce budworm in Canada, estimates of adult dispersal prior to egg-laying for the years 1949–55 suggested values ranging from a loss

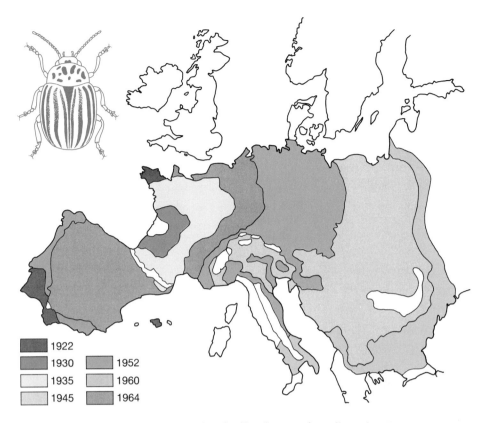

Figure 5.7. Spread of the Colorado beetle (*Leptinotarsa decemlineata*) in Europe (from Johnson, 1967).

of more than 90% by emigration to an increase of 1700% by immigration (Greenbank, 1957). In a population of the Colorado potato beetle in Canada, the average emigration rate of newly emerged adults was 97% (Harcourt, 1971). This can be interestingly compared with the spread of the beetle in Europe in the middle years of this century (Figure 5.7; Johnson, 1967). Dispersal is undoubtedly a crucially important, albeit neglected demographic process.

5.4.4 Passive dispersal on land and in the air: the seed rain

the blurred distinction between active and passive dispersal

Most biological categories are blurred at the edges. One example is the distinction between active and passive dispersers. For most mobile animals there is some element of behaviour involved in movement, whereas in sessile organisms forces outside the organism control the distance and direction of dispersal. Yet passive dispersal in air currents is not a phenomenon restricted to plants. Young spiders that climb to high places and then release a gossamer thread which carries them on the wind are as passively at the mercy of air currents as the winged fruits of a maple or the winged seeds of a pine. Nor is passive dispersal confined to organisms totally lacking in control of their movement (Johnson, 1969). The wings of birds and insects are not only active

Table 5.2. Comparison of the relative abundance of selected groups of Diptera at different heights in the air (data from Glick, 1939). (The italicized figures in column 3 are estimated from column 2.)

Height (feet)	20	(20)	200	1000	2000	3000	5000	Mean of 1000–5000	Mean of catches at 1000–5000 feet as percentage of catch at 200 feet
Flying time (hours)	12	(171)	171	168	163	168	160		
Total insects	1860	*26512*	13396	4749	2357	1364	612	2270	17
Total Diptera	773	*11018*	5175	1979	1024	586	279	967	19
Chloropidae	241	*3435*	1010	459	212	104	74	212	21
Sarcophagidae	4	*57*	17	4	1	2	2		
Muscidae	1	*14*	6	1	2	0	0		
Calliphoridae	0	—	3	0	0	1	1		
Total:	5	*71*	26	5	3	3	3	3.5	13

sources of directed flight; they are also often aids to the passive movement of these organisms in air currents (Table 5.2).

Despite these blurred distinctions, however, the distance and direction that a passive unit is carried can be seen as being determined not by any mobile quality of its own, but by the force that carries it (whether this be wind, water or a moving organism that carries seed on its coat or in its gut). The higher plant has control not of where its seeds will land, but of how they will interact with wind currents and eddies.

The rain of seed that falls from a parent plant is probably never distributed at random. Most seeds fall close to the parent and their density declines with distance. This is the case for wind-dispersed seeds and also for those that are ejected actively by maternal tissue (many legumes and *Geranium*). The density of seeds is often low immediately under the parent, rises to a peak close by, and then falls off steeply with distance. A number of seed dispersal curves are shown in Figure 5.8. The decline in density with distance from the parent often approaches the inverse square law for wind-dispersed seeds, but approaches an inverse cube law in the case of species that ejaculate their seeds from explosive capsules or in which the seeds have no obvious dispersal mechanism. Plant pathologists have recognized that diseases of which the dispersal is approximated by the inverse square law tend to form isolated foci of infection at a distance, whereas diseases with dispersal following the inverse cube law tend to spread as an advancing and quite well marked horizon or front. The same generalizations probably apply to the dispersal of seeds (Harper, 1977).

One consequence of the high density of seeds that tends to fall close to the parent is that the majority of seedlings that develop are likely to meet the parent or close relatives as neighbours. This is just one of the ways in which the nature of the dispersal process determines the way in which a population meets the forces of natural selection.

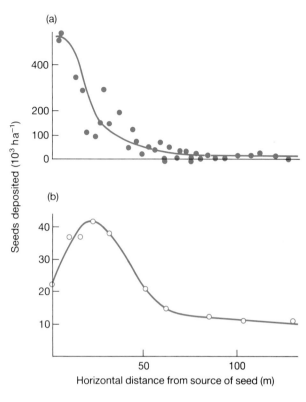

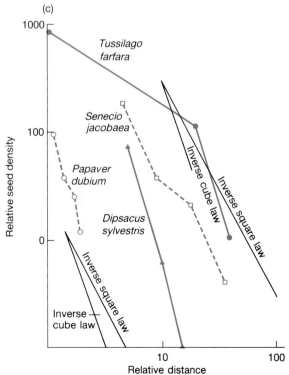

Figure 5.8. (a) The distribution of seeds of *Eucalyptus regnans* from trees *c.* 75 m high from the edge of a dense forest and (b) from an isolated tree. (c) The relationship between relative seed density and relative dispersal distance for a variety of plants: (i) *Senecio jacobaea* from the edge of a dense population in the direction of the prevailing wind; (ii) *Tussilago farfara* from the edge of a dense population; (iii) *Papaver dubium* from an isolated plant; (iv) *Dipsacus sylvestris* from isolated plants. (From Harper, 1977, after a variety of sources.)

5.4.5 Costs and constraints of dispersal

Specializations for dispersal all involve the sacrifice of tissues and/or resources that might have been used in other activities. The wings or plumes that support wind-borne seeds, the fleshy coats of bird-dispersed seeds (see Figure 1.10), and the hooks or spines of seeds that are dispersed on the coats of animals are all always maternal tissue, not part of the dispersed embryonic plant. Their possession is therefore likely to detract from the vitality and growth-potential of the mother herself, because of the resources she has diverted to offspring dispersal. There is also a conflict between the weight of a dispersed unit and its dispersibility. Weight clearly has a major effect on the distance that a seed (or baby spider) is dispersed. But the weight of a seed largely represents the capital of resources that it carries with it to support it during its early phases of establishment. There is therefore, as with a human traveller, a conflict between dispersibility and the advantages of a comprehensive supply of 'luggage'. Moreover, there is yet a further conflict between the weight of resources allocated to individual progeny by a parent and the number of progeny produced. All organisms have limited resources that can be devoted to reproduction; these may be allocated either to a few heavy or many lightweight progeny (Harper *et al.*, 1970).

the cost of dispersal to the 'mother'

dispersibility and capital investment

The evolutionary resolution of conflicts like these is discussed much more fully in Chapter 14. For the present it is worth noting that different species have resolved these conflicts in different ways. For example, tropical trees of the genus *Mora* have seeds that are like cannon-balls. They fall under and at the edge of the tree canopy, and dispersal is so ineffectual that the distribution of these species may still be limited by the dispersal that has occurred since they evolved! Such species are good examples of well-endowed, heavy progeny that appear to have sacrificed dispersibility. By contrast, many parasitic plants (Chapter 12) produce vast numbers of wind-dispersed 'dust seeds', which are almost entirely dependent on their host for the nutrition of even their earliest stages of development.

5.4.6 Passive dispersal by an active agent

Much of the uncertainty in the fate of dispersed progeny is removed if an active agent of dispersal is involved. The seeds of many herbs of the woodland floor carry spines or prickles that increase their chance of being carried passively on the coats of animals. The seeds may then be concentrated in nests or burrows when the animal grooms itself (Sorensen, 1978). The seeds of many species, particularly of shrubs and lower canopy trees, are fleshy and attractive to birds, and the seed coats resist digestion in the gut. Such seeds are then deposited in faeces. Where the seed is dispersed to depends on the defaecating behaviour of the bird.

There are many examples of what might be coevolved relationships between particular plant species and specialist dispersing birds, though there seems to be no bird that is totally specialized on any one species of fruit. A close approach to such specialization, however, is found in the mistletoe bird of Australia (Figure 5.9) which feeds almost exclusively on mistletoe berries. It

birds and seeds

defaecates frequently, and at each defaecation it wriggles itself against the branch of a tree. This has the effect of lodging the seeds in bark crevices where they germinate and establish themselves as parasites on the host. It is dangerous, however, to interpret all bird feeding habits as in some way optimizing the dispersal of seeds, and as examples of coevolution. Figure 5.10 shows the seeds of oak (acorns) stored in holes in fencing posts by the acorn woodpecker (there may be 50 000 holes bored in a single tree). There is little in such behaviour that can be interpreted as increasing the fitness of the oaks!

Much of the ecology of seed dispersal is still in the realm of anecdotal natural history—beautifully described in the classic text of Ridley (1930) and by van der Pijl (1969).

Figure 5.9. The Australian mistletoe bird (*Dicaeum livundinaceum*) feeding young on mistletoe berries.

beetles and mites

There are also important examples in which animals are dispersed by an active agent. For instance, there are many species of mite that are taken very effectively and directly from dung-pat to dung-pat, or from one piece of carrion to another, by attaching themselves to dung beetles or carrion beetles. They usually attach to a newly emerging adult, and leave again when that adult reaches a new patch of dung or carrion. Often the interaction is mutualistic (advantageous to both—see Chapter 13): the mites gain a dispersive agent, but many of them attack and eat the eggs of flies which would otherwise compete with the beetles.

Figure 5.10. Acorns stored in fencing posts by the acorn woodpecker (*Melanerpes formicovorus*) (photograph by J.L. Harper).

5.4.7 Passive dispersal in water

Weight is less of a hindrance to the dispersal of propagules in water. It is striking, however, that in freshwater the passive dispersal of seeds is rare; many aquatic flowering plants produce their seeds out of water. Larvae of freshwater invertebrates in rivers do make use of the flowing column of water as a means of dispersing from hatching sites to appropriate microhabitats (their movements are referred to as 'invertebrate drift'—Townsend 1980), but many freshwater insects depend on flying adults for upstream dispersal and movement from stream to stream. Clearly, passive transport in the water is a useless means of dissemination between ponds and lakes. The dispersal of most freshwater organisms without a free-flying stage depends on resistant wind-blown structures (e.g. gemmules of sponges, cysts of brine shrimps ephippia of cladocerans, eggs of *Hydra* and the dessicated bodies of tardigrades and rotifers).

In the sea, the situation is quite different. Marine habitats are large and continuous, and are more or less accessible to pelagic larvae thanks to rapid diffusion by tidal and residual currents. In marine invertebrates, the pelagic, short-lived larvae are usually the dispersal units; the sedentary adult is usually the stage in the life cycle at which most feeding and growth is done. This is in complete contrast to the freshwater insects (Figure 5.11).

In discussing seed dispersal we made the point that there is a conflict between producing a few large, well-provisioned seeds or a larger number of small ones carrying few reserves. Among the pelagic larvae of the marine habitat there is an alternative—many fend for themselves, and do not depend wholly on yolk reserves provided by the parent. This has been called the planktotrophic strategy. Such pelagic larvae are active swimmers, using energy and actively consuming resources while they are dispersing—very

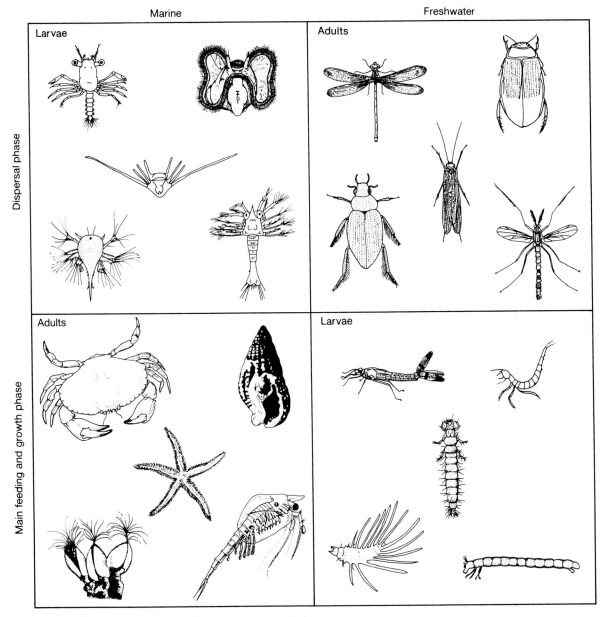

Figure 5.11. Dispersal stages in freshwater and marine habitats.

different from a passively dispersed seed (Crisp, 1976). In such cases settlement is also an active process. The larvae can make a choice of where to settle, and this contrasts dramatically with the impotence of a seed to determine where it will fall.

5.5 Variation in dispersal within and among populations

5.5.1 The genetic dimension

The Japanese geneticist Sakai (1958) studied dispersal and migration in *Drosophila melanogaster* by establishing populations in culture tubes which

183   MIGRATION AND DISPERSAL

were connected to vacant tubes. He then counted the numbers that entered the vacant tubes—the migrants. There were six populations: one with a long history of laboratory culture, and the other five obtained from field samples captured on small islands not far from each other. Random migration (i.e. dispersal) continued at the same rate irrespective of the density of the population in the founder tube. It resulted from the random movements of individual flies and was more common among the wild flies than in those with a history of life in culture. Against this background of dispersal, however, there was also mass migration which was most pronounced at higher densities. Importantly, though, the strains differed. The laboratory strain did not make mass migrations until the density in the founder tubes exceeded 150, whereas in wild strains the critical density that stimulated migration was much lower: 40, 60 or 80 depending on the strain (Figure 5.12). Thus there appeared to be genetic differences among strains suggesting that a tendency to disperse had been lost after a long period of culture in the laboratory.

The technique of electrophoretic analysis of isoenzymes has made it relatively easy to detect genetic differences, not only among strains but among individual organisms. This has made it possible to compare the genotypes of dispersing and non-dispersing individuals in nature. Such a comparison for voles (*Microtus pennsylvanicus*) in southern Indiana showed that they differed at at least two loci, and one rare homozygote was found only among dispersers (Myers & Krebs, 1971). Other studies have obtained similar results with different species of vole. There is, of course, no reason to

strains of Drosophila

are dispersers genetically different?

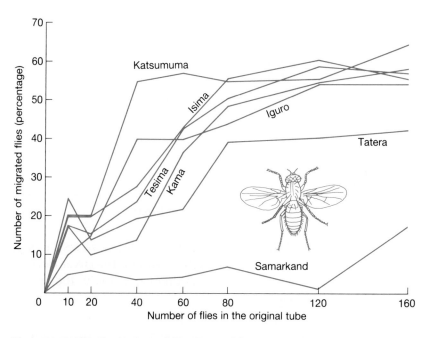

Figure 5.12. The percentage of flies (*Drosophila melanogaster*) that migrated from populations in culture tubes after they were connected to new, unoccupied tubes. The figure shows the effect of the density of the population on the number of flies that migrate. The names refer to islands off the coast of Japan from which the strains of *Drosophila* had been collected. Samarkand is a long-established laboratory strain. (From Sakai, 1958.)

suppose that these particular genetic differences have anything to do with tendency to disperse, as such. But many of the data on small mammals (though not all, see Gaines & McClenaghan, 1980) do suggest that dispersers are not a genetically random subsample of the whole population.

5.5.2 The sexual dimension

Males and females often differ in their liability to disperse. Among species of birds it seems to be usually females that are the main dispersers, but among mammals it is the males (Table 5.3; Greenwood, 1980). This difference between the sexes in the two taxa is linked to a number of other trends: the mating system in mammals is usually based on competition between males for mates rather than for territories; mammals are mainly polygamous whereas birds are mainly monogamous; and male mammals contribute less to the care of their progeny than is commonly the case with birds. Altruism also appears to be more common in the non-migrating sex, i.e. commoner among male than among female birds, but commoner among female than among male mammals.

Table 5.3. Number of species with predominant dispersal... (from Greenwood, 1980).

	. . . by males	. . . by females	No sex difference
Birds	3	21	6
Mammals	45	5	15

Differences in dispersal between the sexes are especially strong in some insects, where it is the male that is usually the more active disperser. The female winter moth, for example, is wingless while the male is free-flying. There are some exceptions to this general rule, such as the beet leaf-hopper (*Circulifer tenellus*), which makes long-distance migrations, crossing mountains thousands of feet high. In this case it is the females that disperse further, and the populations that the migrants leave behind therefore become male dominated.

5.5.3 Dispersal polymorphism: 'bet-hedging' in dispersal

Another source of variability in dispersal within populations is a somatic polymorphism amongst the progeny of a single parent. Environments and habitats are patchy and change with time. Thus the sites in which progeny establish themselves successfully may sometimes be close to home and sometimes far away. Also, the situation may change from season to season and from year to year. There is then no single perfect pattern of dispersal behaviour, and the fittest parents may be those that produce both dispersing and non-dispersing progeny (Figure 5.13).

A classic example is the desert annual plant *Gymnarrhena micrantha*. This bears a very few (one to three) large seeds (achenes) in flowers that remain unopened below the soil surface, and these seeds germinate in the original site of the parent. The root system of the seedling may even grow down

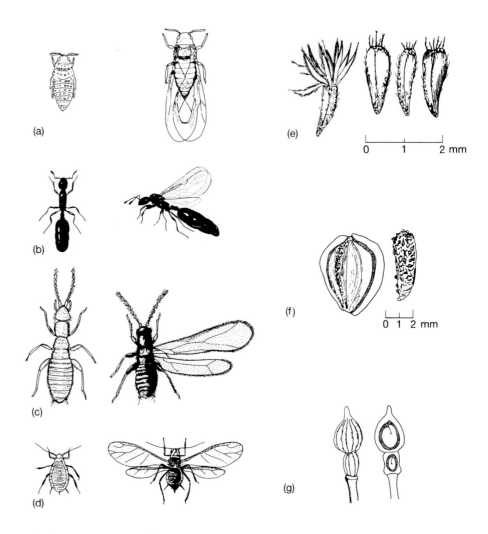

Figure 5.13. Dimorphism with respect to dispersal in animals and plants. Winged and wingless females of: (a) the grape phylloxera (*Viteus vitifoliae*); (b) the little black ant (*Monomorium minimum*); (c) the zorapteran (*Zorotypus hubbardi*); and (d) the apple aphid (*Aphis pomi*). Contrasting sizes and dispersal features of seeds and fruits: (e) *Galinsoga parviflora*, fruits from the centre (on left) and margin (on right) of the capitulum; (f) *Dimorphotheca pluvialis*, again showing central and marginal fruits; (g) *Rapistrum rugosum*, a two-seeded pod with marked difference in seed size. (From a variety of sources.)

through the dead parent's root channel. But the same plants also produce above ground smaller seeds with a feathery pappus, and these are wind-dispersed. In very dry years only the undispersed underground seeds are produced, but in wetter years the plants grow vigorously and produce a large number of seeds above ground, which are released to the hazards of dispersal (Koller & Roth, 1964).

There are very many examples of such seed dimorphism among the flowering plants, especially among grasses, composites and members of the families Chenopodiaceae and Cruciferae (all rather recently evolved families). In these, the difference between seed types is determined by the mother, not

seed polymorphism

by genetic differences between the seeds. Both the dispersed and the 'stay-at-home' seeds will, in their turn, produce both dispersed and stay-at-home progeny.

Commonly in seed polymorphism, the 'stay-at-home' seed is produced from flowers below ground or from unopened flowers. The seeds that are dispersed are more often the product of cross-fertilization. Two phenomena are thus coupled: the tendency to disperse is coupled with the possession of new, recombinant ('experimental') genotypes, whereas the stay-at-home progeny are more likely to be the product of selfing.

A dimorphism of dispersers and non-dispersers is also a common phenomenon among aphids (winged and wingless progeny). As this differentiation occurs during the phase of population growth when reproduction is parthenogenetic, the winged and wingless forms are genetically identical. Whether a mother produces winged or wingless progeny or a mixture of the two seems to be determined by the conditions of crowding and the quality of food that she receives (Harrison, 1980). This situation is closely parallel to that in *Gymnarrhena*.

Often the dispersal polymorphism in insect populations is of long-winged and short-winged forms. Among the water striders (*Gerris*), short-winged forms tend to predominate during phases of population growth, and long-winged forms are developed just before the end of the breeding season. Environmental factors seem to be of prime importance in determining the proportion of long-winged and short-winged individuals in a population. Temperature and photoperiod, the availability and quality of food, and the effect of crowding have all been implicated (Harrison, 1980). These are presumably used as environmental cues that provide some information about the favourableness of the current environment and prospects for the future (e.g. if days are short and growing shorter, then winter is approaching).

5.5.4 Social differences within populations of small mammals

There has been a considerable amount of work carried out on dispersal in small mammals, especially voles and mice, largely because dispersal is believed to be a central element in the determination of their abundance (see Chapter 15). Much of this work has been concerned with the nature of the differences in small mammal populations between those that disperse (i.e. tend to leave their home site) and those that do not (Gaines & McClenaghan, 1980). Four hypotheses have been proposed, with varying levels of confirmation and support from field and laboratory data (Gaines & McClenaghan, 1980).

(1) The social subordination hypothesis (Christian, 1970). This proposes that as population density increases, the resulting shortage of resources leads to increased levels of aggression, and this in turn forces social subordinates to disperse into sub-optimal habitats (i.e. subordinates disperse, dominants do not). Most of the few studies on behavioural differences do suggest that dispersers are subordinate, but in populations of the vole *Microtus pennsylvanicus*, dispersing males were more aggressive than resident males. The data also indicate that dispersers are predominantly young individuals

(i.e. a particular type of subordinate). However, it appears that the proportion of dispersers in the population is highest not when density is highest but when it is rising. A somewhat similar hypothesis suggests that as density increases, individuals are more and more likely to meet others that are not close kin and to react aggressively to them, increasing the urge to escape (Charnov & Finnerty, 1980). On these interpretations it is phenotypic changes that distinguish between dispersers and non-dispersers.

genetic polymorphism?

(2) The genetic–behavioural polymorphism hypothesis (Krebs, 1978). This proposes that individuals tend to be *either* innately aggressive *or* innately capable of a high reproductive output. At low densities natural selection is thought to favour individuals of the latter type. But at higher densities aggressive genotypes are favoured, so that individuals with a high reproductive output are forced to disperse and the stage is set for a decline in density. As with the previous hypothesis, subordinates disperse, dominants do not—but this time the differences are genetic.

The facts that dispersers are generally subordinate, and that there are sometimes genetic differences between dispersers and non-dispersers (p. 183) are both consistent with (though by no means proof of) this hypothesis. So too is the high level of dispersal from increasing populations (as reproducers are replaced by aggressors perhaps). Moreover, dispersers do seem generally to begin reproduction earlier than non-dispersers. However, the data are generally sparse, and the crucial evidence regarding the genetic basis of the difference between reproducers and aggressors is lacking.

saturation–presaturation dispersal?

(3) The saturation–presaturation dispersal hypothesis (Lidicker, 1975). This proposes two distinct categories of dispersers from a population. (i) Those that leave overcrowded or saturated populations. These are the 'social outcasts, juveniles, and very old individuals, those in poor condition and in general those least able to cope' (Lidicker, 1975). (ii) Presaturation dispersers, which are individuals in relatively good condition but are particularly sensitive to population density in the early phases of population growth. This hypothesis clearly contains elements of the previous two hypotheses, and thus receives some support from the available data. However, good evidence for a distinction between the two different types of disperser is lacking.

social cohesion?

(4) The social cohesion hypothesis (Bekoff, 1977). This proposes that dispersers are the individuals that have interacted least with their sibs and have therefore not formed social ties, i.e. dispersers are predominantly asocial individuals rather than the oppressed. There are obvious analogies with human populations, where family ties may hold back an individual from adventurous exploration! The hypothesis also has some support from field observations on canids and marmots.

Overall, it is clear that there are at least some behavioural/social differences between dispersing and non-dispersing individuals within populations of small mammals. But it is not clear how often such differences have a genetic basis, nor whether there are recognizable types of disperser leaving a population under different circumstances. In principle, we should expect individuals to disperse when they have more to gain by leaving than staying. This will frequently be the case for subordinate individuals with a future (e.g.

young but healthy individuals): they will be capable of finding and exploiting new habitats, and can be expected to leave expanding populations. In overcrowded populations, on the other hand, fewer of the subordinates will be healthy. Perhaps the only ones to leave will be those that are so oppressed that they are doomed if they remain.

5.6 Dispersal and outbreeding

If organisms and their progeny remain aggregated together, then there are many opportunities for inbreeding between sibs and half-sibs, progeny and parents, and other close relatives. One consequence of dispersal, on the other hand, is that it favours the mixing of the progeny from different parents and hence favours outbreeding. Whether and when outbreeding results in improved fitness is another question!

It is particularly among populations that normally outbreed that matings between close relatives often result in reduced viability and fertility in the offspring. This point has been beautifully illustrated by Darlington (1960), who examined family histories (e.g. of the Bach and Darwin families), comparing branches where there had and had not been a history of inbreeding. Genealogies are also known in some other long-studied animals, especially birds, because the progeny of ring-marked parents can themselves be marked in the nest. In a long-term study of the great tit in Wytham Woods near Oxford, England, for instance, there were 885 pairings between 1964 and 1975 in which the identities of both male and female were known. Pairs were formed both from birds born in the wood (residents) and from those born outside (immigrants). These 885 pairings consisted of 194 matings between resident males and resident females, 239 between resident males and immigrant females, 158 between immigrant males and resident females and 294 between immigrant males and immigrant females. There was no evidence

inbreeding depression in great tits?

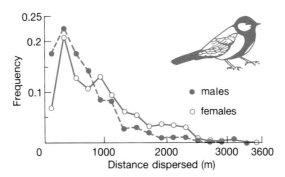

Figure 5.14. The distance dispersed by outbreeding individuals of the great tit (*Parus major*) from the site of birth to the site of first breeding (from Greenwood *et al.*, 1978).

that inbreeding pairs produced smaller clutches of eggs than outbreeding pairs, but nestling mortality was significantly higher among inbreeding (27.7%) than outbreeding (16.2%) pairs (Greenwood *et al.*, 1978). Females dispersed further than males in their first year (Figure 5.14), though subsequently the birds tended to retain their territories and rarely to disperse further. Significantly, many of the cases of inbreeding were associated with less dispersal than might be expected on average. The evidence from this study, therefore, suggests that dispersal contributes significantly to the extent of outbreeding, and that outbreeding produces fitter progeny.

optimal outbreeding in delphiniums?

If dispersal is poor, close neighbours are likely to be related and are also likely to have experienced much the same local selective forces. Outbreeding brings the risk that locally specialized gene combinations may be broken up. Inbreeding might be expected to hold together and congeal locally selected gene combinations. However, inbreeding commonly produces 'inbreeding depression' (a reduction in offspring vitality), especially in populations that have a history of some outbreeding. There are then two forces that act against each other, and it might be expected that the healthiest offspring result from matings between individuals that are neither too similar in genetic make-up nor too dissimilar (Price & Waser, 1979; Bateson, 1978, 1980). An experiment was made to test this hypothesis by hand-pollinating flowers of *Delphinium nelsoni* at various distances from each other (Price & Waser, 1979). There was a clear peak in the number of seeds set per flower at an outcrossing distance of 10 m compared both with the results of selfing and mating close neighbours, and with the results of mating very distant neighbours, 100 and 1000 m apart (Figure 5.15).

Thus, the studies of great tits and of delphiniums are both suggestive of ways in which natural selection might act to influence the distance that individuals disperse. We have, however, a chicken and egg problem: does the degree of in- and outbreeding depression that we observe in the field reflect past dispersal habits, or does the present degree of dispersal that we observe

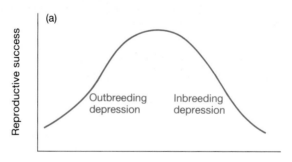

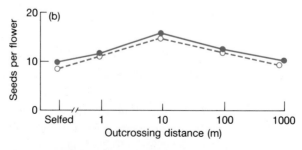

Figure 5.15. (a) A generalized diagram to show the relationship between the similarity of mating individuals and the success of the mating (vigour or number of progeny). (b) The effect of crossing plants of *Delphinium nelsoni* by deliberate hand-pollination of individuals chosen from different distances apart in the community (from Price & Waser, 1979). The symbols ● and ○ show the results of two different experiments.

reflect the selective consequences of different degrees of outbreeding? Has natural selection adjusted the degree of dispersal in a population to optimize the degree of outbreeding, or has the degree of outbreeding been selected to a level appropriate for a particular pattern of dispersal?

Gametes and spores often move further than the individuals that produce them. This means that genes and particular gene combinations of higher plants and some aquatic animals may colonize populations which the adults would never meet. The consequence may be that precise local specialization of populations is hindered by genetic contamination from elsewhere—but also that local populations may be able to draw on genetic variants that were not present within the population itself. The enormous variety of specialized mechanisms that ensure the dispersal of pollen is some measure of the importance that dispersal at this scale has had in the evolution of the lives of higher plants. It is convenient to distinguish such dispersal as *haploid* dispersal. Haploid dispersal in plants is reviewed by Levin and Kerster (1974).

5.7 Dormancy: dispersal in time

An organism gains in fitness by dispersing its progeny as long as they are more likely to leave descendants than they would be if they remained undispersed. Similarly, an organism gains in fitness by delaying its arrival on the scene as long as the delay increases its chances of leaving descendants. This will often be the case when current conditions are unfavourable for survival, growth or reproduction and when future conditions might be better. Thus a delay in the recruitment of an individual to a population may be regarded as *dispersal in time*.

Organisms generally spend their period of delay in a state of *dormancy*. This relatively inactive state confers the benefit of conserving energy, which can then be used during the period following the delay. In addition, the dormant phase of an organism is often more tolerant of the adverse environmental conditions prevailing during the delay (i.e. tolerant of drought, extremes of temperature, lack of light and so on).

predictive and consequential dormancy

Dormancy can be either *predictive* or *consequential* (Müller, 1970). Predictive dormancy is initiated in advance of the adverse conditions, and is most often found in predictable, seasonal environments. It is generally referred to as 'diapause' in animals, and in plants is often called 'innate dormancy' (Harper, 1977) or 'primary dormancy'. Consequential (or 'secondary') dormancy, on the other hand, is initiated in response to the adverse conditions themselves. As we shall see, it may be either 'enforced' or 'induced' (Harper, 1977). Note, though, that all of these distinctions are somewhat blurred at the edges.

5.7.1 Diapause: predictive dormancy in animals

Diapause has been most intensively studied in insects, and examples occur in all developmental stages, throughout the Class Insecta. The common field grasshopper *Chorthippus brunneus* (p. 134) is a fairly typical example. This annual species passes through an *obligatory* diapause in its egg stage, where,

obligatory diapause

in a state of arrested development, it is resistant to the cold winter conditions that would quickly kill the nymphs and adults. In fact, the eggs require a long cold period before development is able to recommence (around 5 weeks at 0°C, or rather longer at a slightly higher temperature; Richards & Waloff, 1954). This ensures that the eggs are not affected by a short, freak period of warm winter weather which may then be followed by normal, dangerous, cold conditions. It also means that there is an enhanced synchronization of subsequent development in the population as a whole. The grasshoppers 'migrate in time' from late summer to the following spring.

Diapause is also commonly shown by species with more than one generation per year. The fruit-fly *Drosophila obscura*, for instance, passes through four generations per year in England, but enters diapause during only one of them (Begon, 1976). This *facultative* diapause shares important features with obligatory diapause; it enhances survivorship during a predictably adverse winter period, and it is experienced by *resistant* diapause adults with arrested gonadal development and large reserves of stored abdominal fat. In this case, though, synchronization is achieved not only during diapause but also prior to it. Emerging adults react to the short day-lengths of autumn by laying down fat and entering the diapause state; they recommence development in response to the longer days of spring. Thus, by relying, like many species, on the utterly predictable *photoperiod* as a cue for seasonal development, *D. obscura* enters a state of predictive diapause which is confined to those generations that inevitably pass through the adverse conditions.

facultative diapause . . .

. . . the importance of photoperiod

5.7.2 Seed dormancy in plants

Seed dormancy is an extremely widespread phenomenon in flowering plants. The young embryo ceases development while still attached to the mother plant and enters a phase of suspended activity, usually losing much of its water and becoming dormant in a desiccated condition. In a few species of higher plants, such as some mangroves, a dormant period is absent, but this is very much the exception—almost all seeds are dormant when they are shed from the parent and require special stimuli to return them to an active state (germination).

We can recognize three types of dormancy-breaking process.

innate dormancy

(a) Innate dormancy is a state in which the embryo or the maternal tissues that enclose it have an absolute requirement for some special external stimulus to re-activate the process of growth and development. The stimulus may be the presence of water, light, photoperiod or an appropriate balance of near- and far-red radiation. Such dormancy-breaking requirements tend to synchronize germination with some phase of the seasons. Seedlings of species with this sort of dormancy tend to appear in sudden flushes of almost simultaneous germination.

enforced dormancy

(b) Enforced dormancy is a state of the seed that has been imposed on it by external conditions (i.e. it is consequential dormancy). The absence of normal requirements for growth (water, an appropriate temperature, a supply of oxygen) or the presence of inhibiting elements (such as high concentrations of carbon dioxide) may hold a seed in a dormant state. It may, of course, then die, but in many groups of higher plants seeds can be held in the dormant state for very long periods. Seeds of *Chenopodium album* collected from archaeological excavations have been shown to be viable when 1700 years old (Ødum, 1965). Seeds that are held in enforced dormancy may be brought to germinate simply by supplying the missing resource or conditions. Many of the seeds in a seedsman's packet are held in enforced dormancy only by the lack of water, and they germinate readily as soon as water is supplied. In nature, enforced dormancy may be broken by the rain falling after a period of drought, or by exposure to the more favourable atmosphere near the soil surface (perhaps as a result of the activities of earthworms or other burrowing animals). The progeny of a single plant with enforced dormancy may be dispersed in time over years, decades or even centuries.

induced dormancy

(c) Induced dormancy is a state produced in a seed by a period of enforced dormancy. In addition though, it acquires some new requirement before it can germinate. The seeds of many agricultural and horticultural weeds will germinate without a light stimulus when they are released from the parent; but after a period of enforced dormancy they require exposure to light before they will germinate. For a long time it was a puzzle that soil samples taken from the field to the laboratory would quickly generate huge crops of seedlings, although these same seeds had failed to germinate in the field. It was a simple idea of genius that prompted Wesson and Wareing (1967) to collect soil samples from the field at night and bring them to the laboratory in darkness. They obtained large crops of seedlings from the soil only when the samples were exposed to light. They also dug pits in the field at night, and covered some with slate to exclude light while they covered others with glass. Seedlings appeared in the pit only under the glass. This type of induced dormancy is responsible for the accumulation of large populations of seeds in the soil. They germinate only when they are brought to the soil surface by earthworms, or by the exposure of soil after a tree falls, or when the soil is disturbed by burrowing animals such as moles or gophers.

seed dormancy and the light beneath a canopy

Seed dormancy may be induced by radiation that contains a relatively high ratio of far-red (730 nm) to near-red (c. 660 nm) wavelengths. Light that has filtered through a leafy canopy has its spectral composition biased in this way, and the seeds of sensitive species can be brought into a state of dormancy by exposing them to light that has filtered through leaves. This must have the effect in the field of holding sensitive seeds in the dormant state when they land on the ground under a canopy, while releasing them into germination only when the overtopping plants have died away. It may also disperse a population of seeds in time, ensuring that individual seeds germinate only when there is a reasonable chance that they are not overtopped by vegetation.

dormancy as a compromise

Seedlings that germinate early, if they survive, usually have an expectation of a longer growth period and a higher fecundity than those that germinate late. Every delay in the life cycle of an organism is a reduction in its potential fecundity and so of its fitness. On the other hand, delayed germination may mean that individuals escape lethal hazards in the environment.

There is obviously a compromise involved in dispersal in time—a compromise between the risk of death by starting growth too early and a loss of fecundity from starting growth too late.

Effective dispersal in time depends not only on seeds being dormant for a period, but also on them being long lived. Such longevity appears to be under genetic control because it has been lost from most long-domesticated cultivars together with much of their innate dormancy. Most of the species of plant with seeds that persist for long in the soil are annuals and biennials, and they are mainly weedy species (which is cause and which is effect?). They are largely species that lack features that will disperse them extensively in space: perhaps dispersal in time and in space are not readily compatible. The seeds of trees usually have very short expectation of life in the soil, and many are extremely difficult to store artificially for more than a year. The seeds of many tropical trees are particularly short lived, a matter of weeks or even days. Among trees, the most striking longevity is seen in those that retain the seeds in cones or pods on the tree until they are released after fire (many *Eucalyptus* species and *Pinus*). This phenomenon of 'serotiny' protects the seeds against risks on the ground until fire creates an environment suitable for their rapid establishment.

the distribution of seed dormancy among plant types

5.7.3 Non-seed dormancy in plants

Dormancy in plants is not confined to seeds. The sand sedge *Carex arenaria*, for example, tends as it grows to accumulate dormant buds along the length of its predominantly linear rhizome. These may remain alive but dormant long after the shoots with which they were produced have died, and they have been found in numbers up to 400–500 per m² (Noble *et al.*, 1979). They play a role analogous to the bank of dormant seeds produced by other species. In a similar vein, Prescott chervill (*Chaerophyllum prescottii*), found in meadows in the U.S.S.R., has no detectable reserve of dormant seed, but can remain as a dormant underground tuber in the soil for over ten years. When meadows are ploughed, or areas of grass are killed, hundreds of dormant chervill tubers are activated into growth (Rabotnov, 1964).

Another type of plant dormancy was described on p. 152: the long-lived individuals in many tree populations that remain stunted until an opening appears in the canopy above them. Finally, the very widespread habit of deciduousness is a form of innate dormancy displayed by many perennial trees and shrubs. Established individuals pass through periods of adversity (usually low temperatures and low light levels) in an energy-conserving and resistant state of low metabolic activity.

5.7.4 Consequential dormancy in animals

Consequential dormancy in animals, like consequential (enforced and induced) dormancy in plants, may be expected to evolve in environments that are relatively unpredictable. In such circumstances there will be a disadvantage in responding to adverse conditions only after they have appeared. But this may very well be outweighed by the advantages (a) of responding to favourable conditions *immediately* they reappear, and (b) of only entering a dormant state if adverse conditions *do* appear.

Thus, when many mammals enter hibernation, they do so (after an obligatory preparatory phase) in direct response to the adverse conditions. And, having achieved 'resistance' by virtue of the energy they conserve at a lowered body temperature, and having periodically emerged and monitored their environment, they eventually cease hibernation whenever the adversity disappears. A particularly remarkable example of consequential dormancy in mammals is provided by those marsupials that are able to hold back the development of a foetus for many months during periods when the mother is short of resources. Such a delaying option is unavailable to placental mammals.

A consequential dormancy similar to that shown by plant seeds is apparent in the lives of many nematode worms, particularly parasites, where dormant cysts may remain in suspended animation for many years, breaking dormancy only after some special stimulus is received that signifies that the environment has become favourable for development (Sunderland, 1960). The signal may be that the cyst has been ingested by a host or, in the case of nematodes parasitic on higher plants, it may be that the growing host has released germination stimulants into the environment of the cyst. Protozoa too may encyst in a dormant state that is broken only by the return of specialized conditions such as the refilling of a dried-up pond.

5.8 Clonal dispersal

By definition, modular growth involves the repetition of units of structure. Thus in almost all clonal organisms, an individual genet spreads its parts around it as it grows. A developing tree or coral disperses its parts into, and therefore explores, the surrounding environment.

When much of the growth is horizontal as a stolon or rhizome (in many higher plants), a laterally spreading clone is formed that commonly roots at the nodes. Often, the interconnections of such a clone decay so that it becomes represented by a number of dispersed parts. In extreme cases, such clonal spread may result in the product of one zygote ultimately being represented by a clone of great age spread over great distances. The Finnish botanist Oinonen estimated that individual clones of the rhizomatous bracken fern (*Pteridium*) had reached an age of 1400 years; and one extended over an area of 474×292 m (nearly 14 ha). A few trees are also capable of clonal spread from root suckers, and the aspen (*Populus tremuloides*) may cover even greater areas than bracken from the product of a single original seed.

Organisms that form clones in this way achieve a form reflecting the disposition of their modules. Among the forms that extend by rhizomes or stolons, we can recognize two extremes in a continuum. At one extreme are types in which the connections between modules are long (often also thin and short-lived), so that the shoots of a clone are widely spaced. These have been called 'guerilla' forms because they give the plant, hydroid or coral a character like that of the military guerilla force—constantly on the move, disappearing from some territories and penetrating into others. They are both fugitive and

guerilla and phalanx growth forms

opportunist. At the other end of the continuum are 'phalanx' forms in which the connections between modules are short (often thick and long-lasting) and the modules are tightly packed (Lovett Doust & Lovett Doust, 1982). The tussock grasses of arid regions are fine examples of phalanx growth forms. Plants and modular animals with phalanx growth forms expand their clones slowly, retain their original site occupancy for long periods, and neither penetrate readily among neighbouring plants nor are easily penetrated by them. The term phalanx is used to describe this growth form by analogy with the units of a Roman army, tightly packed with their shields held around the group. (it is interesting, and ecologically very relevant, that to organize an army of guerillas is no way to fight a phalanx, and an army arranged in phalanxes is no way to fight guerillas.)

Even among non-clonal trees, it is easy to see that the way in which the buds are placed gives them a guerilla or a phalanx type of growth form. The dense packing of shoot modules in species like cypresses (*Cupressus*) produces a relatively undispersed and impenetrable foliage (phalanx canopy), while many loose-structured broad-leaved trees (*Acacia, Betula*) can be seen as guerilla canopies, bearing buds that are widely dispersed and shoots that interweave with the buds and branches of neighbours. The twining or clambering lianes in a forest are guerilla growth forms *par excellence*, dispersing their foliage and buds over immense distances both vertically and laterally.

The ways in which modular organisms disperse and display their modules affect the ways in which they interact with neighbours. Those with guerilla form will continually meet and interact with (compete with) other species and other genets of their own kind. With a phalanx structure, however, most meetings will be between modules of a single genet. For a tussock grass or a cypress tree, competition must occur very largely between the parts of itself.

Clonal growth is most effective as a means of dispersal in aquatic environments. Many aquatic plants fragment easily, and the parts of a single clone become independently dispersed because they are not dependent on the presence of roots to maintain their water relations. The major aquatic weed problems of the world are caused by plants that multiply as clones and fragment and fall to pieces as they grow. Duckweeds (see Figure 4.2), the water hyacinth, Canadian pond weed and the water fern *Salvinia* (see Figure 15.24) are all examples of organisms that disperse their modules, and in which the product of a single zygote may disperse through the water courses of an entire nation.

PART 2

Interactions

Introduction

The activity of any organism changes the environment in which it lives. It may alter conditions—as when the movement of an earthworm through soil aerates the soil, or when the transpiration of a tree cools the atmosphere. Or it may add or subtract resources from the environment that might have been available to other organisms—as when a tree shades what is beneath it, or a cow eats grass. In addition though, organisms will actually interact when individuals, in one way or another, enter into the lives of others. In the following chapters (6–13) we consider the variety of these interactions. We distinguish five main categories: competition, predation, parasitism, mutualism and detritivory. Like most biological categories, these five are not perfect pigeon-holes.

In very broad terms, 'competition' is an interaction in which one organism consumes a resource that would have been available to, and might have been consumed by, another. One organism deprives another, and, as a consequence, the other organism grows more slowly, leaves fewer progeny or is at greater risk of death. The act of deprivation can occur between two members of the same species or between individuals of different species. We choose to consider these situations in separate chapters (Chapters 6 and 7), partly because it seems a reasonable assumption that members of the same species are particularly likely to make the same resource demands on the environment (and to react symmetrically to the presence of each other) while members of different species are likely to differ (and so react asymmetrically to each other). But we need to be very careful not to take this distinction too literally—the very process of natural selection depends on differences between members of the same species, and conspecific individuals can differ, too, in their condition, their stage of development and so on. Moreover, members of two very different species may use and deplete the same resources: a rabbit in a field may be more effectively deprived of food by a sheep than by another rabbit.

Chapters 8, 9 and 10 deal with various aspects of 'predation', though we have defined predation very broadly. We have combined those situations in which one organism eats another and kills it (such as an owl preying on mice), and those in which the consumer takes only part of its prey, which may then regrow to provide another bite another day (grazing). We have also combined herbivory (animals eating plants) and carnivory (animals eating animals). In Chapter 8 we examine the nature of predation, i.e. what happens to the predator and what happens to the prey (be it animal or plant). Particular attention is paid to herbivory because of the subtleties that characterize the response of a plant to attack. In Chapter 9 we discuss the behaviour of predators—an area in which population ecology and behavioural ecology merge. The way in which individual predators behave clearly influences the dynamics of predator and prey populations, and these 'consequences of consumption' form the material of Chapter 10. This is the part of ecology that has the most obvious relevance to those concerned with the management of natural resources—the efficiency of harvesting (whether of fish, whales, grasslands or prairies) and the biological and chemical control of pests and weeds.

Four of the processes in this section—competition, predation, parasitism and mutualism—involve interactions between organisms that generate processes of natural selection. The activities of predators may be expected to influence the evolution of their prey and vice versa; and the activities of parasites may be expected to influence the evolution of their hosts and vice versa. If the process of evolution is largely the consequence of interactions between organisms (as Darwin supposed), it is among predators and prey, parasites and hosts, pairs of mutualists and between competitors that we might expect to find ecological and evolutionary processes most closely intertwined in action. By contrast, dead material leaves no descendants and cannot itself evolve, nor can it multiply. However, it is involved in complex interactions, and the processes by which it is decomposed involve competition, parasitism, predation and mutualism: microcosms of all the major ecological processes (except photosynthesis). Detritivory, the consumption of dead organic matter, is discussed in Chapter 11.

Chapter 12, Parasitism and Disease, deals with a subject that is rarely represented by a full chapter in ecology texts. Yet more than half of the species on the face of the earth are parasites. The category of parasitism has blurred edges, particularly where it merges into predation. But whereas a predator usually takes all or part of many individual prey, a parasite normally takes its resources from one or a very few hosts, and (like many grazing predators) it rarely kills its host immediately if at all. The insect parasitoids (like ichneumon wasps which develop (often singly) within the bodies of their hosts and kill them) fit neatly into neither the predators nor the parasites. We have discussed them, along with predators and grazers, in Chapters 8–10.

Finally, Chapter 13 is concerned with mutualistic interactions between individuals of different species. These are interactions in which both organisms experience a net benefit. The ecology of mutalism has been neglected and, like parasitism, it rarely claims a chapter in an ecology textbook. Yet the greater part of the world's biomass is composed of mutualists.

Interactions between organisms have often been summarized by a simple code which represents each one of the pair of interacting organisms by a '+', '−' or '0', depending on how it is affected by the interaction. Thus, a predator–prey (including a herbivore–plant) interaction, in which the predator benefits and the prey is harmed, is denoted by $+-$; and the parasite–host interaction is also clearly $+-$. Another straightforward case is mutualism, which is obviously $++$; whereas if organisms do not interact at all, we can denote this by 00 (sometimes called 'neutralism'). Detritivory must be denoted by +0, since the detritivore itself benefits, while its food (dead already) is unaffected. The general term applied to +0 interactions is 'commensalism', but paradoxically this term is not usually used for detritivores. Instead, it is reserved for cases, allied to parasitism, in which one organism ('the host') provides resources or a home for another organism, but in which the host itself suffers no tangible ill-effects. Competition is often described as a $--$ interaction, but, as we shall see in Chapters 6 and 7, it is often impossible to establish that both organisms are harmed. Such asymmetric interactions may then approximate to a -0 description, generally referred to as 'amensalism'. True cases of amensalism may occur when one organism produces its ill-effect (for instance a toxin) whether or not the potentially affected organism is present.

The eight chapters of this section are rather different in style. This reflects some very real differences in the various 'states of the art'. Studies of parasitism and mutualism have tended to concentrate on disentangling the intricacies of the relationships between the organisms concerned, rather than dealing with the role of the phenomena in the larger communities in which the organisms live. Thus we know a great deal about the ecology of rhizobial nitrogen-fixing bacteria in the root nodules of legume plants, but we know much less about the ecology of nodulated legumes in natural vegetation. Likewise, we know a great deal about the ecology of parasitic worms in intestines, but we know little about the role of infection in determining the abundance of natural populations. With interspecific competition, on the other hand, the problem is slightly different. We have a very sophisticated science concerned with the experimental and theoretical study of competition, but we are still fearfully ignorant about just how often it occurs and how important it is as a force in natural communities. As another point of contrast, there has been a long history of mathematical modelling of predator–prey interactions, whereas it is only recently that modellers have turned their attention to mutualism. For reasons such as these, and also because of the special interests of the most influential investigators, ecological study proceeds in fits and starts. There are parts that have powerful theory but lack a good underpinning of natural history, and there are other parts that are largely anecdotes from nature, lacking any real theory to give them coherence. Ecology is a science that is full of gaps and areas of ignorance—much like any other science.

Intraspecific Competition

6.1 Introduction: the nature of intraspecific competition

Organisms grow, reproduce, die and migrate (Chapters 4 and 5). They are affected by the conditions in which they live (Chapter 2), and by the resources which they obtain (Chapter 3). Yet no organism lives in isolation. Each, for at least part of its life, is a member of a population composed of individuals of its own species.

Individuals of the same species have very similar requirements for survival, growth and reproduction; but their combined demand for a resource may exceed the immediate supply. The individuals then *compete* for the resource and, not surprisingly, at least some of them become deprived. This chapter is concerned with the nature of such intraspecific competition, and its effects on the competing individuals. It will make sense, though, to begin with a working definition: 'competition is an interaction between individuals, brought about by a shared requirement for a resource in limited supply, and leading to a reduction in the survivorship, growth and/or reproduction of the competing individuals concerned'. We can now look more closely at competition.

a definition of competition

Consider, initially, a simple hypothetical community: a thriving population of grasshoppers (all of one species) feeding on a field of grass (also of one species). In order to live at all, grasshoppers must consume grass to provide themselves with energy and material for body-building. But in the process of finding and consuming the food, they also use energy, and expose themselves to a risk from predators. Each grasshopper will frequently find itself at a spot where there had previously been a blade of grass—before, that is, some other grasshopper ate it. Whenever this happens, the grasshopper must move on; it must expend more energy, and run a greater risk than it would otherwise have done, before it takes in food. And the more grasshoppers there are competing for food, the more often this will happen. Yet an increased energy expenditure, an increased risk of mortality, and a decreased rate of food intake may all decrease a grasshopper's chances of survival; while an increased energy expenditure and a decreased food intake may also leave less energy available for development, and less available for reproduction. Thus, since survival and reproduction determine a grasshopper's contribution to the next generation, the more intraspecific competitors for food a grasshopper has, the less its likely contribution will be.

a hypothetical example

As far as the grass itself is concerned, the contribution of an individual genet to the next generation will depend on the number of its progeny which themselves eventually develop into reproductive adults. An isolated seedling in fertile soil may have a very high chance of surviving to reproductive maturity. It will probably exhibit an extensive amount of modular growth, and will probably therefore produce a large number of offspring. However, a seedling which is closely surrounded by neighbours (shading it with their leaves and depleting its soil with their roots) will be very unlikely to survive, and if it does, will almost certainly be small and simple, and set few seeds. Increases in density will therefore decrease the contribution made by each individual to the next generation.

6.2 Common features of intraspecific competition

the ultimate effect is on fecundity and survivorship

Clearly there are a number of features common to both of these cases of intraspecific competition. The first is that the *ultimate* effect of competition is a decreased contribution to the next generation—a decrease, that is, compared with what would have happened had there been no competitors. Intraspecific competition leads to decreased rates of resource intake per individual, perhaps to decreased rates of individual growth or development, or to decreases in amounts of stored reserves. These may lead, in turn, to decreases in survivorship and/or decreases in fecundity. As we saw in Chapter 4, survivorship and fecundity together determine an individual's reproductive output.

competition is for limited resources

The second feature of intraspecific competition is that the resource for which individuals compete must be in limited supply. Oxygen, for example, although an absolutely essential resource, is not something for which grasshoppers or grass plants compete; the supply exceeds the rate at which the densest populations can consume it. Similarly, light, food, space or any other resource is only competed for if it *is* in limited supply.

In many cases, competing individuals do not interact with one another directly. Instead, individuals respond to the level of a resource, which has been depressed by the presence and activity of other individuals. Thus grasshoppers competing for food are not directly affected by other grasshoppers, but by the reduction in food level and the increased difficulty of finding good food that has been left by the others. Similarly, a competing grass plant is adversely affected by the presence of close neighbours because the zone from which it extracts resources (light, water, nutrients) has been overlapped by the 'resource depletion zones' of the neighbours. In all these cases, competition may be described as *exploitation*, in that each individual is affected by the amount of resource that remains after it has been exploited by the others.

exploitation and interference

In many other cases, however, competition takes another form, known as *interference*. Here individuals interact directly with each other, and one individual will actually prevent another from occupying a portion of the habitat and so from exploiting the resources in it. This is seen, for instance, amongst motile animals that defend *territories* (discussed in more detail in section 6.11): the result is often that the territory itself becomes the resource. Interference can also occur among sessile organisms. The presence of a barnacle on a rock, for example, prevents any other barnacle from occupying that same position, even though their supplies of food at that position may be in excess. Indeed, interference is very widespread amongst the sessile animals and plants that live on rocky shores: they frequently compete through the 'overgrowth' of one individual by another. In such cases the effects of competition tend to be obvious—in many cases of exploitation, the effects are much more subtle. In practice, interference is almost always accompanied by an element of exploitation, though, of course, there are many cases of exploitation without interference.

'one-sided reciprocity'

The third feature of intraspecific competition is that the competing individuals are in essence equivalent—but in practice very much less so. The very fact that they have been classified as 'the same species' implies that they have many fundamental features in common, and they may be expected to use similar resources and react in much the same way to conditions. We must be careful, however, how far we push the idea that the effects between competing individuals are reciprocal. There are many occasions when intraspecific competition is very one-sided: a strong, early seedling will probably shade a stunted, late one, and an older and larger bryozoan on the shore will probably 'overgrow' (or, better still, grow over) a smaller and younger one. Moreover, heritable differences between individuals can certainly ensure that competitive interactions are not reciprocal. Tall genotypes of corn, for instance, will usually shade out and suppress short genotypes of the same species. We therefore cannot say that competing individuals of the same species are entirely equivalent. What we can say is that members of the same species are more likely than members of different species to require the same resource, and they are more likely to react reciprocally to each other's presence.

competition can increase fitness

This lack of exact equivalence means that the ultimate effect of competition is far from being the same on different individuals. Weak competititors may make only a small contribution to the next generation, or no contribution at all. Strong competitors may have their contribution only negligibly affected. Indeed, a strong competitor may actually make a larger *proportional* contribution when there is intense competition than when there is no competition at all (i.e. if he—or she—maintains his contribution while all around him are losing theirs). In other words, although the ultimate effect of competition is a decrease in reproductive output, this does not always mean a decrease in individual fitness (i.e. relative contribution), especially not for the strongest competitors. It would not be correct, therefore, to say that competition 'adversely affects' all competing individuals (Wall & Begon, 1985)

the effects of competition are density-dependent

Finally, the fourth feature of intraspecific competition is that its likely effect on any individual is greater, the more competitors there are. The effects of intraspecific competition are therefore said to be *density-dependent*. In order to look more closely at intraspecific competition, we must examine the effects of population density on individuals, and in particular its effects on death, birth and growth.

6.3 Intraspecific competition, and density-dependent mortality and fecundity

Figure 6.1 shows the pattern of mortality in the flour beetle *Tribolium confusum* when cohorts were reared at a range of densities (Bellows, 1981). Known numbers of eggs were placed in glass tubes with 0.5 g of a flour–yeast mixture, and the number of individuals that survived to become adults in each tube was then scored. The data have been expressed in three complementary ways, and in each case the resultant curve has been divided into three regions. Figure 6.1a describes the effect of density on the per capita mortality rate, i.e. the proportion that died between the egg and adult stages or the chance or probability of an individual dying. Figure 6.1b describes how the *number* that died prior to the adult stage changed with density. (Of course, as density increases, so the number 'available' to die increases also. It is therefore not surprising if the number that die increases with density.) Lastly, Figure 6.1c describes the effect of density on the numbers that survive. (Once again, as density increases, the numbers 'available' to survive increases.)

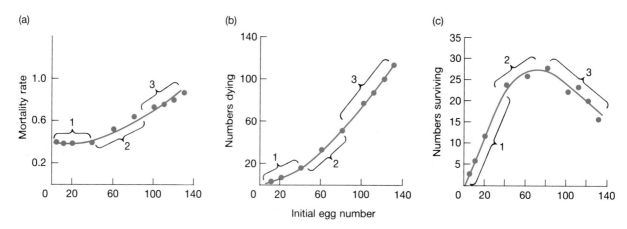

Figure 6.1. Density-dependent mortality in the flour beetle *Tribolium confusum* (a) as it affects mortality rate, (b) as it affects the numbers dying, and (c) as it affects the numbers surviving. In region 1 mortality is density-independent; in region 2 there is undercompensating density-dependent mortality; in region 3 there is overcompensating density-dependent mortality. (After Bellows, 1981.)

density-independent mortality

Throughout region 1 (low density), mortality rate remained constant as density was increased (Figure 6.1a). The numbers dying and the numbers surviving both rose (Figures 6.1b and c), but the *proportion* dying remained the same, which accounts for the straight lines in region 1 of Figures 6.1b and c. Mortality in this region was independent of density, and judged by this, there was apparently no intraspecific competition between the beetles at these densities. Individuals died, but the chance of an individual surviving to become an adult was not changed. Such density-independent deaths affect the population at all densities, and represent a base-line which any density-dependent mortality will exceed.

In region 2, the density-dependent effects of intraspecific competition are apparent. The numbers dying continued to rise with density, but they did

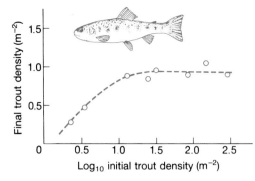

Figure 6.2. Density-dependent mortality amongst trout fry. At high trout densities the increasing mortality rate compensates exactly for increasing trout density and there are a constant number of surviving trout. (From Hassell, 1976, after Le Cren, 1973.)

undercompensating
density-dependence

so more rapidly than in region 1 (Figure 6.1b). The numbers surviving also continued to rise, but they did so less rapidly than in region 1 (Figure 6.1c). The reason for this is that the mortality *rate* increased with density (Figure 6.1a): there was density-dependent mortality. However, although the risk of death increased in region 2, it did so less rapidly than the increase in density. Over this range, increases in egg density continued to lead to increases in the total number of surviving adults. The mortality rate had increased, but it still 'undercompensated' for increases in density.

overcompensating
density-dependence

In region 3, intraspecific competition was even more intense. The increasing mortality rate 'overcompensated' for any increase in density, i.e. an increase in the initial number of eggs led to an even greater proportional increase in the mortality rate. Thus, over this range, the more eggs there were present, the fewer adults survived. Indeed, if the range of densities had been extended, there would have been tubes with no survivors: the developing beetles would have eaten all the available food before any of them reached the adult stage. However, irrespective of these variations in over- and under-compensation, the essential point is a simple one: at appropriate densities, intraspecific competition may lead to density-dependent mortality, which means that mortality rate increases as density increases.

exactly compensating
density-dependence

A slightly different situation is shown in Figure 6.2. This illustrates the effect of density on mortality in young trout (Le Cren, 1973). None of the

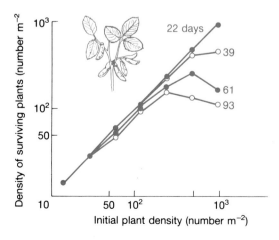

Figure 6.3. Density-dependent mortality in the soybean (*Glycine soja*). After 61 and 93 days the increasing mortality rate overcompensates for increases in sowing density, and the number of surviving plants declines. (After Yoda *et al.*, 1963.)

densities examined were low enough to avoid intraspecific competition entirely; even at the lower densities there was undercompensating density-dependence. At higher densities, however, mortality never overcompensated, but *compensated exactly* for any increase in density: any rise in the number of fry was matched by an exactly equivalent rise in the mortality rate. The number of survivors therefore approached and maintained a constant level, irrespective of initial density.

a plant example

Another example, mortality of soybeans, is illustrated in Figure 6.3. After 22 days almost every seed was represented by a plant; there had been scarcely any mortality and certainly no density-dependent mortality. After 39 days, however, there was some evidence of under-compensating density-dependent mortality; and after 61 and 93 days, the mortality at high density was overcompensating—the more seeds were sown, the fewer plants survived.

intraspecific competition and fecundity

The patterns of density-dependent fecundity that result from intra-

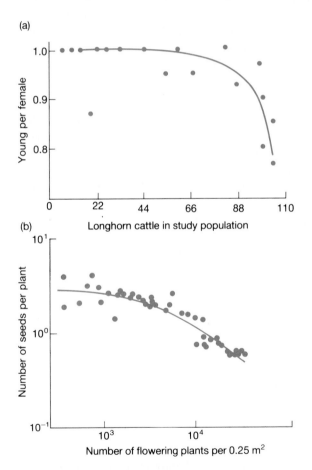

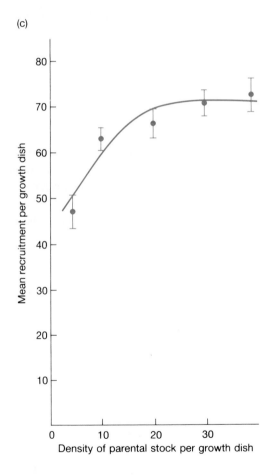

Figure 6.4. Density-dependent fecundity. (a) In longhorn cattle, fecundity changes from density-independence to overcompensating density-dependence as density increases (Fowler, 1981). (b) In the annual dune plant *Vulpia fasciculata,* fecundity changes from approximate density-independence to undercompensating density-dependence (from Watkinson & Davy, 1985, after Watkinson & Harper, 1978). (c) In the fingernail clam, *Musculium securis,* fecundity changes from undercompensating density-dependence to nearly exactly compensating density-dependence (Mackie *et al.*, 1978).

specific competition are, in a sense, a mirror-image of those for density-dependent mortality (Figure 6.4). Here, though, the per capita birth rate *falls* as intraspecific competition intensifies. At the lowest densities, birth rate is density-independent (Figures 6.4a and b); but as density increases, and the effects of intraspecific competition become apparent, birth rate initially shows undercompensating density-dependence (Figures 6.4a, b and c), and may then show exactly compensating (Figure 6.4c) or overcompensating density-dependence (Figure 6.4a).

Density-dependence and intraspecific competition are obviously bound closely together. Whenever there is intraspecific competition, its effect, whether on survival, fecundity or a combination of the two, is density-dependent. However, as subsequent chapters will show, there are processes other than intraspecific competition which also have density-dependent effects.

density or crowding?

Of course, the intensity of intraspecific competition experienced by an individual is not really determined by the density of the population as a whole. An individual is affected, rather, by the extent to which it is crowded and inhibited by its immediate neighbours, and this is particularly true of plants and other sessile organisms. As density increases, so will the typical level of crowding. However, if density varies within the population, many individuals will experience levels of crowding that are far from typical. Indeed, the same is true, though to a lesser extent, with mobile organisms; different individuals meet or suffer from different numbers of competitors. Density is therefore an abstraction which applies to the population as a whole, but need not apply to each individual within it. Density is often, but not always, the most convenient way of expressing the degree to which individuals are crowded.

6.4 Intraspecific competition and the regulation of population size

Despite the variations from example to example, there are obviously typical patterns in the effects of intraspecific competition on death (Figures 6.1–6.3) and birth (Figure 6.4). These generalized patterns are summarized in Figures 6.5 and 6.7.

competition may lead to a stable equilibrium

Figure 6.5 reiterates the fact that as density increases, the per capita birth rate eventually falls and the per capita death rate eventually rises. There must, therefore, be a density at which these curves cross. At densities below this point, the birth rate exceeds the death rate and the population increases in size. At densities above the cross-over point, the death rate exceeds the birth rate and the population declines. At the cross-over density itself, the two rates are equal and there is no net change in population size. This density therefore represents a *stable equilibrium*, in that all other densities will tend to approach it. In other words, intraspecific competition, by acting on birth rates and death rates, can *regulate* populations at a stable density at which the birth rate equals the death rate. This density is known as the *carrying capacity* of the population and is usually denoted by K (Figure 6.5). It is called a carrying capacity because it represents the population size which the resources of the

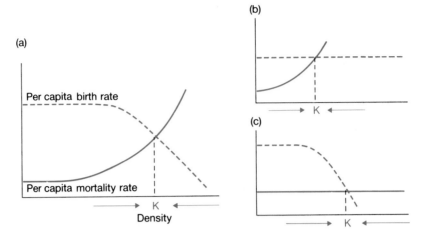

Figure 6.5. Density-dependent birth and mortality rates lead to the regulation of population size. When both are density-dependent (a), or when either of them is (b and c), their two curves cross. The density at which they do so is called the carrying capacity (K). Below this the population increases, above it the population decreases: K is a stable equilibrium. But this figure is a mere caricature of real populations.

environment can just maintain ('carry') without a tendency to either increase or decrease.

natural populations lack simple carrying capacities

However, while hypothetical populations caricatured by line drawings like Figure 6.5 can be characterized by a simple carrying capacity, this is not true of any natural population. There are unpredictable environmental fluctuations; individuals are affected by a whole wealth of factors of which intraspecific competition is only one; and resources not only affect density but respond to density as well. Intraspecific competition does not therefore hold natural populations to a predictable and unchanging level, the carrying capacity. Yet it may act upon a very wide range of starting densities and bring them to a much narrower range of final densities, and it therefore tends to keep density within certain limits. It is in this sense that intraspecific competition may be said typically to be capable of regulating population size. Figure 6.6a, for instance, shows the fluctuations within and between years in populations of the brown trout (*Salmo trutta*) (Elliott, 1984), and Figure 6.6b shows comparable fluctuations in the grasshopper *Chorthippus brunneus* (Richards & Waloff, 1954). There are no simple carrying capacities in these examples, but there are clear tendencies for the 'final' density each year ('late summer numbers' in the first case, 'adults' in the second) to be relatively constant, despite the large fluctuations in density within each year and the obvious potential for increase which both populations possess.

Indeed, the concept of a stable density settling at the carrying capacity, even in caricatured populations, is relevant only to situations in which density-dependence is not strongly overcompensating. Where there is overcompensation, cycles or even chaotic changes in population size may be the result. We will return to this point later (section 6.8).

An alternative general view of intraspecific competition is shown in Figure 6.7a, which deals with numbers rather than rates. The difference between the two curves ('births minus deaths') is the net number of additions expected in the population during the appropriate stage. It is therefore the

the fastest rate of population increase is at intermediate densities

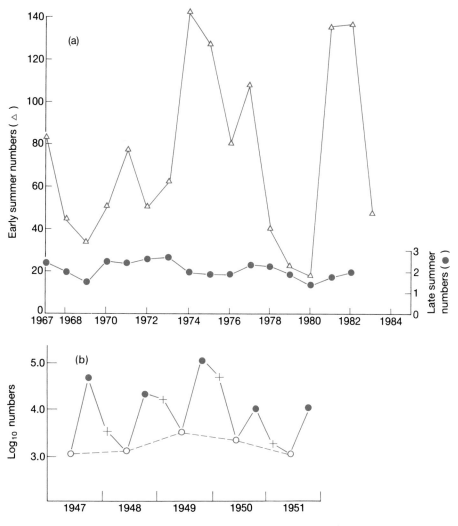

Figure 6.6. Population regulation in practice. (a) Trout (*Salmo trutta*) in an English Lake District stream, where △ = numbers in early summer, including those newly hatched from eggs, and ● = numbers in late summer. Note the difference in vertical scales (after Elliott, 1984.) (b) The grasshopper *Chorthippus brunneus* in southern England, where ● = eggs, + = nymphs and ○ = adults (note the logarithmic scale) (after Richards & Waloff, 1954). There are no definitive carrying capacities, but the 'final' densities each year ('late summer' and 'adults') are relatively constant despite large fluctuations within years.

population's *net rate of recruitment*: the amount by which the population changes in size during one stage or over one interval of time. Because of the shapes of the birth and death curves, the number of additions is small at the lowest densities, increases as density rises, declines again as the carrying capacity is approached, and is then negative (deaths exceed births) when the initial density exceeds K (Figure 6.7b). Thus total recruitment into a population is small when there are few individuals available to give birth, and small when intraspecific competition is intense. It reaches a peak, i.e. the population increases in size most rapidly, at some intermediate density.

The precise nature of the relationship between a population's net rate of recruitment and its density varies with the detailed biology of the species

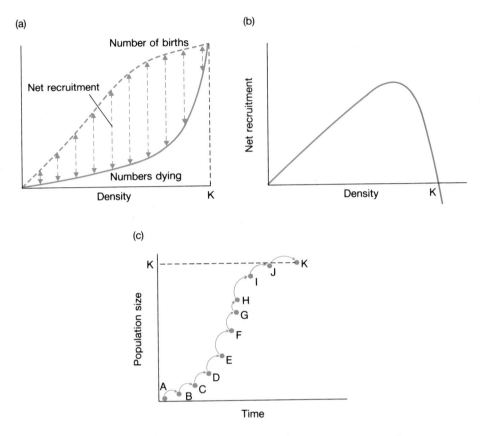

Figure 6.7. Some general aspects of intraspecific competition. (a) Density-dependent effects on the numbers dying and the number of births in a population: net recruitment is 'births minus deaths'. (b) The density-dependent effects of intraspecific competition on net recruitment: a humped or 'n'-shaped curve. (c) A population increasing in size under the influence of the relationships in (a) and (b): an 'S'-shaped or sigmoidal pattern of population increase, approaching the carrying capacity.

populations rise from low density following an 'S'-shaped curve

concerned (for instance, the pheasants, flies and whales in Figure 6.8). Moreover, because recruitment is affected by a whole multiplicity of factors, the data points never fall exactly on any single curve. Yet, in each case in Figure 6.8, an 'n'-shaped curve is apparent. This reflects the general nature of density-dependent birth and death whenever there is intraspecific competition. Note, therefore, that the 'n'-shaped curve describing the relationship between the leaf area index (LAI) of a plant population and the population's growth rate (see Figure 3.7) is of exactly the same type: peak crop growth at an intermediate LAI, and low growth at a high LAI (much mutual shading and competition).

In addition, curves of the type shown in Figures 6.7a and b may be used to suggest the pattern by which a population might increase from an initially very small size (e.g. when a species colonizes a previously unoccupied area). If a succession of time-intervals are taken singly, then each final density can be treated as the initial density for the next time-interval. This is illustrated in Figure 6.7c. Imagine a small population, well below the carrying capacity of its environment (Figure 6.7c, point A). Because the population is small, it

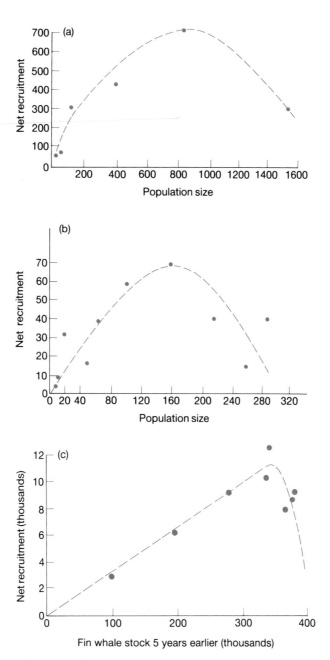

Figure 6.8. Some 'n'-shaped net recruitment curves, drawn by eye through the data points shown. (a) The ring-necked pheasant on Protection Island following its introduction in 1937 (data from Einarsen, 1945). (b) An experimental population of the fruit-fly *Drosophila melanogaster* (data from Pearl, 1927). (c) Estimates for the stock of antarctic fin whales (after Allen, 1972).

increases in size only slightly during one time-interval (Figure 6.7b), and only reaches point B (Figure 6.7c). Now, however, being larger, it increases in size more rapidly during the next time-interval (to point C), and even more during the next (to point D). This process continues until the population passes beyond the peak of its recruitment rate curve (Figure 6.7b). Thereafter, the population increases in size less and less with each time-interval (points G, H, I and J), until the population reaches its carrying capacity (*K*) and ceases completely to increase in size. The population might therefore be expected to follow an 'S'-shaped or 'sigmoidal' curve as it rises from a low density to its carrying capacity. This is a consequence of the hump in its recruitment rate curve, which is itself a consequence of intraspecific competition.

Of course, Figure 6.7c, like the rest of Figure 6.7, is a gross simplification, a caricature. It assumes, apart from anything else, that changes in population size are affected *only* by intraspecific competition. Nevertheless, something akin to sigmoidal population growth can be perceived in many natural and experimental situations (Figure 6.9).

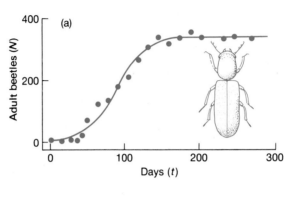

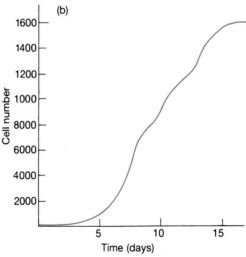

Figure 6.9. Real examples of 'S'-shaped population increase. (a) The beetle *Rhizopertha dominica* in 10 g of wheat grains replenished each week (Crombie, 1945). (b) The alga *Chlorella* in culture (Pearsall & Bengry, 1940).

resumé

Intraspecific competition leads to density-dependent increases in death rate and density-dependent decreases in birth rate; these may under-compensate, overcompensate or exactly compensate for rises in density. Intraspecific competition thus tends to regulate the size of populations, with the population's net rate of increase being greatest at intermediate densities below the carrying capacity; this can lead to a sigmoidal population growth curve. However, while such intraspecific competition will be obvious in certain cases (such as overgrowth competition between sessile organisms on a rocky shore), this will not be true of every population examined. Individuals are affected not only by intraspecific competitors, but also by predators, parasites and prey, competitors from other species, and the many facets of their physical and chemical environment. Any of these may outweigh or obscure the effects of intraspecific competition; or the effect of these other factors at one stage may reduce density to well below the carrying capacity for all subsequent stages. Nevertheless, intraspecific competition probably affects most populations at least sometimes during at least one stage of their life cycle.

6.5 Intraspecific competition and density-dependent growth

competition affects
growth and development
in unitary organisms

Intraspecific competition, then, can have a profound effect on the number of individuals in a population; but it can have an equally profound effect on the individuals themselves. In populations of unitary organisms, rates of growth and rates of development are commonly influenced by intraspecific competition. This necessarily leads to density-dependent effects on the composition of a population. Figures 6.10a and b, for instance, show two examples in which the distribution of sizes within a population has been altered as a result of intraspecific competition. This, in turn, often means that although the numerical size of a population is regulated only approximately by intraspecific competition, the total biomass is regulated much more precisely. This is illustrated in Figure 6.11 from observations on the limpet *Patella cochlear* (Branch, 1975). In the low-density populations the individuals were relatively large, while in the high-density populations they were relatively small. Overall, however, the total biomass was roughly the same at all densities in excess of 400 individuals per m².

modular organisms:
constant final yield

Such effects are particularly marked in modular organisms, one example being an experiment in which seeds of subterranean clover (*Trifolium subterraneum*) were sown at a range of densities (Figure 6.12a; Donald, 1951). As the plants grew, the yield (i.e. the total weight) per unit area increased, and at the first harvest (62 days) the yield was closely related to the density of seeds

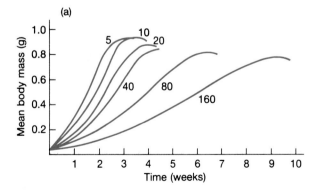
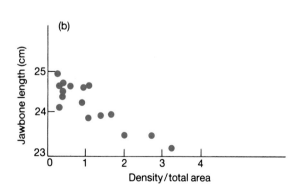

Figure 6.10. The effects of density on growth rate and size. (a) The density-dependent growth rate of the frog *Rana tigrina*, where numbers refer to individuals per 2 litre aquarium (after Dash & Hota, 1980). (b) Mean size (jawbone length) decreases with density in a reindeer population (after Skogland, 1983).

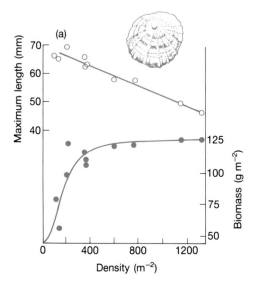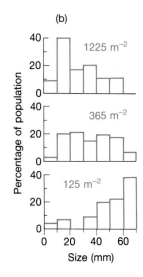

Figure 6.11. Intraspecific competition and growth in populations of the limpet *Patella cochlear*. (a) Individual size declines with density leading to an exact regulation of the population's total biomass. (b) High-density populations have many small and a few large individuals; low-density populations have many large and a few small individuals. (After Branch, 1975.)

sown. After 181 days, however, yield was no longer related to sowing density in this way—it had become the same over a wide range of initial densities. In fact, this is a pattern that has been frequently uncovered by plant ecologists, and it has been called *'the law of constant final yield'* (Kira *et al.*, 1953). The yield becomes constant over a wide range of densities because individuals suffer density-dependent reductions in growth rate and thus in individual plant size. Moreover, in such experiments the reduction in mean plant weight *compensates exactly* for increases in density.

Yield is density (d) multiplied by mean weight per plant ($\bar{w}$). Thus, if yield is constant:

$$d.\bar{w} = c,$$
and so
$$\log d + \log \bar{w} = \log c,$$
and
$$\log \bar{w} = \log c - 1.\log d.$$

Hence, when yield per unit area is independent of density (yield $= c$), a plot of log mean weight against log density should have a slope of -1.

competition affects growth and development in unitary organisms

The data from an experiment on the effects of density on the growth of the grass *Vulpia fasciculata* are shown in Figure 6.12b, and it can be seen that the slope of the curve rises over time and does indeed approach a value of -1. Notice that in this experiment, as in that with the clover, individual plant weight at the first harvest was reduced only at very high densities. As the plants became larger, they interfered with each other at lower and lower densities.

The results of these experiments suggest that there are limited resources

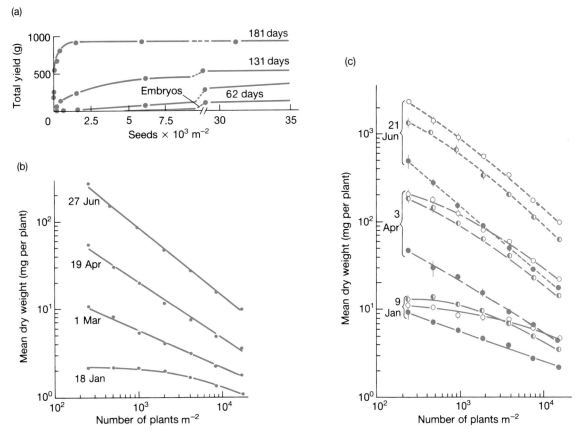

Figure 6.12. The 'constant final yield' of plants sown at a range of densities. This can be illustrated either by a horizontal line when total yield is plotted against density, or by a line of slope−1 when log mean weight is plotted against log density (see text). (a) Subterranean clover, *Trifolium subterraneum* (Donald, 1951). (b) and (c) The dune annual *Vulpia fasciculata*, firstly showing the yields at successive harvests, and secondly showing the additional influence of low (●), medium (◐) and high (○) concentrations of soil nitrogen; vertical lines indicate standard errors (Watkinson, 1984).

available for plant growth, and that at high densities these are shared among the individuals. If this is the case, we would expect the provision of extra resources to allow greater growth of individual plants and greater yield per unit area. This expectation is borne out by the results of an experiment in which three levels of nitrogen fertilizer were given to populations of *Vulpia fasciculata* growing at a range of densities (Figure 6.12c).

competition tends to regulate module number

The constancy of the final yield is a result, to a large extent, of the modularity of plants. This can be seen more explicitly in Figure 6.13, which shows what happened when Kays and Harper (1974) sowed perennial rye grass (*Lolium perenne*) at a 30-fold range of densities. After 180 days, some genets had died; but as a result of density-dependent modular growth, the range of final tiller densities was far narrower than that of genets. In modular organisms, then, the regulatory powers of intraspecific competition frequently operate by affecting the number of modules per genet instead of, or as well as, affecting the number of genets themselves.

It must not be imagined, however, that plants at high densities are typically scaled-down versions of those at low densities, differing only in the

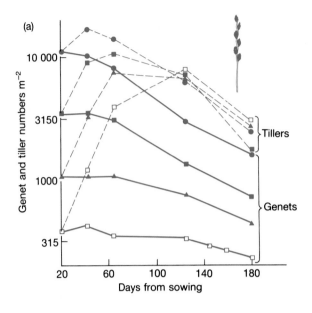

Figure 6.13. When populations of ryegrass, *Lolium perenne*, were sown at a range of densities, the range of final tiller densities was far narrower than that of genets (data from Kays & Harper, 1974).

different plant parts are affected to different extents

possession of fewer modules. Intraspecific competition affects not only rates of growth, but also rates of development and maturation, and it therefore affects the way in which biomass is distributed *within* individual plants. Figure 6.14 shows an example of a very common phenomenon. At high densities the maize plants were not only smaller but devoted a lower proportion of their biomass to seeds; the result was that seed output per unit area declined at the higher densities (Harper, 1961). Indeed, in many examples of density-dependent fecundity, in both unitary and modular organisms, the immediate effect of density is on growth rate and size, and it is this which affects fecundity.

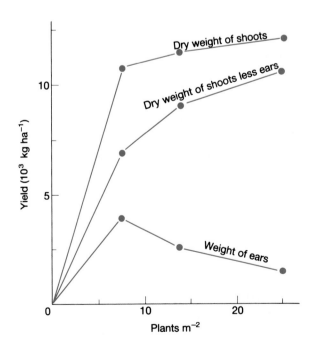

Figure 6.14. Competition in populations of maize (*Zea mays*) affects seed-production (weight of ears) far more than it affects the growth of shoots (Harper, 1961).

6.6 Quantifying intraspecific competition

Every population is in one way or another unique. Nevertheless, as we have already seen, there are general patterns in the action of intraspecific competition which can be clearly discerned. In this section, we take such generalization one stage further. A method will be described which can be used to summarize the effects of intraspecific competition on mortality, fecundity and growth. The method makes use of *k*-values, which were discussed in Chapter 4. Mortality will be dealt with first, and the method will then be extended for use with fecundity and growth.

A *k*-value was defined by the formula:

$$k = \log(\text{initial density}) - \log(\text{final density})$$

or, equivalently,

$$k = \log\left(\frac{\text{initial density}}{\text{final density}}\right).$$

For present purposes, 'initial density' may be denoted by B, standing for 'numbers *before* the action of intraspecific competition', while 'final density' may be denoted by A, standing for 'numbers *after* the action of intraspecific competition'. Thus:

$$k = \log B/A$$

Note that k increases as mortality rate increases, i.e. as the proportion surviving (A/B) decreases.

Some examples of the effects of intraspecific competition on mortality are shown in Figure 6.15, in which k is plotted against $\log B$. In several cases, k is constant at the lowest densities. This is an indication of density-independence: the proportion surviving is unaffected by initial density. At higher densities, k increases with initial density; this indicates density-dependence. Most importantly, however, the way in which k varies with the logarithm of density indicates the precise nature of the density-dependence. Figures 6.15a and b, for example, describe respectively situations in which there is under- and exact compensation at higher densities. The exact compensation in Figure 6.15b is indicated by the slope of the curve (denoted by b) taking a constant value of 1. The undercompensation which preceded this at lower densities, and which is seen in Figure 6.15a even at higher densities, is indicated by the fact that b is less than 1.

For the mathematically inclined: exact compensation means that A is constant. The slope, b, is given by:

$$b = \frac{k_2 - k_1}{\log_{10} B_2 - \log_{10} B_1}$$

$$= \frac{\log_{10} B_2 - \log_{10} A - (\log_{10} B_1 + \log_{10} A)}{\log_{10} B_2 - \log_{10} B_1}$$

$$= \frac{\log_{10} B_2 - \log_{10} B_1}{\log_{10} B_2 - \log_{10} B_1} = 1$$

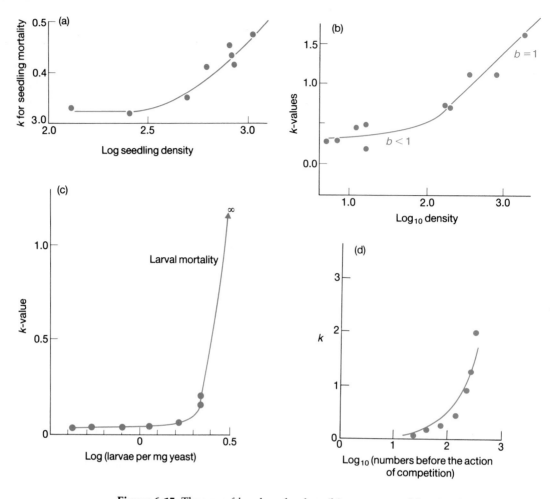

Figure 6.15. The use of k-values for describing patterns of density-dependent mortality. (a) Seedling mortality in the dune annual *Androsace septentrionalis* in Poland (Symonides, 1979a). (b) Egg mortality and larval competition in the almond moth, *Ephestia cautella* (Benson, 1973a). (c) Larval competition in the fruit-fly *Drosophila melanogaster* (Bakker, 1961). (d) Larval mortality in the moth *Plodia interpunctella* (Snyman, 1949).

contest and scramble competition

Exact compensation ($b = 1$) is often referred to as pure *contest* competition, because there are a constant number of winners (survivors) in the competitive process. The term was initially proposed by Nicholson (1954b) who contrasted it with what he called pure *scramble* competition. Pure scramble is the most extreme form of overcompensating density-dependence, in which *all* competing individuals are so adversely affected that none of them survive, i.e. $A = 0$. This would be indicated in Figure 6.15 by a b-value of infinity (a vertical line), and Figure 6.15c is an example in which this is the case. More common, however, are examples in which competition is scramble-*like*, i.e. there is considerable but not total overcompensation ($b \gg 1$). This is shown, for instance, in Figure 6.15d.

Plotting k against $\log B$ is thus an informative way of depicting the effects of intraspecific competition on mortality. Variations in the slope of the curve

k-values can also be used
for fecundity and
growth

(*b*) give a clear indication of the manner in which density-dependence changes with density. The real value of the method, however, is that it can be extended to fecundity and growth.

For fecundity, it is necessary to think of *B* as 'the total number of offspring that *would* have been produced had there been no intraspecific competition', i.e. if each reproducing individual had produced as many offspring as it would have done in a competition-free environment. *A* is then the total number of offspring *actually* produced. (In practice, *B* can be obtained from the population experiencing the least competition—not necessarily competition-free.) For growth, *B* must be thought of as the total biomass, or total number of modules, that would have been produced had all individuals grown as if they were in a competition-free environment (or, in practice, in the environment with least competition). *A* is then the total biomass or total number of modules actually produced.

There is still, in these cases, an underlying comparison of 'before competition' and 'after competition'. Now, however, 'before' must be imagined on the basis of what would have happened had there been no competition. 'After' clearly corresponds with what actually happens.

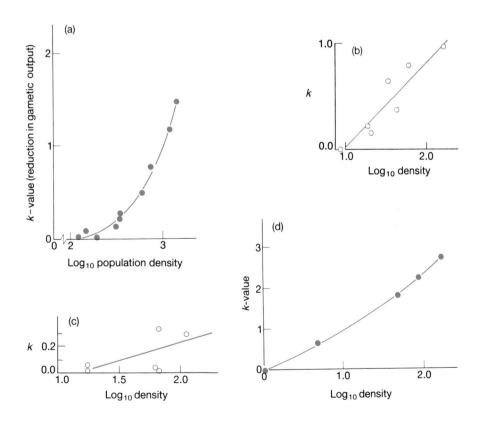

Figure 6.16. The use of *k*-values for describing density-dependent reductions in fecundity as a consequence of competition. (a) The limpet *Patella cochlear* in South Africa (Branch, 1975). (b) The cabbage root fly, *Erioischia brassicae* (Benson, 1973b). (c) The grass mirid *Leptoterna dolabrata* (McNeill, 1973). (d) The plantain *Plantago major* (Palmblad, 1968).

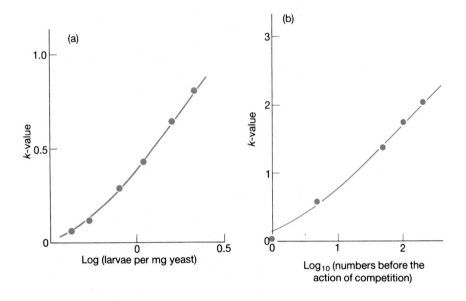

Figure 6.17. The use of *k*-values for describing density-dependent reductions in growth as a consequence of competition. (a) Adult female weight loss as a consequence of larval competition in *Drosophila melanogaster* (Bakker, 1961). (b) Weight loss in the shepherd's purse, *Capsella bursa-pastoris* (Palmblad, 1968).

Figures 6.16 and 6.17 illustrate examples in which *k*-values are used to describe the effects of intraspecific competition on fecundity and growth respectively. The *patterns* are essentially similar to those in Figure 6.15. Each falls somewhere on the continuum ranging between density-independence and pure scramble, and their position along that continuum is immediately apparent. Using *k*-values, all examples of intraspecific competition can be quantified in the same terms. With fecundity and growth, however, the terms 'scramble' and, especially, 'contest' are less appropriate, and it is usually preferable to talk in terms of exact, over- and undercompensation.

6.7 Mathematical models: introduction

The desire to formulate general rules in ecology often finds its expression in the construction of mathematical or graphical models. It may seem surprising that those interested in the natural living world should spend time reconstructing it in an artificial mathematical form; but there are several good reasons why this should be done. The first is that models can crystallize, or at least bring together in terms of a few parameters, the important, shared properties of a wealth of unique examples. This simply makes it easier for the ecologist to think about the problem or process under consideration. The second reason, related to the first, is that a model can provide a 'common language' in which each unique example can be expressed; and if each can be expressed in a common language, then their properties *relative to one another* will be more apparent. The third reason, again related to the first two, is that a model can provide a standard of 'ideal' or idealized behaviour against which reality can be judged and measured. And last but not least, models can actually shed light on the real world of which they are such imperfect mimics. Some ways in which they can do this will become apparent below.

These four reasons for constructing models are also criteria by which any model should be judged. Indeed, a model is only useful (i.e. worth constructing) if it *does* perform one or more of these four functions. Of course, in order to perform them a model must adequately describe real situations and real sets of data, and this 'ability to describe' or 'ability to mimic' is itself a further criterion by which a model can be judged. However, the crucial word is 'adequate': models must *adequately* describe. The only perfect description of the real world is the real world itself. A model is an adequate description, ultimately, as long as it is good enough to satisfy at least one of the four criteria.

In the present case, some simple models of intraspecific competition will be described. They will be built up from a very elementary starting-point, and their properties (i.e. their ability to satisfy the criteria described above) will then be examined. Initially, a model will be constructed for a population with discrete breeding seasons. This will be followed by a model in which breeding is continuous.

6.8 A model with discrete breeding seasons

6.8.1 The basic equations

In section 4.7 we developed a simple model for species with discrete breeding seasons, in which the population size was N_t at time t, and the population altered in size over time under the influence of a fundamental net reproductive rate, R. This model can be summarized in two equations, namely:

$$N_{t+1} = N_t R \qquad (6.1)$$

and

$$N_t = N_0 R^t. \qquad (6.2)$$

no competition: an exponentially increasing population

The model, however, describes a population in which there is no competition, and in which R is constant; and if $R > 1$, the population will continue to increase in size indefinitely ('exponential growth' shown in Figure 6.18). The

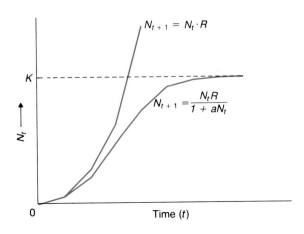

Figure 6.18.
Mathematical models of population increase with time, in populations with discrete generations: exponential increase (left) and sigmoidal increase (right).

first step is therefore to modify the equations by making the net reproductive rate subject to intraspecific competition. This is done in Figure 6.19, which has three components.

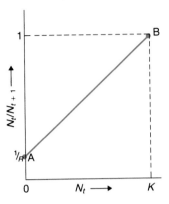

the incorporation of intraspecific competition

Point A describes a situation in which the population size is very small (N_t is virtually zero). Competition is therefore negligible, and the actual net reproductive rate is adequately defined by an unmodified R. Thus at these very low densities, equation 6.1 is still appropriate, or, rearranging the equation:

$$\frac{N_t}{N_{t+1}} = \frac{1}{R}.$$

Point B, by contrast, describes a situation in which the population size (N_t) is very much larger and there is a significant amount of intraspecific competition. In fact, at point B, the actual net reproductive rate has been so modified by competition that the population can collectively do no better than replace itself each generation, because 'births' equal 'deaths'. In other words, N_{t+1} is simply the same as N_t, and N_t/N_{t+1} equals 1; and the population size at which this occurs is, by definition, the carrying capacity, K (Figure 6.5).

The third component of Figure 6.19 is the straight line joining point A to point B and extending beyond it. This describes the progressive modification of the actual net reproductive rate as population size increases; but its straightness is simply an assumption made for the sake of expendiency, since all straight lines are of the simple form: $y = (slope)x + (intercept)$. In Figure 6.19, N_t/N_{t+1} is measured on the y-axis, N_t on the x-axis, the intercept is $1/R$, and the slope, based on the segment between points A and B, is $(1-1/R)/K$. Thus:

$$\frac{N_t}{N_{t+1}} = \frac{1-\dfrac{1}{R}}{K} \cdot N_t + \frac{1}{R}$$

or, rearranging:

$$N_{t+1} = \frac{N_t R}{1 + \dfrac{(R-1)N_t}{K}}.$$

224   CHAPTER 6

For further simplicity, $(R-1)/K$ may be donated by a, giving:

$$N_{t+1} = \frac{N_t R}{(1+aN_t)} \tag{6.3}$$

This is a model of population increase limited by intraspecific competition. Its essence lies in the fact that the unrealistically constant R in equation 6.1 has been replaced by an actual net reproductive rate, $R/(1+aN_t)$, which decreases as population size (N_t) increases.

The properties of this model may be seen by reference to Figure 6.19 (from which the model was derived) and Figure 6.18 (which shows a hypothetical population increasing in size over time in conformity with the model). The population in Figure 6.18 increases exponentially when N_t is very low, but the rate of increase declines progressively as population size rises, until at the carrying capacity the rate is zero; the result is an 'S'-shaped or 'sigmoidal' curve. This is a desirable quality of the model, but it must be recognized that there are many other models which would also generate such a curve. The advantage of equation 6.3 is its simplicity. The behaviour of the model in the vicinity of the carrying capacity can best be seen by reference to Figure 6.19. At population sizes that are less than K the population will increase in size; at population sizes that are greater than K the population size will decline; and at K itself the population neither increases or decreases. The carrying capacity is therefore a stable equilibrium for the population, and the model exhibits the regulatory properties classically characteristic of intraspecific competition.

It is not yet clear, however, just exactly what type or range of competition this model is able to describe; but this can be explored by tracing the relationship between k-values and $\log N_t$ (as in section 6.6). Each generation, the *potential* number of individuals produced (i.e. the number that would be produced if there were no competition) is $N_t R$. The *actual* number produced (i.e. the number that survive the effects of competition) is $N_t R/(1+aN_t)$. Section 6.6 established that:

$k = \log$ (number produced)$-\log$ (number surviving).

Thus, in the present case:

$k = \log N_t R - \log N_t R/(1+aN_t),$

or, simplifying:

$k = \log (1+aN_t).$

Figure 6.20 shows a number of plots of k against $\log_{10} N_t$ with a variety of values of a inserted into the model. In every case, the slope of the graph approaches and then attains a value of 1. In other words, the density-dependence always begins by undercompensating and then compensates perfectly at higher values of N_t. The model is therefore limited in the type of competition that it can produce, and all we have been able to say so far is that *this type* of competition leads to very tightly controlled regulation of populations.

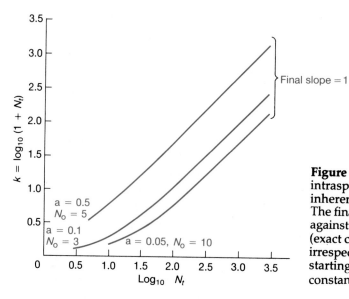

Figure 6.20. The intraspecific competition inherent in equation 6.3. The final slope of k against $\log_{10}N_t$ is unity (exact compensation), irrespective of the starting density N_0, or the constant $a\,(=\,(R-1)/K)$.

6.8.2 Incorporating a range of competition

Fortunately, a far more general model can be produced by incorporating a simple modification of equation 6.3. The modification was originally suggested by Maynard Smith and Slatkin (1973) and has been discussed in detail by Bellows (1981). It alters the equation to:

$$N_{t+1} = \frac{N_t R}{1+(aN_t)^b}.$$

(6.4)

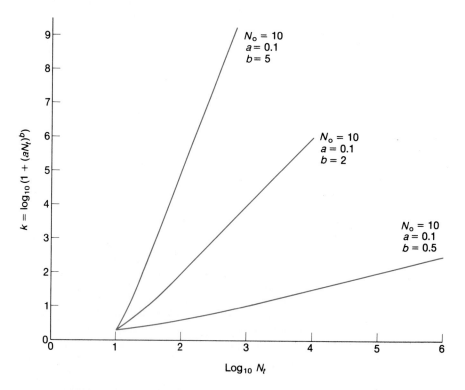

Figure 6.21. The intraspecific competition inherent in equation 6.4. The final slope is equal to the value of b in the equation.

The justification for this modification may be seen by examining some of the properties of the revised model. Figure 6.21 for example, shows plots of k against log N_t analogous to those in Figure 6.20: k is now log $[1+(aN_t)^b]$. It is apparent that the slope of the curve, instead of approaching 1 as it did previously, now approaches the value taken by b in equation 6.4. Thus, by the choice of appropriate values, the model can portray undercompensation ($b < 1$), perfect compensation ($b = 1$), scramble-like overcompensation ($b \gg 1$) or even density-independence ($b = 0$). The model has the generality that equation 6.3 lacks, with the value of b determining the type of density-dependence which is being incorporated.

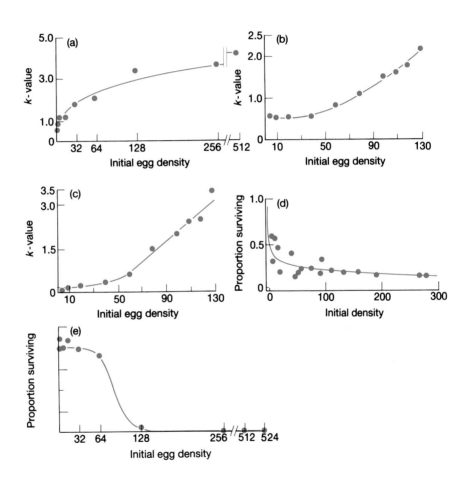

Figure 6.22. Equation 6.4 as a description of density-dependent mortality (the lines) fitted to sets of data (the dots). (a) The beetle *Stegobium panaceum* in the laboratory. (b) The beetle *Tribolium confusum* in the laboratory. (c) The beetle *Tribolium castaneum* in the laboratory. (d) The winter moth *Operophtera brumata* in the field (Varley & Gradwell, 1968). (e) The beetle *Lasioderma serricorne* in the laboratory. (After Bellows, 1981.)

6.8.3 The descriptive powers of the equation

Further evidence of the descriptive powers and generality of equation 6.4 is displayed in Figures 6.22a–e. Each plot in this figure represents a set of field or laboratory data (the dots) to which a curve conforming to the equation has been fitted by computer; the computer programme has selected the combination of values of a, b and R which gives the most closely fitting curve. Equation 6.4 can clearly model situations in which k tends to level off with density (Figure 6.22a) as well as those in which there is an upward sweep to the curve (Figures 6.22b and c); and it can model situations in which the proportion surviving declines monotonically with density (Figure 6.22d) as well as those in which the proportion remains roughly constant at low density, then declines sharply, and then levels out again at higher densities (Figure 6.22e).

the model provides a good description of real data, and a common language

The descriptive properties of equation 6.4 are impressive, and this is a desirable quality for any model to have. In addition, these properties can be utilized as an aid to classifying the enormous variety of density-dependent relationships that have been uncovered. Figure 6.22 indicated that a wide range of data sets could be suitably described by assigning values to three variables of the model: a, b and R. These values, therefore, immediately become a common language which can be used to compare and contrast very disparate situations. The advantages of this should not be underestimated. Ecological data-sets come from an enormous variety of organisms in an enormous variety of circumstances; progress is only possible if the connection between each data-set and all other data-sets can be clearly and simply expressed. Without a common language, ecologists could, for instance, merely point to the general difference in shape between Figures 6.22a and b, and to the fact that Figures 6.22b and c are essentially similar. As it is, each relationship can be precisely encapsulated in a three-figure code, allowing a pattern to emerge from the variation.

6.8.4 Causes of population fluctuations

Another desirable quality which equation 6.4 shares with other good models is an ability to throw fresh light on the real world. By sensible analysis of the population dynamics generated by the equation, it is possible to draw guarded conclusions regarding the dynamics of natural populations. The method by which this and similar equations may be examined has been set out and discussed by May (1975a). It is a method using fairly sophisticated mathematical techniques, but the results of the analysis can be understood and appreciated without dwelling on the analysis itself. These results are shown in Figures 6.23a and b. Figure 6.23a sets out the various patterns of population growth and dynamics which equation 6.4 can generate. Figure 6.23b sets out the conditions under which each of these patterns occurs. There are several points to note. The first is that the pattern of dynamics depends on two things: b, the precise type of competition or density-dependence, and R, the effective net reproductive rate (taking density-independent mortality into

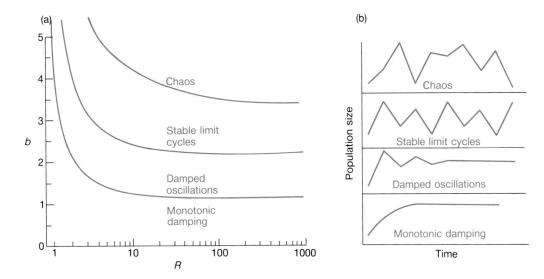

Figure 6.23. (a) The range of population fluctuations (themselves shown in (b)) generated by equation 6.4 with various combinations of b and R inserted. The line dividing stable limit cycles from 'chaos' is approximate. (After May, 1975a, and Bellows, 1981.)

account). By contrast, a determines not the type of pattern but only the level about which any fluctuations occur.

As Figure 6.23b shows, low values of b and/or R lead to populations which approach their equilibrium size without fluctuating at all. This has already been hinted at in Figure 6.18. There, a population behaving in conformity with equation 6.3 approached equilibrium directly, irrespective of the value of R. Equation 6.3 is a special case of equation 6.4 in which $b = 1$ (perfect compensation); Figure 6.23b confirms that for $b = 1$, monotonic damping is the rule whatever the effective net reproductive rate.

the model indicates that intraspecific competition can lead to a wide range of population dynamics

As the values of b and/or R increase, the behaviour of the population changes first to damped oscillations gradually approaching equilibrium, and then to 'stable limit cycles' in which the population fluctuates around an equilibrium level, revisiting the same two, four or even more points time and time again. Finally, with large values of b and R, the population fluctuates in a wholly irregular and chaotic fashion.

Thus a model built around a density-dependent, supposedly regulatory process (intraspecific competition) can lead to a very wide range of population dynamics. If a model population has even a moderate fundamental net reproductive rate (and an individual leaving 100 ($= R$) individuals in the next generation in a competition-free environment is not unreasonable), and if it has a density-dependent reaction which even moderately overcompensates, then *far from being stable*, it may fluctuate widely in numbers without the action of any extrinsic factor. The biological significance of this is the strong suggestion that even in an environment which is wholly constant and predictable, the *intrinsic* qualities of a population and the individuals within it may, by themselves, give rise to population dynamics with large and perhaps even chaotic fluctuations. The consequences of intraspecific competition are clearly not limited to 'tightly controlled regulation'.

An additional analysis of model properties can be carried out using the simpler equation 6.3 as a starting point. In that equation (and the more general equation 6.4) it has been assumed until now that populations respond instantaneously to changes in their own density, i.e. the overall reproductive rate within a population reflects the population size *at that time*. Suppose instead that the reproductive rate is determined by the amount of resource available, but that the amount of resource is determined by the density *one time-interval ago*. (The amount of grass in a field in spring being determined by the level of grazing the previous year is a simple but reasonable example of this.) In such a case, the reproductive rate itself will be dependent on the density one time-interval ago. Thus, since in equations 6.1, 6.3 and 6.4:

$$N_{t+1} = N_t \times \text{reproductive rate},$$

equation 6.3 may be modified to:

$$N_{t+1} = \frac{N_t R}{1 + aN_{t-1}}. \tag{6.5}$$

There is a *time-lag* in the population's response to its own density, caused by a time-lag in the response of its resources; and the behaviour of the modified model is as follows:

$R < 1.33$: monotonic damping
$R > 1.33$: damped oscillations.

In comparison, the original equation, without a time-lag, led to monotonic damping for all values of R. The time-lag has provoked the fluctuations in the model, and it can be assumed to have similar, destabilizing effects on real populations (see also section 10.2.2).

There are two important conclusions to be drawn from the present section. The first is that time-lags, high reproductive rates and over-compensating density-dependence are capable (either alone or in combination) of provoking all types of fluctuations in population density, without invoking any extrinsic cause. The second, equally important, conclusion is that this has been made apparent by the analysis of mathematical models.

6.9 Continuous breeding: the logistic equation

The model derived and discussed in the preceding section was appropriate for populations that have discrete breeding seasons and can therefore be described by equations growing in discrete steps, i.e. by 'difference equations'. Such models are not appropriate, however, for those populations in which birth and death are continuous. These are best described by models of continuous growth, or 'differential' equations, which will be considered next.

The net rate of increase of such a population will be denoted by dN/dt (referred to in speech as 'dN by dt'). This represents the 'speed' at which a population increases in size, N, as time, t, progresses. Each individual in the population will contribute to this increase. Indeed, it is possible to see the increase in size of the population as the sum of the contributions of the various individuals within it. Thus, the average rate of increase *per* individual, or the 'per capita rate of increase' is given by $dN/dt . 1/N$. But we have already seen in section 4.7 that in the absence of competition, this is the definition of the 'intrinsic rate of natural increase', *r*. Thus:

$$\frac{dN}{dt} \cdot \frac{1}{N} = r$$

and

$$\frac{dN}{dt} = rN. \tag{6.6}$$

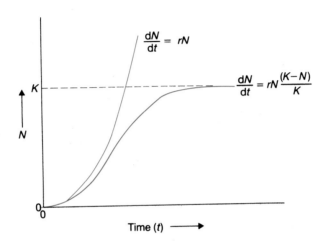

Figure 6.24. Exponential (dashes) and sigmoidal increase in density (*N*) with time for models of continuous breeding. The equation giving sigmoidal increase is the logistic equation.

A population increasing in size under the influence of equation 6.6, with $r > 0$, is shown in Figure 6.24. Not surprisingly, there is unlimited, 'exponential' increase. In fact, equation 6.6. is the continuous form of the exponential difference equation 6.2. Indeed, as discussed in section 4.7, *r* is simply $\log_e R$. (Mathematically adept readers will see that equation 6.6 can be obtained by differentiating equation 6.2.) *R* and *r* are clearly measures of the

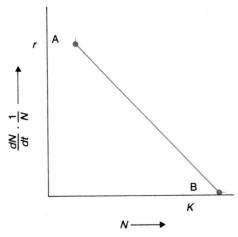

Figure 6.25. The simplest, straight-line way in which the rate of increase per individual ($dN/dt . 1/N$) might fall with density (*N*). For further discussion, see text.

same commodity: 'birth plus survival' or 'birth minus death'; the difference between R and r is merely a change of currency.

For the sake of realism, intraspecific competition must obviously be added to equation 6.6. This can be achieved most simply by the method set out in Figure 6.25, which is exactly equivalent to the one used in Figure 6.19. The net rate of increase per individual is unaffected by competition when N is very close to zero. It is still therefore given by r (point A). When N rises to K (the carrying capacity) the net rate of increase per individual is zero (point B). As before, a straight line between A and B is assumed. Thus:

$$\frac{dN}{dt} \cdot \frac{1}{N} = \frac{-r}{K} \cdot N + r,$$

or

$$\frac{dN}{dt} \cdot \frac{1}{N} = r \left(1 - \frac{N}{K}\right),$$

and

the logistic equation

$$\frac{dN}{dt} = rN \frac{(K-N)}{K}, \tag{6.7}$$

This is known as the *logistic equation* (coined by Verhulst, 1838), and a population increasing in size under its influence is shown in Figure 6.24.

The logistic equation is the continuous equivalent of equation 6.3, and it therefore has all of the essential characteristics of equation 6.3 and all of its shortcomings. It describes a sigmoidal growth curve approaching a stable carrying capacity, but it is only one of many reasonable equations which do this. Its major advantage is its simplicity. Moreover, while it was possible to incorporate a range of competitive intensities into equation 6.3, this is by no means easy with the logistic. The logistic is therefore doomed to be a model of perfectly compensating density-dependence. Nevertheless, in spite of these limitations, the equation will be an integral component of models in Chapters 7 and 10, and it has played a central role in the development of ecology in the past.

6.10 Individual differences: asymmetric competition

competition can lead to skewed weight distributions within populations

Until now, we have focused on what happens to the whole population or the average individual within it. Intraspecific competition, however, can have an important influence on the differences between individuals *within* a population. Figure 6.26 shows the results of an experiment in which flax (*Linum usitatissimum*) was sown at three densities, and harvested at three stages of development, recording the weight of each plant individually (Obeid *et al.*, 1967). This made it possible to monitor the effects of increasing amounts of competition not only as a result of variations in sowing density but also as a result of plant growth (between the first and the last harvests). When intraspecific competition was at its least intense (at the lowest sowing density after only two weeks' growth) the individual plant weights were distributed symmetrically about the mean. When competition was at its most intense,

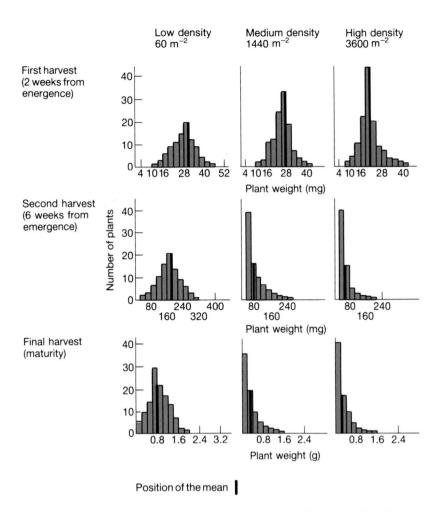

Figure 6.26. Competition and a skewed distribution of plant weights. Frequency distributions of individual plant weights in populations of flax, *Linum usitatissimum*, sown at three densities and harvested at three ages (after Obeid *et al.*, 1967).

however, the distribution was strongly skewed to the left: there were many very small individuals and a few large ones. And as the intensity of competition gradually increased, the degree of skewness increased as well. Rather similar results have been obtained from natural populations of limpets (Figure 6.11; Branch, 1975) and from tadpoles reared under experimental conditions (Figure 6.27; Wilbur & Collins, 1973). In all cases, the populations that experienced the most intense competition had a size distribution in which there were many small and a few large individuals. Characterizing a population by an arbitrary 'average' individual can obviously be very misleading under such circumstances, and can divert attention from some important effects of intraspecific competition.

An indication of the way in which competition can exaggerate or even generate skewness in a population is given in Figure 6.28. This illustrates the results of an experiment in which the grass *Dactylis glomerata* was sown at random on a soil surface (Ross & Harper, 1972). Seedlings were marked as they emerged, so that their time of emergence would be known sub-

individual differences are exaggerated in plants by pre-emption of space

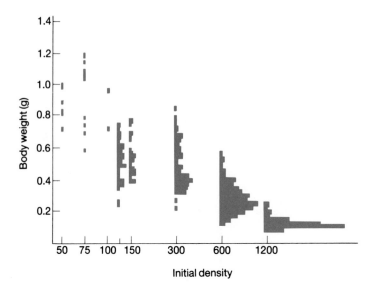

Figure 6.27. Frequency histograms of body weights of *Rana sylvatica* larvae after 50 days of growth at various initial densities (after Wilbur & Collins, 1973).

sequently. Then, 7 weeks after the start of the experiment, all plants were harvested and weighed. Not surprisingly, those that had emerged earliest (and had grown longest) were the largest. However, most plants did not reach the weight they should have attained had weight been determined simply by the length of the growing period (indicated by the dashed line in Figure 6.28). Instead, the later a plant emerged, the further it fell below this expected line (the slower it grew). In other words, the later a plant emerged, the more it was affected by neighbours that had become established earlier. Plants which emerged early *pre-empted* or *'captured' space*, and were little affected by intraspecific competition subsequently. Plants which emerged late entered a universe in which most of the available space had already been

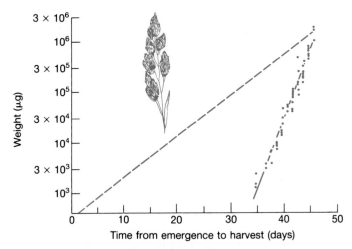

Figure 6.28. Space capture affecting growth amongst individuals in a population of the grass *Dactylis glomerata*, which were harvested simultaneously. Those that emerged later (shorter period from emergence to harvest) were smaller (grew less) than would have been expected had growth been determined by growing period alone (dashed line). (After Ross & Harper, 1972.)

pre-empted; they were therefore greatly affected by intraspecific competition. Such competition is asymmetric: some individuals were affected far more than others, and small initial differences were transformed by competition into a 1000-fold range of plant weights at the time of harvest.

A similar explanation is appropriate for the data on flax in Figure 6.26. All populations contained a range of sizes from a very early stage (this was presumably due to genetic differences, slight differences in emergence time and so on). Then, as competition intensified, the large individuals were affected least and the small individuals affected most. The large individuals therefore grew larger still and were affected even less, while the small individuals lagged further behind and were affected even more. This hierarchy of competition led, at least in part, to the skewed distributions.

In fact, asymmetric intraspecific competition is almost certainly the general rule. The skewed distributions that have been illustrated are one possible manifestation of this, but there are many others. Rubenstein (1981), for instance, studied competition in populations of the Everglades pygmy sunfish (*Elassonia evergladei*). An increase in density led to marked decreases in growth rate, in the fecundity of females and in the reproductive activity of males—or at least it did when 'average' individuals at high and low densities were compared. But the 'best' individuals were only very slightly affected by competition. This emphasizes that intraspecific competition is not only capable of exaggerating individual differences, it is also greatly affected by individual differences.

In particular, there is often a marked asymmetry between different stages or age-classes in the same population. Persson (1983), for example, found that two-year-old perch (*Perca fluviatilis*) were much more affected than three-year-olds when they competed together for the same food. In a similar vein,

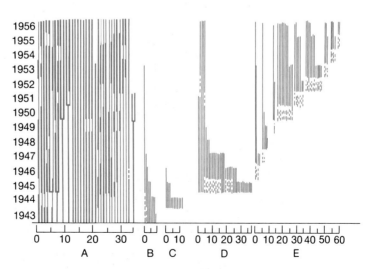

Figure 6.29. Space pre-emption in a perennial: *Anemone hepatica* in a Swedish forest. Each line represents one individual: straight for unramified ones, branched where the plant has ramified, bold where the plant flowered, and broken where the plant was not seen that year. Group A were alive and large in 1943, group B alive and small, group C appeared first in 1944, group D in 1945, and group E thereafter, presumably from seedlings. (After Tamm, 1956.)

235   INTRASPECIFIC COMPETITION

Tamm (1956) examined a population of the herbaceous perennial *Anemone hepatica*, growing in Sweden (Figure 6.29). Despite the crops of seedlings which entered the population between 1943 and 1952, it is quite clear that the most important factor determining which individuals survived to 1956 was whether or not they were established in 1943. Of the 30 individuals that had reached large or intermediate size by 1943, 28 survived until 1956, and some of these had branched. By contrast, of the 112 plants that were either small in 1943 or appeared as seedlings subsequently, only 26 survived to 1956, and not one of these was sufficiently well established to have flowered. Similar patterns can be observed in tree populations. The survival rates, the birth rates and thus the fitnesses of the few established adults are high; those of the many seedlings and saplings are comparatively low.

asymmetries reinforce the regulatory powers of competition

These considerations illustrate a point which is pertinent to asymmetric competition generally: asymmetries tend to reinforce the regulatory powers of intraspecific competition. Tamm's established plants were successful competitors year after year, but his small plants and seedlings were repeatedly unsuccessful. This guaranteed a near-constancy in the number of established plants between 1943 and 1956: each year there was a near-constant number of 'winners' accompanied by a variable number of 'losers'.

A further pattern that can arise out of asymmetric competition can be illustrated using data in which effects were studied at the level of the individual. Mithen *et al.* (1984) sowed seeds of the annual plant *Lapsana communis* by scattering them irregularly on the surface of sterilized compost. The seedlings emerged almost synchronously, and detailed maps were made of their distribution (Figure 6.30a). Then, the position of each individual with respect to its neighbours was described by constructing 'Thiessen polygons'. This was done by drawing the perpendicular bisectors of the lines joining each plant to its neighbours (Figure 6.30b). After a period of 15 weeks the population was mapped again, new polygons were drawn (Figure 6.30c), and the plants were dried and weighed. There were three major findings. First, plants with close neighbours (i.e. with small polygons) had the greatest risk of dying. Secondly, survivors that had had close neighbours were smaller. These were clearly both facets of the asymmetry of competition. In addition, however, although the frequency distributions of plant sizes and polygon areas started highly skewed (Figure 6.30d), they subsequently became distributed symmetrically (Figure 6.30e). Thus, while asymmetric competition can create a skewed distribution in a population, it can also destroy the skew by causing the mortality of the small weaklings.

asymmetries can also destroy skewness when they lead to the mortality of weaklings

The same form of analysis can be applied to the results of an experiment carried out by Connell (1963) on a sessile amphipod crustacean, *Erichthonius braziliensis*. These animals build tubes on solid surfaces, and feed by stretching out from their tubes to clean around the area that they can reach. Connell allowed the amphipods to colonize the freshly cleaned wall of an aquarium. Figures 6.31a and c show the distributions of individuals and the sizes of polygons after 8 days. Figures 6.31b and d show the distributions and polygon sizes two weeks later. Again, we see that it was the smaller individuals that disappeared from the population, and that the polygons therefore became larger but also more even in size.

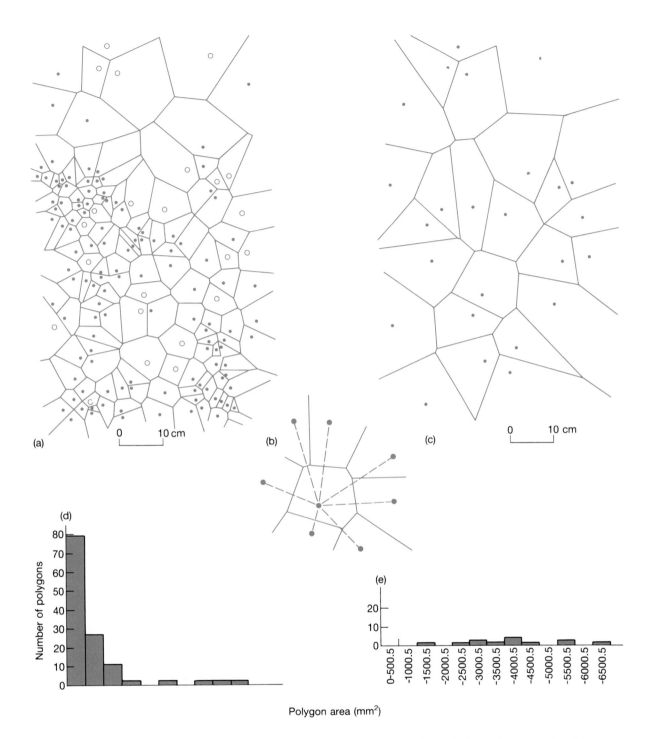

Figure 6.30. (a) The distribution of plants in a seedling population of the annual plant *Lapsana communis* 3 days after seedling emergence. Each seedling is represented by a point and those that survived after 15 weeks are shown as circles. (b) The spatial distribution of sessile organisms (as in (a)) can be described by drawing 'Thiessen polygons', in which polygon sides are drawn at right-angles through the mid-point of lines joining neighbouring organisms. (c) The distribution of survivors after 15 weeks. (d) The frequency distribution of polygon areas after seedling emergence. (e) The frequency distribution of polygon areas 15 weeks after emergence after self-thinning had occurred. (From Mithen *et al.*, 1984.)

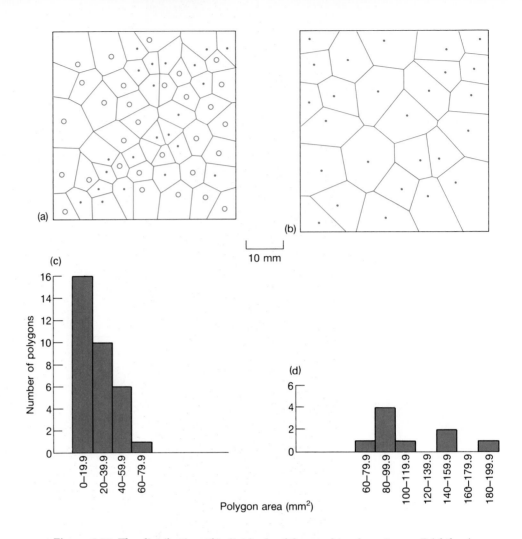

Figure 6.31. The distribution of individuals of the amphipod crustacean *Erichthonius braziliensis* settled on an aquarium wall. (a) Individuals settled on days 1–8. Circled individuals are those that survived to day 9. (b) The distribution of individuals at day 20. There were further deaths and some new recruits between day 9 and day 20. (c) The frequency distribution of polygon areas at day 8. (d) The frequency distribution of polygon areas at day 20. (Calculated from data in Connell, 1963.)

6.11 Territoriality

Territoriality—competition with members of the same species for territories—is one particularly important and widespread form of asymmetric intraspecific competition. Davies (1978) has suggested that a territory may be recognized '. . . whenever [individuals] or groups are spaced out more than would be expected from a random occupation of suitable habitats'. Territoriality would then be apparent not only in conventional cases like breeding great tits (Figure 6.32a), but also, for instance, in barnacles (Figure 6.32b) and in plants (Figure 6.30). It is more usual, however, to reserve the term territoriality for cases in which there is active interference, such that a more or less exclusive area, the territory, is defended against intruders by a recognizable pattern of behaviour.

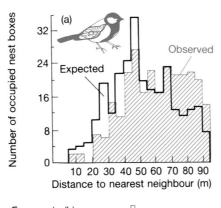

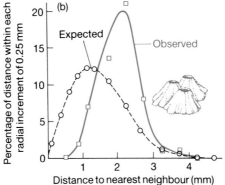

Figure 6.32. Illustrations of territoriality. (a) Great tits (Krebs, 1971) and (b) barnacles (Crisp, 1961) spaced out more than would be expected from a random distribution on the available suitable habitats. (After Davies, 1978.)

population regulation is a consequence of territoriality

The significance of territoriality in the present context lies in the fact that those individuals that hold a territory generally have a far higher fitness than those that do not. In fact, individuals of a territorial species that fail to obtain a territory often make no contribution whatsoever to future generations. The most important *consequence* of territoriality, therefore, is population regulation, or more particularly, the regulation of the number of territory holders. Thus, when territory owners die, or are experimentally removed, their places are often rapidly taken by newcomers. Krebs (1971), for instance, found that in great tit populations, vacated woodland territories were reoccupied by birds coming from hedgerows where reproductive success was noticeably lower; and Watson (1967) found that when territory-holding red grouse were removed, the replacements were non-territorial individuals living in flocks, which would not have bred and would probably have died in the absence of a territory. Territoriality, then, is an extreme form of asymmetric competition, or more evocatively, a 'contest'. There are winners (those that come to hold a territory) and losers (those that do not), and at any one time there can be only a limited number of winners. The exact number of territories (winners) is usually somewhat indeterminate in any one year, and certainly varies from year to year depending on environmental conditions. Nevertheless, the contest nature of territoriality ensures a comparative constancy in the number of surviving, reproducing individuals.

population regulation is *not* the cause of territoriality

Wynne-Edwards (1962) felt that the regulatory consequences of territoriality must themselves be the root causes underlying the evolution of territorial behaviour. He suggested that territoriality was favoured because the population *as a whole* benefited from the rationing effects of territoriality, which guaranteed that the population did not over-exploit its resources. Yet there are powerful and fundamental reasons for rejecting this 'group-selectionist' explanation (essentially, it stretches evolutionary theory beyond reasonable limits), and Wynne-Edwards (1977) has himself subsequently recognized these reasons and accepted the rejection of his ideas. Hence, the *ultimate* cause of territoriality must be sought within the realms of natural selection, in some advantage accruing to the individual.

the advantages of territoriality must outweigh the costs

Any benefit that an individual does gain from territoriality must, of course, be set against the costs of defending the territory. In some animals this defence involves fierce combat between competitors, while in others there is a more subtle mutual recognition by competitors of one another's keep-out signals (e.g. song or scent). Yet even when the chances of physical injury are minimal, territorial animals typically expend energy in patrolling and advertising their territories, and these energetic costs must be exceeded by any benefits if territoriality is to be favoured by natural selection (see Davies & Houston, 1984).

territorial individuals may escape predation

In the great tit, for instance, the male spends time and energy setting up a territory in late winter. This territory is apparently not essential for feeding or mate attraction (Perrins, 1979), but in the summer the spacing-out of nests which results from territoriality greatly reduces the chances of eggs and incubating females being preyed upon (Krebs, 1971), especially by weasels (Dunn, 1977). In fact, the larger the territory, the smaller is the chance of predation, and there is clearly an advantage to the individual. Along similar lines, Sherman (1981) found that in Belding's ground squirrel (*Spermophilus beldingi*) larger territories led to decreased chances of cannibalism of the young by other members of the same species.

territorial individuals may obtain more food

Probably the most common benefit an individual gains from being territorial is an increased rate of food intake. This is difficult to observe directly, but good indirect evidence comes from studies in which territorial behaviour is observed to change in response to changing food levels. An example is the work of Gass *et al.* (1976) on the nectar-feeding rufous humming bird (*Salasphorus rufus*). In north-west California, where the work was carried out, the territories of these birds varied greatly in size; but they also varied in the numbers and species of flowers that they contained, and thus in their overall density of nectar. In fact, there was an inverse relationship between territory size and nectar density, the probable explanation of which is described diagrammatically in Figure 6.33. Defence costs rise rapidly as territory size increases. Benefits also increase with territory size, since they stem from the high consumption rates that arise out of having exclusive rights to flowers. But these benefits reach a plateau when there is more nectar in a territory than a bird can consume. When the density of nectar per unit area is low (because the flowers are few in number and/or poor in nectar), benefits rise slowly with territory size, and *net* advantage (benefit minus cost) should be maximized by a relatively large territory (Figure 6.33); but when nectar density is high (many nectar-rich flowers) benefits rise rapidly and net advantage should be maximized by a relatively small territory. Thus if natural selection has favoured birds that defend a territory of the most favourable

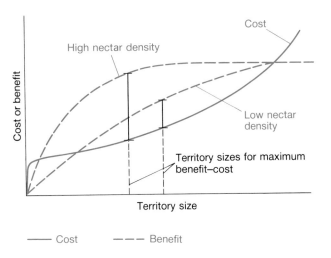

— Cost   — — — Benefit

Figure 6.33. The probable costs, benefits and net benefits (benefits minus costs) of territorial defence. Costs make an initial jump with the move from non-territoriality to territoriality, and thereafter they increase at an accelerating rate as the territory becomes increasingly difficult to defend. Benefits reach a plateau once the territory encompasses all that the organism requires. They therefore do so at a small territory size when there is a high density of resources (in this case nectar), but at a larger territory size for a lower density of resources. As a consequence, net benefits are maximized at lower territory sizes for higher resource densities.

size, an inverse relationship between territory size and nectar-density is to be expected—which is precisely what Gass *et al.* found.

resumé for territoriality   Territoriality can therefore be seen as a flexible and subtle pattern of behaviour which has evolved as a result of the *net* advantages accruing to individual competitors. *As an independent consequence of this*, there is a particularly powerful regulatory influence on the populations concerned. This is just one (albeit extreme) example in which intraspecific competition is recognizably asymmetric.

6.12 Self-thinning

progressive effects in growing cohorts   We have seen that intraspecific competition can, over a period of time, influence the number of deaths, the number of births, the amount of growth and the distribution of biomass within a population. In previous sections this has usually been illustrated by looking at the end-results of competition. But in practice the effects are often progressive. As a cohort ages, the individuals grow in size, their requirements increase, and they therefore compete at a greater and greater intensity. This in turn tends gradually to increase their risk of dying. Thus the number that survive and the growth rate of the survivors are simultaneously influenced by density.

The patterns emerging in ageing, crowded cohorts have been studied in a number of plant populations. An example is illustrated in Figure 6.34a, where both axes have a logarithmic scale. Perennial rye grass, *Lolium perenne,* was sown at a range of densities and samples from each density were harvested

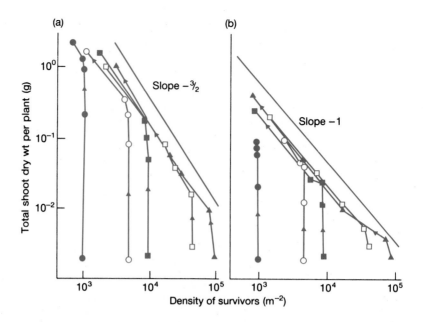

Figure 6.34. Self-thinning in *Lolium perenne* sown at five densities: 1000 (●), 5000 (○), 10 000 (■), 50 000 (□) and 100 000 (▲) 'seeds' m⁻². (a) 0% shade (b) 83% shade. The lines join populations of the five sowing densities harvested on five successive occasions. They therefore indicate the trajectories, over time, that these populations would have followed. The arrows indicate the directions of the trajectories, i.e. the direction of self-thinning. For further discussion, see text. (After Lonsdale & Watkinson, 1982.)

after 14, 35, 76, 104 and 146 days (Lonsdale & Watkinson, 1982). Different lines in the figure represent different sowing densities. Successive points along a single line represent populations of the same initial sowing density at different ages. The lines are therefore trajectories that follow the passage of time. This is indicated by arrows, pointing from many small, young individuals (bottom-right) to fewer larger, older individuals (top-left).

From 35 days onwards, plant growth was slower in the higher density populations: mean plant weight (at a given age) was always greatest in the lowest density populations (Figure 6.34a). It is also clear that the slowest growing, highest density populations were the first to suffer substantial mortality: from 35 days to the last harvest, the two populations sown at the highest density experienced a progressive decline in density. What is most **cohorts approach and** noticeable in these high-density populations, however, is that *density declined* **then follow a thinning** *and plant weight increased in unison*: both populations progressed along a **line** single straight line with a slope of approximately −3/2. The populations are said to have experienced *self-thinning* (i.e a progressive decline in density in a population of growing individuals), and the line with a slope of roughly −3/2 which they approached and then followed is known as the *thinning line*.

In the populations sown at lower densities, self-thinning began rather later. In fact, the lower the density, the later was the onset of self-thinning. In all cases, though, the populations initially followed a trajectory that was almost vertical, i.e. there was little mortality. Then, as they neared the thinning line, the populations suffered increasing amounts of mortality, so that the slopes of all the self-thinning trajectories gradually approached $-3/2$. Finally, on reaching the thinning line, the populations progressed along it.

similar slopes in different thinning lines lead to the $-3/2$ power law

Plant ecologists have repeatedly found that growing, self-thinning plant populations (if sown at sufficiently high densities) approach and then follow a thinning line with a slope of roughly $-3/2$. The relationship is therefore often referred to as *'the $-3/2$ power law'* (Yoda *et al.*, 1963), since density (*d*) is related to mean weight ($\bar{w}$) by the equation:

$$\log \bar{w} = \log c - \frac{3}{2} \log d$$

or

$$\bar{w} = cd^{-3/2},$$

where *c* is a constant. In fact, in many of the cases where the relationship has been revealed, it is not a single cohort which has been followed over time, but a series of similar populations of different ages that have been compared. This has been especially true with populations of trees and other long-lived species.

thinning slopes of -1

A slope of $-3/2$ indicates that in a growing, self-thinning population, mean plant weight *increases* faster than density *decreases*. A population following a $-3/2$ thinning line will therefore steadily increase its *total* weight (or yield). Eventually, of course, this must stop: yield cannot increase indefinitely. Instead, the thinning line might be expected to change from a slope of $-3/2$ to a slope of -1, such that the total weight per unit area remains constant. This actually occurred when populations of *Lolium perenne* (Figure 6.34b) were grown at low light intensities (17% of full light). A slope of -1 means that the further growth of survivors exactly balances the deaths of other individuals, i.e. it seems to occur when total yield reaches a maximum which cannot be exceeded by that species in that environment. In an environment with reduced light intensities, maximum yield is reduced; slopes of -1 are thus apparent at lower densities (Figure 6.34b). In practice though, even in artificial experiments, few self-thinning populations reach these maximum yields; self-thinning slopes of -1 are rare.

different species lie on roughly the same thinning line

Intriguingly, all sorts of plants have a thinning line with a slope of roughly $-3/2$, and they also appear to lie on approximately the *same* thinning line, with intercepts (i.e. values of *c* in the equation) falling within a remarkably narrow range (Figure 6.35). At the right-hand end of the line in Figure 6.35 are high-density populations of small plants (annual herbs and perennials with short-lived shoots), while at the left-hand end are sparse populations of very large plants, mainly trees. The whole spectrum of plant sizes and forms lies along the line between these extremes: the range includes coastal redwoods (*Sequoia sempervirens*), the tallest known trees, and also *Chlorella*, a unicellular alga (J. White, pers. comm.). But each species thins along only a part of the overall line in Figure 6.35, i.e. a given species in a given environment begins its trajectory as a vertical line, which then approaches and ultimately follows a part of the $-3/2$ line. (It would presumably leave the line again to follow a slope of -1 when its maximum yield was attained.) Different species enter and leave the overall line at different points.

Note, in passing, that Figures 6.34 and 6.35 have been drawn, following convention, with log density on the x-axis and log mean weight on the y-axis. This is not meant to imply that density is the independent variable on which mean weight depends. Indeed, it can be argued that mean weight increases naturally during plant growth, and this determines the decrease in density.

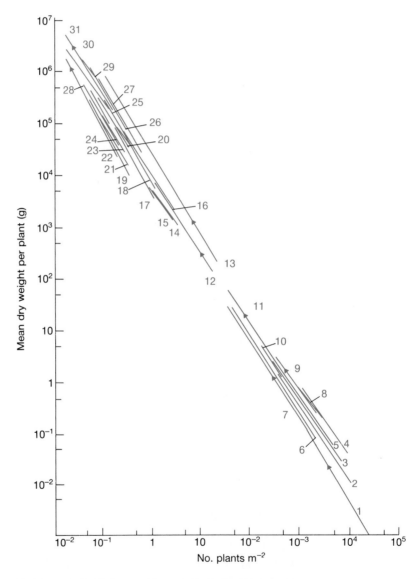

Figure 6.35. Self-thinning in a wide variety of herbs and trees. Each line is a different species, and the line itself indicates the range over which observations were made. The arrows, drawn on representative lines only, indicate the direction of self-thinning over time. The figure is based on Figure 2.9 of White (1980), which also gives the original sources and the species names for the 31 data sets. Note that all lines have a slope approximating to $-3/2$, and that their intercepts also fall within a relatively narrow band.

an explanation for the −3/2 slope

The most satisfactory view is that density and mean weight are wholly inter-dependent: neither is independent of the other.

The precise answer to why thinning lines have a slope of −3/2 is uncertain (see, for instance, White, 1981). However, we know that in a growing cohort, the canopy area per unit area of land (the leaf area index) rapidly becomes independent of plant density—it does not keep on increasing as the mass of the population increases. The −3/2 relationship is usually seen as a reflection of the fact that populations of plants represent weights or volumes (dimensions to the power 3) packed beneath an area of light-capturing canopy (dimensions to the power 2) which remains constant and is thus proportional to the inverse of density.

Looking in more detail at a developing cohort, we see that it is the leaves which overtop their neighbours that are successful in capturing radiation. Thus the successful leaves are those that are supported on longer petioles or stems, or on persistent dead tissue like tree trunks. The population, therefore, becomes composed of progressively fewer and larger individuals, which carry their canopies on an ever-larger proportion of dead or non-photosynthetic support tissue; and it is this dead tissue that a population accumulates as it progresses along the −3/2 thinning line. Indeed, if we look from right to left along the thinning line in Figure 6.35, we pass from herbaceous plants through shrubs to woody trees, of which an increasing proportion is dead. It is necromass not biomass! We cannot, therefore, account for the −3/2 thinning law simply in terms of some present limiting relationship between canopy volume and land area. The explanation for the present has to lie in the past. (This view can be extended to consider not only how thinning occurs within a population of a single species, but also how one species replaces another in the course of vegetational succession—see Chapter 16).

much of the 'biomass' that accumulates in a thinning cohort is necromass

Finally, although we have stressed the apparent universality of the −3/2 thinning law, there are a number of important variants that we must consider. As already discussed, the trees are located to the left in Figure 6.35 because their canopies maintain a very considerable weight of tissue; and they can do this because most of that tissue is dead and requires very little maintenance. Grasses, by contrast, can maintain relatively little tissue, because most of it does indeed require active maintenance. They are therefore located to the right in Figure 6.35. Reduced levels of light induce thinning lines with slopes of −1 (Figure 6.34b), because the reduced level of photosynthesis can maintain only a low weight of tissue. But reduced nutrient levels do not alter the slope of the thinning line. Instead they reduce the rate at which a population progresses along it (Yoda *et al.*, 1963; White & Harper, 1970). This occurs because the unaltered amount of light can still maintain (via photosynthesis) the same weight of tissue, but the plants at low nutrient levels are able to build this tissue only slowly. Lastly, plant shape and type seem to play crucial roles in determining the intercept of the thinning line. Coniferous trees generally have higher intercepts than deciduous trees, and grasses have higher intercepts than dicotyledonous herbs. The significance of this is discussed by Lonsdale and Watkinson (1983).

variations in the thinning line

Animals, whether they be sessile or mobile, must also 'self-thin', insofar as growing individuals within a cohort increasingly compete with one another and reduce their own density. It remains to be seen, however, whether animals have their own self-thinning 'law'. Of course, they do not depend on light in the same way as plants; but crowded sessile animals, at least, can be seen as needing to pack 'volumes' beneath an approximately constant area. On the other hand, a widely applicable self-thinning law for mobile animals is rather less easy to envisage (but see Begon *et al.*, 1986).

7 Interspecific Competition

7.1 Introduction

The essence of interspecific competition is that individuals of one species suffer a reduction in fecundity, survivorship or growth as a result of resource exploitation or interference by individuals of another species. However, behind this simple statement lies a wealth of subtle detail. Interspecific competition can affect the population dynamics of competing species in many ways. The dynamics, in their turn, can influence the species' distributions and their evolution. There are, moreover, numerous ways in which interspecific competition can be investigated. This chapter will therefore cover a range of evidence, a range of end-results and a range of underlying influences. Chapter 18 will examine the role of interspecific competition in shaping the structure of ecological communities. In fact, there are several themes introduced in this chapter that are taken up and discussed more fully in Chapter 18. The two chapters should be read together for a full coverage of interspecific competition.

At this point, it will be as well to emphasize an aspect of the definition of competition that has fundamental significance. Competition occurs when two or more organisms obtain their resources from a supply that is insufficient for all. Competition can *only* occur if a resource is in limiting supply. In other words, two species with very similar resource requirements will, nevertheless, not be in competition if the resources available to them are superabundant. This may be the case, for example, if the species are regulated by their own predators or parasites. In the present chapter, attention is largely restricted to cases where interspecific competition can be assumed to be operating.

7.2 Some examples of interspecific competition

There have been many studies of interspecific competition between pairs (or more) of species of all kinds. We have chosen our examples because they illustrate a number of important ideas particularly clearly.

7.2.1 Competition between salamanders

The first example concerns two species of terrestrial salamanders, *Plethodon glutinosus* and *P. jordani,* which live in the Southern Appalachian Mountains of the United States. Generally, *P. jordani* lives at higher altitudes than *P.*

glutinosus, but in certain areas their altitudinal distributions overlap. Hairston (1980) carried out an experiment at two locations, one in the Great Smoky Mountains, where the distributions overlap over only a small range of altitudes, and the other in the Balsam Mountains, where the species coexist over a very much wider range of altitudes. These sites both had populations of the two salamander species, and they had similar salamander faunas overall; they were at the same elevation and they faced the same direction. At both sites Hairston established seven experimental plots: two from which *P. jordani* was removed, two from which *P. glutinosus* was removed, and three as controls. This work was begun in 1974, and six times in each of the next five years the numbers of both species were estimated in all plots, and individuals were classified into three groups: one-year-olds, two-year-olds, and all those that were older.

In the control plots, and naturally, *P. jordani* was by far the more abundant of the two species; and in the plots from which it was removed, there was a statistically significant increase in the abundance of *P. glutinosus* at both locations. In the plots from which *P. glutinosus* was removed, there was no significant reciprocal increase in the abundance of *P. jordani.* However, there was, at both sites, a statistically significant increase in the proportion of *P. jordani* in the one- and two-year-old age-classes. This was presumably a result of increased fecundity and/or increased survival of young, both of which are crucial components of the basic reproductive rate.

The important point is that individuals of both species must, originally, have been adversely affected by individuals of the other species, since when one species was removed, the remaining species showed a significant increase in abundance and/or fecundity and/or survivorship. It appears, therefore, that in the control plots and in the other zones of overlap generally, these species competed with one another but still coexisted.

7.2.2 Competition between bedstraws (*Galium* spp.)

A second example is provided by an experiment carried out by one of the greatest of the 'founding fathers' of plant ecology, A.G. Tansley, who studied competition between two species of bedstraw (Tansley, 1917). *Galium hercynicum* is a species which grows in Britain at acidic sites, while *Galium pumilum* is confined to more calcareous soils (in Tansley's time they were known as *G. saxatile* and *G. sylvestre*). Tansley found that as long as he grew them alone, *both* species would thrive on both the acidic soil from a *G. hercynicum* site and the calcareous soil from a *G. pumilum* site. Yet if the species were grown together, only *G. hercynicum* grew successfully in the acidic soil and only *G. pumilum* grew successfully in the calcareous soil. It seems, therefore, that when they grow together the species compete, and that one species wins, while the other loses so badly that it is competitively excluded from the site. The outcome depends on the habitat in which the competition occurs.

7.2.3 Competition between barnacles

The third example comes from the work of Connell (1961) on two species of barnacle in Scotland: *Chthamalus stellatus* and *Balanus balanoides* (Figure 7.1). These species are frequently found together on the same Atlantic rocky shores of north-west Europe. However, adult *Chthamalus* generally occur in an intertidal zone which is higher up the shore than that of adult *Balanus*, even though young *Chthamalus* settle in considerable numbers in the *Balanus* zone. In an attempt to understand this zonation, Connell monitored the survival of young *Chthamalus* in the *Balanus* zone. He took successive censuses of mapped individuals over the period of a year and, most important, he ensured at some of his sites that young *Chthamalus* that settled in the *Balanus* zone were kept free from contact with *Balanus*. In contrast with the normal pattern, such individuals survived well, irrespective of the intertidal level. Thus it seemed that the usual cause of mortality in young *Chthamalus* was not the increased submergence times of the lower zones, but competition from *Balanus*. Direct observation confirmed that *Balanus* smothered, undercut or crushed *Chthamalus*, and the greatest *Chthamalus* mortality occurred during the seasons of most rapid *Balanus* growth. Moreover, the few *Chthamalus* individuals that survived a year of *Balanus* crowding were much smaller than uncrowded ones, showing, since smaller barnacles produce fewer offspring, that interspecific competition was also reducing fecundity.

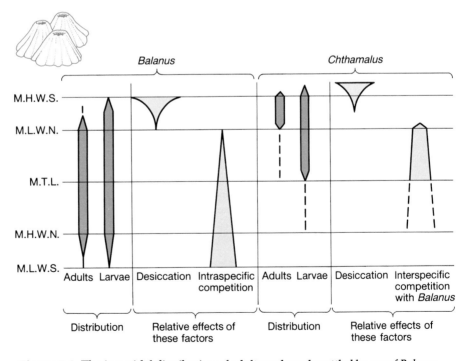

Figure 7.1. The intertidal distribution of adults and newly settled larvae of *Balanus balanoides* and *Chthamalus stellatus*, with a diagrammatic representation of the relative effects of desiccation and competition. Zones are indicated to the left: from M.H.W.S. (mean high water, spring) down to M.L.W.S. (mean low water, spring). (After Connell, 1961.)

Thus *Balanus* and *Chthamalus* compete. They coexist on the same shore, but on a finer scale their distributions overlap very little. *Balanus* outcompetes and excludes *Chthamalus* from the lower zones; but *Chthamalus* can survive in the upper zones where *Balanus*, because of its comparative sensitivity to dessication, cannot.

7.2.4 Competition between *Paramecium* species

The fourth example is provided by the classic work of the great Russian ecologist G.F. Gause, who in a series of experiments studied competition in the laboratory using three species of the protozoan *Paramecium* (Gause, 1934, 1935). All three species grew well alone, reaching stable carrying capacities in tubes of liquid medium in which *Paramecium* consumed bacteria or yeast cells, which themselves lived on regularly replenished oatmeal (Figure 7.2a).

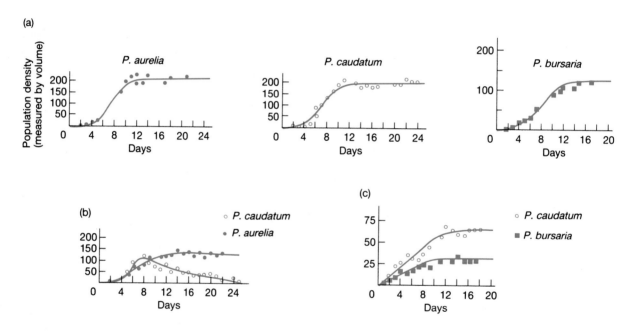

Figure 7.2. Competition in *Paramecium*. (a) *P. aurelia*, *P. caudatum* and *P. bursaria* all established populations when grown alone in culture medium. (b) When grown together, *P. aurelia* drives *P. caudatum* towards extinction. (c) When grown together, *P. caudatum* and *P. bursaria* coexist, though at lower densities than when alone. (Data from Gause, 1934; after Clapham, 1973.)

When Gause grew *Paramecium aurelia* and *P. caudatum* together, *P. caudatum* always declined to the point of extinction, leaving *P. aurelia* as the victor (Figure 7.2b). *P. caudatum* would not normally have starved to death as quickly as implied by the curve to extinction in the figure, but Gause's experimental procedure involved the daily removal of 10% of the culture and animals. Thus, *P. aurelia* was successful in competition because near the point where its population size levelled off, it was still increasing by 10% per day (and able to counteract the enforced mortality) while *P. caudatum* was only increasing by 1.5% per day (Williamson, 1972).

By contrast, when *P. aurelia* and *P. bursaria* were grown together, neither species suffered a decline to the point of extinction. They coexisted, but at stable densities much lower than when grown alone (Figure 7.2c). These coexisting species were therefore still in competition with one another. Yet a closer look revealed that although they lived together in the same tubes, they were, like Connell's barnacles, spatially separated, with *P. aurelia* tending to live and feed on the bacteria suspended in the medium, while *P. bursaria* was concentrated on the yeast cells at the bottom of the tubes.

7.2.5 Competition between diatoms

The final example comes from the work of Tilman *et al.* (1981) on two species of freshwater diatom: *Asterionella formosa* and *Synedra ulna*. This laboratory investigation was unusual because the impact of the species on their limiting resource was recorded at the same time as the population densities were monitored. Both algal species require silicate in the construction of their cell walls. When either species was cultured alone in a liquid medium to which resources were continuously being added, it reached a stable carrying capacity while maintaining the silicate at a constant low concentration (Figure 7.3a and b). In other words, the diatoms *consumed* silicate, and in exploiting

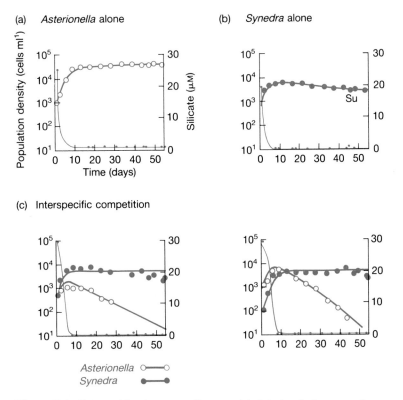

Figure 7.3. Competition between diatoms. (a) *Asterionella formosa*, when grown alone in a culture flask, establishes a stable population and maintains a resource, silicate, at a constant low level. (b) When *Synedra ulna* is grown alone it does the same, but maintains silicate at an even lower level. (c) When grown together, in the replicates, *Synedra* drives *Asterionella* to extinction. (After Tilman *et al.*, 1981.)

this resource they kept its concentration low. However, *Synedra* reduced the silicate concentration to a lower level than *Asterionella*. Thus, when the two species were grown together, *Synedra* maintained the concentration at a level which was too low for *Asterionella* to survive and reproduce. *Synedra* therefore competitively displaced *Asterionella* from mixed cultures (Figure 7.3c).

7.3 Assessment—some general features of interspecific competition

All five examples show that individuals of different species can compete. This is hardly surprising. The salamanders and barnacles also show that different species *do* compete in nature (i.e. there was a measurable interspecific reduction in abundance and/or fecundity and/or survivorship). It seems, moreover, that competing species may either exclude one another from particular habitats so that they do not coexist (as with the bedstraws, the diatoms and the first pair of *Paramecium* species), or may coexist (as with the salamanders), perhaps by utilizing the habitat in slightly different ways (for example, the barnacles and the second pair of *Paramecium* species).

interference and exploitation

As with intraspecific competition, a basic distinction can be made between interference and exploitation competition (although elements of both may be found in a single interaction). In exploitation competition, individuals interact with each other indirectly, responding to a resource level which has been depressed by the activity of competitors. Tilman and his collaborators' experiments with diatoms provide an example of this. When interspecific competition involves resource exploitation, one species consumes and depletes the resource down to a level at which the other species' growth, reproduction or survivorship is affected. By contrast, Connell's barnacles provide a clear example of interference competition. *Balanus*, in particular, directly and physically interfered with the occupation by *Chthamalus* of limited space on the rocky substratum.

interspecific competition is frequently highly asymmetric

A further point is that interspecific competition (like intraspecific competition) is frequently highly asymmetric—the consequences are often not the same for both species. With Connell's barnacles, for instance, *Balanus* excluded *Chthamalus* from their zone of potential overlap, but any effect of *Chthamalus* on *Balanus* was negligible: *Balanus* was limited by its own sensitivity to dessication. Grace and Wetzel (1981) have reported a closely analogous situation involving two species of cattail (reedmace) in Michigan. In the ponds where these aquatic plants were studied, one species, *Typha latifolia*, was mostly found in shallow water, while the other, *Typha angustifolia*, occurred in deeper water (Figure 7.4). Experimental manipulations established that *T. latifolia* normally excludes *T. angustifolia* from shallower water, whereas the distribution of *T. latifolia* is unaffected by competition with *T. angustifolia*.

On a broader front, Lawton and Hassell (1981) reviewed the published field studies of interspecific competition in insects, and found that highly asymmetric cases (where one species was negligibly if at all affected) outnumbered symmetric cases by two to one. The term *amensalism* is generally used to describe an interaction in which one species adversely affects another,

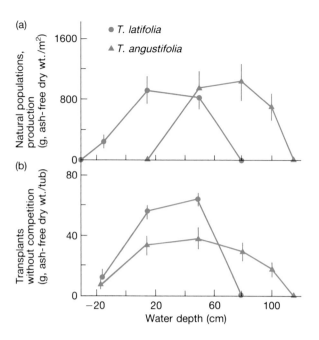

Figure 7.4. Asymmetric competition between cattail species. (a) The natural distributions of coexisting populations of *Typha latifolia* (in shallower water) and *T. angustifolia* (in deeper water). (b) When transplants are grown alone, *T. angustifolia* grows over a much wider range of depths (suggesting it is normally excluded from shallower water), but *T. latifolia* grows at the same range of depths as it did when in competition. (After Grace & Wetzel, 1981.)

but the second species has no effect, good or bad, on the first; and cases of highly asymmetric interspecific competition are clearly candidates for such a description. On the other hand, describing them as amensal rather than competitive obscures their essential continuity with more symmetric cases of interspecific competition. Jackson (1979), for instance, studied the very overt 'overgrowth competition' that occurs amongst bryozoan species (colonial, modular animals) living on the undersurfaces of corals off the coast of Jamaica (overgrowth competition is very widespread among sessile marine organisms generally). He found that for the pair-wise interactions amongst the seven most commonly interacting species, the 'percentage wins' varied more or less continuously from 50% ('symmetry') to 100%.

competition for one resource affects competition for other resources: . . .

Finally, it is worth noting that individuals competing for one resource often have their ability to exploit another resource affected. Buss (1979), for example, showed that in bryozoan overgrowth interactions, there appears to be an interdependence between competition for space and for food. When a colony of one species contacts a colony of another species, it interferes with the self-generated feeding currents on which bryozoans rely—but a colony short of food will, in turn, have a much reduced ability to compete for space (by overgrowth). Comparable examples are found among rooted plants. Suppose, for instance, that an 'aggressive' species invades the canopy of a 'suppressed' species and deprives it of light. The suppressed species will suffer directly from the reduction in light energy that it obtains, but this will also reduce its rate of root growth, and it will therefore be less able to exploit the supply of water and nutrients in the soil. This in turn will reduce its rate of shoot and leaf growth. Thus, when plant species compete, repercussions flow backwards and forwards between roots and shoots. A number of workers have attempted to separate the effects of canopy and root competition by an experimental design in which two species are grown (a) alone, (b) together, (c) in the same soil but with their canopies separated, and (d) in separate soil with their canopies intermingling. One example is the work of Groves and Williams (1975) on subterranean clover (*Trifolium subterraneum*) and skeleton weed (*Chondrilla juncea*). The clover was not significantly affected under any circumstances (another example of asymmetric competition). However, as Figure 7.5 shows, the skeleton weed was affected when the roots intermingled (reduced to 65% of the control value of dry weight) and when the canopies intermingled (47% of the control), and when both intermingled the effect was multiplicative (65%×47% = 30.6%, compared to the 31% reduction in dry weight observed). Clearly, both were important.

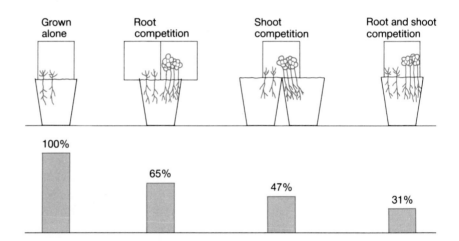

Figure 7.5. Root and shoot competition between subterranean clover (*Trifolium subterraneum*) and skeleton weed (*Chondrilla juncea*). Above are the experimental designs used; below are the dry weights of skeleton weed produced as a percentage of the production when grown alone. (After Groves & Williams, 1975.)

The results of experiments such as those described here highlight a critical question in the study of interspecific competition: what are the general conditions which permit the coexistence of competitors, and what circumstances lead to competitive exclusion? Mathematical models have provided important insights into this question, and these will be discussed in the next section, along with the results of relevant laboratory and field investigations. Consideration of further important questions—how common is interspecific competition in nature, and what kinds of species are likely to compete?—will be reserved for later (Chapter 18).

7.4 Competitive exclusion or coexistence?

7.4.1 A logistic model of interspecific competition

The 'Lotka–Volterra' model of interspecific competition, named in honour of its originators (Lotka, 1925; Volterra, 1926), is an extension of the *logistic equation* described in section 6.9. As such it incorporates all of the logistic's shortcomings, but a useful model can nonetheless be constructed, shedding light on the factors which determine the outcome of a competitive interaction.

The logistic equation:

$$\frac{dN}{dt} = r\,N\,\frac{(K-N)}{K}$$

contains, within the brackets, a term responsible for the incorporation of intraspecific competition. The basis of the Lotka–Volterra model is the replacement of this term by one which incorporates both intra- *and* interspecific competition.

The population size of one species can be denoted by N_1, and that of a second species by N_2. Their carrying capacities and intrinsic rates of increase are K_1, K_2, r_1 and r_2 respectively.

Suppose that ten individuals of species 2 have, between them, the same competitive, inhibitory effect on species 1 as does a single individual of species 1. The *total* competitive effect on species 1 (intra- *and* interspecific) will then be equivalent to the effect of $(N_1+N_2/10)$ species 1 individuals. The constant—1/10 in the present case—is called a *competition coefficient* and is denoted by α_{12} ('alpha one-two'). It measures the per capita competitive effect *on* species 1 *of* species 2. Thus multiplying N_2 by α_{12} converts it to a number of 'N_1-equivalents'. (Note that $\alpha_{12} < 1$ means that species 2 has less inhibitory effect on species 1 than species 1 has on itself, while $\alpha_{12} > 1$ means that species 2 has a greater inhibitory effect than species 1 has on itself.) The crucial element in the model is the replacement of N_1 in the bracket of the logistic equation with a term signifying 'N_1 plus N_1-equivalents', i.e.

α: the competition coefficient

$$\frac{dN_1}{dt} = r_1 N_1 \; \frac{(K_1 - \{N_1 + \alpha_{12}N_2\})}{K_1}$$

or

$$\frac{dN_1}{dt} = \frac{r_1 N_1 (K_1 - N_1 - \alpha_{12}N_2)}{K_1}, \tag{7.1a}$$

the Lotka–Volterra model: a logistic model for two species

and in the case of the second species:

$$\frac{dN_2}{dt} = \frac{r_2 N_2 (K_2 - N_2 - \alpha_{21}N_1)}{K_2}. \tag{7.1b}$$

These two equations constitute the Lotka–Volterra model.

To appreciate the properties of this model, we must ask the question: When (under what circumstances) does each species increase or decrease in abundance? In order to answer this, it is necessary to construct diagrams in which all possible combinations of species 1 and species 2 abundance can be displayed (i.e. all possible combinations of N_1 and N_2). These will be diagrams (Figures 7.6 and 7.8) with N_1 plotted on the horizontal axis and N_2 plotted on the vertical axis, such that there are low numbers of both species towards the bottom-left, high numbers of both species towards the top-right, and so on. Certain combinations of N_1 and N_2 will give rise to increases in species 1 and/or species 2, while other combinations will give rise to decreases in species 1 and/or species 2; and there will be *zero isoclines* for each species (lines along which there is neither increase nor decrease) dividing the combinations leading to increase from those leading to decrease. Moreover, if a zero isocline is drawn first, there will be combinations leading to increase on one side of it, and combinations leading to decrease on the other.

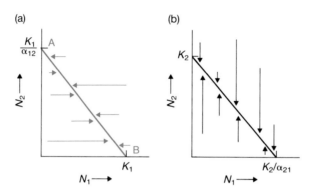

Figure 7.6. The zero isoclines generated by the Lotka–Volterra competition equations. (a) The N_1 zero isocline: species 1 increases below and to the left of it, and decreases above and to the right of it. (b) The equivalent N_2 zero isocline.

In order to draw a zero isocline for species 1, we can use the fact that *on the zero isocline* $dN_1/dt = 0$ (by definition), that is (from equation 7.1a):

$$r_1 N_1 (K_1 - N_1 - \alpha_{12} N_2) = 0.$$

This is true when the intrinsic rate of increase (r_1) is zero, and when the population size (N_1) is zero, but—more importantly in the present context— it is true when

$$K_1 - N_1 - \alpha_{12} N_2 = 0,$$

which can be rearranged as:

$$N_1 = K_1 - \alpha_{12} N_2. \tag{7.2}$$

In other words, everywhere along the straight line which this equation represents, $dN_1/dt = 0$. The line is therefore the zero isocline for species 1; and since it is a straight line it can be drawn by finding two points on it and joining them. Thus, in equation 7.2,

when $N_1 = 0$, $N_2 = \dfrac{K_1}{\alpha_{12}}$ (point A, Figure 7.6a),

when $N_2 = 0$, $N_1 = K$ (point B, Figure 7.6a),

and joining them gives the zero isocline for species 1. Below and to the left of this, numbers of both species are relatively low and species 1, subjected to only weak competition, increases in abundance (arrows from left to right, N_1

on the horizontal axis). Above and to the right of the line, numbers are high and species 1 decreases in abundance (arrows from right to left). Based on an equivalent derivation, Figure 7.6b has combinations leading to increase and decrease in species 2, separated by a species 2 zero isocline, with arrows, like the N_2-axis, running vertically.

Finally, in order to determine the outcome of competition in this model, it is necessary to fuse Figures 7.6a and 7.6b, allowing the behaviour of a joint population to be predicted. In doing this, it should be noted that the arrows in Figure 7.6 are actually vectors—with a strength as well as a direction—and that to determine the behaviour of a joint N_1, N_2 population the normal rules of vector addition should be applied (see Figure 7.7).

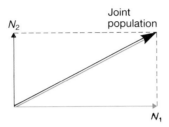

Figure 7.7. Vector addition. When species 1 and 2 increase in the manner indicated by the N_1 and N_2 arrows (vectors), the joint population increase is given by the vector along the diagonal of the rectangle, generated as shown by the N_1 and N_2 vectors.

there are four ways in which the two zero isoclines can be arranged

Figure 7.8a–d shows that there are, in fact, four different ways in which the two zero isoclines can be arranged relative to one another, and the outcome of competition will be different in each case. The different cases can be defined and distinguished by the intercepts of the zero isoclines. In Figure 7.8a for instance:

$$\frac{K_1}{\alpha_{12}} > K_2 \text{ and } K_1 > \frac{K_2}{\alpha_{21}}$$

i.e. $K_1 > K_2\alpha_{12}$ and $K_1\alpha_{21} > K_2$.

strong interspecific competitors out-compete weak interspecific competitors

The first inequality indicates that the inhibitory intraspecific effects that species 1 can exert on itself are greater than the *interspecific* effects that species 2 can exert on species 1. The second inequality, however, indicates that species 1 can exert more of an effect on species 2 than species 2 can on itself. Species 1 is thus a strong interspecific competitor, while species 2 is a weak interspecific competitor; and as the vectors in Figure 7.8a show, species 1 drives species 2 to extinction and attains its own carrying capacity. The situation is reversed in Figure 7.8b. Hence Figures 7.8a and 7.8b describe cases in which the environment is such that one species invariably out-competes the other.

In Figure 7.8c:

$$K_2 > \frac{K_1}{\alpha_{12}} \text{ and } K_1 > \frac{K_2}{\alpha_{21}}$$

when interspecific competition is more important than intraspecific, the outcome depends on the species' densities

i.e. $K_2\alpha_{12} > K_1$ and $K_1\alpha_{21} > K_2$.

Thus both species are stronger competitors on the other species than they are on themselves. This will occur, for instance, when each species produces a substance which is toxic to the other species but harmless to itself, or when

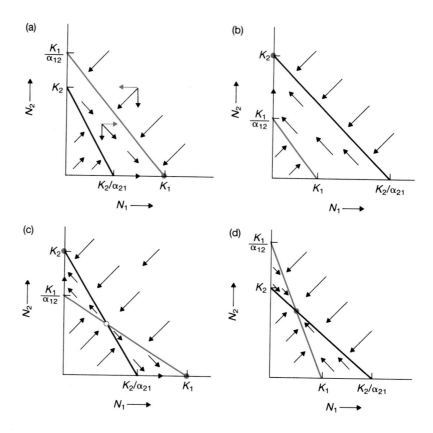

Figure 7.8. The outcomes of competition generated by the Lotka–Volterra competition equations for the four possible arrangements of the N_1 and N_2 zero isoclines. Vectors, generally, refer to joint populations, and are derived as indicated in (a). The solid circles show stable equilibrium points. The open circle in (c) is an unstable equilibrium point. For further discussion, see text.

each species is aggressive towards or even preys upon individuals of the other species, more than individuals of its own species. The consequence, as the figure shows, is an unstable equilibrium combination of N_1 and N_2 (where the isoclines cross), and two stable points. At the first of these stable points, species 1 reaches its carrying capacity with species 2 extinct; while at the second, species 2 reaches its carrying capacity with species 1 extinct. Which of these two outcomes is actually attained is determined by the initial densities: the species which has the initial advantage will drive the other species to extinction.

Finally, in Figure 7.7d:

$$\frac{K_1}{\alpha_{12}} > K_2 \quad \text{and} \quad \frac{K_2}{\alpha_{21}} > K_1$$

i.e. $K_1 > K_2\alpha_{12}$ and $K_2 > K_1\alpha_{21}$.

when inter- is less significant than intra-, the species coexist

In this case, both species have less competitive effect on the other species than they have on themselves. This can also be expressed by noting that:

$$\alpha_{12} \cdot \alpha_{21} < 1,$$

which corresponds very broadly to situations in which competition between species is less significant than competition within species. The outcome, as Figure 7.7d shows, is a stable, equilibrium combination of the two species, which all joint populations tend to approach.

Overall, therefore, the Lotka–Volterra model of interspecific competition is able to generate a range of possible outcomes: the predictable exclusion of one species by another, exclusion dependent on initial densities, and stable coexistence. Each of these possibilities will now be discussed in turn, alongside the results of laboratory and field investigations. We will see that the three outcomes from the model correspond to biologically reasonable circumstances.

7.4.2 The competitive exclusion principle

Figure 7.8a and b describe cases in which a strong interspecific competitor invariably out-competes a weak interspecific competitor. It is useful to consider this situation from the point of view of niche theory (sections 2.12 and 3.7). Recall that the niche of a species in the absence of competitors from other species is its *fundamental* niche (defined by the combination of conditions and resources which allow the species to maintain a viable population). In the presence of competitors, however, the species may be restricted to a *realized* niche, the precise nature of which is determined by which competing species are present. This distinction stresses that interspecific competition reduces fecundity and survival, and that there may be parts of a species' fundamental niche in which, as a result of interspecific competition, the species can no longer survive and reproduce successfully. These parts of its fundamental niche are absent from its realized niche. Thus, returning to Figures 7.8a and b, we can say that the weak interspecific competitor lacks a realized niche when in competition with the stronger competitor. The real examples of interspecific competition previously discussed can now be re-examined in terms of niches.

In the case of Tilman and colleagues' diatom species, the fundamental niches of both species were provided by the laboratory regime (they both thrived when alone). Yet when *Synedra* and *Asterionella* competed, *Synedra* had a realized niche while *Asterionella* did not: there was competitive exclusion of *Asterionella*. The same outcome was recorded when Gause's *P. aurelia* and *P. caudatum* competed; *P. caudatum* lacked a realized niche and was competitively excluded by *P. aurelia*. When *P. aurelia* and *P. bursaria* competed, on the other hand, both species had realized niches, but they were noticeably different—*P. aurelia* living and feeding on the bacteria in the medium, *P. bursaria* concentrating on the yeast cells on the bottom of the tube. Coexistence was therefore associated with a differentiation of realized niches, or a 'partitioning' of resources.

In Tansley's *Galium* experiments, the fundamental niches of both species included both acidic and calcareous soils. In competition with one another, however, the realized niche of *G. hercynicum* was restricted to acidic soils, while that of *G. pumilum* was restricted to calcareous ones: there was reciprocal competitive exclusion. Neither habitat allowed niche differentiation, and neither habitat niche fostered coexistence.

fundamental and realized niches

species deprived of a realized niche by a competitor are driven to extinction

coexisting competitors often exhibit a differentiation of realized niches

Amongst Connell's barnacles, the fundamental niche of *Chthamalus* extended down into the *Balanus* zone; but competition from *Balanus* restricted *Chthamalus* to a realized niche higher up the shore. In other words, *Balanus* competitively excluded *Chthamalus* from the lower zones, but for *Balanus* itself, even its fundamental niche did not extend up into the *Chthamalus* zone—its sensitivity to desiccation prevented it surviving even in the absence of *Chthamalus*. Hence, overall, the coexistence of these species was also associated with a differentiation of realized niches. The pattern that has emerged from these examples has also been uncovered in many others, and has been elevated to the status of a principle: the 'Competitive Exclusion Principle' or 'Gause's Principle'. It can be stated as follows: if two *competing* species coexist in a stable environment, then they do so as a result of niche differentiation, i.e. differentiation of their realized niches. If, however, there is no such differentiation, or if it is precluded by the habitat, then one competing species will eliminate or exclude the other. Thus exclusion occurs when the realized niche of the superior competitor completely fills those parts of the inferior competitor's fundamental niche which the habitat provides.

However, there can be a very real methodological problem in positively establishing the pertinence of the Competitive Exclusion Principle in any particular situation. Consider Hairston's salamanders, for example. The two species competed and coexisted, and the Competitive Exclusion Principle would suggest that this was a result of niche differentiation. This is certainly a reasonable supposition; but until such differentiation is observed and shown to ameliorate the effects of interspecific competition, it must remain no more than a supposition. Thus when two competitors coexist, it is often difficult to establish positively that there is niche differentiation, and it is *impossible* to prove the absence of it. When an ecologist fails to find differentiation, this might simply mean that he has looked in the wrong place or in the wrong way. The Competitive Exclusion Principle has become widely accepted (a) because of the weight of evidence in its favour, (b) because it makes intuitive good sense, and (c) because there are theoretical grounds for believing in it (Lokta–Volterra model). But there will always be cases in which it has not been positively established; and as section 7.6 will make plain, there are many other cases in which it simply does not apply. In short, interspecific competition is a process that is often associated with a particular pattern (niche differentiation); but the pattern can arise through other processes, and the process need not lead to the pattern.

7.4.3 Mutual antagonism

Figure 7.8c, derived from the Lotka–Volterra model, describes a situation in which interspecific competition is, for both species, a more powerful force than intraspecific competition. This is known as mutual antagonism.

An extreme example of such a situation is provided by the work of Park (1962) on two species of flour beetle: *Tribolium confusum* and *Tribolium castaneum*. Park's experiments in the 1940s, '50s and '60s were amongst the most influential in shaping ideas about interspecific competition. He reared

'Competitive Exclusion Principle'

difficult methodological problems in proving and especially *disproving* the Principle

niche differentiation and interspecific competition: a pattern and a process that are not always linked.

the beetles in simple containers of flour, which provided fundamental and
often realized niches for the eggs, larvae, pupae and adults of both species.
There was certainly exploitation of common resources by the different species
and by the different stages; but in the present context, one particular aspect of
Park's results is of special interest. The beetles preyed upon one another.
Larvae and adults ate eggs and pupae, cannibalizing their own species as
well as attacking the other species, and their propensity for doing so is
summarized in Table 7.1 (after Park *et al.*, 1965). The important point is this:
taken overall, beetles of both species ate more individuals of the other species
than they did of their own. Thus a crucial mechanism in the interaction of
these competing species was reciprocal predation (i.e. mutual antagonism),
and it is easy to see that both species were more affected by inter- than
intraspecific action. Reciprocal overt aggression (if not reciprocal predation)
is probably the most common form of mutual antagonism between competing animal species. Among plants, it has often been claimed that
mutual antagonism is displayed through the production of chemicals that

are toxic to other species but not to the producer (known as *allelopathy*).
There is no doubt that chemicals with such properties can be extracted
from plants, but it remains questionable whether they play any part in
interactions in the real world. These problems have been discussed by
Harper (1977).

Table 7.1. Reciprocal predation (a form of mutual antagonism) between two species of flour beetle, *Tribolium confusum* and *Tribolium castaneum*. Both adults and larvae eat both eggs and pupae. In each case, and overall, the preference of each species for its own or the other species is indicated. Interspecific predation is more marked than intraspecific predation. (After Park *et al.*, 1965.)

	'Predator'	'shows a preference for . . .'
Adults eating eggs	*confusum*	*confusum*
	castaneum	*confusum*
Adults eating pupae	*confusum*	*castaneum*
	castaneum	*confusum*
Larvae eating eggs	*confusum*	*castaneum*
	castaneum	*castaneum*
Larvae eating pupae	*confusum*	*castaneum*
	castaneum	*confusum*
Overall	*confusum*	*castaneum*
	castaneum	*confusum*

Figure 7.8c, the Lotka–Volterra model, would suggest that the
consequences of mutual antagonism are essentially the same whatever the
exact mechanism. Because species are affected more by inter- than
intraspecific competition, the outcome is strongly dependent on the relative
abundances of the competing species. The small amount of interspecific
aggression displayed by a rare species will have relatively little effect on an
abundant competitor; but the large amount of aggression displayed by the
abundant species might easily drive the rare species to local extinction.
Moreover, if abundances are finely balanced, a small change in relative
abundance will be sufficient to shift the advantage from one species to the
other. The outcome of competition will then be unpredictable; either species

Table 7.2. Competition between *Tribolium confusum* and *T. castaneum* in a range of climates. One species is always eliminated and climate alters the outcome, but the outcome is nevertheless probable rather than definite. (After Park, 1954.)

| Climate | Percentage wins | |
	confusum	*castaneum*
hot–moist	0	100
temperate–moist	14	86
cold–moist	71	29
hot–dry	90	10
temperate–dry	87	13
cold–dry	100	0

mutual antagonism: the outcome depends on the densities attained

could exclude the other, depending on the exact densities that they start with *or attain*. Table 7.2 shows that this is indeed the case with Park's flour beetles (Park, 1954). There was always only one winner, and the balance between the species changed with climatic conditions. Yet at all intermediate climates *the outcome was probable rather than definite.* Even the inherently inferior competitor occasionally achieved a density at which it could outcompete the other species.

7.5 Coexistence through niche differentiation: the concept of limiting similarity

The Lotka–Volterra model predicts an association between the stable coexistence of competitors and the condition:

$$\alpha_{12} . \alpha_{21} < 1,$$

i.e. situations where interspecific competition is, overall, less significant than intraspecific competition. Niche differentiation will obviously tend to concentrate competitive effects more within species than between species; and the Lotka–Volterra model can therefore be seen to imply that *any* amount of niche differentiation will allow the stable coexistence of competitors. Indeed, the Competitive Exclusion Principle, as usually stated, carries the same implication. But is this implication true? We have already seen a number of examples in which the coexistence of competitors is associated with some degree of niche differentiation. But is there a minimum amount of niche differentiation that has to be exceeded for stable coexistence? And how much differentiation is generally observed? These are amongst the most crucial questions within the study of interspecific competition at present, and the Lotka–Volterra model has been modified and extended in an attempt to answer them. The argument, initiated by MacArthur and Levins (1967) and developed by May (1973), runs as follows.

how much niche differentiation amongst coexisting competitors?

Imagine three species competing for a resource which is unidimensional and distributed continuously; food size is a clear example, but others include food at different heights in a forest canopy or perhaps water along a moisture gradient in the soil. Each species has its own realized niche in this single dimension within which it consumes resources. Moreover, its efficiency, and thus its consumption rate, is assumed to be highest at the centre of its niche and to tail off to zero at either end. Its niche can therefore be visualized as a

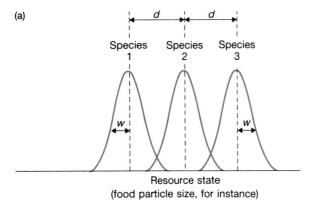

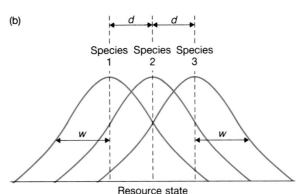

Figure 7.9. Resource utilization curves for three species coexisting along a one-dimensional resource spectrum. *d* is the distance between adjacent curve peaks, *w* is the standard deviation of the curves. (a) Narrow niches with little overlap (*d* > *w*), i.e. relatively little interspecific competition. (b) Broader niches with greater overlap (*d* < *w*), i.e. relatively intense interspecific competition.

a simple model developed to provide an answer

resource utilization curve (Figure 7.9a and b); and the more the utilization curves of adjacent species overlap, the more the species compete. Indeed, by assuming first that the curves are 'normal' distributions (in the statistical sense), and secondly that the different species have similarly shaped curves, the competition coefficient α (applicable to both adjacent species) can be expressed by the following formula:

$$d = e^{-d^2/4w^2}$$

where w is the standard deviation (or, roughly, 'relative width') of the curves, and d is the distance between adjacent peaks. Thus, α is very small when there is considerable separation of adjacent curves ($d/w \gg 1$, Figure 7.9a), and approaches unity as the curves themselves approach one another ($d/w < 1$, Figure 7.9b).

How much overlap of adjacent utilization curves is compatible with stable coexistence? Obviously, if there is little overlap (Figure 7.9a), then there is little interspecific competition and competitors can coexist. On the other hand, if three niches are to be fitted along the resource dimension with little overlap, then at least one of the niches is likely to be narrow. Hence there will be intense *intraspecific* competition within the narrow niche (or niches), and there will also be portions of the resource spectrum that go virtually unutilized by any species. Natural selection is therefore expected to favour an increased utilization of these neglected portions, an increase in niche width, and thus an increase in niche (i.e. curve) overlap. The question is: how much?

MacArthur and Levins (1967) and May (1973) provided an answer by assuming that the two peripheral species had the same carrying capacity (K_1, representing the suitability of the available resources for species 1 and 3), and considering the coexistence between them of another species (carrying capacity K_2). Their results are illustrated in Figure 7.10, which indicates the values of K_1/K_2 that are compatible with stable coexistence for various values of d/w. When d/w is low (α is high and the species are similar) the conditions for coexistence are extremely restrictive in terms of the K_1/K_2 ratio; but these restrictions lift rapidly as d/w approaches and exceeds unity. In other words, coexistence is *possible* when d/w is low, but only if the suitabilities of the environment for the different species are extremely finely balanced. And if the environment is assumed to vary, the conditions become even more restrictive. Environmental fluctuations lead to variations in the K_1/K_2 ratio, and coexistence will now only be possible if, generally, this ratio remains within the stable region of Figure 7.10. For low values of d/w (similar species) and even a moderate amount of variation this is effectively impossible.

coexistence with minimal overlap is possible—but only under prohibitively restrictive conditions

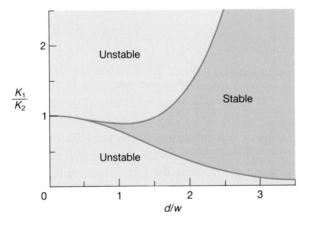

Figure 7.10. The range of habitat favourabilities (indicated by the carrying capacities K_1 and K_2, where $K_1 = K_3$) which permit a three-species equilibrium community with various degrees of niche overlap (d/w). (After May, 1973.)

Thus, high values of d/w allow the stable coexistence of species, but they are unlikely because they lead to intense intraspecific competition and the underexploitation of resources. And when d/w is low, the equilibrium is too fragile to be maintained in the real world. This model, therefore, suggests that the coexistence of competitors (using a unidimensional resource) will be based on niche differentiation in which d/w is approximately equal to, or slightly greater than unity. Indeed, a number of other theoretical approaches lead to the same suggestion (reviewed by May, 1981b). Unfortunately, however, the testing of this suggestion is hindered by two major problems, as follows.

the model suggests that d/w will be approximately equal to or slightly greater than unity

The first is that it applies only to situations in which there is a simple, unidimensional resource (probably quite rare), and in which the utilization curves are at least approximately the same as in the model. Competition in several dimensions, and certain alternative utilization curves (Abrams, 1976), would both lead to lower values of d/w being compatible with robust, stable coexistence.

The second problem is the collection and interpretation of data. In particular, there is a grave danger, when 'testing' the model's predictions, that those field examples that support it are selected, while those that do not support it are ignored. Thus the field evidence persuasively used by May (1973) in support of the model's suggestions was even more persuasively criticized by Abrams (1976). Tests are needed of whether such 'supportive' patterns occur much more frequently than would be expected by chance alone; or, to put it another way, evidence is required that the supportive patterns have arisen through this process and not some other. These difficult problems are addressed in Chapter 18.

Nevertheless, we can see at this stage that the model is far from worthless. It takes us beyond very vague notions of niche differentiation amongst coexisting competitors, by suggesting that there is likely to be a limit to the similarity of competing species; and that this limit represents a balance between, on the one hand, the intensity of intraspecific competition and the underexploitation of resources, and on the other hand, equilibria which are too fragile to withstand the vagaries of the real world. The model, therefore, without necessarily being 'correct', has been immensely instructive.

7.6 Heterogeneity, colonization and pre-emptive competition

At this point it is necessary to sound a loud note of caution. It has been assumed in this chapter until now that the environment is sufficiently constant for the outcome of competition to be determined by the competitive abilities of the competing species. In reality though, such conditions are far from universal. Environments are usually a patchwork of favourable and unfavourable habitats; patches are often only available temporarily; and patches often appear at unpredictable times and in unpredictable places. Even when interspecific competition occurs, it does not necessarily continue to completion. Systems do not necessarily reach equilibrium, and superior competitors do not necessarily have time to exclude their inferiors. Thus an understanding of interspecific competition itself is not always enough. It is often also necessary to consider how interspecific competition is influenced by, and interacts with, an inconstant or unpredictable environment.

7.6.1 Unpredictable gaps: the poorer competitor is a better colonizer

'Gaps' of unoccupied space occur unpredictably in many environments. Fires, landslips and lightning can create gaps in woodlands, storm-force seas can create gaps on the shore, and voracious predators can create gaps almost anywhere. Invariably these gaps are recolonized. But the first species to do so is not necessarily the one that is best able to exclude other species from that area in the longer term. Thus, so long as gaps are created at the appropriate frequency, it is possible for a 'fugitive' species and a highly competitive species to coexist. The fugitive species tends to be the first to colonize gaps; it establishes itself, and it reproduces. The other species tends to be slower to invade the gaps; but having begun to do so, it outcompetes and eventually excludes the fugitive from that particular gap.

coexistence of a
competitive mussel and a
fugitive sea palm

An example of this is provided by the coexistence of the sea palm *Postelsia palmaeformis* (a brown alga) and the mussel *Mytilus californicus* on the coast of Washington, U.S.A. (Paine, 1979). *Postelsia* is an annual which must re-establish itself each year in order to persist at a site. It does so by attaching to the bare rock, usually in gaps in the mussel bed created by wave action. However, the mussels themselves slowly encroach on these gaps, gradually filling them and precluding colonization by *Postelsia*. Paine found that these species coexisted only at sites in which there was a relatively high average rate of gap formation (about 7% of surface area per year), and in which this rate was approximately the same each year. Where the average rate was lower, or where it varied considerably from year to year, there was (either regularly or occasionally) a lack of bare rock for colonization. This led to the overall exclusion of *Postelsia*. At the sites of coexistence, however, although *Postelsia* was eventually excluded from all gaps, these were created with sufficient frequency and regularity for there to be coexistence in the site as a whole.

7.6.2 Unpredictable gaps: the pre-emption of space

When two species compete on equal terms, the result is usually predictable. But in the colonization of unoccupied space, competition is rarely even-handed. Individuals of one species are likely to arrive, or germinate from the seed-bank, in advance of individuals of another species. This, in itself, may be enough to tip the competitive balance in favour of the first species. And if space is pre-empted by different species in different gaps, then this may allow coexistence, even though one species would always exclude the other if they competed 'on equal terms'.

Figure 7.11, for instance, shows the results of a competition experiment between the annual grasses *Bromus madritensis* and *Bromus rigidus* which occur together in Californian rangelands (Harper, 1961). When they were sown simultaneously in an equiproportional mixture, *B. rigidus* contributed overwhelmingly to the biomass of the mixed population. But by delaying the introduction of *B. rigidus* into the mixtures, the balance was tipped decisively in favour of *B. madritensis*. It is therefore quite wrong to think of the outcome of competition as being always determined by the inherent competitive abilities of the competing species. Even an 'inferior' competitor can exclude its superior if it has enough of a head start. And this can foster coexistence when repeated colonization occurs in a changing or unpredictable environment.

7.6.3 Fluctuating environments

Indeed, the balance between competing species can be repeatedly shifted, and coexistence therefore fostered, simply as a result of environmental change. This was the argument used by Hutchinson (1961) to explain 'the paradox of the plankton'—the paradox being that numerous species of planktonic alga frequently coexist in simple environments with little apparent scope for niche differentiation. Hutchinson suggested that the environment, although simple, was continually changing, particularly on a seasonal basis.

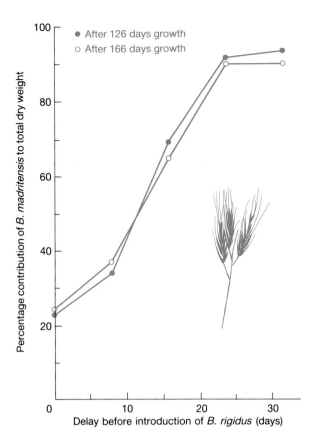

Figure 7.11. The effect of timing on competition. *Bromus rigidus* makes an overwhelming contribution to the total dry weight per pot when sown at the same time as *B. madritensis*. But as the introduction of *B. ridigus* is delayed, its contribution declines. Total yield per pot was unaffected by delaying the introduction of *B. rigidus*. ●, after 126 days growth; ○, after 166 days growth. (After Harper, 1961.)

Thus although the environment at any one time would tend to promote the exclusion of certain species, it would alter and perhaps even favour these same species before exclusion occurred. In other words, the equilibrium outcome of a competitive interaction may not be of paramount importance if the environment typically changes long before the equilibrium can be reached. And since all environments vary, competitive balances must be forever shifting, and coexistence must commonly be associated with niche differences that would promote exclusion in an unvarying world.

7.6.4 Ephemeral patches with variable life-spans

Many environments, by their very nature, are not simply variable but ephemeral. Among the more obvious examples are decaying corpses (carrion), dung, rotting fruit and fungi, and temporary ponds. But note too that a leaf or an annual plant can be seen as an ephemeral patch, especially if it is palatable to its consumer for only a limited period. Often these ephemeral patches have a variable life-span—a piece of fruit and its attendant insects, for instance, may be eaten at any time by a bird. In these cases, it is easy to imagine the coexistence of two species: a superior competitor, and an inferior competitor that nevertheless reproduces early.

coexistence of the good with the fast

An example of this is provided by the work of Brown (1982) on two species of pulmonate snail living in ponds in north-eastern Indiana, U.S.A. Artificially altering the density of one or other species in the field showed that the fecundity of *Physa gyrina* was significantly reduced by interspecific competition from *Lymnaea elodes,* but the effect was not reciprocated. *L. elodes* was clearly the superior competitor when competition continued throughout the summer. Yet *P. gyrina* reproduced earlier and at a smaller size than *L. elodes,* and in the many ponds that dried up by early July it was often the only one of the species to have produced resistant eggs in time. The species therefore coexisted in the area as a whole, in spite of *P. gyrina*'s apparent inferiority.

7.6.5 Aggregated distributions

A more subtle, but more generally applicable model of coexistence between a superior and an inferior competitor on a patchy and ephemeral resource has been investigated through computer simulations by Atkinson and Shorrocks (1981). The crux of their model is that the two species have independent, aggregated (i.e. clumped) distributions over the available patches. This means that the powers of the superior competitor are mostly directed against members of its own species (in the high-density clumps). It means too that the aggregated superior competitor will be absent from many patches within which the inferior competitor, with its independent distribution, can escape competition. Atkinson and Shorrocks found that as a consequence, an inferior competitor was able to coexist with a superior that would have rapidly excluded it from a continuous, homogeneous environment. They also found that the persistence of the coexistence increased with the degree of aggregation, until at high levels of aggregation, coexistence was apparently permanent (Figure 7.12). Since many species have aggregated distributions in nature, these results may be applicable widely.

a clumped superior competitor adversely affects itself and leaves gaps for its inferior

Hanski and Kuusela (1977) provide one example of this phenomenon from a study of carrion flies in Finland. They established 50 small patches of carrion in a very limited area (*c.* 5 m²), and allowed the naturally occurring flies to lay eggs in (i.e. colonize) the patches. Overall, nine species of carrion fly emerged from the patches. Yet the average number of species per patch was only 2.7. This was because all of the species were very highly aggregated in their distribution (with many offspring emerging from a few patches, and a few offspring or none emerging from many others). As a result, particular pairs of species came into contact only rarely, and this is bound to have fostered coexistence overall, even though there may have been exclusion where the species *did* come into contact.

Once again, the heterogeneous nature of the environment can be seen to have fostered coexistence without there being a marked differentiation of niches. A realistic view of interspecific competition, therefore, must acknowledge that it often proceeds not in isolation, but under the influence

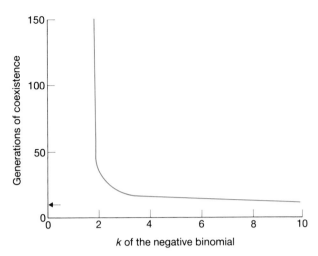

Figure 7.12. When two species compete on a patchy and ephemeral resource, the number of generations of coexistence increases with the degree of aggregation of the competitors as measured by the parameter k of the 'negative binomial' distribution. Values above 5 are effectively random distributions; values below 5 represent increasingly aggregated distributions. On a continuously distributed resource, one species would exclude the other in approximately ten generations. (After Atkinson & Shorrocks, 1981.)

of, and within the constraints of, a patchy, impermanent or unpredictable world.

7.7 Apparent competition: enemy-free space

Another reason for being cautious in our discussion of competition is the existence of what Holt (1977, 1984) has called 'apparent competition' and what others have called 'competition for enemy-free space' (Jeffries & Lawton, 1984, 1985).

Imagine a single species of predator that attacks two species of prey. Both prey species are adversely affected by the predator, and the predator is beneficially affected by both species of prey. Moreover, the increase in abundance that the predator achieves by being able to consume prey species 1 increases its adverse effect on prey species 2. Indirectly, therefore, prey species 1 adversely affects prey species 2 and vice versa. These interactions are summarized in Figure 7.13, which shows that from the point of view of the two prey species, the signs of the interactions are indistinguishable from those that would apply if they were two species competing for a single resource. In the present case, of course, there is no limiting resource. Hence the term 'apparent competition'.

two prey species being attacked by a predator are, in essence, indistinguishable from two consumer species competing for a resource

To be more specific, Holt (1977) has shown, for a simple but illuminating model, that the condition for prey species 1 to coexist with prey species 2 is:

$$r_1 > a_1 . N_{\text{predator}},$$

where r_1 is the intrinsic rate of natural increase of prey species 1, N_{predator} is the equilibrium density of the predator in the absence of prey species 1, and a_1 is

269　INTERSPECIFIC COMPETITION

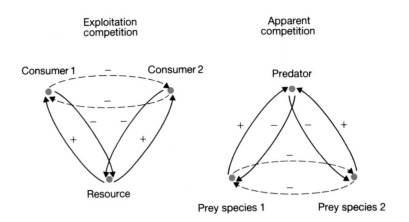

Figure 7.13. In terms of the signs of their interactions, two prey species being attacked by a common predator ('apparent competition') are indistinguishable from two species consuming a common resource (exploitation competition). Solid lines are direct interactions, dotted lines indirect interactions. (After Holt, 1984.)

the 'attack rate' of the predator on prey species 1 (attack rates are discussed in Chapter 9). The crucial point here is that coexistence will be favoured by reducing a_1, i.e. by avoiding attacks from the predator, which we know attacks prey species 2. Clearly, prey species 1 can achieve this by occupying a habitat, or adopting a form or a behavioural pattern, that is sufficiently different from that of prey species 2. In short, 'being different' (i.e. niche differentiation) will favour coexistence—but it will do so because it diminishes apparent competition or competition for enemy-free space. It is easy to see that the pattern (niche differentiation) can result from more than one process.

To take just one example where niche differentiation seems to result from competition for enemy-free space, Gilbert (1984) has suggested that rare *Heliconius* butterfly species are confined to a restricted range of larval host plants because they thereby escape attack from parasitoids that afflict the more common *Heliconius* species that live on other host plants. Overall, the reviews by Holt (1977, 1984) and Jeffries and Lawton (1984) indicate that the relative neglect of apparent competition in the past has been unwarranted.

7.8 Interpreting niche differences in the field

Notwithstanding the important interactions between competition and environmental heterogeneity, and the complications of enemy-free space, a great deal of attention has been focused on competition itself. In this and the next two sections we shall examine some of the more important ways in which information on interspecific competition has been sought.

We begin by noting that the common association between the coexistence of competitors and niche differentiation has led many workers to seek evidence for competition simply by documenting niche differences between species in the field. However, there are profound difficulties in interpreting such differences.

Lack's tits: niche
differentiation amongst
coexisting competitors—
or is it?

The problem can be illustrated by considering some work by Lack (1971). Lack described the coexistence of five species of tit in English broad-leaved woodlands: the blue tit (*Parus caeruleus*), the great tit (*P. major*), the marsh tit (*P. palustris*), the willow tit (*P. montanus*) and the coal tit (*P. ater*) (Figure 7.14). Four of these species weigh between 9.3 g and 11.4 g on average (the great tit weighs 20.0 g); all have short beaks and hunt for food chiefly on leaves and twigs, but at times on the ground; all eat insects throughout the year, and also seed in winter; and all nest in holes, normally in trees. Nevertheless, in Marley Wood, Oxford, all five species breed and the blue, great and marsh tits are common.

All five species feed their young on leaf-eating caterpillars, and all except the willow tit feed on beechmast in those winters when it is plentiful; but both of these foods are temporarily so abundant that competition for them is most unlikely. The small blue tit feeds mostly on oak trees, concentrating on the smaller twigs and leaves of the canopy, to which it is suited by its agility. It also strips bark to feed on the insects underneath, and generally takes insects less than 2 mm in length. It eats hardly any seeds except those of the birch, which it takes from the tree itself. The great tit, by contrast, feeds mainly on the ground, especially in winter. Most of the insects it takes exceed 6 mm in length; it eats more acorns, sweet chestnut and wood sorrel seeds than the other species; and it is the only species to take hazel nuts. The marsh tit has a feeding station between the other two common species. It feeds in the shrub layer, in large trees on twigs and branches below 6 m, or in herbage. It is also intermediate in size between the other two species, and takes insects between

Great tit

Marsh tit

Blue tit

Coal tit

Figure 7.14. The tit species that live in Marley Wood.

Willow tit

3 mm and 4 mm in length. In addition, it takes the fruits and seed of burdock, spindle, honeysuckle, violet and wood sorrel. The coal tit is another small species feeding on oak, but also, later in the year, on ash; and it is also found more often in conifers than the other species. It generally takes insects which are even shorter, on average, than those favoured by the blue tit, and unlike the blue tit it feeds mainly from branches rather than in the canopy. Finally, the willow tit is most like the marsh tit, feeding on birch, to a lesser extent on elder, and in herbage. Unlike the marsh tit, however, the willow tit avoids oak and takes very few seeds.

As Lack concluded, the species are separated at most times of the year by their feeding station, the size of their insect prey, and the hardness of the seeds they take; and this separation is associated with differences in overall size, and in the size and shape of their beaks. Despite their similarities, the species exploit slightly different resources in slightly different fashions. But there are three possible ways in which this can be 'explained'.

coexisting competitors?

The first interpretation is 'current competition': the tits are competing species that coexist by virtue of niche differentiation, but they retain the ability to expand their niches (i.e. occupy the whole of their fundamental niches) in the absence of competitors.

The second interpretation is 'evolutionary avoidance of competition'—what Connell (1980) has called invoking 'the ghost of competition past'. This itself requires some further explanation. When two species compete, individuals of one or both species suffer reductions in fecundity and/or survivorship. Such individuals may then be less fit than certain other individuals, namely those that escape competition because they have a fundamental niche which does not overlap with that of the other species. Natural selection will act on such differences in fitness. It may favour individuals that escape interspecific competition, and eventually the population may consist entirely of such individuals. Of course, this does not always happen: ongoing competition is widespread and important. But it may happen, in which case any niche differentiation would result not from present competition but from the evolutionary avoidance of past competition. To explain niche differentiation in these terms then, is to invoke the ghost of competition past.

or 'the ghost of competition past'?

or, simply, evolution?

The third possible interpretation of the tit data is that the species have, in the course of their evolution, responded to natural selection in different and independent ways. They are distinct species, and they have distinctive features. But they do not compete now, nor have they ever competed; they simply happen to be different.

It is difficult to distinguish between these three interpretations. An experimental manipulation (e.g. the removal of one or more species) could prove the truth of the first interpretation if it led, say, to an increase in the abundance of the remaining species. But negative results would be equally compatible with both the second and the third interpretations. In fact there are no easy or agreed methods of distinguishing between 'evolutionary avoidance of competition' and simple 'evolution' (Chapter 18). And certainly, on the basis of observational data *alone*, it is impossible to reject *any* of the interpretations. There are undoubtedly cases where currently competing

very little can be
concluded from mere
observational evidence

species coexist as a result of niche differentiation; and there are undoubtedly cases in which species' ecologies and morphologies have been moulded by competition in the past. But differences between species are *not*, in themselves, indications of niche differentiation forged by competition; and interspecific competition *cannot* be studied by the mere documentation of interspecific differences.

7.9 Experimental evidence for interspecific competition

The ambiguities of interpreting observational evidence have been avoided in many cases by taking an experimental approach to the study of interspecific competition. In the earlier parts of this chapter, for example, we saw that evidence has been derived from manipulative field experiments involving salamanders (section 7.2.1), barnacles (7.3.2), cattails (7.3) and freshwater pulmonates (7.6.4). In such experiments, the density of one or both species was altered (usually reduced). The fecundity, the survivorship, the abundance or the resource-utilisation of the remaining species was subsequently monitored. It was then compared either with the situation prior to the manipulation, or, far better, with a comparable control plot in which no manipulation had occurred. Such experiments have consistently provided valuable information, but they are typically easier to perform on some types of organism (for instance sessile organisms) than they are on others. Extensive reviews of this type of evidence have been carried out by Schoener (1983) and Connell (1983) (see Chapter 18).

The second type of experimental evidence has come from work carried out under artificial, controlled (often laboratory) conditions. Examples include Tansley's work on bedstraws (section 7.2.2), Gause's on *Paramecium* (7.2.4), Tilman *et al.*'s on diatoms (7.2.5), Groves and Williams's on clover and skeleton weed (7.3) and Park's on flour beetles (7.4.3). Again, the crucial element has been a comparison between the responses of the species living alone and their responses when in combination. Such experiments have the advantages of being comparatively easy to perform and to control, but they have two major disadvantages. The first is that species are examined in environments that are different from those they experience in nature. The second is the simplicity of the environment: it may preclude niche differentiation because niche dimensions are missing that would otherwise be important. Nevertheless, these experiments can provide useful clues to the likely effects of competition in nature.

In short-term experiments with plants, for example, there are two basic designs that have been used: 'substitutive', in which total density is held constant but the proportions of the two species are varied, and 'additive', in which one species is sown at a constant density along with a range of densities of a second species.

7.9.1 Substitutive experiments

In this approach, pioneered by de Wit and his colleagues in the Netherlands (de Wit, 1960), the effect of varying the proportion of each of two species is

explored while keeping overall density constant. Thus at an overall density of
say 200 plants, a series of mixtures would be set up: 100 of species A with 100
of species B, 150A and 50B, zero A and 200 B, and so on. Such *replacement
series* may then be established at a range of total densities, usually chosen to
ensure that competition *does* occur. In practice, however, many workers have
used only a single total density.

A good example of a substitutive experiment is provided by the work of
Marshall and Jain (1969). They used two species of wild oats (*Avena fatua* and
A. barbata) which occur naturally together in the annual grasslands of
California, and grew them together at four densities: 32, 64, 128 and 256
plants per pot. At each of these densities, the two species were sown at the
same range of proportions: 0, 12.5, 50, 87.5 and 100% of the total sown. After
29 weeks' growth in a greenhouse, the yield of each species was assessed in
terms of number of spikelets per pot.

In order to appreciate fully the effects of interspecific competition, it is
necessary first to know the effects of intraspecific competition. Marshall and
Jain, therefore, grew each species alone over the range of densities that they
were to experience in the replacement series. Their results are shown in
Figure 7.15. Effects of intraspecific competition are clearly demonstrated.

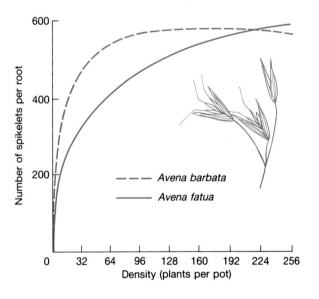

Figure 7.15. Intraspecific
competition affects the
fecundity of two species
of wild oats, *Avena fatua*
and *A. barbata* (after
Marshall & Jain, 1969).

The results of the four replacement series (one at each of four densities)
are shown as *replacement diagrams* in Figure 7.16. In each case, the solid lines
represent the yields from the replacement series, while the dotted lines
represent the yields that would have been expected had the species been
grown alone. Point A (which appears in Figures 7.16a, b and d), for instance,
is the yield obtained from *A. fatua* when it was grown alone at a density of 32
plants per pot. In the complete *absence* of interspecific competition, this yield
would be obtained not only in Figure 7.16a, but also in Figure 7.16b (where
these 32 *A. fatua* were grown with 32 *A. barbata*) and Figure 7.16d (where the
32 *A. fatua* were grown with 224 *A. barbata*). Yet in the latter two cases the

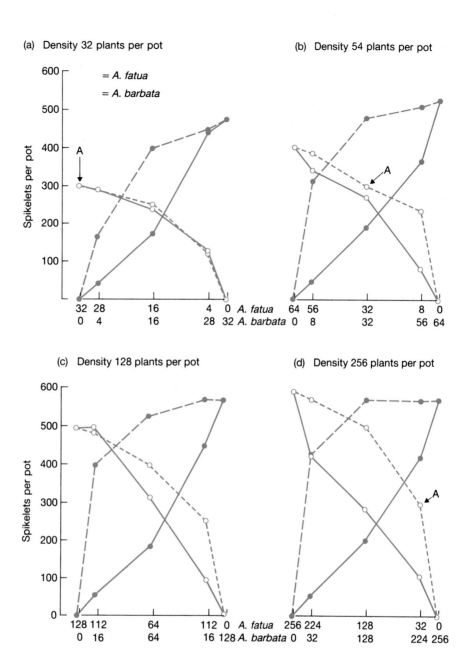

Figure 7.16. Replacement diagrams for *Avena fatua* and *A. barbata* when grown together at four overall densities. Dotted lines show the yield expectations from monoculture (Figure 7.15); point A, for example, appears in (a), (b) and (d). Solid lines show the yield in mixture. (After Marshall & Jain, 1969.)

yield was less than this expectation. It is this depression of yield (the solid line below the dotted line) which indicates the action of interspecific competition.

Interspecific competition is apparent in all cases except that of *A. fatua* grown in the lowest-density replacement series (Figure 7.16a). In fact, for both species, the difference between the yields in mixture and the yields in monoculture increased as overall density increased. This indicates, not surprisingly, that the intensity of competition is dependent on the density at which the interaction takes place. It is also worth noting that *A. fatua* was consistently less strongly affected by interspecific competition than *A. barbata* (competition was asymmetric).

7.9.2 Additive experiments

In additive experiments, one species (typically a crop) is sown at a constant density, along with a range of densities of a second species (typically a weed). The justification for this design is that it mimics the natural situation of a crop infested by a weed, and it therefore provides information on the likely effect on the crop of various levels of infestation. An example is shown in Figure 7.17, describing the effects of two weeds, sicklepod (*Cassia obtusifolia*) and

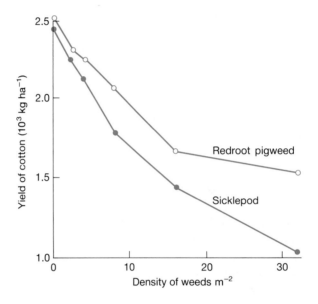

Figure 7.17. An 'additive design' competition experiment: the yield of cotton produced from stands planted at constant density, infested with weeds (either sicklepod or redroot pigweed) at a range of densities (after Buchanan *et al.*, 1980).

redroot pigweed (*Amaranthus retroflexus*), on the yield of cotton grown in Alabama, U.S.A. (Buchanan *et al.*, 1980). As weed density increased, so cotton yield decreased, and this effect of interspecific competition was always more pronounced with sicklepod than with redroot pigweed.

A problem with experiments like this is that overall density and species proportion are changed simultaneously. It has therefore proved difficult to separate the effect of the weed itself on crop yield, from the simple effect of increasing total density (crop plus weed). Note that substitutive experiments avoid this problem by holding overall density constant and only varying proportion. They are more straightforward to analyse, but artificial in that most plant populations that change in proportion over time also change in density. They are therefore limited to predicting the outcome of interspecific competition when density is held constant. Watkinson (1981) has argued that rather than avoid the problems of density, it is preferable to develop models that take account of the variations in both density and proportion. He used a modification of the Lotka–Volterra model to analyse an experiment in which the density of wheat was held constant, while that of the annual weed

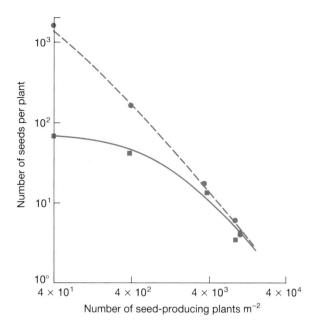

Figure 7.18. The relationship between the density of seed-producing plants of *Agrostemma* and the number of seeds produced per plant in pure stand (●——●) and in the presence of wheat (■——■). The equation of the curve in the presence of wheat is

$$\text{Number of seeds per plant} = 3685\,(1+0.0325\,(N_1+1.5N_2))^{-1.15}$$

where N_1 is the density of surviving *Agrostemma* plants and N_2 is the density of surviving wheat plants. (After Watkinson, 1981.)

Watkinson's model for dealing concurrently with density and proportion

Agrostemma githago was varied. *Agrostemma* was also grown in pure culture. Figure 7.18 shows the relationship between number of seeds produced per *Agrostemma* plant and the density of seed-bearing plants. There is a clear, negative, density-dependent relationship in both the pure and mixed populations. However, seed production was considerably depressed, particularly at low densities, by the presence of wheat. The details of Watkinson's model need not concern us here. It is sufficient to note that it produces a very good fit to the data. It also provides us with a value for the competition coefficient describing the effect of wheat on *Agrostemma* ($\alpha_{AW} = 1.5$), indicating that a single wheat plant (W) is the equivalent of one and a half *Agrostemma* plants (A).

7.10 Natural experiments

the pros and cons of natural experiments

We have seen that interspecific competition is commonly studied by an experimenter comparing species alone and in combination. Nature, too, often provides information of this sort: the distribution of certain potentially competing species are such that they sometimes occur together (*sympatry*) and sometimes occur alone (*allopatry*). These 'natural experiments' can provide additional information about interspecific competition. Their advantage is that they *are* natural: they are concerned with organisms living in their natural habitats and involve no outside influence exerted by the investigator. Their disadvantage stems from the difference between the 'experimental' and the 'control' populations. Ideally there should be only *one* difference: the presence or absence of a competitor species. In natural experiments, though, the populations *may* differ in other ways, simply because they exist under different conditions in different locations. Natural experiments should therefore always be interpreted cautiously.

7.10.1 Competitive release

Evidence for competition often comes from the contraction of a fundamental niche in the presence of a competitor, or from niche expansion in the competitor's absence (known as *competitive release*). Such evidence can sometimes appear to occur naturally rather than as a result of a manipulative experiment. Diamond (1975) has provided a particularly good example of natural competitive release involving ground doves in New Guinea. The New Guinea archipelago comprises one large, several moderately sized and very many small islands. The distributions of species on islands have been extensively studied (see Chapter 19). In the present context, though, it need only be noted that in the New Guinea region, as elsewhere, small islands tend to lack species which are present on large islands and on the mainland. As Figure 7.19 illustrates, there are three similar species of ground dove on New Guinea itself, and moving inland from the coast one encounters them in sequence: *Chalcophaps indica* in the coastal scrub, *Chalcophaps stephani* in the light or second-growth forest and *Gallicolumba rufigula* in the rainforest. However, on the island of Bagabag, where *G. rufigula* is absent, *C. stephani* extends its range inland into rainforest; while on Karkar, Tolokiwa, New Britain and numerous other small islands where *C. indica* is also absent, *C. stephani* expands coastwards to occupy the whole habitat gradient. On Espiritu Santo, on the other hand, *C. indica* is the only species present, and it occupies all three habitats. It is this range or niche expansion in the absence of a presumed potential competitor which is known as competitive release.

ground doves in New Guinea

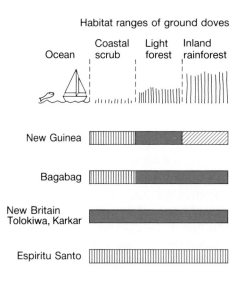

Figure 7.19. Competitive release: the habitats occupied by three species of ground doves on various islands in the New Guinea region: *Chalcophaps indica* (vertical bars), *Chalcophaps stephani* (solid shading), *Gallicollumba rufigula* (diagonal bars). (After Diamond, 1975.)

Direct interspecific competition has not been positively established in this case. But *C. stephani* and *C. indica* only occupy rainforest in the absence of *G. rufigula*; *C. stephani* only occupies coastal scrub in the absence of *C. indica*; and *C. indica* only occupies light forest in the absence of *C. stephani*. The conclusion, therefore, seems more or less unavoidable, that there is competitive exclusion and niche differentiation on New Guinea, and various degrees of competitive release on the other islands.

This conclusion may be tempered, however, by examining a second apparent example of natural competitive release, provided by the work of Abramsky and Sellah (1982) on two gerbilline rodents living in the coastal sand dunes of Israel. In northern Israel, the protrusion of the Mt. Carmel ridge towards the sea separates the narrow coastal strip into two isolated areas, north and south; but in both, the dunes intermingle with a number of other soil types. In biogeographical terms, all of these dunes are of recent origin (between a few hundred and several thousand years old). *Meriones tristrami* is a gerbil that has colonized Israel from the north. It now occurs, associated with the dunes, throughout the length of the coast, including the areas both north and south of Mt. Carmel. *Gerbillus allenbyi* is another gerbil, also associated with the dunes and feeding on similar seeds to *M. tristrami*; but this species has colonized Israel from the south and has *not* crossed the Mt. Carmel ridge. To the north of Mt. Carmel, where *M. tristrami* lives alone, it is found on sand as well as other soil types. However, south of Mt. Carmel it occupies several soil types but *not* the coastal sand dunes. Here, only *G. allenbyi* occurs on dunes.

This appears to be another case of competitive exclusion and competitive release: exclusion of *M. tristrami* from the sand to the south of Mt. Carmel, release of *M. tristrami* to the north. Abramsky and Sellah, however, actually tested this conclusion experimentally. They set up a number of plots south of Mt. Carmel from which *G. allenbyi* was removed, and they compared the densities of *M. tristrami* in these plots with those in a number of similar control plots. They monitored the plots for a year; but the abundance of *M. tristrami* remained essentially unchanged. Invoking the ghost of competition past, it seems that south of Mt. Carmel, *M. tristrami* has evolved to select those habitats in which it avoids competition with *G. allenbyi*, and that even in the absence of *G. allenbyi* it retains this genetically fixed preference. With Lack's tit data (p.271), such an interpretation would have been mere speculation. But here, because a natural experiment appears to have demonstrated competitive release, the interpretation is much sounder and more sensible. We must remember, however, that it is not established fact.

7.10.2 Character displacement

In certain examples, it is not only the realized niche of a species which appears to alter from sympatry to allopatry, but also its morphology. These are referred to as cases of *character displacement*. However, while the idea of character displacement is appealing and plausible, good examples are very few in number (Connell, 1980; Arthur, 1982).

One possible case is illustrated in Figures 7.20 and 7.21. These data come

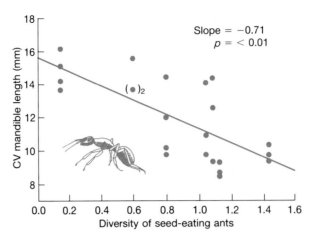

Figure 7.20. Character displacement: the relationship between the within-colony coefficient of variation (CV) in mandible length for *Veromessor pergandei* and the species diversity of seed-eating ants in the community (after Davidson, 1978).

from the work of Davidson (1978) on the seed-eating harvester ant, *Veromessor pergandei*, collected at various desert sites in the south-western United States. Food (i.e. seeds) can be an important limiting resource for these ants (Brown & Davidson, 1977), and different ant species specialize on seeds of different sizes depending on their own size (Davidson, 1977). In the case of *V. pergandei*, Davidson looked at mandible length and variability in foraging workers; and as Figure 7.20 shows, variability decreased significantly as the diversity of potential competitors at a site increased. In other words, *V. pergandei* is more of a size-specialist at those sites in which interspecific competition is most likely. The same picture emerges from Figure 7.21, which

worker ants

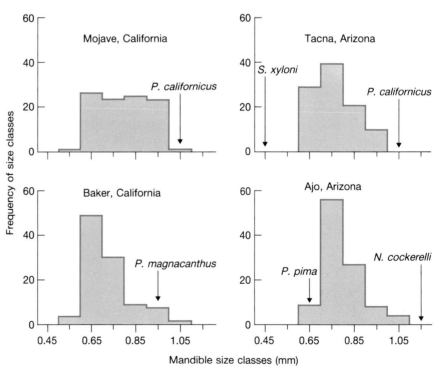

Figure 7.21. Character displacement: some frequency distributions of mandible size classes for *Veromessor pergandei* at different sites, with the mean mandible lengths of competitors most similar in size indicated by arrows (after Davidson, 1978).

suggests, in addition, that worker size itself varies from site to site in a way which tends to make *V. pergandei* different in size from the species with which it coexists. It is this morphological differentiation in the presence of potential competitors that is known as character displacement. The inference is that *V. pergandei* workers have been selected which are significantly different in size from individuals of coexisting species, thus allowing resource partitioning (different-sized ants eat different-sized seeds) and the coexistence of potential competitors.

mud snails in Denmark

Another plausible example of character displacement is provided by the work of Fenchel on the mud snails *Hydrobia ulvae* and *Hydrobia ventrosa* living in the Limfjord, Denmark (Fenchel, 1975; Fenchel & Kofoed, 1976). Fenchel found (Figure 7.22) that when the two species live apart (which they do in a range of habitats), their sizes are more or less identical; but when they coexist there is always a marked difference in their size. He also found that when the similarly sized species lived apart they consumed similarly sized food particles, but when they coexisted, *H. ulvae*, which was larger, tended to consume larger food particles than *H. ventrosa* (Figure 7.23). The data, therefore, strongly suggest character displacement, allowing resource partitioning and coexistence. Fenchel and Kofoed (1976) also found that when similarly sized individuals of the two species grew together in experimental containers, interspecific competition was as intense as intraspecific competition; but this depressant effect was significantly reduced when the individuals of the two species were different in size. If snail species of the same size coexisted at a site, they would exploit and compete for the same food resource. Local extinction of the less competitive species would then seem to be inevitable.

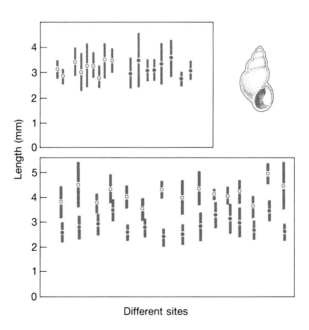

Figure 7.22. Character displacement: average lengths (plus standard deviations) of *Hydrobia ulvae* (open circles) and *H. ventrosa* (filled circles) at various sites. When they live alone (above) their sizes are similar; but when they live together (below) their sizes differ. (After Fenchel, 1975.)

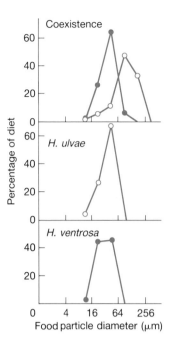

Figure 7.23. Character displacement in *Hydrobia:* distribution of food particle sizes eaten by *H. ulvae* and *H. ventrosa* at typical sites at which they live alone or coexist (after Fenchel, 1975).

Character displacement, on the other hand, would ensure that different species tend to exploit different resources. The intensity of interspecific competition would thus be relaxed, and the species would coexist. Even in this example, however, because of the absence of data on the population dynamics of competition in the field, character displacement can only be suggested rather than proved.

Character displacement, where it is proved to exist, is another example in which at least a partial avoidance of interspecific competition has evolved and become genetically fixed.

7.11 The nature of niche differentiation

In spite of all the difficulties of making a direct connection between interspecific competition and niche differentation, there is no doubt that niche differentiation is often the basis for the coexistence of competitors. The question arises, therefore, as to the precise nature of niche differentiation. An outline will be given here which will be fleshed out in Chapter 18. This will then be followed (section 7.11.1) by a consideration of the special problems of niche differentiation in plants.

There are a number of ways in which niches can be differentiated. The first is resource partitioning, or, more generally, *differential resource utilization*. This can be observed when species living in precisely the same habitat nevertheless utilize different resources. Since the majority of resources for animals are individuals of other species, or parts of individuals (of which there are literally millions of types), there is no difficulty, in principle, in imagining how competing animals might partition resources amongst themselves. Plants, on the other hand, all have very similar requirements for the same potentially limited resources (Chapter 3), and there is much less

differential resource utilization: easy to imagine in animals— less easy in plants

282   CHAPTER 7

apparent scope for resource partitioning (but see below). However, nitrogen is a noteworthy exception. All terrestrial plants utilize fixed nitrogen from the soil, but a number of species, especially legumes, are also able to utilize free nitrogen from the air because of their mutualistic association with nitrogen-fixing bacteria (Chapter 13). Moreover, a few insectivorous plants can obtain nitrogen directly from their prey. This immediately suggests that competing plant species might coexist by differential utilization of the 'total nitrogen' resource, and confirmation that they can do so has been provided by the results of a substitutive experiment carried out by de Wit and his collaborators (de Wit *et al.*, 1966).

spatial and temporal separations based on resources

In many cases, the resources utilized by ecologically similar species are separated spatially. Differential resource utilization will then express itself as either a microhabitat differentiation between the species, or even a difference in geographical distribution. Alternatively, the availability of the different resources may be separated in time, i.e. different resources may become available at different times of the day or in different seasons of the year. Differential resource utilization may then express itself as a temporal separation between the species. There are many examples of segregation of resources in space or time involving both animals and plants. But it is among plants and other sessile organisms, because of their limited scope for differential resource utilization at the same location and instant, that spatial and temporal separation is likely to be of particular significance.

niche differentiation based on conditions

The other major way in which niches can be differentiated is on the basis of conditions. Two species may utilize precisely the same resources; but if their ability to do so is influenced by environmental conditions (as it is bound to be), and if they respond differently to these conditions, then each may be competitively superior in different environments. This too can express itself as either a microhabitat differentiation, or a difference in geographical distribution, or a temporal separation, depending on whether the appropriate conditions vary on a small spatial scale, a large spatial scale or over time. Of course, in a number of cases (especially with plants) it is not easy to distinguish between conditions and resources (Chapter 3). Niches may then be differentiated on the basis of a factor (such as water) which is both a resource and a condition.

7.11.1 Tilman's model of differential resource utilization

Turning to the problem of differential resource utilization in plants, Tilman (1982; see also section 3.6) has made a particularly incisive attempt to see whether this can explain the coexistence of plant species competing for limiting resources (though the method may be applicable to other groups). The crucial point about Tilman's model is that he explicitly considers the dynamics of the resources as well as the dynamics of the species that compete for them. Here, rather than going into the details of his methods, we examine the outlines of his model and his major conclusions.

a model based on the dynamics of competitors *and* their resources

the zero net growth isocline: a niche boundary

Tilman begins by defining the *zero net growth isocline* (ZNGI) for a single species utilizing two essential resources (section 3.6). This is the boundary between the resource combinations which allow the species to survive and

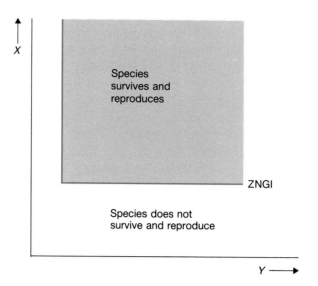

Figure 7.24. The zero net growth isocline (ZNGI) of a species potentially limited by two resources (X and Y) divides resource combinations on which the species can survive and reproduce from those on which it cannot. The ZNGI is rectangular in this case because X and Y are essential resources.

reproduce, and the resource combinations which do not (Figure 7.24). The ZNGI therefore represents the boundary of the species' niche in these two dimensions.

In order to understand interspecific competition in Tilman's terms it is first necessary to understand intraspecific competition. Intraspecific competition should (ideally) bring the population to a stable equilibrium; but in the present case this equilibrium has two components: both population size and the resource levels should remain constant. Population size is constant

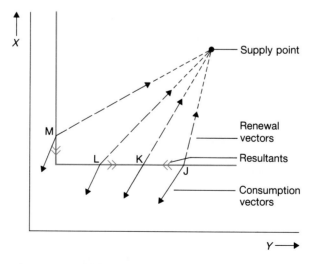

Figure 7.25. The balance between resource consumption and resource renewal. Point K is the only point on the ZNGI (no net population growth) at which there is also no net change in resource concentrations (consumption and renewal are equal and opposite). For further discussion, see text.

intraspecific competition: renewal and consumption in balance

(by definition) at all points on the ZNGI. It is therefore necessary to establish where resource levels are constant.

Any net change in resource levels is the resultant of two opposing forces. A species that consumes and depletes resources draws their levels towards the bottom-left of diagrams like Figure 7.25. But the resources are continually being renewed, i.e. rising towards the top-right. In fact, the resources will tend to rise towards the particular combination of levels that they would attain in the absence of consumption (the *supply point*). Thus in Figure 7.25 there are consumption vectors and renewal vectors ('vectors' because they have a direction *and* a strength). The consumption vectors reflect the species' consumption rates of the two resources; the renewal vectors point towards the supply point. When two vectors are in direct opposition there is no net change in resource level; otherwise the resource levels alter.

Figure 7.25 shows that there is only one point on a ZNGI (constant population size) where resource levels are also constant. At point J, for instance, the consumption rate of resource Y is greater than its renewal rate and its level decreases (towards point K). Only at point K itself are consumption and renewal rates equal; K is therefore the point at which both population size and resource levels are constant.

To move from intra- to interspecific competition, it is necessary to superimpose the ZNGIs of two species on the same diagram (Figures 7.26 and 7.27). The two species will have dissimilar consumption rates and their own consumption vectors, but there will still be a single supply point towards which all renewal vectors will be directed. The consumption and renewal vectors can be put together to determine the outcome of interspecific competition, and as the following paragraphs will explain, the outcome depends on the position of the supply point.

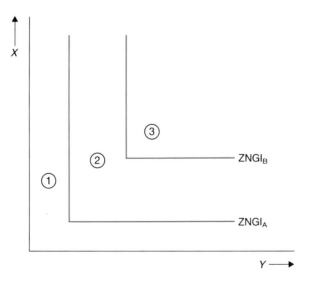

Figure 7.26. Competitive exclusion: the ZNGI of species A lies closer to the resource axes than the ZNGI of species B. If the resource supply point is in region 1, then neither species can exist. But if the resource supply point is in regions 2 or 3, then species A reduces the resource concentrations to a point on its own ZNGI (where species B cannot survive and reproduce): species A excludes species B.

In Figure 7.26, the ZNGI of species A is closer to both axes than the ZNGI of species B. There are three regions in which the supply point might be found. If it was in region 1, below the ZNGIs of both species, then there would never be sufficient resources for either species and neither would survive. If it was in region 2, below the ZNGI of species B but above that of species A, then species B would be unable to survive and the system would equilibrate on the ZNGI of species A. And if the supply point was in region 3, then this system too would equilibrate on the ZNGI of species A. Species A would competitively exclude species B because of its ability to exploit both resources down to levels at which species B could not survive (i.e. this is a typical case of exploitative competition). Of course, the outcome would be reversed if the positions of the ZNGIs were reversed.

In Figure 7.27 the ZNGIs of the two species overlap, and there are six regions in which the supply point might be found. Points in region 1 are below both ZNGIs and would allow neither species to exist; those in region 2 are below the ZNGI of species B and would only allow species A to exist; and those in region 6 are below the ZNGI of species A and would only allow species B to exist. Regions 3, 4 and 5 lie within the fundamental niches of both species. However, the outcome of competition depends on which of these regions the supply point is located in.

coexistence—dependent
on the ratio of resource
levels at the supply point

The most crucial region in Figure 7.27 is region 4. For supply points here, the resource levels are such that species A is more limited by resource X than by resource Y, while species B is more limited by Y than X. However, the consumption vectors are such that species A consumes more X than Y, while species B consumes more Y than X. *And because each species consumes more of the resource that more limits its own growth,* the system equilibrates at the

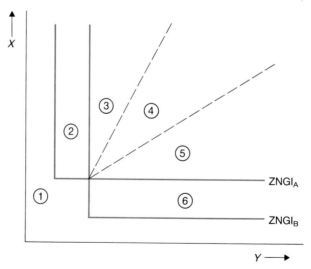

Figure 7.27. Potential coexistence of two competitors limited by two essential resources. The ZNGIs of species A and B overlap, leading to six regions of interest. With supply points in region 1 neither species can exist; with points in regions 2 and 3, species A excludes species B; and with points in regions 5 and 6, species B excludes species A. Region 4 contains supply points lying between the limits defined by the two dashed lines. With supply points in region 4 the two species coexist. For further discussion, see text.

intersection of the two ZNGIs, and this equilibrium is stable: the species coexist.

a subtle niche differentiation: each species consumes more of the resource that more limits its own growth

This is niche differentiation, but of a subtle kind. Rather than the two species exploiting different resources, species A disproportionately limits itself by its exploitation of resource X, while species B disproportionately limits itself by *its* exploitation of resource Y. The result is the coexistence of competitors. By contrast, for supply points in region 3, both species are more limited by Y than X. But species A can reduce the level of Y to a point on its own ZNGI below species B's ZNGI, where species B cannot exist. Conversely, for supply points in region 5, both species are more limited by X than Y, but species B depresses X to a point below species A's ZNGI. Thus, in regions 3 and 5, the supply of resources favours one species or the other, and there is competitive exclusion.

It seems then that two species can compete for two resources and coexist as long as two conditions are met. First, the habitat (i.e. the supply point) must be such that one species is more limited by one resource, and the other species more limited by the other resource. Secondly, each species must consume more of the resource that more limits its own growth. Thus it is possible, in principle, to understand coexistence in competing plants on the basis of differential resource utilization. The key seems to be an explicit consideration of the dynamics of the resources as well as the dynamics of the competing species. As with other cases of coexistence by niche differen-

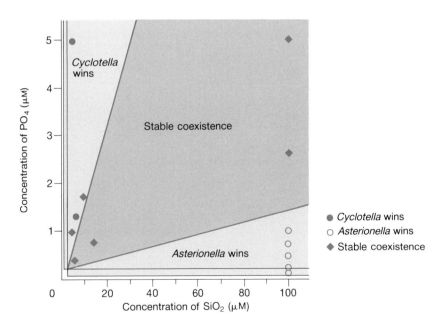

Figure 7.28. The observed ZNGIs and consumption vectors of two diatom species, *Asterionella formosa* and *Cyclotella meneghiniana*, were used to predict the outcome of competition between them for silicate and phosphate. The symbols, representing the three outcomes of the interaction, are explained in the key. Most experiments confirmed the predictions, with the exception of two lying close to the regional boundary. (After Tilman, 1977, 1982.)

tiation, the essence is that intraspecific competition is, for both species, a more powerful force than interspecific competition.

The best evidence for the validity of the model comes from Tilman's own experimental work on competition between the diatoms *Asterionella formosa* and *Cyclotella meneghiniana* (Tilman, 1977). For both species, Tilman observed directly the consumption rates and the ZNGIs for both phosphate and silicate. He then used these to predict the outcome of competition with a range of resource supply points (Figure 7.28). Finally, he ran a number of competition experiments with a variety of supply points, and the results of these are also illustrated in Figure 7.28. In most cases the results confirmed the predictions. In the two that did not, the supply points were very close to the regional boundary. The results are therefore encouraging. But they are in considerable need of consolidation from work on other types of plants and animals. Nevertheless, Tilman's approach is one among several themes that we will be able to pick up again in Chapter 18 when we come to consider the role of interspecific competition in shaping the structure of ecological communities.

The Nature of Predation

8.1 Introduction: the types of predators

There can be no doubt that, in general terms, consumers affect the distribution and abundance of the things they consume and vice versa. Equally, there can be no doubt that these effects are of central importance in ecology. Yet it is never an easy task to determine what the effects are, how they vary, and why they vary. These topics will be dealt with in this and the next few chapters. We begin here by addressing ourselves to the question: what is the nature of predation? In other words we will be looking at the effects of predation on the predators themselves and on their prey. Then, in the next chapter, we examine the behaviour of predators and the way this affects what they consume and how much they consume. In Chapter 10, we turn to the consequences of predation in terms of the dynamics of predator and prey populations.

a definition of predation

Predation, put simply, is consumption of one organism (the prey) by another organism (the predator), in which the prey is alive when the predator first attacks it. This excludes detritivory, the consumption of dead organic matter, which is discussed in its own right in Chapter 11. Nevertheless, it is a definition that encompasses a wide variety of interactions and a wide variety of 'predators'.

taxonomic and functional classifications of predators

There are two main ways in which predators can be classified. Neither is perfect, but both can be useful in certain contexts. We should stress right away that these classifications will not be pursued simply with the aim of constructing a neat and tidy catalogue, nor of settling any issues in semantics. However, by distinguishing various types of predator and establishing what they share and how they differ, we will be able to develop a fuller understanding of the precise nature of predation. The most obvious classification, perhaps, is 'taxonomic': carnivores consume animals, herbivores consume plants, and omnivores consume both. An alternative, however, is a 'functional' classification (Thompson, 1982) of the type already outlined in Chapter 3. Here there are four main types of predator: true predators, grazers, parasitoids and parasites (the last divisible further into microparasites and macroparasites as explained in Chapter 12).

true predators

True predators kill their prey more or less immediately after attacking them, and over the course of their lifetime they kill several or many different prey individuals. Often, they consume prey in their entirety, but some true predators consume only parts of their prey. Most of the more obvious carnivores like tigers, eagles, coccinellid beetles and carnivorous plants are true predators, but so too are seed-eating rodents and ants, plankton-consuming whales and so on.

grazers

Grazers also attack large numbers of prey during their lifetime, but they remove only a part of each prey individual rather than the whole. Their effect on a prey individual is therefore variable, but it is typically harmful. Nevertheless, their attacks are rarely lethal in the short term, and they are certainly never predictably lethal (in which case they would be true predators). Among the more obvious examples are the large vertebrate herbivores like sheep and cattle, but the flies that bite vertebrates, and leeches that suck blood, are also undoubtedly grazers by this definition.

parasites

Parasites, like grazers, consume parts of their prey (their 'host') rather than the whole. Also like grazers, their attacks are typically harmful but rarely lethal in the short term. Unlike grazers, however, their attacks are concentrated on one or a very few individuals during the course of their life. There is, therefore, an intimacy of association between parasites and their hosts which is not seen in true predators and grazers. Tapeworms, liver flukes, the measles virus and the tuberculosis bacterium are all obvious examples of parasites. However, there are considerable numbers of plants, fungi and microorganisms that are parasitic on plants (often called 'plant pathogens'), for example the tobacco mosaic virus, the rusts and smuts, and the mistletoes; and there are also many herbivores that are undoubtedly parasites. Aphids, for example, extract sap from one or a very few individual plants with which they enter into intimate contact, and even caterpillars often rely on a single plant for their development (though the association here is not so intimate). Plant pathogens, and animals parasitic on animals, are dealt with together in Chapter 12. Parasitic herbivores, like aphids and caterpillars, are dealt with here and in the two chapters that follow. In these three chapters they will be grouped together with true predators, grazers and parasitoids under the umbrella term 'predator'.

parasitoids

The parasitoids (Figure 8.1) are a group of insects that are classified as such on the basis of the egg-laying behaviour of the adult female and the subsequent developmental pattern of the larva. They belong mainly to the order Hymenoptera, but also include many Diptera. They are free living as adults, but they lay their eggs in, on or near other insects (or, more rarely, in spiders or woodlice). The larval parasitoid then develops inside (or, more rarely, on) its host individual, which is itself usually a pre-adult. Initially, it does little apparent harm to the host, but eventually it almost totally consumes the host and therefore kills it before or during the pupal stage. Thus it is an adult parasitoid, rather than an adult host, that emerges from what is apparently a host pupa. Often just one parasitoid develops from each host, but in some cases several individuals share a host. In summary then, parasitoids are intimately associated with a single host individual (like parasites), they do not cause immediate death of the host (like parasites and grazers), but their eventual lethality is inevitable (like predators). They may seem to be an unusual group of limited general importance. However, it has been estimated that they account for 25% of the world's species (Price, 1980). This is not surprising when we consider that there are so many species of insects, that most of these support at least one parasitoid, and that even

Figure 8.1. Parasitoids. Left: *Chrysis ignata* (Hymenoptera: Chrysididae), a cuckoo wasp which parasitizes bees. Right: *Aphidius hetricaria* (Hymenoptera: Braconidae) lays its eggs in *Myzus persicae,* an aphid. (Photographs: Heather Angel.)

parasitoids themselves may support parasitoids. Moreover, a number of parasitoid species have been intensively studied by ecologists, and they have provided a wealth of information relevant to predation generally.

For parasitoids, and also for the many herbivorous insects that feed as larvae on plants, the rate of 'predation' is determined very largely by the rate at which adult females oviposit. Each egg laid by the female is an 'attack' on the prey or host, even though it is the larva that develops from the egg that actually does the eating. In most other cases, the rate of predation depends on the rate of physical encounter between the predator itself and its prey, i.e. the predator does indeed 'attack' its prey. In the sections that follow, 'predation' and 'attacks by predators' will refer to both of these types of behaviour.

We are now in a position, in the remainder of this chapter, to examine the nature of predation. We will look at the effects of predation on the prey individual (section 8.2), the effects on the prey population as a whole (section 8.3), and the effects on the predatory individual itself (section 8.4). In the cases of attacks by true predators and parasitoids, the effects on prey individuals are very straightforward: the prey are killed. Attention will therefore be focused in section 8.2 on prey subject to grazing and parasitic attack. In fact, attention will be mainly focused on herbivory. This demands special attention from ecologists firstly because herbivores can act as either true predators, grazers or parasitic consumers, and also because the consumed items may be whole plant individuals, whole plant modules or even parts of modules. Herbivory, therefore, apart from being important in its own right, serves as a useful vehicle for discussing the subtleties and variations in the effects that predators can have on their prey.

8.2 The effects of herbivory on individual plants

The effects of herbivory on a plant depend on precisely which parts are affected, and on the timing of the attack relative to the plant's development. Leaf biting, sap sucking, mining, meristem consumption, flower and fruit damage, and root pruning are all likely to differ in the effect they have on the plant. And the consequences of defoliating a germinating seedling are unlikely to be the same as those of defoliating a plant that is setting its own seed. Moreover, because the plant usually remains alive in the short term, the effects of herbivory are crucially dependent on the response of the plant. Minerals or nutrients may be diverted from one part to another, the overall rate of metabolism may change, the relative rates of root growth, shoot growth and reproduction may alter, and special protective chemicals or tissues may be produced. Overall, the effect of a herbivore may be more drastic than it appears, or less drastic. It is only rarely what it seems.

8.2.1 Plant compensation

herbivory can reduce self-shading

In a variety of ways, individual plants can compensate for the effects of herbivory. In the first place, the removal of leaves from a plant may decrease the shading of other leaves and thereby increase their rate of photosynthesis. Alternatively, if shaded leaves are removed (with normal rates of respiration but low rates of photosynthesis; see Chapter 3), then this may improve the balance between photosynthesis and respiration in the plant as a whole. Thus the weevil *Phyllobius argentatus* has been observed to feed mainly on the lower, shaded leaves towards the centre of a beech tree, and it has little impact on a tree's overall productivity (Nielsen & Ejlerson, 1977).

herbivory can lead to the mobilization of stored carbohydrates

Secondly, in the immediate aftermath of an attack from a herbivore, many plants compensate by utilizing carbohydrates stored in a variety of tissues and organs. For example, when two varieties of Italian ryegrass (*Lolium multiflorum*) were completely defoliated, the variety 'Liscate', which had higher levels of stored carbohydrates in its roots and stubble, exhibited higher initial rates of leaf re-growth than the variety 'S.22' (Figure 8.2; Kigel, 1980). However, within a short time of attack, the role of supplying new tissues usually passes from stored reserves to current photosynthesis.

herbivory can lead to an altered pattern of photosynthate distribution

Herbivory also frequently alters the distribution of photosynthate within the plant, the general rule being apparently that a balanced root/shoot ratio is maintained. When shoots are defoliated, an increased fraction of net production is channelled to the shoots themselves; and when roots are destroyed, the switch is towards the roots (Crawley, 1983). In fact, the defoliation of grasses often stops root growth altogether, and it may even lead to a reduction in root weight when roots that die naturally are not replaced by fresh growth (Ryle, 1970). Thus root-damaged plants, with inadequate uptake of water, nutrients and minerals, have their proportion of roots restored, whereas shoot-damaged plants, with inadequate rates of photosynthesis, have their proportion of shoots and leaves restored. This clearly ameliorates the effects of herbivory. Furthermore, the re-distribution of photosynthate can compensate for the effects of patchy attack *within* a plant.

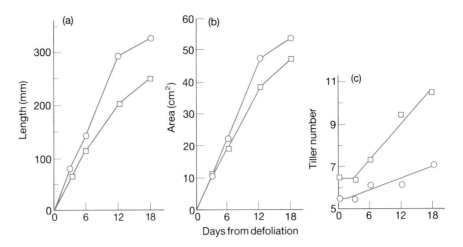

Figure 8.2. After defoliation, the regrowth of the variety 'Liscate' of *Lolium multiflorum* (○) is better than that of the variety 'S.22' (□) in terms of (a) leaf elongation and (b) leaf area expansion, but not (c) tillering. The fraction of water-soluble carbohydrates in the roots and stubble at the time of defoliation was 27% higher in Liscate than in S.22. (After Kigel, 1980.)

Sometimes when tillers are differentially defoliated, the plant responds by redirecting carbohydrates to the damaged tillers themselves (e.g. Italian ryegrass—Marshall & Sagar, 1968); but there are some grass species that cut their losses by redirecting carbohydrates to the *less* badly damaged tillers (Ong *et al.*, 1978).

The redistribution of photosynthate can also play a compensatory role during a plant's reproductive phase. Pod loss in soybeans (*Glycine max*), for example, is compensated by an increase in the weight of the individual seeds produced by the remaining pods (Smith & Bass, 1972).

herbivory can lead to an increase in unit leaf rate

Another method of plant compensation after herbivore attack is an increase in the rate of photosynthesis per unit area of surviving leaf (the 'unit leaf rate' or ULR). To see how this works, we should note that plants are composed of parts that are 'sources' (net exporters of photosynthates, usually leaves) and parts that are 'sinks' (net users of photosynthates, like tubers, expanding leaf buds, roots and so on); and the production by the sources usually just matches the requirements of the sinks, but does not exceed them. However, when sources are removed, the ULRs of the surviving sources often increase, and a rough parity between production and requirement is thereby maintained. Thus, when *Agropyron smithii* was experimentally defoliated, there was a 10% increase in the unit leaf rate over the succeeding 10 days at a time when control plants showed a 10% decrease (Painter & Detling, 1981). ULR can also show a compensatory increase when new 'sinks' are created, as when sap-sucking insects like aphids consume vast quantities of carbohydrate. However, when sinks are destroyed by herbivores, ULR may decline (though it is more usual in these cases for the strength of the sink to be maintained by the production of new tissue).

Often, there is compensatory regrowth of defoliated plants when buds that would otherwise remain dormant are stimulated to develop. There is

herbivory can lead to a
reduced death rate of
plant parts

also, commonly, a reduced subsequent death rate of surviving plant parts. This is especially prevalent in plants that exhibit a high natural rate of flower abortion prior to the production of fruit or seed. The wild parsnip (*Pastinaca sativa*), for example, produces a moderate number of seeds on its primary umbels, a larger number on its secondary umbels, but only a small number on its tertiary umbels because of flower abortion prior to seed production (Figure 8.3). However, when it is attacked by larvae of the parsnip webworm (*Depressaria pastinacella*—a moth), although most of the flowers and fruits on the primary umbels are destroyed, there is little effect on the secondary umbels, and a greatly reduced rate of abortion on the tertiary umbels. Overall, therefore, the quantity of seed produced is largely unaltered (Hendrix, 1979).

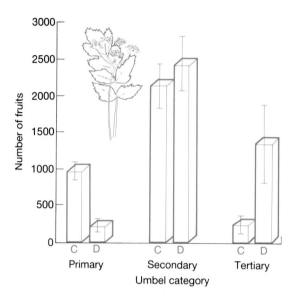

Figure 8.3. Compensation via reduced death rate of flowers. Although most of the flowers and fruits of primary umbels of *Pastinaca sativa* are destroyed by parsnip webworm, damaged plants (D) produce similar numbers of fruits from their secondary umbels and many more fruits from their tertiary umbels than do control plants (C) (means ± s.e.) (From Crawley, 1983, after Hendrix, 1979.)

Clearly, then, there are a number of ways in which individual plants compensate for the effects of herbivory. But *perfect* compensation is rare. Plants are usually harmed by herbivores even though the compensatory reactions tend to counteract the harmful effects. Moreover, as we shall see next, there are many cases where the effects of herbivory are worse than they might at first appear.

8.2.2 Disproportionate effects on plants

One of the most extreme cases where the removal of a small amount of plant has a disproportionately profound effect is ring-barking of trees by goats, ring-barking and
meristem consumption
can kill plants squirrels, rabbits, voles and sheep. The cambial tissues and the phloem are torn away from the woody xylem, and the carbohydrate supply-link between the leaves and the roots is broken. Thus these pests of forestry plantations

often kill young trees while removing very little tissue. Surface-feeding slugs can also do more damage to newly established grass populations than might be expected from the quantity of material they consume (Harper, 1977). The slugs chew through the young shoots at ground level. They leave the felled leaves uneaten on the soil surface, but consume the meristematic region at the base of shoots from which regrowth would occur. They therefore effectively destroy the plant.

herbivores can act as vectors for disease

Herbivores can also have severe effects on plants when they act as vectors for plant pathogens: what they take from the plant is far less important than what they give it! Scolytid beetles feeding on the growing twigs of elm trees, for instance, act as vectors for the fungus that causes Dutch elm disease. This killed vast numbers of elms in the north-eastern United States in the 1960s, and virtually eradicated the elms in southern England in the '70s and early '80s (Strobel & Lanier, 1981). Likewise, the *Cactoblastis* caterpillars which control *Opuntia* cacti in Australia (pp. 585 and 731) exert a large part of their ill-effect by creating feeding scars which are invaded by bacteria that destroy the cactus tissues (Dodd, 1940).

herbivory and plant competition often combine to produce a severe effect

Probably the most widespread reason for herbivory having a more drastic effect than is initially apparent is the interaction between herbivory and plant competition (Whittaker, 1979). Quite moderate levels of grazing may be combined with levels of competition that would otherwise be ineffective to produce a severe, often lethal effect on the target plant. Figure 8.4, for example, shows the results of an experiment in which oats and barley were grown together in two replacement series (Sibma *et al.*, 1964). In one series, the soil was infested with root-feeding nematodes (*Heterodera avenae*) to which the oats were susceptible but the barley resistant; in the other series there was no infestation. In the absence of barley the oats were unaffected by the nematodes; and in the absence of nematodes the oats competed strongly against the barley. But the combination of herbivory from the nematodes and competition from the barley had a profound effect on the oats (Figure 8.4). In a similar vein, it has been observed that when rabbits eat the tops from unshaded young gorse plants (*Ulex europaeus*), the lower buds develop and the plant simply becomes more bushy. But when the gorse is surrounded by dense, tall, competing grasses, the shaded lower buds do not develop and frequently the gorse plant dies (Chater, 1931).

herbivores can have effects that are overtly slight but actually profound

Finally, the effects of herbivory may be underestimated because the herbivores remove sap or xylem without altering the obvious physical structure of the plant. This can be seen, for example, in the effects of the lime aphid (*Eucallipterus tiliae*) on the growth of lime saplings (*Tilia vulgaris*) studied by Dixon (1971). The aphids, which live on the leaves and extract phloem sap through their piercing stylets, can build up extremely heavy infestations: a 14 m lime tree with 58 000 leaves may support more than one million aphids. Yet when Dixon compared uninfested saplings with saplings infested from the time their leaves were half-grown, he found no differences in tree girth, height increment, leaf number or leaf size. However, the aphid infestations led to a virtual cessation of root growth; and this in turn led, over the season, to a total weight increase of infested saplings which was only 8% that of uninfested saplings. Hence the visible effects of the aphids were

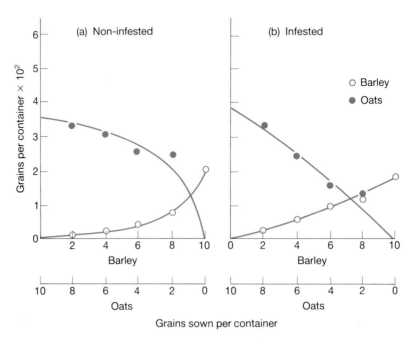

Figure 8.4. The outcome of a replacement series experiment between oats and barley is altered when the soil is infested with a root-feeding nematode, the eelworm *Heterodera avenae,* to which the oats are susceptible (from Sibma *et al.,* 1964).

misleading; infested trees grew normally in volume above the ground, but grew hardly any new roots and put on very little weight. This would doubtless have had repercussions subsequently.

8.2.3 Defensive responses of plants

Plants may also respond to herbivore attack by initiating or increasing their production of defensive structures or chemicals. This production must, to some extent, be costly to the plant, but there are benefits stemming from the lessening of subsequent herbivore attacks. Thus, for example, pines attacked by wood wasps and sawflies show altered phenol metabolism and the appearance of novel defensive chemicals (Thiegles, 1968); artificially wounded potato and tomato plants produce increased levels of protease inhibitors (Green & Ryan, 1972); and the prickles of dewberries on cattle-grazed plants are longer and sharper than those on ungrazed plants nearby (Abrahamson, 1975). Moreover, these responses do generally appear to lessen the effect of subsequent attacks. For instance, when larch trees were defoliated by the larch budmoth, *Zeiraphera diniana,* the survival and adult fecundity of the moths were reduced throughout the succeeding 4 to 5 years as a combined result of delayed leaf production, tougher leaves, higher fibre and resin concentration, and lower nitrogen levels (Baltensweiler *et al.,* 1977). Similarly, snowshoe hares (*Lepus americanus*) show all the adverse signs normally associated with high density (p. 575) when they are fed birch leaves which have regenerated after severe defoliation (Bryant & Kuropat, 1980).

8.2.4 Herbivory and plant survival

repeated defoliation can kill plants

Generally, it is more usual for herbivores to increase a plant's *susceptibility* to mortality than it is for them to kill a plant outright. Repeated defoliation, however, can have a drastic effect. Thus, a single defoliation of oak trees by the gypsy moth (*Lymantria dispar*) led to only a 5% mortality rate (no different from the natural rate in unattacked but crowded woodlands); but three successive heavy defoliations led to mortality rates of up to 80% (Stephens, 1971).

many seedlings are killed by herbivores

However, even a single attack is frequently sufficient to kill a seedling—where the plant is scarcely established and its powers of compensation are least developed. Indeed, as long ago as 1859, Charles Darwin (Figure 8.5) wrote: '. . . on a piece of ground three feet long and two wide, dug and cleared, and where there could be no choking from other plants, I marked all

Figure 8.5. Charles Darwin. (Photograph: Mary Evans Picture Library.)

the seedlings of our native weeds as they came up, and out of the 357 no less than 295 were destroyed, chiefly by slugs and insects.'. Predation of seeds, not surprisingly, has an even more predictably harmful effect on individual plants.

8.2.5 Herbivory and plant growth

Herbivory can stop plant growth; it can have a negligible effect on growth rate; and it can do just about anything in between. When leaves are produced synchronously, the effects of defoliation depend crucially on timing: removal of 75% of the foliage of mature oaks early in the season leads to a 50% loss in wood production, but similar removal later has no noticeable effect on growth (Franklin, 1970; Rafes, 1970). By contrast, in plants where leaf production is continuous, the loss of young leaves may be compensated for by the production of new leaves.

grasses are particularly resistant to grazing

The plants that are most tolerant of grazing are almost certainly the grasses. In most species, the meristem is almost at ground level amongst the basal leaf sheaths, and this major point of growth (and regrowth) is therefore usually protected from grazing. Following defoliation, new leaves are produced either from stored carbohydrates or from the photosynthates of surviving leaves, and new tillers are also often produced.

8.2.6. Herbivory and plant fecundity

smaller plants bear fewer seeds

The effects of herbivory on plant fecundity are, to a considerable extent, reflections of the effects on plant growth: smaller plants bear fewer seeds. But herbivores also influence fecundity in a number of other ways. One of the most common responses to herbivore attack is a delay in flowering. Within a season, this may be particularly detrimental if it leads to a reduced rate of encounter with pollinators late in the season, or an increased chance of being exposed to frosts. For example, regrowth shoots of ragwort (*Senecio jacobaea*) bearing flowers in November are very susceptible to frost damage (Crawley, 1983). Moreover, in longer-lived semelparous species, herbivory frequently delays flowering for a year or more; and this typically increases the longevity of such plants since death almost invariably follows their single burst of reproduction (Chapter 4). Groundsmen can make *Poa annua* on a putting green almost immortal by mowing it at weekly intervals (Crawley, 1983), whereas in natural habitats where it is allowed to flower, it is commonly an annual as its name implies.

herbivory can lead to delayed flowering . . .

. . . which can increase plant longevity

Generally, the timing of defoliation is critical in determining the effect on plant fecundity. If leaves are removed before inflorescences are formed, then the extent to which fecundity is depressed clearly depends on the extent to which the plant is able to compensate. Early defoliation of a plant with sequential leaf production may have a negligible effect on fecundity; but where defoliation takes place later, or where leaf production is synchronous, flowering may be reduced or even inhibited completely. If leaves are removed after the inflorescence has been formed, the effect is usually to increase seed abortion or to reduce the size of individual seeds.

herbivores often destroy reproductive structures directly . . .

The destruction on the plant of the flowers, fruits or seeds themselves can obviously have a much more direct effect on fecundity than mere defoliation. Thus, in a North Wales pasture heavily grazed by cattle, only 15% of the buttercups (*Ranunculus* spp.) that flowered set any seed, whereas 48% set seed in lightly grazed swards (Sarukhán, 1974). And a weevil (*Rhinocyllus conicus*) which feeds on the flowerheads of the nodding thistle (*Carduus nutans*) has been used successfully to control this weed in Virginia, U.S.A., reducing thistle density by 95% (Kok & Surles, 1975).

. . . but much pollen and fruit 'herbivory' is mutualistic

It is important to realize, however, that many cases of 'herbivory' of reproductive tissues are actually mutualistic, i.e. beneficial to both the herbivore *and* the plant (Chapter 13). Animals that 'consume' pollen and nectar usually transfer pollen inadvertently from plant to plant in the process; and there are many fruit-eating animals that also confer a net benefit on both the parent plant and the individual seed within the fruit. Most vertebrate fruit-eaters, in particular, either eat the fruit but discard the seed, or eat the fruit but expel the seed in the faeces. This disperses the seed, rarely harms it, and frequently enhances its ability to germinate. The plant may therefore rely on an animal eating its fruit, and in one case at least this appears to have had profound consequences. Temple (1977) has argued that the tree *Calvaria major* on the island of Mauritius has had no recruitment for the last 300 years because its seeds needed to be processed by the now-extinct dodo (Figure 8.6). Temple fed 17 *Calvaria* seeds to domestic turkeys; and although seven of

Figure 8.6. Dodo, painted by John Savery *c.* 1650, exhibited with remains of a head and foot at University Museum, Oxford. (By permission of the Committee for Scientific Collections, University Museum, Oxford.)

these were crushed in the birds' gizzards, three of the remaining ten subsequently germinated when planted in a nursery. These are probably the first *Calvaria* seeds to germinate in over 300 years.

Insects that attack fruit, on the other hand, are very unlikely to have a beneficial effect on the plant. They do nothing to enhance dispersal, and they may even make the fruit less palatable or unpalatable to vertebrates. However, some large animals that normally kill seeds can also play a part in dispersing them, and they may therefore have at least a partially beneficial effect. There are some 'scatter-hoarding' species, like certain squirrels, that take nuts and bury them at scattered locations; and there are other 'seed-caching' species, like some mice and voles, that collect scattered seeds into a number of hidden caches. In both cases, although many seeds are eaten, the seeds are dispersed, they are hidden from other seed-predators, and a number are never relocated by the hoarder or cacher (Crawley, 1983).

even some seed-eaters benefit plants

Fruit-eating, then, is clearly towards one end of the spectrum of effects that herbivores have on plants. We have seen that with a preponderance of grazers and parasitic consumers, these effects are varied and often subtle—especially so because of the range of responses that the plants display. Yet overall it remains true that herbivores generally harm plants.

yet generally, herbivores are harmful

8.3 The effect of predation on a prey population

Given that the effects of predation on prey individuals are generally harmful (whether the prey be animals or plants), it might seem that the immediate effect of predation on a prey or plant population would also be predictably detrimental. However, these effects are not always so predictable, for one or both of two important reasons: (i) the individuals that are killed (or harmed) are not always a random sample of the population as a whole; and (ii) the individuals that escape predation often exhibit reactions which compensate for the loss of those that are killed.

Errington (1946) made a long and intensive study of populations of the musk-rat (*Ondatra zibethicus*) in the north-central United States. He took censuses, recorded mortalities and movements, followed the fates of individual litters, and was particularly concerned with predation on the musk-rat by the mink (*Mustela vison*). He found that adult musk-rats that were well established in a breeding territory were largely free from mink predation; but those that were wandering without a territory, or exposed by drought, or injured in intraspecific fights, were very frequently preyed upon. Thus, the individuals that were killed were those that were least likely to survive and reproduce. Similar results have been obtained for predation on other vertebrates. Those most likely to succumb are the young, the homeless, the sick and the decrepit. Hence, the effects of predation on the prey population will be far less than might be expected.

predatory attacks are often directed at the weakest prey

Similar patterns may also be found in plant populations. Mortality of mature *Eucalyptus* trees in Australia, resulting from defoliation by the sawfly *Perga affinis affinis,* was restricted almost entirely to weakened trees on poor sites, or to trees that had suffered from root damage or from altered drainage following cultivation (Carne, 1969).

The impact of predation may also be limited by compensatory reactions among the survivors—this being most commonly the result of reduced intraspecific competition. Thus in an experiment in which large numbers of wood pigeons (*Columba palumbus*) were shot, shooting failed to increase the overall level of winter mortality, and stopping the shooting led to no increase in pigeon abundance (Murton *et al.*, 1974). This was because the number of surviving pigeons was determined ultimately not by shooting but by food availability, and so when shooting reduced density, there were compensatory reductions in intraspecific competition and in natural mortality, as well as density-dependent immigration of birds moving in to take advantage of unexploited food.

the effects of predation are often ameliorated by reduced competition . . .

Indeed, whenever density is high enough for intraspecific competition to occur, the effects of predation *on the population* will be ameliorated by the consequent reductions in intraspecific competition. This can be seen very clearly by reconsidering the n-shaped curves of net recruitment or net productivity against density that were discussed in section 6.5. Net recruitment is low when there are few reproducing individuals, just as net plant productivity is low when there are few leaves (a small leaf area index). But net recruitment is also low when there are many, crowded individuals; and plant productivity is low where there is a large leaf area index and a great deal of shading (Figure 8.7). Thus, if a predator or a herbivore attacks a population that lies on the right-hand limb of such a curve, density falls but net recruitment or net productivity rises (Figure 8.7). This hastens the rate at which the population recovers.

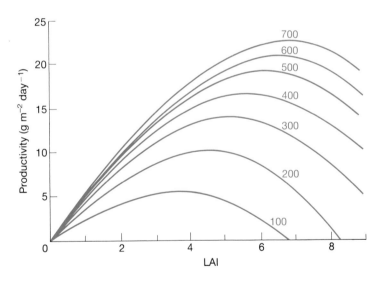

Figure 8.7. Net productivity is an n-shaped function of leaf area index (LAI) in subterranean clover. Optimum LAI increases with light intensity (joules cm^{-2} d^{-1}) since light penetrates deeper into the canopy and a higher fraction of the leaves remain above the compensation point. (From Crawley, 1983, after Black, 1964.)

The effect is probably most important in populations of plants (and especially grasses) where it is surviving plant parts and not only surviving individuals that compensate. Thus even when defoliation has a disastrous effect on an individual tiller or even a whole individual genet, it may have no serious effect on the yield of the sward as a whole. Indeed, if defoliation increases the population's net productivity, then this may increase the quantity of photosynthate available for the production and filling of seeds. It has been observed that autumn grazing of wheat, rye and oats can improve subsequent seed production (Sprague, 1954).

Compensation, however, is by no means always perfect. When 75% of emerging adults were removed each day from experimental populations of blowflies (*Lucilia cuprina*), there was some compensation but nonetheless a 40% reduction in population size (Nicholson, 1954b). And when a subterranean clover population was defoliated below a leaf area index of 4.5 (on the *left*-hand limb of Figure 8.7), there was drastic reduction in the rate of leaf production. Typically, therefore, predation leads to compensatory decreases in intraspecific competition. But equally typically, those powers of compensation are limited (especially in low-density prey or plant populations). These topics will be considered further in section 10.8 in the context of harvesting. For now, though, it is worth noting that man relies on the compensatory powers of populations to allow him to take repeated harvests; but the limitations in these compensatory powers may bring a heavily harvested population to (or beyond) the brink of extinction.

Compensation within a population is not always the result of reduced intraspecific competition. A reduction in one type of predation may lead to

... but compensation is usually not perfect

Table 8.1. The fates, in percentage terms, of Douglas fir seeds sown in open and screened plots. The reductions in bird and rodent predation in screened plots is compensated for by increases in attacks by insects and fungi. (From Lawrence & Rediske, 1962.)

	Plots	
	Open	Screened
PRE-GERMINATION PERIOD		
Losses due to:		
fungi	19.0	20.1
insects	9.5	12.8
rodents	14.0	1.8
birds	4.1	0.9
unknown	6.8	1.8
Total loss, pre-germination	53.4	37.4
Seeds remaining	46.6	62.6
GERMINATION PERIOD		
Non-germination due to:		
fungal attack	13.1	17.3
seed dormancy	12.7	10.5
Total not germinating	25.8	27.8
Seedlings	20.8	34.8
POST-GERMINATION PERIOD (1 YEAR AFTER GERMINATION)		
Mortality due to:		
fungi	5.4	12.8
other causes	7.3	4.6
Total mortality	12.7	17.4
Seedlings surviving	8.2	17.4

a density-dependent, compensatory increase in another type of predation. Table 8.1, for example, shows the results of an experiment in which seeds of Douglas fir (*Pseudotsuga menziesii*) were monitored both in open plots and in plots supposedly screened from vertebrate herbivores (Lawrence & Rediske, 1962). The screens were largely effective in that predation by birds and rodents was greatly reduced. However, there were compensatory increases in the attacks by insects and especially fungi on the seed and seedling stages, and the overall rate of survival was relatively little changed. Once again, there was compensation which lessened, rather than negated, the effects of predation.

8.4 The effects of consumption on consumers

The beneficial effects that food has on individual predators are not difficult to imagine. Generally speaking, an increase in the amount of food consumed leads to increased rates of growth, development and birth, and decreased rates of mortality. This, after all, is implicit in any discussion of intraspecific competition amongst consumers (Chapter 6): high densities, implying small amounts of food per individual, lead to low growth rates, high death rates and so on. Similarly, many of the effects of migration previously considered (Chapter 5) reflect the responses of individual consumers to the distribution of food availability. However, there are a number of ways in which the relationships between consumption rate and consumer benefit can be more complicated than they appear.

In the first place, all animals require a certain amount of food simply for maintenance (Figure 8.8); and unless this threshold is exceeded the animal will be unable to grow or reproduce, and will therefore be unable to contribute to future generations. In other words, low consumption rates, rather than leading to a small benefit to the consumer, simply alter the rate at which the consumer starves to death.

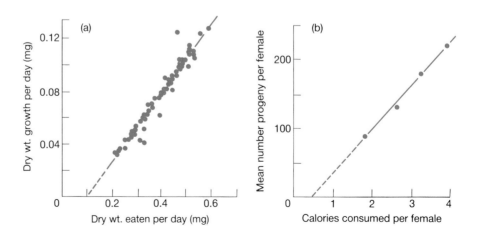

Figure 8.8. Prey thresholds for predators. (a) Growth in the spider *Linyphia triangularis* (Turnbull, 1962). (b) Reproduction in the water flea *Daphnia pulex* var. *pulicaria* (Richman, 1958). (After Hassell, 1978.)

At the other extreme, the birth, growth and survival rates of consumers cannot be expected to rise indefinitely as food availability is increased. Rather, the consumers become *satiated*. Consumption rate eventually reaches a plateau, where it becomes independent of the amount of food available (Figure 8.9), and benefit to consumers therefore also reaches a plateau. Thus there is a limit to the amount that a particular consumer population can eat, a limit to the amount of harm that it can do to its prey population, and a limit to the extent by which the consumer population can increase in size.

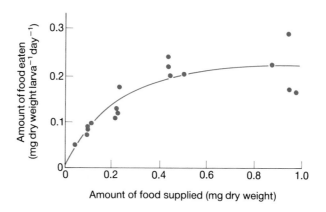

Figure 8.9. As the amount of food (ragwort leaves, *Senecio jacobaea*) supplied to first-instar cinnabar moth caterpillars (*Tyria jacobaeae*) increases, the amount consumed approaches and then reaches a plateau (from Crawley, 1983, after Zahirul Islam, 1981).

One important case in which whole populations of consumers are satiated is provided by the many tree species that have *mast years*—occasional years in which there is the synchronous production of a large volume of seed by most conspecific trees in the same geographic area, with a dearth of seeds produced in the years in between (Figure 8.10). This is seen particularly often in tree species that suffer generally high intensities of seed predation (Silvertown, 1980), and it is therefore especially significant that the chances of escaping seed predation are typically much higher in mast years than in other years. The individual predators of seeds are satiated in mast years, and the populations of predators cannot increase in abundance rapidly enough to exploit the glut. By the time they *have* increased (usually the following year), glut has given way to famine.

mast years and the
satiation of seed
predators

On the other hand, the production of a mast crop makes great demands on the internal resources of a plant. In a mast year a spruce tree averages 38% less annual growth than in other years, and the annual ring increment in forest trees may be reduced by as much during a mast year as by a heavy attack of defoliating caterpillars. The years of seed famine are therefore essentially years of plant recovery. Finally, the synchrony of masting between individuals is obviously important for the satiation of predators, and it is significant that mast years have often been correlated with climatic variables.

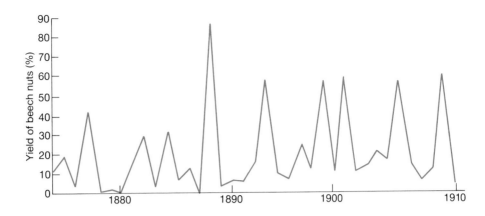

Figure 8.10. Masting: the variation from year to year in the crop of beech nuts (*Fagus sylvatica*) at Rohrbrunn (from Rohmeder, 1967).

Apart from illustrating the potential importance of predator satiation, the example of masting highlights a further point which relates to time-scales. The seed predators are unable to extract the maximum benefit from (or do the maximum harm to) the mast crop because their generation times are too long. A hypothetical seed predator that could pass through several generations during a season would be able to increase exponentially and explosively on the mast crop and destroy it. Generally speaking, consumers with relatively short generation times tend to track fluctuations in the quantity or abundance of their food or prey, whereas consumers with relatively long generation times take longer to respond to increases in prey abundance, and longer to recover when reduced to low densities.

> the numerical response of a consumer is limited by its generation time

Chapter 3 stressed that the abundance or quantity of prey or food consumed may be less important than its quality. In fact, food quality, which has both positive aspects (like the concentrations of nutrients) and negative aspects (like the concentrations of toxins), can only sensibly be defined in terms of the effects of the food on the animal that eats it; and this is particularly pertinent in the case of herbivores. It was mentioned on p. 296, for instance, that the survival and fecundity of larch budmoths and snowshoe hares is as much dependent on the quality of their food as on how much of it they consume. Along similar lines, Sinclair (1975) examined the effects of grass quality (protein content) on the survival of wildebeest in the Serengeti of Tanzania. Despite selecting protein-rich plants and plant parts (Figure 8.11a), the wildebeest consumed food in the dry season which contained well below the level of protein necessary even for maintenance (5–6% crude protein); and to judge by the depleted fat reserves of dead males (Figure 8.11b), this was an important cause of mortality. Moreover, it is highly relevant that the protein requirements of females during late pregnancy and lactation (December–May in the wildebeest) are three to four times higher than the normal (Agricultural Research Council, 1965). It is therefore clear that the shortage of high-quality food (and not just food shortage *per se*) can have a drastic effect on the growth, survival and fecundity of a consumer.

> food quality rather than quantity can be of paramount importance

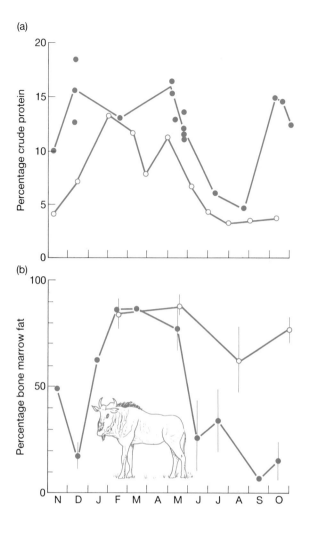

(a)

(b)

Figure 8.11. (a) The quality of food measured as percentage crude protein available to (○) and eaten by (●) wildebeest in the Serengeti during 1971. During the dry season, this fell below the level for maintenance of nitrogen balance (5–6% crude protein), despite selection. (b) The fat content of the bone marrow of the live male population (○) and those found dead from natural causes (●). Vertical lines, where present, show 95% confidence limits. (After Sinclair, 1975.)

In the case of herbivores especially, it is possible for an animal to be apparently surrounded by its food while still experiencing a food shortage. We can see the problem if we imagine that we ourselves are provided with a perfectly balanced diet—diluted in an enormous swimming-pool. The pool contains everything we need, and we can see it there before us, but we may very well starve to death before we can drink enough water to extract enough nutrients to sustain ourselves. In a similar fashion, herbivores may frequently be confronted with a pool of protein that is so dilute that they have difficulty processing enough material to extract what they need. Outbreaks of herbivorous insects may then be associated with rare elevations in the protein concentration of their food plants, perhaps associated with unusually dry or, conversely, unusually waterlogged conditions as suggested by White (1978, 1984). Certainly, it is clear overall, firstly, that the effects of consumption on a consumer are not simply and unqualifiedly beneficial, secondly that the effects on a consumed population are not simply and unqualifiedly harmful, but lastly that consumption is *never* beneficial to the individual consumed.

The Behaviour of Predators

9.1 Introduction

In this chapter, we discuss the behaviour of predators. We shall examine where they feed, what they feed on, how they are affected by other predators and how they are affected by the density of their prey. These topics are of interest in their own right, but they are also relevant in two other, broader contexts. First, foraging is just one aspect of animal behaviour which is being increasingly subjected to the scrutiny of evolutionary biologists, within the general field of what is usually referred to as 'behavioural ecology'. The aim, put simply, is to try to understand how natural selection has favoured particular patterns of behaviour in particular circumstances (how, behaviourally, organisms match their environment). In this chapter, we deal directly with this problem in the sections on 'optimal foraging' (9.3 and 9.11), but the general evolutionary approach will underlie much of what is said throughout the chapter.

Secondly, the various aspects of predatory behaviour can be seen as components that combine to influence the population dynamics of both the predator itself and its prey. The population ecology of predation is dealt with much more fully in the next chapter. Nonetheless, it will be useful at various points in the present chapter to indicate, in general terms, the way in which individual behaviour will tend to affect population dynamics. As Chapter 6 made clear, if predation is such that one or other of the populations is subjected to density-dependent increases in mortality (or decreases in birth rate), then this will tend to regulate the size of that population within certain limits, i.e. it will tend to stabilize the dynamics of the population. This in turn will tend to stabilize the population dynamics of the interaction as a whole (the populations will be persistent and show relatively little variation in abundance). Conversely, if there is 'inverse density-dependence', such that mortality decreases (or birth rate increases) with increasing density, then this will tend to destabilize the dynamics of the interaction. Predatory behaviour can clearly have a significance beyond its effects on the individuals concerned.

9.2 The widths and compositions of diets

Consumers can be classified as either monophagous (feeding on a single prey type), oligophagous (few prey types) or polyphagous (many prey types). Often, an equally useful distinction is between specialists (broadly, mono-

predatory behaviour as behavioural ecology

predatory behaviour and population dynamics

the range and classification of diet widths

phages and oligophages) and generalists (polyphages). Herbivores, parasitoids and true predators can all provide examples of monophagous, oligophagous and polyphagous species. But the distribution of diet widths differs amongst the various types of consumer. True predators with specialized diets do exist; for instance, the Everglades kite (*Rostrahamus sociabilis*) feeds almost entirely on snails of the genus *Pomacea*. But most true predators have relatively broad diets. Parasitoids, on the other hand, are typically specialized and are often monophagous, while herbivores are well represented in all categories. However, while grazing and 'predatory' herbivores typically have broad diets, 'parasitic' herbivores are very often highly specialized. Janzen (1980), for instance, examined 110 species of beetle that feed as larvae inside the seeds of dicotyledonous plants in Costa Rica, and found that 83 attacked only one plant species, 14 attacked only two, nine attacked three, two attacked four, one attacked six and one attacked eight. And this was in spite of there being 975 plant species present in the area.

9.2.1 Food preferences

It must not be imagined that polyphagous and oligophagous species are indiscriminate in what they choose from their acceptable range. On the contrary, some degree of preference is almost always apparent. Technically, an animal is said to exhibit a preference for a particular type of food when the proportion of that type in the animal's diet is higher than its proportion in the animal's environment. To measure food preference in nature, therefore, it is necessary not only to examine the animals' diet (usually by the analysis of gut contents), but also to assess the 'availability' of different food types. And ideally, this should be done not through the eyes of the observer (i.e. not by simply sampling the environment), but through the eyes of the animal itself.

preference is defined by comparing diet with 'availability'

Table 9.1. A feeding preference: the percentages of various planted trees browsed by deer (after Horton 1964).

	White pine	Red pine	Jack pine	White spruce
Winter 1956–57	31	19	84	0
Winter 1958–59	9	1	48	0
Winter 1960–61	17	0	70	0

The exact quantification of preference is therefore fraught with difficulties—but it is generally much easier at least to establish that a preference does exist. For instance, the results of an accidental field experiment, in which deer broke into a tree plantation, provide a nice example. The plantation contained equal numbers of four species arranged at random: white pine, red pine, jack pine and white spruce. As Table 9.1 shows, the deer, with free access to all four species, exhibited a fairly consistent preference for jack pine, followed by white pine, with red pine being only lightly browsed, and white spruce ignored (Horton, 1964).

9.2.2 Ranked and balanced preferences

A food preference can be expressed in two rather different contexts. There can be a preference for items which are *the most valuable* amongst those available, or for items that provide an *integral part* of a mixed and balanced diet. These will be referred to as ranked and balanced preferences respectively.

Ranked preferences are usually seen most clearly amongst carnivores. Figure 9.1, for instance, shows two examples in which carnivores actively

ranked preferences predominate when food items can be classified on a single scale

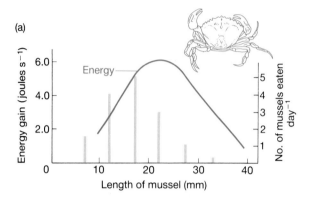

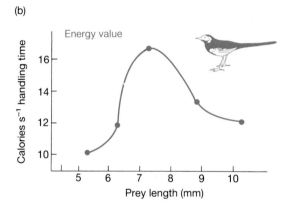

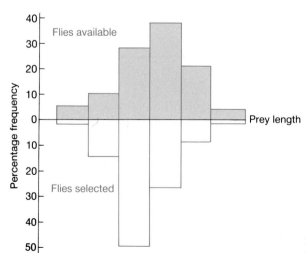

Figure 9.1. Predators eating 'profitable' prey, i.e. predators showing a preponderance in their diet of those prey items which provide them with the most energy. (a) Crabs eating mussels (Elner & Hughes, 1978). (b) Pied wagtails eating flies (Davies, 1977). (After Krebs, 1978.)

selected prey items which were the most profitable in terms of energy intake per unit time spent handling prey. Results such as these reflect the fact that a carnivore's food often varies little in composition, but may vary in size or accessibility. This allows a single measure (like 'energy gained per unit handling time') to be used to characterize food items, and it therefore allows food items to be ranked. Figure 9.1, in other words, shows consumers exhibiting an active preference for food of a high rank.

For many consumers, however, especially herbivores and omnivores, no simple ranking is appropriate, since none of the available food items matches the nutritional requirements of the consumer. These requirements can therefore only be satisfied either by eating large quantities of food, and eliminating much of it in order to get a sufficient quantity of the nutrient in most limited supply, or by eating a combination of food items that between them match the consumer's requirements. In fact, many animals exhibit both sorts of response. They select food which is of generally high quality (so the proportion eliminated is minimized), but they also select items to meet specific requirements. Sheep and cattle, for instance, show a preference for high-quality food. They select leaves in preference to stems, and green matter in preference to dry or old material; and compared to the forage as a whole, the selected material is usually higher in nitrogen, phosphorus, sugars and gross energy, and lower in fibres (Arnold, 1964). In fact, reviews suggest that *all* generalist herbivores show rankings in the rate at which they eat different food plants when given a free choice in experimental tests (Crawley, 1983).

On the other hand, a balanced preference is also frequently apparent. The plate limpet, *Acumaea scutum,* for instance, selects a diet of two species of encrusting microalgae which contains 60% of one species and 40% of the other, almost irrespective of the proportions in which they are offered (Kitting, 1980); while caribou, which survive on lichen through the winter, develop a mineral deficiency by the spring which they overcome by drinking sea-water, eating urine-contaminated snow and gnawing shed antlers (Staaland *et al.,* 1980). Moreover, we have only to look at ourselves to see an example in which 'performance' is far better on a mixed diet than on a pure diet of even the 'best' food.

There are also two other important reasons why a mixed diet may be favoured. In the first place, consumers may accept low-quality items simply because, having encountered them, they have more to gain by eating them (poor as they are) than they would 'gain' by ignoring them and continuing to search. This is discussed in more detail in section 9.3. In the second place, consumers may benefit from a mixed diet because each food-type contains a different undesirable toxic chemical. A mixed diet would then keep the concentrations of all of these chemicals within acceptable limits. It is certainly the case that toxins can play an important role in food preference. One study, for instance, involved the winter food of a variety of arctic animals: three species of ptarmigan, three grouse, capercaillie, two kinds of hare, and the moose (Bryant & Kuropat, 1980). In each case the conclusions were the same: animals ranked their foods on neither energy nor nutrient content. Instead, preference was strongly and negatively correlated with the concentrations of certain toxins.

but many consumers exhibit a combination of ranked and balanced preferences

mixed diets can be favoured for a variety of reasons

9.2.3 Switching

switching involves a preference for food types that are common

The preferences of many consumers are fixed, i.e. they are maintained irrespective of the relative availabilities of alternative food types. But many others *switch* their preference, such that food items are eaten disproportionately often when they are common and are disproportionately ignored when they are rare. The two types of preference are contrasted in Figure 9.2. Figure 9.2a shows the fixed preference exhibited by predatory shore snails when they were presented with two species of mussel prey at a range of proportions (Murdoch, 1969). The line in Figure 9.2a has been drawn on the assumption that they exhibited the same preference at all proportions. This assumption is clearly justified: irrespective of availability, the predatory snails showed the same marked preference for the thin-shelled, less protected *Mytilus edulis*, which they could exploit more effectively. By contrast, Figure 9.2b shows what happened when guppies (a species of fish) were offered a choice between fruit-flies and tubificid worms as prey (Murdoch, Avery & Smith, in Murdoch & Oaten, 1975). The guppies clearly switched their preference, and consumed a disproportionate number of the more abundant prey type.

the circumstances when switching arises

There are a number of situations in which switching can arise. Probably the most common is where different types of prey are found in different microhabitats, and the consumers concentrate on the most profitable microhabitat. This was the case for the guppies in Figure 9.2b: the fruit-flies floated at the water surface while the tubificids were found at the bottom.

Switching can also occur when the consumer becomes more efficient or more successful in dealing with the more abundant food-type. For instance, switching occurs in the carnivorous water bug *Notonecta glauca* (Figure 9.2c) because it becomes more successful in its attacks on freshwater isopods, the more experience it has of them (i.e. the more abundant they are compared to the alternative prey, mayflies) (Lawton *et al.*, 1974). In a similar way, mallard ducks experience changes in their digestive physiology such that their efficiency of utilization of plants improves, the more abundant the plants are (Miller, 1975).

Switching may also be the result of consumers developing 'specific search images' for abundant foods (Tinbergen, 1960). Such search images are thought to result in consumers (particularly vertebrates) concentrating on their 'image' prey to the relative exclusion of their non-image prey; and since these search images develop as a result of previous experience, they are more likely to develop for a common food type than for an uncommon one.

switching and the individual

Interestingly, switching in a population often seems to be a consequence not of individual consumers gradually changing their preference, but of the proportion of specialists changing. This is illustrated by a study of switching in wood-pigeons (*Columba palumbus*) feeding on maple peas and tick beans (Murton, 1971). When the two were equally abundant there was a slight preference for maple peas; but when there were 82% tick beans on offer, the birds switched to an average of 91% tick beans in their diet. This average, however, included two birds that specialized on the rarer maple peas, taking only 5% and 0% tick beans. Figure 9.2d shows a comparable pattern for the guppies fed on fruit-flies and tubificids: when the prey types were equally

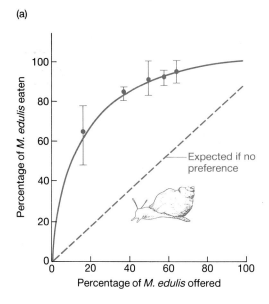

(a)

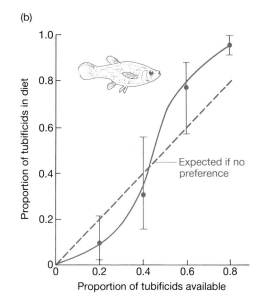

(b)

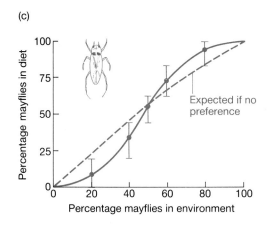

(c)

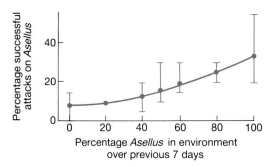

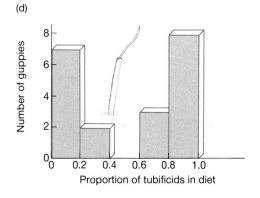

(d)

Figure 9.2. Switching. (a) A lack of switching: snails exhibiting a consistent preference amongst the mussels *Mytilus edulis* and *M. californianus* irrespective of their relative abundance (means±s.e.) (after Murdoch & Oaten, 1975). (b) Switching by guppies fed on tubificids and fruit-flies: they take a supra-proportional amount of whichever prey type is the more available (means and total ranges) (after Murdoch & Oaten, 1975). (c) Switching by *Notonecta* fed on mayfly larvae and

Asellus (above), and also (below) the basis for this switching: the *Notonecta* become more successful in their attacks on *Asellus* the more previous experience they have of them (means and total ranges) (after Lawton *et al.*, 1974). (d) Preferences shown by the individual guppies in (b) when offered equal amounts of the two prey types: individuals were mostly specialists on one or other type.

abundant, individual guppies were not generalists. Rather, there were approximately equal numbers of fruit-fly and tubificid specialists.

9.2.4 Diet width and evolution

A first step towards understanding the variations in diet width can be made by considering the evolution of predators and their prey. The influence that predators and prey have on one another's evolution can be seen in prey attributes like the distasteful or poisonous leaves of many plants, in the spines of hedgehogs and in the camouflage colouration of many insects (Figure 9.3); and it can be seen in such predator attributes as the stout ovipositors of wood wasps, the multi-chambered stomachs of cattle, and the silent approach and sensory excellence of owls. Such specialization makes it clear, however, that no predator can possibly be capable of consuming all types of prey. Simple design constraints prevent shrews from eating owls (even though shrews are carnivores), and they prevent humming-birds from eating seeds. Evolution, therefore, often leads to diet restriction.

Fig. 9.3. Angle shades moth (*Phlogophora meticulosa*) camouflaged against a background of dead oak leaves. (Photograph: Heather Angel.)

the advantages of monophagy

This restriction is typically taken furthest where the consumer lives in intimate association with its prey or host, and where an individual consumer is linked to an individual prey. Parasitoids and parasitic herbivores often specialize on one or a very few host species, because they live in such intimate association with their hosts (and this is even truer of parasites, see Chapter 12). Their whole life-style and life cycle is finely tuned to that of their host, and this precludes their being finely tuned to other host species. In fact, so long as a prey or plant species remains abundant, accessible and predictable, selection will favour finer and finer specialization towards monophagy. This improves the efficiency of the specialist species, and frees it (at least relatively) from interspecific competition.

On the other hand, polyphagy also has definite advantages. A consumer can construct a balanced diet by selecting a range of different foods, and it can maintain this balance by varying its preference to suit altered circumstances. It can also avoid consuming large quantities of a toxin produced by one of its food types. Moreover, for a polyphagous consumer, food is easy to find and search costs are typically low, and an individual is unlikely to starve because of fluctuations in the abundance of one type of food.

Overall, then, evolution may broaden or restrict diets. Where prey exert evolutionary pressures demanding specialized morphological or physiological responses from the consumer, restriction is often taken to extremes. But where consumers feed on items which are individually inaccessible or unpredictable or lacking in certain nutrients, the diet often remains broad.

coevolution: predator–prey arms races?

An appealing and much-discussed idea is that particular pairs of predator and prey species have not only evolved but have *coevolved*. In other words, there has been an evolutionary 'arms race', whereby each improvement in predatory ability has been followed by an improvement in the prey's ability to avoid or resist the predator, which has been followed by a further improvement in predatory ability, and so on. If this occurred, it would certainly be an additional force in favour of diet restriction. At present, however, hard evidence for predator–prey or plant–herbivore coevolution is more or less non-existent (Futuyma & Slatkin, 1983).

9.3 The optimal foraging approach to diet width

Design constraints notwithstanding, most animals have the potential to consume a wider range of foods than they actually choose. In other words, evolution usually gives rise to foraging strategies in which animals consume a narrower range of food types than they are morphologically capable of consuming. In trying to understand what determines a consumer's actual diet within its wide potential range, ecologists have increasingly turned to optimal foraging theory.

certain assumptions are inherent in optimal foraging theory

The aim of optimal foraging theory is to predict the foraging strategy to be expected under specified conditions, and it generally makes such predictions on the basis of a number of assumptions, namely the following.
(i) The pattern of foraging behaviour that will be exhibited by present-day animals will be the one that has been favoured by natural selection—and this will be the one that most enhances an animal's fitness.
(ii) High fitness is achieved by a high net rate of energy intake (i.e. gross energy intake minus the energetic costs of obtaining that energy).
(iii) The animals under consideration are being observed in an environment to which their foraging behaviour is suited, i.e. it is a natural environment very similar to that in which they evolved, or an experimental arena similar in important respects to the natural environment.

These assumptions will not always be acceptable. First, other aspects of an organism's behaviour may influence fitness more than optimal foraging does. For example, there may be such a premium on the avoidance of predators that animals forage at a location and at a time where the likelihood of predation is lower, and in consequence gather their food less efficiently than is theoretically possible. Secondly, and just as important, for many consumers (particularly herbivores and omnivores) the efficient gathering of energy may be less critical than some other dietary constituent (e.g. nitrogen), or it may be of prime importance for the forager to consume a mixed and balanced diet. In such cases, the value of existing optimal foraging theory is limited. However, in circumstances where the energy maximization premise can be expected to apply, optimal foraging theory offers a powerful insight into the significance of the foraging 'decisions' that predators make (for reviews see Krebs, 1978; Townsend & Hughes, 1981; Krebs *et al.*, 1983).

the theoreticians are omniscient mathematicians—but the foragers need not be

Typically, optimal foraging theory makes predictions about foraging behaviour based on mathematical models constructed by ecological theoreticians who are omniscient ('all-knowing') as far as their model ecosystem is concerned. The question therefore arises—is it necessary for a real forager to be equally omniscient and mathematical, if it is to adopt the appropriate, optimal strategy? The answer is 'no'. The theory simply says that if there is a forager that in some way (in *any* way) manages to do the right thing in the right circumstances, then this forager will be favoured by natural selection. It will exploit food and obtain energy economically, and it will therefore have sufficient time and energy for successful reproduction. Ultimately, it should leave more offspring than other foragers; and if its abilities are inherited, these should spread, in evolutionary time, throughout the population.

mechanistic models complement optimal foraging theory

Optimal foraging theory does not specify precisely how the forager should make the right decisions, and it does not require the forager to carry out the same calculations as the modeller. Instead there is another group of 'mechanistic' models designed to account for the behaviour of the forager in terms of fixed action patterns and innate responses to environmental cues. The mechanistic models, therefore, attempt to show how a forager, given that it is *not* omniscient, might nevertheless manage to respond by 'rules of thumb' to limited environmental information and thereby exhibit a strategy which is favoured by natural selection. But it is optimal foraging theory that predicts the nature of the strategy that should be so favoured.

9.3.1 The diet width model—'searching and handling'

MacArthur and Pianka (1966), in the first paper on optimal foraging theory, sought to understand the determination of diet width within a habitat. Subsequently, their model has been developed into a more rigorous algebraic form, notably by Charnov (1976a). MacArthur and Pianka argued that to obtain food, any predator must expend time and energy, first in *searching* for its prey and then in *handling* it (i.e. pursuing, subduing and consuming it). Searching is bound to be directed, to some extent, towards particular prey types, but while searching, a predator is nevertheless likely to encounter a wide variety of food items. MacArthur and Pianka therefore saw diet width as depending on the responses of predators once they had encountered prey. Generalists pursue (and may then subdue and consume) a large proportion of the prey they encounter; specialists continue searching except when they encounter prey of their specifically preferred type.

to pursue or not to pursue?

The 'problem' for any forager is this: if it is a specialist, then it will only pursue profitable prey items, but it may expend a great deal of time and energy searching for them; whereas if it is a generalist, it will spend relatively little time searching, but it will pursue both unprofitable and profitable types of prey. An optimal forager should balance the pros and cons so as to maximize its overall rate of energy intake. MacArthur and Pianka expressed the problem as follows. Given that a predator already includes a certain number of profitable items in its diet, should it expand its diet (and thereby decrease its search time) by including the next-most-profitable item as well?

We can refer to this 'next-most-profitable' item as the ith item. E_i/h_i is then the profitability of the item, where E_i is its energy content, and h_i its handling time. In addition, $\bar{E}/\bar{h}$ is the average profitability of the 'present' diet (i.e. one that includes all prey types that are more profitable than i, but does not include prey type i itself), and $\bar{s}$ is the average search time for the present diet. If a predator does pursue a prey item of type i, then its expected rate of energy intake is E_i/h_i. But if it ignores this prey item, while pursuing all those that are more profitable, then it can expect to search for a further $\bar{s}$, following which its expected rate of energy intake is $\bar{E}/\bar{h}$. The total time spent in this latter case is $\bar{s}+\bar{h}$, and so the overall expected rate of energy intake is $\bar{E}/(\bar{s}+\bar{h})$. The most profitable, optimal strategy for a predator will be to pursue the ith item if, and only if,

$$E_i/h_i \geq \bar{E}/(\bar{s}+\bar{h}). \tag{9.1}$$

In other words, a predator should continue to add increasingly less profitable items to its diet as long as condition 9.1 is satisfied (i.e. as long as this increases its overall rate of energy intake). This will serve to maximize its overall rate of energy intake, $\bar{E}/(\bar{s}+\bar{h})$.

This optimal diet model leads to a number of predictions.

searchers should be generalists

(1) Predators with handling times that are typically short compared to their search times should be generalists, because in the short time it takes them to handle a prey item that has already been found, they can barely begin to search for another prey item. (In other words, h_i is small and E_i/h_i therefore large for a wide range of prey types, whereas $\bar{s}$ is large and $\bar{E}/(\bar{s}+\bar{h})$ therefore small even for broad diets). This prediction seems to be supported by the broad diets of many insectivorous birds that 'glean' foliage. Searching is always moderately time-consuming: but handling the minute, stationary insects takes negligible time and is almost always successful. A gleaning bird, therefore, has something to gain but virtually nothing to lose by consuming an item once found, and overall profitability is maximized by a broad diet.

handlers should be specialists

(2) By contrast, predators with handling times that are long relative to their search times should be specialists. This can be seen by noting that if $\bar{s}$ is always small, then $\bar{E}/(\bar{s}+\bar{h})$ is not much smaller than $\bar{E}/\bar{h}$. Thus maximizing $\bar{E}/(\bar{s}+\bar{h})$ is the same as maximizing $\bar{E}/\bar{h}$, which is achieved, clearly, by including only the most profitable items in the diet. Lions, for instance, live more or less constantly in sight of their prey so that search time is negligible:

handling time, on the other hand, and particularly pursuit time, can be long (and very energy-consuming). Lions consequently specialize on those prey that can be pursued most profitably: the immature, the lame and the old.

specialization should be greater in more productive environments

(3) Other things being equal, a predator should have a broader diet in an unproductive environment (where prey items are relatively rare and $\bar{s}$ is relatively large) than in a productive environment (where $\bar{s}$ is smaller). This prediction is broadly supported by the two examples shown in Figure 9.4: in experimental arenas, both bluegill sunfish (*Lepomis macrochirus*) and great tits (*Parus major*) had more specialized diets when prey density was higher.

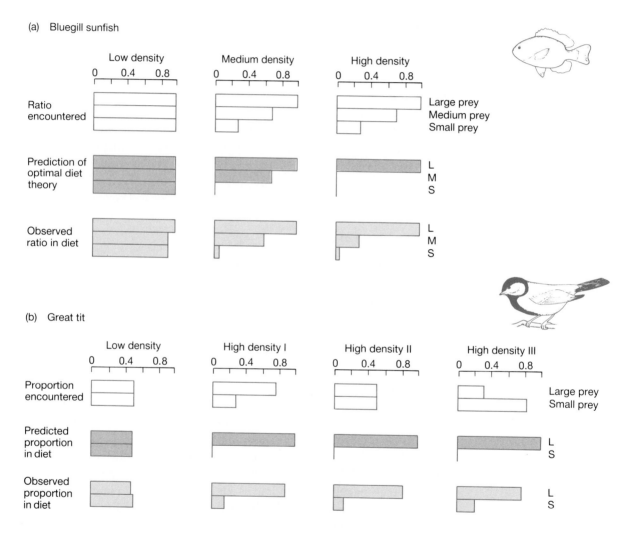

Figure 9.4. Two studies of optimal diet choice showing a clear but limited correspondence with the predictions of Charnov's (1976a) optimal diet model. Diets are more specialized at high prey densities, but more low-profitability items are included than predicted by the theory. (a) Bluegill sunfish (Werner & Hall, 1974) preying on different size classes of *Daphnia*: the histograms show ratios of encounter rates with each size class at three different densities, together with the predicted and observed ratios in the diet. (b) Great tits preying on large and small pieces of mealworm (Krebs *et al.*, 1977). The histograms in this case refer to the proportions of the two types of item taken. (After Krebs, 1978.)

(4) Perhaps the most interesting prediction arising from the model can be seen by re-considering the condition (9.1) for including the ith item in the diet. It depends on the profitability of the ith item (E_i/h_i), and it depends on the profitabilities of the items already in the diet ($\bar{E}/\bar{h}$). It depends too on the search times for items already in the diet ($\bar{s}$), and thus on their abundance. But it does *not* depend on the search time for the ith item, s_i, nor therefore on its abundance or the rate at which the predator encounters it. In other words, predators should specialize when profitable food-types are common and/or differences in profitability are great; and they should be indiscriminate when profitable types are rare and/or differences in profitability are slight. But they should ignore insufficiently profitable food-types *irrespective* of their abundance.

the abundance of
unprofitable prey types
is irrelevant

Re-examining the examples in Figure 9.4, we can see that these both refer to circumstances in which the optimal diet model predicts that the least profitable items should indeed be ignored completely. The foraging behaviour was very similar to this prediction, but in both cases the animals consistently took slightly more than expected of the less profitable food types. In fact, this sort of discrepancy has been uncovered consistently, and there are a number of reasons why it may occur (Krebs & McCleery, 1984). We can summarize these reasons crudely by noting that the animals are not omniscient. It is important to realize, however, that the optimal diet model does not necessarily predict a perfect correspondence between observation and expectation. It predicts the sort of strategy that will be favoured by natural selection, and says that the animals that come closest to this strategy will be most favoured. From this more realistic point of view, the correspondence between data and theory in Figure 9.4 seems much more satisfactory.

9.3.2 Switching and optimal diets

switching complements
the optimal diet model

There may seem, at first sight, to be a contradiction between the predictions of the optimal diet model and switching. In the latter, a consumer switches from one prey-type to another as their relative densities change. But the optimal diet model suggests that the more profitable prey-type should always be taken, irrespective of its density or the density of any alternative. The contradiction can be resolved, however, by noting that switching is presumed to occur in circumstances to which the optimal diet model does not apply. Specifically, switching often occurs when the different prey-types occupy different microhabitats, whereas the optimal diet model predicts behaviour *within* a microhabitat. Moreover, most other cases of switching involve a change in the profitabilities of items of prey as their density changes, whereas in the optimal diet model these are constants. Of course, the changes associated with switching typically mean that the more abundant prey-type is the more profitable, and in such a case the optimal diet model predicts specialization on that prey-type (i.e. switching). Certainly, these arguments provide a plausible reconciliation, though we know of no instance where the ideas have been experimentally tested and verified. Such experiments would help us understand switching and optimal foraging models, and would provide an important basis for understanding variations in diet width and predatory behaviour generally.

9.4 Foraging in a broader context

It is worth stressing that foraging strategies will not always be strategies for simply maximizing feeding efficiency. On the contrary, natural selection will favour foragers that maximize their *net* benefits, and strategies will therefore often be modified by other, conflicting demands on the individuals concerned. In particular, the need to avoid predators will frequently affect an animal's foraging behaviour.

backswimmers forage 'sub-optimally' but avoid being preyed upon . . .

One example in which this has been shown is the work of Sih (1982) on foraging by nymphs of an aquatic insect predator, the backswimmer *Notonecta hoffmanni*. These animals pass through five nymphal instars (with 'I' being the smallest and youngest, and 'V' the oldest), and Sih showed in the laboratory that the first three instars are liable to be preyed upon by adults of the same species. In fact, the relative risk of predation from adults was:

$$I > II > III > IV = V \simeq \text{no risk.}$$

These risks appear to modify the behaviour of the nymphs, in that they tend (both in the laboratory and in the field) to avoid the central areas of water bodies, where the concentration of adults is greatest. Indeed, Sih found that the relative degree of avoidance was:

$$I > II > III > IV = V \simeq \text{no avoidance.}$$

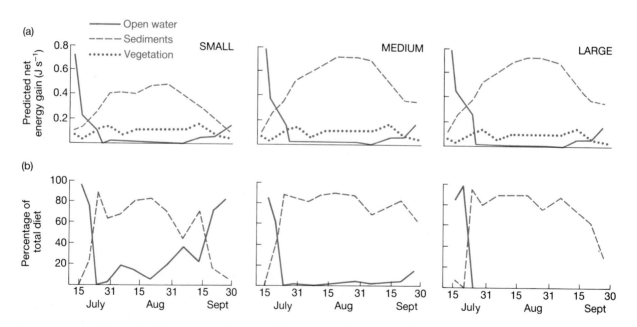

Figure 9.5. Seasonal patterns in (a) the predicted habitat profitabilities (net rate of energy gain) and (b) the actual percentage of the diet originating from each habitat, for three size classes of bluegill sunfish (*Lepomis macrochirus*). Piscivores were absent. (The 'vegetation' habitat is omitted from (b) for the sake of clarity—only 8–13% of the diet originated from this habitat for all size classes of fish.) There is good correspondence between the patterns in (a) and (b). (After Werner *et al.*, 1983a.)

Yet these central areas also contain the greatest concentration of prey items for the nymphs. Hence, as a consequence of their predator-avoidance behaviour, nymphs of instars I and II showed a reduction in feeding rate in the presence of adults (though those of instar III did not). In other words, these young nymphs displayed a less than maximal feeding rate but an increased survivorship as a result of their avoidance of predation.

. . . as do certain fish . . .

The modifying influence of predators on foraging behaviour has also been studied by Werner and his colleagues working on bluegill sunfish. They estimated the net energy returns from foraging in three contrasting laboratory habitats—in open water, amongst water weeds and on bare sediment—and they examined how prey densities varied in comparable natural habitats in a lake through the seasons (Werner *et al.*, 1983a). They were then able to predict the time at which the sunfish should switch between different lake habitats so as to maximize their overall net energy returns. In the absence of predators, three sizes of sunfish behaved as predicted (Figure 9.5). But in a further field experiment, this time in the presence of predatory largemouth bass, the small sunfish restricted their foraging to the water-weed habitat (Figure 9.6; Werner *et al.*, 1983b). Here they were relatively safe from predation, though they could only achieve a markedly sub-maximal rate of energy intake. By contrast, the larger sunfish are more or less safe from predation by bass, and they continued to forage according to the optimal foraging predictions. In a similar vein, several species of zooplanktivorous fish largely restrict their feeding to the hours of darkness, a time when their feeding rates are likely to be relatively low but when the risk of predation by piscivorous fish and birds is much reduced (Townsend & Winfield, 1985). And in plovers studied in pasture fields, feeding efficiency is reduced when flocks are joined by

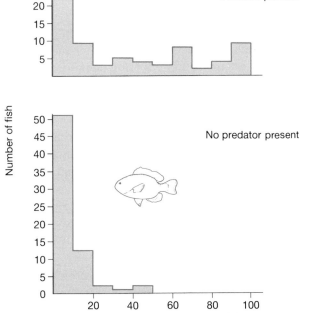

Figure 9.6. In contrast to Figure 9.5, when largemouth bass (which prey on small bluegill sunfish) are present, many sunfish take prey from the vegetation habitat where they are relatively protected from predation. (After Werner *et al.*, 1983b.)

kleptoparasitic gulls (i.e. gulls that steal food from the plovers). The risk of kleptoparasitism appears to constrain the amount of time plovers can devote to searching for and handling profitable prey (Thompson, 1983; Thompson & Barnard, 1984).

Taken together, the work on *Notonecta*, fish and birds illustrates some points of very general importance. A foraging strategy is an integral part of an animal's overall pattern of behaviour. It is strongly influenced by the selective pressures favouring the maximization of feeding efficiency, but it may also be influenced by other, possibly conflicting demands.

Finally, it is worth pointing out one other thing. The places where animals occur, where they are maximally abundant, and where they choose to feed are all key components of their 'realized niches'. We saw in Chapter 7 that realized niches can be highly constrained by competitors. Here we see that they can also be highly constrained by predators.

9.5 Functional responses: consumption rate and food density

a functional response
relates consumption rate
to food density

One of the most obvious things of crucial importance to a consumer is the density of its food since, generally, the greater the density of food, the more the consumer eats. The relationship between an individual's consumption rate and food density is known as the consumer's *functional response* (Solomon, 1949). Not surprisingly, the detailed nature of the response varies. It has been classified into three 'types' by Holling (1959).

9.5.1 The type 2 functional response

The most frequently observed functional response is the 'type 2' response, in which consumption rate rises with prey density but gradually decelerates, until a plateau is reached at which consumption rate remains constant irrespective of prey density. This is shown for a carnivore and a herbivore in Figure 9.7. Holling's (1959) explanation for the type 2 response can be stated as follows. A consumer has to devote a certain handling time to each prey item it consumes (i.e. pursuing, subduing and consuming the prey item, and then preparing itself for further search). As prey density increases, finding prey becomes increasingly easy. Handling a prey item, however, still takes the same length of time, and handling overall therefore takes up an increasing proportion of the consumer's time—until at high prey densities the consumer is effectively spending all of its time handling prey. Consumption rate therefore approaches and then reaches a maximum (the plateau), determined by the maximum number of handling times that can be fitted into the total time available.

This view of the type 2 functional response can be examined by considering a study of the ichneumonid parasitoid *Pleolophus basizonus* attacking cocoons of the European pine sawfly, *Neodiprion sertifer* (Griffiths, 1969). Griffiths plotted the number of ovipositions per parasitoid over a range of host densities, dealing separately with parasitoids of different ages. He also calculated the actual maximum oviposition rate by presenting other parasitoids of the same ages with a super-abundant supply of host cocoons. Figure

9.8 shows that the type 2 curves did indeed approach their appropriate maxima. However, while these maxima (of around 3.5 ovipositions per day) suggest a handling time of about 7 hours, further direct observation appeared to indicate that oviposition takes, on average, only 0.36 hours. In fact, this type of discrepancy is extremely common. For example, when oystercatchers (*Haematopus ostralegus*) fed on cockles in North Wales, the observed handling time of a cockle varied from 19 to 29 seconds; but the value estimated from a functional response curve was 75 seconds (Sutherland, 1982).

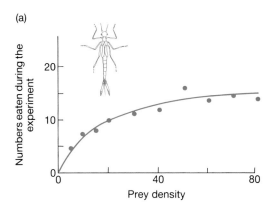

(a)

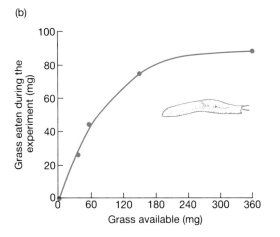

(b)

Figure 9.7. Type 2 functional responses. (a) Tenth-instar damselfly nymphs eating *Daphnia* of approximately constant size (after Thompson, 1975). (b) Slugs eating the grass *Lolium perenne* (after Hatto & Harper, 1969).

the meaning of handling times

In the case of Griffiths's work, the discrepancy is accounted for by the existence of a 'refractory period' following oviposition, during which there are no eggs ready to be laid. 'Handling time', therefore, includes not only the time actually taken in oviposition, but also the time taken preparing for the next oviposition. Likewise, the true handling times of Sutherland's oyster-catchers, and of the animals in Figure 9.7, includes time devoted to feeding-related activities other than the direct manipulation of food items. The damselfly larvae in Figure 9.7a, for example, are affected by a factor known as 'gut limitation', whereby their maximum rate of consumption is determined by the capacity of their gut, and by the speed with which it can make space available for further food items. This, however, can be thought of as the speed with which their gut can 'handle' food, and the gut thus determines handling time. 'Handling time' includes all such phenomena.

A further point to note from Figure 9.8 is that while the plateau level (and thus the handling time) was approximately the same for parasitoids of different ages, the rate of approach to that plateau was much more gradual in the younger parasitoids. This indicates that the younger parasitoids have a lower '*searching efficiency*', or, synonymously, a lower '*attack rate*'. At low host densities they oviposit less often than the older parasitoids, but at high host densities there is such a ready supply of hosts that even they are limited only by their handling time. The form taken by a type 2 functional response curve, therefore, can be characterized simply in terms of a handling time (determining the level of the plateau) and an attack rate (determining the rate of approach to the plateau).

a type 2 response can be defined by a handling time and a searching efficiency

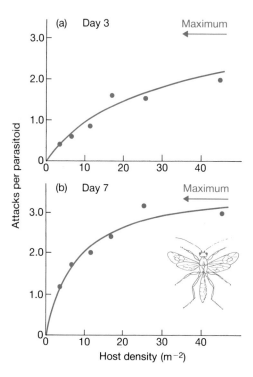

Figure 9.8. The type 2 functional responses of the ichneumonid parasitoid *Pleolophus basizonus* to changes in the density of its host *Neodiprion sertifer*. Arrows indicate maxima observed in the presence of excess hosts. (a) Parasitoids on their third day of adult life. (b) Parasitoids on their seventh day. (After Griffiths, 1969.)

More specifically, we can derive a relationship between P_e (the number of prey items eaten by a predator during a period of searching time, T_s) and N, the density of those prey items (Holling, 1959). In fact, P_e increases with the time available for searching, it increases with prey density, and it increases with the searching efficiency or attack rate of the predator, a'. Thus:

$$P_e = a'T_sN. \tag{9.2}$$

However, the time available for searching will be less than the total time, T, because of time spent handling prey. Hence, if T_h is the handling time of each prey item, then T_hP_e is the total time spent handling prey, and

$$T_s = T - T_hP_e.$$

Substituting this into equation 9.2 we have:

$$P_e = a'(T - T_hP_e)N,$$

or, rearranging,

Holling's disc equation

$$P_e = \frac{a'NT}{1 + a'T_hN}. \tag{9.3}$$

This equation describes a type 2 functional response, and is known as Holling's 'disc equation' because Holling first generated type 2 responses experimentally by getting a blindfolded assistant to pick up ('prey upon') sandpaper discs. Note that the equation describes the amount eaten during a specified period of time, T; and that the density of prey, N, is assumed to remain constant throughout that period. In experiments, this can sometimes be guaranteed by replacing any prey that are eaten; but more sophisticated models are required if prey density is depleted by the predator. Such models are described by Hassell (1978), who also discusses methods of estimating attack rates and handling times from a set of data.

there are alternative reasons for a type 2 response

It would be wrong however, to imagine, that the existence of a handling time is the only or the complete explanation for all type 2 functional responses. For instance, if the prey items are actually of variable profitability, then at high densities the diet may tend towards a decelerating number of items which are nevertheless of high profitability (Krebs *et al.*, 1983). Or a predator may become confused and less efficient at high prey densities.

9.5.2 The type 1 functional response

An example of a 'type 1' functional response is illustrated in Figure 9.9, which shows the rate at which *Daphnia magna* consumed yeast cells when the density of cells varied (Rigler, 1961). The consumption rate rose linearly to a maximum as density increased, and then remained at that maximum irrespective of further increases. This occurred because the yeast cells were extracted by the *Daphnia* from a constant volume of water washed over their filtering apparatus, and the amount extracted therefore rose in line with food concentration. Above 10^5 cells ml^{-1}, however, the *Daphnia* were unable to swallow (i.e. handle) all the food they filtered, and they therefore ingested

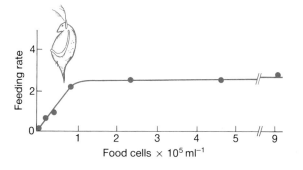

Figure 9.9. The type 1 functional response of *Daphnia magna* to different concentrations of the yeast *Saccharomyces cerevisiae* (After Rigler, 1961).

food at their maximum (plateau) rate irrespective of its concentration. In other words, in a type 1 functional response, below the plateau, the handling time is zero and $T_s = T$. Equation 9.2 therefore applies, and the slope of the response is a' (the attack rate or searching efficiency), i.e. the proportion of prey eaten per unit time.

(a)

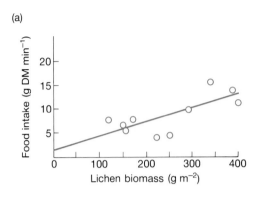

(b)

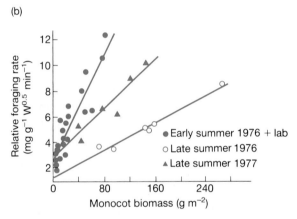

Figure 9.10. Linear functional responses of two herbivores. (a) Reindeer feeding on lichen (Batzli *et al.*, 1981). (b) Brown lemmings feeding in cotton grass/sedge communities (White *et al.*, 1981). (After Crawley, 1983.)

Fully described type 1 responses with a linear rise and a plateau like that in Figure 9.9 are rather rare. But there are many examples, especially amongst herbivores, where, over the range of food densities examined, the consumption rate rises linearly (Figure 9.10). In other words, there is no evidence in these cases for the deceleration which characterizes type 2 response.

9.5.3 The type 3 functional response

A type 3 functional response is illustrated in Figure 9.11a. At high food densities it is similar to a type 2 response, and the explanations for the two are the same. At low food densities, however, the type 3 response has an accelerating phase where an increase in density leads to a more-than-linear increase in consumption rate. Overall, therefore, a type 3 response is 'S-shaped' or 'sigmoidal'.

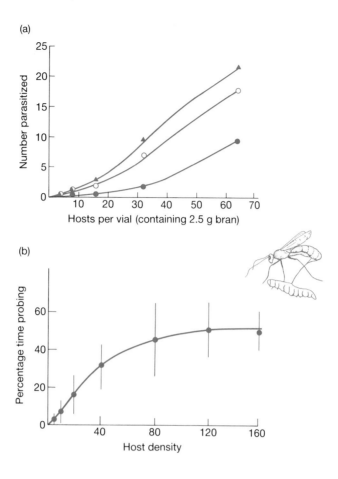

Figure 9.11. (a) A type 3 (sigmoidal) functional response of the ichneumonid *Venturia canescens* parasitizing *Cadra* larvae of second (●), third (○) and fourth (▲) instars (Takahashi, 1968). (b) The basis of this response: the relationship between the time spent probing by *Venturia canescens* (as a percentage of total observation time) and the density of its host larvae, *Plodia interpunctella* (means and 95% confidence limits). (After Hassell *et al.*, 1977.)

type 3 responses—and switching . . .

One important way in which a type 3 response can be generated is by switching on the part of the consumer (section 9.2.3). Indeed the similarities between Figures 9.11a and 9.2 are readily apparent. The difference between them is that discussions of switching focus on the density of a prey-type *relative* to the densities of alternatives, whereas functional responses are based on only the *absolute* density of the single prey-type being considered. Nevertheless, in practice, absolute and relative densities are likely to be closely correlated, and switching is therefore likely to lead frequently to a type 3 functional response.

. . . and changes in handling time or searching efficiency

More generally, a type 3 functional response will arise whenever an increase in food density leads to an increase in the consumer's searching efficiency or a decrease in its handling time. Between them, these two determine consumption rate, and so an increase in a' or a decrease in T_h will make consumption rate rise faster than would be expected from the increase in food density alone. As an example, Figure 9.11b shows that the parasitoid

Venturia canescens spends an increasing proportion of its time searching for hosts as host density increases. It therefore becomes more efficient and effective as host density increases, and its consumption rate not only increases but actually accelerates initially. A type 3 functional response is the result.

At one time, type 2 functional responses were referred to as invertebrate functional responses, and type 3 responses as vertebrate responses, the implication being that the alterations in behaviour associated with type 3 responses were limited largely to vertebrates. Now, however, as the section on switching would tend to confirm, it seems likely that type 3 functional responses are widely displayed by both vertebrates and invertebrates.

9.5.4 The consequences of functional responses for the dynamics of populations

The different types of functional response have different effects on the population dynamics of the consumers and the food-species concerned. If the rise in consumption rate *decelerates* as food density increases (type 2 responses, type 1 responses as the plateau is reached, and type 3 responses at higher densities), then prey in higher density populations will have less chance of being affected than prey in lower density populations. This is inverse density-dependence, and it will tend to have a destabilizing effect on the dynamics of the populations. On the other hand, if the rise in consumption rate *accelerates* as food density increases (type 3 responses at lower densities), then individuals at higher densities have more chance of being affected than individuals at lower densities, and this effect, being density-dependent, will tend to stabilize the population dynamics of the interaction.

It is important to realize, however, that functional responses are only one element in an array of factors affecting the dynamics of these interacting populations. They have *tendencies* to stabilize or destabilize, but these tendencies may be overridden (or supported) by forces arising from other components of the interaction. In particular, the powers that functional responses have to affect population dynamics depend on the extent to which consumption rate accelerates or decelerates over the range of densities normally experienced by the 'prey' population.

9.6 The effects of consumer density: mutual interference

No consumer lives in isolation: all are affected by other consumers. The most obvious effects are competitive; many consumers experience exploitation competition for limited amounts of food when their density is high or the amount of food is small, and this results in a reduction in the consumption rate per individual as consumer density increases. However, even when food is not limited, the consumption rate per individual can be reduced by increases in consumer density by a number of processes known collectively as *mutual interference*. For example, many consumers interact behaviourally with other members of their population, leaving less time for feeding and therefore depressing the overall feeding rate. Humming-birds, for instance, actively and aggressively defend rich sources of nectar; badgers patrol and visit the 'latrines' around the boundaries between their territories and those of their neighbours; and females of *Rhyssa persuasoria* (an ichneumonid parasitoid of wood wasp larvae) will threaten and, if need be, fiercely drive away an intruding female from their own area of tree trunk (Spradbery, 1970).

mutual interference leads to reductions in consumption rate

Mutual interference can also arise when an increase in consumer density leads to an increased rate of emigration (Figure 9.12), or when consumers steal food from one another (as do many gulls), or when prey respond to the presence of consumers and become less available for capture. In all these cases though, the essential effect is the same: the consumption rate of the average consumer is depressed by the presence of other consumers, and the impact of this increases with consumer density.

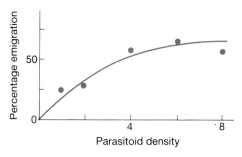

Figure 9.12. Mutual interference: there is an increase in emigration of females of the ichneumonid parasitoid *Diadromus pulchellus* from an experimental cage as parasitoid density increases (from Hassell, 1978, after Noyes, 1974).

Hassell and Varley (1969; see Hassell, 1978) have shown how various cases of mutual interference can all be reduced to a common form by calculating the searching efficiency of the consumer and plotting this against consumer density on logarithmic scales (Figure 9.13). (The searching efficiency of the consumer, a', is the same as that in Holling's disc equation.) As expected, the slopes of the graphs are negative: searching efficiency, and thus the consumption rate per individual, declines with density. The slope in each case is said to take the value $-m$, and m is known as the *coefficient of interference*. At the very lowest densities (e.g. Figure 9.13a) the effect of

the coefficient of interference

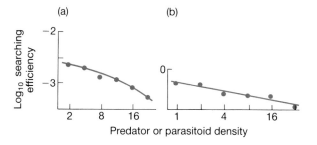

Figure 9.13. Mutual interference: the relationship between searching efficiency and the density of searching parasitoids or predators (log scales). (a) *Encarsia formosa* parasitizing the whitefly *Trialeurodes vaporariorum* (Burnett, 1958). (b) *Phytoseiulus persimilis* feeding on nymphs of the mite *Tetranchyus urticae* (Fernando, 1977). (After Hassell, 1978.)

329   BEHAVIOUR OF PREDATORS

interference is negligible. But at moderate and high densities m tends to remain constant (Figure 9.13a and b); and m is actually often constant over the entire range of densities examined (e.g. constant slope in Figure 9.13b).

It is important to remember of course that the consumption rate per individual does not always decline with increasing consumer density. Sometimes, especially at low densities, there can be an increase in consumption rate which occurs because of *social facilitation* amongst the consumers. In many herds, flocks or schools of vertebrates, for instance, an increase in density can lead to a reduction in the time each individual spends on the look-out for its own predators, and this leaves more time for feeding (see Barnard & Thompson, 1985, for examples of this in birds).

the reverse of interference: facilitation

Without doubt, though, the more general pattern is for individual consumption rate to *decrease* with consumer density. This reduction is likely to have an adverse effect on the fecundity, growth and mortality of individual consumers, and the adverse effect will increase as consumer density increases. The consumer population is thus subject to density-dependent control, and mutual interference therefore tends to stabilize the dynamics of predatory populations, and to stabilize predator–prey dynamics generally.

mutual interference tends to stabilize predator–prey dynamics

9.7 Consumers and food patches

Although it has not been discussed explicitly so far, it is indisputable that for all consumers the world is heterogeneous—their food is distributed patchily. The patches may be natural and discrete physical objects: a bush laden with berries is a patch for a fruit-eating bird, while a leaf covered with aphids is a patch for a predatory ladybird. Alternatively, a 'patch' may only exist as an arbitrarily defined area in an apparently uniform environment; for a wading bird feeding on a sandy beach, different 10m² areas may be thought of as patches that contain different densities of worms. In either case though, a patch must be defined with a particular consumer in mind. One leaf is an appropriate patch for a ladybird, but for a larger and more active insectivorous bird, a square-metre of canopy or even a whole tree may represent a more appropriate patch.

the meaning of 'patch'

Consumers select those habitats that contain their food. But even within these broad habitats, food is distributed patchily, and consumers typically show a preference for particular patches. The basis for this preference may be the quality of the food; mountain hares, for instance, preferentially graze patches of heather that are rich in nitrogen (Moss *et al.*, 1981). Alternatively, food patches may be chosen so as to avoid predators, or parasites, or severe weather conditions (Crawley, 1983).

a preference for particular patch types is typical

9.7.1 Aggregative responses and partial refuges

The best-documented basis for patch preference, however, is when patches vary in the *density* of the food or prey items they contain. Figure 9.14a–d illustrates a number of examples in which individual consumers spend most time in patches containing the greatest densities of prey. This is known as an *aggregative response* on the part of the consumers, because one of its consequences is that consumers tend to aggregate in the patches of high prey density. Figure 9.14e–f shows this explicitly for a further pair of examples.

an aggregative response typically leads to a parallel distribution of ill-effects . . .

Aggregative responses represent a preference by consumers for patches in which the density of food is high, and thus the rate of food intake is high. In addition, however, the patches in which the consumer spends most time are usually the patches in which a 'prey' individual has the greatest chance of being attacked. This can be seen, for instance, for the parasitoids in Figure 9.15. The parasitoids spent longest in patches with high densities of hosts (i.e.

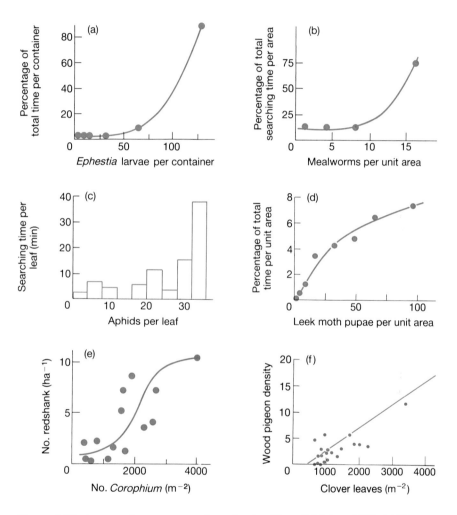

Figure 9.14. Aggregative responses (mostly after Hassell & May, 1974). (a) The parasitoid *Venturia canescens* spends more time in containers with its host *Ephestia cautella* at high densities (after Hassell, 1971). (b) Great tits (*Parus major*) spend more searching time at higher densities of mealworms (*Tenebrio mollitor*) (after Smith & Dawkins, 1971). (c) Coccinellid larvae (*Coccinella septempunctata*) spend more time on leaves with high densities of their aphid prey (*Brevicoryne brassicae*) (after Hassell & May, 1974). (d) The parasitoid *Diadromus pulchellus* spends more time at higher densities of leek moth pupae (*Acrolepia assectella*) (after Noyes, 1974). (e) Redshank (*Tringa totanus*) aggregate in patches with higher densities of their amphipod prey *Corophium volutator* (after Goss-Custard, 1970). (f) Wood-pigeons (*Columba palumbus*) aggregate in areas with higher densities of clover leaves (after Murton *et al.*, 1966).

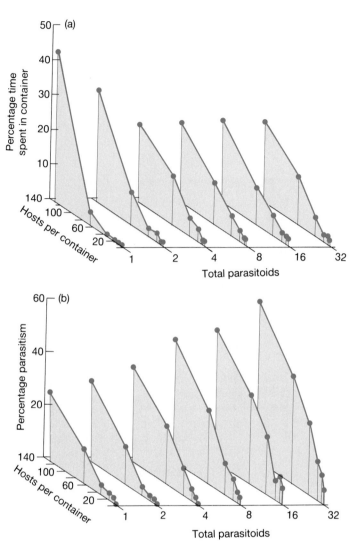

Figure 9.15. (a) The aggregative response of *Venturia canescens,* which spends more time in high-density containers of its host *Ephestia cautella.* (b) The resultant distribution of ill-effects: hosts in low-density containers are in a partial refuge—they are least likely to be parasitized. (After Hassell, 1982.)

there was an aggregative response); they therefore searched these patches most thoroughly, and found and parasitized a large percentage of the hosts. As a consequence, the hosts in the low-density patches had a very low likelihood of being attacked, and the low-density patches were thus a *partial refuge* for the host population. In other words, a portion of the prey population was less susceptible to predation than it would have been had the predators distributed themselves over prey patches at random.

On the other hand, this parallel between an aggregative response and the distribution of ill-effects does not always occur. Figure 9.16, for example, shows the aggregative response of another parasitoid species, but this time the consequences are that hosts in *low*-density patches (where the parasitoids

... but not always

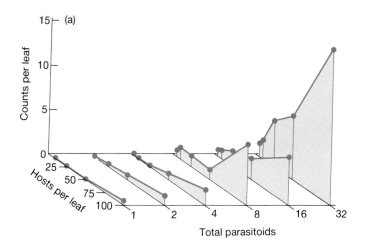

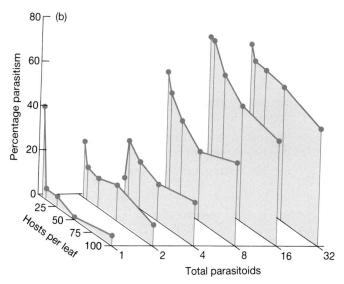

Figure 9.16. (a) The aggregative response of *Trichogramma pretiosum*, which aggregates on leaves with high densities of its host *Plodia interpunctella*. (b) The resultant distribution of ill-effects: hosts on *high* density leaves are in a partial refuge—they are least likely to be parasitized. (After Hassell, 1982.)

spend *least* time) have the greatest chance of being attacked. This seems surprising at first sight, but there are a number of reasons why it may occur (Morrison & Strong, 1981; Hassell, 1982). For instance, the parasitoids may have a handling time which is so long that they can only attack a very small proportion of the hosts in patches of high host density; or there may be higher rates of encounter with already parasitized hosts in these patches, causing the parasitoids to depart from there prematurely. Even in these cases, however, the prey population has a partial refuge. The difference is that the refuge lies in the high-density rather than the low-density patches. Thus partial refuges occur (and partial protection arises) whenever the ill-effects of predators are distributed in a consistently aggregated manner amongst patches of prey. The relatively protected prey are most commonly (but not always) those in low-density patches.

partial refuges arise from a patchy distribution of ill-effects

9.7.2 The consequences of aggregated distributions for the dynamics of populations

partial refuges tend to stabilize predator–prey dynamics

Partial refuges tend to *stabilize* the population dynamics of a predator–prey interaction. A number of prey remain unattacked even at high predator densities, and this buffers the prey population from the most drastic effects of its predator. This in turn tends to ensure that predators always have a population of prey on which to feed. Note, however, that this stabilizing effect is a by-product of the consumers' preference for high-profitability patches; the predators are *not* behaving in this way 'in order to' stabilize the predator–prey interaction.

aggregation and pseudo-interference

The stabilizing effect can also be explained by considering the effect of the aggregated distribution on consumption rates. When consumer density is low, the consumers concentrate on patches of high prey density, and they therefore have a much higher consumption rate than they would if they distributed themselves at random. But at high consumer densities the patches with a high prey density are rapidly depleted. The consumers are thus unable to match the high consumption rates they achieve when their density is low. In fact, consumption rate declines as consumer density increases; and this density-dependent effect is bound to stabilize the dynamics of the interaction. The effect is known as *pseudo-interference* (Free *et al.*, 1977). This is because it stems from a pattern (consumption rate declining as consumer density increases) which has conventionally been attributed to mutual interference amongst consumers (section 9.6).

9.7.3 Aggregations of herbivores

Simple aggregative responses are far less common amongst herbivores than amongst carnivores and parasitoids (but see Figure 9.14f). Thus, for example, different species of herbivorous insect peak in abundance at different densities of *Brassica* plants: larvae of the butterfly *Pieris rapae* are most abundant in low-density patches of the plants; the beetle *Phyllotreta striolata* is most abundant in intermediate-density plots; while *Phyllotreta cruciferae* is most abundant in the densest *Brassica* populations (Cromartie, 1975). The likely explanation for this is that many herbivores treat each plant as a patch, and select the plants that are most profitable. Plants at lower density may be chosen because they are bigger or of higher quality (since they suffer less competition), or because there is less predator pressure on the herbivores on these plants.

herbivores often aggregate without this being an 'aggregative response'

On the other hand, many herbivores display a marked tendency to aggregate without this being an aggregative response (i.e. without it being a response to 'prey density'). The cabbage aphid (*Brevicoryne brassicae*) forms aggregates at two separate levels (Way & Cammell, 1970). Nymphs quickly form large groups when isolated on the surface of a single leaf; and populations on a single plant tend to be restricted to particular leaves. At the level of the whole plant, aggregations can be seen when the moth *Cactoblastis cactorum* attacks the prickly pear cacti *Opuntia inermis* and *Opuntia stricta* (Monro, 1967). The female moths deposit egg-sticks on the plants, each

Table 9.2. The observed distribution of *Cactoblastis* egg-sticks on *Opuntia* plants is aggregated compared to a random, Poisson distribution (in which the variance and mean are equal) (after Monro, 1967).

Site	Mean density (egg-sticks per segment)	Egg-sticks per plant		Comparison by χ^2 test of distributions with Poisson distributions of the same mean
		Mean	Variance	
1	0.398	2.42	6.16	†P < 0.001
2	0.265	2.09	22.40	†P < 0.001
3	0.084	1.24	5.09	†P < 0.001
4	0.031	0.167	0.247	0 P > 0.05
5	0.112	0.53	1.47	†P < 0.01
6	0.137	1.97	18.76	†P < 0.001
7	0.175	0.62	3.55	†P < 0.001
8	0.112	0.34	0.90	†P < 0.05

†Egg-sticks more clumped than expected for random oviposition
0 Egg-sticks not distributed differently from random

containing 70–90 eggs; and as Table 9.2 shows, the distribution of these egg sticks over the plants is highly aggregated.

herbivore aggregations create partial refuges for plants

Herbivore aggregations create 'partial refuges' for the plants in essentially the same way as aggregative responses do, i.e. by leading to an aggregated distribution of ill-effects. When cabbage aphids attack only one leaf of a four-leaved cabbage plant (as they do naturally), the other three leaves survive, but if the same number of aphids are evenly spread over the four leaves, then all four leaves are destroyed (Way & Cammell, 1970). And when *Cactoblastis* attacks a prickly pear population, there are an 'unexpectedly' large number of plants that escape attack altogether (Monro, 1967; Table 9.2). The aggregative behaviour of the herbivores affords protection to a number of cabbage leaves and a number of prickly pear plants. Such partial refuges also tend to lend stability to the population dynamics of the interaction.

9.8 The ideal free distribution: aggregation and interference

It seems, then, that consumers tend to aggregate in profitable (often high prey density) patches where their expected rate of food consumption is highest. We also saw (in section 9.6) that consumers tend to compete and interfere with one another, thereby reducing their per capita consumption rate. It follows from this that patches that are initially most profitable become immediately less profitable because they attract most consumers. Indeed these patches of high prey density may assume a lower profitability than patches of lower prey density which initially attracted fewer consumers. We might therefore expect the consumers to redistribute themselves.

It has been proposed that if a consumer forages optimally, the process of redistribution will continue until the profitabilities of all patches are equal (Fretwell & Lucas, 1970; Parker, 1970). This will happen because as long as there are dissimilar profitabilities, consumers should leave less profitable patches and be attracted to more profitable ones. Fretwell and Lucas called the consequent distribution the *ideal free distribution*: the consumers are 'ideal' in their judgement of profitability, and 'free' to move from patch to patch. Note that in the ideal free distribution, because all patches have the same profitability, all consumers have the same consumption rate. Note too, however, that the distribution of individuals will vary in detail from case to case, since it depends on the precise differences in profitability from patch to patch, and on the precise nature and strength of the competition and interference amongst consumers.

the ideal free distribution: a balance between attractive and repellent forces

There are some simple cases where consumers appear to conform to an ideal free distribution (Figure 9.17). There are also many cases where the correspondence is extremely weak, presumably because the consumers are not both 'ideal' and 'free'. The important, more general point, however, is that the distributions of consumers amongst food patches should be seen as the resultant of attractive and repellent forces. The aggregative responses and aggregated distributions of section 9.7 are thus the combined results of on the one hand a tendency for consumers to aggregate where the density of prey is highest, and on the other hand a tendency for consumers to avoid patches with high levels of competition and interference.

This balance between an underlying aggregation and the avoidance of

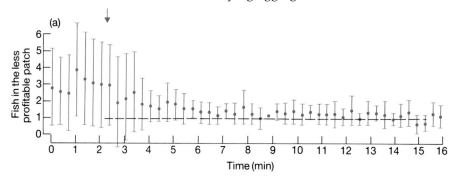

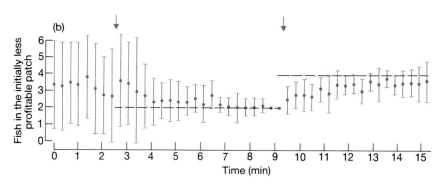

Figure 9.17. Sticklebacks (a small fish) conforming to the ideal free distribution. Six fish were used in each case, and were fed at 'more profitable' and 'less profitable' ends of a fish tank. Dots and bars are means and standard deviations of 8 trials in (a) and 11 trials in (b). (a) Number of fish feeding at the less profitable end with a profitability ratio of 5:1. Feeding started at the arrow, and the dashed line is the ideal free expectation. (b) Similar to (a), but an initial profitability ratio of 2:1, followed by a reversal of profitabilities. (After Milinski, 1979.)

competition and interference, however, does not always express itself as an aggregated distribution. For some consumers, the pressures to avoid competition and interference exceed the attractions of 'better' patches; and the individuals, far from aggregating, actually distribute themselves such that they are more *evenly* dispersed than would be expected from a random occupation of patches. For both examples in Figure 9.18 there are large numbers of patches occupied by just a single consumer. This simply means, of course, that the forces of attraction and repulsion are balanced in a different way from the more usual balance of an aggregated distribution.

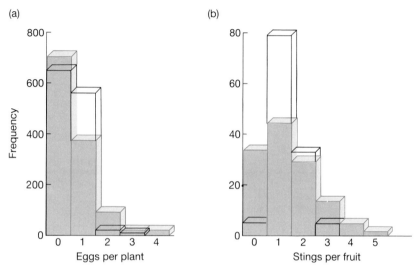

Figure 9.18. Regular or evenly distributed attacks by insects. (a) Orange tip butterfly eggs on plants of *Cardamine pratensis* (Wiklund & Ahrberg, 1978). (b) *Dacus tryoni* egg-laying 'stings' on loquat fruits (Monro, 1967). (Observed distribution ——; expected Poisson distribution – – –.) These patterns arise by avoidance of plant parts already attacked. (After Crawley, 1983.)

9.9 Patchiness and time: hide-and-seek

The responses of consumers to their food patches often have a temporal as well as spatial component. When this is the case, the protagonists can appear to play 'hide-and-seek'. The most famous example is the experimental work of Huffaker (Huffaker, 1958 and Huffaker *et al.*, 1963), who studied a system in which the predatory mite *Typhlodromus occidentalis* fed on the herbivorous mite *Eotetranychus sexmaculatus,* which fed on oranges interspersed amongst rubber balls in a tray. In the absence of its predator, *Eotetranychus* maintained a fluctuating but persistent population (Figure 9.19a); but if *Typhlodromus* was added during the early stages of prey population growth, it rapidly increased its own population size, consumed all of its prey and then became extinct itself (Figure 9.19b).

The interaction was altered, however, when Huffaker made his microcosm more 'patchy'. He spread the oranges further apart, and partially isolated each one by placing a complex arrangement of vaseline barriers in the

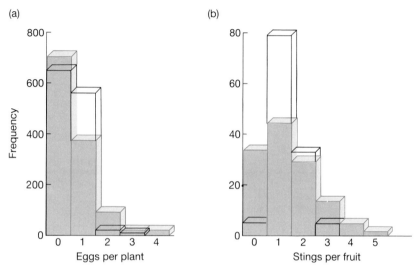

'even' distributions

Huffaker's mites and oranges

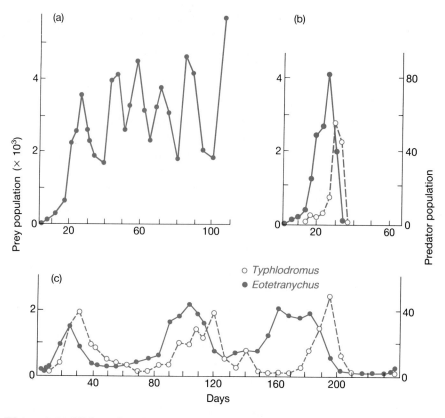

Figure 9.19. Hide and seek. Predator–prey interactions between the mite *Eotetranychus sexmaculatus* (●) and its predator, the mite *Typhlodromus occidentalis* (○). (a) Population fluctuations of *Eotetranychus* without its predator. (b) A single oscillation of the predator and prey in a simple system. (c) Sustained oscillations in a more complex system. (After Huffaker, 1958.)

tray, which the mites could not cross. But he facilitated the dispersal of *Eotetranychus* by inserting a number of upright sticks from which they could launch themselves on silken strands carried by air currents. Dispersal between patches was therefore much easier for prey than it was for predators. In a patch occupied by both *Eotetranychus* and *Typhlodromus,* the predators consumed all the prey and then either became extinct themselves or dispersed (with a low rate of success) to a new patch. But in patches occupied by prey alone, there was rapid, unhampered growth accompanied by successful dispersal to new patches. In a patch occupied by predators alone, there was usually death of the predators before their food arrived. Each patch was therefore ultimately doomed to the extinction of both predators and prey; but overall, at any one time, there was a mosaic of unoccupied patches, prey–predator patches heading for extinction, and thriving prey patches; and this mosaic was capable of maintaining persistent populations of both predators and prey (Figure 9.19c). Patchiness or heterogeneity, therefore, appears to confer stability on the interaction by providing the prey with a succession of temporary 'refuges in time'.

A similar example, from a natural population, is provided by work off the coast of Southern California on the predation by starfish of clumps of mussels

(Landenberger, 1973, in Murdoch & Oaten, 1975). Clumps which are heavily preyed upon are liable to be dislodged by heavy seas so that the mussels die. The starfish are continually driving patches of their mussel prey to extinction. Yet the mussels have planktonic larvae which are continually colonizing new locations and initiating new clumps. The starfish, however, disperse much less readily. They aggregate at the larger clumps (they have an aggregative response), and there is a time-lag before they leave an area when the food is gone. Thus patches of mussels are continually becoming extinct, but other clumps are growing prior to the arrival of the starfish. The aggregative behaviour of the starfish allows initially small clumps of mussels to become large and profitable. 'Hide-and-seek' and aggregative responses are therefore essentially indistinguishable. They certainly share a tendency to stabilize predator–prey interactions.

aggregation, hide-and-seek and stability

9.10 Behaviour that leads to aggregated distributions

patch location

There are various types of behaviour underlying the aggregative responses of consumers (and their responses to patches generally) but they fall into two broad categories: those involved with the location of profitable patches, and those that represent the response of a consumer once within a patch. The first category includes all examples in which consumers perceive, at a distance, the existence of heterogeneity in the distribution of their prey. For instance, the parasitoid *Callaspidia defonscolombei* is attracted to concentrations of its syrphid (fly larva) host by the odours produced by the syrphids' own prey, which are various species of aphid (Rotheray, 1979).

Within the second category—responses of consumers within patches—there are two main aspects of behaviour. The first is a change in the consumer's pattern of searching in response to encounters with items of food. In particular, there is often a slowing-down of movement and an increased rate of turning immediately following the intake of food, both of which lead to the consumer remaining in the vicinity of its last food item ('area-restricted

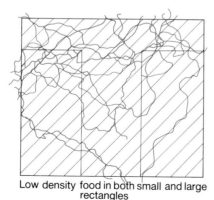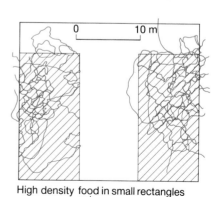

Low density food in both small and large rectangles

High density food in small rectangles only

Figure 9.20. The search paths of thrushes (the song thrush, *Turdus philomelos,* and the European blackbird, *T. merula*). The birds tend to remain within areas of high prey density (right), because in these, compared with low-density areas (left), they turn more often and more sharply. (After Smith, 1974.)

search'). Consumers therefore tend to remain in high-density patches of food (where the high rate of encounter leads to a slowing down and a high turning rate), and to leave low-density patches (where they move faster and turn less). Figure 9.20 illustrates this sort of behaviour for birds feeding on a lawn. Alternatively, consumers may simply abandon unprofitable patches more rapidly than they abandon profitable ones. This is what happens when the carnivorous, net-spinning larva of the caddis-fly *Plectrocnemia conspersa* feeds on chironomid larvae in a laboratory stream. Caddis in their nets were provided with one prey item at the beginning of the experiment and then fed

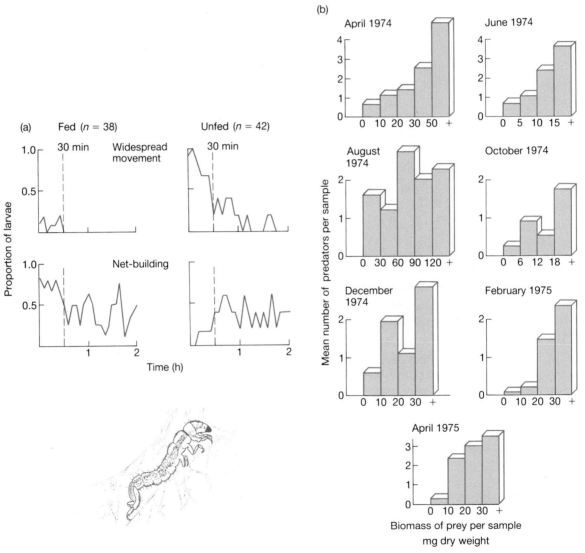

Figure 9.21. (a) On arrival in a patch, fifth-instar *Plectrocnemia conspersa* larvae which encounter and eat a chironomid prey item quickly cease wandering and commence net-building. Predators that fail to encounter a prey item exhibit much more widespread movement during the first 30 minutes of the experiment, and are significantly more likely to move out of the patch.

(b) Aggregative responses of fifth instar larvae on seven occasions through the year expressed as mean number of predators against combined biomass of chironomid and stonefly prey per 0.0625 m² sample of stream bed (*n* = 40), in a number of prey biomass categories. (After Townsend & Hildrew, 1980, and Hildrew & Townsend, 1980).

daily rations of zero, one or three prey. The average time to abandoning of the net was greatest at the high feeding rates (Hildrew & Townsend, 1980; Townsend & Hildrew, 1980).

Plectrocnemia's behaviour in relation to prey patches also has an element of area-restricted search; the likelihood that it will spin a net in the first place depends on whether it happens to encounter a food item (which it can consume even without a net). Figure 9.21a shows that larvae that have fed begin net-building immediately, whereas unfed larvae continue wandering and are more likely to leave the patch. Overall, therefore, a prey item is more likely to be encountered in a rich patch (leading to construction of a net), and a rich patch will provide high feeding rates (reducing the likelihood of abandoning a net). These two behaviours account for an aggregative response in the natural stream environment observed for much of the year (Figure 9.21b).

thresholds and
giving-up times

The difference in the rates of abandonment of patches of high and low profitability can be achieved in a number of ways, but two are especially easy to envisage. A consumer might leave a patch when its feeding rate drops below a threshhold level, or a consumer might have a giving-up time—it might abandon its patch whenever a particular time interval passes without the successful capture of food. Whichever mechanism is used, or indeed if the consumer simply uses area-restricted search, the consequences will be the same: individual consumers will spend longer in more profitable patches, and these patches will therefore contain more consumers.

9.11 The optimal foraging approach to patch use

The advantages to a consumer of spending more time in higher profitability patches are easy to see. However, the *detailed* allocation of time to different patches is a subtle problem, since it depends on the precise differentials in profitability, the average profitability of the environment as a whole, the distance between patches, and so on. The problem has been a considerable focus of attention for optimal foraging theory. In particular, a great deal of interest has been directed at the very common situation in which foragers themselves deplete the resources of a patch, causing its profitability to decline with time. Amongst the many examples of this are insectivorous insects removing prey from a leaf, and bees consuming nectar from a flower.

9.11.1 The marginal value theorem

Charnov (1976b) and Parker and Stuart (1976) produced similar models to predict the behaviour of an optimal forager in such situations. They found that the optimum stay-time in a patch should be defined in terms of the rate of energy extraction experienced by the forager at the moment it leaves a patch (the 'marginal value' of the patch). Charnov called the results 'the marginal value theorem'. The models were formulated mathematically, but their salient features are shown in graphical form in Figure 9.22.

The primary assumption of the model is that an optimal forager will maximize its overall intake of energy during a bout of foraging. In so doing it

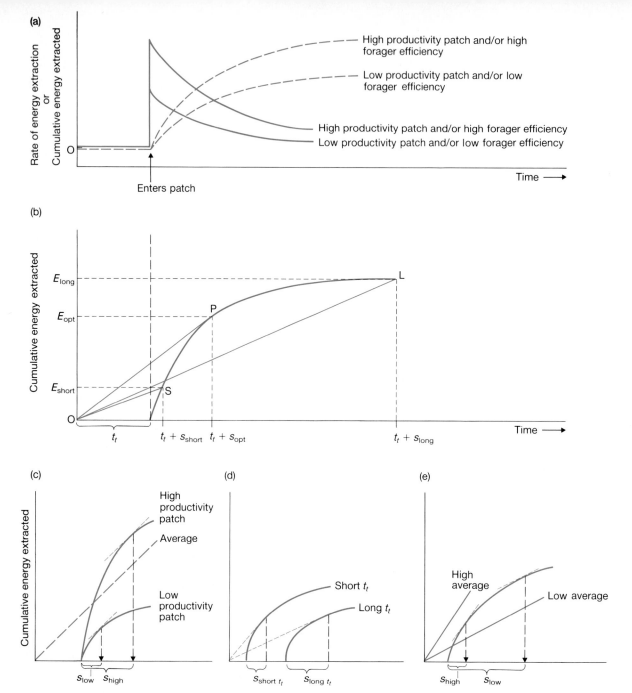

Figure 9.22. The marginal value theorem. (a) When a forager enters a patch, its rate of energy extraction is initially high (especially in a highly productive patch or where the forager has a high foraging efficiency), but this rate declines with time as the patch becomes depleted. The cumulative energy intake approaches an asymptote. (b) The options for a forager. The solid curve is cumulative energy extracted from an average patch, and t_t is the average travelling time between patches. The rate of energy extraction (which should be maximized) is energy extracted divided by total time, i.e. the slope of a straight line from origin to curve. Short stays in the patch (slope $= E_{short}/(t_t+s_{short})$) and long stays (slope $= E_{long}(t_t+s_{long})$) both have lower rates of energy extraction (shallower slopes) than a stay (s_{opt}) which leads to a line just tangential to the curve. s_{opt} is therefore the optimum stay-time, giving the maximum overall rate of energy extraction. *All* patches should be abandoned at the *same* rate of energy extraction (the slope of the line OP). (c) Low-productivity patches should be abandoned after shorter stays than high-productivity patches. (d) Patches should be abandoned more quickly when travelling time is short than when it is long. (e) Patches should be abandoned more quickly when the average overall productivity is high than when it is low.

will maximize the average rate of energy intake for the bout *as a whole*. Energy will, in fact, be extracted in bursts because the food is distributed patchily, and the forager will therefore sometimes move between patches, during which time its intake of energy will be zero. But once in a patch, the forager will extract energy in a manner described by the curves in Figure 9.22a. Its initial rate of extraction will be high; but as time progresses and the resources are depleted, the rate of extraction will steadily decline. Of course, the rate will itself depend on the initial contents of the patch (i.e. its productivity or profitability) and on the forager's efficiency and motivation (Figure 9.22a).

when should a forager leave a patch that it is depleting?

The problem under consideration is this: at what point should a forager leave a patch? If it left all patches immediately after reaching them, then it would spend most of its time travelling between patches, and its overall rate of intake would be low. If it stayed in all patches for considerable lengths of time, then it would spend little time travelling, but it would spend extended periods in depleted patches, and its overall rate of intake would again be low. Some intermediate stay-time is therefore optimal. In addition, though, the optimal stay-time must clearly be greater for profitable patches than for unprofitable ones, and it must depend on the profitability of the environment as a whole.

Consider, in particular, the forager in Figure 9.22b. It is foraging in an environment where food is distributed patchily and where some patches are more valuable than others. The average travelling time between patches is t_t. This is therefore also the length of time the forager can expect to spend on average after leaving one patch before it finds another. The forager in Figure 9.22b has arrived at an *average* patch for its particular environment, and it therefore follows an average extraction curve. In order to forage optimally it must maximize its rate of energy intake not merely for its period in the patch but for the whole period since its departure from the last patch (i.e. for the period $t_t + s$, where s is the stay-time in the patch).

the way to maximize overall energy intake

If it leaves the patch rapidly then this period will be short ($t_t + s_{short}$ in Figure 9.22b). But, by the same token, little energy will be extracted (E_{short}). The *rate* of extraction (for the whole period $t_t + s$) will be given by the slope of the line OS (i.e. $E_{short}/(t_t + s_{short})$). On the other hand, if the forager remains for a long period (s_{long}) then far more energy will be extracted (E_{long}); but the overall rate of extraction (the slope of OL) will be little changed. To maximize the rate of extraction over the period $t_t + s$, it is necessasry to maximize the slope of the line from O to the extraction curve. This is achieved simply by making the line a *tangent* to the curve (OP in Figure 9.22b). No line from O to the curve can be steeper, and the stay-time associated with it is therefore optimal (s_{opt}).

The optimal solution for the forager in Figure 9.22b, therefore, is to leave that patch when its extraction rate is equal to (tangential to) the slope of OP, i.e. it should leave at point P. In fact, Charnov, and Parker and Stuart, found that the optimal solution for the forager is to leave all patches, irrespective of their profitability, at the *same* extraction rate (i.e. the same 'marginal value'). This extraction rate is given by the slope of the tangent to the average extraction curve (e.g. in Figure 9.22b), and it is therefore the maximum average overall rate for that environment as a whole.

The model therefore confirms that the optimal stay-time should be greater in more productive patches than in less productive patches (Figure 9.22c). Moreover, for the least productive patches (where the extraction rate is never as high as OP) the stay-time should be zero. The model also predicts that all patches should be depleted such that the final extraction rate from each is the same (i.e. the 'marginal value' of each is the same); and it predicts that stay-times should be longer in environments where the travelling time between patches is longer (Figure 9.22d), and that stay-times should be longer where the environment as a whole is less profitable (Figure 9.22e).

9.11.2 Experimental tests of the marginal value theorem

Encouragingly, there is evidence from a number of cases that lends support to the marginal value theorem. Cowie (1977), for instance, has quantitatively tested the prediction set out in Figure 9.22d: that a forager should spend longer in each patch when the travelling time is longer. He used captive great tits in a large indoor aviary, and got the birds to forage for small pieces of mealworm hidden in sawdust-filled plastic cups—the cups were 'patches'. All patches on all occasions contained the same number of prey; but travelling time was manipulated by covering the cups with cardboard lids that varied in their tightness and therefore varied in the time needed to prize them off. Birds foraged alone, and Cowie used six in all, subjecting each to two habitats. One of these habitats always had longer travelling times (tighter lids) than the other.

For each bird in each habitat (12 in all) Cowie measured the average travelling time and the curve of cumulative food intake within a patch. He

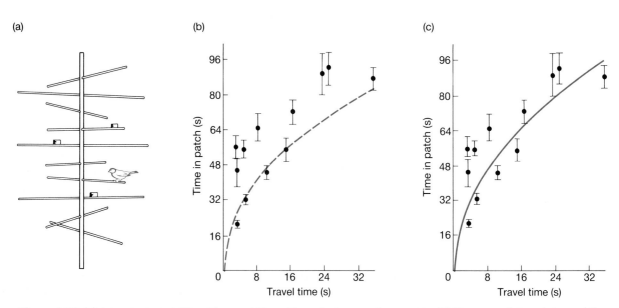

Figure 9.23. (a) An experimental 'tree' for great tits, with three patches. (b) Predicted optimal time in a patch plotted against travelling time (dashed line), together with the observed mean points (±s.e.) for six birds, each in two environments. (c) The same data points, and the predicted time taking into account the energetic costs of travelling between patches. (From Krebs, 1978, after Cowie, 1977).

then used the marginal value theorem to predict the optimal stay-time in habitats with different travelling-times, and compared these predictions with the stay-times he actually observed. As Figure 9.23 shows, the correspondence was quite close. It was closer still when he took account of the fact that there was a net *loss* of energy when the birds were travelling between patches.

Hubbard and Cook's experiments with *Venturia*

The predictions of the marginal value theorem have also been examined using the ichneumon parasitoid *Venturia canescens* attacking the flour moth *Ephestia cautella* (Hubbard & Cook, 1978). Individual *Venturia* searched in an arena containing patches of prey concealed in plastic dishes covered with wheat bran, and patches contained varying numbers of prey. Hubbard and Cook made a number of observations broadly consistent with the marginal value theorem, amongst which was the one shown in Figure 9.24. This indicates that the encounter rate with unparasitized hosts at the end of the experiment (the marginal value) was very similar in patches of different initial density. By contrast, the initial encounter rate was, as expected, higher in the more profitable patches. The *Venturia* appeared to deplete the patches until all profitabilities were similar.

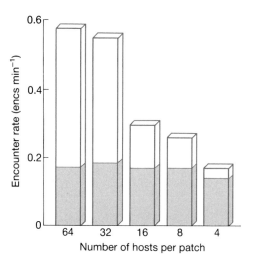

Figure 9.24. The estimated terminal encounter rates of *Venturia canescens* with its host in patches of different host density (shaded areas) compared with the initial rates in these patches (open areas). (After Hubbard & Cook, 1978.)

9.11.3 Mechanistic explanations for 'marginal value behaviour'

Much fuller reviews of the tests of the marginal value theorem are provided by Townsend and Hughes (1981) and Krebs and McCleery (1984). The picture these convey is one of encouraging but not perfect correspondence. The main reason for the imperfection is that the animals, unlike the modellers, are not omniscient. They may need to spend time learning about and sampling their environment, and they may need to spend time doing things other than foraging (e.g. looking for predators). Nevertheless, they often seem to come remarkably close to the predicted strategy. Ollason (1980) has developed a mechanistic model to account for this in the great tits studied by Cowie, while Waage (1979) has developed a mechanistic model for the foraging behaviour of *Venturia*.

Ollason's mechanistic model for the great tit

Ollason's is a memory model. It assumes that an animal has a 'remembrance of past food' which Ollason likens to a bath of water without a plug.

Fresh remembrance flows in every time the animal feeds. But remembrance is also draining away continuously. The rate of input depends on the animal's feeding efficiency and the productivity of the current feeding area. The rate of out-flow depends on the animal's ability to memorize and the amount of remembrance. Remembrance drains away quickly, for instance, when the amount is large (high water level) or the memorizing ability is poor (tall, narrow bath). Ollason's model simply proposes that an animal should stay in a patch until remembrance ceases to rise; i.e. an animal should leave a patch when its rate of input from feeding is slower than its rate of declining remembrance.

An animal foraging consistently with Ollason's model behaves in a way very similar to that predicted by the marginal value theorem. This is shown for the case of Cowie's great tits in Figure 9.25 (and Ollason argues that the discrepancy can be explained in terms of Cowie's experimental procedure). As Ollason himself remarks, this shows that to forage in a patchy environment in a way that approximates closely to optimality, an animal need not be omniscient, it does not need to sample, and it does not need to perform numerical analyses to find the maxima of functions of many variables. All it needs to do is to remember, and to leave each patch if it is not feeding as fast as it remembers doing.

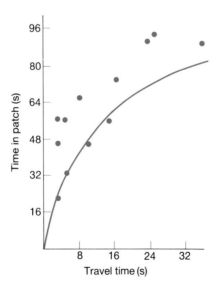

Figure 9.25. Cowie's (1977) great tit data (see Figure 9.23) compared to the predictions of Ollason's (1980) mechanistic memory model.

Waage's mechanistic model for *Venturia*

Waage's (1979) model for *Venturia* is based very largely on his own and others' observations of the parasitoid's behaviour. Host larvae produce a contact chemical which elicits a sharp U-turn by *Venturia* when contact is lost (i.e. when the edge of the patch is reached). This effect is enhanced by increases in the concentration of the chemical (i.e. with increases in host density). However, *Venturia* gradually becomes habituated to this chemical and will eventually leave a patch (no U-turn) despite the loss of contact. A successful oviposition in a patch, on the other hand, temporarily reverses the effects of habituation. Frequent oviposition will therefore prolong the stay-time in a patch. In addition, it seems that low oviposition rates over several days may increase responsiveness to patch stimuli, thereby increasing average stay-time in low-productivity habitats. Waage argues that these patterns of behaviour, taken together, can account for the observed foraging pattern of the parasitoids.

For both the great tit and *Venturia*, therefore, optimal foraging and mechanistic models are seen to be compatible and complementary in explaining *how* a predator has achieved its observed foraging pattern, and *why* that pattern has been favoured by natural selection. Between them they provide grounds for hoping that a comprehensive understanding of patch use by foragers may emerge.

10 The Population Dynamics of Predation

10.1 Introduction: patterns of abundance and the need for their explanation

We turn now to the effects of predation on the population dynamics of the predator and its prey, where even a limited survey of the data reveals a varied array of patterns that we might wish to understand. There are certainly cases where predation has a profoundly detrimental effect on the prey. For example, the 'vedalia' ladybird beetle (*Rodolia cardinalis*) is famous for having virtually eradicated the cottony-cushion scale insect (*Icerya purchasi*), a pest that threatened the California citrus industry in the late 1880s (DeBach, 1964). On the other hand, there are many cases where predators and herbivores have no apparent effect on their prey's dynamics or abundance. The weevil *Apion ulicis,* for example, was introduced into New Zealand in an attempt to control the abundance of gorse bushes (*Ulex europaeus*), and it has become one of the most abundant insects in New Zealand. Yet despite eating up to 95% of the gorse seeds every year, it has had no appreciable impact on the numbers of the plant (Miller, 1970).

There are also many examples in which a predator retains a fairly constant density in spite of fluctuations in the abundance of its prey (tawny owls and small mammals in Figure 10.1a); and there are cases in which a predator or herbivore population tracks the abundance of its prey, although the prey itself varies in density as a result of some other factor (cinnabar moth larvae and ragwort plants in Figure 10.1b). There are studies, too, that appear to show predator and prey populations linked together by coupled oscillations in abundance (Figure 10.1c). And finally, of course, there are many examples in which predator and prey populations fluctuate in abundance apparently independently of one another.

It is clearly a major task for ecologists to develop an understanding of the patterns of predator–prey abundance, and to account for the differences from one example to the next. Equally clearly though, there is a limited extent to which predator–prey abundance can be understood by considering predators and prey in isolation. They always exist as parts of multi-species systems, and all these species are affected by environmental conditions. These more complex problems are addressed in Chapter 15. However, as with any complex process in science, it is valid, interesting and instructive to see how much we can learn by ignoring the complexities, and focusing instead on interactions between conceptually isolated components—in this case, populations of

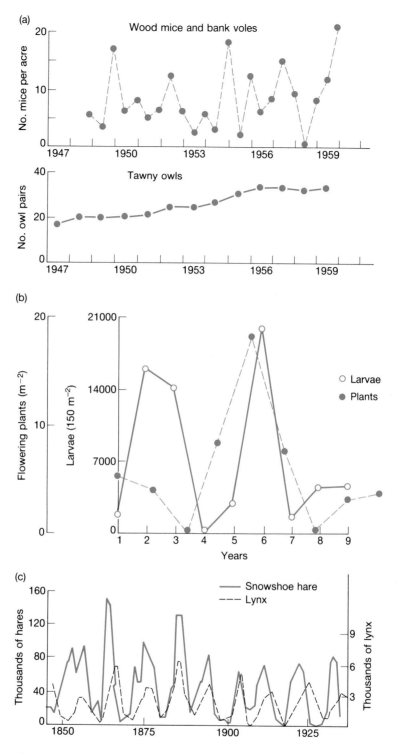

Figure 10.1. The various patterns of predator–prey abundance. (a) Tawny owls (*Strix aluco*) near Oxford maintain a constancy of abundance despite fluctuations in the numbers of their small mammal prey (Southern, 1970). (b) The number of cinnabar moth larvae (*Tyria jacobaeae*) in one year at an East Anglian site is determined largely by the number of flowering ragwort plants (*Senecio jacobaea*) in the previous year—but changes in plant abundance are due mainly to changes in germination conditions. The insect is food limited, but the plant is not herbivore limited (from Crawley, 1983, after Dempster & Lakhani, 1979). (c) The apparently coupled oscillations in abundance of the snowshoe hare (*Lepus americanus*) and Canada lynx (*Lynx canadensis*) as determined from numbers of pelts lodged with the Hudson Bay Co. (MacLulick, 1937).

predators and prey. Hence, this chapter will deal with the consequences of predator–prey interactions for the dynamics of the populations concerned.

The approach will be firstly to use simple models to deduce the effects produced by different components of the interactions. Then, in each case, field data will be examined to see whether the deductions appear to be supported or refuted. In fact, simple models are of their greatest utility when their predictions are *not* supported by real data—as long as the reason for the discrepancy can subsequently be discovered. Confirmation of a model's predictions constitutes consolidation; refutation with subsequent explanation constitutes progress.

In the final section of the chapter (10.8) we will turn from predator–prey dynamics in general to the particular problems surrounding activities in which man himself acts as a predator of a population that he wishes to exploit. We will look at the problems of harvesting or culling natural populations.

10.2 The basic dynamics of predator–prey and plant–herbivore systems: the tendency towards cycles

There have been two main series of models developed as attempts to understand predator–prey dynamics. One, based originally on the work of Nicholson and Bailey (1935), uses difference equations to model host–parasitoid interactions with discrete generations. These models have been subjected to a relatively rigorous mathematical treatment, and are reviewed by Hassell (1978), and Begon and Mortimer (1981). The second series, which will be examined here, is based on differential equations and relies heavily on simple graphical models (following Rosenzweig & MacArthur, 1963).

10.2.1 The Lotka–Volterra model

The simplest differential equation model is known (like the model of interspecific competition) by the name of its originators: Lotka–Volterra (Lotka, 1925; Volterra, 1926). It is extremely naïve, but it will serve as a useful point of departure. The model has two components: C, the numbers present in a consumer (= predator) population, and N, the numbers or biomass present in a prey or plant population.

It can be assumed initially that in the absence of consumers the prey population increases exponentially:

$$\frac{dN}{dt} = rN.$$

But prey individuals are removed by predators, and this occurs at a rate that depends on the frequency of predator–prey encounters. Encounters will increase as the numbers of predators (C) increase and as the numbers of prey (N) increase. However, the exact number encountered and successfully consumed will depend on the searching and attacking efficiencies of the predator, i.e. on a', the 'searching efficiency' or 'attack rate' (Chapter 9). The frequency of 'successful' predator–prey encounters, and hence the consumption rate of prey, will thus be $a'CN$, and overall:

the Lotka–Volterra
prey equation

$$\frac{dN}{dt} = rN - a'CN \qquad (10.1)$$

As we saw in Chapter 8, in the absence of food, individual predators lose weight and starve to death. Thus, in the model, predator numbers are assumed to decline exponentially through starvation in the absence of prey:

$$\frac{dC}{dt} = -qC,$$

where q is their mortality rate. This is counteracted by predator birth, the rate of which is assumed to depend on only two things: (i) the rate at which food is consumed, $a'CN$, and (ii) the predator's efficiency, f, at turning this food into predator offspring. Predator birth rate is therefore $fa'CN$, and overall:

the Lotka–Volterra
predator equation

$$\frac{dC}{dt} = fa'CN - qC \qquad (10.2)$$

Equations 10.1 and 10.2 constitute the Lotka–Volterra model.

The properties of this model can be investigated by finding zero isoclines (on which a population is just maintained, neither increasing nor decreasing) and using these to determine the behaviour of joint predator–prey populations. In the case of the prey (equation 10.1):

$$\text{when } \frac{dN}{dt} = 0, \; rN = a'CN$$

the properties of the
model are revealed by
zero isoclines

$$\text{or} \qquad C = \frac{r}{a'}.$$

Thus, since r and a' are constants, the prey zero isocline is a line for which C itself is a constant (Figure 10.2a).

Likewise, for the consumers (equation 10.2):

$$\text{when } \frac{dC}{dt} = 0, \; fa'CN = qC$$

$$\text{or} \qquad N = \frac{q}{fa'}.$$

The predator zero isocline is therefore a line along which N is constant (Figure 10.2b).

Putting the two isoclines together (Figure 10.2c) shows the behaviour of joint populations: they undergo indefinite, coupled oscillations in abundance. Predators increase in abundance when there are large numbers of prey. But this leads to an increased predation pressure on the prey and thus to a decrease in prey abundance. This, though, leads to a food shortage for predators and a decrease in predator abundance, which leads to a relaxation of predation pressure and an increase in prey abundance, which leads to an increase in predator abundance, and so on (Figure 10.2d).

The behaviour of the model, however, should not be taken too seriously.

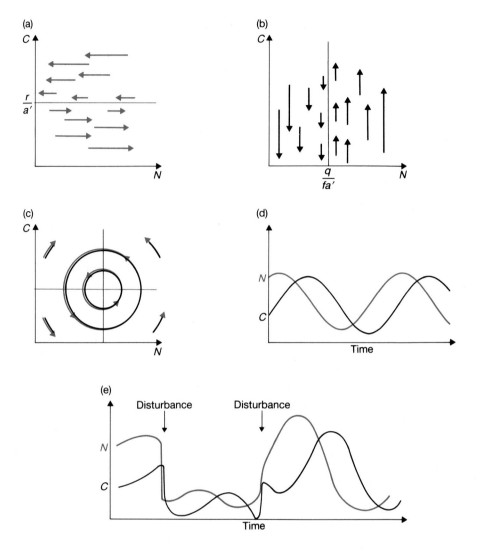

Figure 10.2. The Lotka–Volterra predator–prey model. (a) The prey zero isocline, with prey (N) increasing in abundance at lower predator densities (low C) and decreasing at higher predator densities. (b) The predator zero isocline, with predators increasing in abundance at higher prey densities and decreasing at lower prey densities. (c) When the zero isoclines are combined, the outcome is indefinite coupled cycles of predator–prey abundance like those shown in (d). However, as shown in (e), these cycles exhibit neutral stability: they continue indefinitely if undisturbed, but each disturbance to a new abundance initiates a new series of neutrally stable cycles.

the model exhibits indefinite, neutrally stable fluctuations . . .

It exhibits 'neutral stability', which means that the populations follow precisely the same cycles indefinitely unless some external influence shifts them to new values, after which they follow new cycles indefinitely (Figure 10.2e). In practice, of course, environments *are* continually changing, and populations would continually be 'shifted to new values'. A population following the Lotka–Volterra model would, therefore, fluctuate erratically. No sooner would it start one cycle than it would be diverted to a new one. For a population to exhibit regular and recognizable cycles, those cycles must themselves be stable: when an external influence changes the population level, there must be *a tendency to return to the original cycle.* Such cycles, in contrast to the neutrally stable Lotka–Volterra fluctuations, are known as *stable limit cycles.*

... which reveals an underlying tendency for predator–prey populations to undergo coupled oscillations

Nevertheless, the Lotka–Volterra model is useful in pointing to an underlying tendency for predator–prey interactions to generate fluctuations in the prey population tracked by fluctuations in the predator population (i.e. coupled oscillations); and the basic mechanism is the time-delay inherent in the sequence from many prey to many predators to few prey to few predators to many prey and so on (Table 10.1).

10.2.2 The time-delay logistic

the time-delay logistic can be applied to predator–prey interactions ...

This recognition of a time-delay provides a clue towards an alternative way of modelling a simple predator–prey system. This alternative is based on the logistic equation (section 6.9):

$$\frac{dC}{dt} = rC \left(\frac{K-C}{K} \right).$$

Here, the rate of predator population growth depends on the number present (C) and the per capita rate of increase, $r(K-C)/K$, where K is the carrying capacity of the predator population. This in turn depends on the extent to which the environment is unsaturated $(K-C)$, which in the case of a predator population can be thought of as the extent to which prey availability exceeds the predators' requirements. However, prey availability, and thus the per capita rate of predator increase, often reflects the density of predators at some *previous* time (section 6.8.4). In other words, there may be a *time-lag* in the response of the predator to its own density:

$$\frac{dC}{dt} = rC_{\text{now}} \left(\frac{K-C_{\text{now}-\text{lag}}}{K} \right)$$

If the time-lag is only small or the predator reproduces only slowly (i.e. r is small), then the dynamics of this population are not very different from the

Table 10.1. The essential similarity between the patterns of abundance generated by the Lotka–Volterra model (and predator–prey models generally) on the one hand, and the time-delay logistic model on the other hand. In both cases there is a four-phase cycle, with peaks (and troughs) of predator abundance one phase behind the peaks (and troughs) of prey abundance.

Lotka–Volterra and predator–prey in general			Time-delay logistic			
Time	Predator numbers	Prey numbers	Time	Predator numbers	Predator per capita rate of increase	$K-C_{\text{now}-\text{lag}}$
0		High	0	High		
1	High		Lag×1		Low	Low
2		Low	Lag×2	Low		
3	Low		Lag×3		High	High
4		High	Lag×4	High		
5	High		Lag×5		Low	Low

simple logistic (see May, 1981a). But for time-lags and reproductive rates that
are moderate or large, the population follows stable limit cycles. Moreover,
these stable limit cycles, once they arise in equations like the time-delay
logistic, have a length (or 'period') which is roughly equal to four times the
time-lag (May, 1981a). The essential similarity between this time-delay logistic
and the Lotka–Volterra predator–prey model can be seen in Table 10.1.

10.2.3 Delayed density-dependence

Varley (1947) used the term 'delayed density-dependence' to describe the
time-delayed regulatory effect that a predator has on a prey population when
their abundances are linked closely together. Compared to other regulatory
effects, however, those of delayed density-dependence are relatively difficult
to demonstrate.

The coupled oscillations produced by a particular predator–prey
(actually parasitoid–host) model are shown in Figure 10.3a (Hassell, 1985).
The model population is more stable than those of the previous sections, in
that the oscillations are damped, but the details of the model need not concern

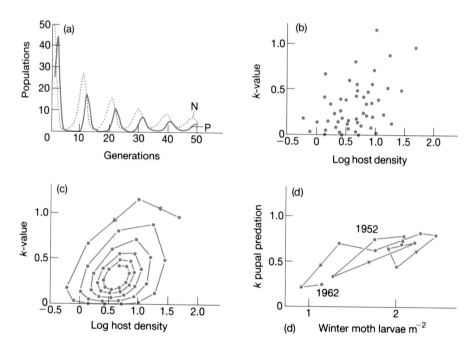

Figure 10.3. Delayed density-dependence. (a) A parasitoid–host model followed over
50 generations (Hassell, 1985): despite oscillations, the parasitoid has a regulatory
effect on the host population. (b) For the same model, the k-value of generation
mortality against the log of host density: no clear density-dependent relationship is
apparent. (c) The points from (b) linked serially from generation to generation: they
spiral in an anti-clockwise direction (Hassell, 1985). (d) Field data: the k-value for
predation (mainly by beetles) of pupae of the winter moth (*Operophthera brumata*)
plotted against the log of density. This emerges unequivocally as a density-dependent
factor in spite of a certain amount of delayed density-dependence (anti-clockwise
spiralling) (Varley *et al.*, 1973).

us. What is important is that the prey population, subject to delayed density-dependence, is regulated in size by the predator. Yet, when we plot the k-values of predator-induced mortality over a generation against the log of prey density (Figure 10.3b), no clear relationship is apparent, in spite of the fact that this is the conventional way of revealing density-dependence (Chapter 6). On the other hand, when the same points are linked serially from generation to generation (Figure 10.3c), they can be seen to spiral in an anti-clockwise direction. It is this spiralling that is characteristic of delayed density-dependence; and here, because the oscillations are damped, the points actually spiral inwards.

the regulatory effects of predators are frequently difficult to demonstrate

The model population of Figure 10.3 is not subject to the fluctuations of a natural environment; it is not subject to the density-dependent attacks of any other predator; and it is not subject to the inaccuracies of sampling error. Yet, even here, serial linking is necessary to reveal the regulatory effects of delayed density-dependence. It is not surprising, therefore, that there have been difficulties in determining the role of predators in the regulation of natural populations (Dempster, 1983; Hassell, 1985)—difficulties that have been made worse when it has not been possible to follow the same population over extended periods (which would allow serial linking). In some cases, the oscillations are damped rapidly enough for a positive density-dependent relationship to be apparent in spite of a tendency to spiral (e.g. Figure 10.3d). But we must certainly be wary of underestimating the regulatory effects of predators simply because intrinsic time-delays and natural variability combine to cloud clear density-dependent relationships.

10.2.4 Predator–prey cycles—or are they?

The inherent tendency for predator–prey interactions to generate coupled oscillations in abundance has, at times in the past, produced an 'expectation' of such oscillations in real populations. This expectation, however, should immediately be tempered by two thoughts. First, there are many important aspects of predator and prey ecology that have not been considered in the models derived so far; and as subsequent sections will show, these can greatly modify any expectations. Secondly, even if a population exhibits regular oscillations, this does *not* necessarily provide support for the Lotka–Volterra, the time-delay logistic or any other simple model. If a herbivore population fluctuates in abundance, that may reflect its interaction with its food *or* with its predators. And if a prey population cycles inherently, then the abundance of its predators may still track these cycles, even though the predator–prey interaction itself did not generate them. Thus predator–prey interactions *can* generate regular cycles in the abundance of both interacting populations, and they can reinforce such cycles if they exist for some other reason; but attributing a cause to regular cycles in nature is generally a difficult task (see section 15.4).

despite the underlying tendencies, predator–prey cycles are not necessarily seen—nor are they to be 'expected'

It has been possible in some cases to generate coupled predator–prey oscillations in the laboratory. Figure 10.4, for instance, shows this for a parasitoid–host system—the azuki bean weevil, *Callosobruchus chinensis*, and a braconid parasitoid, *Heterospilus prosopidis* (Utida, 1957). A more

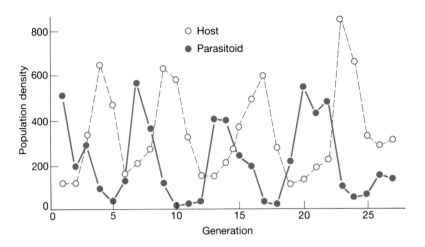

Figure 10.4. Coupled predator–prey oscillations in the laboratory: the azuki bean weevil (*Callosobruchus chinensis*) and its braconid parasitoid *Heterospilus prosopidis* (after Utida, 1957).

typical example, however, was discussed in section 9.9 (Figure 9.19): Huffaker's predatory mites and their prey were only able to sustain coupled oscillations in a heterogeneous environment. In the homogeneous type of environment that the simple models envisage, both mite populations rapidly fluctuated to extinction. Even under controlled laboratory conditions, therefore, it is necessary to take account of many aspects of predator and prey ecology before cycles of abundance can be understood.

the complex natural plant–hare–lynx–grouse cycles . . .

Amongst field populations, there are a number of examples in which regular cycles of prey and predator abundance can be discerned. These are discussed in section 15.4, but for now a single example should prove illuminating (see Keith, 1983). Cycles in hare populations have been discussed by ecologists since the 1920s, and were recognized by fur trappers more than 100 years earlier. Most famous of all is the snowshoe hare, *Lepus americanus*, which in the boreal forests of North America follows a '10-year cycle' (though in reality this varies in length between 8 and 11 years—Figure 10.1c). The snowshoe hare is the dominant herbivore of the region, feeding on the terminal twigs of numerous shrubs and small trees. Its cycle is tracked by a number of predators, including the Canada lynx (*Lynx canadensis*). There are also 10-year cycles of certain other herbivores, notably the ruffed grouse and the spruce grouse. The hare cycles often involve 10- to 30-fold changes in abundance, and 100-fold changes can occur in favourable habitats. They are made all the more spectacular by being virtually synchronous over a vast area from Alaska to Newfoundland.

The declines in snowshoe hare abundance are accompanied by low birth rates, low juvenile survivorship, high weight loss and low growth rates, all of which can be induced experimentally by food shortages. Moreover, direct measurements do indeed suggest that there is low availability of food during periods of peak hare abundance. Perhaps more important, though, is the fact that the plants respond to heavy grazing by hares by producing shoots with high levels of toxins, making them unpalatable to the hares. And particularly

... which tend to support
the predictions of the
time-delay logistic ...

significant is the fact that the plants remain protected by these chemicals for
2–3 years after heavy grazing, leading to a time-lag of roughly 2.5 years
between the decline in hare abundance and the recovery of the hares' food
supply. This, of course, is a time-lag which is one-quarter the length of the
cycle—precisely what the simple models predicted. Overall, therefore, it
seems to be the hare–plant interaction that generates the hare's decline, and
the time-lag that generates the cycles.

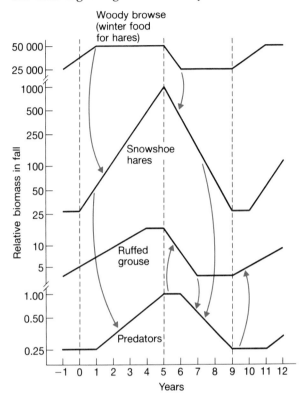

Figure 10.5. Fluctuations in the relative biomass of the major components of 'wildlife's
10-year cycle' in Alberta, Canada. The arrows indicate the major causative influences.
(Modified from Keith, 1983.)

... as part of a more
complex picture

The predators, on the other hand, *track* the hare cycles rather than
generating them. Yet the cycles are probably accentuated by the high ratio of
predators to hares as the hare population declines, and also by the low ratio
following a hare 'trough', as the hares recover before their predators (Figure
10.5). Moreover, the predators eat a large number of grouse when the
predator-to-hare ratio is high, but only a small number when the ratio is low.
It appears to be this that generates the cycles in these subsidiary herbivores
(Figure 10.5). Thus the hare–plant interaction generates cycles, the predators
track the hares, and the grouse cycles are generated by variations in predation
pressure. The simple models are clearly helpful in understanding natural
cycles in abundance, but they are far from being able to provide a complete
explanation.

10.3 The effects of self-limitation

cycles in the laboratory

To gain further insights into cycles, or indeed any other patterns of abundance in predator–prey or plant–herbivore systems, it is necessary to look beyond simple models. Other components of the interaction must be considered, and their effects investigated. To begin with, most predator–prey interactions are influenced by *self*-limitation in either one or both of the populations, i.e. by intraspecific competition or mutual interference. The effects of these processes can be investigated by incorporating them into the Lotka–Volterra predator and prey zero isoclines. The predators will be dealt with first.

10.3.1 Self-limitation in the model

a predator zero isocline with predator self-limitation

The vertical predator zero isocline in the Lotka–Volterra model expresses the idea that a constant number of prey (q/fa') is sufficient just to maintain *any* number of predators (Figure 10.6, curve A). This can immediately be improved by assuming that larger populations of predators require larger populations of prey to maintain them (Figure 10.6, curve B). In addition, it is likely that at some density, mutual interference will reduce individual consumption rates (see Chapter 9). This is turn will increase the number of prey required just to maintain a given number of predators. Thus, as predator density increases, so too will mutual interference and the requirement for additional prey (Figure 10.6, curve C). Finally, it seems likely that at high densities, even in the presence of excess food, most predator populations will be limited by availability of some other resource: nesting sites perhaps, or safe refuges of their own. This will put an upper limit on the predator population irrespective of prey numbers (Figure 10.6, curve D).

Overall then, it seems reasonable to assume that consumers generally have a zero isocline resembling curve D in Figure 10.6. Note that this is a line along which the predators neither increase nor decrease in abundance. At

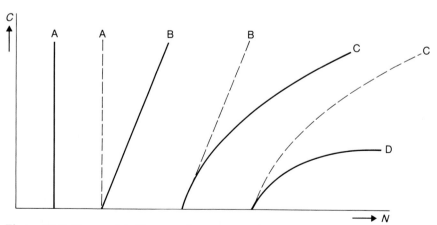

Figure 10.6. Curves A to D are predator zero isoclines of increasing complexity. A: The Lotka–Volterra isocline. B: more predators require more prey. C: consumption rate is progressively reduced by mutual interference amongst the predators. D: the predators are limited by something other than their food.

Box 10.1

The solid line in the upper figure describes variation in prey recruitment rate or plant productivity with density (see Chapter 6). The dashed lines in the upper figure describe the removal or consumption of prey by predators. In this simplest case, no distinction is made between type 1, type 2 and type 3 functional responses (section 8.5): an individual predator's consumption rate is assumed to rise linearly over the relevant density range. There is a family of dashed curves because the *total* rate of consumption depends on predator density: increasingly steep dashed curves reflect these increasing densities.

At the points where a consumption curve crosses the recruitment curve the *net* rate of prey increase is zero (consumption equal recruitment). Each of these points is characterized by a prey density and a predator density, and these pairs of densities therefore represent joint populations lying on the prey zero isocline. This has been used to construct the prey zero isocline in the lower figure (and in Figure 10.7): the shape is characteristic of a self-limited prey population and an approximately linear functional response. The arrows in the lower figure show the direction of change in prey abundance.

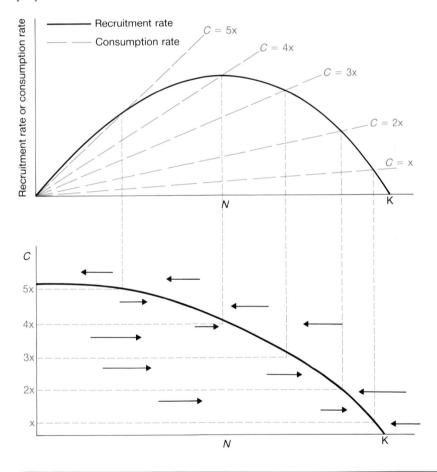

predator–prey combinations above and to the left of it they decrease, while at combinations below and to the right of it they increase. There will be considerable variation from predator to predator in the location of the isocline on the axes, and in the densities at which particular effects become apparent.

The detailed methods of incorporating intraspecific competition into the prey zero isoclines are described in Box 10.1. The end-result shown in Figure 10.7, however, can be understood without reference to these details. At low prey densities there is no intraspecific competition, and the prey isocline is the same as in the Lotka–Volterra model. But as density and intraspecific competition increase, the isocline is increasingly depressed until at the carrying capacity (K_N) it actually reaches the prey axis; in other words, at a density of K_N, the prey population can only just maintain itself even in the absence of predators.

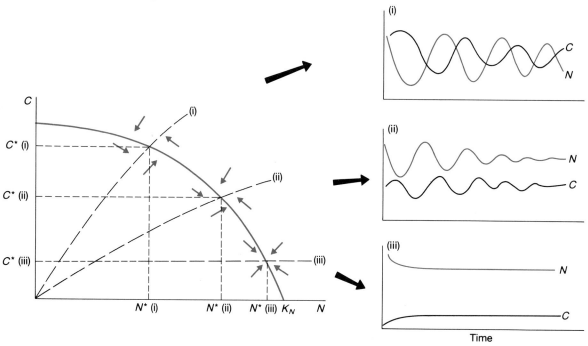

Figure 10.7. A prey zero isocline with self-limitation, combined with predator zero isoclines with increasing levels of self-limitation: (i), (ii) and (iii). C^* is the equilibrium abundance of predators, and N^* the equilibrium abundance of prey. Combination (i) is least stable (most persistent oscillations), and has most predators and least prey: the predators are relatively efficient. Less efficient predators (ii) give rise to a lowered predator abundance, an increased prey abundance and less persistent oscillations. Strong predator self-limitation (iii) can eliminate oscillations altogether, but C^* is low and N^* close to K_N.

The likely effects of self-limitation in either population can now be deduced by incorporating the isoclines from Figure 10.6 in Figure 10.7. Oscillations are still apparent for the most part, but these are no longer neutrally stable. Instead they are either stable limit cycles or are actually damped so that they converge to a stable equilibrium. Predator–prey inter-actions in which either or both populations are substantially self-limited, therefore, are likely to exhibit patterns of abundance which are relatively stable, i.e. in which fluctuations in abundance are relatively slight.

More particularly, when the predator is relatively inefficient, that is when many prey are needed to maintain a small population (curve (ii) in Figure 10.7), the oscillations are damped quickly but the equilibrium prey abundance (N^*) is not much less that the equilibrium in the absence of

... but this is most
marked for inefficient
predators where prey
abundance is therefore
little affected

predators (K_N). By contrast, when the predators are more efficient (curve (i)), N^* is lower and the equilibrium density of predators, C^*, is higher—but the interaction is less stable (the oscillations are more persistent). Moreover, if the predators are very strongly self-limited, perhaps by mutual interference, then stability is relatively high and abundance may not oscillate at all (curve (iii)); but C^* will tend to be low, while N^* will tend to be not much less than K_N. Hence, for interactions with self-limitation, there appears to be a contrast between stable patterns of abundance in which predator density is low and prey abundance is little affected, and less stable patterns in which predator density is higher and prey abundance is more drastically reduced.

10.3.2 Self-limitation in practice

ground squirrels provide
support for the
stabilizing effects of self-
limitation ...

There are certainly examples that appear to confirm the stabilizing effects of self-limitation in predator–prey interactions. For instance, there are two groups of primarily herbivorous rodents that are widespread in the Arctic: the microtine rodents (lemmings and voles) and the ground squirrels. The microtines are renowned for their dramatic, cyclic fluctuations in abundance (see section 15.4.2), but the ground squirrels have populations that remain remarkably constant from year to year. Significantly, the ground squirrels are strongly self-limited by their aggressive territorial defence of burrows used for breeding and hibernating, and it is to this that their stability has been convincingly attributed (Batzil, 1983).

An example that lends more specific support to Figure 10.7 is provided by work on an avian herbivore, the red grouse (*Lagopus lagopus scoticus*) feeding on heather (*Calluna vulgaris*) on Scottish moorlands (Watson & Moss, 1972; see also Caughley & Lawton, 1981). Heather comprises at least 90% of the red grouse's diet over most of the year, and it is the dominant (and sometimes virtually the only) higher plant on the moors where the grouse live. The grouse themselves are strongly territorial, with the size of the spring breeding population being determined by the number of territories established by cocks in the previous autumn. The 'surplus' birds are then forced from the moor, and account for the bulk of the overwinter mortality. Some of the grouse populations are fairly constant in size, though others fluctuate over perhaps a three-fold range of densities (Jenkins *et al.*, 1967); but in line with Figure 10.7 (curves (ii) and (iii)) the moors support a great deal of heather with relatively few red grouse. In fact the grouse eat only about 2% of the total annual heather production. This may, of course, be a higher percentage of the *nutritious* heather, but the predicted effects of mutual interference on an interaction are, at the very least, not refuted by these populations.

10.4 Heterogeneity, aggregation and partial refuges

Probably the most important shortcoming in the models discussed so far has been an assumption of homogeneity—homogeneity in the environment, and homogeneity amongst the organisms themselves. We turn now to consider the consequences of relaxing this assumption.

We can do so within the context of the Lotka–Volterra isoclines by noting that various types of heterogeneity all lead to a disproportionately low predation rate at low prey densities. This is true, in particular, when prey are distributed patchily, the predators have an aggregative response, and a certain number of prey therefore have a 'partial refuge' (section 9.9). It is also true where the prey have an actual refuge, within which a number escape predation altogether. In such cases, the predation rate is low at low prey densities because a number of prey are relatively or totally immune from predation. The 'heterogeneity', therefore, rests in a variation in the prey's susceptibility to predation.

partial refuges lead to vertical prey isoclines at low prey densities . . .

Box 10.2 describes the derivation of a prey zero isocline in these circumstances. The end-result is shown in Figure 10.8; the isocline rises vertically or near-vertically at low prey densities, indicating that prey at low densities can increase in abundance irrespective of predator density, since they are largely unaffected by the predators.

Box 10.2

The figures in this box are analogous to those in Box 10.1, but here the total rate of consumption at low prey densities is low (left) or zero (right), irrespective of predator abundance. This leads to a vertical prey zero isocline at low densities.

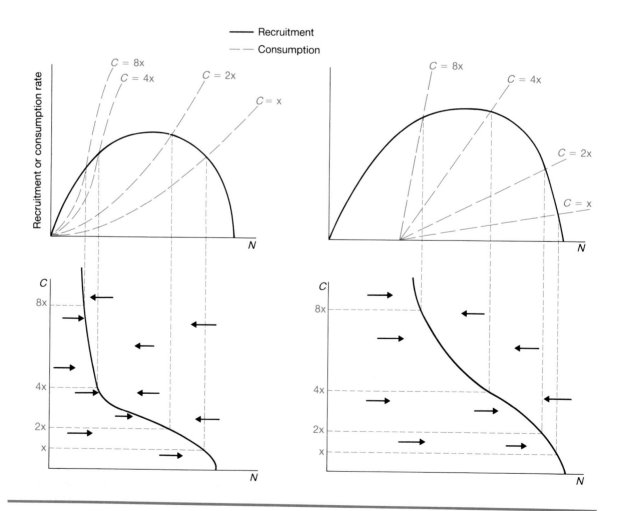

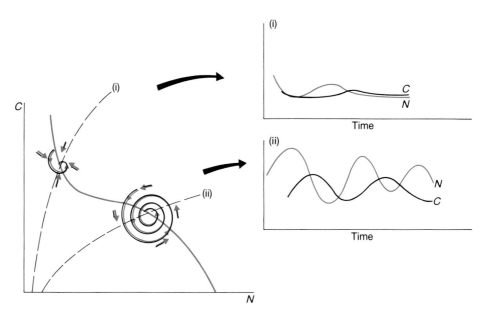

Figure 10.8. The prey zero isocline is that which is appropriate when consumption rate is particularly low at low prey densities because of an aggregative response (and partial refuge), or because of an actual refuge, or because of a reserve of plant material that is not palatable. With a relatively inefficient predator, predator zero isocline (ii) is appropriate and the outcome is not dissimilar from Figure 10.7. However, a relatively efficient predator will still be able to maintain itself at low prey densities. Predator zero isocline (i) will therefore be appropriate, leading to a stable pattern of abundance in which prey density is well below the carrying capacity and predator density is relatively high.

The predicted outcome from such an isocline depends on where along its length it crosses the predator isocline. If the two cross towards the right of the prey isocline (curve (ii), Figure 10.8) then the outcome is more or less the same as has previously been described (Figure 10.7). But if they cross towards the left (curve (i)), in the vertical section of the isocline, then the outcome is a highly stable pattern of abundance with N^* much less than K_N. Until now, no other model has predicted such an outcome. The curves cross in this way when the predator has a high searching efficiency relative to the reproductive rate of the prey (pushing its isocline to the left).

. . . which can lead to low, stable prey abundances when the predator is also efficient

On balance, therefore, this type of heterogeneity is clearly a stabilizing influence. More specifically, a predator will have the ability to depress prey abundance to a stable level well below K_N if it combines an aggregative response (leading to stability) with a high searching efficiency (leading to low prey abundance).

heterogeneity is therefore stabilizing— as we have seen before

Such a conclusion is clearly a reiteration of a theme that has occurred several times already in previous chapters: heterogeneity leads to stability. We saw this for populations of single species in Chapter 6, where contest competition and asymmetric competition generally (heterogeneity between winners and losers) led to particularly powerful regulation; and we saw it for interspecific competition in Chapter 7 (where species coexisted in hetero-geneous environments though they would have excluded one another in homogeneous environments). In the present case, moreover, it is important to realize that stability is conferred by heterogeneity (i.e. differential suscept-ibility to predation) not only when it arises out of an aggregative response by the predator (leading to a partial refuge). It can also occur where there is a mosaic environment giving the prey a 'refuge in time' (section 9.9), it can occur where plants maintain a reserve of material that cannot be grazed possibly lying close to or under the ground (section 8.2), and it can occur where the predators have a type 3 functional response (see next section). It can even occur if the prey themselves are simply intrinsically variable in their susceptibility to predation, perhaps because of genetic differences or because they are members of a competitive hierarchy (Chapter 6). Overall, therefore, heterogeneity in its broadest sense has a strong *prima facie* case for being considered an important stabilizing force in predator–prey interactions. But what is its role in practice?

10.4.1 Aggregation and heterogeneity in practice

The stabilizing effects of heterogeneity have already been described for the case of Huffaker's mites (section 9.9). In a similar vein, it is significant that populations of the 'cyclic' snowshoe hare (p. 349) never show cyclic be-haviour in habitats that are a mosaic of habitable and uninhabitable areas. In mountainous regions, and in areas fragmented by the incursions of agri-culture, the snowshoe hare maintains relatively stable, non-cyclic popu-lations (Keith, 1983). The effects of aggregative responses, on the other hand, are perhaps best appreciated by considering the properties and nature of biological control agents.

theoretically, aggregative responses are important for successful biological control . . .

A biological control agent is a natural enemy of a pest species (often a parasitoid of the pest, though sometimes a predator or a herbivorous insect of a weed) that is imported into an area in order to control that pest. The ability to depress prey abundance to a stable level well below K_N' is precisely what a good biological control agent should have, and this can be seen in Figure 10.9a (Beddington *et al.*, 1978). In each of the six cases of successful biological control by a parasitoid where an estimate of N^*/K_N could be made, there was a marked depression in abundance; but in four laboratory studies of host–parasitoid interactions, this depression was one or two orders of magnitude less dramatic. Beddington and his colleagues examined the behaviour of a number of simple but reasonable mathematical host–parasitoid models, and they came to a conclusion that has already been reached in a less formal and exact way here: a low, stable value of N^*/K_N can only be obtained from models which incorporate an aggregated distribution of parasitoid-induced mortality (Figure 10.9b). This is turn leads to two further conclusions: (i) an aggregative response, leading to an aggregated distribution of ill-effects, is a desirable, and perhaps even an essential property of a successful biological control agent; and (ii) the contrast in Figure 10.9a probably reflects the simplicity and spatial homogeneity of most laboratory environments, and the lack of oppor-tunity this gives for aggregative responses.

. . . and this appears to be borne out by work on the holly leaf-miner . . .

Arising directly from this work, Heads and Lawton (1983) examined some of the natural enemies of the holly leaf-miner (*Phytomyza ilicis*, a small agromyzid fly) which is a pest species in Canada. The production of faultless

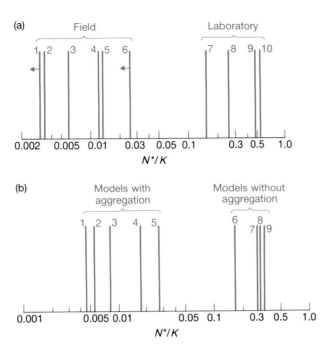

Figure 10.9. The reduction in prey (i.e. host) abundance in a number of parasitoid–host studies and a number of parasitoid–host models. (a) In six biological control studies in the field, host abundance was reduced to the order of 1% of carrying capacity or less, whereas in four laboratory studies the reduced level varied from 50% to 15%. (b) In four models without an aggregative response on the part of the parasitoid, the maximally reduced level varied from 30% to 15%, whereas in models which included an aggregative response the reduced levels varied from 3% to less than 0.5%. (From Beddington *et al.*, 1978.)

ornamental holly requires the abundance of *P. ilicis* to be reduced to a stable low level. Of the parasitoids and predators considered by Heads and Lawton, one, the larval parasitoid *Chryoscharis gemma,* produced an aggregated distribution of host mortality. Significantly, this distribution became most marked at the lowest host densities. On the other hand, there was no evidence of aggregative responses from any other of the natural enemies, including the pupal parasitoids *Sphegigaster pallicornis* and *Chrysocharis pubicornis.* Heads and Lawton were working not in Canada but in York in northern England. Nonetheless, it is noteworthy that whereas neither *S. pallicornis* nor *C. pubicornis* was apparently successful in combating the leaf-miner in Canada, *C. gemma* was reported by the entomologists working on the project to be an effective control agent. It does indeed appear that in biological control, and more generally too, spatial heterogeneity and aggregative responses may be crucial in producing stable low-density patterns of predator–prey abundance.

On the other hand, a case of successful biological control (of the olive scale, *Parlatoria oleae*) has been described in which no aggregative responses were apparent even though they were searched for (Murdoch *et al.,* 1984). Undoubtedly this is an area of theoretical and experimental population dynamics that would repay further study.

Despite these uncertainties, the importance of heterogeneity appears to be reaffirmed at the next trophic level down in the control of the prickly-pear cacti *Opuntia inermis* and *Opuntia stricta* in Australia (Monro, 1967; see p. 585). Prior to the introduction of the moth *Cactoblastis cactorum* in 1925, vast areas of otherwise useful land in Australia were covered with dense stands of these cacti. *Cactoblastis* was then introduced precisely because it was a natural consumer of the prickly-pear in South America, where they both originated. As the moths spread to more and more areas between 1928 and 1935, the dense stands of cacti were virtually wiped out. A population crash usually took about two years, and the prickly-pear has remained at low (and economically acceptable) levels ever since ($N^*/K_N \simeq 0.002$, according to Caughley and Lawton (1981)).

The *Cactoblastis* larvae are doubly aggregated: eggs are laid 80 at a time in egg-sticks, and the egg-sticks themselves have a clumped distribution amongst the plants. In addition, the larvae have a very limited mobility, and they are likely to perish along with the plant itself when the plant carries too many egg-sticks (roughly, more than two per plant). Thus, in this case, the interaction is not only stabilized at low densities by the partial refuge which the cacti have as a result of the moth's aggregations, but it is also stabilized by the high death rate of larvae on overloaded plants at high larval densities.

On balance then, it appears that a good case can be made for spatial and other heterogeneities frequently being important when predators exert profound regulatory effects on prey populations. However, as explained in section 10.2.3, this can be difficult to demonstrate simply on the basis of generation-to-generation mortality. The combination of delayed density-dependence and the fluctuations characteristic of all natural environments can lead to patterns of generation mortality which do not readily suggest population regulation. These may nonetheless hide powerful regulatory forces that act within each generation. As Hassell (1985) has remarked, the only real solution is a 'two-dimensional' sampling programme that follows a population over several generations but also documents the spatial and other variations occurring within generations. Without such a dual approach, there is a very real risk of failing to identify the true reasons for the relative stability of many natural populations.

populations must be
examined both 'spatially'
and 'temporally' in the
search for regulatory
processes

10.5 Functional responses and the Allee effect

Prey zero isoclines can be further modified to take account of the various types of functional response (section 9.5), and also to incorporate the 'Allee effect' (where the prey have a disproportionately low rate of recruitment when their own density is low, perhaps because mates are difficult to find or because a 'critical number' must be exceeded before a resource can be properly exploited). However, rather than go into details once again, we will limit ourselves largely to the conclusions that can be derived from such modifications.

As already explained, a type 3 response will in principle have the same effect as a partial refuge: a low predation rate where prey density is low. In a homogeneous environment, however, there is a contradiction in the idea of a

the 'stabilizing' effects of
type 3 responses are
probably of little
importance in
practice . . .

. . . but switching type 3
responses can stabilize
prey abundance

predator having a type 3 response (ignoring prey at low densities) and also being highly efficient at low prey densities. Curve (ii) in Figure 10.8 will therefore apply, and the stabilizing influence of a type 3 response (section 9.5.4) will in practice be of little importance.

On the other hand, a predator may have a type 3 function response to one particular type of prey because it *switches* its attacks amongst various prey types depending on which are most abundant. In such a case, the population dynamics of the predator would be independent of the abundance of any particular prey type, and the position of its zero isocline would therefore be the same at all prey densities. As Figure 10.10 shows, this can lead potentially to the predators regulating the prey at a low and stable level of abundance.

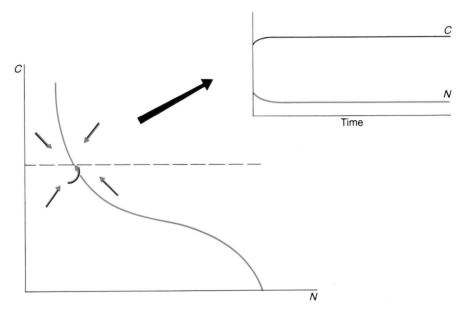

Figure 10.10. When a type 3 functional response arises because the predator exhibits switching behaviour, the predator's abundance may be independent of the density of any particular prey type (main figure), and the predator zero isocline may therefore be horizontal (unchanging with prey density). This can lead to a stable pattern of abundance (inset) with prey density well below the carrying capacity.

An apparent example of this is provided by studies on the field vole (*Microtus agrestis*) in southern Sweden (Erlinge *et al.*, 1984). Vole populations there were preyed upon by a number of generalists: badgers, red foxes, domestic cats, weasels, buzzards and tawny owls. But these predators switched to alternative prey (notably rabbits, other rodents, insectivores, earthworms and insects) when the field voles were relatively scarce. Significantly, the field vole populations remained fairly stable from year to year. By contrast, in northern Sweden and elsewhere, where rich supplies of alternative prey to the voles are not present, the field vole populations exhibit marked cycles. Hence, it is possible, though by no means proven, that the switching of the predators in southern Sweden is responsible for stabilization in the manner predicted by the simple model.

If the predator has a type 2 response that reaches its plateau well below

Box 10.3

The figures in this box are analogous to those in Boxes 10.1 and 10.2. Those on the left-hand side describe the effects of a type 2 functional response which reaches a plateau at a prey density well below K_N. This means that at intermediate predator densities the consumption rate curves cross the recruitment curve twice, leading to a prey zero isocline with a hump. The figures on the right-hand side describe the effects of the prey having a recruitment rate which is particularly low or even zero at the lowest prey densities. This may come about, for instance, because of difficulties in finding mates or because the prey need to cooperate with one another in order to obtain food, and it is known as the 'Allee effect' (Allee, 1931). Once again the consumption curves cross the recruitment curve twice, and once again the prey zero isocline has a hump.

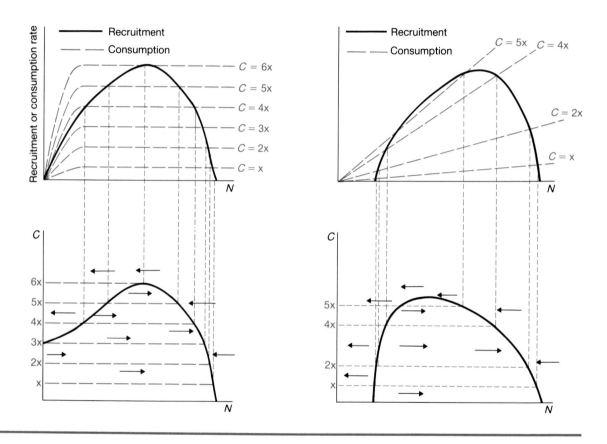

the 'destabilizing' effects of type 2 responses are also probably of little importance in practice

K_N, or if there is an Allee effect, then the prey zero isocline has a 'hump' (Box 10.3). And if the predator zero isocline crosses this to the left of the hump, then the outcome will be *divergent* oscillations, i.e. the interaction will be destabilized. However, for a type 2 response to have this effect, handling-time must be very long, whereas in practice handling-times are relatively short (Hassell, 1981). The potentially destabilizing effects of type 2 responses (section 9.5.4) are therefore likely to be of little practical importance.

The destabilizing role of the Allee effect has apparently been neither supported nor refuted quantitatively. However, the idea of a prey population declining towards extinction because it is 'too small' (i.e. below some critical density) is familiar, at least, in the context of exploited whale and fish populations (section 10.8). Decline to extinction, of course, is the ultimate in instability.

10.6 Multiple equilibria: an explanation for outbreaks

Ecologists working in a variety of fields have come to realize that when predator and prey populations interact, there is not necessarily just one equilibrium combination of the two (about which there may or may not be oscillations). There can, instead, be 'multiple equilibria' or 'alternative stable states'. The idea has been pursued most often in cases where the predator or herbivore numbers are not determined by the abundance of their prey (see May, 1977 for a review). Good examples of this would be a herd or flock of grazing mammals (where the flock size is determined by the farmer—see Noy-Meir, 1975), or a fleet of trawlers exploiting fish populations (where the fleet size is determined by the trawler owners—or perhaps by international treaty). Situations of this type will be considered in the sections on harvesting (below), but multiple equilibria can also emerge when predators and prey *interact* with one another.

models can be constructed with multiple equilibria...

Figure 10.11 is a model with multiple equilibria. The prey zero isocline has both a vertical section at low densities *and* a hump. This could reflect a

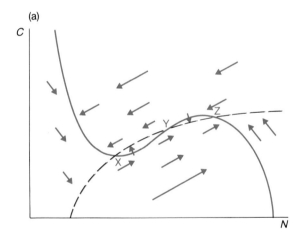

(a)

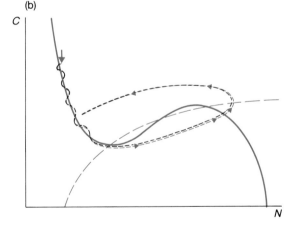

(b)

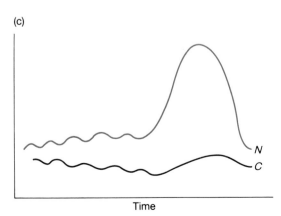

(c)

Figure 10.11. A predator–prey zero isocline model with multiple equilibria. (a) The prey zero isocline has a vertical section at low densities and a hump; the predator zero isocline can therefore cross it three times. Intersections X and Z are stable equilibria, but intersection Y is an unstable 'break-point' from which the joint abundances move towards either intersection X or intersection Z. (b) A feasible path that the joint abundances might take when subject to the forces shown in (a). (c) The same joint abundances plotted as numbers against time, showing that an intersection with characteristics that do not change can lead to apparent 'outbreaks' in abundance.

type 3 functional response of a predator which also has a long handling-time, or perhaps the combination of an aggregative response and a decline in prey recruitment at low density. As a consequence, the predator zero isocline crosses the prey zero isocline three times. The strengths and directions of the arrows in Figure 10.11a indicate that two of these points (X and Z) are fairly stable equilibria (though some oscillations around each are to be expected). The third point however (point Y) is unstable: populations near here will move towards either point X or point Z. Moreover, there are joint populations close to point X where the arrows lead to the zone around point Z, and joint populations close to point Z where the arrows lead back to the zone around point X. Even small environmental perturbations could put a population near point X on a path towards point Z and vice versa.

. . . leading to 'outbreak' or 'eruptive' patterns of abundance

The behaviour of a hypothetical population, consistent with the arrows in Figure 10.11a, is plotted in Figure 10.11b on a joint abundance diagram, and in Figure 10.11c as a graph of numbers against time. The prey population, in particular, displays an 'outbreak' or 'eruption' in abundance, as it moves from a low-density equilibrium to a high-density equilibrium and back again. And this eruption is in no sense a reflection of an equally marked change in the environment. It is, on the contrary, a pattern of abundance generated by the interaction itself (plus a small amount of environmental 'noise'), and in particular it reflects the existence of multiple equilibria. Similar explanations may be invoked to explain apparently complicated patterns of abundance in nature.

10.6.1 Multiple equilibria in nature?

There are certainly examples of natural populations exhibiting outbreaks of abundance from levels that are otherwise low and apparently stable (Figure 10.12a,b,c); and there are other examples in which populations appear to alternate between two stable densities (Figure 10.12d). But it does not follow that each of these examples is necessarily an interaction with multiple equilibria.

In some cases a plausible argument for multiple equilibria can be put forward. This is true, for instance, of Clark's (1964) work in Australia on the eucalyptus psyllid (*Cardiaspina albitextura*), a homopteran bug (Figure 10.12a). These insects appear to have a low-density equilibrium maintained by their natural predators (especially birds), and a much less stable high-density equilibrium reflecting intraspecific competition (the destruction of host-tree foliage leading to reductions in fecundity and survivorship). Outbreaks from one to the other can occur when there is just a short-term failure of the predators to react to an increase in the density of adult psyllids.

Figure 10.12. Possible examples of outbreaks and multiple equilibria. (a) Mean ratings of relative abundance of the eucalyptus psyllid, *Cardiaspina albitextura,* in three study areas in Australia (A5, A7 and A9) (after Clark, 1962). (b) Observed changes in numbers of the spruce budworm, *Choristoneura fumiferana,* in Canada (after Morris, 1963).
(c) Counts of pupae of the pine beauty moth, *Panolis flammae,* whose larvae eat needles of the Scots pine in Germany (after Schwerdtfeger, 1941). (d) Density fluctuations of the bobwhite quail, *Colinus virginianus,* in Wisconsin, U.S.A. (after Errington, 1945).

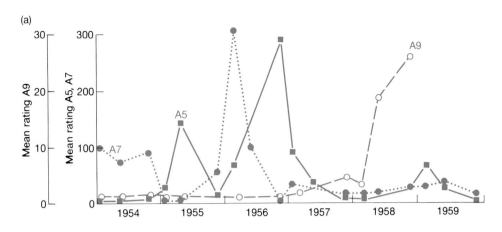

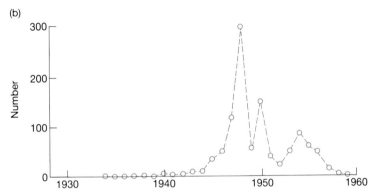

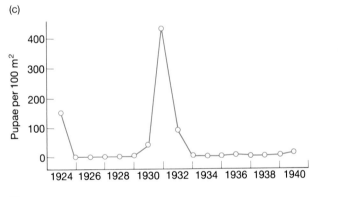

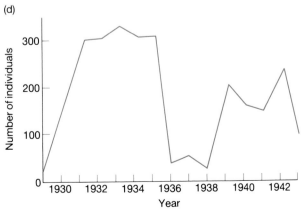

Spruce budworm populations (*Choristoneura fumiferana*; Figure 10.12b) also appear to exhibit multiple equilibria, though probably not of the exact type described in Figure 10.11 (see Peterman *et al.*, 1979). In an immature balsam fir and white spruce forest, or in a mixed spruce–fir–hardwood stand, the quality of food for the budworm is low and their rate of recruitment is low. It has therefore been hypothesized that they are held in check at a low-density stable equilibrium by their predators (as in Figure 10.8, curve (i)), though there is no experimental proof of this important assumption. However, as a spruce and fir forest gradually matures, food suitability improves, and there is an increase in the rate of budworm recruitment and in the carrying capacity of the forest. The spruce budworm zero isocline may then rise above that of its predators such that the isoclines only cross at a high-density equilibrium (as in Figure 10.7). This outbreak, however, quickly destroys the mature trees. The forest therefore reverts to relative immaturity, and the isoclines and equilibria revert to those in Figure 10.8.

The important point is that there can be a large change in abundance (i.e. an outbreak and subsequent crash) as a result of either a small change in carrying capacity (i.e. maturity) or a small environmental perturbation when maturity is close to a critical threshold value.

... but sudden changes in abundance can also result from sudden changes in the environment

On the other hand, there are many cases in which sudden changes in abundance are fairly accurate reflections of sudden changes in the environment or a food source. The number of herons nesting in England and Wales, for instance, normally fluctuates around 4000–4500 pairs, but the population declines markedly after particularly severe winters (Figure 10.13; Stafford, 1971). This fish-eating bird is unable to find sufficient food when inland waters become frozen for long periods, but there is no suggestion of the lower

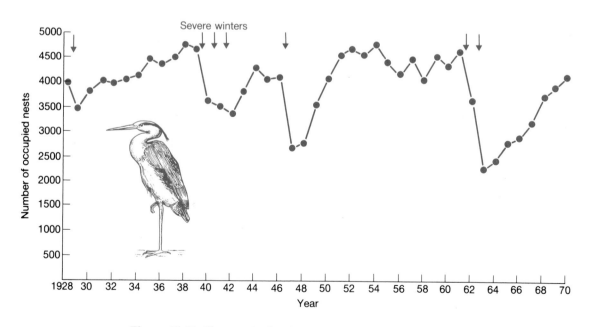

Figure 10.13. Changes in the abundance of herons (*Ardea cinerea*) in England and Wales (measured by the numbers of nests occupied) are readily attributable to changes in environmental conditions (particularly severe winters) (from Stafford, 1971).

population levels (2000–3000 pairs) being an alternative equilibrium. The population crashes are simply the result of density-independent mortality from which the herons rapidly recover.

10.7 Resumé

The simplest mathematical models of predator–prey interactions produce coupled oscillations that are highly unstable. However, by adding various elements of realism to these models it has been possible to reveal the features of real predator–prey relationships that are likely to contribute to their stability. Self-limitation in either population (intraspecific competition among prey, mutual interference among predators) and heterogeneity/ aggregative responses can be particularly powerful stabilizing influences. A further insight provided by models is that predator–prey systems may exist in more than one stable state.

We have seen that a variety of patterns in the abundance of predators and prey, both in nature and in the laboratory, are consistent with the conclusions derived from models. Unfortunately, we are rarely in a position to apply specific explanations to particular sets of data, because the critical experiments and observations to test the models have rarely been made. Moreover, natural populations are affected not just by their predators or their prey, but also by many other environmental factors that serve to 'muddy the waters' when direct comparisons are attempted with simple models. We return to the problems of 'abundance' in a broader context in Chapter 15.

10.8 Harvesting, fishing, shooting and culling

Of all the predator–prey interactions in nature, the ones that probably concern us most are those in which we ourselves are the predators. There are two main types of interaction. In the first, we remove all or part of a pest population because it is harmful to some other population which we wish to protect. In the second, we 'crop' or 'harvest' a population, utilizing the individuals we remove, and leaving other individuals to maintain a population for future harvests. This section deals with harvested populations.

In an ideal world, all harvesting operations would have the same basic aim, whether they involved the removal of whales or fish from the sea, the removal of deer from a moorland or the removal of timber from a forest. This aim would be to follow the narrow path between overexploitation and underexploitation. In a continuously overexploited population, too many individuals are removed and ultimately the population is driven to extinction; in an underexploited population there are fewer individuals removed than the population can bear, and a crop is therefore produced which may be smaller than necessary.

Unfortunately, however, this reasonable biological aim ignores the fact that harvesting is usually a commercial undertaking. Economic factors must be taken into account. In particular, from an economic point of view, *current* profits (which can be invested at a favourable rate of interest) are likely to be more valuable than future profits (which have to be waited for). In many

harvesting aims to avoid both overexploitation and underexploitation

economic considerations influence harvesting strategies

cases, therefore, it may make sense *economically* to overexploit a population, since this increases current profits at the expense of future ones. Of course, this is ecologically short-sighted, and it is a distainful (and distasteful) way of treating the hungry mouths to be fed in future generations. But profit is nonetheless an important factor that harvesters have to consider. The detailed interactions and conflicts between economic and purely biological aims are beyond the scope of this book (but for further discussions see Clark, 1976, 1981). Attention is focused instead on the central biological issues.

Table 10.2. Effects produced in populations of the blowfly *Lucilia cuprina* by the destruction of different constant percentages of emerging adults. (After Nicholson 1954b.)

Exploitation rate of emerging adults	Pupae produced per day (a)	Adults emerged per day (b)	Mean adult population (c)	Mean birth rate (per individual per day) (a/c)	Natural adult deaths per day	Adults destroyed per day (d)	Accessions of adults per day (e = b−d)	Mean adult life-span (days) (c/e)
0%	624	573	2520	0.25	573	0	573	4.4
50%	782	712	2335	0.33	356	356	356	6.6
75%	948	878	1588	0.60	220	658	229	7.2
90%	1361	1260	878	1.55	125	1134	126	7.0

A good impression of the effects of harvesting can be gained by examining the results obtained by Nicholson (1954b) when he harvested Australian sheep blowflies (*Lucilia cuprina*) from cultures maintained in the laboratory (Table 10.2). As the exploitation or harvesting rate of emerging adults increased, both pupal production and adult emergence increased, and the rate of natural adult mortality decreased (leading to an increased average life-span for the non-harvested adults). In other words, harvesting reduced density and therefore reduced intraspecific competition. This increased the survivorship and life-time fecundity of those that remained. However, although the adult population size was depressed to well below the apparent carrying capacity (down to 878, compared with 2520 at a zero exploitation rate), the *yield* from the harvest increased as the exploitation rate rose to 90%. High yields are thus obtained from populations held below (and in this case well below) the carrying capacity. The crucial point though is that the responses of a harvested population, and the yield obtained from it, depend critically on the reductions in intraspecific competition which harvesting induces (see section 8.3).

blowflies stress the importance of intraspecific competition in harvesting

10.8.1 A simple model of harvesting: fixed quotas

These conclusions can be expanded upon by considering the model population shown in Figure 10.14. There, net recruitment (or net productivity) is described by an n-shaped curve (section 6.4). Recruitment rate is low when there are few individuals and low when there is intense intraspecific competition. It is zero at the carrying capacity (K). The density giving the *highest* net recruitment rate depends on the exact form of intraspecific competion; this density is $K/2$ in the logistic equation (section 6.9), much less than $K/2$ in

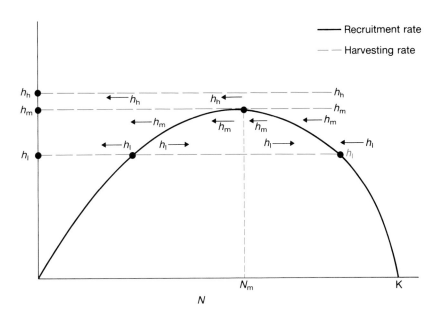

Figure 10.14. Fixed quota harvesting. The figure shows a single recruitment curve (solid line) and three fixed quota harvesting curves (broken lines); high quota (h_h), medium quota (h_m) and low quota (h_l). Arrows in the figure refer to changes to be expected in abundance under the influence of the harvesting rate specified. Dots (●) are equilibria. At h_h the only 'equilibrium' is when the population is driven to extinction. At h_l there is a stable equilibrium at a relatively high density, and also an unstable break-point at a relatively low density. The maximum sustainable yield is obtained at h_m because it just touches the peak of the recruitment curve (at a density N_m): populations greater than N_m are reduced to N_m, but populations smaller than N_m are driven to extinction.

Nicholson's blowflies (Table 10.2) and only slightly less than K in many large mammals (Figure 6.8c). It is, though, always less than K.

Figure 10.14 also illustrates three possible harvesting strategies, though in each case there is a constant harvesting *rate*, i.e. a constant number of individuals removed during a given period of time. (In an actual harvesting operation, this would be described as a *fixed quota*.) When the harvesting and recruitment lines cross, the harvesting and recruitment rates are equal and opposite; the number removed per unit time by the harvester equals the number recruited per unit time by the population. Of particular interest is the harvesting rate h_m, the line which crosses (or, in fact, just touches) the recruitment rate curve at its peak. This is the highest harvesting rate that the population can match with its own recruitment. It is therefore known as the *maximum sustainable yield* or MSY. (The maximum yield would be obtained by harvesting the whole population—but this could only be done once, not sustained.) The MSY, then, represents the desired balance between underexploitation and overexploitation. It is equal to the maximum rate of recruitment, and it is obtained from the population by depressing it to the density at which the recruitment rate curve peaks (always below K).

the 'maximum sustainable yield' is obtained from the peak of the net recruitment curve

The MSY concept is central to much of the theory and practice of harvesting. This makes the recognition of its shortcomings all the more essential. In the first place, by treating the population as a number of similar individuals, or as an undifferentiated 'biomass', it ignores all aspects of population structure such as size- or age-classes and their differential rates of growth, survival and reproduction; alternatives which incorporate structure are considered below. Secondly, by being based on a single recruitment curve, it treats the environment as unvarying. This, however, is a problem common to virtually all harvesting models (but see Iles, 1973), and usually the only possible response is to make allowances for bad years by building safety factors into recommendations. Even ignoring these problems though, there is a third difficulty: obtaining an estimate of MSY. In order to do this, it is necessary to have estimates of both population sizes and recruitment rates. However, many actual data sets are far from perfect, and obtaining the relevant data may be impractical or judged too expensive. The biologist then has no alternative but to 'make do' with what is available. It is not unusual, for example, for the shape of the recruitment curve simply to be assumed. Thus the MSY level is often taken to be half the carrying capacity (as it is in the logistic). Sadly, when real recruitment curves are examined, such assumptions are frequently found to be unjustified.

the MSY concept has severe shortcomings . . .

Despite all these difficulties, the MSY concept has dominated resource management for many years in such fields as fisheries, forestry and wildlife exploitation. Prior to recent changes, for example, there were 39 agencies for the management of marine fisheries, every one of which was required by its establishing convention to manage on the basis of an MSY objective (Clark, 1981). In many other areas the MSY concept is still the guiding principle. The approach is clearly risky but not entirely without value; MSY can be an important *component* of a management scheme—so long as it is estimated carefully and utilized cautiously (see below).

. . . but it is frequently used

Returning to Figure 10.14, we can see that the MSY density (N_m) is an equilibrium (gains = losses), but because harvesting is based on the removal of a fixed quota, it is a very fragile equilibrium. If the density exceeds the MSY density, then h_m exceeds the recruitment rate and the population declines towards N_m. This, in itself, is satisfactory. But if, by chance, the density is even slightly less than N_m, then h_m will once again exceed the recruitment rate. Density will then decline even further (Figure 10.14); and if, when this happens, a fixed quota at the MSY level is maintained, the population will decline until it is extinct. Furthermore, if the MSY is even slightly over-estimated, the harvesting rate will always exceed the recruitment rate (h_h in Figure 10.14). Extinction will then follow, whatever the initial density. In short, a fixed quota at the MSY level might be desirable and reasonable in a wholly predictable world about which we had perfect knowledge. But in the real world of fluctuating environments and imperfect data sets, these fixed quotas are open invitations to disaster. Regrettably, it has sometimes proved difficult for those involved in fishing and other industries to understand that an estimate of MSY is not a catch that can be taken from the population every year regardless of the prevailing conditions (Clark, 1981).

fixed quota MSY harvesting is extremely risky

10.8.2 Fixed quota harvesting in practice

Nevertheless, a fixed quota strategy has frequently been used. A management agency typically formulates an estimate of the MSY, and this is then

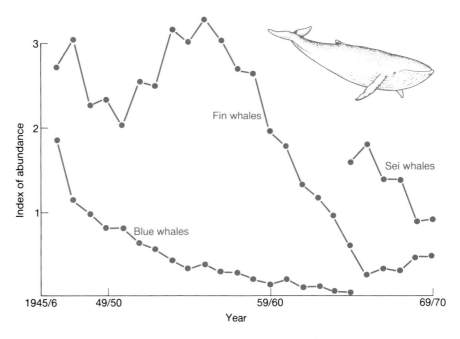

Figure 10.15. The declines in the abundance of antarctic baleen whales under the influence of human harvesting (after Gulland, 1971).

the dangers of fixed quota MSY harvesting are illustrated by whaling . . .

adopted as the annual quota. On a specified day in the year, the fishery (or hunting season or whatever) is opened and the accumulated catch is logged. Then, when the quota has been taken, the fishery is closed for the rest of the year. This obviously encourages fishermen to compete; and when annual quotas were used by the International Whaling Commission from 1949 to 1960, the ensuing scramble was commonly referred to as 'the whaling Olympics' (Clark, 1981). The whaling nations eventually agreed to allocate quotas to one another before the season opened; but even so, the effects of fixed quotas on whale populations have not been encouraging (Figure 10.15).

. . . and by the Peruvian anchoveta fishery

Another, fairly typical example of the use of fixed quotas is provided by the Peruvian anchoveta (*Engraulis ringens*) fishery. From 1960 to 1972 this was the world's largest single fishery, and it constituted a major sector of the Peruvian economy. Fisheries experts advised that the MSY was around 10 million tons annually, and catches were limited accordingly. But the fishing capacity of the fleet expanded and in 1972 the catch crashed. Over-fishing seems to have been the main cause (Murphy, 1977). A moratorium on fishing might have allowed the stocks to recover, but this was not politically feasible: 20 000 people were dependent on the anchoveta industry for employment. The Peruvian government has therefore allowed fishing to continue each year. The poor catches have continued to decline.

Of course, it is not necessary to fix quotas at the MSY level itself, or to leave quotas unchanged from year to year. The management procedure instituted by the International Whaling Commission in 1975, for instance, bases the regulation of each stock on the current size of the stock relative to the MSY level. It then fixes quotas as a proportion of the estimated MSY. Along similar lines, the quota from the North Sea plaice fishery is 112 000 tonnes annually, but this is reduced if there is poor recruitment (which is monitored). This is clearly more enlightened than waiting for low catches before changing strategy. But it also demands good population dynamics data, and in many fisheries these are non-existent.

10.8.3 The regulation of harvesting effort

The risk associated with fixed quotas can be reduced if instead there is regulation of the harvesting *effort*. The yield from a harvest (h) can be thought of, simply, as being dependent on three things:

$$h = g.E.N.$$

Yield h increases with the size of the harvested population N; it increases with the level of harvesting effort, E (for instance, the number of 'trawler–days' in a fishery or the number of 'gun–days' with a hunted population); and it increases with harvesting efficiency, g. On the assumption that this efficiency remains constant, Figure 10.16 depicts an exploited population subjected to three potential harvesting strategies differing in harvesting effort.

Choosing an effort that should lead to the MSY (E_m) is a much safer strategy than fixing an MSY quota. Now, in contrast to Figure 10.14, if density drops below N_m, recruitment exceeds the harvesting rate and the population recovers. In fact, there needs to be a considerable overestimate of E_m before the population is driven to extinction (E_o in Figure 10.16). However, because there is a constant effort, the yield varies with population size. In particular, the yield will be less than the MSY whenever the population size drops below N_m. The appropriate reaction to this would be to reduce effort slightly or at

regulating harvesting effort is less risky, but it leads to a more variable catch

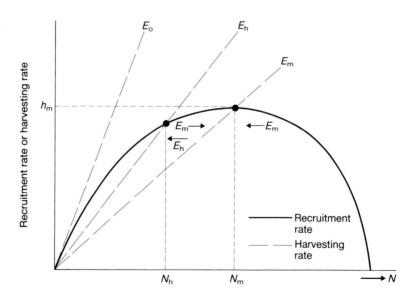

Figure 10.16. Constant-effort harvesting. Curves, arrows and dots as in Figure 10.14. The maximum sustainable yield is obtained with an effort of E_m, leading to a stable equilibrium at a density of N_m with a yield of h_m. At a somewhat higher effort (E_h), the equilibrium density and the yield are both lower than with E_m. Only at a much higher effort (E_o) is the population driven to extinction.

least hold it steady while the population recovers. But an understandable (albeit misguided) reaction would be to compensate by increasing the effort. This, however, might depress population size further (E_h in Figure 10.16); and it is therefore easy to imagine the population being driven to extinction as very gradual increases in effort chase an ever-diminishing yield.

There are many examples of harvests being managed by legislative regulation of effort, and this occurs in spite of the fact that effort usually defies precise measurement and control. Issuing a number of gun licences, for instance, leaves the accuracy of the hunters uncontrolled; and regulating the size and composition of a fishing fleet leaves the weather to chance. Nevertheless, the harvesting of mule deer, pronghorn antelope and elks in Colorado are all controlled by issuing a limited but varying number of hunting permits (Pojar, 1981). And in the management of the important Pacific halibut stock, effort is limited by seasonal closures and sanctuary zones—though a heavy investment in fishery-protection vessels is needed to make this work (Pitcher & Hart, 1982).

10.8.4 The instability of harvested populations—multiple equilibria

Even with the regulation of effort, however, harvesting near the MSY level may be courting disaster. Recruitment rate may be particularly low in the smallest populations (Figure 10.17a, Box 10.3), or harvesting efficiency may decline in large populations (Figure 10.17b) perhaps because the species expands its range. In either case, small overestimates of E_m are liable to lead to overexploitation or even eventual extinction (Figure 10.17).

Such responses, like those in Figure 10.14, are simply further illustrations of 'multiple equilibria'. As quotas or harvesting efforts are gradually increased from low levels, the equilibrium population size gradually declines.

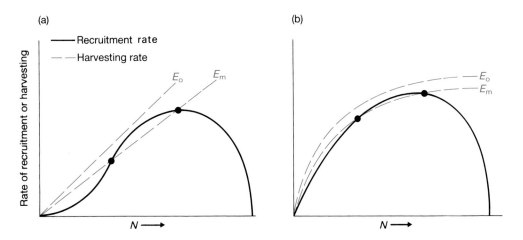

Figure 10.17. Multiple equilibria in harvesting. (a) When recruitment rate is particularly low at low densities, the harvesting effort giving the maximum sustainable yield (E_m) has not only a stable equilibrium but also an unstable break-point at a density below which the population declines to extinction. The population can also be driven to extinction by harvesting efforts (E_o) not much greater than E_m. (b) When harvesting efficiency declines at high densities, comments similar to those in (a) are appropriate.

many harvesting
operations have multiple
equilibria, and are
therefore susceptible to
dramatic, irreversible
crashes

But as soon as the quota exceeds the MSY (Figure 10.14) or the effort exceeds some threshold (Figure 10.17), the situation suddenly alters: the only 'equilibrium' is a population size of zero and a group of harvesters with nothing to harvest. Note too that even fixed quota levels below the MSY have alternative equilibria (h_1 in Figure 10.14). So long as density exceeds the lower equilibrium, the population will approach the upper, stable equilibrium. But if density drops below the lower equilibrium the population will decline to extinction.

Regrettably, there are a number of examples where stocks have 'crashed' suddenly, and where they have failed to recover even when harvesting efforts have been reduced. This has been seen most importantly in a number of clupeid fisheries, including the Peruvian anchoveta, the North Sea herring (Figure 10.18) and the Pacific sardine (Murphy, 1977); and it has also been seen in certain Pacific salmon fisheries (Peterman *et al.*, 1979). Interestingly, multiple equilibria like those in Figure 10.17 have been proposed in each case. Many clupeids, for instance, are especially prone to capture at low densities, because they form a small number of large schools which follow stereotyped migratory paths that the trawlers can interrupt (Figure 10.17b); and recruitment of young salmon is low at low densities because of intense predation from larger fish (Figure 10.17a). It is obviously important for stock-managers to appreciate the consequences of multiple equilibria: drastic changes in stock abundance can result from only small changes in harvesting strategy or small changes in the environment.

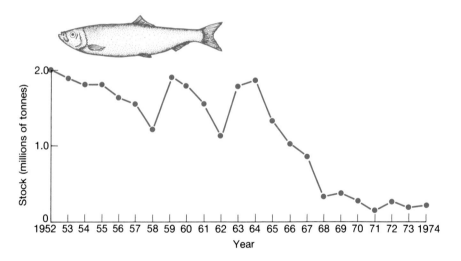

Figure 10.18. The decline in the stock of North Sea herring (measured in millions of tonnes) (from Iles, 1981).

10.8.5 Regulated-percentage and regulated-escapement harvesting

Figure 10.16 is also an appropriate way of depicting a third harvesting strategy: removing a constant *percentage* of the population. This, not surprisingly, has the same safety properties as the simple constant-effort strategy.

Finally, a fourth way of regulating harvests is to hold neither yield, nor effort, nor percentage harvest constant, but to let a constant number 'escape'. This is the safest strategy, because it is the most sensitive to changing densities. But yield and income vary even more than with a regulated-effort (or percentage) strategy; and when population size is less than the desired escapement, harvesting must be suspended completely. Moreover, with both the regulated-escapement and the regulated-percentage strategies, a variable effort must be achievable and enforceable, and the population size must be measurable at a time which allows yield and effort to be adjusted accordingly.

regulated escapement is the safest but least practicable strategy

In spite of these practical difficulties though, there are many examples where a regulated percentage or a regulated escapement is at least the management *objective*. Amongst ungulates in Arizona and New Mexico, for example, the peccary is harvested at an annual rate of 15% and mule deer at 17%; but a species with a low rate of natural recruitment, like the desert bighorn sheep, is harvested at an annual rate of less than 2% (van Dyne *et al.*, 1980).

A constant-escapement strategy can be seen most clearly in the rather special but very widespread case of harvesting seed-crop plants. For instance, a constant quantity of wheat grain or rape seed is retained (i.e. 'allowed to escape') each year, to be sown for the following season: it is the excess which is then harvested. Constant-escapement policies are also currently being used in the management of certain fisheries, including those of the Pacific salmon. There is substantial variation in the annual catches, and the fishermen's incomes fluctuate accordingly—but this seems to be accepted as part of the fisherman's lot (Clark, 1981).

10.8.6 Recognizing structure in harvested population: dynamic pool models

The simple models of harvesting that have been described so far are known as 'surplus yield' models. They are useful as a means of establishing some basic principles (like MSY), and they are good for investigating the possible consequences of different types of harvesting strategy. But they ignore population structure, and this is a bad fault for two reasons. The first is that 'recruitment' is, in practice, a complex process incorporating adult survival, adult fecundity, juvenile survival, juvenile growth and so on, each of which may respond in its own way to changes in density and harvesting strategy. The second reason is that most harvesting practices are primarily interested in only a portion of the harvested population (for instance baby seals, mature trees, or fish that are large enough to be saleable). The approach that attempts to take these complications into account involves the construction of what are called 'dynamic pool' models.

'dynamic pool' (as opposed to 'surplus yield') models recognize population structure

The general structure of a dynamic pool model is illustrated in Figure 10.19. In its simplest form it is not much more sophisticated than the surplus-yield models. Thus, for example, in constructing the first, classic dynamic pool model, Beverton and Holt (1957) made a number of simplifying assumptions. They assumed that recruitment rate was constant, that both natural and harvesting mortality rates were constant, that both of these affected an age-class either completely or not at all, and that individual growth followed

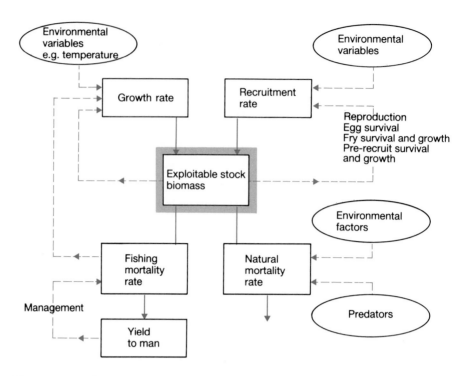

Figure 10.19. The dynamic pool approach to fishery harvesting and management, illustrated as a flow diagram. There are four main 'sub-models': the growth rate of individuals and the recruitment rate into the population (which add to the exploitable biomass), and the natural mortality rate and the fishing mortality rate (which deplete the exploitable biomass). Solid lines and arrows refer to changes in biomass under the influence of these sub-models. Dashed lines and arrows refer to influences either of one sub-model on another, or of the level of biomass on a sub-model, or of environmental factors on a sub-model. Each of the sub-models can itself be broken down into more complex and realistic systems. Yield to man is estimated under various regimes characterized by particular values inserted into the sub-models. These values may be derived theoretically (in which case they are 'assumptions') or from field data. (Modified from Pitcher & Hart, 1982.)

the trajectory of a theoretically derived curve (the von Bertalanffy curve). However, they did take two important considerations into account. First, the biomass yield to man depended not only on the number of individuals caught but also on their size (*past growth*); and secondly, the quantity of exploitable (i.e. catchable) biomass depended not just on 'net recruitment' but on an explicit combination of natural mortality, harvesting mortality, individual growth, and recruitment into the catchable age-classes.

dynamic pool models may themselves be simple or complex

More modern dynamic pool models have improved on Beverton and Holt's work in a number of ways. In the first place, growth, natural mortality, susceptibility to capture, etc., have been dealt with separately in each of the age-classes. For example, recruitment into the catchable part of the population does not need to be sudden: there can be non-catchable age-classes, gradations of poorly catchable age-classes and so on. In the second place, the four 'sub-models' (growth, recruitment, natural mortality and harvesting mortality) have, in different cases, incorporated as much or as little information as is available or desirable. Thus growth curves have been obtained directly from field samples (ageing fish, for instance, by examining their scales or their otoliths); and density-dependence in either recruitment or natural mortality has been assumed and modelled or (better still) actually monitored in the field.

These more modern dynamic pool models are reviewed very clearly (along with harvesting strategies generally) by Pitcher and Hart (1982). The basic approach is always the same. Available information (both theoretical and empirical) is incorporated into a form which reflects the dynamics of the structured population. This then allows the yield and the response of the population to different harvesting strategies to be estimated. This in turn should allow a recommendation to the stock-manager to be formulated. The crucial point is that in the case of the dynamic pool approach, a harvesting strategy includes not only a harvesting intensity—it also involves a decision as to how effort should be partitioned amongst the various age-classes.

a dynamic pool harvesting strategy includes a differential harvesting effort within the population

As soon as population structure is recognized, it has also to be recognized that harvesting is likely to alter structure; and if age-classes are affected differentially, then harvesting is certain to affect structure. This can be seen for example, in the case of Nicholson's blowflies (Table 10.2). As the rate of harvesting of adults increased, the number of adults declined; but the number of pupae (and larvae) rose. By contrast, when Slobodkin and Richman (1956) increased the exploitation rate of the youngest age-class in an experimental *Daphnia* (water flea) population, the proportion of that class in the pre-harvest populations actually rose (Figure 10.20). The precise effects of harvesting on population structure, therefore, depend on the differing extents to

harvesting itself alters population structure

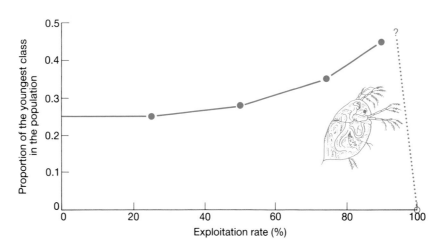

Figure 10.20. As the exploitation rate of the youngest age-class in an experimental *Daphnia* population rises, so the proportion of that class in the population rises (after Slobodkin & Richman, 1956).

which various processes experience a reduction in intraspecific competition. In the blowflies, exploitation led to an increase in pre-adult survival; in the *Daphnia* there was probably a sharp increase in adult fecundity.

One commercially exploited population for which a dynamic pool model has been produced is the Arcto-Norwegian cod fishery, the most northerly of the Atlantic stocks (Garrod & Jones, 1974; see also Pitcher & Hart, 1982). Garrod and Jones took the age-class structure of the late 1960s, and used this to predict the medium-term effects on yield of different fishing intensities and different mesh sizes in the trawl. Some of their results are shown in Figure 10.21. The temporary peak after five or so years is a result of the very large 1969 year-class working through the population. Overall, however, it is clear that the best longer-term prospects are predicted with a low fishing intensity and a large mesh size. Both of these give the fish more opportunity to grow (and reproduce) before they are caught, which is important because yield is measured in biomass, not simply in numbers. Higher fishing intensities and

dynamic pool models can lead to valuable recommendations . . .

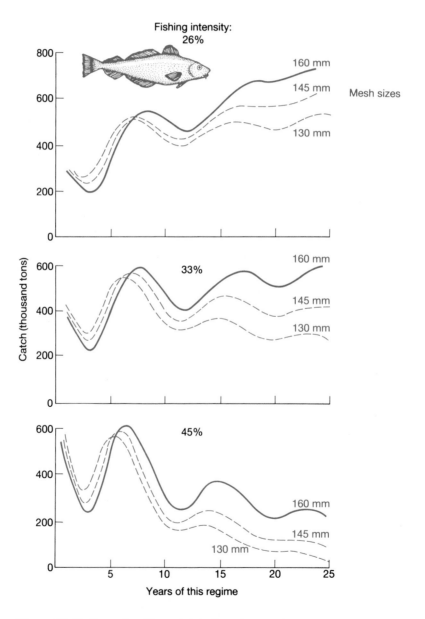

Figure 10.21. Garrod and Jones's (1974) predictions for the Arctic cod stock under three fishing intensities and with three different mesh sizes (after Pitcher & Hart, 1982).

mesh sizes of 130 mm were predicted to lead to overexploitation of the stock.

. . . but these may still be ignored

Sadly, Garrod and Jones's clear recommendations were ignored by those with the power to determine fishing strategies. Mesh sizes were not increased until 1979, and then only from 120 to 125 mm. Fishing intensity never dropped below 45% and catches of 900 000 tons were taken in the late 1970s. Not surprisingly perhaps, surveys late in 1980 showed that these and other north Atlantic cod stocks were very seriously depleted as a result of over-fishing. Garrod and Jones's work provides encouragement for the use of dynamic pool models. But the response to their work has been somewhat less encouraging.

10.8.7 Finalé

In conclusion, some words of caution are appropriate. In the real world, as the cod example suggests, there are many populations that are not harvested scientifically or ecologically. Instead, they are harvested on the basis of folklore, guess-work or sheer greed; and there are many other populations that are exploited with no management or restraint whatever. The examples in which harvesting models have actually been used successfully are rather few, and the examples in which good data have been collected are fewer still. In the case of a river fishery in a remote area where access is limited, unrestrained harvesting by anglers or sports fishermen may be acceptable. But for most exploited populations, the future can only be guaranteed (or even realistically predicted) by detailed understanding and enlightened management.

Of course, there will be many cases in which the management objective itself is a matter of dispute. Long-term ecological and short-term economic aims may often be in conflict. Moreover, in the management of a woodland, for instance, the strategy aimed at high yield (a single population with a high growth rate) is not the same as the strategy for aesthetics and recreation (a mixture of populations of mature trees close to the carrying capacity). Yet if there are contradictions to be resolved, a firm foundation of knowledge is even more essential.

Many natural exploited populations are either being pushed towards the brink of collapse, or have already passed beyond the brink. Most are under-stood in only the scantiest of detail. Harvesting theories usually outstrip the database; but the theories themselves still face many daunting challenges. Politicians, economists, and the fishermen and hunters themselves will all have to play their part if the exploitation of these populations is to continue. But ultimately and crucially, controlled and continued exploitation will depend on the progress made by the ecologists of the future.

enlightened harvesting requires enlightened ecologists!

11 Decomposers and Detritivores

11.1 Introduction

When plants and animals die, their bodies become resources for other organisms. Of course, in a sense, all consumers live on dead material—the carnivore catches and kills its prey, and the living leaf taken by a herbivore is dead by the time digestion starts. The critical distinction between the organisms that we are concerned with in this chapter, and the categories of herbivore, carnivore and parasite, is that the latter all directly affect the rate at which their resources are produced. Whether it is lions eating gazelles, gazelles eating grass or grass parasitized by a rust fungus, the act of taking the resource harms its ability to regenerate new resource (more gazelles or grass leaves). Some categories of mutualist, on the other hand, can *increase* the supply of resources provided by their partner (Chapter 13). In contrast with these groups, decomposers (bacteria and fungi) and detritivores (animal consumers of dead matter) do not control the rate at which their resources are made available or regenerate; they are totally dependent on the rate at which some other force (senescence, illness, fighting, the shedding of leaves by trees) releases the resource on which they live.

decomposers and detritivores do not control their supply of resources : . . .

Mathematically, we can express these effects on rates of resource (R) renewal as follows. For carnivores, herbivores and parasites, the rate of renewal is given by:

$$\frac{dR}{dt} = F(R) - aP$$

where $F(R)$ means 'a function of the amount of the resource R', P is the number of predators and a is an expression of the efficiency with which individuals find and capture their food resource.

For mutualists:

$$\frac{dR}{dt} = F(R) + \delta M$$

where M is the number of mutualists and δ is an expression of the beneficial effect exerted by individual mutualists on the rate at which resources are made available by their partners.

For decomposers and detritivores:

$$\frac{dR}{dt} = F(R). \tag{11.1}$$

In this case there is no expression involving the consumers, because they have no direct effect on the rate at which their resources become available.

Pimm (1982) describes the relationship between decomposers or detritivores and their food as *donor-controlled*: the donor (prey) controls the density of the recipient (predator) but not the reverse. The dynamics of donor-controlled models differ in a number of ways from the traditional Lotka–Volterra models of predator–prey interactions (Chapter 10). One important distinction is that interacting assemblages of species which possess donor-controlled dynamics are expected to be particularly stable, and further, that stability is independent of, or actually increases with, increased species diversity and food web complexity. This is in complete contrast to the situation where Lotka–Volterra dynamics apply. Detailed discussion of the important topic of food web complexity and community stability is reserved for Chapter 21.

Equation 11.1 indicates that there is no direct feedback between the consumer population and the resource. Nevertheless, it is possible to see an indirect effect, for example via the release of nutrients from decomposing litter, which may ultimately affect the rate at which trees produce more litter. In fact, it is in nutrient recycling that decomposers and detritivores play their most fundamental role.

Immobilization occurs when an inorganic nutrient element is incorporated into organic form—primarily during the growth of green plants. Conversely, decomposition involves the release of energy and the *mineralization* of chemical nutrients—conversion of elements from organic to inorganic form. Decomposition is defined as the gradual disintegration of dead organic matter and is brought about by both physical and biological agencies. It culminates with complex energy-rich molecules being broken down by their consumers (decomposers and detritivores) into carbon dioxide, water and inorganic nutrients. Some of the chemical elements will have been locked up for a time as part of the body structure of the decomposer organisms, and the energy present in the organic matter will have been used to do work and is eventually lost as heat. Ultimately, the incorporation of solar energy in photosynthesis, and the immobilization of inorganic nutrients into biomass, is balanced by the loss of heat energy and organic nutrients when the organic matter is mineralized. Thus a given nutrient molecule may be successively immobilized and mineralized in a repeated round of nutrient cycling. We discuss the overall role played by decomposers and detritivores in the fluxes of energy and nutrients at the community level in Chapter 17. In the present chapter, we introduce the organisms involved and look in detail at the ways they deal with their resources.

It is not only the bodies of dead animals and plants that serve as resources for decomposers and detritivores. Dead organic matter is continually produced during the life of both animals and plants and can be a major resource. Unitary organisms shed dead parts as they develop and grow—the larval skins of arthropods, the skins of snakes, the skin, hair, feathers and horn of other vertebrates. Specialist feeders are often associated with these cast-off resources. Among the fungi there are specialist decomposers of feathers and of horn, and there are arthropods that specialize on sloughed off skin. Human skin is a resource for the household mites that are omnipresent inhabitants of house dust and cause real problems for many asthma sufferers.

The continual shedding of dead parts is even more characteristic of modular organisms. The older polyps on a colonial hydroid or coral die and decompose, while other parts of the same genet continue to regenerate new polyps. Most plants shed old leaves and grow new ones; the seasonal litter fall onto a forest floor is the most dramatic of all the sources of resource for decomposers and detritivores, but the producers do not die in the process. Higher plants also continually slough off cells from the root caps, and root cortical cells die as a root grows through the soil. This supply of organic material produces the very resource-rich *rhizosphere*. Plant tissues are generally leaky, and soluble sugars and nitrogenous compounds also become available on the surface of leaves, supporting the growth of bacteria and fungi in the *phyllosphere*.

Finally, animal faeces, whether produced by detritivores, herbivores, carnivores or parasites, are a further category of resource for decomposers and detritivores. They are composed of dead organic material that is chemically related to what their producers have been eating.

The remainder of this chapter is in two parts. In section 11.2, we describe the 'actors' in the decomposition 'play', and consider the relative roles of the bacteria and fungi on the one hand, and the detritivores on the other. Then, in section 11.3, we consider, in turn, problems and processes involved in the consumption by detritivores of plant detritus, faeces and carrion.

11.2 The organisms

11.2.1 The decomposers: bacteria and fungi

If a scavenger does not take a dead resource immediately, the process of decomposition usually starts with colonization by bacteria and fungi. Other changes may occur at the same time: enzymes in the dead tissue may start to autolyse it and break down carbohydrates and proteins to simpler, soluble forms. The dead material may also become leached by rainfall or, in an aquatic environment, may lose minerals and soluble organic compounds as they are washed out in solution.

bacteria and fungi are early colonists of newly dead material

Bacteria and fungal spores are omnipresent in the air and the water, and are usually present on (and often in) dead material before it is dead. They usually have first access to a resource, together with the necrotrophic parasites (see Chapter 12). These early colonists tend to use soluble materials, mainly amino acids and sugars, that are freely diffusible. They lack the array of enzymes necessary for digesting structural materials such as cellulose, lignin, chitin and keratin. Many species of *Penicillium*, *Mucor* and *Rhizopus*, the so-called 'sugar fungi' in soil, grow fast in the early phases of decomposition; together with bacteria having similar opportunistic physiologies, they tend to undergo population explosions on newly dead substrates. As the freely available resources are consumed, these populations collapse, leaving very high densities of resting stages from which new population explosions may develop when another freshly dead resource becomes available. They may be thought of as the opportunist 'r-selected species' among the decomposers (see Chapter 14), and can be compared with the 'tramps' and 'supertramps' among the bird colonists of islands (see pp. 174–5).

The early phases of the decomposition process are encountered in a variety of domestic and industrial situations.

domestic and industrial decomposition

(a) The early colonizers of staling bread are the opportunistic starch and sugar fungi such as *Penicillium* and *Mucor*.

(b) The early colonizers of the nectar in flowers are predominantly yeasts (simple sugar fungi), and these may spread to the ripe fruit, or explode as populations in juice to give the well known staling products that are wine and beer.

(c) Making silage from green fodder for conservation as a livestock feed, or sauerkraut from cabbage for human consumption, depends on primary colonizers that metabolize soluble or readily solubilized carbohydrates. The organisms involved are bacteria, mainly lactobacilli, which respire soluble sugars and produce organic acids (mainly lactic acid) when the fermentation is anaerobic. The acidity may 'pickle' the plant material by reducing the pH below a level at which other decomposers can act.

(d) Plants that are used as sources of stem fibres (e.g. flax and jute) are 'retted' microbially to free the fibres from the rest of the plant tissue. Bacteria are primarily responsible for this process, and they enter through cuts and damage points and grow up through the soft, easily decomposed cambial tissues, attacking pectins and hemicellulose cell walls that have not yet become thickened with cellulose and lignins. Again we see simple, opportunistic invaders as the first decomposers.

In nature, as in the various industrial and domestic processes, the activity of the early colonizers is dominated by the metabolism of sugars and is strongly influenced by aeration. When oxygen is in free supply, sugars are metabolized to carbon dioxide by growing microbes. Under anaerobic conditions, fermentations produce a less efficient breakdown of sugars to byproducts such as alcohol and organic acids that change the nature of the environment for subsequent colonizers. In particular, the lowering of the pH by the production of acids has the effect of favouring fungal as opposed to bacterial activity.

aerobic and anaerobic decomposition in nature

A strong element of chance determines which species are the first to colonize newly dead material, but in some environments there are specialists with properties that enhance their chances of arriving early. Litter that falls into streams or ponds is often colonized by aquatic fungi (e.g. Hyphomycetes), which bear spores with sticky tips (Figure 11.1a), and are often of a curious form that seems to maximize their chance of being carried to and sticking to leaf litter. They may spread by growing from cell to cell within the tissues (Figure 11.2).

after the initial phase, decomposition of more resistant tissues proceeds more slowly

After colonization of dead material by the 'sugar' fungi and bacteria, and perhaps also after leaching by rain or in the water, the residual resources are not diffusible and are more resistant to attack. In broad terms, the major components of dead organic matter are, in a sequence of increasing resistance to decomposition: sugars < (less resistant than) starch < hemicelluloses, pectins and proteins < cellulose < lignins < suberins < cutins. Hence, after

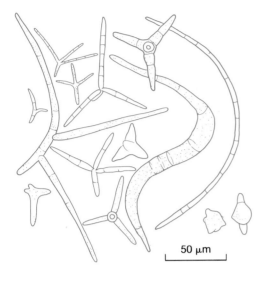

Figure 11.1. Left: spores (conidia) of aquatic hyphomycete fungi from river foam (after Webster, 1970). Right: A quartz grain in the soil, colonized by rod-shaped bacteria (from Rovua & Campbell, 1974).

an initial rapid breakdown of sugar, decomposition proceeds more slowly, and involves microbial specialists that can use celluloses and lignins and break down the more complex proteins, suberin (cork) and cuticle. These are structural compounds, and their breakdown and metabolism depend on very intimate contact with the decomposers (most cellulases are surface enzymes requiring actual physical contact between the decomposer organism and its resource). The processes of decomposition may now depend on the rate at which fungal hyphae can penetrate from cell to cell through lignified cell walls. In the decomposition of wood by fungi, two major categories of specialist decomposers can be recognized: the brown rots which can decompose cellulose but leave a predominantly lignin-based brown residue, and the white rots which decompose mainly the lignin and leave a white cellulosic residue.

there is a natural succession of decomposing microorganisms

The organisms capable of dealing with progressively more refractory compounds represent a natural succession starting with simple sugar fungi (mainly Phycomycetes and Fungi Imperfecti) and usually followed by septate

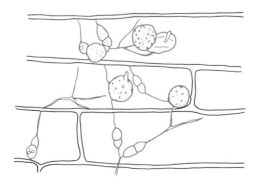

Figure 11.2. Rhizomycelium of the aquatic fungus *Cladochytrium replicatum* within the epidermis of an aquatic plant. The circular bodies are zoosporangia. (From Webster, 1970.)

fungi (Ascomycetes, Basidiomycetes and Actinomycetes) which are slower growing, spore less freely, make intimate contact with their substrate and have more specialized metabolism (Pugh, 1980). The diversity of the microflora that decomposes a fallen leaf tends to decrease as fewer but more highly specialized species are concerned with the last and most resistant remains. An example of the succession of fungi involved in the decomposition of forest litter is given in Chapter 16, p. 616.

The nature of a decomposing resource changes during the course of the succession; indeed the changes are largely responsible for driving the succession. The changing nature of a resource is illustrated in Figure 11.3 for oak leaf litter on a woodland floor and in a small stream running through woodland. The patterns are remarkably similar. Soluble carbohydrates quickly disappeared, mainly as a result of leaching in water (and comparatively more quickly in the aquatic environment). The resistant structural cellulose and hemicellulose components were reduced through enzymatic degradation more slowly, and lignin (only measured in the woodland floor study) most slowly of all. Lipid (only measured in the steam study) was degraded at an intermediate rate.

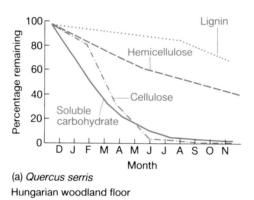

(a) *Quercus serris*
Hungarian woodland floor

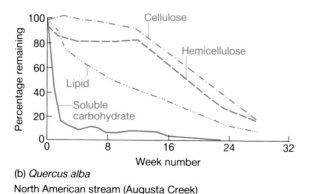

(b) *Quercus alba*
North American stream (Augusta Creek)

Figure 11.3. Changes in composition of oak leaf litter during decomposition in contrasting situations: (a) leaves of *Quercus cerris* on a woodland floor in Hungary, through the year; (b) leaves of *Quercus alba* in a small stream in North America, during a 28 week experiment. Amounts are expressed as percentages of the starting quantities. (Respectively from Toth *et al.*, 1975; and Suberkropp *et al.*, 1976.)

Individual species of microbial decomposer are not biochemically very versatile; most of them can cope with only a limited number of substrates. It is the diversity of species involved that allows the structurally and chemically complex tissues of a plant or animal corpse to be decomposed. Between them, a varied flora of bacteria and fungi can accomplish the complete degradation of dead material of both plants and animals; but in practice they seldom act alone, and the process would be much slower if they did so. The major factor that delays the decomposition of organic residues is the compartmentalized cellular structure of plant remains—an invading decomposer meets far fewer barriers in an animal body. The process of plant decomposition is enormously speeded up by any activity that grinds up and fragments the tissues, such as the chewing action of detritivores. This breaks open cells and exposes the contents and the surfaces of cell walls to attack.

11.2.2 The detritivores and specialist microbivores

The microbivores are a group of animals that operate alongside the detritivores, and which can be difficult to distinguish from the latter. The name microbivore is reserved for the minute animals that specialize at feeding on microflora, and are able to ingest bacteria or fungi but to exclude detritus from their guts. Exploitation of the two major groups of microflora requires quite different feeding techniques, principally because of differences in growth form. Bacteria (and yeasts) show a colonial growth form arising by the division of unicells, usually on the surface of small particles (Figure 11.1b).
Specialist consumers of bacteria are inevitably very small; they include free-living protozoans such as amoebae, in both soil and aquatic environments, and the terrestrial nematode worm *Pelodera,* which does not consume whole sediment particles but grazes among them consuming the bacteria on their surface. Fungi, in contrast to bacteria, produce extensively branching, filamentous hyphae, which in many species are capable of penetrating organic matter. Some specialist consumers of fungi possess piercing, sucking stylets (e.g. the nematode *Ditylenchus*) which they insert into individual fungal hyphae. However, most fungivorous animals graze on the hyphae and consume them whole. In some cases, close mutualistic relationships exist between fungivorous beetles, ants and termites and characteristic species of fungi. These mutualisms are discussed in Chapter 13. Note that microbivores consume a living resource and cannot truly be described as being donor-controlled (p. 386).

most detritivores
consume both detritus
and its microflora

The larger the animal, the less able it is to distinguish between microflora as food and the plant or animal detritus on which these are growing. In fact, the majority of the detritivorous animals involved in the decomposition of dead organic matter are generalist consumers, of both the detritus itself and the associated microfloral populations.

The invertebrates that take part in the decomposition of dead plant and animal materials are a taxonomically diverse group. In terrestrial environments they are usually classified according to their size. This is not an arbitrary basis for classification, because size is an important feature for organisms that reach their resources by burrowing or crawling among cracks

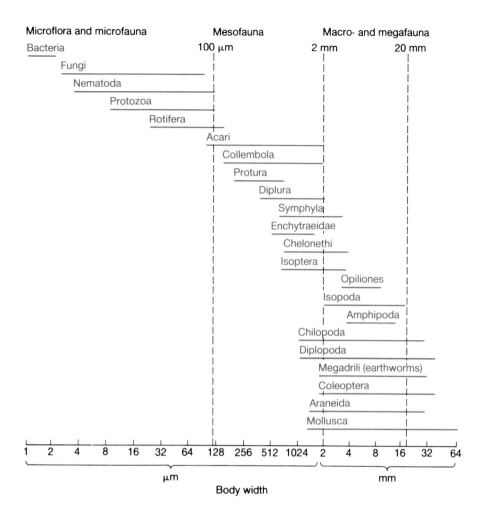

Figure 11.4. Size classification by body width of organisms in terrestrial decomposer food webs. The following groups are wholly carnivorous: Opiliones (harvest spiders), Chilopoda (centipedes), Araneida (spiders). (From Swift *et al.*, 1979.)

terrestrial detritivores are usually classified according to their size

and crevices of litter or soil. The *microfauna* (including the specialist microbivores) includes protozoans, nematode worms and rotifers (Figure 11.4). The principle groups of the *mesofauna* (animals with a body width between 100 μm and 2 mm) are litter mites (Acari), springtails (Collembola) and pot worms (Enchytraeidae). The *macrofauna* (2 mm to 20 mm body width) and, lastly, the *megafauna* (> 20 mm) include woodlice (Isopoda), millipedes (Diplopoda), earthworms (Megadrili), snails and slugs (Mollusca) and the larvae of certain flies (Diptera) and beetles (Coleoptera). These animals are mainly responsible for the initial shredding of plant remains. By their action, they may bring about a large-scale redistribution of detritus and thus contribute directly to the development of soil structure. A selection of organisms that dwell in soil and litter is illustrated in Figure 11.5.

Darwin (1888) estimated that earthworms in some pastures close to his house formed a new layer of soil 18 cm deep in 30 years, bringing about 50 tons ha^{-1} to the soil surface each year as worm casts. Figures of this order of

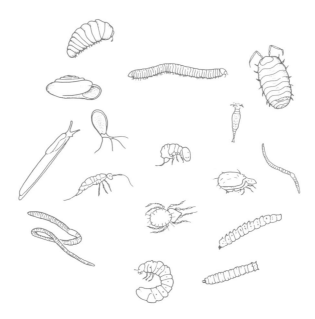

Figure 11.5.
Representatives of the most important groups of detritivores in litter and soil (not drawn to same scale).

the important role of earthworms

magnitude have since been confirmed on a number of occasions. Even higher values have been reported from a pasture in western Nigeria: 170 tons ha^{-1}, all in the space of 2–6 months of the rainy season (Madge, 1969). Moreover, not all species of earthworm put their casts above ground, so the total amount of soil and organic matter that they move may be much greater than this. Where earthworms are abundant, they bury litter, mix it with the soil (and so expose it to other decomposers and detritivores), create burrows (so increasing soil aeration and drainage) and deposit faeces rich in organic matter. It is not surprising that agricultural ecologists become worried about the consequences to ecosystems of practices that reduce worm populations.

Detritivores occur in all types of terrestrial habitat and are often found at remarkable species richness and in very great numbers. Thus, for example, a square metre of temperate woodland soil may contain 1000 species of animals, in populations exceeding 10 million for nematode worms and Protozoa, 100 000 for springtails (Collembola) and soil mites (Acari), and 50 000 or so for other invertebrates (Anderson, 1978). The relative importance of microfauna, mesofauna and macrofauna in terrestrial communities varies along a lati-

the relative importance of micro-, meso- and macrofauna varies latitudinally

tudinal gradient (Figure 11.6). The microfauna is relatively more important in the organic soils in boreal forest, tundra and polar desert. Here the plentiful organic matter stabilizes the moisture regime in the soil and provides suitable microhabitats for protozoans, nematodes and rotifers which live in interstitial water films. The hot, dry, mineral soils of the tropics have few of these animals. The deep organic soils of temperate forests are intermediate in character; they maintain the highest mesofaunal populations of litter mites, springtails and pot worms. The majority of the other soil animal groups decline in numbers towards the drier tropics, where they are replaced by termites.

On a more local scale too, the nature and activity of the decomposer community depends on the conditions in which the organisms live. Temp-

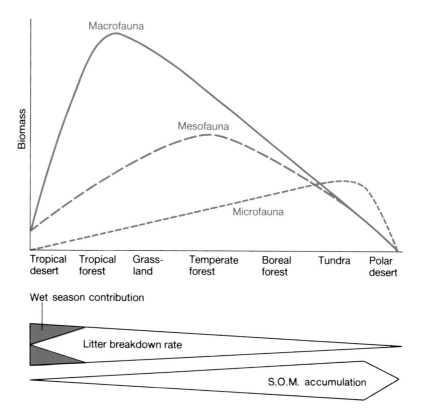

Figure 11.6. Patterns of latitudinal variation in the contribution of the macro-, meso- and microfauna to decomposition in terrestrial ecosystems. Soil organic matter accumulation (SOM) (inversely related to litter breakdown rate) is promoted by low temperatures and waterlogging, where microbial activity is impaired. (From Swift *et al.*, 1979.)

the terrestrial detritivore community also varies with dryness

erature has a fundamental role in determining the rate of decomposition (Swift *et al.*, 1979), and moreover, the thickness of water films on decomposing material places absolute limits on mobile microfauna and microflora (protozoa, nematode worms, rotifers and those fungi that have motile stages in their life cycles; see Figure 3.8). In dry soils, such organisms are virtually absent. A continuum can be recognized from dry conditions through waterlogged soils to true aquatic environments. In the former, the amount of water and thickness of water films are of paramount importance, but as we move along the continuum, conditions change to resemble more and more closely those of the bed of an open-water community, where oxygen shortage, rather than water availability, may dominate the lives of the organisms.

In contrast to terrestrial studies, the emphasis in freshwater ecology has been concerned less with size classification than with the functional role of the consumer. Cummins (1974) devised a scheme which recognizes four main categories of invertebrate consumer in streams. *Shredders* are detritivores which feed on coarse particulate organic matter (particles > 2 mm in size), and during feeding these serve to fragment the material. Very often in streams, the shredders, such as cased caddis-fly larvae of *Stenophylax* spp., freshwater shrimps (*Gammarus* spp.) and isopods (e.g. *Asellus* spp.), feed on

aquatic detritivores are usually classified according to their functional role

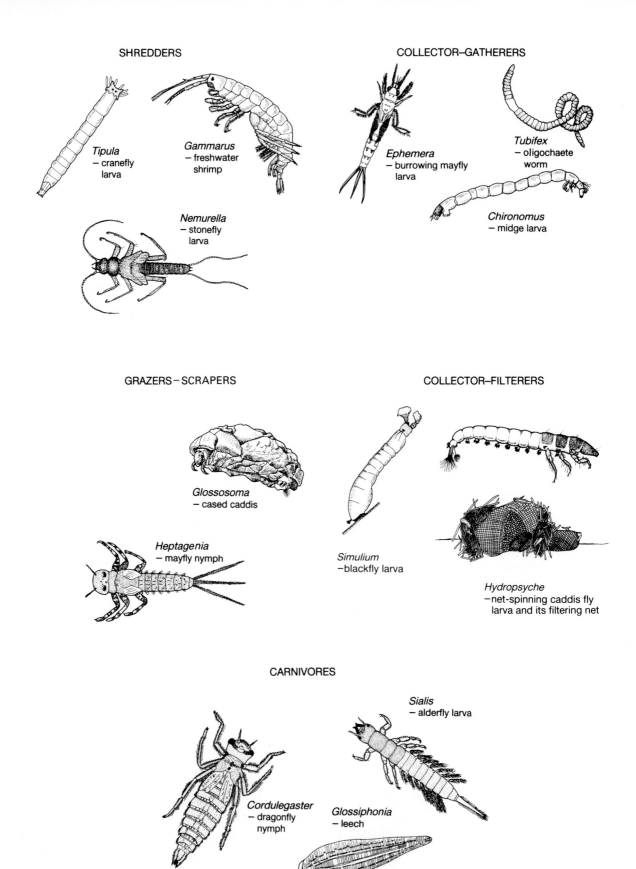

Figure 11.7. Examples of the various categories of invertebrate consumer in freshwater environments.

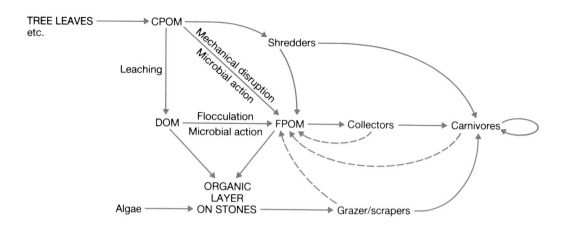

Figure 11.8. A general model of energy flow in a stream. A fraction of coarse particulate organic matter (CPOM) is quickly lost to the dissolved organic matter (DOM) compartment by leaching. The remainder is converted by three processes to fine particulate organic matter (FPOM): (i) mechanical disruption by battering; (ii) processing by microorganisms causing gradual break up; (iii) fragmentation by the shredders. Note also that all animal groups contribute to FPOM by producing faeces (dashed lines). DOM is also converted into FPOM by a physical process of flocculation or via uptake by microorganisms. The organic layer attached to stones on the stream bed derives from algae, DOM and FPOM adsorbed onto an organic matrix.

tree leaves which fall into the stream. *Collectors* feed on fine particulate organic matter (< 2 mm). Two subcategories of collectors are defined. *Collector–gatherers* obtain dead organic particles from the debris and sediments on the bed of the stream, whereas *collector–filterers* sift small particles from the flowing column of water. Some examples are shown in Figure 11.7. *Grazer–scrapers* have mouth parts appropriate for scraping off and consuming the organic layer attached to rocks and stones and comprised of attached algae and dead organic matter adsorbed to the substrate surface. The final invertebrate category is *carnivores*. Figure 11.8 shows the relationships amongst these invertebrate feeding groups and three categories of dead organic matter. This scheme, developed for stream communities, obviously has parallels in terrestrial ecosystems as well as in other aquatic ecosystems. Earthworms are important shredders in soils, while a variety of crustaceans perform the same role on the sea bed. On the other hand, filtering is common among marine but not terrestrial organisms.

most aquatic detritivores are markedly generalized

The faeces and bodies of aquatic invertebrates are generally processed along with dead organic matter from other sources by shredders and collectors. Even the large faeces of aquatic vertebrates do not appear to possess a characteristic fauna, probably because such faeces are likely to become fragmented quickly and dispersed as a result of water movement. Carrion also lacks a specialized fauna—many aquatic invertebrates are omnivorous, feeding for much of the time on plant detritus and faeces with their associated microorganisms, but ever ready to tackle a piece of dead invertebrate or fish when this is available. This contrasts very strongly with the situation in the terrestrial environment, where both faeces and carrion have specialized detritivore faunas (see later).

Some animal communities are composed almost exclusively of detriti-vores and their predators. This is true not only of the forest floor, but also of shaded streams, the depths of oceans and lakes, and the permanent residents of caves: in short, wherever there is insufficient light for appreciable photo-synthesis but nevertheless an input of organic matter from nearby plant communities. The forest floor and shaded streams receive most of their organic matter as dead leaves from the trees. The beds of oceans and lakes are subject to a continuous settlement of detritus from above. Caves receive dissolved and particulate organic matter percolating down through soil and rock, together with windblown material and debris of migrating animals. In other habitats, detritivores may be numerically less important than herbi-vores, carnivores and parasites, but invariably some are present.

11.2.3 The relative roles of microflora and detritivores

The roles of the microflora and the detritivores in decomposing dead organic matter can be compared in a variety of ways. A comparison of numbers will reveal a predominance of bacteria. This is almost inevitable because we are counting individual cells. A comparison of biomass gives a quite different picture. Figure 11.9 shows the relative amounts of biomass represented in different groups involved in the decomposition of litter on a forest floor (expressed as the relative amounts of nitrogen present). For most of the year, microorganisms accounted for five to ten times as much of the biomass as the

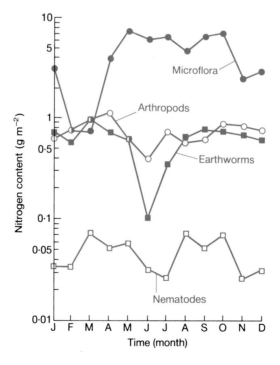

Figure 11.9. The relative importance in forest litter decomposition of microflora in comparison to arthropods, earthworms and nematodes, expressed in terms of their relative contents of nitrogen. Microbial activity is much greater than that of detritivores but the latter is more constant through the year. (After Ausmus *et al.*, 1972.)

biomass: an interesting
but imperfect measure of
importance

detritivores. The detritivore biomass varied less through the year because these are less sensitive to climatic change, and during a period in the winter they were actually predominant.

Unfortunately, the biomass present in different groups of decomposers is itself a poor measure of their relative importance in the process of decomposition. Populations of organisms with short lives and high activity may contribute more to the activities in the community than larger, long-lived, sluggish species (e.g. slugs!) that make a greater contribution to biomass. All in all, it is very difficult to assess the relative contribution of microorganisms and detritivores to the process of decomposition. This is particularly the case because their lives are so intimately interwoven.

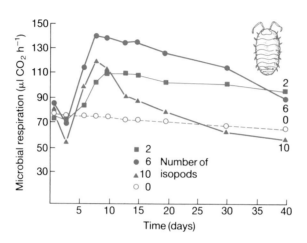

Figure 11.10. The influence of woodlice (Isopoda) activity on microbial breakdown of leaf litter in laboratory microcosms (from Hanlon & Anderson, 1980).

the crucial importance of
interactions between
detritivores and microbes

The decomposition of dead materials is not simply due to the sum of the activities of microbes and detritivores: it is largely the result of interaction between the two. The nature of such interaction is seen in Figure 11.10, which gives the results of a very simple experiment in which up to six isopods were added to microcosms containing 1 g of oak leaf litter. The presence of the isopods caused increased microbial respiration (an index of rate of decomposition), which continued for more than 40 days. A higher density of ten animals per experiment also stimulated microbial respiration, at least initially.

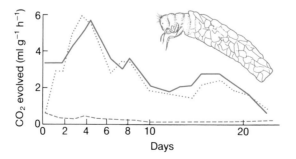

Figure 11.11. Microbial respiration on food and faeces of the terrestrial caddis *Enoicyla pusilla*. The similarity in activity on faecal pellets (——) and on ground oak leaves of similar fragment size (· · ·) contrasts with that on intact leaves (–––). (From Drift & Witkamp, 1959.)

The shredding action of these detritivores produced smaller particles with a larger surface area (per unit volume of litter) and thus increased the area of substrate available for microorganism growth. However, there were further, more subtle consequences, since counts of bacterial cells and lengths of fungal hyphae indicated that the animals had stimulated bacterial growth but inhibited the fungi. In another experiment, also involving oak leaf litter, microbial respiration was significantly higher on the faeces produced by the terrestrial larva of the caddis-fly *Enoicyla pusilla* than on its food, dried oak leaves (Figure 11.11). Close similarity in rates of respiration on faeces and on oak leaves ground to the same fragment size as faeces indicates that enhancement of microbial activity was due mainly to the stimulus of increased surface area.

Enhancement of microbial respiration by the action of detritivores has also been reported in the decomposition of small mammal carcasses. Two sets of insect-free rodent carcasses weighing 25 g were exposed under experimental conditions in the field. In one set the carcasses were left intact. In the other, the bodies were artificially riddled with tunnels by repeated piercing of the material with a dissecting needle to simulate the action of blowfly larvae in the carcass. The results of the experiment, carried out in an English grassland in autumn, showed that the tunnels enhanced microbial activity (Figure 11.12). Tunnelling by blowfly larvae disseminates the microflora as well as increasing the aeration of the carcass.

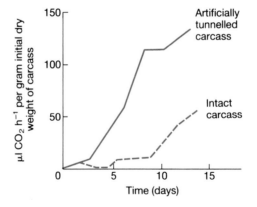

Figure 11.12. The evolution of CO_2, a measure of microbial activity, from carcasses of small mammals placed in 'respiration' cylinders and screened from insect attack. One set of carcasses was left intact, while the second set was pierced repeatedly with a dissecting needle to simulate the action of tunnelling by blowfly larvae. (From Putman, 1978a.)

the life (after death) of a plant cell wall

The ways in which the microflora and the detritivores interact might be studied by following a leaf fragment through the process of decomposition, focusing attention on a part of the wall of a single cell. Initially, when the leaf falls to the ground, the piece of cell wall is protected from microbial attack because it lies within the plant tissue. The leaf is now chewed and the fragment enters the gut of an isopod. Here it meets a new microbial flora in the gut and is acted on by the digestive enzymes of the isopod. The fragment emerges, changed by passage through the gut. It is now part of the isopod's faeces and is much more easily attacked by microorganisms, because it has been fragmented and partially digested. While microorganisms are colonizing, it may again be eaten, perhaps by a coprophagous springtail, and pass through the new environment of the springtail's gut. Incompletely digested fragments may again appear, this time in springtail faeces, yet more easily accessible to microorganisms. The fragment may pass through several other guts in its progress from being a piece of dead tissue to its inevitable fate of becoming carbon dioxide and minerals.

This 'reductionist' way of thinking about the process of decomposition emphasizes the manner in which microflora and detritivores interact. There is, however, an even more dramatic way of visualizing the process, when we remember that the gut of the animal is just a tube of the external environment that passes through its body. The particle of dead leaf passes through a succession of different environments in the course of its decomposition; some of these environments are the controlled environments of the gut, others are those within the faeces, or on the soil particles with which they become mixed. Looked at this way, we can see that the process of decomposition often begins within the digestive tract of a herbivore: the decomposition is largely microbial and it is the by-products of microbial decomposition that are the primary source of nutrition for the herbivore. The process will be discussed in detail under the heading of mutualism (between ruminant herbivores and microorganisms; Chapter 13) but could equally well figure here as the start of a process of microbial decomposition.

11.2.4 The chemical composition of decomposers, detritivores and their resources

Table 11.1 Major organic components (as percentage of the total dry weight) of detritivore resources. (Compiled by Swift *et al.*, 1979, who give the original references.)

There is a great contrast between the composition of dead plant tissue (primary resource) and that of the tissues of the heterotrophic organisms that consume and decompose it. This is illustrated in Table 11.1 for major organic components and in Table 11.2 for nutrient elements (see also Figure 3.13). While the major components of plant tissues, particularly cell walls, are structural polysaccharides, these are only of minor significance in the bodies of microorganisms and detritivores. However, being harder to digest than

	Lipid	Storage carbo-hydrate	Cellulose	Other cell wall polysac-charides	Lignin	Protein	Ash
Deciduous leaf	8	22	16	13	21	9	6
Deciduous wood	2–6	1–2	45–48	19–24	17–26	—	0.3–1.1
Bacteria	10–35	5–30	—	4–32	—	50–60	5–15
Fungi	1–42	8–60	—	2–15	—	14–52	5–12
Invertebrate detritivores	2–26	11–31	—	5–9	—	38–72	9–23
Invertebrate faeces (millipede)	—	2	38	—	?	11	8
Vertebrate faeces (horse)	2	5	28	24	14	7	9
Vertebrate carcass (steer)	50	?	—	—	—	39	11

Table 11.2. Nutrient
element composition
of resources used by
detritivores (percentage
dry weight). (Compiled
by Swift *et al.*, 1979, who
give the original
references.)

	N	P	K	Ca	Mg
Deciduous leaf	0.56	0.15	0.60	2.35	?
Deciduous wood	0.1–0.5	0.1–0.17	0.06–0.4	0.06–1.3	0.01–0.15
Bacteria	4–15	1–6	1–2	1	0.15–1.0
Fungi	1.3–3.6	0.1–0.7	0.1–2.9	0.1–3.3	0.1–0.2
Invertebrate detritivores	5.8–10.5	0.8–6.9	0.1–0.7	0.3–10.3	0.2–0.3

storage carbohydrates and protein, the structural chemicals still form a significant component of detritivore faeces. Detritivore faeces and plant tissue have much in common chemically, but the protein and lipid contents of detritivores and microflora are significantly higher than those of plants and faeces.

The rate at which dead organic matter decomposes is strongly dependent on its content of available nitrogen, or on nitrogen (ammonium or nitrate) that is available from outside. This is because microbial tissue has a ratio of carbon to nitrogen of approximately 10:1. In other words, a microbial population of 11 g can only develop if there is 1 g of nitrogen available. Most plant material has a carbon-to-nitrogen ratio of 40–80:1. Consequently, it can support only a limited biomass of decomposer organisms and the whole pace of the decomposition process will itself be limited. A greater microbial biomass can be supported, and decomposition proceeds faster, if nitrogen is absorbed from elsewhere. One important consequence is that after plant material is added to soil, the level of soil nitrogen tends to fall rapidly as it is incorporated into microbial biomass. The effect is particularly evident in agriculture, where the ploughing in of stubble can result in nitrogen deficiency of the subsequent crop. The same process occurs when leaf litter falls on a forest floor. When nitrogen is not available, the whole process of decomposition is slowed down, and undecomposed litter accumulates. In contrast to plants, the bodies of animals have C/N ratios that are of the same order as those of microbial biomass; thus their decomposition is not limited by the availability of nitrogen, and animal bodies tend to decompose much faster than plant material.

the small proportion of nitrogen in plant material limits the rate of its decomposition

When dead organisms or their parts decompose in or on soil, they begin to acquire the carbon-to-nitrogen ratio of the decomposers. This largely accounts for the fact that the C/N ratios of soils are so similar. The decomposer system is in this way remarkably homeostatic. On the whole, if material of nitrogen content less than 1.2–1.3% is added to soil, any available ammonium ions are absorbed. If the material has a nitrogen content greater than 1.8%, ammonium ions tend to be released. One consequence is that the C/N ratios of soils tend to be rather constant around values of 10—though in extreme situations, where the soil is very acid or waterlogged, it may rise to 17 (an indication that decomposition is slow).

the C/N ratio in soil is remarkably constant

Other minerals, most notably phosphorous, are also at higher concentrations in the bodies of detritivores and microorganisms than in the detritus on which they feed. They, like nitrogen, may place limits on the size of the decomposer populations and the rate at which decomposition occurs. The idea that the availability of mineral nutrients may commonly limit the decomposition process is discussed by Swift *et al.* (1979).

It should not be thought that the only activity of the microbial decomposers of dead material is to respire away the carbon and mineralize the remainder. A major consequence of microbial growth is the accumulation of microbial by-products, particularly fungal cellulose and microbial polysaccharides, which may themselves be slow to decompose and contribute to maintaining soil structure.

11.3 Detritivore–resource interactions

11.3.1 Consumption of plant detritus

Two of the major organic components of dead leaves and wood are cellulose and lignin (Table 11.1). These pose considerable digestive problems for animal consumers, most of which are not capable of manufacturing the enzymatic machinery to deal with them. Cellulose catabolism (cellulolysis) requires *cellulase* enzymes. Without these, detritivores are unable to digest the cellulose component of detritus, and so cannot derive from it either energy to do work or the simpler chemical modules to use in their own tissue synthesis.

only a few detritivores have their own cellulase

Cellulases of animal origin have been definitely identified in remarkably few species, be they detritivorous or herbivorous invertebrates. Animal cellulase has been found in the herbivorous snail *Helix pomatia* (Koopmans, 1970) and seems also to occur in several other types of mollusc, some fly larvae and a few earthworms. In these organisms, cellulolysis poses no special problems.

The majority of detritivores, lacking their own cellulases, often rely on

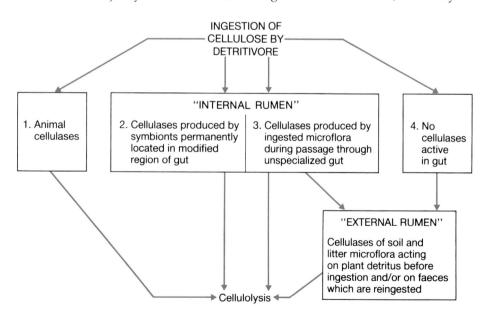

Figure 11.13. The range of mechanisms which detritivores adopt for digesting cellulose (cellulolysis) (after Swift *et al.*, 1979).

the production of cellulases by associated microflora or, in some cases, protozoa. The interactions range from *obligate mutualism* between a detritivore and a specific and permanent gut-microflora or microfauna, through *facultative mutualism,* where the animals make use of cellulases produced by a microflora which is ingested with detritus as it passes through an unspecialized gut, to the 'external rumen', where animals assimilate the metabolic products of the cellulase-producing microflora associated with decomposing plant remains or faeces (Figure 11.13).

> **most detritivores rely on microbial cellulases**

Clear examples of obligate mutualism are found amongst certain species of cockroach and termite which rely on symbiotic bacteria or protozoa for digestion of structural plant polysaccharides. In lower termites, such as *Eutermes,* protozoa may make up more than 60% of the insect's body weight. The protozoa are located in the hindgut, which is dilated to form a rectal pouch. They ingest fine particles of wood, and are responsible for extensive cellulolytic activity, though bacteria are also implicated. Termites feeding on wood made up of 54.6% cellulose, 18.0% pentosans and 27.4% lignin, produced faeces containing 18% cellulose, 8.5% pentosans and 75.5% lignin, indicating effective digestion of cellulose but not of lignin (Lee & Wood, 1971). This pattern is apparently generally the case for termites, but *Reticulitermes* has been reported to digest 80% or more of the lignin present in its food. This topic is dealt with in detail in Chapter 13.

> **termites rely on protozoa and bacteria**

An extraordinary symbiosis has recently been reported which involves the shipworm (*Lyrodus pedicellatus*). Despite its common name, this is a species of shellfish which, historically, was the scourge of all wooden navies. It rasps away at the wood, but makes use of a colony of bacteria, unknown until recently, to do the work of digesting cellulose. These symbionts live in a special gland linking the shipworm's gills with its oesophagus. The bacteria obtain their cellulose from the shipworm's gut, but they also obtain another resource, dissolved molecular nitrogen, from the water which flows over the gills. They are remarkable because they digest cellulose and fix nitrogen at the same time. Figure 11.14 shows how an isolated population of bacteria can grow when cellulose is present but in the absence of any source of nitrogen other than dissolved nitrogen molecules from the atmosphere. Genetic engineers would be delighted to produce such a creature, because it could serve to convert source materials to high-quality protein food.

> **shipworms rely on very special bacteria**

Woodlice provide a good example of detritivores that indulge in a facultative symbiotic association in an unspecialized gut. Cellulase activity has been recorded in isopod guts, but careful experimentation has shown this to be of microbial origin (Hassall & Jennings, 1975). When excised guts of *Philoscia muscorum* were incubated in contact with a film of carboxy-methyl-cellulose and then stained to determine whether cellulolysis had occurred, it was possible to demonstrate the presence of cellulase at particular places in the gut. After the addition of anti-microbial agents to kill any microflora in the gut, no cellulolytic activity could be detected, but this was restored when the animal was allowed to ingest leaf litter, with its associated microflora.

> **woodlice rely on an ingested microflora**

Finally, a wide range of detritivores appear to have to rely on the exogenous microflora to digest cellulose. The invertebrates then consume the partially digested plant detritus along with its associated bacteria and fungi,

> **many species rely on an exogenous microflora**

no doubt obtaining a significant proportion of the necessary energy and nutrients by digesting the microflora itself. These animals, such as the spring-tail *Tomocerus,* can be said to be making use of an 'external rumen' in the provision of assimilable materials from indigestible plant remains. The 'external rumen' phenomenon reaches a pinnacle of specialization in 'ambrosia' beetles and in certain species of ants and termites which 'farm' fungus in specially excavated gardens (section 13.2.2).

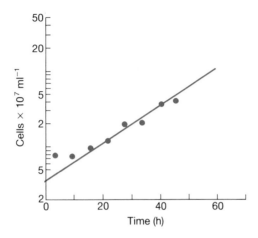

Figure 11.14. Growth of bacteria isolated from a shipworm's 'gland of Deshayes' at 35°C. in a liquid medium under an atmosphere of nitrogen, oxygen and carbon dioxide. The culture contained powdered cellulose but no additional source of nitrogen. (From Waterbury *et al.,* 1983.)

why no cellulases?

Given the versatility apparent in the evolutionary process, it is surprising that so few animals that consume plants can produce their own cellulase enzymes. Janzen (1981) has argued that cellulose has been chosen as the master construction material of plants 'for the same reason that we construct houses of concrete in areas of high termite activity'. He views the use of cellulose, therefore, as a defence against attack, since higher organisms can rarely digest it unaided. But if microflora have evolved the ability to produce cellulases, why have animals not? Perhaps the answer is partly that the diets of plant-eaters generally suffer from a limited supply of critical nutrients, such as nitrogen and phosphorus, rather than of energy. This imposes the need for processing large volumes of material to extract the required quantities of nutrients. In fact, the energy yield from cellulolysis, if this were routinely employed by animals, could conceivably create serious problems in disposal of waste energy (Swift *et al.,* 1979).

Of course, not all plant detritus is so difficult for detritivores to digest. Fallen fruit, for example, is readily exploited by many kinds of opportunist feeders, including insects, birds and mammals. However, like all detritus, decaying fruits have associated with them a microflora, in this case mainly dominated by yeasts. Fruit-flies (*Drosophila* spp.) specialize at feeding on these yeasts and their by-products; and in fruit-laden domestic compost heaps in Australia, five species of fruit-fly show preferences for particular categories of rotting fruit and vegetables (Oakshott *et al.,* 1982). *Drosophila*

fruit-flies and rotten fruit

hydei and *D. immigrans* prefer melons, *D. busckii* specializes on rotting vegetables, while *D. simulans* is catholic in its tastes for a variety of fruits. The common *D. melanogaster*, however, shows a clear preference for rotting grapes and pears. Note that rotting fruits can be highly alcoholic. Yeasts are commonly the early colonists and the fruit sugars are fermented to alcohol

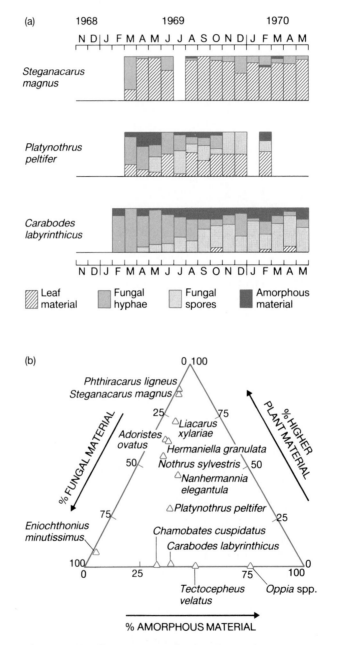

Figure 11.15. Gut contents of oribatid mites from a woodland in Kent, England. (a) Temporal variation in feeding by three species from *Castanea* litter showing the average composition of gut contents on each sampling occasion. (b) Mean proportions of the three major food items in the species which were most abundant during a 20 month study. Specialized feeders occur near the apices of the triangle, with more generalist species nearer the centre. (From Anderson, 1975.)

which is normally toxic, eventually even to the yeasts themselves. *D. melanogaster* tolerates such high levels of alcohol because it produces large quantities of the enzyme alcohol dehydrogenase (ADH) which breaks down ethanol to harmless metabolites. More surprisingly, it has been shown that this fruit-fly can use ethanol as a resource (Parsons & Spence, 1981). Decaying vegetables produce little alcohol, and *D. busckii,* which is associated with them, produces very little ADH. Intermediate levels of ADH were produced by the species preferring moderately alcoholic melons. The boozy *D. melanogaster* is also associated with winery wastes!

plant detritus and the microflora are typically consumed together...

Because of the intimate intermeshing of microfloral and plant detrital tissues there are inevitably many generalist consumers which ingest both resources. These animals simply cannot manage to take a mouthful of one without the other. Thus, for example, Figures 11.15a and b illustrate the various components of the diet of several oribatid mites occurring in the litter of a woodland in Kent, England. There are clear interspecific differences in the relative importance of fungal material, higher plant material and amorphous organic matter, and some species are more specialized feeders than others, at least at some times of year. All consume more than one of the potential diet components and most are remarkably generalist.

Microfloral biomass seems to constitute a better food resource than plant detritus. A clue to its relative importance in diet has been provided by studies on the efficiency with which ingested pure fungus or unconditioned tree leaf is assimilated. For example, a freshwater shredder, *Gammarus pseudolimnaeus,* assimilated pure sterile leaf discs of elm or maple with an efficiency of only 17–19% of the energy content, the remainder being lost in faeces. On the other hand, when the detritivores were provided with a pure culture of aquatic fungi, normally associated with tree leaves in streams, the assimilation efficiency was 74–83% (Barlocher & Kendrick, 1975).

... but the microflora is the more nutritious

Such results have led Cummins (1974) to liken microbial tissue to peanut butter—a nutritious food which occurs on a nutritionally unsuitable cracker biscuit, the plant detritus. While this analogy is quite apt, it perhaps over-stresses the role of microbial biomass. By far the largest organic component ingested by detritivores is generally plant detritus, so although it is assimilated with low efficiency, it may still provide an appreciable portion of the animal's requirements. It is perhaps more accurate to compare what detritivores ingest to a very thick cracker biscuit spread only thinly with peanut butter. Nevertheless, there is no doubt that an important part of the diet of terrestrial and aquatic detritivores derives from fungi and bacteria, and there is considerable evidence, among freshwater detritivores for instance, that the animals prefer microbially conditioned detritus and grow better when fed on it.

11.3.2 Feeding on faeces

Invertebrate faeces

A large proportion of dead organic matter in soils and aquatic sediments may consist of invertebrate faeces. Detritivores with generalist feeding habits will usually include these in their diet. In many cases, re-ingestion of faeces is critically important. In the laboratory, the litter-dwelling millipede *Apheloria montana* showed only a small weight increase if it was fed solely on leaf detritus (1.1% in 30 days). A second group of animals, which were allowed access to their faeces, increased their ingestion rates by more than 60% and their growth rates by 16% compared to the controls (McBrayer, 1973). It seems that the conversion of plant remains to digestible microbial biomass and also to partially degraded plant compounds is of great importance to many detritivores. In one sense, the detritivores' guts act as factory conveyor belts producing cracker biscuits (faeces) to be spread with peanut butter (microflora) over and over again.

MacLachlan and his students (MacLachlan *et al.*, 1979) have unravelled a remarkable story of coprophagy (eating faeces) that occurs in small bog lakes in north-east England. These murky water bodies have restricted light penetration because of dissolved humic substances derived from the surrounding sphagnum peat, and they are characteristically poor in plant nutrients. Primary production is insignificant. The main organic input consists of poor-quality peat particles resulting from wave erosion of the banks. By the time the peat has settled from suspension it has been colonized, mainly by bacteria, and its caloric and protein contents have increased by 23% and 200% respectively. These small particles are consumed by *Chironomus lugubris* larvae, the detritivorous young of a non-biting chironomid midge. The faeces the larvae produce become quite richly colonized by fungi, microbial activity is enhanced, and they would seem to constitute a high quality food resource. But they are not re-ingested by *Chironomus* larvae, mainly because they are too large and too tough for its mouthparts to deal with. However, another common inhabitant of the lake, the cladoceran *Chydorus sphaericus*, finds chironomid faeces very attractive. It seems always to be associated with them and probably depends on them for food. *Chydorus* clasps the chironomid faecal pellet just inside the valve of its carapace and rotates it while grazing the surface, causing gradual disintegration. In the laboratory, the presence of chydorids has been shown to speed up dramatically the breakdown of large *Chironomus* pellets to smaller particles. The final and most intriguing twist to the story is that the fragmented chironomid faeces (mixed probably with chydorid faeces) are now small enough to be used again by *Chironomus*. It is probable that *Chironomus lugubris* larvae grow faster when in the presence of *Chydorus sphaericus* because of the availability of suitable faecal material to eat. The interaction benefits both participants.

Vertebrate faeces

The dung of carnivorous vertebrates is relatively poor quality stuff. Carnivores assimilate their food with high efficiency (usually 80% or more is digested) and their faeces retain only the least digestible components. In addition, carnivores are necessarily much less common than herbivores, and their dung is probably not sufficiently abundant to support a specialist detritivore fauna. What little research has been done suggests that decay is effected almost entirely by bacteria and fungi.

the highly efficient
decomposition of
elephant dung by dung
beetles in the wet
season . . .

In contrast, herbivore dung still contains an abundance of organic matter and is sufficiently thickly spread in the environment to support its own characteristic fauna, consisting of many occasional visitors but with several specific dung feeders. A good example is provided by elephant dung. Two main patterns of decay can be recognized, related to the wet and dry seasons in the tropics (Kingston, 1977). During the rains, within a few minutes of dung deposition the area is alive with beetles. The adult dung beetles feed on the dung but they also bury large quantities along with their eggs to provide food for developing larvae. For example, the large African dung beetle, *Heliocopris dilloni,* carves a lump out of fresh dung and rolls this away for burying several metres from the original dung pile. Each beetle buries sufficient dung for several eggs. Once underground, a small quantity of dung is shaped into a cup, and lined with soil; a single egg is laid and then more dung is added to produce a sphere which is almost entirely covered with a thin layer of soil. A small area at the top of the ball, close to the location of the egg, is left clear of soil, possibly to facilitate gas exchange. After hatching, the larva feeds

Figure 11.16. (a) African dung beetle rolling a ball of dung (photograph: Heather Angel). (b) The larvae of the dung beetle *Heliocopris* excavates a hollow as it feeds within the dung ball. (From Kingston & Coe, 1977.)

by a rotating action in the dung ball, excavating a hollow, and, incidentally, feeding on its own faeces as well as the elephant's (Figure 11.16). When all the food supplied by its parents is used up, the larva covers the inside of its cell with a paste of its own faeces, and pupates.

The full range of tropical dung beetles in the family Scarabeidae vary in size from a few millimetres in length up to the 6 cm *Heliocopris*. Not all remove dung and bury it at a distance from the dung pile. Some excavate their nests at various depths immediately below the pile, while others build nest chambers within the dung pile itself. Species in other families do not construct chambers but simply lay their eggs in the dung, and their larvae feed and grow within the dung mass until fully developed, when they move away to pupate in the soil. The beetles associated with elephant dung in the wet season may remove 100% of the dung pile. Any left may be processed by other detritivores such as flies and termites, as well as by microflora.

. . . but a very different picture emerges in the dry season

Dung which is deposited in the dry season is colonized by relatively few beetles (adults emerge only in the rains). Some microbial activity is evident but this soon declines as the faeces dry out. Rewetting during the rains stimulates more microbial activity but beetles do not exploit old dung. In fact a dung pile deposited in the dry season may persist for longer than two years, compared with 24 hours or less for one deposited during the rains.

The breakdown of ungulate dung in temperate environments has also been much studied. Cow dung in Britain is attacked by many invertebrates (amongst which earthworms are dominant) as well as by bacteria and fungi (Denholm Young, 1978). Very few species of dung-burying beetles operate in Britain but one, *Geotrupes spiniger*, may remove up to 13% by dry weight of a cow pat. The maggots of many species of fly, including the specialist yellow dungfly *Scatophaga stercoraria*, also play a role. Adult *Scatophaga* are predatory or nectar-feeding and take only small quantities of liquid dung. However, they congregate to mate around fresh droppings and the female lays her eggs on the dungpile. Larvae hatch out on the surface and immediately burrow down into the dung. A new generation of adults emerges within 3–4 weeks.

An Australian dilemma—bovine dung but no bovine dung beetles

Bovine dung has provided an extraordinary and economically very important problem in Australia. During the past two hundred years the cow population has risen from just seven (brought over by the first English colonists in 1788) to 30 million or so. These produce some 300 million cow pats per day, covering as much as six million acres per year with dung. Deposition of bovine dung poses no particular problem elsewhere in the world, where bovines have existed for millions of years and have an associated fauna which exploits the faecal resources. However, the largest herbivorous animals in Australia until European colonization were marsupials such as kangaroos. The native detritivores which deal with the dry, fibrous dung pellets that these leave cannot cope with cow dung, and the loss of pasture under dung has imposed a huge economic burden on Australian agriculture. In addition, Australia is plagued by native flies which deposit eggs on dungs pats. The decision was therefore made in 1963 to establish in Australia beetles of African origin, able to dispose of bovine dung in the most important places and under the most prevalent conditions where cattle are raised (Waterhouse, 1974). This huge project has been a comparative success in some areas.

11.3.3 Consumption of carrion

many carnivores are opportunistic carrion feeders

The chemical composition of the diet of carrion-feeders is quite distinct from that of other detritivores, and this is reflected in their complement of enzymes. Carbohydrase activity is weak or absent, but protease and lipase activity is vigorous. Carrion-feeding detritivores possess basically the same enzymatic machinery as carnivores, reflecting the chemical identity of their food. In fact, many species of carnivore are also opportunistic carrion-feeders.

scavenging vertebrates often remove whole carcasses . . .

When considering the decomposition of dead bodies, it is helpful to distinguish three categories of organisms that attack carcasses. As before, both microflora and invertebrate detritivores have a role to play, but, in addition, scavenging vertebrates are often of considerable importance. Many carcasses of a size to make a single meal for one or a few of these scavenging detritivores will be removed completely within a very short time of death, leaving nothing for bacteria, fungi or invertebrates. This role is played, for example, by arctic foxes and skuas in polar regions, by crows, gluttons and badgers in temperate areas, and by a wide variety of birds and mammals, including kites, jackals and hyenas, in the tropics.

. . . though this can vary seasonally in a subtle way

The relative roles played by microflora, invertebrates and vertebrates are influenced by factors that affect the speed with which carcasses are discovered by scavengers in relation to the rate at which they disappear through microbial and invertebrate activity. This is illustrated graphically for small rodent carcasses whose disappearance/decomposition was monitored in the Oxfordshire countryside in both summer–autumn and winter–spring periods (Figure 11.17). There are two points to note. First, the rate at which carcasses were removed was faster during summer and autumn, reflecting greater scavenger activity at this time (presumably because of higher

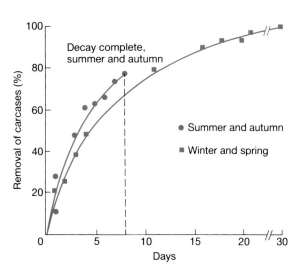

Figure 11.17. Rate of removal of small mammal corpses in the Oxfordshire countryside in two periods: summer–autumn and winter–spring (from Putman, 1983).

scavenger population densities and/or higher feeding rates—these were not monitored in the study). Secondly, a greater percentage of the rodent bodies were removed in the winter–spring period, albeit over a longer time-scale. At a time when microbial decay proceeds most slowly, all the carcasses persisted for long enough to be found by scavengers. During summer and autumn, decomposition was much more rapid and any carcass which was undiscovered for seven or eight days would have been largely decomposed and removed by bacteria, fungi and invertebrate detritivores.

Larger carcasses, such as those of dogs, sheep and antelopes, may also be dealt with predominantly by vertebrate scavengers. Sheep carcasses in temperate regions are often stripped completely by foxes or eagles within a short time of death, while vultures and hyenas remove as many as 90% of the bodies of ungulates that die in southern Africa. In a study of decomposition of the largest terrestrial carcasses of all, those of elephants, Coe (1978) found that although scavengers played a major role, some carcasses went unexploited by vertebrates, and in these cases invertebrate detritivores and particularly the microflora played an important role. However, this study was carried out in a period of unusually high elephant mortality in Kenya, and the scavenger populations may have been unable to exploit the exceptional availability of food as fully as usual.

seasonal variations in invertebrate and microbial activity

Carrion undiscovered by large scavengers is available for processing by microflora and invertebrates, the relative roles of which are influenced by prevailing conditions. Invertebrates such as the larvae of blowflies (which are characteristic of decaying carrion, and not surprisingly pose a major problem by contaminating meat intended for human consumption) may remove a significant portion of carrion while conditions are favourable. For example, Putman (1978b) estimates that in Britain blowfly maggots consume as much as 80% of small mammal carcasses available during the summer. In the winter, however, decomposition is largely a microbial process. The activity of blowflies is seasonal in tropical environments too. In wet seasons they play an important role in decomposition, but they are unable to colonize in dry seasons, and carrion is likely to shrivel and dry, any decomposition being attributable to microorganisms. Tropical dry season mummification parallels the winter mummification of carrion often observed in temperate regions.

the specialist consumers of bone, hair and feathers

Certain components of animal corpses are particularly resistant to attack and are the slowest to disappear. However, some consumer species possess the enzymes to deal with them. For example, the blowfly larvae of *Lucilia* species produce a collagenase which can digest the collagen and elastin present in tendons and soft bones. The chief constituent of hair and feathers, keratin, forms the basis of the diet of species characteristic of the later stages of carrion decomposition, in particular tineid moths and dermestid beetles. The midgut of these insects secretes strong reducing agents which break the resistant covalent links binding together peptide chains in the keratin. Hydrolytic enzymes then deal with the residues. Fungi in the family Onygenaceae are specialist consumers of horn and feathers.

the remarkable burying beetles—*Necrophorus* spp.

One group of carrion-feeding invertebrates deserves special attention— the burying beetles (*Necrophorus* spp.). These species live exclusively on carrion and require the medium to play out their extraordinary life-history.

Adult *Necrophorus,* presumably orientating by smell, arrive at the carcass of a small mammal or bird within an hour or two of death. The beetle may tear flesh from the corpse and eat it or, if decomposition is sufficiently advanced, consume blowfly larvae instead. However, should a burying beetle arrive at a completely fresh corpse it sets about burying it where it lies; or if the substrate is not suitable, it may drag the body (many times its own weight) for several metres before starting to dig. It works beneath the corpse, painstakingly excavating and dragging the small mammal down little by little until it is completely underground (Figure 11.18). Some species, such as *N. vespilloides,* only just cover the corpse, while others, including *N. germanicus,* may bury it to a depth of 20 cm, a remarkable feat for so small an animal.

Figure 11.18. Burial of a mouse by a pair of *Necrophorus* beetles (from Milne & Milne, 1976).

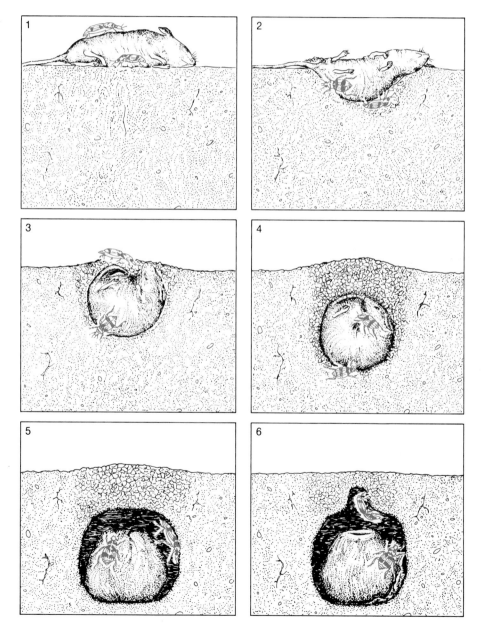

During the excavation, other burying beetles are likely to arrive. Competing individuals of the same or other species are fiercely repulsed, sometimes leading to the death of one combatant. A prospective mate, on the other hand, is accepted and the male and female work on together.

The buried corpse is much less susceptible to attack by other invertebrates than it was while on the surface. Additional protection seems to be afforded by virtue of a mutualistic relationship between *Necrophorus* and a species of mite, *Poecilochirus necrophori,* which invariably infests adult burying beetles, hitching a ride to a suitable carrion source. These mites are believed to help keep the carcass free of potentially competing detritivores, such as fly larvae, by removing and feeding upon any eggs deposited on the surface of the corpse.

When excavation of the burial chamber is complete, the beetles set about removing fur or feathers, and they work the mass of flesh into a compact ball. Now the beetles copulate, and eggs are laid in a blind passage leading to the roof of the chamber. The eggs are not abandoned, but rather both parents, or sometimes just the female, remain in the chamber to provide parental care for the larvae. A conical depression is prepared in the top of the meat-ball, into which droplets of partially digested meat are regurgitated. When the eggs hatch, the adult calls the larvae to the feeding location by stridulation, and the larvae are fed droplets from the pool of liquid food by the parent or parents. Older larvae are able to feed themselves from the pool or to remove strips of meat from the carcass. Only when their offspring are ready to pupate do the adults force their way out through the soil and fly away.

We have already noted that in freshwater environments carrion lacks a specialized fauna. However, specialist carrion-feeders are found on the sea bed in very deep parts of the oceans. The great depth through which detritus has to sink usually means that all but the largest particles of organic matter are completely decomposed on their way down. In contrast, the occasional body of a fish, mammal or large invertebrate does settle on the sea bed. A remarkable diversity of scavengers exist there, though at low density, and these possess several characteristics which match a way of life in which meals are well spread out in space and time. For example, Dahl (1979) describes several genera of deep sea gammarid crustaceans which, unlike their relatives at shallower depths and in freshwater, possess dense bundles of exposed chemosensory hairs for food location, and sharp mandibles that can take large bites from carrion. These animals also have the capacity to gorge themselves far beyond what is normal in amphipods. Thus *Paralicella* possesses a soft body wall which can be stretched when feeding on a large meal so that the animal swells to two or three times its normal size, and *Hirondella* has a midgut which expands to fill almost the entire abdominal cavity and in which it can store meat.

carrion feeders on the sea bed

11.4 Conclusion

Decomposer communities are, in their composition and activities, as diverse as or more diverse than any of the communities more commonly studied by ecologists. Generalizing about them is unusually difficult because the range of conditions experienced in their lives is so varied. As in all natural communities, the inhabitants not only have specialized requirements for resources and conditions, but their activities change the resources and conditions available for others. Most of this happens hidden from the view of the observer, in the crevices and recesses of soil and litter and in the depths of water bodies.

Despite these difficulties, some broad generalizations may be made.

(i) Decomposers and detritivores tend to have low population densities and to have low levels of activity when temperatures are low, aeration is poor, soil water is scarce and conditions are acid.

(ii) The structure and porosity of the environment (soil or litter) is of crucial importance, not only because it affects the factors listed in (i) but because most of the organisms responsible for decomposition must swim, creep, grow or force their way through the medium in which their resources are dispersed.

(iii) The activities of the microflora and detritivores are intimately interlocked and tend to be synergistic. For this reason it is very difficult, and perhaps not sensible, to try to discover their relative importance to the group.

(iv) Many of the decomposers and detritivores are specialists and the decay of dead organic matter results from the combined activities of organisms with widely different structure, form and feeding habits.

(v) Individual particles of organic matter may cycle repeatedly through a succession of microhabitats within and outside the guts and faeces of different organisms, as they are degraded from highly organized structures to their eventual fate as carbon dioxide and mineral nutrients.

(vi) The activity of decomposers unlocks the mineral resources such as phosphorus and nitrogen that are fixed in dead organic matter. The speed of decomposition will determine the rate at which such resources are released to growing plants (or become free to diffuse and thus to be lost from the ecosystem). This topic is taken up and discussed in Chapter 17.

(vii) Many dead resources are patchily distributed in space and time. A strong element of chance operates in the process of their colonization; the first to arrive have a rich resource to exploit, but the successful species vary from dung pat to dung pat, and from corpse to corpse. The dynamics of competition between exploiters of such patchy resources require their own particular mathematical models (Atkinson & Shorrocks, 1981; see p. 268). Because detritus is often an 'island' in a sea of unused habitat, its study is conceptually very similar to that discussed under the heading of island biology in Chapter 20.

(viii) Finally, it may be instructive at this point to switch the emphasis away from the success with which decomposers and detritivores deal with their resources. It is easy to underestimate the importance of the barriers to decomposition that enable so many communities to retain their structure. It is, after all, the failure of organisms to decompose wood efficiently and rapidly that makes the existence of forests possible! Similar inefficiencies have resulted in the build-up of peat and the formation of coal and oil deposits.

Parasitism and Disease

12.1 Introduction

In Chapter 8 we defined a parasite as an organism that obtains its nutrients from one or a very few host individuals, normally causing harm but not causing death immediately. However, there are many definitions of parasitism. Parasitologists tend to emphasise (a) the intimacy of the association between parasite and host and (b) the dependence of the parasite on the host for the regulation of its environment; but they may omit any mention of 'harm' to the host. It is important for the ecologist to stress this intimacy of association in a full definition of parasitism, but if there is really neither benefit *nor* harm accruing to the host, the association must be seen as commensal (p. 201). For this reason it seems sensible to distinguish between *parasitic* interactions, in which at least some harm is predictably caused, though perhaps only in appropriate circumstances (e.g. a sufficient number of parasites or a host in poor condition), and *commensal* interactions, in which the commensal organism benefits but the host is consistently neither benefitted nor harmed. We also need an operational definition of 'harm', if such definitions are to be useful. In practice we may measure the harm that a parasite does to its host as the reduction in the intrinsic growth rate of the host and/or its populations.

Parasites and pathogens (disease-causing agents) are an extraordinarily important group of organisms. Millions of people are killed each year by various types of infection, and many millions more are debilitated or deformed (250 million cases of elephantiasis at present, over 200 million cases of bilharzia, and the list goes on and on). When the effects of pathogens on domesticated animals and crops are added to this, the cost in terms of human misery and economic loss becomes incalculable. Of course humans make things easy for the parasites and pathogens (see below) by living in dense and aggregated populations and forcing their domesticated animals and crops to do the same. But it is nonetheless reasonable to believe that animals and plants *in general* are harmed and killed in vast numbers by parasitism and disease. Pathogens are certain to represent an important source of mortality and unfulfilled fecundity in many, probably most, natural populations.

Finally, in terms of numbers of individuals and species, we have only to consider two points. Firstly, a free-living organism that does *not* harbour several parasitic individuals of a number of species is a rarity; and secondly, many parasites and pathogens are host-specific or at least have a limited range of hosts. Together, these two points would seem to make the conclusion unavoidable that more than half the species on the earth, and many more than half the individuals, are parasites or pathogens. Most bacterial and viral parasites probably still remain unidentified!

In the light of this richness, a survey of the diversity of parasites and pathogens might seem impracticable. On the other hand, an outline survey, especially of the types that have been most intensively studied and are of most importance to man, will provide a valuable background for the remainder of this chapter.

12.2 The diversity of parasites

The language and jargon used by plant pathologists and animal parasitologists are often so different as to give the impression that the subjects have nothing in common. For the ecologist, however, such distinction is superficial. There are important differences in the ways in which living animals and plants serve as habitats for parasites, and important differences in their responses to attack; but the differences are less striking than the resemblances. As ecologists, therefore, we will deal with the two together.

It is convenient from the outset to recognize two major categories among parasites: the *microparasites* and the *macroparasites*.

The distinction between micro- and macroparasites has been made relatively recently (May & Anderson, 1979). Microparasites multiply directly within their host (usually within the host cells). Macroparasites *grow* in their host, but multiply by producing infective stages which are released from the host to infect new hosts. They are often intercellular (in plants) or live in body cavities, rather than within the host cells.

12.2.1 Microparasites

the range of microparasites

Probably the most obvious microparasites are the bacteria and viruses that infect animals, such as the measles virus and the typhoid bacterium. There are also many virus diseases (though rather fewer bacterial diseases) of plants (e.g. the beet and tomato yellow net viruses, the cauliflower, radish and pea mosaic viruses, and the bacterial crown gall disease). The other major group of microparasites affecting animals are the protozoa (e.g. the trypanosomes that cause sleeping sickness and the *Plasmodium* species that cause malaria—Figure 12.1); while for plants the other major group are the simple fungi (e.g. the slime mould *Plasmodiophora brassicae* which causes club root of brassicas, and species of *Synchytrium*, such as *S. endobioticum*, which causes wart of potato—Figure 12.2).

Microparasites can be divided into those that are transmitted directly from host to host, and those that are transmitted indirectly via some other species, the vector.

directly transmitted microparasites

Direct transmission between hosts can occur almost instantaneously, as in the cases of venereal disease and the short-lived infective agents carried in the water-droplets of coughs and sneezes (influenza, measles, etc.). Alternatively, the parasite may spend an extended dormant period 'waiting' for its

417 PARASITISM AND DISEASE

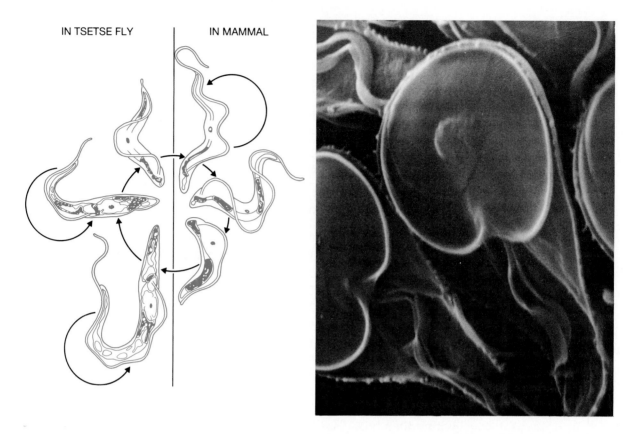

IN TSETSE FLY   IN MAMMAL

Figure 12.1. Microparasites of animals. Left: *Trypanosoma brucei.* This is the causal agent of sleeping sickness which is transmitted to mammals by a tsetse fly injecting the parasite into the blood of mammals when it feeds. The parasite multiplies by fission during its life in the mammalian blood. It also undergoes division in the gut of the fly and later in the salivary glands. Right: scanning electromicrograph of *Giardia intestinalis,* a protozoan gut parasite of man, baboons and monkeys throughout the world. ×9000. (Photograph by courtesy of K. Vickerman.)

new host. This is the case with the ingestion of food or water contaminated with the protozoan *Entamoeba histolytica* which causes amoebic dysentry, and with the plant pathogens that leave resting spores in the soil, where infection of a new host depends on physical contact between the host and the spores. Resting spores of *Plasmodiophora brassicae* (club root), for instance, can apparently remain viable in a dormant state for decades. But in the presence of a suitable host they germinate and penetrate the root hairs, after which a cycle of growth and multiplication follows, leading to an enlarged gall on the root. Within this, many resting spores are ultimately formed which are released back into the soil as the gall tissue decays.

vector-transmitted microparasites: sleeping sickness and malaria . . .

    The two most economically important groups of vector-transmitted protozoan parasites of animals are first the trypanosomes, which cause sleeping sickness in man and Nagana in domesticated (and wild) mammals and are transmitted by tsetse flies (*Glossina*); and secondly the various species of *Plasmodium,* which cause malaria and are transmitted by anopheline mosquitoes. Typical life cycles from the two groups are illustrated and

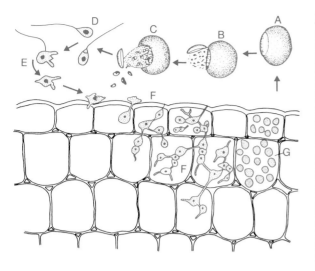
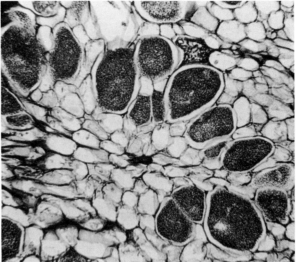

Figure 12.2. Microparasites of plants. These are usually intracellular and multiply within the host releasing dispersal spores on the death of the host or its parts. Left: *Physoderma zeae-maydis* a pathogen infecting the leaves of corn (*Zea mays*) (from Tisdale 1919). Right: *Plasmodiophora brassicae*, the causal agent of club root disease of crucifers. The pathogen causes the cortical cells of the root to divide and enlarge. (Photograph: Heather Angel.)

described in Figures 12.1 and 12.3. Note that in both cases the flies act not only as vectors but also as intermediate hosts within which there is multiplication of the parasites.

... plant viruses

Many plant viruses are transmitted by aphids. In some, 'non-persistent' species (e.g. cauliflower mosaic virus), the virus is only viable in the vector for an hour or so and is often only borne *on* the aphid's mouth parts. In other, 'circulative' species (e.g. lettuce necrotic yellow virus) the viruses pass from the aphid's gut to its circulatory system and thence to its salivary glands. There is therefore a latent period before the vector becomes infective, but it then remains infective for an extended period (often its entire life). Finally, there are also 'propagative' viruses (e.g. the potato leaf roll virus) which multiply within the aphid. Nematode worms are also widespread vectors of plant viruses.

12.2.2 Macroparasites

the range of macroparasites

Amongst the major macroparasites of animals (Figure 12.4) are the parasitic helminth worms, which include the platyhelminths (e.g. the tapeworms and the trematode schistosomes and flukes), the acanthocephalans and the roundworms (nematodes). In addition, there are lice, fleas, ticks and mites, and also the fungi that attack animals. Plant macroparasites (Figure 12.6) include the higher fungi that give rise to the powdery mildews, downy mildews, rusts and smuts, as well as the gall-forming and mining insects, and such flowering plants as the dodders and broomrapes that are themselves parasitic on other higher plants.

Like the microparasites, the macroparasites can be sub-divided into

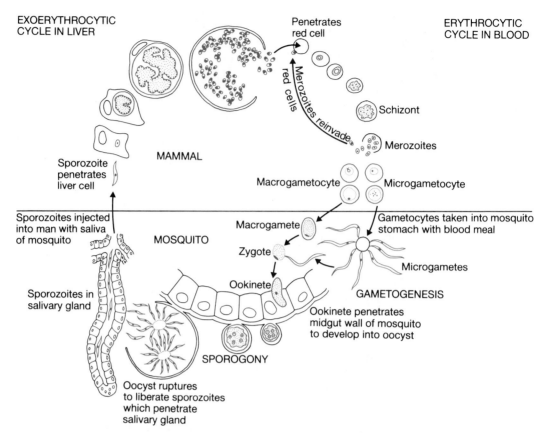

EXOERYTHROCYTIC
CYCLE IN LIVER

Penetrates
red cell

ERYTHROCYTIC
CYCLE IN BLOOD

Merozoites reinvade red cells

Schizont

Merozoites

MAMMAL

Sporozoite
penetrates
liver cell

Macrogametocyte

Microgametocyte

Sporozoites injected
into man with saliva
of mosquito

Gametocytes taken into mosquito
stomach with blood meal

Macrogamete

MOSQUITO

Zygote

Microgametes

Sporozoites in
salivary gland

Ookinete

GAMETOGENESIS

Ookinete penetrates
midgut wall of mosquito
to develop into oocyst

SPOROGONY

Oocyst ruptures
to liberate sporozoites
which penetrate
salivary gland

Figure 12.3. The life cycle of the malarial parasite *Plasmodium*. The parasite is injected directly into the bloodstream of the host by a mosquito. Multiplication occurs in the liver cells of the host and the products enter the red blood cells where they multiply again and form sexual stages which are taken up by a feeding mosquito. Fertilization occurs in the mosquito gut and after yet a further phase of multiplication the salivary glands of the mosquito become infected and the cycle is ready to start again. (After Vickerman & Cox, 1967.)

those that are transmitted directly and those that require a vector or intermediate host for transmission and therefore have an indirect life cycle.

directly-transmitted macroparasites: monogeneans, . . .

The monogeneans are ectoparasitic platyhelminth worms that feed from the skin or gills of fish particularly, but also from amphibians, reptiles, cetaceans and cephalopods. They maintain their position by means of a specialized posterior attachment organ. New hosts are actively located by free-swimming larvae or by the adults themselves.

. . . intestinal nematodes,

The intestinal nematodes of man, all of which are transmitted directly, are perhaps the most important human intestinal parasites, both in terms of the number of people infected and their potential for causing ill health. The life cycle of the blood-feeding and anaemia-inducing hookworm is shown in Figure 12.5. In this, and in most other intestinal nematodes, the eggs and larvae which are responsible for transmission normally only become infective after a period of development in the soil.

. . . lice and fleas . . .

Lice spend all stages of their life cycle on their host (either a mammal or a bird), and transmission is usually by direct physical contact between host individuals, often between mother and offspring. Fleas, by contrast, lay their eggs and spend their larval lives in the 'home' (usually the nest) of their host (again, a mammal or a bird). The emerging adult then actively locates a new

Figure 12.4. Macroparasites of animals. Top left: Hedgehog ticks (*Ixodes hexagonus*) *in situ* with heads buried in the host's skin. Bottom: Three-spined stickleback with cestode worm parasites (the pleurocercoid stage of *Schistocephalus solidus*). (Photographs: Heather Angel.) Top right: Fungal parasite of *Campanotus* ant. (Photograph: BPS.)

host individual, often jumping and walking considerable distances in order to do so.

... mildews and other plant-infecting fungi ...

Direct transmission is common amongst the fungal macroparasites of plants. For example, in the development of an infection by a leaf-infecting fungus, such as mildew on a crop of wheat, infection involves contact between a spore (usually wind-dispersed) and a leaf surface, followed by penetration of the fungus into or between host cells, where it begins to grow. Sooner or later this infection, if successful, will become apparent as a lesion of altered host tissue. This phase of invasion and colonization (the lag or juvenile phase) precedes an infective stage when the lesion matures and becomes a spore-producing area. Subsequently, the lesion may cease to produce spores, and the host tissue is then either dead or incapable of being reinfected.

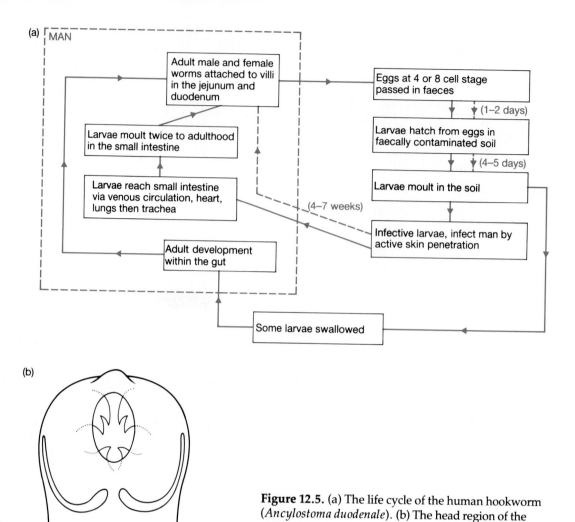

(a) MAN

Adult male and female worms attached to villi in the jejunum and duodenum

Larvae moult twice to adulthood in the small intestine

Larvae reach small intestine via venous circulation, heart, lungs then trachea

Adult development within the gut

Eggs at 4 or 8 cell stage passed in faeces

(1–2 days)

Larvae hatch from eggs in faecally contaminated soil

(4–5 days)

Larvae moult in the soil

Infective larvae, infect man by active skin penetration

(4–7 weeks)

Some larvae swallowed

Figure 12.5. (a) The life cycle of the human hookworm (*Ancylostoma duodenale*). (b) The head region of the worm showing the cutting teeth in the buccal cavity. (After Cox, 1982.)

... parasitic flowering plants: ...

One major and fairly separate group of plant macroparasites with direct transmission are the parasitic flowering plants (Figure 12.6). These are of two quite distinct types: *holoparasites*, which lack chlorophyll and are wholly dependent on the host plant for the supply of water, nutrients and fixed carbon, and *hemiparasites*, which are photosynthetic but form connections with the roots or stems of other species and draw most or all of their water and mineral nutrient resources from the host. These hemiparasites have a poorly developed root system of their own or none at all.

The most extreme holoparasitism is found in *Rafflesia arnoldii*, in which the entire vegetative body of the plant is a network within the host; the only external structure is the flower, which is borne on the surface of the host roots and is the biggest known flower, reaching a metre in diameter. All of the resources used in growth are derived directly from the host. At the other end of the spectrum, the hemiparasitic species of Scrophulariaceae, such as *Odontites verna, Rhinanthus minor* and *Euphrasia* spp., are capable of an autotrophic but stunted existence in the absence of a host.

Figure 12.6. Macroparasites of plants. Top left: *Orobanche gracilis,* a broomrape parasitic on legumes. Top right: Bladder cytinus (*Cytinus hypocistis*) parasitic on the roots of rock roses (*Cistus*). Bottom: Lesser dodder (*Cuscuta epithymum*) growing on ling (*Calluna vulgaris*). (Photographs: Heather Angel.)

The more complete the dependence of plant parasites on their hosts, the more limited is the range of hosts that they parasitize. Thus *Odontites verna* has a wide range of potential hosts, but the holoparasitic species of the genus *Orobanche* (the broomrapes) have a range which is often restricted to a single host species. Also, the more wholly parasitic the parasite, the smaller tend to be its seeds, and the more numerous. It is as if these species have gained access to a food resource for their early development that is an alternative to parentally supplied seed reserves.

. . . mistletoes,
A number of hemiparasites develop on the branches of trees. Like other hemiparasites, these mistletoes and their allies have rather large seeds. In most (perhaps all) of the species, the seeds are dispersed by birds which eat the fleshy single-seeded fruits (Figure 5.9). They either wipe off the seed by rubbing it against the bark of the tree, or they swallow it and deposit the seed with the faeces.

The hemiparasites on trees tap the sapwood of their host either by sending in one primary haustorium that branches in the cortex of the host and may encircle a branch from within, or by developing a root system on the outside of the branch from which haustoria tap into the sapwood of the host at intervals. There is clear evidence that the presence of branch hemiparasites reduces the growth of the host. Stands of silver fir in France that carried mistletoe populations were reduced in stem volume by 19% (Klepak, 1955) and dwarf mistletoe has produced losses of one-third in extractable timber in stands of lodgepole pine in Colorado (Gill, 1957).

indirectly transmitted macroparasites: tapeworms, . . .
There are many types of medically important animal macroparasites with indirect life cycles. The tapeworms, for example, are intestinal parasites as adults, absorbing host nutrients directly across their body wall, and consisting of a large number of segments (or 'proglottids') proliferating behind an attached head. The most distal proglottids contain masses of eggs and are voided via the host's faeces. The larval stages of the life cycle then proceed through one or two intermediate hosts before the definitive host is reinfected. Figure 12.7 outlines the life cycles of two human intestinal tapeworms. There are other tapeworms, however, that use man as an intermediate (larval) host (e.g. *Echinococcus granulosus,* which causes hydatid disease when it infects a variety of deep tissue sites, especially the liver and lungs).

. . . schistosomes, . . .
The schistosomes causing human schistosomiasis (bilharzia) are representative of a much larger group of trematode worms, in that their life cycle involves sexual reproduction in a terrestrial vertebrate host and asexual multiplication in a snail (Figure 12.8). After asexual multiplication of *Schistosoma mansoni* in a snail, for example, free-living aquatic larvae are released which infect man by penetrating the skin. The larvae migrate within the blood vessels to the liver, where they mature and pair. They then migrate to the blood vessels of the intestine, where they may live and continue to produce eggs for many years. The eggs reach the outside world via the faeces after disrupting host tissues on their way out. Schistosomes are therefore macroparasites of humans but microparasites of snails. Schistosomiasis affects the gut wall because eggs become lodged there, and also affects the blood vessels of the liver and lungs when eggs become trapped there too.

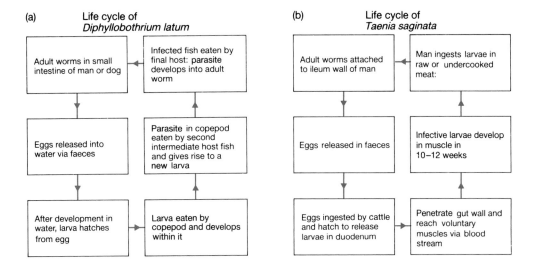

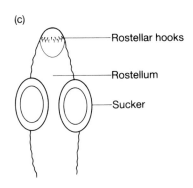

Figure 12.7. (a and b) The life cycle of two cestode parasites that infect man. (a) *Diphyllobothrium latum,* which is transmitted in incompletely cooked fish and (b) *Taenia saginata* which is transmitted in raw or undercooked beef. (c) The four muscular suckers and the recurved hooks that anchor a tapeworm in the gut—in this case *Echinococcus granulosus,* a parasite of dogs. (After Whitfield, 1982.)

... filarial nematodes, ...

Filarial nematodes are long-lived parasites of man that all require a period of larval development in a blood-sucking insect vector (Figure 12.9). *Wuchereria bancrofti,* which causes Bancroftian filariasis, does its damage by the accumulation of adults in the lymphatic system (classically but only rarely leading to elephantiasis). Larvae (microfilariae) are released into the blood and are ingested by culicine mosquitoes which also transmit more developed, infective larvae back into the host. *Onchocerca volvulus* which causes 'river blindness' is transmitted by blackflies (with river-inhabiting larvae, hence the name). Here though it is the microfilariae that do the major damage when they are released into skin tissue and reach the eyes.

... rust fungi

Indirect transmission via an intermediate host is relatively rare amongst plant macroparasites, but it is not uncommon amongst the rust fungi. In black stem rust (Figure 12.10), for example, two phases of wind dispersal are involved in the transmission of the infection from the cereal host (e.g. wheat) to the barberry shrub (*Berberis vulgaris*) and other host species, and from *Berberis* back to wheat. Infections on the cereal are polycyclic, i.e. within a season spores may infect, form lesions on the leaves and stem, and release spores which disperse in turn to infect further cereal plants. This cycle may continue throughout most of the season, and is essentially an intense multi-

(a)

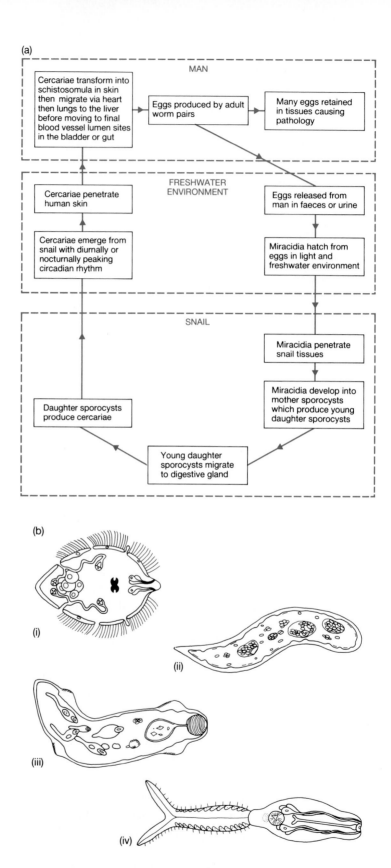

MAN

Cercariae transform into schistosomula in skin then migrate via heart then lungs to the liver before moving to final blood vessel lumen sites in the bladder or gut → Eggs produced by adult worm pairs → Many eggs retained in tissues causing pathology

FRESHWATER ENVIRONMENT

Cercariae penetrate human skin

Cercariae emerge from snail with diurnally or nocturnally peaking circadian rhythm

Eggs released from man in faeces or urine

Miracidia hatch from eggs in light and freshwater environment

SNAIL

Daughter sporocysts produce cercariae

Miracidia penetrate snail tissues

Miracidia develop into mother sporocysts which produce young daughter sporocysts

Young daughter sporocysts migrate to digestive gland

(b)

(i)

(ii)

(iii)

(iv)

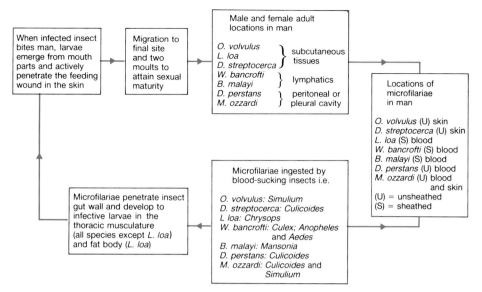

Figure 12.9. The life cycle of the seven most important filarial parasites of man. The filariae are long-lived nematode worms that are transmitted by blood-sucking hosts. (After Whitfield, 1982.)

plication of clones of the fungus—but it is on the alternate host that a proper sexual process occurs, generating recombinants.

The barberry is infected by secondary haploid spores formed from over-wintering resting spores (teliospores) that had developed on the cereal at the end of the season. Haploid infections on the barberry are 'spermatized' by the insect transmission of specialized spores that are, in effect, gametes. The parasite on the barberry then forms lesions from which yet a further type of spore is released, and this can infect only the cereal. The barberry is a long-lived shrub and the rust is persistent within it. Infected barberry plants may therefore serve as primary and persistent foci for the spread of the rust into crops of the shorter-lived annual cereal.

12.3 Transmission and distribution

The study of the behaviour of diseases within populations of hosts (i.e. the population dynamics of disease) is referred to as *epidemiology*. The way in which we study these populations in practice, however, varies with the type of parasite. It is usually possible to count or at least estimate the numbers of macroparasites in or on a host (e.g. worms in an intestine or lesions on a plant), especially as they do not multiply in the host. The macroparasite itself is then the appropriate unit of study. Microparasites, on the other hand, are

the appropriate units of study: macroparasites, and hosts infected with microparasites

Figure 12.8. (a) The life cycle of human schistosomes. (b) Larval stages in digenean worms (the schistosomes are a class of digenean worms): (i) the ciliated larva which infects molluscs (miracidia); (ii) the sporocyst, which develops in the mollusc; (iii) rediae, which migrate to the molluscs digestive gland and may multiply there eventually releasing (iv) larval stages that leave the mollusc—these cercariae may in some cases infect the host directly or may encyst as a dormant stage that is eaten by the host. (After Whitfield, 1982.)

small, often extremely numerous, and possess the ability to multiply rapidly and directly within an individual host. Thus it is always difficult and usually impossible to count the number of microparasites in a host, and the infected host is the appropriate unit for practical study. In other words, the epidemiologist studying microparasites counts numbers of infected hosts rather than numbers of parasites. Of course, an epidemiologist working with a more complex life cycle, such as that of a schistosome, will use the infected snail as

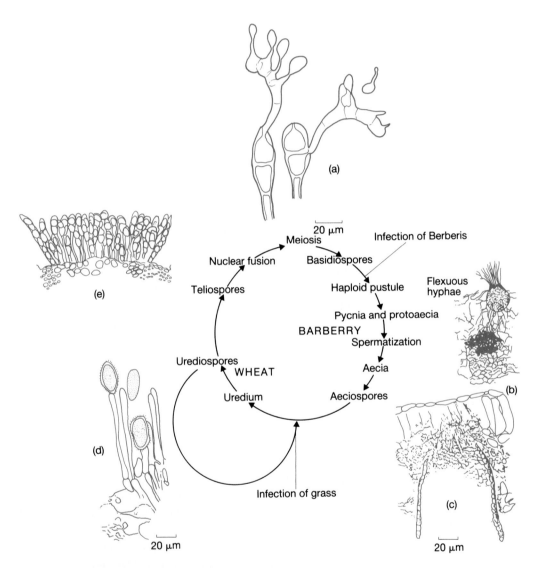

Figure 12.10. The life cycle of the black stem rust disease of cereals (*Puccinia graminis*). (a) Resting spores (teliospores) germinate to produce wind-dispersed small basidiospores which infect the barberry (*Berberis vulgaris*). The fungus develops in the leaf of the barberry and forms pustules that are involved in a sexual process (b), and that lead after fertilization to the development of new pustules (aecia) that release binucleate aeciospores (c). The aeciospores then infect grasses, including cereals, where they develop to produce a quite different type of pustule (uredia) (d). These in turn release urediospores that can reinfect the grass. Towards the end of the season, resting teliospores develop, which recommence the cycle (e). (After Webster, 1970.)

one unit of study (because the parasite multiplies within the snail), but in an infected human, where the parasite grows but does not multiply, he will estimate the number of parasites directly. However, the critical difference between microparasites and macroparasites is operational rather than fundamental: its effect is on the way we study them.

12.3.1 Hosts as islands: transmission

A crucial element in the epidemiology of any disease is the transmission of parasites between hosts. Janzen (1968, 1973) introduced the idea that we could usefully think of hosts as islands that are colonized by parasites. The further an individual plant (or leaf), for instance, is isolated from others of its own sort, the more remote are the chances that it will be colonized by a parasite. If it is isolated from others of its own kind by a mixture of other species, the transmission stages of parasites are likely to end up stranded on alien 'non-hosts'. This is probably one reason why there are seldom widespread epidemics of plant disease in nature: monotonous stands of a single host species are rare in natural vegetation. The major disease epidemics known among plants have occurred in crops which are not islands in a sea of other vegetation but 'continents'—large areas of land occupied by one single species (and often by one single variety of that species).

The analogy of animal hosts with islands is not so obvious, because islands are not usually mobile; but elements of the analogy are still real. A man colonized by the malarial parasite is an island. The malarial parasite can pass from one island to another only by means of a transmission agent, the mosquito. Limitations on the flight path of the mosquito represent the distance between islands from one host to another.

transmission is affected by contact rates and therefore densities

Not surprisingly, the factors affecting the rate of parasite transmission vary between different types of parasite. For directly transmitted microparasites, where infection is effected either by physical contact between hosts or by means of a very short-lived infective agent, the net rate of transmission is usually directly proportional to the frequency of encounters between infected hosts and susceptible (uninfected) hosts. The net rate of parasite transmission is therefore always greater in dense populations than in sparse ones. In fact, the transmission rate often varies seasonally, due to the influence of climatic factors on the frequency of host contact or the life-expectancy of the short-lived infective stages. Seasonal contact rates are important in the epidemiology of many common viral infections of man, such as measles, mumps and chicken-pox (Anderson, 1982).

Where infection is effected by a longer-lived infective agent, the rate of parasite acquisition is usually directly proportional to the frequency of contact between hosts and infective stages (and is therefore dependent on the densities of both). For vector-transmitted microparasites, on the other hand, the transmission rate from infected vectors to hosts is proportional to the 'host biting rate'. This itself is dependent on the overall frequency of vector 'bites' and the fraction of the host population that are susceptible. The transmission rate from infected hosts back to susceptible vectors depends on the frequency of vector bites and the fraction of hosts that are infected.

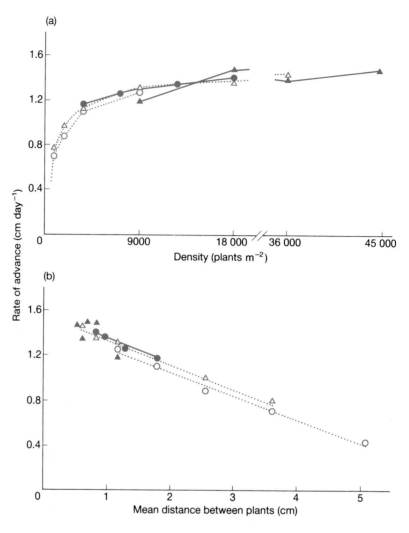

Figure 12.11. The rate of advance of the pathogen *Pythium irregulare* into populations of cress (*Lepidium sativum*), from infected seedlings placed at the edge of the plant populations. The spread of the disease is shown (a) in relation to the density of cress plants, and (b) in relation to the mean distance between plants. ○, ● and ▲ represent replicates. (After Burdon & Chilvers, 1975.)

the spread of a disease is affected by the distance between hosts

Many soil-borne diseases of plants are spread from one host plant to another by root contacts, or by the growth of the fungus through the soil from a 'food-base' established on one plant which gives it the resources from which to attack another. The honey fungus *Armillaria mellea*, for example, spreads through soil as a bootlace-like 'rhizomorph', and can infect another host (usually a woody tree or shrub) where their roots meet. Such a parasite can spread through a plantation or along a hedge, and can be controlled by digging a trench across its path; its passage 'from island to island' can be prevented by a purely physical barrier. On a much finer scale, fungi of the genus *Pythium*, which parasitize and kill the seedlings of many higher plants, depend for rapid spread on the proximity of the seedlings to each other (Figure 12.11). In both commercial horticulture and forestry, the risk of epidemic damage from *Pythium* is controlled by sowing seeds at low density. It is very characteristic of such soil-borne diseases that epidemics spread as 'invading fronts' or 'horizons of infection'. Swollen shoot disease of cacao is a similar disease of the above-ground parts of a plant. It is a virus infection transmitted by mealy bugs, but these are flightless. The main spread of disease is therefore between trees that make physical contact with each other, and cacao is grown under just such conditions—as 'continents' of continuous inter-plant contacts. In the period from 1922 to 1930 this disease spread to destroy plantations over wide areas of West Africa.

wind-borne diseases: leptokurtic transmission

For diseases that are spread by wind, foci of infection may become established at great distances from the origin; but the rate of spread of an epidemic locally is strongly dependent on the distance between plants. It is characteristic of wind-dispersed propagules (spores, pollen and seeds) that the distribution achieved by dispersal is usually strongly 'leptokurtic': a few propagules go a very long way but the majority are deposited close to the origin. An extreme example of a wind-borne parasite of animals is the virus causing foot-and-mouth disease in farm livestock (Davidson, 1983). The disease is normally absent from Great Britain, but occasionally a case is reported involving long-distance spread, for instance from the Netherlands to eastern England. Foci of infection very rarely become established at this sort of distance, but there can be extremely rapid spread within herds once they have, especially in the crowded conditions of cattle markets. In Britain, because it is an island, the effective control of the disease is by the compulsory slaughter of all infected animals and likely contacts. On the continent of Europe such a policy could not be sustained, and vaccination is adopted.

12.3.2 Disease in mixtures of species and genotypes

We have seen in the chapters on the interactions between predators and prey that there is often a high degree of specialization of a particular predator species on a particular species of prey (monophagy). The specificity is often very much more strongly developed in the relationship between parasites and hosts. Not only is there very commonly a restriction of one species of parasite to one species of host, but there is also a detailed correspondence between specific genotypes of parasite and specific host genotypes that they may parasitize. In its most extreme form, there is a strict one-for-one correspondence between genes that confer pathogenicity on a parasite and genes conferring resistance on a host. Natural populations of both hosts and parasites may contain many different host and parasite genotypes. Under these circumstances, the effective density of a host (to a particular race of parasite) is the density of susceptible hosts, not the total density of the host species.

mixtures dilute densities

In agriculture and forestry, some control of disease may be obtained by growing mixtures of species, or mixture of genotypes of a single species. In mixtures, the effective density of each species is reduced—diluted among the others. Each disease-resistant form in a mixture may serve as a barrier to the spread of infection from one susceptible to another.

In agricultural practice, resistant cultivars offer a challenge to evolving parasites: mutants that can attack the resistant strain have an immediate gain in fitness. New, disease-resistant crop varieties therefore tend to be widely adopted into commercial practice; but they then often succumb, rather suddenly, to a different race of the pathogen. A new resistant strain of crop is then used, and in due course a new race of pathogen emerges. There is a 'boom and bust' cycle which is repeated endlessly. This keeps the pathogen in a continually evolving condition, and plant breeders in continual employment (see p. 458 and Figure 12.28.)

(see p. 458 and Figure 12.28.)

mixtures of susceptibles and immunes

An escape from this boom and bust cycle can be gained by the deliberate mixing of varieties so that the crop is dominated neither by one virulent race of the pathogen nor by one susceptible form of the crop itself. Various types of mixture have been used, including random mixtures of different genotypes, and multilines (populations that differ within themselves only with respect to the resistance genes that they carry). These procedures have indeed been shown to control the rate of disease development. Figure 12.12 (and Plate III) for example, illustrates the effect of varietal mixing on the development of epidemics of late blight of potatoes.

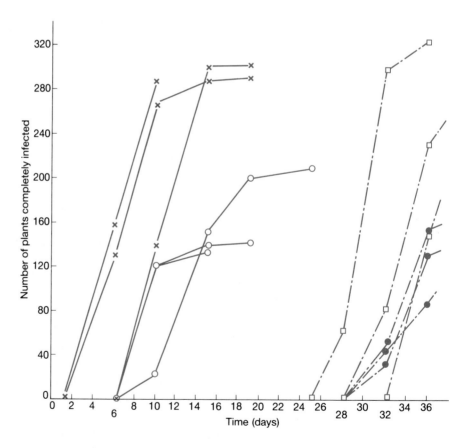

Figure 12.12. The rate of spread of infections by the pathogen causing late blight of potatoes (*Phytophthora infestans*) in pure populations of the potato varieties Pentland Crown and Pentland Dell and in a mixed population of the two varieties. × = pure populations of Pentland Crown, □ = pure populations of Pentland Dell, O = Pentland Crown in mixed populations, ● = Pentland Dell in mixed populations. (After Skidmore, 1983.)

After the spread of an epidemic through a population of higher animals, those that have contracted the disease but recovered may have acquired immunity (see p. 441). They then act in the same way as resistant forms in a population, diluting the density or contact rate of susceptibles, and slowing down the multiplication of the parasite and the spread of the disease. In medical practice, a population of immunes can be created by vaccination. The whole population need not be immunized—only a sufficient proportion to provide a mixture in which susceptibles are sufficiently diluted to prevent effective transmission. In the case of measles in western societies, a rather high proportion of children (92%–94%) would have to be vaccinated to give an appropriate dilution to prevent an epidemic occurring (Anderson, 1982).

12.3.3 The distribution of parasites and infected hosts

parasites are usually aggregated

The distribution of parasites within populations of hosts is rarely random—parasites are usually aggregated or clumped, i.e. many hosts harbour a few or no parasites, and a few hosts harbour large numbers of parasites (Figure 12.13). Some of the aggregation results directly from the leptokurtic dispersal of infection units, when patches of disease develop within a crop from foci of infection. Also, when parasites multiply within their host, randomly infected individuals may quickly generate gross differences, between zero populations of parasites in some hosts and dense populations in others. When dispersal is poor or slow (e.g. multiplication of the wingless form of aphids), very dense, very local aggregations of parasites can occur. But aggregations probably arise most frequently because hosts vary in their susceptibility to infection (whether due to genetic, behavioural, or environmental factors).

the prevalence, intensity and mean intensity of infection

In such locally aggregated populations, the mean density of parasites often has rather little useful meaning. In a population of humans in which only one person is infected with anthrax, the mean density of *Bacillus anthracis* is a particularly useless piece of information! The most widely used epidemiological statistic for microparasites is the *prevalence* of infection: the proportion or percentage of a host population that is infected with a specific parasite. This is a particularly convenient measure in the case of microparasites, where the number of individual parasites cannot be counted anyway (the infected host is the unit of study). On the other hand, the severity of infection is often clearly related to the number of parasites harboured, particularly in the case of macroparasites; and the number of parasites in or on a particular host is referred to as the *intensity* of infection. The *mean intensity* of infection is then the mean number of parasites per host in a population (including those hosts that are not infected).

distribution, prevalence and mean intensity

The relationships between prevalence and mean intensity for various types of frequency distribution are shown in Figure 12.14. If parasites are evenly distributed, prevalence tends to be relatively high and mean intensity relatively low. This could arise because of density-dependent mortality amongst the parasites (see below), or acquired resistance to reinfection, or because hosts with high intensities of infection are killed. In the more usual case of aggregation, mean intensity is relatively high but prevalence is relatively low. The epidemiological consequences of aggregation are dis-

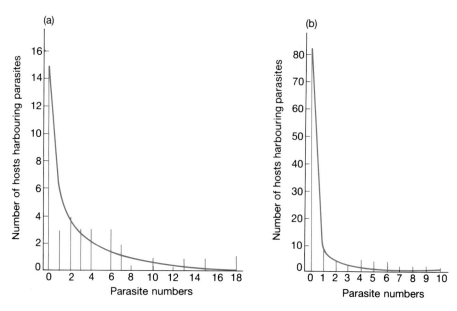

Figure 12.13. Examples of aggregated distributions of parasite numbers per host for which the negative binomial is a good empirical model. The vertical bars represent the observed frequencies and the solid line is the fit of the negative bionomial distribution. The distributions of (a) *Toxicara canis*, a gut nematode parasite of foxes and (b) lice (*Pediculus humanis capitis*) on a population of man. (After Watkins & Harvey, 1942; and Williams, 1944.)

cussed in subsequent sections; and we return to the concepts of prevalence and intensity in the more general context of 'abundance' and 'rarity' in Chapter 15.

12.4 Hosts as habitats

the habitat of a parasite is alive

The essential difference between the ecology of parasites and that of free-living organisms is that the habitats of parasites are themselves living organisms. A habitat that is alive is capable of growth (in numbers and/or size); it is potentially reactive, i.e. it can respond *actively* to the presence of a parasite by changing its nature, developing immune reactions to the parasite, digesting it, isolating or imprisoning it; it is able to evolve (every host is a descendant of previous hosts and a product of natural selection); and in the case of many animal parasites it is mobile and has patterns of movement that dramatically affect dispersal (transmission) from one habitable host to another.

parasites and hosts vary in the intimacy of their association

Parasites (like mutualists—see Chapter 13) vary in the intimacy with which they enter into the environment that is their host. At one extreme there are ectoparasites, which feed on host material from outside. Examples amongst animal parasites include fleas, lice, ticks, monogeneans and even some fungi; while amongst plant parasites there are aphids and powdery mildews which live on leaf surfaces and send stylets (aphids) or haustoria (mildews) into host tissue. Such parasites share some experience of the outside world with their host; they are relatively uninsulated against cold,

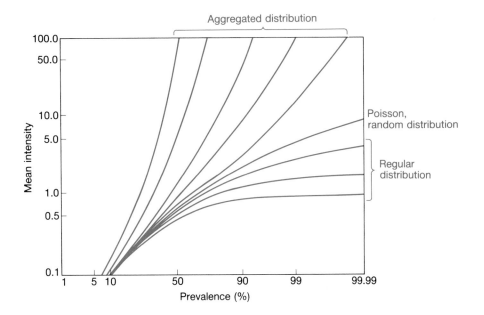

Figure 12.14. The relationship between the prevalence and mean intensity of infection for aggregated, random and regular distributions of parasites per host. (After Anderson, 1982.)

hosts are heterogeneous environments

evaporation, etc.. An intermediate group of parasites develop within cavities in the host; for example worms in the alimentary canal, lungs and Eustachian tubes of animals, or fungi as mycelia between the cells of a plant host. Such parasites may not enter the host cells at all; they are in a strict sense 'outside' the cellular body of the host, though inside its environment. At the other end of the spectrum are parasites that enter directly into host tissues and cells. Microparasites are often truly intracellular. The malarial parasites and the piroplasms that cause babesiosis in domesticated animals live in the red blood cells of vertebrates. *Theileria* parasites of cattle, sheep and goats live in the lymphocytes of the mammal and in the epithelial cells and later the salivary glands cells of the tick which transmits the disease. Among the parasites of higher plants, viruses, bacteria and primitive fungi typically grow and multiply within host cells.

Every organism is a heterogeneous environment—a matrix of potential habitats. Even a bacterial cell is a heterogeneous environment to an infective viral particle. Gut, blood, liver, brain, pancreas and eyes are different environments for the parasites of a mammal. Leaves, stems, roots, fruits and seeds offer different resources, are differently protected, and provide different physical environments for the parasites of plants (Figure 12.15). Different species of host, even when they are closely related, differ in properties (the food they eat, their life cycles, the way they react to infection) that may prevent them being invaded or make them uninhabitable to a parasite.

The distribution of parasites on the gills of fishes is a particularly beautiful example of habitat specialization within a host. The teleost fishes have four pairs of gills. Each gill is formed from a bony arch which carries primary lamellae that in turn carry tightly packed secondary lamellae. They provide an enormous surface for respiratory exchange of gases and for occupancy by parasites. There is a continuous flow of blood close to the surface of the gills and the flow of water over them keeps the microenvironment well oxygenated. The gills are parasitized by fungi, protozoa, monogenean worms, metacercarial larvae of trematodes, copepods, glochidia larvae of bivalve molluscs, leeches and mites. Hanek (1972) surveyed the parasites on the fish *Lepomis globosus* in West Lake, Ontario. The monogenean parasites were most abundant on the anterior face and medial sections of the gills. The copepods were concentrated on the dorsal and ventral regions. The monogeneans and glochidia were found most often on gills II and III, and the copepod *Ergasilus caeruleus* was rather evenly distributed among the different gills.

Perhaps the most remarkable example of habitat specificity known among parasites is that shown by fungi on the bodies of insects. Within the fungal genus *Laboulbenia* are species which are specialists on beetles. On one particular host species, the beetle *Bembidion picipes*, different species of fungus colonize different parts of the body. It is possible to identify many of the species of fungus with precision simply from the part and the sex of the beetle that they parasitize (Figure 12.16).

Many parasites are limited in their sites within a host to the places that they reach first, but in others the specific sites are sought out by migration after the host has been entered. Helminth worms are particularly good examples of parasites that search for specific habitats within their hosts. The nematode *Angiostrongylus cantonensis*, for example, has a remarkable circuit of habitats in the rat (Mackerras & Saunders, 1955; Alicata & Jindrak, 1970). The parasite penetrates the wall of the gut and most of the invaders enter the small veins and pass to the liver in the hepatic portal system. They then travel to the right atrium of the heart via the posterior vena cava, and from there to the lungs through the pulmonary artery and back to the main arterial circulation from the heart. These routes disperse the parasite through the body, and the young adults then pass into cerebral veins, return to the heart and lungs, and then become established in the pulmonary arteries.

The digenean worm *Halipegus eccentricus*, which infects frogs, is eaten by tadpoles and settles at the cardiac end of the stomach. As the tadpoles develop into adult frogs, the parasite moves up the oesophagus and enters the Eustachian tubes of the ear. Such homing of parasites on to target habitats can be demonstrated by deliberately transplanting parasites from one part of the host's body to another. Alphey (1970) transplanted a nematode worm (*Nippostongylus brasiliensis*) into the anterior and posterior parts of the small intestine of rats. The nematodes migrated back to their normal habitat in the jejunum. Another nematode (*Spirocera lupi*) was transplanted into the thoracic cavity of a dog. The parasites migrated back to their normal site on the wall of the oesophagus (Bailey, 1972).

The examples above are of parasites that move bodily, but in other cases habitat search may involve growth (for example, of a fungus) from the point at which infection occurs to a quite different part of the body where spore production takes place. Loose smut of wheat, *Ustilago tritici*, infects the exposed stigmas of the flowers of wheat, and then grows as an extending

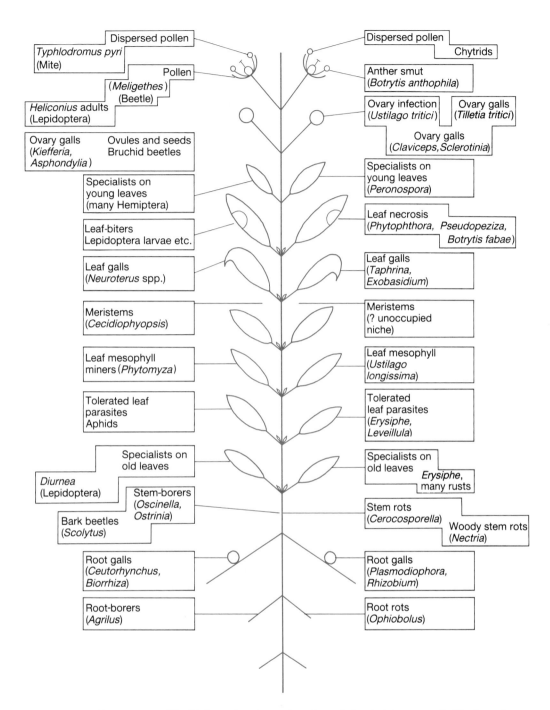

Figure 12.15. The higher plant is a series of specialized niches occupied by fungal pathogens and by insect predators and parasites. Comparable, though rather less diverse, diagrams could be drawn for nematodes and for pathogenic bacteria. The subtlety of niche specialization in even greater than is implied by the diagram—for example, *Neuroterus* species which produced leaf galls on oak are believed to specialize at a subspecific level with respect to the tip, middle and base of a leaf, as well as to leaf age (Askew, 1962; Hough, 1953). Some species may alternate between two highly specific niches on the same plant, e.g. *Biorhiza pallida* forms galls on the roots and shoot meristems in different seasons. (From Harper, 1977.)

filamentous system into the young embryo. Growth continues in the seed-ling, and the fungus mycelium keeps pace with the growth of the shoot. Ultimately the fungus grows rapidly within the developing flowers and converts them into masses of spores. In bunt of wheat, *Tilletia tritici*, it is the developing seedling that is infected, but again the mycelium grows through the host and makes its major development within the flowers, where it turns the ovary into a 'puff-ball' of fungal spores.

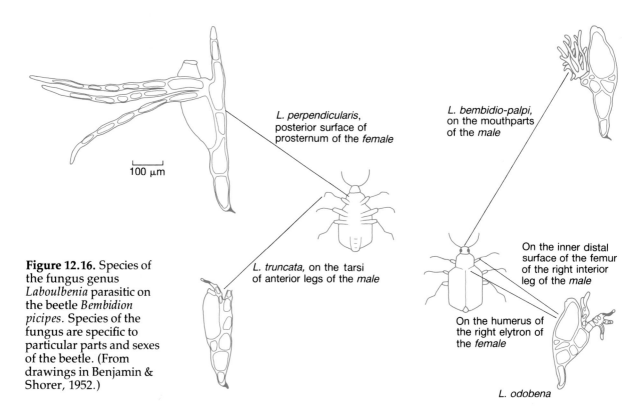

Figure 12.16. Species of the fungus genus *Laboulbenia* parasitic on the beetle *Bembidion picipes*. Species of the fungus are specific to particular parts and sexes of the beetle. (From drawings in Benjamin & Shorer, 1952.)

L. perpendicularis, posterior surface of prosternum of the *female*

L. truncata, on the tarsi of anterior legs of the *male*

L. bembidio-palpi, on the mouthparts of the *male*

On the inner distal surface of the femur of the right interior leg of the *male*

On the humerus of the right elytron of the *female*

L. odobena

100 μm

12.4.1 Density-dependence within hosts

Because populations of parasites occupy specific sites within their host, it is to be expected that individuals will sometimes interfere with each other's activities—that there will be competition between parasites and density-dependence in their growth, birth or death rates. It is extremely difficult to determine the way in which populations of microparasites are regulated within the cells of their host. However, it is relatively easy to manipulate populations of the larger parasites like the helminths, and the experimenter can control the density of populations that are present. Responses of the parasite to its own density can then be shown quite clearly (Figure 12.17). In experimental infections of mice with the tapeworm *Hymenolepis microstoma*, the total weight of worm per mouse rises asymptotically, as if to some ceiling value, and the weight of individual parasites decreases. The total number of eggs produced *per host* is much the same at all levels of infection. These

competition between parasites increases with density . . .

patterns are clearly reminiscent of many of those in Chapter 7, especially the 'constant final yield' in many populations of higher plants.

A similiar crowding effect can be seen in the liver fluke (*Fasciola hepatica*), and two additional effects now come into play. As density increases, the percentage of infections that develop into established worms declines, and, even more important, the development of a worm to the egg-producing stage is greatly delayed. This has the direct effect of slowing down the potential rate of increase of the population of worms.

. . . but so may the host's immune response

We have, however, to be careful when concluding that the effects of parasite density are a result of competition between parasites. The intensity of the immune reaction elicited from the host (see below) invariably depends on the density of parasites. The effect of density on the performance of parasites is not just the result of competition between the parasites themselves; it reflects active reaction by the parasite's environment—its host.

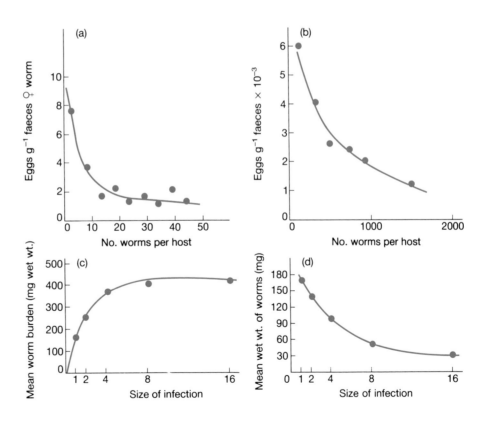

Figure 12.17. Density-dependent reproduction of parasite populations.
(a) Egg production by the roundworm *Ascaris* in man (after Croll *et al.*, 1982). (b) Egg production in *Ancylostoma* (a tapeworm) in man (after Hill, 1926). (c) The mean weight of worms per infected mouse after deliberate infection with different levels of infection by the tapeworm *Hymenolepis microstoma*, and (d) the mean weight of individual worms from the same experiment (after Moss, 1971).

12.5 The responses of hosts

necrotrophs and
biotrophs

The presence of a parasite in a host appears always to elicit a response. This must be so as a matter of definition—if there were no response we would describe the organism as a commensal not a parasite. The most obvious response is for the host to die, either as a whole or in its infected parts. We recognize a major difference among parasites between those that kill and can continue to live on the dead host (*necrotrophic* parasites) and those that cannot do so (*biotrophic* parasites). For a biotroph, the death of its host (or its infected parts) spells the end of the active phase of its life; the dead host is uninhabitable. For the necrotrophic parasites, the resources of the host may be more available when the host is dead. Most parasitic worms, lice and fleas, protozoa and parasitoids and the gall-forming parasites of higher plants are biotrophic, as are the rusts, smuts and mildews among the fungal parasites of plants. Biotrophs have specialized nutritional requirements, and until recently few could be cultured outside their host.

12.5.1 Necrotrophic parasites

Lucilia cuprina, the blowfly of sheep, is a necroparasite on an animal host. The preliminary attack is on the living host on which eggs are laid. Larvae (maggots) develop, and damage may lead to death. Further colonization of the carcase by *Lucilia* is detritivory rather than parasitism. Necroparasites on plants include many that attack the vulnerable seedling stage (especially species of the genus *Pythium*) and cause the 'damping off' of seedlings. They are rarely host-specific and attack a wide range of species.

Botrytis fabi is a typical necroparasite. It develops in the leaves of the bean *Vicia faba* and the cells are killed, usually in advance of penetration. Spots and blotches of dead tissues form on the leaves and the pods. The fungus may continue to develop and form spores on the dead tissue. It does not do so on the living substrate.

When modular organisms are the host, a necroparasite will often kill a part rather than the whole host. Modular organisms grow by continually producing new parts as older parts die. A necroparasite may simply speed up the natural death rate of the modules of its host (e.g. leaves of a plant) leaving the remainder of the plant alive and iterating new parts.

necrotrophs as
pioneering detritivores

Most necroparasites can be regarded as ecological pioneers. They are detritivores that are 'one jump ahead' of competitors because they can kill the host (or its parts); and by being there first they have first access to the resources of its dead body. Their resource requirements are typically simple; they can usually be grown on simple artificial media.

The response of the host to necroparasites is rather limited. The very fact that necroparasites kill host tissues is a demonstration that host defences and tolerances are weak. Among plant hosts, the most common responses are the dropping off of the infected leaves (alfalfa sheds leaves that are attacked by *Pseudopeziza medicaginis*) or the formation of a specialized barrier that isolates the infection. The corky scabs formed on potatoes as a reaction to *Actinomyces scabies* isolate and localize the invading colonies.

12.5.2 The immune response

Any reaction by an organism to the presence of another depends on its recognizing a difference between what is 'self' and what is 'not-self'. In invertebrate animals, populations of phagocytic cells are responsible for most of a host's response to invaders, even to inanimate particles. Phagocytes may engulf and digest small alien bodies, and encapsulate and isolate larger ones. Presumably the recognition that invading particles are not 'self' depends on surface properties of the invader, but the process is still largely obscure. In vertebrates there is also a phagocytic response to 'not-self' materials, but their armoury is considerably extended by a much more elaborate process: the immune response.

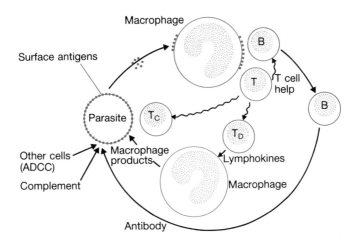

Figure 12.18. A vertebrate host is a reactive environment in which the immune response of the host converts it from being a habitable site into one that is uninhabitable. This sort of reaction distinguishes living organisms as habitats from the sorts of environment more normally considered by ecologists.

The response begins when the immune system is stimulated by an antigen that is taken up and processed by a macrophage. The antigen is a part of the parasite such as a surface molecule. The processed antigen is presented to T and B lymphocytes. T lymphocytes respond by stimulating various clones of cells, some of which are cytotoxic, others of which secrete lymphokines and others stimulate B lymphocytes to produce antibodies. The parasite that bears the antigen can now be attacked in a variety of ways. (After Cox, 1982.)

The ability of vertebrates to recognize and reject tissue-grafts from unrelated members of their own species is just one example of the immune response in action (see Figure 12.18). From the point of view of the ecology of parasites, there are two vital features of the immune response. First, it may enable a host to recover from infection. Secondly, it can give a once-infected host (= habitat) a 'memory' that may change its reaction if the parasite (= colonist) strikes again, i.e. the host may become immune to reinfection. In mammals, the transmission of immunoglobulins to the offspring can sometimes even extend protection to the next generation. The response of vertebrates to infection therefore protects the individual host and increases its potential for survival to reproduction. The response of invertebrates gives poorer protection to the individual, and the recovery of populations after diseases depends more on the high reproductive potential of the survivors than on the recovery of those that have been infected (Tripp, 1974).

transient and persistent infections

For most viral and bacterial infections of higher animals, the colonization of the host is a brief (i.e. transient) episode in the host's life. The parasites multiply within the host and elicit a strong immunological response. By contrast, the immune responses elicited by many of the macroparasites and protozoan microparasites tend to be rather short-lived. The infections themselves, therefore, tend to be persistent, and hosts are subject to continual reinfection.

12.5.3 Hypersensitivity and phytoalexins in plants

The modular structure of plants, the presence of cell walls, and the absence of a true circulating system (such as blood or lymph) would all make any form of immunological response an inefficient protection. There is no migratory population of phagocytes in plants that can be mobilized to deal with invaders. Many plants, however, respond to even the earliest stages of infection by hypersensitive reactions. The infected cells die, and they and the immediately surrounding cells produce 'phytoalexins'. These are low-molecular-weight, antimicrobial products of secondary plant biosynthesis that are induced to accumulate to inhibitory levels in the localized region of pathogen infection (Bailey & Mansfield, 1982). Phytoalexins can prevent the spread of parasites from the initial site of infection, but they are quite unlike an immune reaction: they are strictly localized, rather non-specific in their action, and are produced as a reaction to a variety of stimuli.

12.5.4 Responses to biotrophic parasites: tolerance, morphogenesis and behaviour

many biotrophs are tolerated

The most highly specialized parasites are those whose presence is tolerated by the host. This must involve some behaviour on the part of a parasite that prevents it being recognized. A most remarkable example of a parasite using host resources without stimulating a strong host reaction is seen in the parasitism of higher plants by aphids. Aphids gain access to plant resources by penetrating with their stylets into the phloem and drawing phloem sap from the sieve tubes while these are still actively functional (Figure 3.18). Phloem cells are extremely sensitive to disturbance and extremely difficult to study experimentally. Yet aphid stylets are inserted into phloem with so little disruption that plant physiologists, interested in the nature of phloem sap, have used the inserted stylets as the most effective way of sampling un-disturbed sieve tubes.

Biotrophic fungi, such as the rusts and powdery mildews, penetrate the cells of the host, which usually tolerate the presence of the parasite and remain alive. In some cases, the infections seem to produce no major change in the behaviour of the host apart from reducing the expectation of life of the infected cells and changing the pace of metabolic processes and water loss. In other situations, the presence of a parasite may induce quite profound changes in the metabolism of the infected host. Many of the biotrophic parasites that infect living leaf tissue stimulate the formation of 'green islands' around the infection which are not only sites of increased photosynthetic activity but also serve as 'sinks' to which metabolites flow from uninfected parts of the same plant.

Figure 12.19. Agromyzid flies that parasitize and form galls on oak—see Figure 3.19.

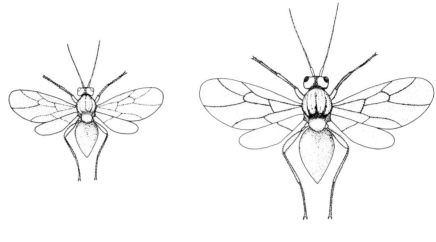

Cynips quercus-folii,   Cynips quercus-folii,

many biotrophs alter the pattern of host morphogenesis

Many of the biotrophic parasites induce a new *programmed* change in the development of the host. The agromyzid flies and cecidomyid and cynipid wasps that form galls on higher plants are remarkable examples (Figure 12.19). These insects lay eggs in host tissue, which responds by renewed growth. The galls that are produced are the result of a morphogenetic response by the host that is quite different from any structure that the plant normally produces. Just the presence for a time of the parasite egg may be sufficient to start a morphogenetic process in the host which can continue even in the absence of the developing larva. There is clearly a recognition process involved in which specific alien material (the insect egg) cues the process of gall formation. Among the gall-formers attacking the common English oak (*Quercus rober*), each elicits a unique morphogenetic response from the host (Askew, 1961, 1971). Fungal parasites can also induce morphogenetic change. Parasites such as *Plasmodiophora brassicae* stimulate cell multiplication and enlargement, increasing the number of habitable sites (root cells) in which the parasite multiplies (Figure 12.2). Nematode parasites may also elicit a morphogenetic response from the host, in the form of enormously enlarged cells and the formation of nodules and other 'deformations'. Perhaps the most profound intervention by a parasite in controlling the growth of the host is caused by the bacterium *Agrobacterium tumefaciens*, which induces the formation of galls on plant tissues. After infection, gall tissue can be recovered that lacks the parasite but has now been set in its new morphogenetic pattern of behaviour; it continues to produce gall tissue. In this case the parasite has induced a genetic transformation of the host cells.

The process involves plasmid transmission, and something similar may be involved in the formation of the insect-induced galls and the nodules of nitrogen-fixing mutualisms (Chapter 13).

A particular type of control of host growth is accomplished by parasitoids. The host remains 'alive' (though its body becomes increasingly the body of its parasite), and the host has no continued life after the parasitic larvae mature. Such parasites kill the host, but use the resources of its body only while it is alive. They kill but are not necroparasites. Some of the more specialized parasitic fungi also 'take control' of their host and 'castrate' or 'sterilize' it. The fungus *Epichloe typhina,* which parasitizes grasses, prevents them from flowering and setting seed—the grass remains a vegetatively vigorous 'eunuch' leaving descendant parasites but no descendants of its own. This 'take-over' of the host is a common phenomenon among systemic parasites of plants, and these vegetative eunuchs may actually have greater vegetative vigour than unparasitized plants (Bradshaw, 1959).

some parasites alter host behaviour

Most of the responses of modular organisms to environmental stimuli involve changes in growth and form. In unitary organisms the responses to parasites may involve a change in behaviour of the host that increases the chance of transmission of the parasite. Irritation of the anus of worm-infected hosts may cause scratching and transmission of the parasite eggs from the fingers or claws to the mouth. Sometimes the behaviour of infected hosts seems to maximize the chance of the parasite reaching a secondary host or vector. Praying mantises have been observed walking to the edge of a river and apparently throwing themselves in. Such suicidal individuals were seen in two successive years by the River Hérault in southern France, and within a minute of entering the water a gordian worm (*Gordius*) had emerged from the anus. This worm is a parasite of terrestrial insects but depends on an aquatic host for part of its life cycle. It seems that an infected host develops a 'hydrophilia' that ensures that the parasite reaches a watery habitat. One author (J.L.H.) has observed suicidal mantises that were rescued. They discharged their worms and immediately returned to the river bank and threw themselves in again!

12.5.5 The survivorship, growth and fecundity of hosts

The success of a biotrophic parasite must depend on it being an effective competitor for resources against the parts of the host that it parasitizes. It is such internal competition that is responsible, at least in part, for the reduced survival, fecundity, growth or competitive ability of the host. The influence of parasite burden on host death rate is shown in Figure 12.20 for mosquitoes infected with nematodes, sheep infected with liver flukes, and humans infected with a variety of diseases in relatively poorly developed countries. The adverse effect of a root-infecting nematode on the competitive ability of oats was described on p. 296, and the adverse effects of mistletoes on tree growth were described on p. 423. Infection with the gut-inhabiting nematode *Trichostrongylus tenuis* has a significantly adverse effect on the reproductive output of female red grouse (Potts *et al.,* 1984); while infection of the water bug *Hydrometra myrae* with the mite *Hydryphantes tenuabilis* has a detrimental effect on survival, reproduction and rate of development (Figure 12.21). More subtly, infection may make hosts more susceptible to predation (Anderson, 1979). For example, cormorants capture a disproportionately large number of roach infected with the tapeworm *Ligula intestinalis*, compared to the prevalence of infection in the fish population as a whole (Van Dobben, 1952).

Of these examples, the ones most instructive about natural populations

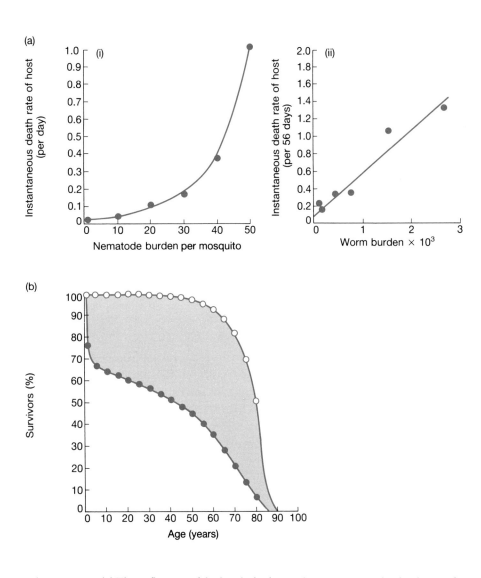

Figure 12.20. (a) The influence of the level of infection by parasites on the death rate of the host. (i) The influence of the burden of the nematode parasite *Dirofilaria immitis* on the death rate of its host mosquito (*Aedes trivittatus*) (after Christensen, 1978). (ii) The influence of the worm burden of the digenean liver fluke *Fasciola hepatica* on the death rate of sheep (after Boray, 1969). (b) The percentage of individuals surviving to a given age in a prosperous industrial country (open circles) and in a poor developing country (after Bradley, 1977).

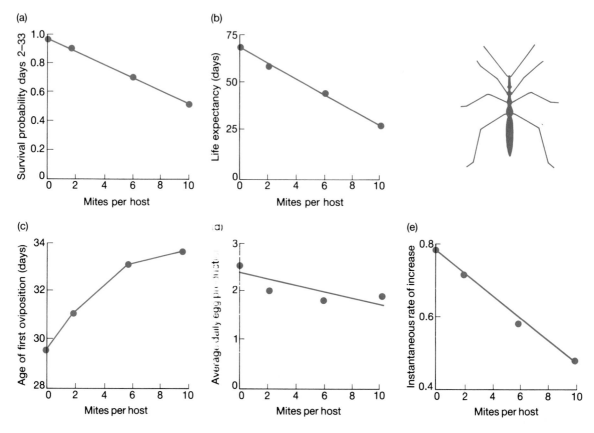

Figure 12.21. The relationships between the rate of infection of the water bug *Hydrometra myrae* by the parasitic mite *Hydraphantes tenuabilis* and (a) host survival, (b) the life expectancy of the host, (c) host maturity, (d) lost fecundity and (e) the rate of increase of the host population. (After Lanciani, 1975.)

parasitism and other factors often interact to harm the host

in general are probably the humans affected by disease, the roach infected with tapeworms, and the oats affected by nematodes. In all cases, the adverse effects reflect an interaction between disease *per se* and some other factor or factors. With the oats there was an interaction between nematode infection and competition from barley; with the roach the interaction was between infection and predation; while with the humans there was an interplay between disease and nutritional deficiency: the majority of infant deaths in under-developed countries appear to be due to childhood infections that are not lethal in the better nourished populations of developed countries. Such interactions between diseases and other factors are certain to be widespread. We must never forget the relevance of genetic variation in these interactions. Diseases such as measles have only recently been introduced into Africa and India. The diseases strike populations of individuals that are not only ill-nourished but whose ancestors have never experienced selection for resistance or tolerance to the disease.

12.6 The population dynamics of parasitism

Mathematical rigour has been brought to the study of the population dynamics of parasitism more effectively perhaps than to any other branch of ecology. Indeed, the theory of the dynamics of infectious disease is the oldest branch of biomathematics. There are also many close parallels between the dynamics of parasite–host interactions and those of the interactions of predators and their prey. For example, cycles arising in the outbreaks of measles and other childhood diseases are exactly equivalent to Lotka–Volterra cycles for predators and prey (section 10.2.1), since the acquisition of immunity by infected individuals is equivalent, in its effects on parasite populations, to the effects of the removal of prey on predator populations.

12.6.1 Directly transmitted microparasites

the basic reproductive rate and the transmission threshold

In all studies of the dynamics of parasite populations or the spread of a disease, the most important parameter is the *basic reproductive rate*, R_p (see May, 1981d, and Anderson, 1982, for clear and general reviews). We have already met the same concept of basic reproductive rate R_0 in Chapter 4 (section 4.5.1) in the more general context of the population dynamics of animals and plants. For parasites we use R_p because there are some subtleties that creep into the use of the term. In particular, in the case of microparasites, where it is the number of infected hosts that is the unit of study (rather than the number of parasites), R_p is the average number of new cases of the disease that arise from each infected host. The *transmission threshold*, which must be crossed if a disease is to spread, is therefore given by the condition $R_p = 1$. An infection will die out for $R_p < 1$ and spread for $R_p > 1$.

Some insights into the dynamics of infection can be gained by considering the various determinants of the basic reproductive rate. R_p increases with the density of susceptibles in the population, N (because higher densities offer more opportunities for transmission of the parasite); it increases with the transmission rate of the disease, β (which itself increases both with the frequency of host contact *and* the infectiousness of the disease—the probability that contact leads to transmission); it increases with the fraction of hosts that survive long enough to become infectious themselves, f; and it increases with the average period of time over which the infected host remains infectious, L. Thus, overall:

the determinants of R_p

$$R_p = \beta N f L \qquad (12.1)$$

Note that for most biotrophic parasites L is the period of the host's *life* when it is infectious; but for necrotrophic parasites and some biotrophs, the hosts may remain infectious long after they are dead (and indeed decomposed). They may leave a residue of resting spores where a diseased corpse or root system had been, and these may then germinate and produce a new infection when they make contact with a new host (resting spores of *Plasmodiophora* after a plant has suffered from clubroot, or the resting spores of *Bacillus anthracis* after a host has died from anthrax).

In terms of the size of the population, the transmission threshold is a *critical threshold density*, N_T, where, because $R_p = 1$,

$$N_T = \frac{1}{\beta f L} \tag{12.2}$$

critical threshold densities

Thus, if diseases are highly infectious (large β's), or are unlikely to kill their host (large f's), or give rise to long periods of infectiousness (large L's), they will have high R_p values and can persist in small populations (N_T is small). Some diseases of plants, with very long periods of infectiousness, are able to persist in small or low-density populations (e.g. clubroot). Conversely, if diseases are of low infectivity, or are likely to kill their hosts, or have short periods of infectiousness, they will have small R_p values and can only persist in large populations (Anderson & May, 1978). This provides an explanation for a number of patterns.

these equations help us to understand a number of patterns.

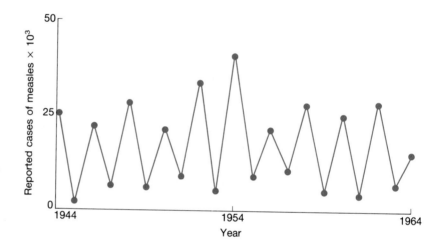

Figure 12.22. Reported cases of measles in New York City from 1944 to 1964 (after Yorke & London, 1973).

Many protozoan diseases of vertebrates are persistent within individual hosts (large L) because the immune response to them is short-lived, and they are able to infect even previously infected individuals. They can therefore survive endemically in small host populations. On the other hand, the immune responses to many viral and bacterial infections mean that they are transient in individual hosts (small L), and they often induce lasting immunity. Thus a disease like measles, for example, only occurs endemically in human populations larger than 500 000.

N in equation 12.1 is the number of *susceptibles* in the population. The immunity induced by bacterial and viral infections therefore reduces N and reduces R_p for the disease itself. This will often lead to a decline in the incidence of the disease, which will only increase again following the influx of new susceptibles into the population (as a result of new births or perhaps immigration). A sequence from 'high incidence' to 'few susceptibles' to 'low incidence' to 'many susceptibles' to 'high incidence', etc., underlies the cyclic behaviour of many diseases like measles (Figure 12.22). By contrast, the prevalence of protozoan infections is characteristically more stable through time.

12.6.2 Vector-transmitted microparasites

For parasites that are spread from one host to another by a vector, the life cycle characteristics of both host and vector enter into the calculation of R_p. The appropriate equation is correspondingly more complicated:

$$R_p = \beta^2 \, \frac{N_v}{N_h} \, f_v f_h L_v L_h \qquad (12.3)$$

Here, N_v and N_h represent the densities of vector and host respectively (mosquito and man, or aphid and sugar beet plant). f_v and f_h represent the fractions of infected vectors and hosts that survive to become infectious themselves and L_v and L_h represent the periods of time over which vectors and hosts remain infectious. β is the effective transmission rate: the rate of mosquito biting or aphid penetrating that effectively leads to infection. It appears as a squared term because the act of biting or phloem penetration transfers the infection *both to and from* the host.

A particularly important feature of the dynamics of vector-borne microparasites is that the transmission threshold ($R_p = 1$) is dependent on a ratio:

the vector-to-host ratio is crucially important with vector-transmitted microparasites

$$\frac{N_v}{N_h} = \frac{1}{\beta^2 f_v f_h L_v L_h} \qquad (12.4; \text{cf. equation } 12.2)$$

For a disease to establish itself and spread, the ratio of vectors to hosts must exceed a critical level—hence disease control measures are usually aimed directly at reducing the numbers of vectors, and are aimed only indirectly at the parasite. Many virus diseases of crops, and vector-transmitted diseases of man and his livestock, are controlled by insecticides rather than chemicals directed at the parasite; and the control of all such diseases is of course crucially dependent on a thorough understanding of the vector's ecology.

As a matter of parasite natural history, it is worth noting that the prevalence of infection of many plant and animal pathogens within their vector populations is characteristically low, even when the prevalence within the host population is high. For example, even in regions of endemic malaria, where more than 50% of the human population are infected, the prevalence within the mosquito vector is typically 1–2%. Prevalence in hosts is high because the survival rate of hosts is high and the pathogen tends to accumulate. This is particularly so with virus diseases of plants, where the plant usually remains infected and infectious for the rest of its life (L_h relatively large). On the other hand, the prevalence in vectors is low because their intrinsic survival rate tends to be low and because the incubation period of the pathogen within the vector is often long relative to the vector's life expectancy. For example, the incubation period of the malarial *Plasmodium* within a mosquito is 10–12 days, while the life expectancy of the mosquito is typically of the order of one week. As a consequence of this pattern, probably millions of uninfected vectors must be killed if any control programme is to be successful.

12.6.3 Directly transmitted macroparasites

Because the unit of study for macroparasites is the individual parasite, the basic reproductive rate is the number of offspring produced by an adult parasite through the course of its reproductive life that themselves survive to produce offspring. The transmission threshold is again defined by the condition $R_p = 1$. For a leaf-infecting parasite of a higher plant, the comparable term is the effective daily multiplication factor R_c: the number of progeny lesions produced per mother lesion per day.

For macroparasites of animals:

$$R_p = (\lambda L_a f_a) \times (\beta N L_i f_i). \tag{12.5}$$

The terms in the two pairs of brackets represent the reproductive contributions of the adult parasite and the infective stage respectively. λ is the rate of egg production per adult, β is the transmission rate, and N is the density of hosts. L_a is the expected life-span of the mature adult parasite within the host (dependent on both host and parasite mortality rates), and L_i is the expected life-span of the infective stage outside the host (dependent on both its mortality rate and its rate of contact with new hosts). The term f_a is the proportion of parasites in the host that attain sexual maturity, and f_i is the proportion of the transmission stage that become infective. For many of the parasitic intestinal worms, the transmission rates and the life expectancies of infective stages depend on the patterns of movement and behaviour of the host, since these affect the chances of an egg entering food or drinking-water or being transferred to the mouth by hand.

density-dependence within the host is crucially important with macroparasites

The effective reproductive rate of a macroparasite is typically affected by the density-dependent constraints on its survival, maturation and reproduction within the host, i.e. it is affected by the density-dependent effects on λ, L_a and f_a. These can arise because of competition between the parasites, or commonly because of the host's immune response. The intensity of the constraints varies with the distribution of the parasite population between its hosts (p. 433), and as we have seen, aggregation of the parasites is the most common condition. This means that a very large proportion of the parasites exist at high densities where the constraints are most intense, and this tightly controlled density-dependence undoubtedly goes a long way towards explaining the observed stability of many helminth infections (such as hookworms and roundworms) even in the face of perturbations induced by climatic change or man's intervention (Anderson, 1982).

Most directly transmitted helminths have an enormous reproductive capability (λ is very large). The female of the human hookworm *Necator*, for instance, produces roughly 15 000 eggs per worm per day, while *Ascaris* can produce in excess of 200 000 eggs per worm per day. The critical threshold densities for these parasites are therefore very low, and they occur and persist endemically in low-density human populations, such as hunter–gatherer communities.

plant macroparasites: the importance of a latent period

When modelling the dynamics of plant macroparasites like mildew on wheat, a further factor must be taken explicitly into account, namely the lag phase or latent period between the time when a lesion is initiated and the time when it becomes spore-forming and infective itself. The rate of increase in the proportion of a plant population affected by lesions (Zadoks & Schein, 1979) is given by:

$$\frac{dx_T}{dt} = R_c \cdot x_{T-P} \cdot (1-x_T) \tag{12.6}$$

Here, x_T is the proportion affected by lesions at time T, and P is the length of the latent period. The rate of increase therefore depends on the multiplication rate (R_c), the proportion of the population affected by *infective* lesions (x_{T-P}), and the proportion of the population susceptible to infection ($1-x_T$).

three phases in the development of a plant macroparasite epidemic

In the development of an epidemic, plant pathologists recognize three phases, as follows.

(a) The exponential phase, when the most rapid multiplication of the parasite occurs. The disease is rarely detectable in this phase: 'epidemiologically significant multiplication escapes observation'. This is also the phase in which

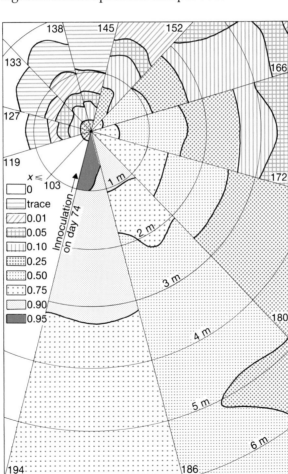

Figure 12.23. Diagram showing the development of an epidemic of the yellow stripe rust of wheat (*Puccinia striiformis*) from artificially established foci of infection. The circles show the distance of infection from the original focus and the shading indicates the intensity of infection. The progress of the epidemic can be followed by following the 'spider's web' clockwise. The figures around the margin indicate time in days from 1st January. (After F.H. Rijsdijk & S. Hoekstra, in Zadoks & Schein, 1979.)

chemical control would be most effective, but in practice it is usually applied in the next phase. The exponential phase is usually considered arbitrarily to end at $x = 0.05$, and this is about the level of infection at which a non-specialist might detect that an epidemic was developing (the perception threshold).

(b) The second phase, which extends to $x = 0.5$. (This is sometimes confusingly called the 'logistic' phase, although the logistic curve describes all of (a), (b) and (c).)

(c) The terminal phase, which continues until x approaches 1.0. In this phase chemical treatment is virtually useless—yet it is at this stage that the greatest damage is done to the yield of a crop.

The population dynamics of epidemic foliage parasites tends to be dominated by three factors:

(i) the initial dispersal of parasites to the host population (epidemics may be initiated by a single spore landing within a stand of a crop (Plate III, p. 84), or a spore cloud being deposited from a heavily infected area elsewhere (Figure 12.23));

(ii) the speed with which infectivity develops on infected hosts (i.e. the value of P); and

(iii) the proximity of susceptible hosts to each other in relation to the direction of air movement.

12.6.4 Macroparasites with indirect transmission

The basic reproductive rate for macroparasites with indirect transmission depends on a large number of parameters, but the formula is readily comparable with those derived for other types of parasite:

$$R_p = (\lambda_1 L_{a1} f_{a1})(\beta_1 N_1 L_{i1} f_{i1})(\lambda_2 L_{a2} f_{a2})(\beta_2 N_2 L_{i2} f_{i2}). \tag{12.7}$$

In terms of the schistosome life cycle, for example, alternating between man and snail (p. 424 and Figure 12.8), λ_1 and λ_2 are the rate of egg production per adult female worm and the rate of cercarial production per infected snail, N_1 and N_2 represent human and snail densities, and β_1 and β_2 denote the transmission rates from cercariae to man and from miracidia to snails. L_{a1}, f_{a1} and so on are the expected life-spans and proportions surviving to infectivity of the adult parasite, the miracidium, the infected snail and the cercaria.

theory throws light on the epidemiology of schistosomiasis

Once again, the formula for R_p is instructive in understanding the epidemiology of the disease. For example, the spread of schistosomiasis is limited because f_{a2} is small (i.e. the proportion of infected snails which survive to release cercariae). This is the case because there is a developmental delay of roughly 28–30 days before cercariae are released, but the snails themselves have a short life-span (roughly 14–54 days).

As with directly transmitted macroparasites, density-dependence within hosts plays a crucial role in the epidemiology of schistosomes (and other indirectly transmitted species). In this case, however, the regulatory constraints can occur in either or both of the hosts: adult worm survival and egg production are influenced in a density-dependent manner in the human host, but production of cercariae is virtually independent of the number of miracidia that penetrate the snail. Thus, levels of schistosome prevalence tend to be stable and resistant to the perturbations of outside influences.

The threshold density for the spread of infection depends directly on the abundance of both humans and snails (i.e. a product, N_1N_2, as opposed to the ratio which was appropriate for vector-transmitted microparasites). This is because transmission in both directions is by means of free-living infective stages. Thus, since it is inappropriate to reduce human densities, shistosomiasis is often controlled by reducing snail densities with molluscicides in an attempt to depress R_p below unity (the transmission threshold). The difficulty with this approach, however, is that the snails have an enormous reproductive capacity, enabling them to recolonize aquatic habitats once molluscicide treatment ceases. The limitations imposed by low snail densities, moreover, are offset to an important extent by the long life-span of the parasite in man (L_{a1} large): the disease can remain endemic despite wide fluctuations in snail density. Finally, the sexual reproduction of the schistosome in its human host depends on colonization by both male and female worms, reducing the frequency of successful infections in the same way that the colonization of islands is more risky for dioecious plants (those with separate sexes).

the complex determination of a basic reproductive rate for black stem rust

The determinants of the basic reproductive rate for a plant macroparasite like black stem rust (p. 428) are even more complex. Forgetting for the moment the existence of the sexual stage on the barberry, the parasite has a basic reproductive rate during the growing season of the cereal. Lesions give rise to daughter lesions, and there is a local multiplication of lesions within a crop and among local crops. But spore clouds of the parasite are carried for long distances, and epidemics often develop far more quickly than would be possible from purely local multiplication. A proper calculation of R_p for black stem rust would therefore need to include the number of daughter lesions produced per lesion along such a path of disease migration. R_p calculated in this way would be the basic reproductive rate of clones, spreading and multiplying without sexual recombination. But there is a quite different R_p that could be calculated—the basic *sexual* reproductive rate that takes into account the fact that new clones (new recombinants) arise only on the barberry. For this we would need to measure the number of daughter lesions produced per mother lesion on the barberry.

The control of a disease with such a complex life-history and mass long-distance dispersal poses interesting problems. If the barberry was the crop that required protection, the eradication of the cereal would eliminate the disease on a time-scale determined solely by the length of life of the barberry (spores released from the barberry cannot infect barberry). Similarly, the elimination of the barberry on a local scale would prevent the local perpetuation of infections on the cereal where it does not overwinter; but epidemics could still arise from the mass influx of spores in the annual 'migration' of spore clouds. In milder climates, the asexual cycle continues on grasses and cereals with overwintering foliage. Some control may then be achieved by preventive spraying with fungicides, but this is expensive. Thus the major means of control of such a disease, whether or not there is an alternate host in the life cycle, is by the use of genetic resistance bred into the crop. An alternate host on which sexual stages occur then has a quite different

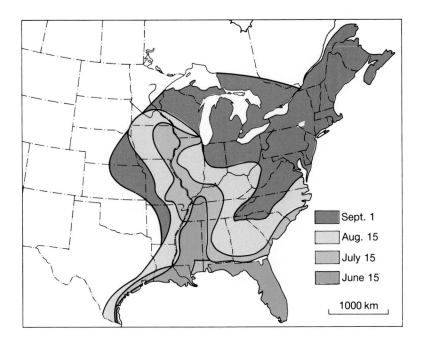

Figure 12.24. The development of an epidemic of southern corn leaf blight (*Helminthosporium maydis*) in the United States in 1970 (After Zadoks & Schein, 1979).

significance in epidemiology. It is the place where new 'races' of the parasite may arise through recombination, breaking through the resistance barriers developed by the plant breeder.

12.6.5 Parasites and the population dynamics of hosts

In the previous parts of this section we have looked at the dynamics of disease itself; but we now turn to the effects of parasites on the population dynamics of their hosts. In principle, the sorts of conclusions that were drawn in Chapter 9 regarding predator–prey and herbivore–plant interactions can be extended here to parasites and hosts. Parasites harm individual hosts with an intensity that varies with the densities of both. In addition, infected and uninfected hosts exhibit compensatory reactions which may greatly reduce the effects on the host population as a whole. Theoretically a range of outcomes can be predicted: varying degrees of reduction in host population density, varying degrees of persistence and amplitude in the fluctuations in abundance, and even multiple equilibria (Anderson, 1979, 1981).

Epidemiology is now one of the most sophisticated parts of ecological science. Medical records provide some of the most extensive data sets, and there are comparable long-term records of epidemics in diseases of crop plants. There is probably no other area of ecological study in which theory, long-term records and the results of experimental study are so tightly interwoven. There are, however, plenty of problems.

applying theory to practice

One difficulty is that parasites often cause a reduction in the 'health' of their host rather than its immediate death, and it is therefore usually difficult to disentangle the effects of the parasite from those of some other factor or factors with which they interact (section 12.5.5). Another problem is that even when parasites cause a death, this may not be obvious without a detailed post-mortem examination (especially in the case of microparasites). In addition, the biologists describing themselves as parasitologists have in the past tended to study the biology of their chosen parasite without much consideration of the effects on whole host populations (and ecologists have tended to ignore parasites). On the other hand, plant pathologists and medical and veterinary parasitologists have, for obvious reasons, chosen to study parasites with known severe effects, living typically in dense and aggregated populations of hosts. They too have therefore paid little attention to the more typical effects of parasites in populations of hosts less strongly influenced by man. Perhaps the most certain thing that can be said about the role of parasites in host population dynamics is that the elucidation of this role is one of the major challenges facing ecology.

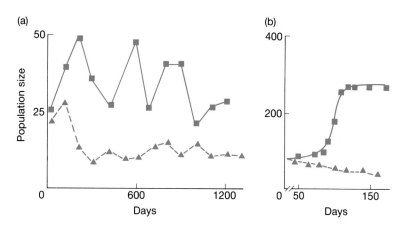

Figure 12.25. The depression of the population growth of insects by protozoan parasites in laboratory culture. (a) The flour beetle *Tribolium castaneum* infected with *Adelina triboli* (after Park, 1948). (b) *Laemophloeus minutus* infected with *Mattesia dispora* (after data in Finlayson, 1949). The solid lines represent uninfected populations and the dashed lines infected populations.

parasites certainly can reduce host densities

There are certainly cases where a parasite can be seen to reduce the population density of its host. The widespread and intensive use of sprays, injections and medicines in agricultural and veterinary practice all bear witness to the disease-induced loss of yield that would result in their absence. (In human medicine, there is often more emphasis on saving the individual rather than increasing the size or yield of the population as a whole.) Reductions of host density by parasites can also be seen in controlled laboratory environments (Figure 12.25). However, such direct comparative evidence is extraordinarily difficult to obtain from natural populations. Even when a parasite is present in one population but absent in another, the parasite-free population is certain to live in an environment that is different from that of the infected population; and it is likely also to be infected with some other parasite that is absent from or of low prevalence in the first population.

the level of host-density reduction in different circumstances can be predicted

The degree of reduction in host density to be *expected* in various circumstances can be seen in Figure 12.26 (Anderson, 1979), based on models similar to those described in the previous parts of this section. For transient infections, where infected hosts do not recover from infection (e.g. many microparasites of plants and invertebrates), the greatest depression in host density is at low to moderate levels of pathogenicity (Figure 12.26a and b). High pathogenicity leads to many hosts dying before they transmit the disease: the parasite becomes extinct and the host attains the carrying capacity of its environment. At the other extreme, host density drops rapidly as pathogenicity rises from zero to even low levels. In addition, host density is

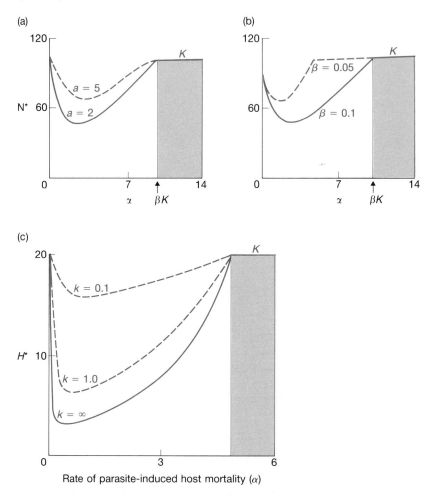

Figure 12.26. The effects of the pathogenicity of a parasite on the degree of depression produced in the host population. (a) Transient-infection models in which the infected hosts do not recover from infection. In the shaded region the parasite becomes extinct and the host population attains its full carrying capacity K. a = the rate of host reproduction, β = the rate of infection and α = the rate of parasite-induced host mortality. In this figure two different rates of host reproduction are shown $a = 2$ and $a = 5$ and other values are held constant ($b = 1$, $\beta = 0.1$ and $K = 100$). (b) As (a) but two different rates of host infection are shown with $\beta = 0.1$, and $\beta = 0.05$, and other values are held constant ($b = 1$, $a = 2$ and $K = 100$). (c) A persistent-infection model showing the relationship between the population equilibrium of the host (H^*) and the parameter α (the severity of the parasite's influence on host survival). The effects of the parasite on the host population are greatly influenced by the degree of aggregation of the parasite and the three curves show the relationship for random distributions ($k = \infty$), and two cases of overdispersion (negative binomial distributions where $k = 0.1$ and 1.0). (After Anderson, 1979.)

depressed more when the host's reproductive capacity is low (Figure 12.26a), than when the parasite's rate of infection is high (Figure 12.26b). A similar picture (Figure 12.26c) emerges for persistent infections (e.g. most macroparasites). It is also apparent that the effect on host density declines as the distribution of the parasites becomes more aggregated. Remember, though, that the reduced densities are more stable when the parasites are aggregated as a result of the enhanced density-dependent control.

Overall, we can see that parasites should be expected to reduce the densities of their host, and that this reduction will be greatest for parasites with low to moderate pathogenicity. This is a lesson worth learning when parasites are used, as they are increasingly, in attempts to control populations of pests.

disease can provide an explanation for complex patterns of host population dynamics

Finally, there is growing evidence that disease can explain (or help explain) more complicated patterns of host population dynamics, especially cycles. Figure 12.27, for example, shows the observed fluctuations in the abundance of the larch bud moth in the European Alps, together with the observed fluctuations in the prevalence of its infection with a granulosis virus. The figure also shows the encouragingly similar fluctuations predicted by a model of the interaction (Anderson & May, 1980). Similar encouragement may be gained, for instance, from models of the fox–rabies interaction, where the foxes exhibit a three- to five-year cycle of density even in areas lacking a more widespread wildlife cycle (Anderson et al., 1981), and also

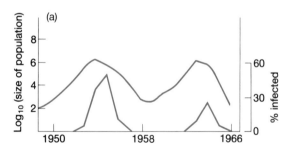

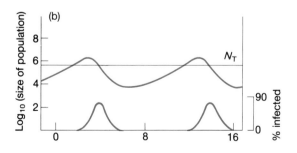

Figure 12.27. (a) Observed changes in the abundance of the larch bud moth (*Zeiraphera diniana*) in the European Alps (plotted logarithmically) and the prevalence of a granulosis virus (plotted linearly as a percentage). (b) Results from a theoretical model. (After Anderson & May, 1980.)

from models of red grouse in northern England infected with the nematode *Trichostrongylus tenuis* (Potts *et al.*, 1984; Hudson, 1986; see p. 567).

The possible importance of disease in population dynamics is increasingly being confirmed.

12.7 Polymorphism and genetic change in parasites and their hosts

It is almost always wrong to describe 'a species' as having 'an ecology'. Species are composed of populations that show genetic variation both within populations and between them. This is nowhere more true than among species of parasites and their hosts. We have seen that in agricultural practice, 'boom and bust' cycles occur as new resistant races of a crop are attacked by new virulent races of pathogen. The presence of disease in a population reduces the fitness of infected individuals. Hosts that are resistant are therefore at an advantage, and resistance genes spread in the population. But the presence of resistant hosts reduces the fitness of disease organisms, and so pathogens with genes that confer virulence are at an advantage and 'virulent' genes spread. In most studies of host resistance and pathogen virulence, a precise, one-for-one, *gene-for-gene* correspondence between resistance in the host and virulence in the pathogen has been shown. This was demonstrated first for rust disease of flax (*Melampsora lini*) by Flor (1960, 1971) and is now known for very many other diseases. There are other sorts of resistance (such as those which distinguish hosts from non-hosts) which are usually controlled by a multiplicity of genes. But it is with gene-for-gene resistance that polymorphism and rapid evolutionary change are most obvious.

When populations of host and parasite evolve together without the intervention of man as a plant breeder, the multiplication of a particular race of pathogen is followed by an increase in the proportion of resistant hosts.

the gene-for-gene relationship between virulence and resistance

pathogens and dynamic polymorphisms in natural populations

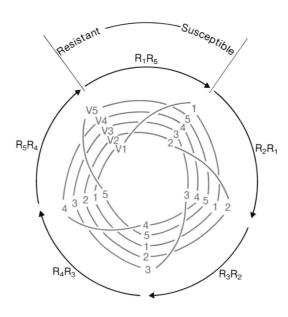

Figure 12.28. Diagram illustrating how five resistance genes (*R*) in the host may interact with five related virulence genes in the parasite in a system of cyclical balanced polymorphism. (After Person, 1966.)

This increase in the representation of resistance genes leads to a decline in the representation of the appropriate virulence gene in the pathogen. A further race of the pathogen then gains ascendancy, and this in turn leads to an increase in the representation of genes for resistance to that race of pathogens. Race after race of the pathogen becomes ascendant and declines, and the result is a maintained dynamic polymorphism in the genetic structure of both pathogen and host (Figure 12.28). In fact, the polymorphisms that develop naturally between interacting and evolving hosts and parasites produce effectively the same result as the deliberate mixing of strains of a crop in, for example, the multilines that are being developed by agriculturists for control of disease.

Genetic polymorphisms are actually only one of the evolved consequences of the reciprocating selection pressures exerted between parasites and their hosts. A cyclical phenotypic response that does not involve genetic change occurs during recurrent bouts of sleeping sickness in an infected human host. The trypanosome parasite has a changing repertoire of antigens that are activated in turn by subtle molecular mechanisms as the host mounts successive immune defences (Figure 12.29). The result is violent fluctuations in the trypanosome population *within* a host.

Finally, in the evolution of myxoma virus and its rabbit host, violent oscillations in the density of infected rabbit populations have been followed by an apparently stable situation in which the virus has settled at an intermediate level of virulence (Figure 12.30; Levin & Pimentel, 1981; May & Anderson, 1983). This process, involving decreasing virulence of the parasite and increasing tolerance by the host, may parallel the long-term evolutionary processes that have led in nature to the closely integrated life-styles of biotrophic pathogens and their hosts (e.g. the rust fungi).

A very large body of recent literature on the ecology of parasites focuses

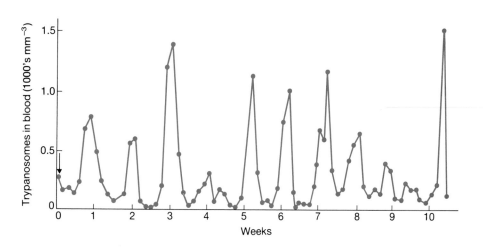

Figure 12.29. Fluctuations in the populations of trypanosomes in a patient with trypanosomiasis (sleeping sickness). The trypanosome population at each peak is antigenically different from those at preceding or succeeding peaks. (After Ross & Thomson, 1910.)

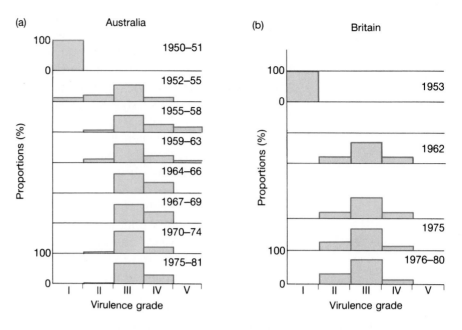

Figure 12.30. (a) The proportions in which various grades of myxoma virus have been found in wild populations of rabbits in Australia at different times from 1951 to 1981. (b) Similar data for wild populations of rabbits in Britain from 1953 to 1980. Grade I is the most and Grade V the least virulent. (After May & Anderson, 1983, from data collected by Fenner, 1983.)

on the importance of genetic variation in both hosts and parasites. Nowhere else in ecology have we become so sensitive to the forces of natural selection that are driven by the interactions between organisms. Sexual recombination and assortment of genetic material are of the greatest importance in the evolving interactions of host and parasites. Indeed, the unending battle between the two may be the major force favouring the evolution and maintenance of sex itself. In the chapter on mutualisms (Chapter 13) we will see that here, in contrast, sexual processes are often absent or rare; host and 'recipient' specificities are less well developed; and epidemics seem not to occur. The evolutionary consequences of battle and of partnership are quite different.

13 Mutualism

13.1 Introduction

Mutualism is the name given to associations between pairs of species that bring mutual benefit $(+, +)$; the individuals in a population of each mutualist species grow and/or survive and/or reproduce at a higher rate when in the presence of individuals of the other. Each mutualist gains one of a variety of kinds of advantage. Most often this involves food resources for at least one of the parties and frequently, for the other, protection from enemies or provision of a favourable environment in which to grow and reproduce. In other cases, the species that gets the food provides a 'service' by ridding its partner of parasites (e.g. cleaner fish) or by bringing about pollination or seed dispersal. Despite the advantages that each partner gains it is important to avoid thinking in terms of a 'cosy' relationship between the mutualists. Each is acting in an essentially 'selfish' manner; mutualistic relationships evolve simply because the benefits to each partner outweigh any costs that might be involved. A detailed review is given by Boucher *et al.* (1984).

the world's biomass is largely composed of mutualists

Previous ecology texts have generally underemphasized or ignored mutualism. Yet it is an extremely widespread phenomenon. A very significant proportion of the world's biomass depends on it—for example, the dominant organisms of all grasslands and forests (plant roots intimately associated with fungi in mycorrhizae), and corals (animal polyps containing unicellular algae) are mutualists. Most rooting plants have mutualistic mycorrhizae; many flowering plants depend on insect pollinators; and a very great number of animals possess guts which contain a mutualistic community of microorganisms.

Mutualism has been subject to very little of the mathematical modelling that has characterized the study of competitive, predator–prey and parasite–host interactions (May, 1982), although there are some important developments in this area (Vandermeer & Boucher, 1978; Vandermeer, 1980). Instead, questions related to the general biology of mutualisms (physiology, morphology and behaviour), rather than to their population dynamics, have dominated the study of their ecology. The structure of this chapter reflects such a bias.

mutualists may be facultative or obligate

First, consideration is given to examples of mutualistic links between individuals of different species that range from those which are *facultative* (where each partner (symbiont) gains a benefit from but is not dependent on the other), through links that are *obligate for one partner* but facultative for the other, finally to links that are *obligate for both partners*. Later we consider

461 MUTUALISM

mutualisms in which one species inhabits the surface or body cavity of the other and also situations in which one partner (now called an endobiont) lives within the cells of another and where the two components begin to show an integrated physiology (Figure 13.1). Towards the end of the chapter we will discuss the view that it has been the evolution of mutualistic relations

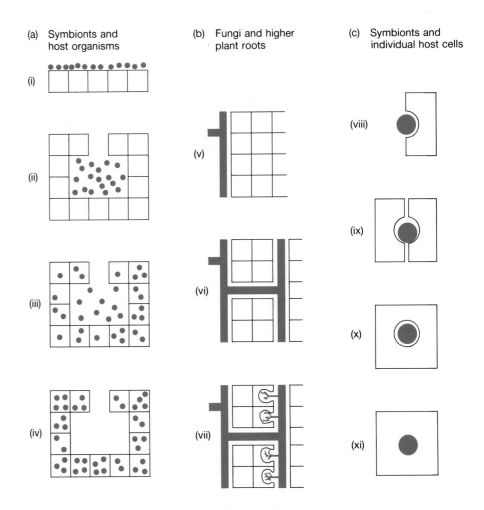

Figure 13.1. Types of morphological integration between symbionts and hosts. (a) Symbionts (coloured circles) and host organisms (squares): (i) symbionts entirely on surface (e.g. epiphytic or epizootic microbes); (ii) symbionts within organism (e.g. digestive-tract microorganisms such as rumen microbes); (iii) symbionts within organism, partly extracellular, partly intracellular, as occurs with blue-green algae in tropical marine sponges; (iv) symbionts entirely intracellular. (b) Fungal hyphae and higher plant roots: (v) rhizosphere fungi which do not penetrate roots; (vi) ectotrophic mycorrhizal fungi of forest trees, which penetrate between but not into root cells; (vii) fungi which penetrate the root and into cells by haustoria as in vesicular/arbuscular mycorrhizas; the haustoria are enclosed by a host membrane. (c) Symbionts and individual host cells: (viii) symbionts in close physical contact with surface of host cells, but also in contact with external environment (e.g. lichens, *Prochloron* on ascidians); (ix) symbionts enclosed by host cells, but not within them; (x) symbiont within host cells, but enclosed in a host vacuole (most intracellular symbionts); (xi) symbionts lying free in cytoplasm (e.g. endonuclear bacteria). (From Smith, 1979.)

between organisms that has been responsible for the major steps in the evolution of higher animals and plants. Some tentative attempts will also be made to compare the ecologies of parasites and mutualists.

13.2 Mutualisms that involve reciprocal links in behaviour

It is not easy to categorize the benefits gained by participants in the various mutualisms that involve intricate behavioural links. Their diversity is illustrated in the following examples.

13.2.1 The honey guide and the honey badger

An African bird, the honey guide, has formed a remarkable relationship with the ratel or honey badger. A honey guide that has located a bees nest leads the honey badger to it. The mammal tears open the nest and feeds on honey and bee larvae, and later the honey guide gets a meal of beeswax and larvae. The honey guide can locate a bees nest but not break it open, while the honey badger is in just the opposite situation. The reciprocal link in their behaviours brings mutual benefit.

13.2.2 Shrimps and gobiid fish

Shrimps of the genus *Alphaeus* dig burrows, and goby fish (*Cryptocentrus*) use these as safe sites in an environment that otherwise provides little or no shelter. The shrimp is almost completely blind and, when it leaves the

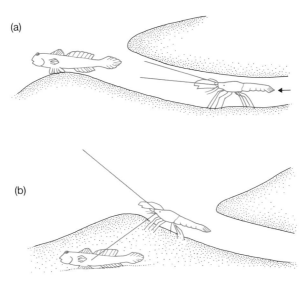

Figure 13.2. The sequence of behaviour during emergence from the burrow in the mutualism of the shrimp *Alpheus djiboutensis* and the fish *Cryptocentrus cryptocentrus*. (a) The shrimp moves head-first toward the entrance of the burrow; antennae are somewhat raised at an oblique forward angle with the tips 2–3 cm apart. (b) Outside the burrow the shrimp keeps one antenna in contact with the fish. (After Karplus *et al.*, 1972.)

burrows, keeps its antenna in contact with the fish, thereby getting warning of any disturbance (Figure 13.2). The goby gains a place to live in an environment of sediment containing abundant food. The shrimp, for its part, benefits by acquiring an optical warning system, allowing it to leave its burrow safely for short periods to feed on sediment outside (Fricke, 1975).

13.2.3 Clown fish and anemones

A variety of mutualistic patterns of behaviour is found among the inhabitants of tropical coral reefs (where the corals themselves are mutualists—see later). The clownfish (*Amphiprion*) lives close to a sea anemone (e.g. *Physobrachia, Radianthus*) and retreats among the anemone's tentacles whenever danger threatens. While within the anemone, the fish gains from it a covering of mucus that protects it from the anemone's stinging nematocysts (the anemone slime is a substance whose normal function is to prevent discharge of nematocysts when neighbouring tentacles touch). The fish certainly derives protection from this relationship, but the anemone also benefits because clownfish attack other fish which come near, including species which normally eat sea anemones (Fricke, 1975).

13.2.4 Cleaner fish and customers

Another apparently mutualistic association exists between 'cleaner' and 'customer' fish, where the former feeds on ectoparasites, bacteria and necrotic tissue from the body surface of the latter. Forty-five species of cleaner fish have been recognized and these often hold territories while their customers migrate to the 'cleaning stations'; indeed, customers visit cleaning stations more often when they carry many parasites. Both the cleaner and the customer have patterns of behaviour in which they display to and recognize each other, and the relationship appears to have become tightly coevolved. The cleaners gain a food source and the customers remain clean. In an experiment in the Bahamas, Limbaugh (1961) removed all the cleaner fish from patches of reef. The 'customers' developed skin diseases and their populations declined within two weeks. A similar experiment made in Hawaii gave negative results but it seems likely that the 'cleaner–customer' relationship is often truly mutual. Six species of shrimp have also been shown to act as cleaners of fish in similar relationships.

13.2.5 Ants and acacia

Another relationship which might be described as 'cleaner–customer' in nature has been reported by Janzen (1967). The bull's horn acacia (*Acacia cornigera*) has a tight mutualistic relationship with the ant *Pseudomyrmex ferruginea* in Central America. The plant bears hollow thorns which are used by the ants as nesting sites; it has finely pinnate leaves bearing protein-rich 'Beltian bodies' at their tips (Figure 13.3), which the ants collect and use for food; and it bears sugar-secreting nectaries on its vegetative parts which attract the ants. The ants, for their part, protect these small trees from

(a)

(b)

Figure 13.3. Structures of the bull's horn acacia (*Acacia cornigera*) that attract its ant mutualist. (a) Protein-rich Beltian bodies at the tips of the leaves. (b) Hollow thorns used by the ants as nesting sites. (Photographs by L.E. Gilbert.)

competitors by actively snipping off shoots that enter the acacia canopy, and also protect the plant from herbivores.

Janzen's demonstration of this ant–plant mutualism lays open the way for new interpretations of other plant structures. Nectaries are present on the vegetative parts of plants of at least 39 families in many ecosystems throughout the world. While nectaries on or in flowers are easily interpreted as attractants for pollinators, the role of extra-floral nectaries on vegetative parts is less obvious. They clearly attract ants, sometimes in vast numbers, but there has been a controversy between the 'protectionists' (who have supported the idea that the ants visiting the nectaries protect the plants from herbivorous animals), and the 'exploitationists' who have felt that the plants have no more use of ants than dogs have of fleas—a sceptical view that is proper in all scientific enquiry. Bentley (1977) provides a review of this controversy. Such disagreement needs to be resolved by carefully designed and controlled experiments such as the one carried out by Inouye and Taylor (1979). They excluded ants from plants of the aspen sunflower (*Helianthella quinquenervis*) which bears extra-floral nectaries. This plant can suffer heavy attacks from a terphritid fly whose larvae may destroy more than 85% of the developing seeds. The plants carrying ants escaped much of this predation.

13.3 Mutualisms that involve the culture of crops or livestock

13.3.1 *Homo sapiens* is a mutualist in relation to crops and livestock

The most dramatic ecological mutualisms are those of human agriculture. The numbers of individual plants of wheat, barley, oats, corn and rice, and the areas these crops occupy, vastly exceed those that would have been present if they had not been brought into cultivation. The increase in the human population since the time of hunter–gatherers is some measure of the reciprocal advantage to *Homo sapiens*. We may not do the experiment, but we can easily imagine the consequence that the extinction of man would cause to the world population of rice plants or the effect that the extinction of rice plants would have on the population of man. It is worth noting that man has also chosen to cultivate plants such as coffee, tobacco and opium poppies, which contain powerful chemical defences (caffeine, nicotine, etc.). An obvious advantage is gained by the plant species, but whether man benefits depends on how the drugs are used. The domestication of cattle, sheep and other mammals also involves a mutualism—the numbers of such animals would fall sharply in the absence of man and the diet of part of mankind would change dramatically in the absence of livestock.

Similar mutualisms have developed in termite and ant societies.

13.3.2 The 'farming' of caterpillars by ants

A classic example of livestock mutualism is that between the large blue butterfly (*Lycaena avion*) and ants. The butterfly lays its eggs on buds of wild thyme and the larvae closely resemble the flowers on which they feed. After about 20 days (the third moult) the larvae leave the plants and wander—they never eat plant food again. The larva bears a honey gland and when an ant finds a larva it strokes it and droplets of secretion are released which the ant drinks. After a complicated mutual signalling sequence, the ant carries the larva into its underground colony where it remains for about 11 months. For part of this time it hibernates or is present as a pupa, but while active it feeds upon young ant larvae. The butterfly emerges in June, completing the life cycle.

The life-history of the large blue butterfly appears so bizarre that it might seem incredible if it were not part of a complex series of ant–butterfly interactions found amongst many members of the family Lycaenidae (to which the large blue belongs) (see, for example, Ford, 1945). Other species of this family are also farmed by ants (though not all are carried into the colony), and the larvae of most species in the family bear active honey glands that are milked by ants. Ants 'farm' and 'milk' many species of Homoptera in much the same way, in return for sugary secretions. The ants defend these insects against predators and parasitoids in the same way that ants attracted by extra-floral nectaries may defend against herbivores the plant that produces them (Strong *et al.*, 1984.)

The large blue butterfly is a rare and endangered species in Europe. It recently became extinct in Britain. Its life cycle traps it in a pattern of behaviour through which it is wholly dependent for food on the presence of flowering thyme and subsequently on ant larvae. Only two species of ant appear to be acceptable as its hosts (*Myrmica scabrinoides* and *M. laevonoides*), and the calcareous grasslands that serve as habitat for both the thyme and the ants are rapidly disappearing. This life cycle is thus a classic example of a species caught in 'the ever narrowing rut of specialization', with its particular risk of extinction.

13.3.3 The farming of fungi by beetles

Much of plant tissue, including wood, is unavailable as a direct source of food to animals because they lack the enzymes capable of digesting cellulose and

(a)

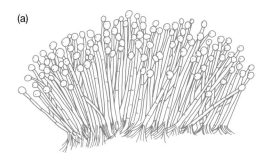

(b)

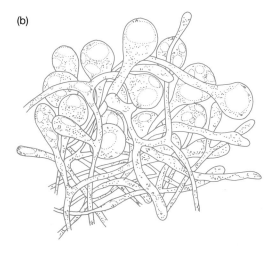

Figure 13.4. (a) The ambrosia fungus which is cultivated in the burrows of the beetle *Xyleborus xylographus* (redrawn from Hubbard). (b) A fraction of the mycelium in the fungus garden of the South American ant *Moellerius*, showing the swollen 'kohl-rabi' bodies that are used by ants as food (redrawn from Bruch).

lignins (see Chapters 3 and 11). However, many fungi possess the appropriate enzymes and an animal that eats fungi gains access to an energy-rich food resource indirectly. In a number of cases very specialized mutualisms have developed between an animal and a fungal decomposer. Beetles in the group *Scolytidae* tunnel deeply in the wood of dead and dying trees, and fungi that are specific for particular species of beetle grow in these burrows and are continually grazed by the beetle larvae. These 'ambrosia' beetles may carry inocula of the fungus in their digestive tract but in some species there are specialized brushes of hairs on the head of the beetle which serve to transport the spores. The fungi serve as food for the beetle and depend on it for dispersal to new tunnels—the relationship is wholly mutualistic (Figure 13.4).

13.3.4 The farming of fungi by ants

In some species of ants (and termites) the subtlety of the fungus-cropping system is developed further. Members of the genus *Atta* are the most remarkable of the fungus-farming ants. They excavate cavities in the soil which are two or three litres in volume. The fungus is cultured in these cavities on leaves that are cut from neighbouring vegetation, and the whole ant colony may depend absolutely on the fungus for its food supply. Again, the fungus gains from the association: it is both fed and dispersed. The reproductive female fills a pouch in her throat with inoculum when she leaves to found a new colony.

13.4 Pollination mutualisms

Most animal-pollinated flowers offer nectar or pollen or both as a reward to their visitors. Nectar seems to have no value to the plant other than as an attractant to animals and it represents a cost, mainly in carbohydrates that might presumably have been used by the plant in the growth of a tissue or in some other activity. The benefit to the plant, however, is that pollination is effected. (A similarly costly investment is made by many plant species in fruit structures that attract and provide food for animals which help to disperse the seeds contained within.)

Pollination involves the transfer of pollen from an anther of one plant to a stigma of the same or another plant. A great many flowering plants, particularly annuals, are normally self-pollinated and thus self-fertilized, without the intervention of a pollen vector. But others are intimately associated with animal pollen vectors so that cross-pollination and cross-fertilization generally result. Presumably, the evolution of specialized flowers and the attraction of animals for pollination has occurred because outbreeding *within* some species, and/or the avoidance of outbreeding *between* species has been strongly favoured. This is not the place to discuss the possible benefits to some plants which have led to the evolution of outbreeding systems (which appear, at the same time, to bring both genetic and material costs). As with the evolution of sex itself, the forces responsible for the evolution of outbreeding are not yet understood (see, for example, Williams, 1975; Maynard Smith, 1978). However, the forces have been so strong that they have led to much of the diversification within the plant kingdom and involve some of the most subtle mutualisms.

A range of animals have entered into pollination liaisons with flowering plants, including humming-birds, bats and even small rodents and marsupials (Figure 13.5). However, the pollinators *par excellence* are, without doubt, the insects (Grant, 1963).

Figure 13.5. Pollinators. Left: Honeybee (*Apis mellifera*) on raspberry flower. Right: Cape sugarbird (*Promerops cafer*) feeding on *Protea eximia* (Photographs: Heather Angel.)

insect-pollinated flowers may be generalists or specialists

In the most simple examples of insect pollination systems, nectar (or pollen) is offered in abundance and freely exposed to all and sundry visitors. This is the case in the bramble (*Rubus fruticosus*), where nectar may be produced in such abundance that a local glut is provided for a wide range of nectar seekers. In other plants, variously specialized flower structures protect the nectar from all but a few insect species that have appropriately specialized mouthparts. Within the family Ranunculaceae a sequence of specialization in the form of the nectaries ranges from the open, unprotected type of *Ranunculus ficaria*, through simple flaps covering the nectar secreting region of the petals, to the complex nectar-filled spurs of *Aquilegia* (Plate IV). Floral specialization restricts the variety of species that may harvest the nectar and thus serve as pollinators. Such restriction has three main consequences: (i) the tendency of the associated insect to move from individual to individual of the same plant species (perhaps because it has formed a specific 'search image') enhances the opportunities for outbreeding within the plant species; (ii) less pollen is wasted on the stigmas of unrelated plant species; and (iii) the insect may develop, by learning or by evolution, specialized handling abilities that make it a more efficient harvester as a floral specialist than it would be as a generalist 'jack of all trades'.

the benefits of specialization

It is in the pollination processes of the orchids that the closest, apparently coevolved, mutualisms are found (Figure 13.6); these have led to obligate mutualism (dependence) for at least one of the partners. Other finely tuned mutualisms may, in fact, not have involved coevolution (reciprocal selection of each mutualist on the other). In the genus *Asclepias,* for example, different plant species are pollinated by the same species of bumblebee but the structure of each flower is such that, in each case, a different part of the bee's body picks up the pollen mass and transfers it to the stigmatic chamber of another individual of the appropriate species. The mechanism prevents or hinders cross-pollination between *Asclepias* species, but all the specialization occurs in the species of *Asclepias* while the bee remains unspecialized.

Flowering is a seasonal event in most plants and this places strict limits on the degree to which a pollinator can become an obligate specialist. The life cycle of a pollinator can only become completely dependent on specific flowers as a source of food if it fits the flowering season of the plant involved. This is feasible for insects which use flowers only for a short period, for example as a maintenance diet for the adult stage in the case of butterflies and moths. Longer-lived pollinators such as bats and rodents, or bees with their long-lived colonies, are more likely to be generalists, turning to the flowers of different species through the seasons or to quite different foods when nectar is unavailable.

the pollination mutualism of the fig and the fig wasp

A remarkable and quite different type of mutualism exists between a plant, the fig (*Ficus*, family Moraceae), and a pollinator, the fig wasp, in which the interaction has become obligate for both (Wiebes, 1979; Janzen, 1979).

The flowers of the fig are borne within swollen receptacles (fleshy stem tissue) and male and female flowers are borne in separate regions within the plant. The female flowers are of two types: long-styled and short-styled. Female wasps enter the fig through an apical pore and lay their eggs in the

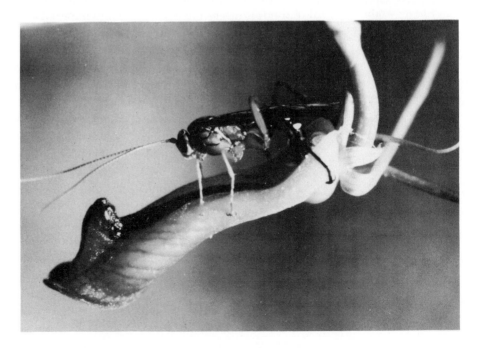

Figure 13.6. Male of the ichneumonid wasp *Lissapimpla excelsa* pseudocopulating with orchid *Cryptostylis subulata.* (Photograph: Mantis Wildlife Films/ OSF.)

Figure 13.7. Fig wasps clustered in and on an opened fig. Also shown are parasitoids of the fig wasps, distinguished by their long ovipositors. (Photograph by W.D. Hamilton.)

short-styled ovaries (their ovipositers cannot reach down the long-styled ovaries). The parasitized short-styled ovaries develop into galls in which the fig wasp larvae complete their development. Each species of fig has its own more-or-less specific species of wasp. Males bore their way out of the fig and the females follow them, but only after passing the male flowers and collecting pollen from these. The pollen is carried in specialized sacs, and when the females enter a new fig they have been observed shaking pollen out of these sacs and placing it on the stigmas of long-styled ovaries, which after pollination develop seeds (Figure 13.7).

Such a specialized sequence of mutualistic events must have involved reciprocating evolutionary pressures in which each stage in the evolution of the fig has evoked its counterpart in the behaviour pattern of the wasp (Wiebes, 1982). The wasp has become not only pollinator but also parasite. The evolution of this complex ecosystem has progressed even further. In most species of fig there are two different species of wasp involved, each specialized on particular flowers within the fruit (in addition, there are usually two species of parasitic wasp which specialize on the two pollinating species). A further wasp is often present which lays eggs in the female flowers of the fig but plays no part in the pollination process. It is not a mutualist but is in a real sense a cuckoo in the nest, reared by the fig but paying no price or fee for the privilege.

This pattern of interactions is repeated time and again in the various wild fig species, with different species of wasps as gall former, parasite and cuckoo. The situation represents one of the great classics of coevolution *and* of parallel evolution of mutualists *and* of parasitism. (Note that the cultivated fig is parthenogenetic, requiring no act of pollination for the production of a ripened fruit. It has therefore escaped from dependence on the wasp.)

13.5 Mutualisms involving gut inhabitants

Most of the mutualisms so far discussed have depended on the patterns of behaviour of the animals involved. In all cases they involve acts of search, and usually (but not always) they have food as a reward. Both partners in such mutualisms are highly evolved complex organisms, and each spends a significant part of its life on its own. In many other mutualisms one of the partners is a unicellular eukaryote or prokaryote and is integrated into the life of its multicellular partner as a more or less permanent part of itself. Animal guts and their living inhabitants provide important examples of this phenomenon.

In most animals the gut is a microcosm of microbial life. In many herbivores the gut microflora plays a major role in the digestion of cellulose and perhaps also in the synthesis of vitamins. The gut receives a more or less steady flow of substrates in the form of food eaten, chewed and partly homogenized by the animal.

the gut as a culture chamber

In the gut the pH of the substrate is regulated, its aeration is controlled and the mixture is stirred. In warm-blooded animals the temperature of the culture chamber is closely regulated. Waste substrate continually flows out of the culture chamber, which therefore does not grow stale. The system is closely analogous to the continuous fermentation vat used in the mass production of beer, where precisely the same set of environmental conditions is regulated by the brewer.

13.5.1 The ecosystem of the rumen

The stomach of ruminants (which include deer, cattle and antelope) is four-chambered and food that is swallowed passes first into the reticulum chamber. The first chewing reduces the size of food particles to volumes varying from 1 to 1000 μl and the particles may be as long as 10 cm. Only particles with a volume of about 5 μl or less can pass out of the reticulum into the next compartment, the omasum; the animal regurgitates and rechews the larger particles (the process of rumination). Dense populations of bacteria (10^{10}–10^{11} ml^{-1}) and of protozoa (10^5–10^6 ml^{-1}) are present in the rumen, the pH of which is regulated by the ruminant through secretion of 100–140 mM bicarbonate and 10–50 mM phosphate from the salivary glands. Thus it is the activities of the host that provide the continual supply of substrates *and* the controlled conditions for fermentation by the microbes, while the products of the microbial fermentation form the principle food of the host (Figure 13.8).

the flora and fauna of the rumen . . .

The bacterial communities of the rumen are composed almost wholly of obligate anaerobes—many of the species are killed instantly by exposure to oxygen. They require carbohydrate and most also need acetic, iso-butyric, isovaleric and 2-methylbutyric acids and ammonia, many of which are provided by other species of bacteria in the rumen. The rumen flora is by no

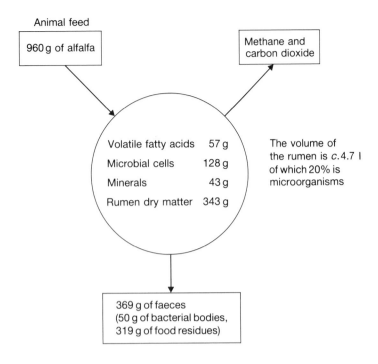

Figure 13.8. Microbial digestion within the rumen of the sheep. The data were obtained by continual feeding of alfalfa pellets. The empirical formula for materials used and produced daily by the rumen system is:

$$C_{20.03}H_{36.99}O_{17.40}N_{1.34} + 5.6\,H_2O \rightarrow C_{12}H_{24}O_{10.1} + 0.83\,CH_4 + 2.76\,CO_2 + 0.50\,NH_3 + C_{4.44}H_{8.88}O_{2.35}N_{0.78}$$

difference between + water → volatile + methane + carbon + ammonia + microbial cells
food and faeces   fatty acids   dioxide

(From data in Hungate *et al.*, 1971.)

means a mixture of microbial generalists. Many have metabolic specialisms that constrain them to life in the unique environment that is provided by the rumen.

Cellulose and other fibres are the main constituents of the ruminant's diet but the ruminant lacks the enzymes that can digest these. The cellulolytic activities of the rumen microflora are therefore of crucial importance. *Bacteroides succinogenes* sticks closely to its fibre substrate and digests cellulose; *Ruminococcus* digests only cellulose, cellobiose and xylose; while others, like *Clostridium locheadii,* can digest not only cellulose but also starch. The rumen flora contains other resource specialists such as *Methanobacterium ruminantium* which can use only hydrogen or formate as an energy source, and *Bacteroides amylophilus* which can use only starch and its derivatives as substrates. These species occupy specialized 'metabolic niches' in the environments of the rumen.

The protozoa in the gut are also a complex mixture of specialized forms. Most are ciliates, holotrichs and entodiniomorphs (a group that is known only from rumen biotas: Figure 13.9). The free-living protozoa of the rumen exist in a constant environment in which they are probably subject to intense com-

. . . which attack cellulose

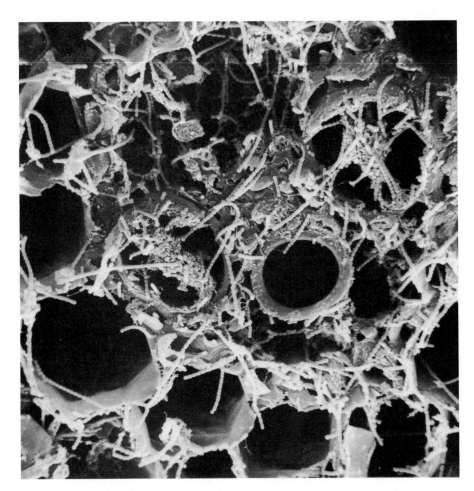

Figure 13.9. Attachment of *Ruminococcus flavefaciens* to the cell walls of perennial ryegrass (*Lolium perenne*). Scanning electron micrograph, magnification×1000. (Photograph by courtesy of B.E. Booker.)

petition from myriads of accompanying microbes. Their morphological diversity and complexity has been compared to that of tropical ecosystems, where favourable environments support great productivity and species diversity (Hungate, 1975).

Only a very few of the protozoa can digest cellulose (and even these may do so only with the aid of their own bacterial symbionts). Many of the protozoa ingest bacteria and if they are removed the numbers of bacteria rise. On the other hand, some of the entodiniomorphs are carnivores, preying on other protozoa. Thus all of the processes of competition, predation, mutualism, and the food chains that characterize terrestrial and aquatic communities in nature are present within the rumen microcosm. To add to the complexity, the species composition of the rumen microflora varies from one host species to another. It can also change dramatically if the diet is suddenly changed.

The microbial population of the rumen is continually multiplying and being depleted as the rumen contents pass into the intestine. The ruminant's own enzymes are responsible for further digestive processes in the intestine, where some microbial bodies are themselves broken down. Within the rumen the main products of digestion are volatile fatty acids (acetic, propionic and

the rumen contains a complete ecological community

butyric), ammonia, carbon dioxide and methane. The fatty acids are absorbed by the ruminant, and are its primary source of carbon nutrition. Propionic acid is particularly important since this is the only one of the volatile fatty acids that can be converted into carbohydrate by the ruminant and is essential for its metabolism, especially during lactation.

The mutualistic character of the association of ruminants with a rumen microflora is clear: the microbial population gains a continuous supply of food and a rather stable environment; the ruminant gains digestible resources from a diet which its own enzymes cannot handle.

13.5.2 The ecosystem of the termite gut

Termites are colonial social insects of the order Isoptera. The supposedly more advanced members of the group Macrotermitineae cultivate fungi (see above). The 'lower' groups feed directly on wood, and the cellulose, hemi-celluloses and lignins are digested in the gut. Some preliminary digestion occurs in the fore- and mid-gut as a consequence of the action of the termite's own enzymes, but the bulk of the food passes into a paunch (part of the segmented caecum), which is a microbial fermentation chamber (Figure 13.10). Termites refecate; they eat their own faeces so that food material passes at least twice through the gut, and microbial bodies produced in the

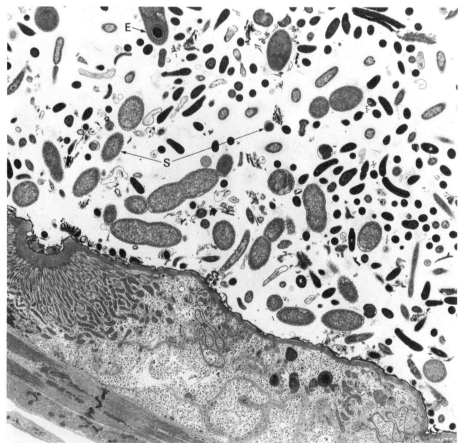

Figure 13.10. Electron micrograph of a thin section of the paunch of the termite *Reticulitermes flavipes*. Much of the flora is composed of aggregates of bacteria. Among them can be seen endospore-forming bacteria, (E) spirochaetes (S) and protozoa. (From Breznak, 1975.)

first passage may be digested the second time. The major group of micro-organisms in the termite's paunch are protozoans, consisting of anaerobic flagellates, such as *Trichomonas termopsidis,* and representing unique genera that are found only in termites and in a closely related species of wood-eating cockroach (*Cryptocercus*). Bacteria are also present but in the termite it is the protozoa themselves that are responsible for most of the cellulose decom-position. Wood particles are engulfed by protozoa and the cellulose is fer-mented intracellularly, releasing carbon dioxide and hydrogen. The principle products are (as in the rumen) volatile fatty acids, but in termites it is primarily acetic acid which is absorbed through the hindgut. In the absence of protozoa termites cannot digest wood.

cellulose decomposition mainly by protozoa in the termite paunch

The bacterial population of the termite gut is less conspicuous than that of the rumen, but appears to play a part in two distinct mutualisms.
(i) Spirochaetes are important members of the bacterial flora and they, together with rod-shaped bacteria, tend to be concentrated at the surface of the flagellates. In the guts of one species (*Mastoterma paradoxa*) the spiro-chaetes have been observed in synchronized movement actually propelling the flagellates. The spirochaetes are so conspicuous that the flagellates were once thought to be ciliates (Figure 13.11). The association of spirochaete and flagellate is mutualistic—the spirochaete receiving nutrient from the pro-tozoan and the protozoan gaining mobility from the spirochaete; so here we have a pair of mutualists living mutualistically within a third species.

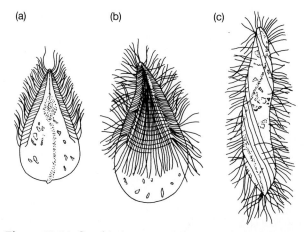

(a)   (b)   (c)

Figure 13.11. Symbiotic protozoa from the intestinal tract of termites. These drawings (from Cupp & Kirby) were made at a time when these organisms were thought to be ciliates. The cilia are now recognized as spirochaetes living in a mutualistic association with their protozoan host.

nitrogen is fixed in the termite gut.

(ii) Bacteria in the termite gut are capable of fixing gaseous nitrogen (di-nitrogen). This was suspected by Cleveland as early as 1925, but he was unable to prove it. More recently, Breznak (1975) has shown that nitrogen fixation does indeed occur in the termite gut, and that the process stops when antibacterial antibiotics are fed. The rate of nitrogen fixation also falls off sharply if the nitrogen content of the diet is increased.

13.6 Symbionts living within animal tissues or cells

There are a great many careful records establishing the presence of symbionts (especially bacteria) within the cells of animals (especially of insects). For example, bacterial cells are invariably present, massed together in enormous numbers in the fat bodies of cockroaches. They are transmitted through the mother, first aggregating around the oocytes and then entering the cytoplasm of the egg. The symbiosis is therefore truly hereditary, by maternal inheritance. Similarly, yeast-like symbionts are conspicuous in specialized bodies (mycetomes) in aphids and again are passed to offspring via the eggs. In most cases it has proved impossible to culture such symbionts and their metabolic activities are not understood. However, it has sometimes been possible to remove the microbial symbionts by using antibiotics and the host may then reveal deficiency symptoms, presumably for some metabolic need that was previously supplied by the symbiont. For example, the blood-sucking bug *Rhodnius prolixus*, in the absence of its symbionts, suffers vitamin B deficiency. Cockroaches (*Blatella germanica*) are slowed down in their development if they are deprived of their symbionts, but their normal growth rate can be restored by adding certain polypeptides to their diet. Again it is presumed these had been provided by the symbiont, and that the relationship is mutualistic.

13.7 Mutualisms involving higher plants and fungi—mycorrhizae

Most higher plants do not have roots, they have mycorrhizae—an intimate mutualism of fungus and root tissue in which the fungus helps to supply mineral nutrients to its host and gains from the plant part of its organic carbon resource (see Harley & Smith, 1983, for a recent survey). Plants of a few families like the Cruciferae are very much the exception in not forming mycorrhizae. Most mosses, ferns, lycopods, gymnosperms and angiosperms contain tissues that are more or less intimately interwoven with fungal mycelium and all the dominant plants of the major vegetation types of the world—forest trees, grasses and heaths—are conspicuous mycorrhiza-formers. (The fossil record of the earliest land plants suggests that they too were heavily infected. These species lacked root hairs, even roots in some cases, and the colonization of the land may have depended on the presence of the fungi.)

We can recognize distinct forms of mycorrhizae.

13.7.1 Sheathing forms

Sheathing forms (Ectomycorrhizae) occur most often on the roots of trees and involve Basidiomycete or Ascomycete fungi, many of which can be cultured in isolation from the host. Infected roots are usually concentrated in the litter layer of the soil and the mycelium extends through the litter and produces large fruiting bodies above ground which release enormous numbers of wind-borne spores. The fungus forms a 'tissue' sheathing the root and usually induces morphogenetic changes. Infected roots cease to grow apically and remain stubby and are often dichotomously branched, while the fungal mycelium extends between the cells of the root cortex providing an intimate cell-to-cell contact between fungus and host (Figure 13.12). In contrast, those roots that penetrate deeper into the less organically rich layers of the soil remain uninfected; they retain normal apical growth and continue to elongate.

Figure 13.12. Mycorrhiza of pine (*Pinus sylvestris*). The swollen, much branched structure is the modified rootlet enveloped in a thick sheath of fungal tissue. (Courtesy of Miss Judith Whiting. Photograph by S. Barber.)

mycorrhizal sheathing fungi require soluble carbohydrates

The fungi that form mycorrhizal sheaths require soluble carbohydrates as their carbon resource, and in this respect differ from most of their non-symbiotic, free-living relatives which are cellulose decomposers. The mycorrhizal fungi gain part at least of their carbon resources from the host. Mineral nutrients are absorbed by the mycelium from the soil and there is now no question that the fungus is active in supplying mineral resources to the host. Studies with radioactive tracers have shown that phosphorus, nitrogen and calcium may all move along the fungal hyphae into the host roots and then into the shoot system. Surprisingly, the mycorrhizae appear to function equally effectively without the hyphae that extend out from the sheath. The sheath itself must have some very remarkable properties as an absorber of nutrients and, subsequently, equally remarkable properties as a releaser of them to the host.

13.7.2 Vesicular arbuscular mycorrhizae

Vesicular arbuscular (V.A.) mycorrhizae have been reported from an exceedingly wide range of host plants. They do not form a sheath but penetrate *within* the cells of the host, and they do not produce morphogenetic effects. These features distinguish them sharply from the sheath formers. Moreover, the fungi that are responsible appear to belong to a single genus, *Endogone* (or *Glomus*), which has so far defied attempts to culture it in the absence of the host. It produces very large spores in small numbers—in striking contrast to sheath formers.

Roots are infected from mycelium present in the soil or from germ tubes that develop from the large spores (Figure 13.13). Initially the fungus grows between host cells but then enters them and forms a finely branched 'arbuscule'. The main advantage to the host of V.A. mycorrhizae seems to be that the fungal mycelium can obtain phosphate from the soil at distances greater than can be reached by an uninfected root and root hairs. Experiments using tracer ^{32}P show that the fungus can transport phosphate over distances of 1–2.7 cm. When phosphate is deficient in the soil, infection with V.A. mycorrhizae increases the growth rate and phosphate content of the host (Figure 13.14).

phosphate is transported by the fungus

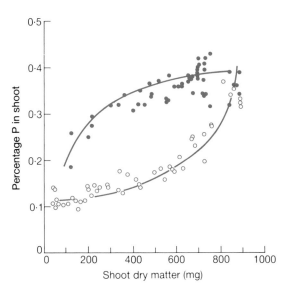

Figure 13.14. The relationship between phosphate concentration in the shoots of leeks and the dry weight of the shoots in non-mycorrhizal plants○ and plants infected with the endophyte *Glomus mosseae*●. (From Stribley *et al.*, 1980.)

The host plant certainly provides some carbon resources to the fungus (^{14}C assimilated by the host can be found in the fungal hyphae within 24 hours). How much of a cost this represents to the plant is difficult to measure. In onions the volume of fungus present in the roots is only 1–2% of the total root volume. There are, however, intriguing reports that the presence of V.A. mycorrhizae reduces the growth of the host when phosphate is readily available and this suggests that the fungus makes real demands on the host. It is quite possible that what we observe in V.A. mycorrhizae is a parasitic root fungus in which the virulence of the parasite has been largely controlled. The transfer of phosphate to the host under conditions of deficiency may be a

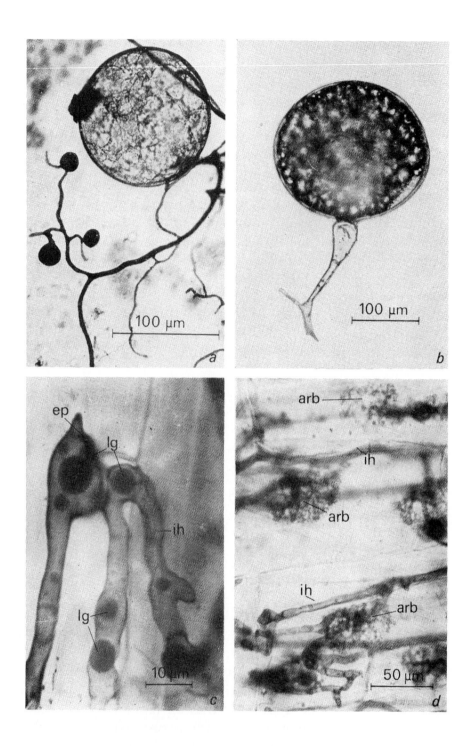

Figure 13.13. Vesicular arbuscular mycorrhizae. (a) A spore of the 'yellow vacuolate endophyte' *Glomus mosseae*. The endophytes have never been cultured successfully and their complete life-history is unknown. The various forms are described as species based on the different forms of their spores. (b) A spore of the 'bulbous reticulate' type, *Gigaspora calospora*. (c) The entry point of the fungal filament into the host tissues. (d) A squashed preparation of an onion root showing the internal filaments of the fungus and the arbuscules (arb). (From Tinker, 1975.)

simple consequence of the juxtaposition of fungal tissue between host and soil, with no special coevolved qualities at all.

13.7.3 Other mycorrhizae

Specialized associations of fungi with host roots occur in situations other than those where sheaths or vesicular arbuscules are formed. For example, there is a characteristic mycorrhizal form in the heaths (Ericaceae) and here there is some evidence that nitrogen may be transferred to the host by the fungus. The orchids also have characteristic fungal associations but the way in which these interact with the host plant is still obscure. They are clearly involved in seed germination, and mycorrhizal fungi are used in commercial practice to ensure good seedling establishment. The seeds of orchids are minute, little more than packets of dried cells almost wholly lacking in food reserves. It seems highly probable that fungal infection provides fixed carbon and other resources to the developing seedings. What it does afterwards is still a puzzle although various authors have described an apparent digestion of the fungus by the host.

Some orchids (e.g. *Neottia*) and members of other families (e.g. *Monotropa*, Ericaceae) are wholly lacking in chlorophyll. They are heavily infected with mycorrhizal fungi which act as a link transporting materials from the roots of a second, photosynthetic host plant. There is no evidence that such relationships are mutualistic. Indeed, we could regard them as true parasites on the fungi (Lewis, 1974). The gametophyte generation in the life cycle of some ferns and lycopods also lack chlorophyll. Their life is spent underground and they are heavily infected with mycorrhizae. Again, they are perhaps parasitic on the fungi.

13.8 Mutualisms of algae with animals

Algae are found within the tissues of a variety of animals, particularly coelenterates.

An especially detailed study of an animal–alga mutualism has been made on *Hydra viridis,* a species that can be easily cultured in the laboratory. In this animal, algal cells of the genus *Chlorella* are present in large numbers (1.5×10^5 per hydroid) *within* the digestive cells of the endoderm. The animal can be grown without its symbionts (it is then called aposymbiotic) but the endosymbiont (the alga) has not been cultured successfully alone. When a suspension of algae is injected into the coelenteron of an aposymbiont, some of the cells are ingested but there is a recognition process involved. Free-living *Chlorella* cells are treated like food particles—only *Chlorella* isolated from symbiotic *Hydra* are retained. These are sequestered in individual vacuoles, and migrate to specialized positions at the base of the digestive cells, where they multiply. In such an endocellular mutualism there must be regulating processes that harmonize the growth of the endosymbiont and its host (Douglas & Smith, 1983). If this were not the case, the symbionts would either overgrow and kill the host or fail to keep pace and become diluted as the host multiplied.

regulatory processes must harmonize the growth of the endosymbiont and its host

Some such regulation presumably occurs in all mutualisms maintaining an equilibrium between the partners. Even when *Hydra* are maintained in darkness and fed daily with organic food, a population of algae is maintained in the cells for at least six months and quickly returns to normal within two days of exposure to light (Muscatine & Poole, 1979). There is no doubt that in the light, *Hydra* receives fixed carbon products of photosynthesis from the algae and also 50–100% of its oxygen needs. It can, however, also use organic food. The dual nutrition of the mutualistic *Hydra* is remarkable. They possess the ability to behave both as autotrophs and as heterotrophs.

mutualisms with dinoflagellates

In marine environments it is most commonly members of the algal family Dinophyceae that form mutualistic associations with invertebrates. These algae have a very distinctive morphology and physiology. Most of them are nutritional opportunists capable of photosynthesis but also of using organic foods. However, their greatest ecological importance comes from their mutualism with corals. Not only do the algae provide photosynthates to their hosts, but the active photosynthesis has the side effect of precipitating calcium carbonate and so permits the formation of the coraline structure—coral reefs would not exist otherwise!

an experimental demonstration of an alga–animal mutualism

Some of the flagellate Dinophyceae can be cultured free from their host, and Taylor (1975) has been able to set up remarkable ecological model systems in which the endosymbiont and its host are linked in circulating illuminated cultures, though kept physically separate by dialysis membranes or by hollow fibre devices (Figure 13.15). The growth rate of the algae in such cultures is much increased (nearly doubled) as a result of linking with the 'host'. Note that in these models the metabolic products of both host and alga will be diluted in the culture medium; the opportunities for mutual exchange of material will be much greater in the true symbiosis. In fact, there is evidence that the export of photosynthate from the alga is considerably increased by

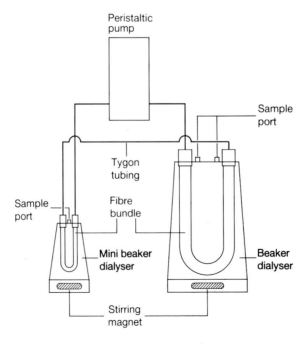

Figure 13.15. A device used to study the interaction between mutualists in which each partner is maintained in a separate chamber, connected by a circulating medium. The two populations are prevented from mixing by fibre bundles and dialysis membranes. (From Taylor, 1975.)

contact with the host. Just how this comes about is quite obscure but it seems characteristic of many mutualistic systems that the host can elicit a higher than normal rate of release (and perhaps also production) of metabolites from its endosymbiont.

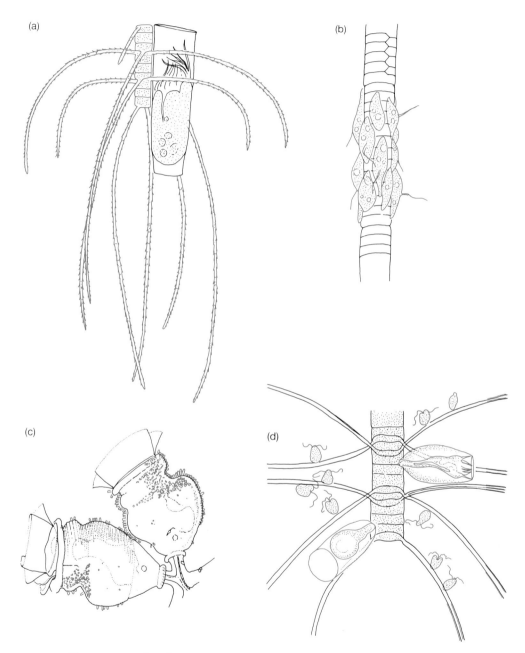

Figure 13.16. Symbiotic associations (presumed mutualisms) from tropical marine microplankton. (a) The filamentous diatom *Chaetoceros tetrastichon* (D) attached to the tintinnid *Eutintinnus pinguis* (T). (b) *Solenicola setigera* aggregated on the diatom *Leptocylindricus mediterraneus*. (c) Two zooids of *Zoothamnium pelagicum* bearing epiphytic blue-green bacteria. (d) A chain of cells of the diatom *Chaetoceros lorenzianum* infected with the flagellate *Ruttnera pringsheimii* and the ciliate *Vaginicola* sp. (From Taylor, 1982, and other sources.)

Figure 13.17. Underwater photographs of molluscs which have been shown to use the photosynthetic products from the chloroplasts of ingested algae. (a) *Tridachia crispata,* (b) *Elysia viridis,* (c) *Placida dendritica,* (d) *Plachobranchus ianthobapsus.* (From Trench, 1975.)

There are now many records of close associations between algae and protozoa in the plankton of tropical seas. Some of the associations are intracellular and the alga is tightly integrated into the morphology of its host. In the ciliate *Mesodinum rubrum*, for example, chloroplasts are present which appear to be symbiotic algae. The mutualistic consortium of animals and algae forms dense populations, called blooms, in up-welling water currents and can fix carbon dioxide and take up mineral nutrients. Extraordinarily high production rates have been recorded from such populations (in excess of 2 g of C m^{-3} hr^{-1}), and these appear to be the highest levels of primary productivity ever recorded for populations of micro-organisms. Other, less integrated associations are common, involving radiolarians carrying algal cells within their radiating pseudopods, and protozoa with individual algal cells (often dinophyceae but also blue-green algae and diatoms) growing on their surface (Figure 13.16).

A very odd situation is found in relationships between algae and some invertebrates, falling somewhere between the conditions usually regarded as mutualism, parasitism and predation. In quite separate groups of invertebrates there has developed a system in which algae are digested but their chloroplasts are retained in a photosynthetically active form within the tissues of the animal. The photosynthetic activity is in a sense 'kidnapped' from the algae. It is only one group of algae, the Siphonales, that can provide chloroplasts used in this way, and principally a group of gastropod molluscs that use them (Figures 13.17 and 13.18). The chloroplasts may remain active in the animal for more than two months and are protected against digestion by the host until they lose activity. Within the host cells they fix carbon and evolve oxygen and the host uses the products.

13.9 The mutualism of fungus and alga—the lichens

A number of fungi have escaped from their normal life habit in a mutualistic association with algae. These 'lichenized' fungi include within their filamentous body a thin layer of algal cells near the surface (Figure 13.19). The algae form only 3–10% by weight of the thallus body. Of the 70 000 or so species of fungus that are known, approximately 25% are lichenized. Fungi that have become lichenized belong to diverse taxonomic groups and the algae that have been identified within them belong to 27 different genera. Presumably the lichen habit has evolved many times. Fungi that are not lichenized occupy a restricted range of habitats, either as parasites of plants or animals, or as decomposers of dead materials. The lichenization extends the ecological range enormously, both onto substrates (rock surfaces, tree trunks) and into arid deserts and arctic and alpine regions that are barred to many other life forms.

The lichens derive photosynthates from their algal symbiont. The advantage to the algae, if any, has not been established so clearly. The fact that isolated algal cells are commonly found exposed on the surface of the lichen suggests that they can survive without the protection of the fungal thallus. It may be that the algae are 'captured' by the fungus and exploited without any recompense. However, some of the species (e.g. of the genus

blooms of ciliates with mutualistic algae

kidnapping of chloroplasts

the 'lichen habit' seems to have evolved many times

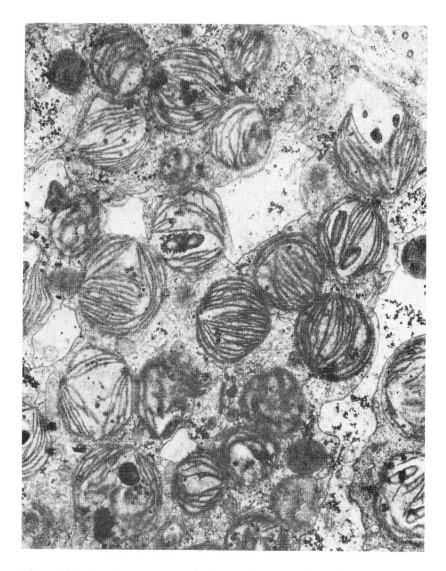

Figure 13.18. Electron micrograph showing the chloroplasts of the alga *Codium fragile* in a digestive cell of the gastropod mollusc *Elysia atro-viridis* (from Trench, 1975).

Trebouxia) are not known in a free-living form, and this suggests that there is something special about life in a fungal thallus that they need for their existence.

The most remarkable feature in the life of the lichenized fungi is that the growth form of the fungus is profoundly changed when the algae are present. When the fungi are cultured in isolation from the algae they grow slowly in compact colonies, much like related free-living fungi; but in the presence of the algal symbionts they take on a variety of morphologies that are characteristic of specific alga–fungus partnerships. In fact, the algae stimulate morphological responses in the fungus that are so precise that the lichens can be classified as distinct species, and different species of algae stimulate development of quite different morphologies from the same fungus.

lichen mutualisms
produce 'new species'

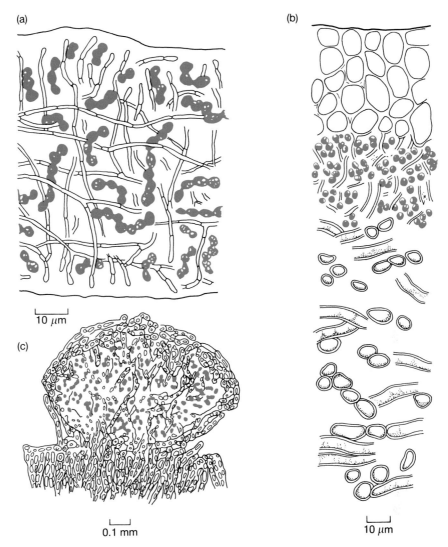

Figure 13.19. The structure of some lichen bodies. (a) A simple two-membered lichen, e.g. *Collema* with algal cells distributed throughout the fungal body. (b) A more structured system in which the algae are confined to a central layer. (c) A section through a specialized dispersal structure (cephalopodium) containing blue-green algal cells.

lichens are slow growers . . .

All lichens are slow growing; the colonizers of rock surfaces rarely extend faster than 1–5 mm per annum. They are very efficient accumulators of mineral cations, which they presumably extract from rain water and from the flow and drip down the branches of trees. Their ability to accumulate minerals makes them particularly sensitive to environmental contamination from pollution by heavy metals and fluoride, and for this reason they are among the most sensitive indicators of environmental pollution. The 'quality' of an environment can be judged rather accurately from the presence or absence of lichen growth on grave stones.

. . . and sensitive indicators of environmental pollution

Ninety per cent of the lichenized fungi are associated with green algae and these cannot fix atmospheric nitrogen. The remaining 10% contain blue-green algae of the genera *Nostoc, Scytonema, Stigonema, Dichothrix* and

Calothrix, which can fix atmospheric nitrogen. Surprisingly, these lichens are not especially characteristic of nitrogen-deficient habitats, though the algae certainly fix nitrogen within the host and leak substantial quantities to it.

13.10 Nitrogen fixation in mutualisms

The inability of most plants and animals to fix atmospheric nitrogen, (conventionally referred to as dinitrogen, N_2) is one of the great puzzles in the process of evolution. Dinitrogen can be fixed by only a small group of prokaryotes which includes a few bacteria, actinomycetes and blue-green algae. Many of these have been caught up in tight mutualisms with systematically quite different groups of eukaryotes. Presumably such symbioses have evolved a number of times independently. They are of enormous ecological importance because nitrogen is in crucial limiting supply in many habitats. Gibson and Jordan (1983) give a detailed review of the ecology of nitrogen-fixing systems. The prokaryotes that have been found in symbiosis (not necessarily mutualistic) are members of the following taxa:

(i) Azotobacteriaceae, which can fix nitrogen aerobically and are commonly found on leaf and root surfaces, e.g. *Azotobacter, Azotococcus*;
(ii) Rhizobiaceae, which fix nitrogen in the root nodules of leguminous plants (and have been persuaded to do so outside the host in laboratory culture);
(iii) Bacillaceae, such as *Clostridium* spp. which occur in ruminant faeces and *Desulfotomaculum* sp., which occurs in the rumen;
(iv) Enterobacteriaceae, which occur almost exclusively in intestinal floras, though occasionally on leaf surfaces and on root nodules;
(v) Spirillaceae, such as *Spirillum lipiferum*, which is an obligate aerobe found on grass roots; and
(vi) Actinomycetes of the genus *Frankia*, which fix nitrogen in the nodules of a number of non-leguminous plants such as the alder (*Alnus*) and sweet gale (*Myrica*).

Of all these symbioses, that of rhizobial bacteria with legumes is the most thoroughly studied, because of the huge importance of legume crops to man.

13.10.1 The mutualism of *Rhizobium* and leguminous plants

The establishment of the liaison between *Rhizobium* and the legume plant proceeds by a series of reciprocating steps. The bacteria occur in a free-living state in the soil (Figure 13.20) and multiply when the root hair of a host plant is close by—some stimulus or resource is apparently provided by the host. A bacterial colony develops on the root hair which then begins to curl, and the cell is penetrated by the bacteria. The host responds by laying down a wall that encloses the multiplying bacteria and forms an infection thread that is part host, part bacteria. This may then grow from cell to cell of the host root cortex, and *in advance of it* the host cells start to divide, beginning to form a nodule. As the host cells become colonized the bacteria change their form to become non-dividing swollen bacteroids. A special vascular system develops in the host that supplies the products of photosynthesis to the nodule tissue and carries away fixed nitrogen compounds (mainly asparagine) to other

nitrogen fixation is of crucial importance because nitrogen is often in limiting supply

parts of the plant (Figure 13.21). A special compound, leghaemoglobin, is formed, giving the active nodule a pink colour. The haem for this molecule is apparently provided by the bacteria and the globin by the host plant—a biochemical mutualism. The leghaemoglobin is partly responsible for maintaining the low oxygen tension in the nodule that is an absolute requirement for dinitrogen fixation. Up to 40% of the nodule weight may be leghaemoglobin and the nodules themselves may represent 5% of the total weight of the plant.

The costs and benefits of this symbiosis have been measured but the picture is still not entirely clear. We need to compare the energetic costs of the

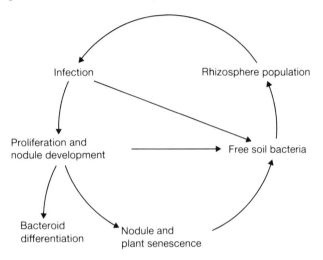

Figure 13.20. An idealized life cycle for *Rhizobium* (from Beringer *et al.*, 1979). Within the nodule of the leguminous host the bacteria develop to a nitrogen-fixing form—the bacteroid—which is no longer viable. The 'rhizosphere' population is that found in the immediate vicinity of the root surface. Surprisingly, the populations of rhizobia are greatest in the rhizosphere. Individual nodules probably contain about 10^6 viable rhizobia, yet the populations in the rhizosphere may be as high as 10^{11} per gram of soil (cf. figure of $< 10^2$ to 10^5 per gram of soil which has not carried legumes recently). Beringer *et al.* speculate that: 'From the point of view of the rhizobium, the colonization of the plant root is merely a device to divert plant metabolites to support large rhizosphere populations of bacteria.' (Beringer *et al.*, 1979.)

the energy costs of dinitrogen fixation

alternative processes by which a plant may obtain its supplies of fixed nitrogen. The normal route is direct from the soil as nitrate or ammonium ions. The metabolically cheapest route is the use of ammonium ions, which do not require to be reduced before their incorporation into proteins, etc. However, in most soils ammonium ions are rapidly converted to nitrates by microbial activity (nitrification) and it is mainly in anaerobic and acid soils that ammonium ions are readily available. The costs of the mutualistic process (including the maintenance costs of the bacteroids) are *c.* 13.5 moles of ATP consumed per mole of ammonia formed. This is slightly more energetically expensive than the 12 moles of ATP involved in reducing nitrate to ammonia. To the cost of nitrogen fixation must be added the costs of forming and maintaining the nodules themselves. A recent estimate puts this at about 12% of the plants total photosynthetic output. It is this enormous cost of nodule

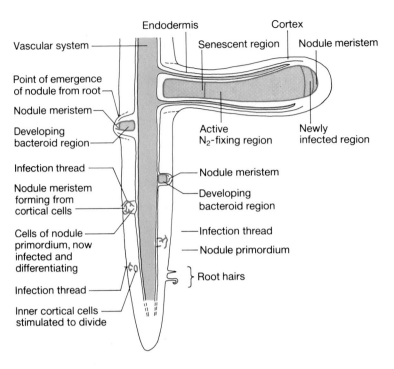

Figure 13.21. The development of the root nodule structure during the course of development of infection of a legume root by rhizobial bacteria (from Sprent, 1979).

construction and maintenance that makes dinitrogen fixation an energetically inefficient process. There is, however, no obvious reason why the efficiency of energy usage should be a necessary constraint on the evolution of higher plants. If green plants are 'pathological overproducers of carbohydrate' (see Chapter 3), they may have energy resources in superabundance but be limited by other resources in short supply, for instance nitrogen. A rare and valuable commodity bought with a cheap currency is no bad bargain. Dinitrogen fixation declines rapidly when a nodulated legume is provided with nitrates and it is in environments where fixed nitrogen is a rare commodity that the pay-off in dinitrogen-fixing mutualism is likely to be greatest. Sprent (1979) gives a good overview of mutualistic nitrogen fixation.

dinitrogen-fixing mutualisms in their ecological context

The symbiosis of rhizobium legumes must not be seen just as an isolated ecosystem of bacteria with a host plant. The ecological status of legumes is profoundly changed when they carry nitrogen-fixing symbionts. In nature, legumes normally form mixed stands in association with non-legumes. These are potential competitors with the legumes for fixed nitrogen (nitrates or ammonium ions in the soil). The nodulated legume sidesteps this competition by its access to a unique source of nitrogen. It is in this ecological context that nitrogen-fixing mutualisms gain their main advantage. When we look for the ecological consequences of symbiotic nitrogen fixation and its evolution we need to consider the performance of legumes in the natural competitive environment in which they are normally found and have presumably evolved.

Figure 13.22 shows the results of an experiment in which soybeans (a

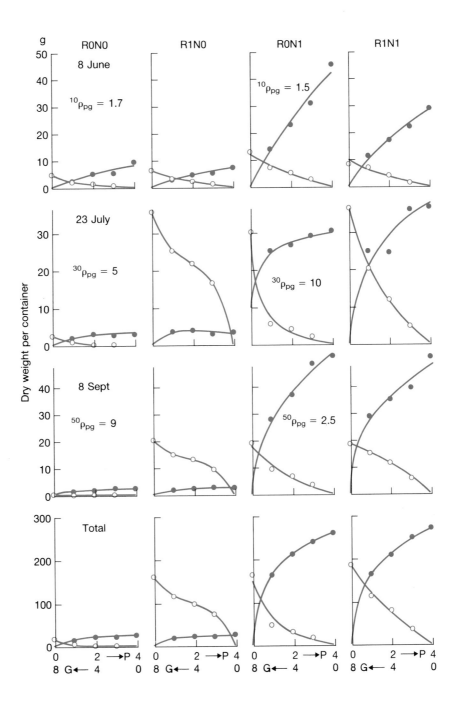

Figure 13.22. The growth of soybeans (*Glycine soja*) and the grass *Paspalum* grown separately and in mixtures with and without nitrogen fertilizer and inoculation with nitrogen-fixing *Rhizobium*. The plants were grown in containers containing 0–4 plants of the grass and 0–8 plants of *Glycine*. The horizontal scale on each figure shows the proportions of plants of the two species in each container. R0N0 = no inoculation, no fertilizer; R1N0 = inoculated with *Rhizobium*, no fertilizer; R0N1 = no inoculation, nitrate fertilizer applied; R1N1 = inoculated with *Rhizobium*, nitrate fertilizer applied. (From de Wit *et al.*, 1966.)

legume) were grown in mixtures with *Paspalum,* a grass. The mixtures either received mineral nitrogen or were inoculated with *Rhizobium.* The experiment was designed as a 'replacement series' (see section 7.9.1), which allows us to compare the growth of pure populations of the grass and the legume with their performance in the presence of each other. If we look first at the pure stands of soybean we see that *either* inoculation with *Rhizobium, or* application of fertilizer nitrogen increased its yield by at least 10-fold; the two treatments had much the same effect. This shows that the legumes can use either source of nitrogen as a substitute for the other. The grass responded only to the fertilizer. In mixtures of soybean and grass and in the presence of *Rhizobium,* the legume contributed about five times as much to the yield of the mixture as did the grass. With fertilizer nitrogen the legume and the grass contributed about equally to the mixture yield. Quite clearly in such eco-systems, it is under conditions of resource deficiency that the pay-offs from the mutualism are strongest—the creators of deficiency (the depleters) will be the species (e.g. *Paspalum*) that lack the mutualism.

There is a sense in which organisms that can fix atmospheric nitrogen are suicidal. If their activity raises the level of fixed nitrogen in their environment so that it becomes available to other species, the advantage to the nitrogen fixer is lost and its neighbours may now dominate it and crowd it out. This is one reason why it is very difficult to grow repeated crops of pure legumes in agricultural practice without aggressive grass weeds invading the nitrogen-enriched environment. It may also be the explanation for the failure of leguminous herbs or trees to form dominant stands in nature.

Legume tissues usually have a high nitrogen content and in mixed clover–grass pastures the legumes make a large contribution to the protein intake of grazing herbivores. Moreover, after death, legumes augment the level of soil nitrogen on a very local scale with a 6–12 months delay as they decompose. A major consequence of this is that the growth of associated grasses will be favoured in these local patches. The grazing animal con-tinually removes the grass foliage and the nitrogen status of the patches again declines to a stage at which the legume may once more be at a competitive advantage. In a stoloniferous legume, such as white clover, the plant is continually 'wandering' through the sward leaving behind it local grass-dominated patches while invading and enriching with nitrogen new patches where the nitrogen status has become low. (The grazing animal itself adds to the process by which patchiness is generated. It grazes over extensive areas of a sward and its dung and urine concentrate excreted nitrogen in local patches.) The symbiotic legume in such an ecosystem is not only the driver of its nitrogen economy but also of some of the cycles that occur within the species patchwork of the community.

13.10.2 Nitrogen fixation in mutualisms with plants that are not legumes

Other organisms that can fix dinitrogen are involved in mutualistic associa-tions with non-legumes. Two main types of organism are involved; blue-green algae and a curious symbiont, *Frankia,* which is still not properly understood taxonomically and is usually classed as an actinomycete. The distribution of these organisms in mutualisms with higher plants is very odd, and does not seem to make much evolutionary sense. The blue-green algae form symbioses with three genera of liverwort (*Anthoceros, Blasia* and *Clavicularia*), with some mosses (e.g. *Sphagnum*), with one fern (the free-floating aquatic *Azolla*), with many cycads (e.g. *Encephalartos*) and with all 40 species of the flowering plant genus *Gunnera*. In the liverworts the blue-green alga *Nostoc* lives in mucilaginous cavities in the thallus and the plant reacts to the presence of the alga by developing fine filaments that maximize contact with it. The alga provides nitrogen to the host and the host provides fixed carbon to the alga.

nitrogen fixation based on *Frankia*

Frankia forms mutualisms with members of eight families of flowering plants, almost all of which are shrubs or trees. The nodules are usually hard and woody. The best known hosts are the alder (*Alnus*), the sea buckthorn (*Hippophae*), sweet gale (*Myrica*) and the arctic–alpine sub-shrubs *Arctostaphylos* and *Dryas*. Plants of *Caeonothus* which form extensive stands in Californian chaparral also develop *Frankia* nodules. All the examples of mutualism with *Frankia* can be interpreted as allowing the host to enter a habitat that is nitrogen deficient—the peaty soils of *Alnus* and *Myrica,* the sand dunes of *Hippophae* or the arctic and alpine habitats of *Dryas* and *Arctostaphylos.*

13.10.3 The evolution of nitrogen-fixing mutualisms

The process of biological nitrogen fixation may have evolved relatively recently. The gene complex responsible for the coordinated metabolic processes of dinitrogen reduction is essentially the same in all the organisms that can perform the reaction. The whole gene complex (the *nif* system) has been experimentally transferred as a plasmid from the bacterium *Klebsiella pneumoniae*, in which it occurs naturally, to *Escherichia coli*, in which it does not. The same kind of transfer could have occurred repeatedly in nature. This would be the easiest way to explain its presence in such diverse groups of prokaryotes. If the ability to fix dinitrogen occurred late in evolutionary history it would also explain its absence from eukaryotes—it developed after prokaryotes and eukaryotes had diverged.

The diversity of the groups with which nitrogen-fixing organisms have entered into mutualisms can most easily be explained as a series of quite separate evolutionary developments originating in different nitrogen-deficient habitats. The ecological consequences are profound.

13.11 The evolution of subcellular structures from symbioses

We have seen that there is remarkable variety in the type of association that could be regarded as mutualistic symbioses in both plant and animal kingdoms. They extend from associations in which behaviour links two very different organisms that spend parts of their lives apart and retain their individual identities (gobiid fish with shrimps, the large blue butterfly with ants), through the chemostat ecosystems (strictly external to the body tissues) of the rumen and the caecum, to the intercellular sheathing mycorrhizae and the intracellular zooxanthellae of coelenterates. The stages can be read as a sequence from integration of separate parts of a community to integration within what becomes in most respects parts of a single 'organism'.

mutualism may have led to the evolution of the eukaryotes

It is not difficult to envisage that some of what are now thought of as single organisms have arisen from symbioses in which the identities of the participants have become inextricably merged. This is the view pressed particularly by Margulis (1975). She argues that critical steps in the evolution of the major groups of higher plants and animals have involved the incorporation of prokaryotes. This is the 'Serial Endosymbiosis Theory of the Evolution of the Eukaryote Cell'. 'The first step in the origin of eukaryotes from prokaryotes occurred when a fermentative anaerobe was invaded by a Krebs-cycle-containing eubacteria (promitochondria) ... giving the mitochondria-containing amoeboids from which all other eukaryotes were derived. These in turn acquired motile surface bacteria giving flagellates and ciliates.' (We have already seen that the flagellates in the termite gut acquire motility from spirochaetes that become attached to them.) At a later stage the ingestion of blue-green algal cells would provide the autotrophic 'organism' from which the plant kingdom might evolve. This sequence is highly imaginative and there are many unbelievers, but it may well be true, at least in part. All the appropriate biota still exist, including the bacterium *Paracoccus denitrificans* which has characteristics resembling the hypothetical free-living ancestor of the mitochondrion. If Margulis's theory is correct, most of this chapter has been about a second stage in the development of mutualisms as further pairs of species become more closely integrated, each into the life of another.

13.12 Models of mutualisms

At first sight it would appear that it is easy to make mathematical models of mutualistic interactions. The Lotka–Volterra equation describing the growth of a population of a single species, $dN_1/dt = r_1N_1(K_1-N_1/K_1)$ is modified by introducing a factor $(+\alpha N_2)/K_1$ which determines how much members of species 2 contribute to increasing the growth rate of species 1. The same equation can be written for the growth of species 2, with $(+\beta N_1)/K_2$ introduced. This allows each species to have its number increased by the presence of the other—the characteristic of a truly mutual relationship. However, it produces an explosive situation in which there are no limits to the population size of both mutualists. It is obviously unrealistic. Other attempts to model mutualisms (see May, 1982) lead us to expect mutualisms to be rather unstable. But most of the evidence from field situations is that they have very high stability and resilience. Almost certainly the models fail because they treat mutualism as a mirror image of competition; the models allow the constraints that limited the species in isolation to be relaxed in mutualism. In fact, most mutualisms probably act not by *relaxing* a constraint but by *removing* it and substituting another. For example, when a coelenterate gains a photosynthetic endosymbiont, carbon nutrition ceases to be a constraint on its growth and some other resource limit presumably comes into play. Such a condition may well be stable.

13.13 Some general features of the lives of mutualists

There are many features of the biology of mutualists (particularly those that enter into a close physical relationship) that set them apart from most other organisms. They contrast very strongly with aspects of the biology of parasites and with free-living relatives.

1. The life cycles of close mutualists are remarkably simple (contrasting particularly with the life cycles of most parasites).

2. Sexuality appears to be suppressed in endosymbiotic mutualists, especially in comparison with parasites and with free-living close relatives (see Law & Lewis, 1983).

3. There is no conspicuous dispersal phase in endosymbionts. When dispersal occurs the two partners are often dispersed together (as when young queen ants take fungal inoculum with them to found a new colony, or when fungus and alga are combined in the dispersal unit of many lichens). Spores from the fruiting bodies of the sheath-forming mycorrhizal fungi are exceptions to this general rule, but these fungi may spend most of their lives in a free-living condition (or as parasites). This contrast between mutualists and parasites is particularly strong: dispersal rules the population dynamics of most parasites.

4. There seems to be nothing in the life of mutualistic organisms comparable to epidemics among parasites. Populations of mutualists seem to have great stability when compared to those of parasites.

5. Within populations of mutualistic organisms, the numbers of endosymbionts per host seem to be remarkably constant.

6. The ecological range (and niche breadth) of organisms in mutualistic symbioses appears usually (perhaps always) to be greater than that of either species when living alone. This again contrasts with parasitic symbioses where the host's ecological range is probably usually reduced by the presence of parasites.

7. Surprisingly, host specificity is often quite flexible—ants and nectaries, algae and fungi in lichens, plants and pollinators, etc., often involve pairs of species that can live mutualistically with several, sometimes many, other species. Strict specificity is *not* the rule.

There is no doubt that mutualisms have been seriously neglected by ecologists—even more so than parasitisms. Attention has perhaps been overly concentrated on the ecology of competitive and predator–prey relationships. One reason may be that many of the mutualistic relationships among organisms have much of the 'stand back and wonder' quality of natural history that is despised by theoreticians. Mutualisms lend themselves easily to the story-telling approach to the explanation of nature, in which everything is seen as a piece of evolved perfection.

Much of the literature concerning mutualisms is a stronghold of anecdotes—the structure of this chapter largely mirrors this truth. But the study of mutualisms attacks one of the most fundamental problems in ecology. Do whole communities of organisms in nature represent more or less tightly coevolved relationships? Do 'holist' properties emerge from evolutionary interactions? Many of the examples in this chapter seem to support the view that there is a tendency for groups of two or more species to become tied together in mutually beneficial associations that come to represent systems within systems—superorganisms. We still know almost nothing about how far such evolved mutualisms extend, perhaps in weaker form, to integrate the activities of whole communities in nature. Are they extreme examples of a widespread phenomenon or just odd and sometimes bizarre curiosities—important freaks? One intriguing suggestion is that mutualisms in temperate zones are rarely obligate, in contrast to those in tropical habitats. The contribution of facultative mutualisms to community or guild structure may not be as negligible as previously presumed (Bristow, cited in May, 1982).

PART 3

Two Overviews

Introduction

This short section has only two chapters. Both deal with topics that could, in principle, have been discussed earlier in the book; but in both cases, delay has brought with it advantages. In Chapter 14 we survey, and seek to understand, the range of life-history patterns or 'life-history strategies' exhibited by living organisms; while in Chapter 15 we survey, and seek to understand, the range of patterns of abundance exhibited by living organisms. Hence, the chapters are, in a sense, overviews of the two levels of the ecological hierarchy that we have covered so far: Chapter 14 deals with individual organisms, and Chapter 15 with populations.

The two chapters have been placed at this point in the book, because we will need to draw on information from the preceding chapters in order to make sense of the topics addressed in the present section. In Chapter 14 we demonstrate that the life-history of an organism can often be seen to reflect the environment in which it lives—yet this environment comprises, to a very large extent, the many other organisms with which it interacts. Likewise, the abundance of an organism is, essentially, a reflection of the combined effects of the many interactions in which it takes part. Thus, if we wish to disentangle these effects, we must first develop an understanding of them individually. Hence, Parts 1 and 2 of the book can be seen as a necessary preparation for Chapters 14 and 15.

Notwithstanding this argument, it may seem regrettable that the present section interrupts the apparently logical flow from individual to population to interaction to community. But an alternative point of view would be to find it remarkable that this is the *only* interruption in the sequence. Every aspect of ecology can be fully understood only by reference to every other aspect—above it, below it and beside it in the hierarchy. To take just one example: we dealt with the responses of individuals to conditions, very early in the book, and these responses set the scene for the interactions of the organisms with other organisms, and for the construction of the communities of which they are part. But the response of an organism to environmental conditions is itself crucially influenced by the presence of other, interacting organisms, and by the physical and biological structure of the surrounding community. Thus the 'logical' structure of the book is artificial. It has been imposed *by us*, for convenience, on a science that would more properly be described as a multi-dimensional array of topics with arrows of influence flowing backwards and forwards in every direction. In the present section, we collect together some of the arrows that have been passing in the opposite direction to the one we have been following up the hierarchy.

14 Life-History Variation

14.1 Introduction

Most natural history museums present us with a collection of mature adult organisms, as if the variety of nature could be properly seen in this one selected moment from the life of each species. But there is no such characteristic moment. Every organism can be truly represented only by its whole life-history. An organism's life-history is its lifetime pattern of growth, differentiation, storage and, especially, reproduction. Different organisms spend varying proportions of their lives in a phase of growth and differentiation prior to reproducing; reproduction itself may occur once or repeatedly; and it may more or less coincide with the end of growth, or an individual may continue to grow during its reproductive life. Trying to understand the similarities and differences in life-histories is one of the fundamental challenges of modern ecology.

what is a life-history?

Every life-history is, to some extent at least, unique. In this chapter, therefore, we consider ways in which life-histories might be grouped, classified and compared. This search for patterns amongst life-histories involves first deciding what are the most effective ways of describing them. Only then can we search for correlations between life-history traits and features of the habitats in which the life-histories are found. We can ask, too, whether particular traits are commonly associated with each other—for example, do all organisms that have long lives start reproducing late? Such questions must be asked though, in an evolutionary context. The questions must be: would natural selection be expected to favour the co-occurrence of particular life-history traits—and does it? And would natural selection be expected to favour particular traits in particular habitats—and does it?

life-history patterns reflect rules imposed within constraints

We will develop a set of rules describing when particular traits might be favoured. It is important to realize, however, that these rules work within constraints: the possession of one life-history trait may limit the possible range of some other trait, and the general morphology of an organism may limit the possible range of *all* its life-history traits. The most that natural selection can do, therefore, is to favour the life-history that is best (not 'perfectly') suited, overall, to the many, various and often conflicting demands of an organism's environment.

Of course, the life-history of an organism is not immutable. It is fixed within limits by the genotype of the individual, but the experiences of the individual during its life may modify this: there is plasticity in any life-history, determined by the way in which the genotype of the organism interacts with its environment. Thus, in our concern with the ecological significance of life-histories, we are dealing with (a) the products of the evolutionary process and (b) the ways in which these products interact with the immediate environment of individuals. In addition, we may need to consider (c) whether the interaction between the evolutionary product and the present environment has itself evolved. In the life-history of an annual plant, for instance, we may find that (a) there is a number of seeds per pod that is characteristic of the species (genotypic); (b) the number is reduced during drought (plasticity); and (c) the extent to which seed number is reduced by drought is itself under genotypic control (some species may reduce the number of seeds set during drought, while others reduce the size of the seeds but keep the number the same).

It is important to realise too that the study of life-history patterns deals typically with comparisons, not absolutes. Life-history studies cannot explain why a particular lizard produces eight eggs, each weighing 10 g, after a period of immaturity lasting seven months. But they may be able to explain how it comes about that the eggs are smaller or larger than those of other lizard species; or why they are produced after a longer or shorter period of immaturity; or why eight 10 g eggs are produced when a related species produces four 20 g eggs. Life-history studies, therefore, deal in currencies of 'more', 'larger', 'longer' and so on. Even when we say an elephant is large, we mean only that it is large compared to animals in general. And if we attempt to explain *why* an elephant is large, the best we can do is explain why it is not smaller (or larger still).

Finally, it may be asked why 'patterns' should be sought. Following Southwood (1977), the search can be seen as an attempt by ecologists to construct their equivalent of the chemists' Periodic Table. For the eighteenth-century chemist '. . . each fact had to be discovered by itself and each had to be remembered in isolation'. A similar fate will await ecologists unless they can discern, in outline at least, some patterns in the ways that organisms live out their lives.

14.2 The components of life-histories and their potential benefits

In order to recognize life-history patterns, it is necessary first to specify the various components of life-histories, and to be aware of the benefits that these components can convey.

14.2.1 Size

Individual size is perhaps the most apparent aspect of an organism's life-history. It varies from taxon to taxon, from population to population, and from individual to individual; and as Chapter 4 made clear, it is particularly variable in organisms with a modular construction. Large size may increase an organism's competitive ability, or increase its success as a predator, or decrease its vulnerability to predation. Large organisms are also often better able to maintain a constancy of body function in the face of environmental

life-histories reflect the genotype, the environment, and the interaction between the two

life-history studies involve comparisons not absolutes

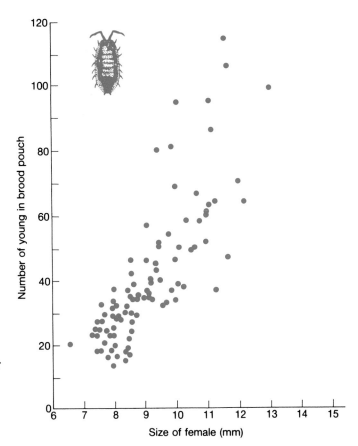

Figure 14.1. Larger females produce a greater number of young in the terrestrial isopod *Armadillidium vulgare* (after Paris & Pitelka, 1962).

variation (because their smaller surface-to-volume ratio makes them less 'exposed' to the environment). All of these factors tend to increase the survivorship of larger organisms. In addition, larger individuals within a species usually produce more offspring (e.g. Figure 14.1). Size, however, can increase some risks. A larger tree is more likely to be felled in a gale or struck by lightning, and there have been many studies in which predators exhibit a preference for larger prey (Chapter 9).

14.2.2 Rates of growth and development

All organisms increase their size by growth—every organism is larger than the zygote from which it developed. A particular size may be achieved by starting large, growing fast, growing for a long time, or any combination of these three.

Development, on the other hand, is the progressive differentiation of parts, enabling an organism to do different things (for instance reproduce) at different stages in its life-history. In many organisms, growth and development occur simultaneously. Development is quite separate from growth, however, in the sense that a given stage of development can be represented by a range of sizes, and a given size can be achieved by individuals at various stages of development. Rapid development can be beneficial because it leads to the rapid initiation of reproduction, to short generation lengths, and thus to high rates of increase (Chapter 4). On the other hand, arrested development (i.e. dormancy or diapause) may be beneficial if the organism passes through a highly unfavourable period during its lifetime (Chapter 5).

14.2.3 Reproduction

A great deal of the variation among life-histories relates to aspects of reproduction, and reproduction is thus the central issue in this chapter.

precocity and delay

(a) The length of the pre-reproductive period is often used as a straightforward measure of the rate of pre-reproductive development. Conversely, organisms can be classified according to the extent that they *delay* reproduction. This has the advantage of stressing that many individuals postpone reproduction to beyond the size or stage at which reproduction occurs in more precocious individuals of the same species. The straightforward benefits of precocity (short-generation length, etc.), and the rather less obvious benefits of delay, are discussed in section 14.6.

itero- and semelparity

(b) Organisms can either produce all of their offspring simultaneously in a single reproductive event (semelparity), or they can produce them in a series of separate events during and after each of which the organisms maintain themselves in a condition that favours survival to reproduce again subsequently (iteroparity).

clutch size, clutch number and offspring size

(c) Among iteroparous organisms, the number of separate clutches of offspring can vary. Clearly, all other things being equal, fecundity will increase as the number of clutches increases.

(d) Variation is obviously possible in the number of offspring in a clutch. Other things being equal, fecundity will increase as clutch size increases.

(e) Individual offspring can vary in size, i.e. the quantity of resources allocated to them by their parent can vary. This variation may result from the provisioning of eggs with food reserves. Here the cost is exclusively to the mother and is totally wasted unless the eggs are fertilized. (Note that the cost is usually incurred before they are fertilized, and so may be a risky investment.) Alternatively, the variation can arise during the early growth and development of embryos provided with resources supplied directly by the mother via a placenta or its equivalent. This is the situation in humans, other mammals and all seed plants. Large newly emerged or newly germinated offspring are often better competitors, better at obtaining nutrients and better at withstanding environmental stress. The number of *successful* offspring produced by a parent will therefore often increase with offspring size.

(f) Finally, life-histories are often described in terms of a composite measure of reproductive activity known as 'reproductive allocation' (also often called 'reproductive effort'). This is sometimes seen as the absolute amount of resources allocated by an organism to reproduction, but it is better defined as the *proportion* of the available resource input that is allocated to reproduction *over a defined period of time*.

reproductive allocation: often measured imperfectly

However, reproductive allocation is far easier to define than it is to measure. Ideally, the appropriate, limited resource should be identified, and its allocation over a period of time to the various parts and physiological processes of an organism should be followed. In practice, even the better

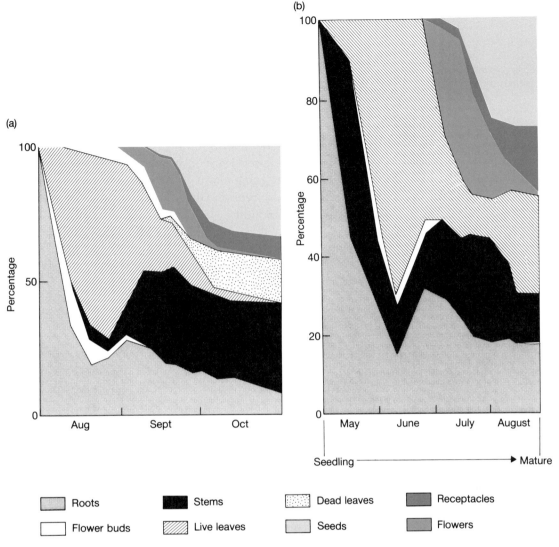

(a)

(b)

| Roots | Stems | Dead leaves | Receptacles |
| Flower buds | Live leaves | Seeds | Flowers |

Figure 14.2. (a) Percentage allocation of calories to different structures throughout the life cycle of the annual weed *Senecio vulgaris* (after Harper & Ogden, 1970). (b) Percentage allocation of dry weight to different structures throughout the life cycle of another annual, *Chrysanthemum segetum* (after Howarth & Williams, 1972).

studies monitor only the allocation of energy (Figure 14.2a) or just dry weight (Figure 14.2b) to various structures at a number of stages in the organism's life cycle (Garper & Ogden, 1970; Howarth & Williams, 1972). The most common way of assessing reproductive allocation, however, has been to measure such ratios as 'gonad weight to body weight', 'seed crop weight to plant weight' or 'clutch volume to body volume'. These obviously fall far short of the ideal, but they can nonetheless be useful approximations when the intention is only to compare reproductive allocations (see discussion in Hart & Begon, 1982).

14.2.4 The uses of soma

Reproduction, however, is not the only component of a life-history. The characteristics of an organism's somatic tissues can also be of crucial importance.

(a) There is considerable variation between species, and indeed within

505 LIFE-HISTORY VARIATION

parental care, longevity, dispersal, storage, resource capture and protection

species, in the amount of resources allocated by parents to the care of off-spring after birth: protecting them, suckling them or foraging on their behalf.

(b) Longevity (length of life) is not important in its own right, but it can certainly be important in the many cases where an increase in longevity leads to an increased number of clutches and/or an increased period of parental care.

(c) Dispersal can affect both fecundity and survivorship and is an integral part of any organism's life-history (see Chapter 5).

(d) Storage of energy and/or resources will be of benefit to those organisms that pass through periods of reduced or irregular nutrient supply (probably most organisms at some time). It may be utilized later for metabolism, growth, defence or reproduction, and in general for enhanced future survivorship and fecundity.

(e) Organisms will also generally benefit from the allocation of energy and/or resources to structures or activities that increase their own rate of resource capture or protect them from their enemies. The reason, of course, is that these in turn may enhance growth, survivorship and fecundity.

14.3 Reproductive value

It seems almost too much to hope that any one measure might encapsulate many of the features of diverse life-histories, and do so in a way that is relevant to the way that evolution occurs. However, the concept of 'reproductive value' introduced by Fisher (1930) does just this.

what is reproductive value?

Natural selection favours those individuals with the greatest fitness, i.e. those that make the greatest *proportionate* contribution to the future of the population to which they belong. All life-history components affect this contribution, and they do so through the media of fecundity and survivorship. Reproductive value is a measure of the combined effects of fecundity and survivorship that also takes into account an individual's proportionate (rather than absolute) contribution to the future of the population. Reproductive value is defined and described in Box 14.1

The box shows that *reproductive value* is a term with a detailed and specific meaning. For most purposes though, these details can be ignored as long as it is remembered that (a) reproductive value is the sum of the current reproductive output and the residual (i.e. future) reproductive value; (b) residual reproductive value combines expected future survivorship and

Figure 14.3.
Reproductive value generally rises and then falls with age. (a) The annual plant *Phlox drummondii* (Leverich & Levin, 1979). (b) Female grey squirrels (Charlesworth, 1980, after Barkalow *et al.*, 1970).

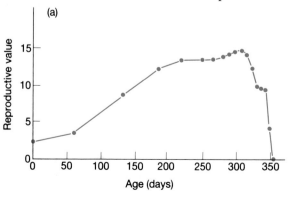

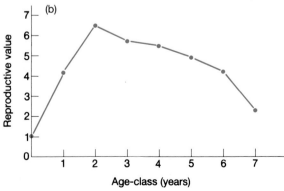

Box 14.1: A formal definition of reproductive value

The reproductive value of an individual of age or stage x (RV_x) is defined in terms of the life-table statistics discussed in Chapter 4. Specifically:

$$RV_x = \sum_{t=x}^{t=\infty} m_t \cdot S_{x \to t} \cdot \frac{N_{T(x)}}{N_{T(t)}}$$

where m_t is the birth rate of the individual in stage or age-class t; $S_{x \to t}$ is the probability that the individual will survive from stage x to stage t, and is therefore equal to l_t/l_x (Chapter 4); $N_{T(t)}$ is the size of the population when the individual is in stage or age-class t; and Σ means 'the sum of'.

To understand this equation, it is easiest to split RV_x into its two components:

$$\text{reproductive value} = \text{contemporary reproductive output} + \text{residual reproductive value}$$

or

$$RV_x = m_x + \sum_{t=x+1}^{t=\infty} m_t \cdot S_{x \to t} \cdot \frac{N_{T(x)}}{N_{T(t)}}.$$

The contemporary reproductive output is simply the individual's birth rate in its current stage (m_x). The *residual reproductive value* (Williams, 1966) is the sum of the 'expectations of reproduction' of *all* subsequent stages, modified for each stage by the change in the number of individuals in the population (i.e. modified by $N_{T(x)}/N_{T(t)}$ for the reasons described below). The 'expectation of reproduction' for stage t is $m_t \cdot S_{x \to t}$, i.e. it is the birth rate of the individual should it reach that stage, reduced by the probability of it doing so.

Reproductive value takes on its simplest form in the many cases where *on average* the population size remains approximately constant, i.e. there is no long-standing tendency for the population to either increase or decrease. In such cases $N_{T(x)} = N_{T(t)}$ and $N_{T(x)}/N_{T(t)}$ can be ignored. The reproductive value of an individual is then simply its total lifetime expectation of reproductive output (from its current age-class and from all subsequent age-classes).

However, when the population consistently increases or decreases, this must be taken into account. If the population increases, then $N_{T(x)}/N_{T(t)} < 1$, and future (i.e. 'residual') reproduction adds relatively little to RV_x (because the *proportionate* contribution is small). Conversely, if the population decreases, then $N_{T(x)}/N_{T(t)} > 1$, and future reproduction can be of greater value than present reproduction because of its greater proportionate contribution.

expected future fecundity; (c) this is done in a way which takes account of the proportionate contribution of an individual to future generations; and (d) reproductive value is the currency in which the worth of a life-history in the hands of natural selection should be calculated. The life-history favoured by natural selection from amongst those available in the population will be the one which has the highest *total* reproductive value (i.e. for which the sum of contemporary and residual reproductive values is highest).

The way in which reproductive value changes with age in two contrasting populations is illustrated in Figure 14.3. It is low for young individuals when they have only a low probability of surviving to reproductive maturity, and low for old individuals because either fecundity or survivorship or both decline in old age. The detailed rise and fall, however, varies with the detailed age- or stage-specific birth or mortality schedules of the species concerned.

14.4 Life-history compromise

real life-histories are a compromise allocation of resources

It is not difficult to describe a hypothetical organism that has all the attributes that confer high reproductive value. It reproduces almost immediately after its own birth; it produces large clutches of large, protected offspring on which it lavishes parental care; it does this repeatedly and frequently throughout an infinitely long life; and it outcompetes its competitors, escapes predation and catches its prey with ease. But while easy to describe, this organism is difficult to imagine, simply because if it puts all of its resources into reproduction it cannot also put all of them into survivorship; and if it takes the greatest care of its offspring it cannot also continue to produce large numbers of them. In short, common sense alone suggests that a real organism's life-style and life-history must be a compromise allocation of the resources that are available to it.

14.4.1 Trade-offs

negative correlations between life-history traits

Evidence that life-histories are compromises comes from examples in which individuals allocate more than normal to one structure or activity and appear to have less available for others. 'Trade-offs' are benefits from one process that are bought at the expense of another. Female fruit-flies (*Drosophila subobscura*), for instance, if subjected to a period of simulated migratory flight, have their subsequent fecundity immediately depressed, and this never quite returns to the levels exhibited by non-migrant controls (Figure 14.4a; Inglesfield & Begon, 1983); Douglas fir trees (*Pseudotsuga menziesii*) grow less the greater the number of cones they produce (Figure 14.4b; Eis *et al.*, 1965); and predatory rotifers (*Asplancha brightwelli*) have a lowered probability of future survivorship the more fecund they are (Figure 14.4c; Snell & King, 1977). Each aspect of the life-history that might be expected to increase reproductive value is offset by an associated change that tends to reduce reproductive value.

resource input differences can obscure trade-offs and lead to positive correlations

Hence trade-offs tend to generate negative correlations: more migration associated with less reproduction; more reproduction associated with less growth; and so on. Yet there is quite often a *positive* correlation between two apparently alternative processes. In Figure 14.5, for instance, the grasshoppers that reach the adult stage first are also the largest (Wall, 1985)—but the explanation is quite simple. These grasshoppers comprise a competitive hierarchy, in which the best competitors develop rapidly *and* grow large because they consume large amounts of food. The worst competitors obtain little food and are therefore small and slow to mature. This does not, however, indicate a lack of compromise (individuals could presumably have been bigger still but rather slower developing, or even smaller but somewhat faster). It indicates simply that many trade-offs are obscured by the effects of the competitive hierarchy. Positive correlations between beneficial processes are therefore to be expected whenever different individuals (or different populations or different species) obtain substantially different amounts of resources.

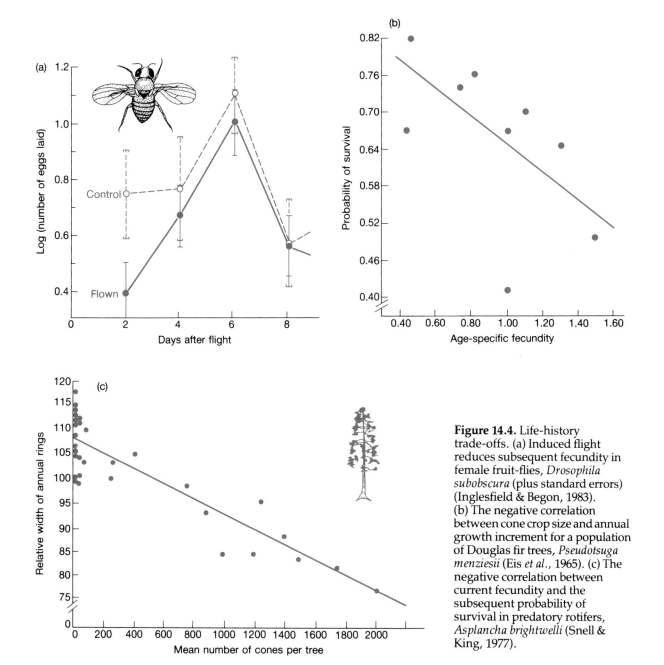

Figure 14.4. Life-history trade-offs. (a) Induced flight reduces subsequent fecundity in female fruit-flies, *Drosophila subobscura* (plus standard errors) (Inglesfield & Begon, 1983). (b) The negative correlation between cone crop size and annual growth increment for a population of Douglas fir trees, *Pseudotsuga menziesii* (Eis *et al.*, 1965). (c) The negative correlation between current fecundity and the subsequent probability of survival in predatory rotifers, *Asplancha brightwelli* (Snell & King, 1977).

14.4.2 The cost of reproduction

current reproduction often leads to reductions in survivorship, growth or future reproduction . .

The examples of the fir trees and rotifers in Figure 14.4 show an apparent '*cost of reproduction*'. Here, 'cost' is used in a particular way to indicate that an individual, by increasing its current allocation to reproduction, is likely to decrease its survivorship and/or its rate of growth, and therefore decrease its potential for reproduction in future (i.e. its residual reproductive value). This is also shown by the data in Figure 14.6 from a long-term study of red deer on the island of Rhum in Scotland (Clutton-Brock *et al.*, 1983). The figure

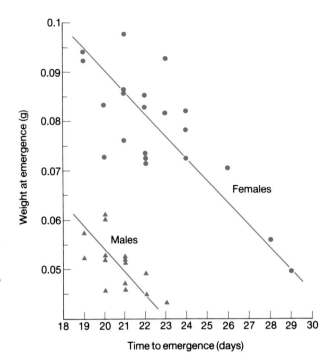

Figure 14.5. Differences in resource input obscuring possible trade-offs: the individuals that are largest at adult emergence are also first to emerge in male and female grasshoppers *Chorthippus brunneus,* reared in the laboratory (after Wall, 1985).

compares females that had produced calves in the previous season and reared them through the summer ('milk' hinds), with potentially reproductive females of a similar age that had failed to conceive in the previous season or had lost their calves ('yeld' hinds). The milk hinds had a higher risk of mortality (Figure 14.6a) and a lower expectation of future reproduction (Figure 14.6b). The body condition of milk hinds after reproduction was depressed in comparison with yeld hinds. This presumably was the physiological cost of bearing and rearing young. It was manifested as a reduction in their residual reproductive value.

The costs of reproduction can be shown even more easily with plants. Semelparous plants such as the foxglove (*Digitalis purpurea*) flower when they reach a critical size in the second or later years after germination. They normally die after they have set seed. However, if seed set fails—for example if the inflorescence is removed or damaged—regrowth occurs from the basal rosette, and the plant flowers again the next year, when it is often even bigger. Gardeners who are interested in prolonging the life of perennial flowering herbs should remove ripening seed heads, since these compete for

Table 14.1. Estimated energy allocated to growth and reproduction by reproductive and non-reproductive *Armadillidium vulgare* females. Values are in calories expended during one complete moult cycle. (After Lawlor, 1976.)

	Size class (♀) (in mg)			
	Reproductive		Non-reproductive	
	25–59	60–100	20–59	60–100
Growth	10.0	11.9	24.1	30.5
Reproduction	16.0	26.4	—	—
Total production	26.0	38.3	24.1	30.5

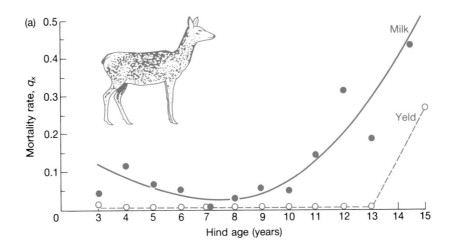

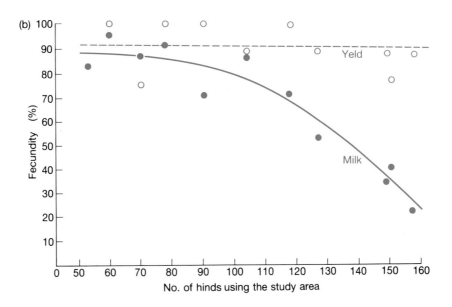

Figure 14.6. The cost of reproduction in female red deer, *Cervus elaphus*. (a) Milk hinds have a higher rate of mortality than yeld hinds at all ages. (b) Yeld hinds have a higher probability of calving in the next year than milk hinds, especially at high densities. (After Clutton-Brock *et al.*, 1983.)

resources that may be available for improved survivorship and even better flowering in the following year.

. . . because the resources which organisms have at their disposal are limited

The cost of reproduction is observable because individuals have only limited resources at their disposal. This is emphasized by the work of Lawlor (1976) on a terrestrial isopod, the common pill-bug *Armadillidium vulgare*. Table 14.1 shows that despite a minor increase in total energy allocation by reproductive females, the major effect of reproduction was to *divert* a considerable amount of energy away from growth.

Thus individuals that delay reproduction, or restrain their reproduction to a level less than the maximum, will typically grow faster, grow larger and have an increased quantity of resources available for maintenance and storage. These restrained reproducers may be said to have a greater 'accumulated somatic investment', but as a shorthand and a simplification we can think of them having a greater size. ('Size' therefore includes not only size *per se*, but also stored reserves, body condition and so on.) In this sense, an increase in contemporary reproductive output is likely to lead to a smaller size than would otherwise have been possible, which will often mean a decrease in residual reproductive value. The 'cost' of contemporary reproduction, therefore, is ultimately a decrease in residual reproductive value. And since natural selection favours the life-history with the highest available total reproductive value (p. 506), it must favour the *compromise* life-history for which the sum of contemporary reproductive output plus residual reproductive value is greatest at all stages of the life-history. This is the most fundamental statement that can be made about the evolution of life-histories.

the 'cost' of reproduction is a reduction in RRV often mediated by a reduction in 'size'

natural selection favours life-histories with the greatest sum of contemporary reproductive output plus RRV.

14.4.3 Compromises and optima

The allocation of limited resources is not the only compromise inherent in an organism's life-history. For example, animals that have large storage organs must 'pay' for this by a decreased ease and speed of movement, and hence presumably a decreased ability to avoid predators or catch prey (Pond, 1981). Similarly, plants that produce numerous large fruits must pay for this either by also producing somatic structures capable of supporting them, or by an increased instability in the face of violent weather. In other words, increased benefits often bring inevitable increased costs along with them; and natural selection can therefore be expected to favour not a maximum gross benefit but a compromise, optimum gross benefit which maximizes the *net* benefit (benefit minus cost). Thus an animal's storage organs can be expected to maximize not its total storage capability but the difference between the benefits of storage and the costs of supporting and carrying round the enlarged organs.

life-histories comprise optima which maximize net rather than gross benefits

A similar argument can be applied to the number of eggs produced in a clutch by a bird. This will be subject to the forces of compromise in resource

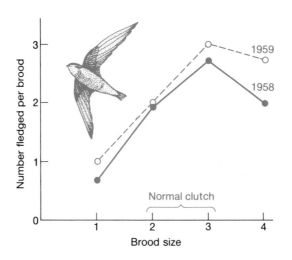

Figure 14.7. Brood size and number fledging in the swift, *Apus apus*. When the brood size was artificially increased to four, this led to a decrease in the number successfully fledged. (After Perrins, 1964.)

allocation (e.g. a larger clutch will leave the bird with fewer resources for predator avoidance, future clutches, etc.). But clutch size will also often be affected by the fact that the more nestlings there are, the more difficult it will be to feed them adequately. Perrins (1964), for instance, manipulated clutch size in the swift (*Apus apus*). He wanted to see whether clutch sizes in excess of the normal two or three would bring with them an increase in net benefit. They did not (Figure 14.7). When a fourth chick was added to the nest at hatching, the survival of *all* chicks in the nest was depressed and the number surviving to fledge (i.e. the number of descendants) was decreased. Swifts apparently cannot feed four nestlings adequately. Natural selection, therefore, appears to have favoured a compromise, optimum clutch size through which net benefits (not total young but total surviving young) are maximized.

14.4.4 Which resources are traded off?

The resources whose allocation matters to the reproductive value of the organism will vary from case to case. The physiological requirements for making different organs are not the same—the anther of a flower contains a different proportion of N, P, K and C from the pistil. In practice, however, for ease of comparison, resource costs are usually expressed in energy terms. Yet the justification for this is usually, a faint-hearted one: measuring energy is technically much easier than measuring most other resources.

There are, however, clear cases in which a particular resource can be identified as being crucial. The growth and reproduction of moose, for instance, may be limited by the amount of sodium they can collect during a short season of aquatic grazing (Botkin *et al.*, 1973). Their life-history is, therefore, likely to be influenced by the optimal allocation of sodium between parent and offspring and between various sodium-demanding activities (Harper, 1977). Fortunately though, the search for life-history patterns and their determinants can usually proceed by assuming that organisms partition 'limited resources', but without specifying what these resources are.

14.5 Habitats and their classification

habitats must be seen
from the organism's
point of view

The exact life-history favoured by natural selection depends on the habitat of the organism concerned. 'Habitat' therefore plays a crucial role in moulding life-histories. Yet each organism's habitat, like each organism's life-history, is unique. Thus if a pattern linking habitats and life-histories is to be established, habitats must be classified in terms that apply to them all. Moreover, these terms must be described and classified from the point of view of the organism concerned. For instance, whether or not an annual plant will in its lifetime encounter an unusually cold winter or hot summer is difficult to predict, since it lives through only one year. But the number of cool, moderate and hot summers experienced by a long-lived tree is much more predictable. Similarly, a small woodland might be a heterogeneous habitat for a beetle attacking aggregations of aphids, a homogeneous habitat for a rodent collecting seeds from the woodland floor, and part of a heterogeneous habitat for a large, predatory buzzard ranging over a wider area of ground. Habitat-type is therefore undoubtedly in the eye and the response of the beholder. So, on what basis can or should we try to classify habitat?

14.5.1 Habitats classified in time and space

One generalized classification of habitats has been suggested by Southwood (1977), who considered the ways in which habitats vary in time and space. Over time, from the organism's point of view, a habitat can be *constant* (in which case conditions remain favourable or unfavourable indefinitely); it can be predictably *seasonal* (in which case there is a regular alternation of favourable and unfavourable periods); it can be *unpredictable* (in which case favourable periods of variable duration are interspersed with equally variable, unfavourable periods); or it can be *ephemeral* (in which case there is a favourable period of predictably short duration followed by an unfavourable period of indefinite duration).

> habitats can be constant, seasonal, unpredictable or ephemeral, . . .

In space, a habitat can be *continuous* (the favourable area is larger than the organism can cover even using specialized dispersive mechanisms); it can be *patchy* (favourable and unfavourable areas are interspersed, but the organism is easily able to disperse from one favourable area to another); or it can be *isolated* (a restricted favourable area too far from other favourable areas for an organism to disperse between them, except rarely and by chance).

> . . .continuous, patchy or isolated

These classifications of habitats in time and space can be combined to give twelve habitat types of which ten could support life. The two which, for obvious reasons, are very unlikely to do so, are the continuous–ephemeral and the isolated–ephemeral habitats.

14.5.2 Habitats classified by their demographic effects

An alternative or, rather, complementary way of classifying habitats is to focus on the effects of the 'size' of the organism (used as a shorthand as on p. 512) on survivorship and future fecundity, i.e. on residual reproductive value (RRV) (Begon, 1985).

> habitats can be classified by the effects of 'size' on RRV . . .

A number of different habitat types can be recognized.

(1) *Size-beneficial* habitats, in which, for established individuals, residual reproductive value increases rapidly with individual size (Figure 14.8a). Here there is a significant cost of reproduction, because present reproduction leads to a smaller-than-possible size which in turn leads to a reduced residual reproductive value.

(2) *Size-neutral* or *size-detrimental* habitats, in which, by contrast, the residual reproductive value of established individual is little affected by or actually decreases with size (Figure 14.8a). Here, therefore, there is a negligible cost of reproduction.

(3) *Offspring-size-beneficial* habitats, in which, *for offspring*, reproductive value increases rapidly with offspring size (Figure 14.8b). (For pre-reproductive individuals, reproductive value and residual reproductive value are the same.)

(4) *Offspring-size-neutral* or *offspring-size-detrimental* habitats (Figure 14.8b).

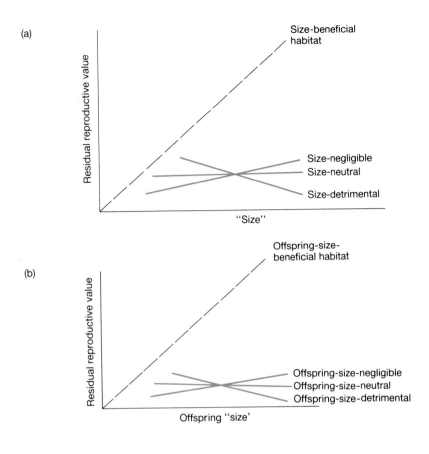

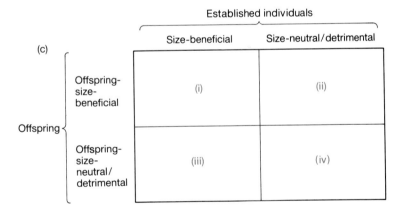

Figure 14.8. A demographic classification of habitats. (a) Habitats of established individuals (as opposed to recently produced offspring) can be *either* relatively size-beneficial (residual reproductive value rises sharply with 'size' *or* relatively size-neutral/size-negligible/size-detrimental. (b) Habitats of recently produced offspring can be either relatively offspring-size-beneficial or relatively offspring-size-neutral/offspring-size-negligible/offspring-size-detrimental. (c) By combining these two contrasting pairs, habitat of an organism over its whole life can be of four basic types, arbitrarily referred to as (i)–(iv) in the figure.

515   LIFE-HISTORY VARIATION

Thus, any one population experiences a habitat which is either size-beneficial of size-neutral (or detrimental), but which is *also* offspring-size-beneficial or offspring-size-neutral (or deterimental). Together, therefore, these habitats can be combined into four types (Figure 14.8c). Perhaps the most important point about this classification of habitats, however, is that, like so many other aspects of the study of life-histories, it is comparative. A habitat can only be described as size-beneficial *relative* to some other habitat which is, comparatively, size-neutral. The purpose of the classification is to contrast habitats with one another, rather than to describe them in absolute terms. But this will in no way diminish the importance of the classification in our search for life-history patterns.

It is also of crucial importance to realise that a habitat can be of a particular type for a variety of different reasons. Habitats are probably size-beneficial for two main reasons. In the first place, large size will be advantageous when there is intense competition amongst established individuals with only the best, 'largest' competitors surviving and reproducing. Red deer stags, where only the best competitors can hold a harem of females, are a good example of this. On the other hand, large size will also be advantageous wherever diminutive adults are particularly susceptible to an important source of predatory or abiotically induced mortality. Large mussels, for instance, may outgrow predation by crabs and eider ducks.

habitats of different
types can arise for a
variety of reasons

Habitats are either size-neutral or size-detrimental for probably three main reasons. First, much mortality may be indiscriminate and unavoidable, irrespective of size. For example, when temporary ponds dry out, most individuals die irrespective of their previous investment in soma. Secondly, conditions may be so benign and competition-free for established individuals that all of them have a high probability of surviving and a large reproductive output, irrespective of their previous investment in soma (i.e. again 'size does not matter'). This is true, at least temporarily, for the first colonists to arrive in a newly arisen habitat. And lastly, a habitat may be size-detrimental simply because there are important sources of mortality to which the largest individuals are especially prone. For instance, in the Amazon, avian predators preferentially prey upon the largest individuals of certain cyprinodont fish species.

Analogous comments apply to offspring-size-beneficial, offspring-size-neutral and offspring-size-detrimental habitats.

14.6 Semelparity or iteroparity; precocity or delay

Armed with this demographic classification of habitats, we can begin to search for pattern by asking what the difference might be between habitats favouring semelparity and those favouring iteroparity, and by asking where relatively delayed or relatively precocious reproduction might be favoured. Consider the four hypothetical organisms in Figure 14.9. They live in a habitat which is size-neutral. There is 100% winter survivorship for all (irrespective of size), and there is no increase in subsequent fecundity resulting from the increased size that arises out of reproductive restraint. Of the four organisms, the first is precocious and semelparous: it reproduces after one year, puts

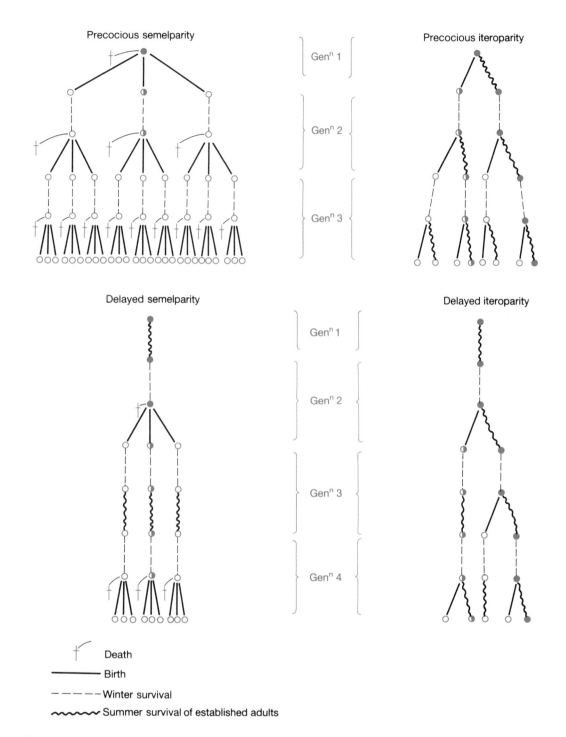

Figure 14.9. Four types of organism in a size-neutral habitat. There is 100% winter survivorship for all; 100% summer survivorship of all established iteroparous adults; semelparous adults produce three offspring and die; iteroparous adults produce one offspring and survive. The symbols ● and ◑ refer to individuals whose passage through one or more generations is followed. In this type of habitat, precocity and semelparity are favoured (i.e. they lead to the production of most descendants) because all individuals pay no cost for reproduction.

maximal investment into its three offspring, but dies itself. The second is precocious but iteroparous; it reproduces after one year and every year thereafter, but it invests in its own further growth and produces only one offspring each year. The third organism is semelparous but delays reproduction until its second year; it then produces three offspring but dies. Finally, the fourth organism is iteroparous and delays its initial reproduction until its second year; it then produces one offspring each year forever.

precocity and semelparity are favoured in size-neutral habitats

From Figure 14.9 it is clear that in size-neutral habitats, precocity is far more successful than delay, and semelparity is far more successful than iteroparity. Precocious individuals have a shorter generation-length, and their rate of population increase is therefore greater. Semelparous individuals, because they make a large investment in reproduction, have a greater basic reproductive rate (Chapter 4). Precocity and semelparity are only favoured, however, because the delayed and iteroparous organisms experience no increase in their residual reproductive value as a consequence of their reproductive restraint (i.e. the precocious and semelparous individuals pay a negligible cost for their reproduction.

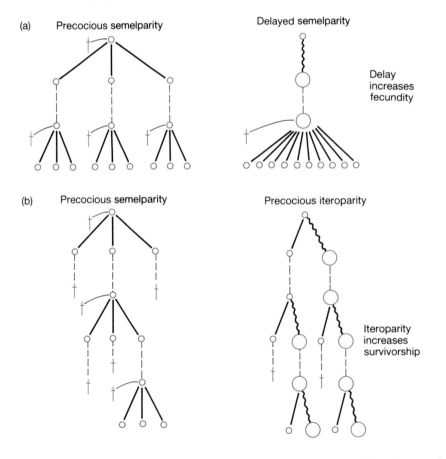

Figure 14.10. Similar types of organism to those in Figure 14.9, but in a size-beneficial habitat. (a) Reproductive delay leads to an increase in size which markedly increases fecundity. Delay rather than precocity is favoured. (b) The reproductive restraint of iteroparity leads to an increase in size which markedly increases winter survivorship. Iteroparity rather than semelparity is favoured.

By contrast, Figure 14.10 portrays a size-beneficial habitat. In Figure 14.10a, for instance, precocious semelparous and delayed semelparous individuals are compared. But this time the delayed individuals, *because of their increased size,* produce ten (not three) offspring when they reproduce. This is one *more* than the number produced in two generations by their precocious counterparts ($3 \times 3 = 9$), and delay would therefore be favoured by natural selection. In a similar vein, in Figure 14.10b, precocious semelparous and precocious iteroparous organisms are compared. But this time winter conditions are harsh, and while the iteroparous parents have a survivorship of 100% (because of their large size), the small, semelparous individuals have a survivorship of only 33%. As the figure shows, the iteroparous individuals leave more descendants to subsequent generations and would be favoured by natural selection.

Thus, in size-beneficial habitats, delay and iteroparity should lead to increases in size which in turn lead to increases in RRV *greater* than the accompanying decreases in contemporary output. Delay and iteroparity should therefore confer a higher total reproductive value than precocity and semelparity, and they should be favoured by natural selection.

By contrast, in size-neutral and size-detrimental habitats delay and iteroparity again lead to increases in size, but then confer negligible increases, or even decreases, in RRV. Precocity and semelparity should therefore be favoured. However, precocity and semelparity are themselves facets of a life-history over which an organism must compromise. A semelparous organism develops all of its offspring simultaneously and releases none until it releases them all. If it produced and released them one at a time, it could produce the first one earlier. In other words, semelparous organisms must 'pay' for their semelparity with a certain amount of reproductive delay. Thus, size-neutral habitats will tend to favour *either* semelparity *or* highly precocious iteroparity.

Arable land is a good example of a habitat which is size-neutral because adult mortality is indiscriminate: it is ploughed each year and established plants are killed irrespective of their size. The characteristic weed flora is made up of precocious and semelparous cornfield annuals (together with some iteroparous perennials for which ploughing is not lethal).

14.7 Reproductive allocation and the cost of reproduction

We have already noted on p. 514 that reproductive allocation carries a significant reproductive cost in size-beneficial habitats but not in size-neutral habitats. Simply by analogy with the previous section, therefore, we can conclude that the reproductive restraint of a low reproductive allocation should be favoured in size-neutral or size-detrimental habitats.

This can be seen, for example, in a study of different 'biotypes' in three populations of the dandelion *Taraxacum officinale* (Gadgil & Solbrig, 1972; Solbrig & Simpson, 1974, 1977). The dandelion populations were composed of a number of distinct clones which could be identified by electrophoresis; these clones were found to belong to one or other of four biotypes (A–D). The habitats of the populations varied from a footpath (the habitat in which adult

Margin notes:

delay and iteroparity are favoured in size-beneficial habitats

precocity and semelparity must be compromised

reproductive allocation in size-beneficial and size-neutral habitats

. . . illustrated by dandelions

mortality was most indiscriminate) to an old, stable pasture (the habitat with most adult competition which was therefore size-beneficial); the third site was intermediate between the other two. In line with predictions, the biotype that predominated in the footpath site (A) made the greatest reproductive allocation (whichever site it was obtained from), while the biotype that predominated in the old pasture (D) made the lowest reproductive allocation (Figure 14.11). Biotypes B and C were appropriately intermediate with respect to their site occupancies and their reproductive allocations.

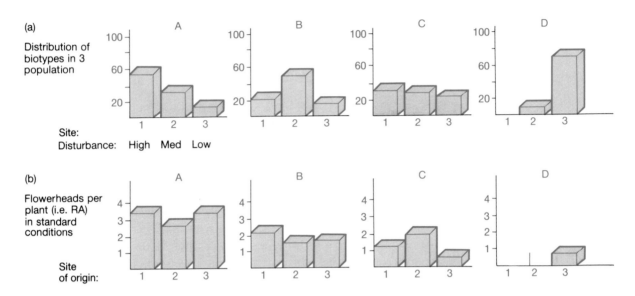

Figure 14.11. (a) The distribution of four biotypes (A–D) of the dandelion *Taraxacum officinale* amongst three populations subject to high, medium and low levels of disturbance (i.e. habitats ranging from relatively size-neutral to relatively size-beneficial). (b) The reproductive allocations (RA's) of the different biotypes from the different sites of origin, showing that the biotype A, which predominates in the relatively size-neutral habitat, has a relatively large RA, and so on. (After Solbrig & Simpson, 1974.)

The cost of reproduction may also be low (or even zero) when organisms experience a superabundance of resources. Such organisms clearly occupy a relatively size-neutral habitat (insofar as the habitat is indiscriminately benign), and they might be expected to have a relatively large reproductive allocation. This may explain why tapeworms and other endoparasites, lying bathed in nutrients, apparently have exceedingly high reproductive allocations but low reproductive costs (Calow, 1981).

However, the arguments about reproductive allocation and the cost of reproduction have until now been based entirely on resource allocation, and have ignored the fact that reproduction sometimes involves 'risks' over and above the simple diversion of resources. Reproducing individuals, for instance, may be more vulnerable to predation or less efficient at food capture because of the increased weight and bulk of their reproductive organs. In such a case the extra costs rise as reproductive allocation rises, and high allocations will have a much worse effect on residual reproductive value than they would otherwise have. We might expect, therefore, that natural selection should favour a relatively low reproductive allocation. In support of this,

the life-histories of tapeworms

'cost' can exceed 'allocation'

the life-histories of lizards

lizards that chase their prey and flee from their predators, and thus pay an extra cost for reproduction in terms of decreased speed and mobility, have been found to have a relatively low reproductive allocation, whereas those that sit and wait for their prey, and use crypsis or armour to avoid their predators, have a high reproductive allocation (Vit & Congdon, 1978).

Alternatively, however, reproduction may demand a dangerous and effortful pattern of behaviour (as when salmon make their upstream migration from the sea to their spawning grounds). In such a case the risks and extra costs are associated with the 'act' of reproduction and are largely independent of the magnitude of the reproductive allocation. Here it is presumably advantageous to have a large reproductive allocation or none at all (it must be disadvantageous to pay a large cost for a small allocation). This would explain why most salmon display suicidal semelparous reproduction.

the suicidal life-histories of salmon

Thus, as judged by the tapeworm, lizard and salmon examples, the relationship between reproductive allocation and the cost of reproduction has a valuable role to play in understanding life-history patterns.

14.8 More, smaller or fewer, larger offspring

'More, smaller' and 'fewer, larger' offspring can be thought of as alternative ways of partitioning the same total reproductive investment, though of course there are many ways of partitioning it: 'more, smaller offspring', 'slightly fewer, slightly larger offspring', and so on. The most favoured alternative is the one in which the offspring, taken together, have the greatest *summed* reproductive value. This may come from a few offspring each with a high reproductive value, or from many offspring each with a somewhat lower their reproductive value.

the combination of offspring size and number is favoured which has the greatest summed reproductive value . . .

Clearly, the production of larger offspring (and thus fewer of them) is favoured in offspring-size-beneficial habitats, while the production of more, smaller offspring is favoured in offspring-size-neutral or offspring-size-detrimental habitats. (In addition, in offspring-size-beneficial habitats, where a large investment in individual offspring brings a high return in terms of offspring reproductive value, a large investment in the parental care of offspring is liable to be favoured.)

In support of this, within the Californian flora there is a positive correlation between seed weight and the dryness of the habitat, which is apparent for herb species within the same genus, for whole herb communities, and for Californian trees (Baker, 1972). The likely explanation is that drier habitats are habitats in which small seeds are particularly prone to mortality: a larger seed enables a seedling to produce a more extensive root system than a small seed, and so allows it to obtain water more rapidly and efficiently in a dry environment, reducing its probability of death.

. . . illustrated by the Californian flora

14.9 *r* and *K* selection

Some of the predictions of the previous sections can be brought together in a concept which has been particularly influential in the search for life-history patterns. This is the concept of *r* and *K* selection, originally propounded by MacArthur and Wilson (1967) and elaborated by Pianka (1970). The letters refer to parameters of the logistic equation (section 6.13). They are used to indicate that r-selected individuals have been favoured for their ability to reproduce rapidly (i.e. have a high r-value), while K-selected individuals have been favoured for their ability to make a large proportional contribution to a population which remains at its carrying capacity (K). The concept is therefore based on there being two contrasting types of habitat: r-selecting and K-selecting. Like all generalizations, this dichotomy is an over-simplification. Yet there is enough truth in it to make it worthy of careful examination.

K-selected individuals in K-selecting environments

A K-selected population is thought of as living in a habitat which is either *constant* or *predictably seasonal* in time (p. 514). It therefore experiences very little by way of random environmental fluctuations (Figure 14.12a). As a consequence, a crowded population of fairly constant size is established. There is intense competition amongst the adults, and the results of this competition largely determine the adults' rates of survivorship and fecundity. The young also have to compete for survival in this crowded environment, and there are few opportunities for the young to become established as breeding adults themselves. In short, the population lives in a habitat which, because of intense competition, in both size-beneficial and offspring-size beneficial.

The predicted characteristics of these K-selected individuals are therefore larger size, delayed reproduction, iteroparity (i.e. more extended reproduction), a lower reproductive allocation, and larger (and thus fewer) offspring with more parental care (Figure 14.12a). The individuals will generally invest in increased survivorship (as opposed to reproduction); but in practice (because of the intense competition) many of them will have very short lives.

By contrast, an r-selected population is thought of as living in a habitat which is either *unpredictable* in time or *ephemeral* (p. 514). Intermittently, the population experiences benign periods of rapid population growth, free from competition (either when the environment fluctuates into a favourable period, or when an ephemeral site has been newly colonized). However, these benign periods are interspersed with malevolent periods of unavoidable mortality (either in an unpredictable, unfavourable phase, or when an ephemeral site has been fully exploited or disappears). The mortality rates of both adults and juveniles are therefore highly variable and unpredictable, and they are frequently independent of population density and of the size and condition of the individual concerned (Figure 14.12b). In short, the habitat is both size-neutral and offspring-size-neutral.

r-selecting individuals in *r*-selecting environments

The predicted characteristics of r-selected individuals are, not surprisingly, smaller size, earlier maturity, possibly semelparity, a larger reproductive allocation and smaller (and thus more) offspring (Figure 14.12b). The individuals will invest little in survivorship, but their actual survivorship will vary considerably depending on the (unpredictable) environment in which they find themselves.

The 'r/K concept', then, envisages two contrasting types of individual (or population or species), and predicts the association of r-type individuals with r-selecting environments and K-type individuals with K-selecting environments. The concept originally emerged (MacArthur & Wilson, 1967) out of the contrast between species that were good at rapidly colonizing relatively 'empty' islands (r species), and species that were good at maintaining them-

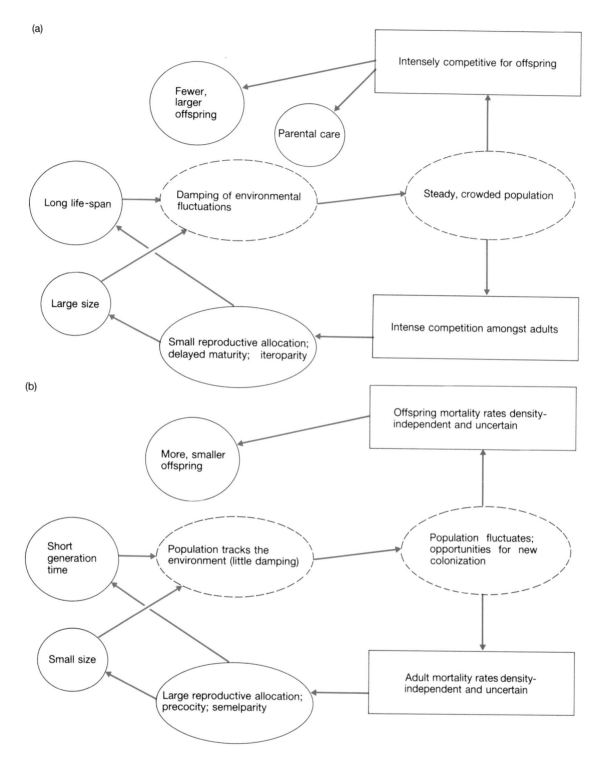

Figure 14.12. The chains of causality presumed to give rise to (a) *K*-selected individuals in *K*-selecting environments, and (b) *r*-selected individuals in *r*-selecting environments. Green = life-history traits, grey = attributes of the population, black = mortality factors acting on individuals. The type of population dynamics (middle-right) leads to a particular type and pattern of mortality, which selects for particular life-history traits. But the long life-span and large size of *K*-selected individuals make the environment 'seem', to them, to be even less variable; while the small size and short generation time of *r*-selected individuals lead to their responding rapidly to environmental change. The chains of causality have moved full circle; the original pressures are reinforced, and contrasting *r* or *K* strategies are selected for. (Modified from Horn, 1978.)

selves on islands once many colonizers had reached there (*K* species). Subsequently, the concept was applied much more generally. It is easy to see, though, that it is really only a special case of the general demographic classification of habitats (Figure 14.8c). There are two important points to note about this. First, adult and offspring habitats need not be linked in this way. Secondly, the life-history characteristics associated with the *r*/*K* scheme can arise for reasons beyond the scope of the *r*/*K* scheme itself (for instance: predation of diminutive adults as opposed to intense competition amongst adults; mortality of large offspring as opposed to a fluctuation between benign periods and indiscriminate mortality; and so on). These points are taken up later.

*the *r*/*K* scheme is merely a special case of a more general classification*

14.10 Evidence for the *r*/*K* concept

14.10.1 Broad comparisons across taxa

*certain broad comparisons conform to the *r*/*K* scheme*

The *r*/*K* concept can certainly be useful in describing some of the general differences amongst taxa. For instance, higher vertebrates with their relatively large size, long life, low reproductive output and high degree of homeostatic control can be thought of as comparatively *K*-selected; while insects with their relatively small size, short life, high reproductive output and so on, may be thought of as comparatively *r*-selected. Similarly, amongst plants it is possible to draw up a number of very broad and general relationships (Figure 14.13a–c), and to find that for trees in relatively *K*-selecting, woodland habitats, there is long life, delayed maturity, large seed size, low reproductive allocation, large individual size and a very high frequency of iteroparity; while in more *r*-selecting habitats, the plants tend to conform to the general syndrome of *r* characteristics. There is, however, one important respect in which plants and other modular organisms stand apart. Modular growth gives the genet the potential to increase in size exponentially by increasing its number of modules (branching corals, trees or clonal herbs). Delayed reproduction does not then necessarily delay population growth. Indeed, the number of descendant progeny produced by a zygote may well be greater when there is exponential growth followed by a burst of reproduction than when there are repeated short generations. This may account for the rather high frequency of species with clonal growth (and potentially infinite life) in *r*-selecting habitats.

14.10.2 Comparison between closely related taxa

*cattails conform to the *r*/*K* scheme*

There are also many cases in which populations of a species or of closely related species have been compared, and the correspondence with the *r*/*K* scheme has been good. This is true, for instance, of McNaughton's (1975) study of *Typha* (cattail or reed mace) populations (Table 14.2). NcNaughton was particularly interested in a southerly species, *Typha domingensis*, and a northerly species, *T. angustifolia*. He took these from sites in Texas and North Dakota respectively, and grew them side-by-side under the same conditions. He also quantified certain aspects of the habitats with long and short growing

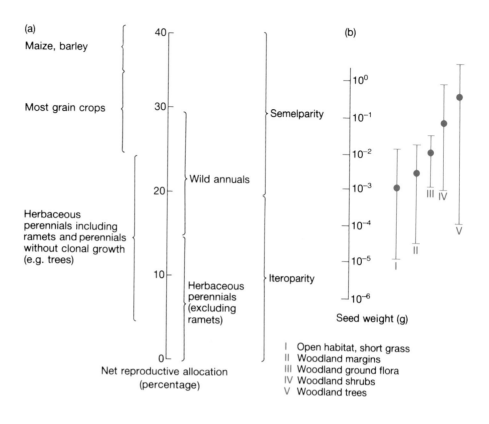

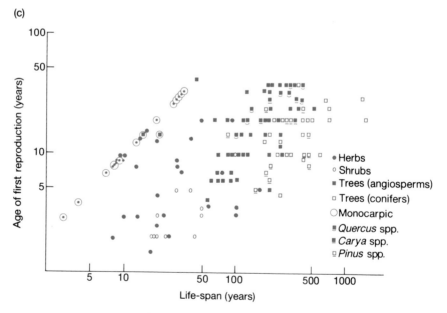

Figure 14.13. Broadly speaking, plants show some conformity with the *r/K* scheme. For example, trees in relatively *K*-selecting woodland habitats: (a) have a relatively high probability of being iteroparous and a relatively small reproductive allocation, (b) have relatively large seeds, and (c) are relatively long-lived with relatively delayed reproduction. (After Harper, 1977; data from Ogden, 1968; Salisbury, 1942; Harper & White, 1974.)

Table 14.2. Life-history traits of two *Typha* (cattail) species, along with properties of the habitats in which they grow. '$s^2/\bar{x}$' refers to the variance-to-mean ratio, a measure of variability. The cattails conform to the *r/K* scheme. (After McNaughton, 1975.)

		Growing season	
Ecosystem property	measured by	Short	Long
Climatic variability	$s^2/\bar{x}$ frost-free days per year	3.05	1.56
Competition	biomass above ground (g m^{-2})	404	1336
Annual recolonization	winter rhizome mortality (%)	74	5
Annual density variation	$s^2/\bar{x}$ shoot numbers m^{-2}	2.75	1.51
Plant traits		*T. angustifolia*	*T. domingensis*
1. Days before flowering		44	70
2. Mean foliage height (cm)		162	186
3. Mean genet weight (g)		12.64	14.34
4. Mean number of fruits per genet		41	8
5. Mean weights of fruits (g)		11.8	21.4
6. Mean total weight of fruits (g)		483	171

seasons in which these species are found, and, as is clear from Table 14.2, the former were relatively *K*-selecting and the latter relatively *r*-selecting. It is equally clear from the data that the species inhabitating these sites conform to the *r/K* scheme. *T. angustifolia* (which naturally has a short growing season) matures earlier (trait 1), is smaller (traits 2 and 3), makes a larger reproductive allocation (traits 3 and 6) and produces more and smaller offspring (traits 4 and 5) than does *T. domingensis* (long growing season).

In a similar vein, Law *et al.* (1977), working in north-west England and North Wales, compared two types of populations within one species: 'annual' meadow grass (*Poa annua*). One type ('opportunist') contained individuals at

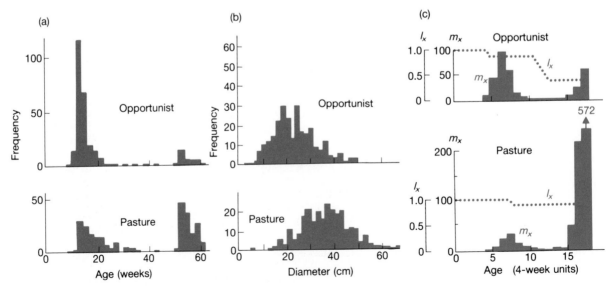

Figure 14.14. *Poa annua* populations conform to the *r/K* scheme. (a) The distribution of pre-reproductive periods in plants from opportunistic and pasture environments. (b) The distribution of sizes in plants from opportunistic and pasture environments. (c) Rates of reproduction (m_x) and survival (l_x) in samples from an opportunistic and a pasture environment. (After Law *et al.*, 1977.)

low density with large areas of bare ground between them, and had been subject to repeated disturbance for some time. The other type contained individuals at high density, and had generally existed as permanent 'pasture' for some time. It was assumed that opportunist habitats were characterized predominantly by density-independent limitation, while pasture habitats were characterized predominantly by density-dependent regulation. They can therefore be thought of as *r*-selecting and *K*-selecting respectively. Samples from each population were grown from seed under controlled, uncrowded conditions, and the life-histories of individuals were monitored. Some of the results are illustrated in Figure 14.14.

annual meadow grass conforms to the *r/K* scheme

The *P. annua* populations conform to the *r/K* scheme. Those from the *r*-selecting environment mature earlier (Figure 14.4a), are smaller (Figure 14.14b) and make a greater reproductive allocation earlier, leading to a shorter life-span (Figure 14.14c). The reverse is true of those from the *K*-selecting environment.

14.10.3 Assessment of the *r/K* concept

There are, then, examples that fit the *r/K* scheme. Yet Stearns (1977), in an extensive review of the available data, found that of 35 thorough studies, 18 conformed to the scheme while 17 did not. This can be seen as a damning criticism of the *r/K* concept, since it certainly shows that the explanatory powers of the scheme are limited. On the other hand, it is equally possible to regard it as very satisfactory that a relatively simple concept can help make sense of a large proportion of the multiplicity of life-histories. Whatever one's point of view, however, much variation certainly remains to be explained.

the *r/K* explains much but leaves as much unexplained

There are a number of very good reasons why particular life-histories might not fit the *r/K* scheme. Indeed, two of these reasons have been discussed already: reproductive cost may be far greater than its corresponding reproductive allocation (section 14.7); and demographic forces beyond the *r/K* scheme may be important (sections 14.5.2 and 14.9). The final sections of the chapter will discuss this and a number of other important considerations.

14.11 'Alternatives' related to the *r/K* concept

14.11.1 'Bet-hedging'

Notwithstanding the usual description of a *K*-selecting habitat, it is indisputable that *all* environments fluctuate and all populations suffer some random variations in mortality rate. Schaffer (1974) focused on this, and distinguished between habitats in which the major effect is on adult mortality, and those in which it is the juveniles that are affected most (and adults might therefore be expected to 'hedge their bets' (Stearns, 1976) and not release all of their offspring into the same environment). Yet unpredictable, unavoidable mortality of established adults (one arm of this dichotomy) is probably the most distinctive features of *r*-selecting environments; while random variations in mortality rates that particularly affect juveniles are likely to be a characteristic of *K*-selecting environments (where the unestablished, weakly competing juveniles are bound to be more affected by environmental fluctuations than the established, strongly competing, low-mortality-rate adults). Schaffer's classification, therefore, is complementary to the r/K concept (Horn, 1978). Encouragingly, Schaffer's mathematical calculations predict essentially r-type traits where adult mortality rates fluctuate most, and K-type traits where juvenile rates fluctuate most.

14.11.2 Grime's classification

Grime (1974, 1979) has developed a classification of habitats and plant life-histories which is outlined in Figure 14.15. Habitats are seen as varying in their level of disturbance (brought about by herbivores, pathogens, trampling, environmental disasters, etc.), and in the extent to which they experience shortages of light, water, minerals, etc. (though a habitat's level of disturbance or shortage will depend on which species is experiencing them).

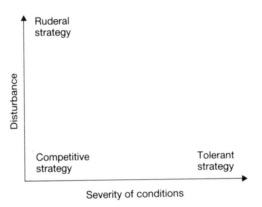

Figure 14.15. Grime's classification of habitats and plant life-histories. Habitats vary in the severity of conditions and the amount of disturbance. With low severity and little disturbance a competitive strategy is appropriate. With severe conditions but little disturbance a tolerant strategy is appropriate. And in disturbed habitats where conditions are benign a ruderal strategy is appropriate.

Grime suggests that when disturbance is rare and resources are abundant, a crowded population is established and a *competitive strategy* is appropriate. A *tolerant strategy* is appropriate when resources are scarce or conditions severe, but disturbance is uncommon. A *ruderal strategy* is appropriate when disturbance levels are high but conditions are benign and resources are abundant. Grime also suggests that there are intermediate strategies like 'competitive–ruderal', and so on.

Ruderal organisms and disturbed habitats correspond closely to r-type organisms and habitats; and competitive organisms and resource-rich, low-disturbance habitats would seem to correspond to K-type organisms and habitats. Grime's classification therefore serves to highlight the fact that there are many organisms that live in habitats that are predictable, but predictably hostile (e.g. arid deserts, polar and infertile habitats). The most important difference between Grime's scheme and the *r/K* concept, however, is in their aims. Grime's is a classification in which all plants can be assigned a category which describes them in absolute terms, whereas the *r/K* scheme is essentially comparative. Grime (1979) has had some success in fulfilling his aim. But the simpler, comparative *r/K* scheme is the one more commonly used.

14.12 Demographic forces beyond the *r/K* scheme

Perhaps the most obvious inadequacy of the *r/K* scheme is the fact that it is merely a special case of a more general demographic theory (p. 524). One example where this inadequacy is apparent concerns two species of the shore snails known as winkles: *Littorina rudis* (Hart & Begon, 1982) and *Littorina nigrolineata* (Naylor & Begon, 1982). At a single location on the exposed west coast of Holy Island, off Anglesey in Wales, both species occupy two rather different habitats. The first is a gently sloping rocky shore covered with medium-sized to large boulders. The second is a large, near-vertical, rocky stack indented with many narrow crevices within which the winkles are almost entirely confined. Females of both species were collected from both habitats (winkles have separate sexes), and the pertinent results are summarized in Table 14.3. Note that the study involves comparisons of populations of each species from the two habitats, not a comparison of the two species.

winkles reflect the demographic classification of habitats but do not conform to the r/K scheme

Table 14.3. A summary of the traits of two species of winkle, *Littorina rudis* and *L. nigrolineata,* from two contrasting habitats: 'crevice' and 'boulders'. Where appropriate, the relative 'r-ness' or 'K-ness' of the traits is indicated. Individuals combine supposedly *r* and *K* traits. (From Hart & Begon, 1982; Naylor & Begon, 1982.)

Crevice population	Boulders population	Species
1. Thinner shells	1. Thicker shells	*rudis* and *nigrolineata*
2. Average size smaller (*r*)	2. Average size larger (*K*)	*rudis* and *nigrolineata*
3. Maturity at smaller size (*r*)	3. Maturity relatively delayed to larger size (*K*)	*rudis* and *nigrolineata*
4. Larger RA (*r*)	4. Smaller RA (*K*)	*rudis* and *nigrolineata*
5. Fewer, larger offspring (*K*)	5. More, smaller offspring (*r*)	*rudis*

Table 14.3 shows that both species have different life-histories in the different habitats, and that they vary 'in parallel' with one another. It also shows that the life-histories of both *L. rudis* populations are compounded of what might be called *r*-type and *K*-type characteristics. These patterns can be understood, though, by looking at the ecology of the winkles and especially at the mortality factors that affect them.

In crevice habitats, the population size of the winkles is regulated by crevice availability. If more and larger crevices are artificially added to a habitat, this increases the size of the population and also the size of the largest individuals (Emson & Faller-Fritsch, 1976; Raffaelli & Hughes, 1978). Crevices provide protection against abiotic mortality factors (wave action, desiccation) and biotic mortality factors (bird predation), but they do so only for those winkles small enough to inhabit them. There is intense competition for limited space, but it is the largest individuals that suffer the most intense competition. The relationship between residual reproductive value and size, therefore, takes the form shown in Figure 14.16. Amongst the newly liberated juveniles, there is a large premium on large size because of the intense intraspecific competition, i.e. the habitat is offspring-size-beneficial. For the same reason, the habitat is also size-beneficial for the smaller established individuals. However, at larger sizes (where there are few suitable crevices) the habitat becomes more size-neutral and eventually size-detrimental.

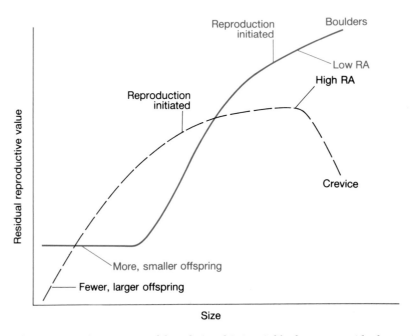

Figure 14.16. A summary of the relationship in winkles between residual reproductive value and size in the boulders and crevice habitats. Habitat types are indicated by the gradient of the relationship, as discussed in the text (reproduction is initiated in both habitats at the same gradient). The life-history traits appropriate to the habitat types are stated, and these conform to what is found in the winkles (Table 14.3).

In the 'boulders' population, the winkles are subject to two very important mortality factors against which thick shells and large size are undoubtedly a protection (Raffaelli & Hughes, 1987; Elner & Raffaelli, 1980). One of these is predation. But the other, which is generally density-independent and catastrophic, is crushing by the moving boulders. As a consequence, the relationship between residual reproductive value and size takes the form shown in Figure 14.16. Amongst the juveniles, the catastrophic mortality is unavoidable and unrelated to size: the habitat is offspring-size-neutral. As size increases, however, large size becomes increasingly a protection against mortality (especially if 'size' includes shell thickness): the habitat becomes increasingly size-beneficial. Finally, though, among the larger, more protected winkles, the habitat returns to being more size-neutral.

If the relationships in Figure 14.16 (reflecting the mortality schedules of the two populations) are compared with the life-history data in Table 14.3, it will be seen that they correspond completely. For the 'crevice' individuals, their fewer, larger offspring reflect the initial rise in the graph, while their smaller size, earlier maturity and larger reproductive allocation reflect the graph's subsequent plateau and fall. For the 'boulder' individuals, their more numerous, smaller offspring reflect the initial size-independence of the graph, while their larger size, delayed maturity and smaller reproductive allocation reflect the graph's subsequent rise.

This single example, therefore, illustrates and reinforces a point of considerable general importance. *Demographic forces can be powerful in their ability to explain life-history patterns, but these forces need not be limited to the r/K dichotomy.*

14.13 Short-term responses to the environment

a life-history is frequently an organism's immediate response to the present environment

Of course, a life-history is not a fixed property that an organism exhibits irrespective of the prevailing environmental conditions. An observed life-history is the result of long-term evolutionary forces, but also of the more immediate responses of an organism to the environment in which it is and has been living.

There were several examples in Chapter 6 which showed that the relative allocations to different plant parts, and to different reproductive parts, changed as population density increased; Figure 14.17 illustrates a comparable difference when the umbellifer alexanders (*Smyrnium olusatrum*) is grown under normal and low-nutrient conditions (Lovett-Doust, 1980). The concurrent effect of competitive ability (and thus food level) on the development rate and size of individual grasshoppers was shown in Figure 14.5; while Figure 14.18 illustrates the contrasting responses of two flatworm species to various levels of food supply (Woolhead, 1983). All these examples

Figure 14.17. The allocation of dry matter to various plant parts by alexanders, *Smyrnium olusatrum,* in control and low-nutrient treatments (after Lovett-Doust, 1980).

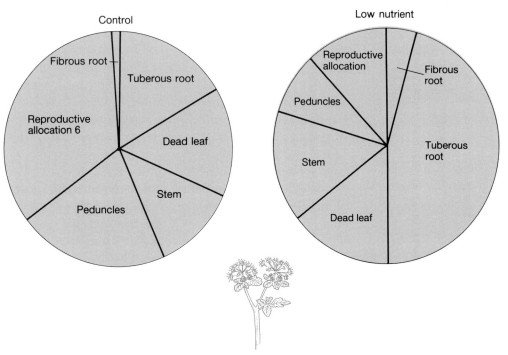

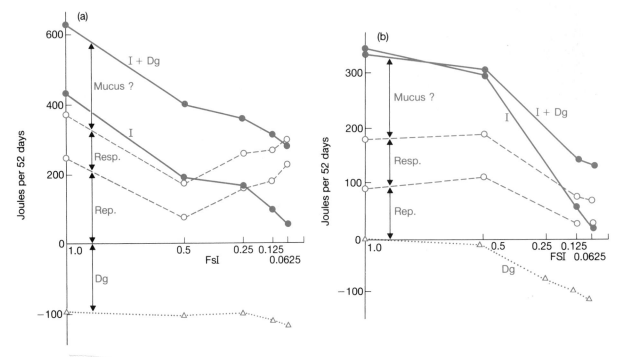

Figure 14.18. (a) Energy allocation by the flatworm *Bdellocephala punctata* to respiration (Resp), reproduction (Rep) and secretion/excretion (Mucus?) of food ingested (I) at different levels (indices) of food supply (FSI). Also included are levels of degrowth (Dg) and total available energy (I + Dg). Degrowth is the mobilization of somatic tissue, making energy available for other processes. (b) Comparable data for the flatworm *Dugesia lugubris*. (After Woolhead, 1983.) The life-histories are affected by the food supply, but the effect is different in the two species, as discussed in the text.

illustrate that life-histories frequently reflect an organism's immediate response to its past and present environments.

immediate responses may be selected, or they may simply be imposed by the environment

In some cases, these responses may themselves be seen as properties of the life-history that are favoured in environments subject to variability and change. Of the flatworms in Figure 14.18, for instance, *Bdellocephala punctata* makes a high reproductive allocation at low food levels, and has hatchlings that are well able to cope with nutritive stress. *Dugesia lugubris*, on the other hand, makes a low reproductive allocation at low food levels and has hatchlings that are far more susceptible to nutritive stress. There are many cases, however, in which the response seems imposed rather than selected: organisms grown at low nutrient levels are small, etc.

Whatever the nature of the responses though, it is nonetheless clear that all organisms are capable of exhibiting a range of life-histories. An understanding of these responses is therefore essential for a true understanding of life-history patterns. Regrettably, this is a relatively undeveloped area in the study of life-histories.

More specifically, an organism's life-history will almost certainly vary with the level of input of resources. And yet there are many studies that ignore input levels when life-histories are compared. Those studies are therefore in danger of attributing to differences in habitats, features that should be attributed to differences in level of input *within* habitats.

14.14 Interactions with physiological demands

In fact, the problems posed by an organism's environment will often be primarily physiological problems which only affect life-history traits indirectly. Stearns (1980), for example, examined two neighbouring populations of the mosquito fish (*Gambusia affinis*) in Texas, one of which lived in freshwater while the other lived in brackish water. He found a higher reproductive allocation in individuals from the brackish water population, but concluded that selection was not acting on reproductive allocation directly. It seemed instead that the individuals in the freshwater population had osmoregulatory problems, and they used extra energy to maintain their body fluids in an acceptable condition. They therefore had less 'disposable' energy, and it was this that led to their smaller reproductive allocation. McClure and Randolph (1980) have reached an analogous conclusion in considering the energy required for homeothermy by the eastern wood rat (*Neotoma floridana*) and the hispid cotton rat (*Sigmodon hispidus*). These examples serve to emphasize that a life-history is a response to the *whole* of an organism's environment.

mosquito fish life-histories reflect physiologial problems

14.15 Phylogenetic and allometric constraints

The life-histories that natural selection favours (and we observe) are not selected from an unlimited supply. The favoured life-history for a buttercup is selected from amongst those available to buttercups; and the favoured butterfly life-history is selected from amongst those available to butterflies. An ecologist might, quite reasonably, try to explain how it comes about that buttercup species differ in their life-histories—but it is no part of an ecologist's job to explain why buttercups are not butterflies.

organisms are, to some extent, prisoners of their evolutionary past

In other words, an organism's life-history is constrained by the developmental possibilities available to that organism, and it is therefore constrained by the phylogenetic position that the organism occupies. For example, in the entire order Procellariiformes (albatrosses, petrels, fulmars) the clutch size is one, and the birds are 'prepared' for this morphologically by having only a single brood patch with which they can incubate this one egg (Ashmole, 1971). A bird might produce a larger clutch, but this is bound to be a waste unless it exhibits concurrent changes in all the processes in development of the brood patch. Albatrosses are therefore prisoners of their evolutionary past, as are all organisms. Their life-histories can evolve to only a limited number of options, and the organisms are therefore confined to a limited range of habitats.

life-histories reflect both habitat and phylogeny

It follows from the existence of these constraints that caution must be exercised when life-histories are compared. The albatrosses, as a group, may be compared with other types of birds in an attempt to discern a link between the typical albatross life-history and the typical albatross habitat. And the life-histories and habitats of two albatross species might reasonably be compared. But if an albatross species is compared with a distantly related bird species, then care must be taken to distinguish between differences attributable to habitat (if any) and those attributable to phylogenetic constraints.

Disentangling these two is a central (but often unresolved) problem in the study of life-histories.

14.15.1 The effects of size

A related, but in many ways more subtle, constraint is illustrated in Figure 14.19a. This shows the relationship between two life-history components in a wide range of organisms from viruses to whales, the components being time to maturity and size (weight) (Blueweiss *et al.*, 1978). The first important point to note about this figure is that particular groups of organisms are confined to particular size ranges. For instance, unicellular organisms cannot exceed a certain size because of their reliance on simple diffusion for the transfer of oxygen from their cell surface to their internal organelles. Insects cannot exceed a certain size because of their reliance on unventilated tracheae for the transfer of gases to and from their interiors. Mammals, being endothermic,

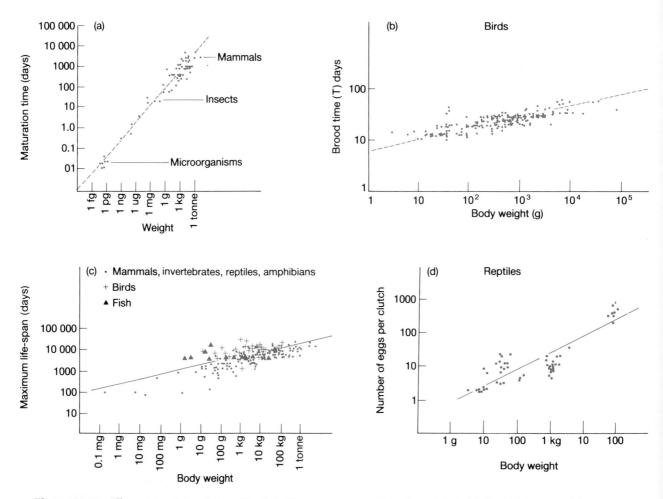

Figure 14.19. Allometric relationships, all collated by Blueweiss *et al.* (1978), and all plotted on log scales. (a) Maturation time as a function of body weight for a broad range of animals. (b) Brood time as a function of maternal body weight in birds. (c) Maximum life-span as a function of adult body weight for a broad range of animals. (d) The number of eggs per clutch as a function of maternal body weight in reptiles.

must exceed a certain size, because at smaller sizes the relatively large body surface would dissipate heat faster than the animal could produce it. And other groups are also confined to their own size range in other ways.

phylogeny affects size affects life-history

The second point to note is that time to maturity and size are strongly correlated. In fact, as Figures 14.19b–d show, size is strongly correlated with many life-history components. Since the sizes of organisms are confined within limits, these other life-history components will be confined too. The life-history of an organism can therefore once again be seen to be constrained by the phylogenetic position of that organism.

14.15.2 Allometric relationships

allometry defined

Closely connected with the relationships in Figure 14.19 are those in Figures 14.20 and 14.21 since they are all *allometric* relationships (Gould, 1966). An allometric relationship is one in which a physical or physiological property of an organism varies with organism size, such that there is a change in the physical or physiological property *relative* to the size of the organism. For example, in Figure 14.20a an increase in size (actually volume) amongst salamander species leads to a decrease in the *proportion* of that volume which is allocated to a clutch of young. Likewise, in Figure 14.19b an increase in weight amongst bird species is associated with a *decrease* in brood time *per unit body weight*.

Such allometric relationships can be *ontogenetic* (changes occurring as an organism develops) or *phylogenetic* (changes which are apparent when related taxa of different size are compared). It is allometric relationships of the latter type that are particularly important in the study of life-histories and are illustrated in Figures 14.19–14.21. It is apparent from these figures that allometric relationships exist at many (probably all) taxonomic levels; and in the context of life-history studies there are a number of crucial points that must be understood from them.

the properties of allometries and their consequences for life-histories

(1) Allometry, by definition, involves *a lack of geometric or physiological similarity* amongst organisms of different size. For instance, the ratios 'life-span:body weight' in Figure 14.19c, 'eggs per female:body weight' in Figure 14.19d and 'tree height:trunk diameter' in Figure 14.21, all *change* as size increases.

(2) The typical, straight-line relationship on a log–log plot means that the ratio changes at a constant rate as size increases.

(3) An allometric relationship at a higher taxonomic level can (and usually does) hide allometric relationships at lower taxonomic levels which have *different* slopes. This can be seen, for instance, for the salamanders in Figure 14.20b.

(4) Although allometric relationships are usually good descriptions of taxonomic assemblages as a whole, individual points (i.e. individuals or individual species or whatever) typically *deviate* to a greater or lesser extent from the relationship in question. All of the allometric relationships in Figures 14.19–14.21 show this to be so.

(5) Finally, a life-history component involved in an allometric relationship affects other life-history components (because of trade-offs and because

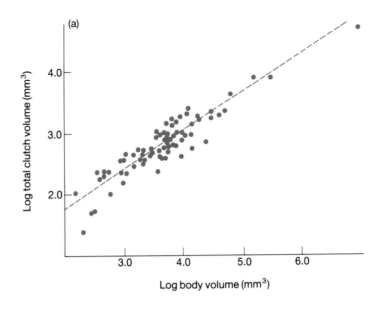

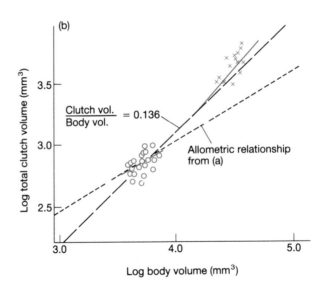

Figure 14.20. Allometric relationships between total clutch volume and body volume in female salamanders (after Kaplan & Salthe, 1979). (a) The overall relationship for 74 salamander species, using one mean value per species ($P < 0.01$). (b) The relationships within a population of *Ambystoma tigrinum*, crosses ($P < 0.01$), and within a population of *A. opacum*, open circles ($P < 0.05$). The allometric relationship from (a) is shown as a dotted line: *A. opacum* conforms closely to it; *A. tigrinum* does not. However, they both lie on an isometric line along which clutch volume is 13.6% of body volume (dashed line).

organisms function as integrated wholes). Allometries can therefore lead, *in themselves*, to life-history variations affecting all life-history components at all taxonomic levels. They are a particularly potent source of phylogenetic constraint.

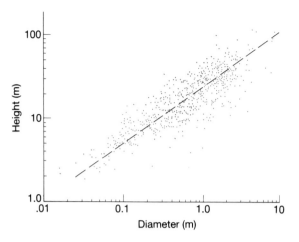

Figure 14.21. The allometric relationship between tree height and trunk diameter 1.525 metres from the ground, for 576 individual 'record' trees representing nearly every American species (after McMahon, 1973).

14.15.3 Why are there allometric relationships?

It is obviously desirable that allometric relationships should be understood. However, while this is a fascinating topic, it is more the province of physiologists than ecologists (for discussion see Gould, 1966; Schmidt-Nielsen, 1984; and, in a more ecological context, Peters, 1983).

without allometry, physiological efficiency would change with size

Briefly, allometric relationships are presumed to exist for the following reason. If similar organisms that differed in size retained a geometric similarity (i.e. if they were *isometric*), then all surface areas would increase as the *square* of linear size, while all volumes and weights increased as the cube. An increase in size would then lead to decreases in length-to-area ratios, decreases in length-to-volume ratios, and, most important, decreases in area-to-volume ratios. Almost every bodily function, however, depends for its efficiency on one of these ratios (or a ratio related to them). A change in size amongst isometric organisms would therefore lead to a change in efficiency.

For example, the transfer of heat, or water, or gases, or nutrients, either within an organism or between an organism and its environment, takes place across a surface, which has an area. The amount of heat produced or water required, however, depends on the volume of the organ or organism concerned. Hence, changes in area-to-volume ratios resulting from changes in size are bound to lead to changes in the efficiency of transfer per unit volume; and if efficiency is to be maintained, this must be done by allometric alterations. Similarly, the weight of any part of an organism depends on its volume, while the strength of a structure supporting that weight depends on its cross-sectional area. Increases in size will therefore lead to problems of strength unless there are allometric alterations to compensate.

In most examples, these simple area-to-volume relationships interact with one another and depend on the detailed morphology and physiology of the organisms concerned. Allometric slopes therefore vary from system to system and from taxon to taxon. From the ecological point of view, however, the important facts are that these allometric relationships exist, and that they commonly involve life-history characteristics.

The usual approach to the ecological study of life-histories is to compare the life-histories of two or more populations (or species or groups), and to seek to understand the differences between them by reference to their environments. It must be clear by now, however, that taxa can also differ because they lie at *different* points on the *same* allometric relationship, or because they are subject to different phylogenetic constraints generally. It is therefore important to disentangle 'ecological' differences from allometric and phylogenetic differences.

Figure 14.20a shows the allometric relationship between clutch volume and body volume for salamanders generally. Figure 14.20b shows the same relationships in outline; but superimposed upon it are the allometric relationships *within* populations for two salamander species, *Ambystoma tigrinum* and *A. opacum* (Kaplan & Salthe, 1979). If the species' *means* are compared without reference to the general salamander allometry, then the species are seen to have the same ratio of clutch volume to body volume (0.136). This seems to suggest that the species' life-histories 'do not differ', and that there is therefore 'nothing to explain'—but any such suggestion would be wrong. *A. opacum* conforms closely to the general salamander relationship. *A. tigrinum*, on the other hand, has a clutch volume which is almost twice as large as would be expected from that relationship. Within the allometric constraints of being a salamander, *A. tigrinum* is making a much greater reproductive allocation than *A. opacum*; and it would be reasonable for an ecologist to look at their respective habitats and seek to understand why this might be so.

In other words, it is reasonable to compare taxa from an 'ecological' point of view as long as the allometric relationship linking them at a higher taxonomic level is known (Clutton-Brock & Harvey, 1979). It will then be their respective deviations from the relationship that form the basis for the comparison. Salamander species have been compared against the background of the general salamander relationship; and individual salamanders from within the *A. trigrinum* population can be compared against the background of the population-specific allometric relationship. The problem comes, though, when allometric relationships are unknown (or ignored). Without the general salamander allometry in Figure 14.20b, the two species would have seemed similar when in fact they are different. Conversely, two other species might have seemed different when in fact they were simply conforming to the same allometric relationship. Comparisons oblivious to allometries are therefore perilous.

In practice, regrettably, ecologists *are* frequently oblivious to allometries. The more general phylogenetic constraints are usually taken into account, but the subtler constraints of allometry are not. Typically, life-histories are compared, and an attempt is made to explain the differences between them in terms of habitat differences. As previous sections have shown, such attempts have often been successful. But they have also often been unsuccessful, and unrecognized allometries undoubtedly go some way towards explaining this.

15 Abundance

15.1 Introduction—the interpretation of census data

A significant part of the science of ecology is concerned with trying to understand what determines the abundance of organisms. Why are some species rare and others common? Why does a species occur at low population densities in some places and at high densities in others? What factors cause fluctuations in a species' abundance? These are crucial questions. To provide complete answers for even a single species in a single location, ideally we would need a knowledge of physicochemical conditions, the level of resources available, the organism's life cycle and the influence of competitors, predators, parasites, etc., and an understanding of how all these things influence the population's rates of birth, death and migration. In previous chapters, each of these topics has been explored more or less in isolation, but in reality populations are likely to be significantly influenced by more than one process, and to some extent by all of them. The aim of the present chapter is to consider the determination of abundance from the multifactorial point of view.

abundance is affected by a range of factors acting in concert

The raw material for the study of these questions is often a census. In its crudest form, this consists of a list of presences and absences in defined sample areas. The most detailed censuses, on the other hand, involve counting individuals (and their parts, in the case of modular organisms), recognizing individuals of different age, sex, size and dominance, and even distinguishing genetic variants.

simple census information may hide vital details . . .

If a census (over an area or over time) simply records the numbers of individuals present, we can attempt to 'explain' variations by correlating them with factors outside the population—weather, soil conditions, numbers of predators, and so on—but vital information, hidden within the population, may be lost. As an example, picture three human populations, shown by census to contain identical numbers of individuals. One of these is an old people's residential area, the second is a population of young children, and the third is a population of mixed age and sex. No amount of attempted correlation with factors outside the population would reveal that the first was doomed to extinction (unless maintained by immigration), the second would grow fast but only after a delay, and the third would continue to grow steadily.

In practice, the ecologist usually has to deal with census data that *are* grossly deficient in detail, and for this reason most of what is known about changes in the abundance of organisms over time and space has had to take the form of correlations with external factors. It must nevertheless be borne strongly in mind that the answers to many questions about the abundance and distribution of organisms lie in the heterogeneities within the populations themselves, and in dispersal into and out of them.

It is not surprising that most censuses lack the detail that we would like. The reasons for this are as follows.

(i) It is usually a technically formidable task to follow individuals in a population throughout their lives, though it may be relatively easy at some stages (it is often easy to count birds at nesting time or frogs at mating sites, but harder at other seasons). Often, a crucial stage in the life cycle of an organism is hidden from view—rabbits within their warrens, butterflies and moths as buried pupae, and seeds in the soil. It is possible to mark birds with numbered leg rings, young fish with dyes or metal tags, roving carnivores with radio transmitters, or seeds with radioactive isotopes, and to recognize marked individuals on resampling; but the species and the numbers that can be censused in this way are severely limited. Only plants and immobile animals stay put and wait to be counted; but even then, censusing the dispersal phases in the life-history poses enormous problems to the investigator.

(ii) The results of a census will be misleading unless sampling is adequate over both space and time, and adequacy of either usually requires great commitment of time and money. The lifetime of investigators, the hurry to produce publishable work, and the short tenure of most research programmes all deter individuals from even starting to make a census over an extended period of time.

... or because techniques
are not always
appropriate (with
hindsight)

(iii) As knowledge about populations grows, so the number of attributes that we hope to find recorded in a census grows and changes. Every census procedure is likely to be out of date almost as soon as it is started. New approaches or techniques emerge, and these should be incorporated to make a census fully interpretable in the light of modern knowledge. Again and again, the analysis of census records shows that the next time it is done it should be done differently.

All the censuses described in this chapter are in one way or another less than ideal, judged in the light of what we now know is needed to understand them fully. Furthermore, most of the really long-term or geographically extensive censuses have been made of organisms of economic importance—fur-bearing animals, game birds, pest and disease organisms—or the furry and, especially, feathered favourites of amateur naturalists. The detailed censuses have been very biased. Insofar as generalizations emerge, they must be treated with great caution.

15.2 Fluctuation or stability

All populations are in a continuous state of flux, as new individuals are born, or arrive as immigrants, and older ones die, or leave as emigrants. Despite this, fluctuations in population size are not unbounded; no population increases without limit and species only occasionally become extinct. One of the central features of population dynamics, therefore, is the simultaneous occurrence of flux and relative constancy. The extent to which populations fluctuate, however, differs dramatically from species to species.

Perhaps the record of a local population which covers the greatest time-span is that of the swifts in the village of Selborne in southern England (Lawton & May, 1983). In one of the very earliest published works on ecology, Gilbert White, who lived in Selborne, wrote (in 1778) of the swifts:

the swifts of Selborne: stability from 1778 to 1983

'I am now confirmed in the opinion that we have every year the same number of pairs invariably; at least, the result of my inquiry has been exactly the same for a long time past. The number that I constantly find are eight pairs, about half of which reside in the church, and the rest in some of the lowest and meanest thatched cottages. Now, as these eight pairs—allowance being made for accidents—breed yearly eight pairs more, what becomes annually of this increase?'.

Lawton and May visited the village in 1983, and found major changes in the 200 years since White described it. It is unlikely that swifts have nested in the church tower for 50 years, and the thatched cottages have disappeared or been covered with wire. Yet the number of breeding pairs of swifts regularly to be found in the village is now 12. In view of the many unquantified changes that have taken place in the intervening centuries, this number is remarkably close to the eight pairs so consistently found by White.

The long-term study of nesting herons in the British Isles (where there is a large population of devoted ornithologists) reveals the same picture of a population that has remained remarkably constant over a long period (see Figure 9.19). This census included seasons of severe weather when the population declined temporarily, but it subsequently recovered.

British herons: stability despite catastrophe

Table 15.1. Population flux of *Ranunculus repens* in three sites of 1 m² in a grazed grassland.

	A	B	C
(a) Number of plants m⁻², April 1969	385	117	148
(b) Number of plants m⁻², April 1971	157	139	222
(c) Net change (b−a)	−228	+22	+74
(d) Rate of increase (b/a)	0.41	1.19	1.50
(e) Number of plants arrived between April 1969 and April 1971	344	244	466
(f) Total number of plants lost between April 1969 and April 1971	577	222	390
(g) Plants present April 1969, alive by April 1971	25	13	22
(h) Percentage survival of plants in (a) (g/a×100)	6.5	11.1	7.4
(i) Expected time for complete turnover (years) 2/(100−h)×100	2.14	2.25	2.16
(j) Total plants recorded during study	729	361	612
(k) Percentage annual mortality of all individuals (f/j)×100	79.1	61.5	63.7

In their census of a population of creeping buttercups (*Ranunculus repens*) in an old permanent pasture in North Wales, Sarukhán and Harper (1973; Sarukhán, 1974) made detailed maps of the distribution of plants and seedlings and were able to follow the fate of every individual (an approach that is rarely possible for mobile animals). Death occurred throughout the year, with seasonal lethal periods occurring at the same time as the survivors were growing most rapidly. Additions to the population were made through two distinct processes: the germination of seeds in annual flushes, and the clonal multiplication of rosettes. In clonal growth, the plants form rosettes of leaves and buds then grow out from the leaf axils to form elongated stolons; a new rooted rosette is formed at the stolon tip. A rosette may form several clonal daughters and these may, in turn, produce further stolons and rosettes in the course of a single season. In this way, a single genet (arising from a seedling) may come to be represented by a family of rosettes all of the same genotype—each plant is now a sub-population of parts which may lose their interconnections but remain modules of the same genetic individual.

The population dynamics of the creeping buttercup, derived from three separate 1 m² sites, are summarized in Table 15.1. From such accumulated data it is possible to see in detail the ways in which the births and deaths of individuals contribute to the overall changes in the size of populations (Figure 15.1). The behaviour of the population can be seen to depend on the behaviour of individual organisms; it is individuals that are born or die, that do or do not produce progeny, and that do or do not leave descendants. Despite the

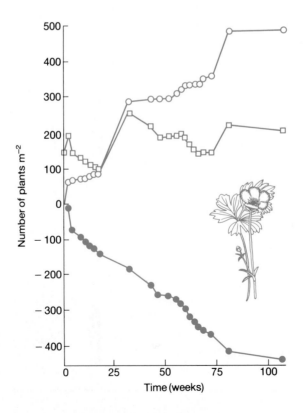

Figure 15.1. Changes in population size of the creeping buttercup (*Ranunculus repens*) at site C (see Table 15.1). O, cumulative gains from seed germination and clonal growth; ●, cumulative losses; □, net population size. (From Sarukhán & Harper, 1973.)

operation of these processes, which underlie population change, the population of creeping buttercups remained remarkably constant over the two years of the study.

Another example of a population showing relatively little change in adult numbers from year to year is seen in an 8-year study in Poland of the small, annual sand dune plant *Androsace septentrionalis* (Symonides, 1979a; Figure 15.2, and see p. 559). Each year there were between 150 and 1000 seedlings per square metre, and each year mortality reduced the population by between

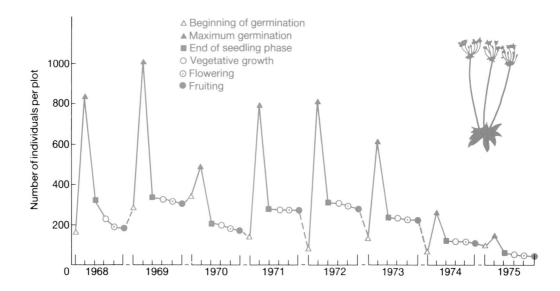

Figure 15.2. The population dynamics of *Androsace septentrionalis* during an eight year study (from Symonides, 1979a).

30% and 70%. However, the population appears to be kept within bounds—at least 50 plants always survived to fruit and produce seeds for the next season.

the apple blossom thrips: a fluctuating population

As a final example of the relationship between population flux and relative constancy we take the detailed study by Davidson and Andrewartha (1948a and b) of the apple blossom thrips (*Thrips imaginis*), a small insect (*c.* 1 mm long) found in the flowers of rosebushes, fruit trees, garden plants and weeds in southern Australia. They censused the thrips population in 20 roses picked at random from a long hedge every day, except Sundays and certain holidays, for 81 consecutive months. The variety of rose sampled (Cecile Brunner) lacks stamens and the thrips require pollen if they are to breed. The roses therefore served as 'traps' for the census—they were not themselves breeding sites. Examples of the census records are shown in Figure 15.3 and mean monthly population counts are summarized in Figure 15.4. Like many insects the thrips underwent very large fluctuations in density. Are these fluctuations determined entirely by randomly acting factors or are there

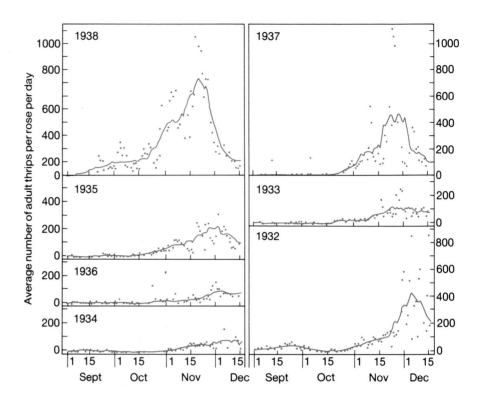

Figure 15.3. The numbers of *Thrips imaginis* per rose during the spring each year for seven consecutive years. The points represent daily records; the curve is a 15-point moving average. (After Davidson & Andrewartha, 1948a.)

processes at work which tend to keep them within bounds? We can begin to answer this question by looking at theories of species abundance in general.

15.2.1 Theories of species abundance

how is abundance determined?—a mid-twentieth century controversy

There have in the past been contrasting theories to explain the abundance of populations of animals and plants. Some investigators have emphasized the apparent stability of the populations they have studied, while others have emphasized the scale of fluctuations, and have sought the causes of the variation rather than the nature of the ultimate limits. Indeed, the same census data have often been used by different authors either to indicate that populations remain within narrow limits or to stress that they change dramatically. The interest has been so great, and the disagreement often so marked, that the subject has been a dominant focus of attention in population ecology throughout much of this century. The present view is that the controversy has not so much been resolved, as recognized for what it is: a product of protagonists taking up extreme positions and arguing at cross purposes. Since most of the arguments were propounded between 1933 and 1958, when understanding of population ecology was less sophisticated than

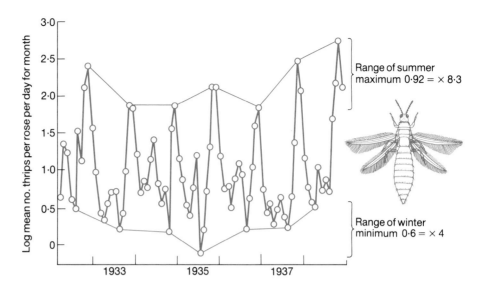

Figure 15.4. Mean monthly population counts per rose of *Thrips imaginis* (log scale) (from Davidson & Andrewartha, 1948a, after Varley *et al.*, 1973).

now, it would be unwise (and unfair) to analyse the contrasting views in too great a detail. However, consideration of the old viewpoints will make it easier to appreciate the details of the modern consensus.

A.J. Nicholson, an Australian theoretical and laboratory animal ecologist, is usually credited as the major proponent of the view that *density-dependent*, biotic interactions (which Nicholson called 'density-governing reactions') play the main role in determining population size (Nicholson, 1933, 1954a and b, 1957, 1958). In his own words: 'Governing reaction induced by density change holds populations in a state of balance in their environments,' and 'The mechanism of density governance is almost always intraspecific competition, either amongst the animals for a critically important requisite, or amongst natural enemies for which the animals concerned are requisites'. Moreover, although he recognized that 'factors which are uninfluenced by density may produce profound effects upon density', he considered that they only did so by 'modifying the properties of the animals, or those of their environment, so influencing the level at which governing reaction adjusts population densities'. Even under the extreme influence of density-independent factors 'density governance is merely relaxed from time to time and subsequently resumed, and it remains the influence which adjusts population densities in relation to environmental favourability,' (Nicholson, 1954b). In other words, Nicholson may be taken to represent the view among ecologists (along with others, notably Haldane, Lack, Varley and Solomon) that density-dependent processes play a crucial role in determining the abundance of species by operating as stabilizing or regulating mechanisms.

By contrast, two other Australian ecologists, Andrewartha and Birch, have taken the view that density-dependent processes are generally of minor

Andrewartha and Birch's
view: the crucial
importance of *r*

or secondary importance and play no part in determining population numbers in some species. Note that in contrast to Nicholson, their research was concerned more with the control of insect pests in the wild. Andrewartha and Birch's (1954) view can be summarized as follows:

'The numbers of animals in a natural population may be limited in three ways: (a) by shortage of material resources, such as food, places in which to make nests, etc; (b) by inaccessability of these material resources relative to the animals' capacities for dispersal and searching; and (c) by shortage of time when the rate of increase *r* is positive. Of these three ways, the first is probably the least, and the last is probably the most important in nature. Concerning (c), the fluctuations in the value of *r* may be caused by weather, predators, or any other component of environment which influences the rate of increase.'.

Andrewartha and Birch, therefore, 'rejected the traditional subdivision of environment into physical and biotic factors and "density-dependent" and "density-independent" factors on the grounds that these were neither a precise nor a useful framework within which to discuss problems of population ecology' (Andrewartha & Birch, 1960).

The views of Andrewartha and Birch can be made more explicit by considering in some detail an example to which they attached great weight (already referred to in Figures 15.3 and 15.4). We have noted already that in the census of *Thrips imaginis*, estimates of abundance were obtained for 81 consecutive months. For a further seven years, estimates were also obtained for spring and early summer only. In addition, local temperature and rainfall were monitored throughout the period. The census data were analysed by a multiple regression technique (see for instance, Poole, 1978) to determine how much of the population's variation from year to year could be 'explained' by a relationship with weather. For the analysis, the peak population each year was taken to be the mean logarithm of the numbers in the 30 days preceding the maximum. These peak values became the dependent variate in the regression analysis (log Y).

In seeking quantities which might be associated with the numbers of thrips in the spring, they looked first for one which would represent the opportunity for growth during autumn and winter afforded to the annual plants which were chiefly important in the ecology of *Thrips imaginis*. They took as the start of the season the date on which the seeds of these annuals started to germinate. As a measure of the temperature in the following period, they summed the 'effective temperature', T, from the start of the season until 31 August, where:

$$T = \frac{\text{maximal daily temperature} - 48°F}{2}.$$

The values of T become the first chosen variate (x_1) for the regression. The second variate x_2 was the total rainfall in September and October (Australian spring), and x_3 was the daily 'effective temperature' in September and October. They recognized that the size of the population that carried over from the winter might be determined by conditions in the previous year, and a further variate was introduced to allow for this—the value of x_1 from the previous year was taken as x_4.

The regression equation obtained from this analysis is

$$\log Y = -2.390 + 0.125x_1 + 0.2019x_2 + 0.1866x_3 + 0.0850x_4.$$

. . . where weather accounts for 78% of the variation in numbers . . .

The real values and those calculated from the regression are shown in Figure 15.5. By far the most important variate was x_1 ($P < 0.001$) and x_2 was the next ($P < 0.01$). Of the variance in population maxima, 78% was accounted for by the regression 'calculated entirely from meteorological records'. 'This left virtually no chance of finding any other systematic cause for variation, because 22 per cent is a rather small residuum to be left as due to random sampling errors . . . All the variation in maximal numbers from year to year may therefore be attributed to causes that are not related to density. Not only did we fail to find a 'density-dependent factor', but we also showed that there was no room for one.' (Andrewartha & Birch, 1954).

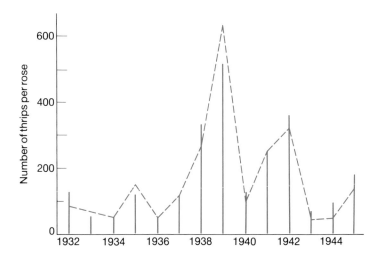

Figure 15.5. Observed and predicted population sizes of *Thrips imaginis*. The columns represent the geometric means of the daily counts of thrips per rose during the population peak. The curve represents the theoretical values calculated from the expression given above. (After Davidson & Andrewartha 1948b.)

The census of *Thrips imaginis* is a very remarkable study both in its intensity and its duration—some 6 000 000 thrips had been recorded by the end of the study. The work also provides an excellent example of what can be achieved by regression analysis. Clearly, the weather (as represented by the four factors) plays a central and crucial role in the determination of thrips numbers at their peak. Yet Andrewartha and Birch (1954) used this, and the fact that no density-dependent factor had been found, to conclude that there was 'no room' for a density-dependent factor as a determinant of peak thrips numbers.

As Varley *et al.* (1975) point out, however, the multiple regression model is not designed to reveal directly the presence of a density-dependent factor. And when Smith (1961) applied to the same data methods that were so

... but does that mean
that density-dependent
processes are
unimportant?

designed, he obtained excellent evidence of density-dependent population
growth prior to the peaks. There were significant negative correlations
between population *change* and the population size immediately preceding
the spring peak. In addition, a fairly strong density-dependent factor acting
between the summer peak and the winter trough can be inferred from Figure
15.3, since the maxima are spread over an eight-fold range and the minima
cover only a four-fold range. In fact, Andrewartha and Davidson themselves
(1948b) felt that weather acted as a density-dependent component of the
environment during the winter, by killing the proportion of the population
inhabiting less favourable 'situations'. (If the number of safe sites is limited
and remains roughly constant from year to year, then the number of
individuals outside these sites killed by the weather will increase with
density.) They also considered, however, that this did not fit the general,
density-dependent theory 'since Nicholson (1933, pp. 135–6) clearly excludes
climate from the list of possible "density-dependent factors".'. Forty years
on, it is not easy to see the validity of this objection to Nicholsonian density-
dependence operating, in this case, via climate.

15.2.2 The determination of abundance and its regulation

It is of critical importance to distinguish clearly between the determination of
a population's abundance and its regulation. *Regulation* refers to the tendency
of a population to decrease in size when it is above a particular level, but to
increase in size when below that level. In other words, regulation of a
population, by definition, can only occur as a result of one or more density-
dependent processes acting on the rates of birth (and/or immigration) and/or
death (and/or emigration, Figure 15.6). Various potential density-dependent
processes have been discussed in earlier chapters on competition, predation
and parasitism. On the other hand, abundance will be *determined* by the
combined effects of all the factors and all the processes that impinge on a
population, be they dependent or independent of density. The argument that
there must be regulating factors that explain the relative stability of natural
populations was well put by J.B.S. Haldane (1953):

Haldane's bordered
whites: regulated but
highly variable

'Clearly a population can only be changed by birth (or hatching), death, or
migration. The easiest population to consider is one with strictly defined
generations, such as an annual insect. Suppose P_n is the population in an area
at a definite date in year n, for example P_{1952} is the number of Bordered White
chrysalises in an isolated pine wood on 1 January 1952. Let $P_{n+1} = R_n P_n$
where R_n is the net rate of increase or decrease of the population. R_n can be
greater or less than one. But since $P_{n+2} = R_{n+1} R_n P_n$, and so on, the product of
R_n over a number of years must be very close to one, or the sum of the
logarithms of R_n very close to zero. How close they must lie is clear if we
suppose that $R_n = 1.01$ over 1000 years, which is a very short time if we are
considering evolution. The population would increase by 21000 times.
Similarly, if R_n were 0.99 it would decrease to 0.000043 of its original number.
    Since no population increases without limit, and species only occasion-
ally become extinct, there must be some regulating factors which, on the

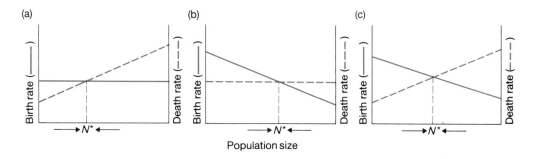

Figure 15.6. Population regulation with (a) density-independent birth and density-dependent death, (b) density-dependent birth and density-independent death, (c) density-dependent birth and death. Population size increases when birth rate exceeds death rate, and decreases when death rate exceeds birth rate. N^* is therefore a stable equilibrium population size. The actual value of the equilibrium population size is seen to depend on both the magnitude of the density-independent rate *and* the magnitude and slope of any density-dependent process.

whole, cause the density of an animal or plant species in a given area to increase when it is small, and to decrease when it is large.' (Haldane 1953).

The example that Haldane used to illustrate his argument is very different from that of the herons and swifts that we have quoted. His illustration was a census of the numbers of the bordered white moth in a German pine forest from 1882 to 1940. During this period the highest density (in 1928) was at least 30000 times the lowest density (in 1911). Clearly, if this was a regulated population, the regulation was not very precise! However, during the period of the study there were phases of extremely rapid population growth (e.g. 1922 to 1928), but these were not sustained—the population had the potential for extremely rapid growth but this was never realized for long. Hence, it can be argued that some regulating factor must have operated that prevented the populations from growing beyond bounds.

regulation seems unavoidable logically . . .

Andrewartha and Birch's view that density-dependent processes play no part in determining the abundance of some species clearly implies that the populations of such species are not regulated. However, it is logically unreasonable to suppose that any population is absolutely free from regulation. Even the bordered white population suffered neither extinction nor unrestrained growth, and the fluctuations in almost all populations are at least limited enough for us to be able to describe the species as 'common', 'rare', and so on.

On the other hand we must remember that in Davidson and Andrewartha's work, the weather accounted for 78% of the variation in peak number of thrips. If we wished to predict abundance, or decide why, in a particular year, one level of abundance was attained rather than another, then weather would undoubtedly be of major importance (by implication, therefore, density-dependent processes would be of secondary importance in this respect).

. . . but many account for little of the variation in population size

Thus it would be unwise to go along with Nicholson altogether. Although density-dependent processes are an absolute necessity as a means of regulating populations, their importance in *determining* abundance depends very much on the species and environment in question. Indeed, much of the argument about regulation and determination now seems rather stale, as a reading of later texts such as *The Theory of Island Biogeography* by MacArthur and Wilson (1967) makes plain. These authors emphasize that populations differ in behaviour; that some species (or local populations) spend most of their time recovering from past crashes or in phases of invasion of new territories, and that others spend much of their time 'bumping up against' the limits of their environmental resources—or in other ways suffering from the effects of overcrowding (density-dependent forces). These two conditions represent the ends of a continuum. At one extreme the size of a population usually reflects (a) the level to which it had last been reduced, (b) the time elapsed for it to regrow, and (c) its intrinsic rate of population increase during that time. At the other end of the continuum, the size of populations reflects the availability of some limiting resource that constrains the further expansion of the population by limiting the birth rate, increasing the death rate or stimulating emigration. The ends of this continuum are obviously MacArthur and Wilson's *r* selection and *K* selection, discussed in Chapter 14.

the *r/K* continuum joins the opposing poles . . .

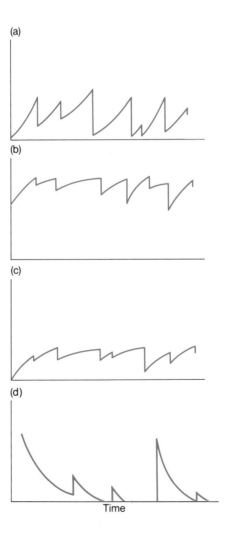

Figure 15.7. Idealized diagrams of population dynamics: (a) dynamics dominated by phases of population growth after disasters (*r* species); (b) dynamics dominated by limitations on environmental carrying capacity— carrying capacity high (*K* species); (c) same as (b) but carrying capacity low (*K* species); (d) dynamics within a habitable site dominated by population decay after more or less sudden episodes of colonization or recruitment, e.g. from the seed bank in the soil. (Harper, 1981.)

Idealized forms are shown in Fig. 15.7. Reconsidering them here emphasizes that all abundances reflect both density-dependent and density-independent factors, but that the relative importance and frequency of action of the two can vary greatly.

Figure 15.8 shows diagrammatically how the equilibrium level of abundance may be very greatly modified by the intensity of a density-independent process. Consider a population in which the birth rate is density-dependent, while the death rate is density-independent but depends on physical conditions which differ in three locations. The figure shows three equilibrium populations (N_1, N_2, N_3) which correspond to the three death rates which in turn correspond to the physical conditions in the three environments. In this context, Watkinson and Harper (1978) showed how variations in density-independent mortality were primarily responsible for differences in the abundance of the annual plant *Vulpia fasciculata* on different parts of a sand dune environment in North Wales. Reproduction was density-dependent and regulatory, but varied little with physical conditions from site to site (as indicated diagrammatically in Figure 15.8).

. . . which are therefore both important— as illustrated by *Vulpia*

Finally, it is important to note that because all environments are variable, the position of any 'balance-point' is continually changing. Thus, in spite of the ubiquity of density-dependent, regulating processes, there seems little value in a view based on universal balance with rare non-equilibrium interludes. On the contrary, it is likely that *no* natural population is *ever* truly at equilibrium.

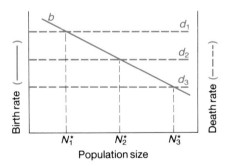

Figure 15.8. Population regulation with density-dependent birth b and density-independent death d. Death rates are determined by physical conditions which differ in three sites (death rates d_1, d_2 and d_3). Equilibrium population size varies as a result (N_1^*, N_2^*, N_3^*).

We have quite a clear idea of what determines the abundance of *Vulpia* plants *and* what regulates their population about its mean density. In the case of *Thrips*, we know much about what determines its abundance but only have a hazy idea of what regulates it—because the study was not designed to answer both questions and the data are therefore inadequate in this respect. In the next section, we describe a technique that can throw light both on determination and regulation of abundance.

15.3 Key-factor analysis

The number of individuals of a species present in an area at a point in time is the number present at some previous time plus births, minus deaths, plus immigrants and minus emigrants. Fluctuations in a population must be accountable in terms of these four processes. If we could break down the contributions of each of these to the changes that occur, it is likely that we could focus research more precisely on the stages that are responsible for changes in density and also for stability.

The more frequently a census can be made and the more completely a life cycle can be exposed, the more likely it is that crucial phases that determine population size may be discovered. This is not at all easy with some organisms—lemmings, migratory birds, plants with widely dispersed seeds—because their life-style may prevent an observer from making a census at a critical stage. But the approach, known as *key-factor analysis,* is applicable for many univoltine insects (for which it was developed) and other animals and plants. It is an approach based on the use of *k*-values (see also Chapter 4, p. 136; Chapter 6, p. 219; and Chapter 10, p. 355.

15.3.1 The Colorado potato beetle

In key-factor analysis, data obtained from a series of censuses are compiled in the form of a life-table, such as that for Canadian population of the Colorado potato beetle (*Leptinotarsa decemlineata*) in Table 15.2 (Harcourt, 1971). In this species, 'spring adults' emerge from hibernation around the middle of June, when potato plants are breaking through the ground. Within three or four days oviposition begins, continuing for about a month and reaching its peak in early July. The eggs are laid in clusters (average size, 34 eggs) on the lower leaf surface, and the larvae crawl to the top of the plant where they feed throughout their development, passing through four instars. When mature, they drop to the ground and form pupal cells in the soil. The 'summer adults' emerge in early August, feed, and then re-enter the soil at the beginning of September to hibernate and become the next season's 'spring adults'.

The sampling programme provided estimates for seven age-intervals: eggs, early larvae, late larvae, pupal cells, summer adults, hibernating adults and spring adults. In addition, one further category was included, 'females ×2', to take account of any unequal sex ratios amongst the summer adults.

Table 15.2 lists these age-intervals and the numbers within them for a single season, and also gives the major 'mortality factors' to which the deaths between successive intervals can be attributed. The *k*-values have been computed for each source of mortality and their mean values over ten seasons for a single population are presented in the first column of Table 15.3. These indicate the relative strengths of the various mortality factors as contributors to the total rate of mortality within a generation. Thus the emigration of summer adults has by far the greatest proportional effect, while the starvation of older larvae, the frost-induced mortality of hibernating adults, the 'non-

key-factor analysis is based on the use of k-values . . .

Table 15.2. Typical set of
life-table data collected
by Harcourt (1971) for the
Colorado potato beetle
(in this case for Merivale,
1961–62).

Age interval	Numbers per 96 potato hills	Numbers 'dying'	'Mortality factor'	$\log_{10}N$	k-value
Eggs	11 799			4.072	
	9268	2531	Not deposited	3.967	0.105 (k_{1a})
	8823	445	Infertile	3.946	0.021 (k_{1b})
	8415	408	Rainfall	3.925	0.021 (k_{1c})
	7268	1147	Cannibalism	3.861	0.064 (k_{1d})
Early larvae	6892	376	Predators	3.838	0.024 (k_{1e})
Late larvae	6892	0	Rainfall	3.838	0 (k_2)
Pupal cells	3170	3722	Starvation	3.501	0.337 (k_3)
Summer adults	3154	16	*D. doryphorae*	3.499	0.002 (k_4)
♀ ×2	3280	−126	Sex (52% ♀)	3.516	−0.017 (k_5)
Hibernating adults	16	3264	Emigration	1.204	2.312 (k_6)
Spring adults	14	2	Frost	1.146	0.058 (k_7)
					2.926(k_{total})

. . . which indicate the
average strengths of
various mortality
factors . . .

deposition' of eggs, the effects of rainfall on young larvae and the cannibal-ization of eggs all play substantial roles.

What the first column of Table 15.3 does not tell us, however, is the relative importance of these factors as determinants of the year-to-year fluctuations in mortality. We can easily imagine, for instance, a factor that repeatedly takes a significant toll from a population, but which, by remaining constant in its effects, plays little part in determining the particular rate of mortality (and thus the particular population size) in any one year. In other words, such a factor may, in a sense, be important in determining population size, but it is certainly not important in determining *changes* in population size, and it cannot help us understand why the population is of a particular size in a particular year. This can be assessed, however, from the second column of Table 15.3, which gives the regression coefficient of each individual

Table 15.3. Summary of
the life-table analysis for
Canadian Colorado
beetle populations (data
from Harcourt, 1971). b
and a are, respectively,
the slope and intercept of
the regression of each k-
factor on the logarithm of
the numbers preceding
its action; r^2 is the
coefficient of deter-
mination. (See text for
further explanation.)

		Mean	Coefficient of regression on k_{total}	b	a	r^2
Eggs not deposited	k_{1a}	0.095	−0.020	−0.05	0.27	0.27
Eggs infertile	k_{1b}	0.026	−0.005	−0.01	0.07	0.86
Rainfall on eggs	k_{1c}	0.006	0.000	0.00	0.00	0.00
Eggs cannibalized	k_{1d}	0.090	−0.002	−0.01	0.12	0.02
Egg predation	k_{1e}	0.036	−0.011	−0.03	0.15	0.41
Larvae 1 (rainfall)	k_2	0.091	0.010	0.03	−0.02	0.05
Larvae 2 (starvation)	k_3	0.185	0.136	0.37	−1.05	0.66
Pupae (*D. doryphorae*)	k_4	0.033	−0.029	−0.11	0.37	0.83
Unequal sex ratio	k_5	−0.012	0.004	0.01	−0.04	0.04
Emigration	k_6	1.543	0.906	2.65	−6.79	0.89
Frost	k_7	0.170	0.010	0.002	0.13	0.02

$$k_{total} = 2.263$$

k-value on the total generation value, k_{total}. Podoler and Rogers (1975) have pointed out that a mortality factor that is important in determining population changes will have a regression coefficient close to unity, because its k-value will tend to fluctuate in line with k_{total} in terms of both size and direction. A mortality factor with a k-value that varies quite randomly with respect to k_{total}, however, will have a regression coefficient close to zero. Moreover, the sum of all the regression coefficients within a generation will always be unity. Their values will, therefore, indicate their relative importance as determinants of fluctuations in mortality, and the largest regression coefficient will be associated with the *key*-factor causing population change (Morris, 1959; Varley & Gradwell, 1960).

... and which is the 'key-factor causing population change' ...

In the present example, it is clear that the emigration of summer adults, with a regression coefficient of 0.906, is the key-factor; and other factors (with the possible exception of larval starvation) have a negligible effect on the changes in generation mortality, even though some have reasonably high mean k-values. A similar conclusion can be drawn, in a more arbitrary fashion, from a simple examination of the fluctuations in k-values with time (Figure 15.9). (Note that Podoler and Rogers's method, even though it is less arbitrary than this graphical alternative, still does not allow us to assess the statistical significance of the regression coefficients, because the two variables are not independent of one another.)

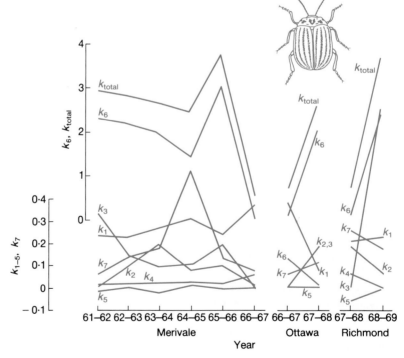

Figure 15.9. The changes with time of the various k-values of Colorado beetle populations in Canada. (Data from Harcourt, 1971.)

Thus, while mean k-values indicate the average strengths of various factors as causes of mortality each generation, key-factor analysis indicates their relative strengths as causes of yearly *changes* in generation mortality, and thus measures their importance as determinants of population size.

... and which factors
regulate rather than
simply determine
abundance

We must now consider the role of these factors in the *regulation* of the Colorado beetle population. In other words, we must examine the density-dependence of each. This can be achieved most easily by plotting k-values for each factor against the common logarithm of the numbers present before the factor acted. Thus columns 3, 4 and 5 in Table 15.3 contain, respectively, the slopes, intercepts and coefficients of determination of the various regressions of k-values on their appropriate '$\log_{10}$ initial densities'. Three factors seem worthy of close examination.

The emigration of summer adults (the key-factor) appears to act in an overcompensating density-dependent fashion, since the slope of the regression (2.65) is considerably in excess of unity. Thus, the key-factor, though density-dependent, does not so much regulate the population as lead, because of overcompensation, to violent fluctuations in abundance. Indeed, the Colorado potato beetle–potato system is only maintained in existence by humans, who by replanting prevent the extinction of the potato population (Harcourt, 1971).

The rate of pupal parasitism by *Doryphorophaga doryphorae*, a tabanid fly (Figure 15.10b) is apparently inversely density-dependent (though not significantly so, statistically), but because the mortality rates are small, any destabilizing effects this may have on the population are negligible.

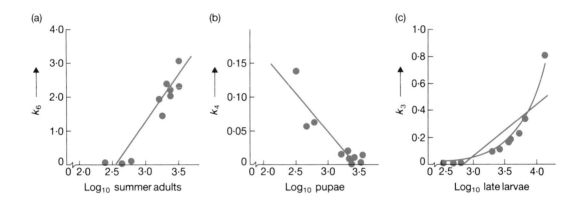

Figure 15.10. (a) Density-dependent emigration by Colorado beetle 'summer' adults (slope 2.65). (b) Inverse density-dependence in parasitization of pupae (slope −0.11). (c) Density-dependence in starvation of larvae (straight line slope 0.37, final slope of curve 30.95). (Data from Harcourt, 1971.)

Finally, the rate of larval starvation appears to exhibit undercompensating density-dependence (though statistically this is not significant). An examination of Figure 15.10c, however, indicates that the relationship would be far better reflected not by a linear regression but by a curve. If such a curve is fitted to the data, then the coefficient of determination rises from 0.66 to 0.97, and the slope (b-value) achieved at high densities is 30.95 (though it is, of course, much less than this in the range of densities observed). Hence, it is quite possible that larval starvation plays an important part in regulating the population, prior to the destabilizing effects of pupal parasitism and adult emigration.

15.3.2 Further examples of key-factor analysis

Key-factor analysis has been applied to a great many insect populations, but to far fewer bird, mammal or plant populations. An example of each is given in Figure 15.11, which in each case summarizes the changes through time of the total population mortality (k_{total}) and of the k-values of various other factors. The key-factor, identified by Podoler and Rogers's regression technique, is given as a dashed line. Shown alongside these diagrams are factors which act in a density-dependent manner.

The winter moth

The key-factor in the life cycle of the winter moth (*Operophthera brumata*) in Wytham Wood near Oxford is easily identified as k_1, the overwintering loss of eggs and larvae before the first larval census each spring. This loss is mainly associated with larvae hatching before opening of the leaf buds of oak, on which the larvae feed. When this happens, the winter moth caterpillars emigrate by spinning a length of silk on which they are blown away from the trees. In this study, the key factor (responsible for determining year-to-year changes in population size) was not the regulating factor that maintains the population within bounds. This is very often the case. In fact, predation in the soil of winter moth pupae (k_5) by beetles and small mammals was found to be a density-dependent factor (Figure 15.11a). The slope of the relationship between k_5 and pupal density is 0.35, showing that this density-dependent factor undercompensates (it is less than 1.0) for changes in density. Note also that k_3, parasitism by insects (other than *Cyzenis*), acted in a slight, but statistically significant, negatively density-dependent way (Figure 10.3).

Tawny owls

A long-term study of the tawny owl (*Strix aluco*) population, again in Wytham Wood, revealed that the failure of birds to breed each year (k_1) was the key-factor. Although the number of territory-holding adults did not vary

Figure 15.11. Key-factor analysis of four contrasting populations. In each case, a graph of total generation mortality (k_{total}) and of various k-factors is presented. The values of regression coefficients of each individual k-factor on k_{total} are given in brackets (values from Podoler & Rogers, 1975, except in (d) which were calculated from a graph given in Silvertown, 1982). The k-factor with the largest regression coefficient is the key-factor and is shown as a dashed line. Alongside each key-factor graph are shown k-factors which act in a statistically significant, density-dependent manner. (a) Winter moth, *Operophthera brumata* (data from Varley & Gradwell, 1968). (b) Tawny owl, *Strix aluco* (data from Southern, 1970). (c) Buffalo, *Syncerus caffer* (data from Sinclair, 1973). (d) The sand dune annual plant *Androsace septentrionalis* (data from Symonides, 1979a, analysis in Silvertown, 1982).

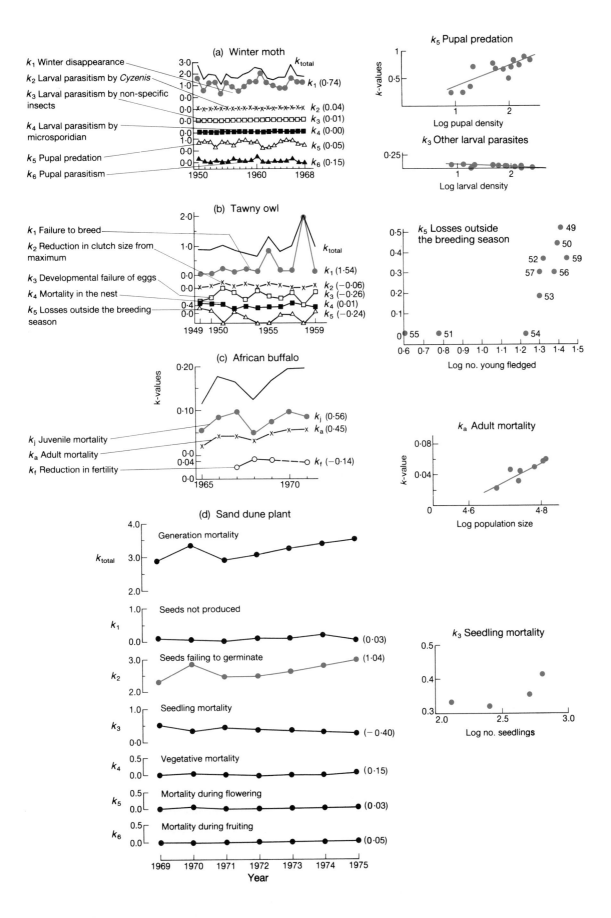

(a) Winter moth

k_1 Winter disappearance
k_2 Larval parasitism by *Cyzenis*
k_3 Larval parasitism by non-specific insects
k_4 Larval parasitism by microsporidian
k_5 Pupal predation
k_6 Pupal parasitism

k_{total}
k_1 (0.74)
k_2 (0.04)
k_3 (0.01)
k_4 (0.00)
k_5 (0.05)
k_6 (0.15)

1950   1960   1968

k_5 Pupal predation
k-values
Log pupal density

k_3 Other larval parasites
Log larval density

(b) Tawny owl

k_1 Failure to breed
k_2 Reduction in clutch size from maximum
k_3 Developmental failure of eggs
k_4 Mortality in the nest
k_5 Losses outside the breeding season

k_{total}
k_1 (1.54)
k_2 (−0.06)
k_3 (−0.26)
k_4 (0.01)
k_5 (−0.24)

1949 1950   1955   1959

k_5 Losses outside the breeding season
Log no. young fledged

(c) African buffalo

k-values

k_j Juvenile mortality
k_a Adult mortality
k_f Reduction in fertility

k_j (0.56)
k_a (0.45)
k_f (−0.14)

1965   1970

k_a Adult mortality
k-value
Log population size

(d) Sand dune plant

k_{total} Generation mortality

k_1 Seeds not produced (0.03)

k_2 Seeds failing to germinate (1.04)

k_3 Seedling mortality (−0.40)

k_4 Vegetative mortality (0.15)

k_5 Mortality during flowering (0.03)

k_6 Mortality during fruiting (0.05)

1969 1970 1971 1972 1973 1974 1975
Year

k_3 Seedling mortality
Log no. seedlings

much (17–32 pairs), there was a great deal of variation in the number that attempted to breed, ranging from 22 pairs in 1959 to none in 1958. Years in which few attempted to breed were those when numbers of the owl's prey of mice and voles were particularly low (see Figure 9.8a). Prey availability did not depend on owl density, and thus k_1 did not operate as a density-dependent factor. In contrast, k_5, which represents death or migration of young owlets after they leave the nest and before the start of the next season, was found to be density-dependent (Figure 15.11b). Losses became density-dependent only above a certain density (when the slope is 1.6, and over-compensating); they were probably brought about via intraspecific competition for territories.

African buffalo

Juvenile mortality (k_j) is the obvious key-factor for an African buffalo (*Syncerus caffer*) population in the Serengeti region of East Africa, and once again it does not operate in a density-dependent way. Juveniles suffered more than adults from a variety of endemic diseases and parasites, but calf mortality, although heavy, was random with respect to density. On the other hand, adult mortality (k_a) was found to be density-dependent (under-compensating slope of 0.24; Figure 15.11c). Undernutrition appears to have been a primary agent.

Can any generalizations be made about density-dependence in different groups of animals? In her analysis of 30 published accounts, Stubbs (1977) reached the following conclusions. Animals from more permanent habitats (vertebrates and some insects, including winter moth) tend to show under-compensating or exactly compensating mortalities (slope less than or equal to 1, never much more than 1). On the other hand, animals from more temporary habitats (many insects, including the Colorado potato beetle) tend to have small, undercompensating mortalities at low densities, rising sharply to overcompensating density-dependent losses as numbers increase. In addition, 86% of density-dependent factors acted on young stages for temporary habitat animals (*r* species), whereas the comparable figure for permanent habitat animals (*K* species) was only 15%. Instead, parasitism and predation (30%) and reduced fecundity (35%) seem to act more frequently in these animals.

An annual plant

Finally, the key-factor acting on a Polish population of the sand dune annual plant *Androsace septentrionalis* (dynamics shown in Figure 15.2) was found to be k_1, i.e. seed mortality in the soil. Again, this key-factor did not operate in a density-dependent manner, whereas the seedling mortality (k_2), which was not a key-factor, was found to be density-dependent (undercompensating slope of 0.2). Seedlings which emerge first in the season stand a much greater chance of surviving (Symonides, 1977), suggesting that competition for resources may be intense (and density-dependent) and/or that the later-emerging, smaller plants are more susceptible to various mortality factors.

15.4.3 Assessment of key-factor analysis

At this point, it is necessary to sound a note of warning about the limitations of analysing k-values. The analysis of *Androsace* failed to show up a strong negative correlation between density and plant size (i.e. a density-dependent relationship) paralleled by larger numbers of fruits on the bigger, low-density plants, as reported by Symonides (1979a). The likely reason for this is that such a relationship was evident apparently only in years with an early spring, when vegetative growth could proceed for longer. The correlation was almost completely missing in years when the life cycle only lasted a short time, and all individuals were small (Symonides, 1979b). Such an irregularly acting factor, regardless of any potential regulating role, is unlikely to be revealed by a simple analysis. Secondly, in a sense key-factor analysis is misnamed. In fact, it picks out *phases* in the life cycle when mortality (or failure to achieve reproductive potential) is important. It cannot by itself indicate what factors are responsible (e.g. competition for nitrate or breeding sites, the action of a specific predator or parasite, etc.) unless only one factor acts in each phase.

the 'factors' are actually phases

Thus, it is not necessarily always desirable to carry out a key-factor analysis of a run of census data to discover the factors underlying a population's abundance. Indeed, the more illuminating studies usually involve the experimental manipulation of density, resources or enemies (see section 15.6) and/or the development of mathematical models (see accounts of *Cakile edentula* in section 15.5 and *Opuntia–Cactoblastis* in section 15.6.5). Nevertheless, key-factor analysis offers a useful technique for unravelling census and life-table data, and in particular makes clear the crucial distinction between determination and regulation of abundance.

experiments are often more revealing than observations

All the animal studies referred to above have been performed on unitary organisms, and *Androsace* was treated from the point of view of the whole plant, rather than its parts. However, higher plants, and modular organisms in general, die in parts, rarely as wholes. For example, predation on corals by *Acanthaster* is more sensibly censused by the number of polyps eaten than by the number of individual corals destroyed, and a student of defoliating forest insects will have reason to be more interested in the population of leaves than in the number of trees in a forest. A beginning has been made to the study of the population dynamics of plant parts; they can, like whole organisms, be individually marked and life-tables can be constructed, age structures determined (Figure 15.12a) and survivorship curves drawn (Figure 15.12b). Key-factor analysis could easily be performed on such data, though, to date, this has not been attempted.

15.4 Population cycles and their analysis

The existence of regular cycles in animal abundance was first observed in the long-term records of fur-trading companies, and of gamekeepers. More recently, cycles have been reported from many studies of voles and lemmings and in certain forest Lepidoptera. Cycles may be driven by periodic fluctuations in the environment. Alternatively, they may arise as a result of internal demographic processes which involve delayed, or over-compensating

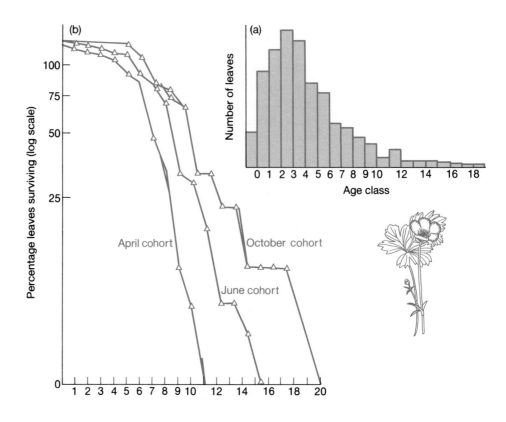

Figure 15.12. (a) Age-structure and (b) survivorship curves of populations of individual leaves of the creeping buttercup, *Ranunculus repens* (from Peters, 1980).

density-dependence (see sections 10.2.2 and 10.2.3). However, such density-dependent processes need to be particularly pronounced if they are not to be masked by a background of random fluctuations in weather.

15.4.1 Cycles and quasi-cycles.

The great classic among long-term censuses is the record of furs received by the Hudson Bay Company from 1821 to 1934. Figure 15.13a shows the number of Canadian lynx trapped for the company and the peaks appear to the eye to be rather evenly spaced and to have much the same amplitude (see also Chapter 10, p. 356).

margin: are they cycles?— correlograms

A statistical procedure has been delayed by Moran (1952) to check whether such cycles are real or simply a series of random fluctuations. This method correlates the number of animals counted in each year with the number counted in each succeeding year at increasing time intervals. In a cyclic series, high correlations occur when the intervals in years match corresponding phases of the cycle. For example, in a time-series which peaks every four years, high positive correlations occur at years 4, 8, 12, etc., with negative correlations (but of similar strength) at 2, 6, 10, etc. The correlations for each time-series are known as *auto-correlations*, and a graph showing the strengths of correlations at different time intervals is a *correlogram*. If a

560   CHAPTER 15

population does not oscillate in a regular periodic fashion, the correlogram damps down quickly to low and insignificant levels of correlation (although purely chance fluctuations prevent it from ever damping down completely). A description of the procedure is given by Poole (1978).

lynx show cycles

The correlogram for lynx (Figure 15.13b) shows no evidence of damping down—the fluctuations in the population are truly cyclic with a period of 10 years. As discussed in section 10.2.4, these arise as a result of cyclic interaction between snowshoe hares (the lynx's prey) and the plant food of the hares. The lynx population simply tracks that of its prey.

(a)

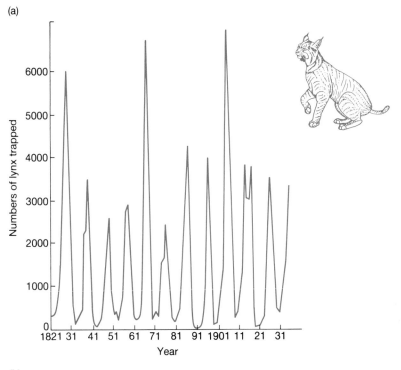

(b)

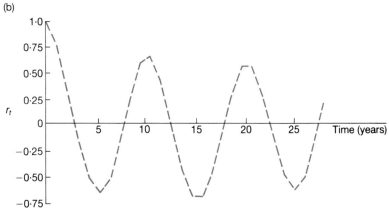

Figure 15.13. (a) Number of Canadian lynx (*Lynx canadensis*) trapped for the Hudson Bay Company (data from Elton & Nicholson, 1942). (b) Correlogram for Canadian lynx. The correlogram does not damp down, showing clearly a true cycle with a period of 10 years (after Moran, 1953).

Other populations do not show such perfect cycling, but may demonstrate a *tendency* towards a cyclic type of population change, which is evident when the correlogram does not damp down as fast as it would do in the complete absence of periodic fluctuation. Such *quasi-cycles* behave as if there is a wave that dies away after a disturbance (or a shout that leaves dying echoes). Analysing the records of red grouse (*Lagopus lagopus scoticus*) shot on a large number of moors in the north of England (each of at least 20 consecutive years during the period 1870 to 1977), Potts *et al.* (1984) found that the majority showed significant *negative* auto-correlation peaks at 2 or 3 years, indicating a cycling period of between 4 and 6 years (in fact, the average for all studies was 4.8 years). Figure 15.14 presents the original population data and correlograms for two of the moors, one with and the other without a quasi-cycle. Using a special statistical procedure, Potts *et al.* determined whether any of the auto-correlation peaks were statistically significant. Two auto-

red grouse show quasi-cycles

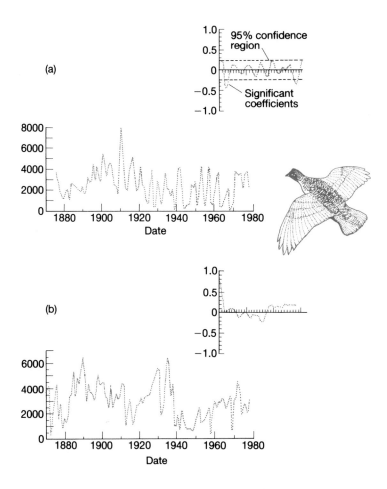

Figure 15.14. Analysis for serial correlations in red grouse (*Lagopus lagopus scoticus*) records from (a) a typical quasi-cyclic English moor and (b) a non-quasi-cyclic moor. Above the time series of original data of birds shot in each year is the correlogram produced by the auto-correlation technique. Significant *negative* coefficients occur in (a) at 2 and 3 years, indicating a quasi-cycle corresponding to between 4 and 6 years. No significant correlations occur in (b). (From Potts, *et al.*, 1984.)

correlation coefficients were significant in the data from the moor shown in Figure 15.14a, but none in Figure 15.14b.

The causes of quasi-cycles in red grouse are still obscure and indeed disputed (Watson & Moss, 1980). The cycle periods appear to be different on different moors, and, for example, seem to be significantly longer in Scotland than in England. Almost certainly, cycles at different places and times will have different causes. Potts *et al.* (1984) have suggested that the quasi-cycles that they describe result from an interaction between grouse and a parasite burden. A model that incorporates an interaction with parasitic nematodes gives a good fit to the cycles observed in the field, but only after random elements are introduced to simulate perturbations due to weather. Potts and his colleagues concluded that the grouse cycles could be caused by the effects of the parasite working together with stochastic elements (weather) and a time-delay arising from the uptake of the worms.

The best-documented cycles and strong quasi-cycles occur in three high-latitude northern vegetation zones. On open tundra, lemmings, voles and willow ptarmigan undergo a 4-year cycle. In the transitional birch forest zone between tundra and boreal forest, the snowshoe hare exhibits a 4-year cycle; while in the boreal forest itself, species such as snowshoe hare, spruce grouse, ruffed grouse and lynx follow a 10-year cycle (see Chapter 10, p. 357). The other important vertebrate herbivores in these environments are the larger caribou or reindeer, which make long-distance annual migrations. Their populations appear not to cycle.

15.4.2 Population changes in microtine rodents

In many northern habitats, the populations of small microtine rodents (particularly voles, *Microtus* spp., and lemmings, *Lemmus* spp.) show enormous changes from year to year, exploding and crashing (e.g. Figure 15.15). These microtine cycles have a typical periodicity of three or four years, though there are some populations that regularly or occasionally display a 2- or a 5-year cycle. There are also many populations and many species, particularly in less northerly regions, that never display multi-annual cycles. Despite massive research effort (see, for example, Krebs *et al.*, 1973; Krebs & Myers, 1974; Krebs, 1985), no clear explanation has yet appeared for these patterns of abundance. We will consider them in some detail here, because their study illustrates well the joys, frustrations and difficulties of wrestling with a conspicuous empirical phenomenon that we ought to be able to explain, but cannot.

Microtine cycles pose, in a particularly striking way, the more general question of how the sizes of populations are determined. The voles and lemmings, like other species, are affected by their food, by their predators, by disease and by the weather, and, as with many species, the characteristics of the individuals themselves vary with density and with the phase of the cycle. There are therefore problems of disentangling cause from effect, of distinguishing factors which change density from those that merely vary with density, and of distinguishing those that affect density from those that actually impose a pattern of cycles.

an important focus of interest, without a clear explanation ...

... posing questions pertinent to populations generally

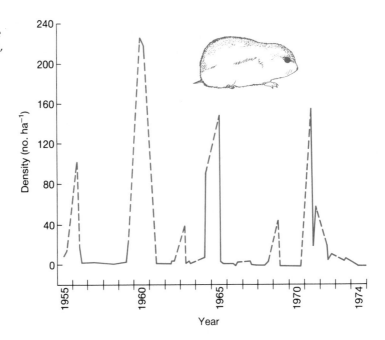

Figure 15.15. Estimated lemming densities in the coastal tundra at Barrow, Alaska for a 20-year period (after Batzli *et al.*, 1980).

the biology of lemmings in Alaska

An impression of microtine biology and the nature of the microtine cycle can be gained by considering an intensive study of the brown lemming (*Lemmus sibericus*) which was carried out in the coastal tundra at Barrow, Alaska, where it is the dominant herbivore (Batzli *et al.*, 1980). A second species, *Dicrostonyx torquatus*, is present but usually scarce (0.1 ha^{-1}), whereas population of the brown lemming may reach mean densities of 225 ha^{-1} over the entire tundra. The behaviour of the animals puts many hindrances in the way of answering the key questions. Much of their life (in particular, their breeding) takes place underground or under snow. Only with the very greatest difficulty can censuses be made, except of animals moving above ground, when trapping is the most effective way of estimating numbers.

The density of lemmings is patchy in both space and time. Much of the studied area is composed of a repeated pattern of polygons, formed by deeply penetrating wedges of ice. The polygons range from a diameter of a few metres to 30 m (average 12 m) and are outlined by narrow troughs that overlie the ice wedges. The patterns of snow accumulation, drainage and vegetation vary across a polygon and between polygons (Brown *et al.*, 1980). The life of the lemmings is strongly related to this pattern of land-form and changes with the season (Table 15.4). Mean values of lemming density greatly underestimate the real extent to which animals are concentrated in the patches. In 1974, for example, 93% of nests were in polygon troughs, and it was here also that the lemmings clipped the greatest amount of vegetation.

a lemming increase and crash

One of the periodic increases in the lemming population starts when the lemmings reproduce in nests of grass and sedges at the base of the snow pack, and the population grows rapidly during the winter to reach a peak in late spring. Breeding ceases in May. Signs of overcrowding appear before the snow melts, when many lemmings burrow to the surface and wander about, some to die. During snow-melt, massive clippings of grasses and sedges and the disruption of moss and lichen carpets are revealed, and lemmings scurry everywhere. Large numbers of predators attack the exposed lemmings. During the summer, lemming survival is low, and the population crashes to a low

Table 15.4. Summary of indicators of brown lemming activity in four habitats at Barrow. Note that densities are seasonal extremes. (Brown *et al.*, 1980.)

	1972	1973	1974
Winter nest density (no. ha^{-1})			
High-centred polygons	—	24.9	18.5
Low-centred polygons	—	18.2	12.9
Meadow	—	17.2	4.2
Polygons and ponds	—	2.7	0.4
Summer population density (no. ha^{-1})			
High-centred polygons	3.4–14.7	3.1–4.0	0.3–0.6
Low-centred polygons	12.9–46.1	3.1–4.9	0.3–0.6
Meadow	—	2.5–3.1	0.9–1.2
Polygons and ponds	1.6–7.7	0.3–1.5	0.0–0.6
Percentage of graminoid tillers clipped			
High-centred polygons	14.8	12.3	2.4
Low-centred polygons	—	25.1	6.7
Meadow	—	24.3	1.9
Polygons and ponds	—	5.4	0.3

level, where it remains for one to three years (Batzli *et al.*, 1980). This is a general picture, but each irruption of a population has its own characteristics. In 1963 and 1969 there was particularly heavy predation by weasels under the snow. In 1964 very high densities preceded the 1965 peak (the population doubled from 1964 to 1965), whereas in 1970 a very low density preceded the 1971 peak (the population increased by a factor of 250 during the intervening winter).

No single theory satisfactorily explains the microtine cycles, but any future theory will have to account for the following features (Warkowska-Dratnal & Stenseth, 1985).

ten features that a theory of microtine cycles must explain . . .

(1) A species is not necessarily cyclic throughout its range.
(2) Not all microtine species in a region are necessarily cyclic.
(3) The cyclic species in a region need not necessarily be in phase with each other.
(4) The ratio of maximum to minimum density can be in the order of thousands.
(5) The period of the cycle seems to be more regular than its amplitude.
(6) During periods of early population increase, most individuals have a high reproductive rate and typically a high dispersal rate, whereas during periods of high but declining densities, most individuals have a low reproductive rate and a low dispersal rate.
(7) The level of individual aggressiveness varies throughout the cycle (see below).
(8) The specific dispersal rate depends on population density as well as on the rate of density change (i.e. the position during the cycle).

Two further features can also be added to Warkowska-Dratnal and Stenseth's list.

(9) The cycle is not symmetrical. Rather, there tends to be a rapid explosion succeeded by a rapid crash, following which there are a number of low-density years.

. . . including the importance of dispersal

(10) Dispersal, and especially immigration, seems to be necessary for the normal pattern of population change (e.g. Krebs *et al.*, 1969; Tamarin, 1978; Gaines *et al.*, 1979; see also Abramsky & Tracy, 1979).

It will be useful, before we proceed, to clarify the various types of theory that have been proposed to explain the cycles. The main distinction is between those theories stressing extrinsic factors and those stressing intrinsic factors. The possible extrinsic factors are weather, food, predators and parasites. In each case, cycles may simply emerge from an interaction without the individual organisms altering in any marked way (see Chapter 6 for time-lags and overcompensation, Chapter 10 for predation and alternative stable states, and Chapter 12 for parasitism). Alternatively, the interactions may lead to significant alterations in the individuals as the cycle progresses. By contrast, the theories that rely on an intrinsic cause see these changes in individual animals as the force *driving* the cycles. Two types of intrinsic cause have received attention: hormonal change and behavioural change. The latter, in its turn, may be either a genotypic change (one type of individual being replaced over time by another) or a phenotypic change (a given individual changing in response to its environment). There is a third type of theory that combines elements from all or some of the others, both intrinsic and extrinsic.

the various types of theory: extrinsic and intrinsic causes

Many workers have considered that food plays an important part in microtine cycles. In Alaska, the vegetation can be devastated by a peak population of lemmings and many dead individuals are seen, but in other studies such extreme starvation has been rare. Nonetheless, the rodents may be affected by varying concentrations of both nutrients and toxins in their plant food (see, for example, Batzli, 1983). Amongst Finnish workers on microtine cycles, food quality has traditionally been seen as important. Laine and Henttonen (1983), following Kalela (1962), have argued that the nutritional state of food plants (as indicated by flowering intensity in particular) varies from year to year with meteorological conditions, and that the microtines respond to this nutritional state. Rodents increase to a peak density on high-quality food. A combination of flowering and microtine consumption, however, leads to the exhaustion of plant nutrients and energy resources, a collapse in food quality, and a decline in the microtine population (the decline being then reinforced by predators and disease). The plants typically take a number of years to recover in the northern-latitude habitats with their short summer, and it is only when they have done so, and when meteorological conditions are favourable, that their nutritional state and the microtine density once again rise. This argument is summarized in Figure 15.16. Despite its plausibility, and some evidence in its favour, this model has not been favoured by many microtine ecologists. At least part of the reason has been the failure of food supplementation to prevent population declines in a number of studies, especially in less northerly habitats. Yet

an underlying cyclicity in food quality and quantity in northern latitudes?

the equivocal results of food supplementation experiments

this may simply be due to the 'pantry effect': the attraction of predators from nearby non-supplemented populations to counteract any microtine increase or lack of decline that there might otherwise have been. Certainly, when Ford and Pitelka (1984) provided supplementary food and water to penned populations of voles in California, *and controlled the number of predators*, they found that experimental populations only declined modestly during a summer when control populations crashed.

Figure 15.16. A conceptual model for the generation of microtine cycles in northern Fennoscandia (after Laine & Henttonen, 1983).

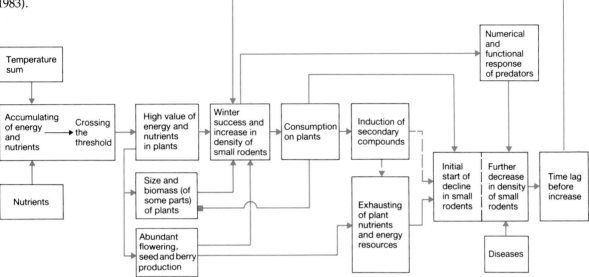

food: the nutrient-recovery hypothesis

A rather more complicated interaction between lemmings and their food was proposed by Schultz (1964, 1969): the 'nutrient-recovery hypothesis' (Figure 15.17). Schultz recognized that shortage of mineral nutrients, particularly phosphate, could place serious limitations on the breeding success of lemmings. He also realized that the growth of vegetation was nutrient-limited and that the lemmings, after eating and digesting plant material, released nutrients in faeces and urine. A lemming 'low' and a lemming 'high' would therefore have quite different effects on the availability of mineral nutrients and the growth rate of vegetation. These effects can be combined to produce conditions that might initiate both growth and declines in the lemming populations. Climatic variations between seasons and years could explain why the peaks and crashes repeated but did not follow a wholly rhythmic periodic cycle. Elements in the hypothesis have been criticized (Batzli *et al.*, 1980), but it has stood up to a partial experimental test. Areas of the tundra were treated with fertilizer (Schultz, 1969) to raise the protein, calcium and phosphorus levels of grasses and sedges (the main food of lemmings) well above that in unfertilized areas. In the fertilized areas the density of winter nests of lemmings was 75 ha^{-1}, whereas there were none in the control areas. The effect persisted, though less dramatically, in the following year.

To gain an impression of the role played by predators, we can again turn to Alaskan lemmings (Batzli *et al.*, 1980). At the time of year when the major increases take place (a 'high' winter) the main predators are the arctic fox

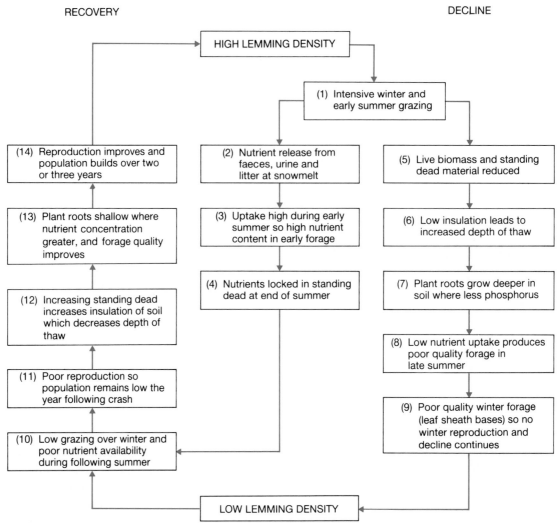

Figure 15.17. Summary of steps in the nutrient-recovery hypothesis (after Schultz, 1964, 1969).

the role of predators: Alaskan lemmings as a typical example

(*Alopax lagopus*) and two species of weasel, the least weasel (*Mustela nivalis*) and the ermine (*M. erminea*). These were absent from the study site at Barrow during periods when the lemming populations were low, but they recolonized and multiplied fast during lemming peaks, reaching densities of up to 25 km^{-2}. The other major predators are migrant birds, especially snowy owls, pomarine jaegars (skuas), glaucous gulls and sometimes short-eared owls. Their arrival precedes (snowy owl) or coincides with the onset of snow-melt, the time at which the lemming populations are exposed. If the density of lemmings is low, the jaegars and owls move on, but if lemming density is high the birds stay and breed. The birds have smaller territories (occur at a higher density), lay larger clutches, and rear more young in years when lemmings are abundant. In other words the density of predators is closely linked to that of the lemmings—both by immigration and by rapid reproduction during lemming population peaks (Figures 15.18 and 15.19).

Populations of the lemmings may be seriously reduced by the various predators (Figure 15.19) and '. . . predation contributes to population declines

and may be sufficient to prevent increases at low densities, but it is not
sufficient to account for summer declines following a peak. Furthermore,
relaxation of winter predation will not necessarily lead to population increases' (Batzli *et al.*, 1980). And this seems to be a conclusion that applies to
microtine cycles generally: predators track the cycles to a certain extent, and
they account for a large number of deaths; but many of those that are killed are
probably doomed anyway, and the predators seem to play little part in
generating the cycles or in actually causing the major changes in density.

Figure 15.18. The
estimated densities of
predators during the
course of a standard
lemming cycle at Barrow.
Periods of snowmelt and
freeze are indicated
between summer and
winter (After Batzli *et al.*,
1980.)

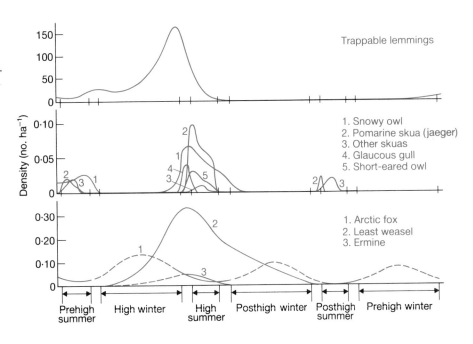

In similar vein, parasites have been known to cause considerable microtine mortality (e.g. Hentonnen *et al.*, 1981), but they too seem to affect
populations of individuals in poor condition that were probably susceptible to
many other sources of mortality as well.

We can now turn to the 'intrinsic' theories. Changes in hormonal or
physiological function (Christian, 1950) may be correlated with cycles (e.g.
Andrews *et al.*, 1975), but no role as a causative agent (rather than a consequence) has been demonstrated.

By contrast, a great deal of attention has been paid to the role of behavioural changes, of which there are two of major importance. (i) Dispersal
rate is high in the increase phase of the cycle, but decreases towards the peak
and is low during the decline phase (Stenseth, 1983). (ii) Aggressiveness
varies throughout the cycle, though the manner in which it does so is disputed. Krebs (1985) concludes that aggressiveness is highest in the increase
phase but low in the decline phase, whereas Warkowska-Dratnal and
Stenseth (1985) conclude that it is lowest during periods of early population
increase but highest when densities are high but declining.

Such disagreements aside, the essence of these views is that changes in
the proportions of aggressive and docile individuals, strong and weak

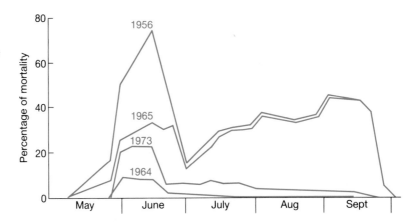

Figure 15.19. The impact of avian predators on lemming populations in four years as indicated by the percentage of mortality accounted for by predators (after Osborn, 1975).

. . . but are the changes genotypic or the result of kin selection?

dispersers, and strong and weak reproducers play a *causative* role in changing density, which in turn leads to a change in the proportions of different types of individual. Following Chitty (1960), many have believed these changes to be genotypic, i.e. aggressive genotypes are favoured at one phase in the cycle and they increase in frequency, but this leads to, or is accompanied by, a change in density which favours docile genotypes. On the other hand, Charnov and Finerty (1980) have offered an alternative to this, based on the theory of kin selection, in which all individuals are docile towards close relatives but aggressive towards those that are unrelated, and aggressive encounters are therefore most common when there are high rates of dispersal (i.e. mixing of families) and high densities.

A final pertinent point is that Rosenzweig and Abramsky (1980), impressed by the apparent necessity of immigration for cycling (see above), were able to generate regular density cycles in mathematical models which incorporated immigration, as long as they also incorporated a significant degree of habitat heterogeneity.

a tentative attempt to put the disparate elements together

Can we, therefore, fit all these elements together? The answer is that we can do so tentatively, where angels fear to tread, in one hypothesis that future studies will, at the very least, refine considerably. Microtine density appears to rise rapidly on a plant resource that is high in nutrients and low in toxins. This rising density is accompanied by a high rate of dispersal, and together these lead to a mixing of unrelated individuals and thus to a high frequency of aggressive encounters. Individuals suffer as a consequence: their reproductive capacity is depressed and they may be afflicted by physiological or hormonal disorders. At the same time, the plants, as a result both of flowering and of grazing, have declined in quantity and in quality (both nutritionally and toxically). We thus have a microtine population at peak density, on a depleted food resource, having suffered from aggressive encounters. Such a population will begin to decline, but will then crash precipitously as the susceptible individuals are attacked by predators (and perhaps pathogens) that have responded to the peak. The microtine population will then only rise again after the predators have declined and dispersed, and the plants have recovered and once again accumulated reserves. If dispersal of the microtines is prevented by either an experimenter or the habitat, then there is less aggression, less decline in individual condition, and thus a tendency to deplete the plants totally (as found by Krebs *et al.* (1969)). In the absence of immigration, there is also less aggression, but a much lower population peak, and the population therefore neither crashes nor cycles (as found by Gaines *et al.* (1979)).

15.5 Abundance determined by dispersal

One element in the population dynamics of microtines is undoubtedly the extent to which individuals disperse. We have already referred to this in Chapter 5, and we turn to it again now.

In many studies of abundance, the assumption has been made that the major events occur within the census area, and that immigrants and emigrants can be safely ignored in the equation

$$N_{t+1} = N_t + \text{Births} - \text{Deaths} + \text{Immigrants} - \text{Emigrants}$$

Of course, this is by no means invariably true; in many populations migration is a vital factor in determining and/or regulating abundance. We have already seen that emigration of summer adults of the Colorado potato beetle is both the key-factor in determining population fluctuations and an overcompensating density-dependent factor. Moreover, the winter moth population was subject to an overwintering loss, due partly to larval emigration, which acted as the key-factor but was not density-dependent; while in the life cycle of the tawny owl, emigration (and death) of young owlets was shown to be density-dependent though not, in this case, the key-factor.

When dispersal is a major event in determining population size, it poses significant additional problems for the investigator. Census *within* an area will usually miss such events entirely, and devising methods that account accurately for losses due to emigration and gains due to immigration demands a far more subtle censusing procedure. Even in fixed organisms that cannot themselves disperse, there is an inevitable dispersal phase amongst their progeny (planktonic larvae, seeds) and these may be difficult to monitor.

the crucial effects of dispersal on the dynamics of *Cakile* populations

In a study of *Cakile edentula,* a summer annual plant growing on the sand dunes at Martinique Bay, Nova Scotia, Keddy (1981) found that population density was greatest in the middle of the dunes and declined at both seaward and landward ends of its distribution along this environmental gradient. The data gathered on seed production and mortality in the three parts of its range show convincingly that, in theory, an equilibrium population should be found only towards the seaward limit, where seed production is high and strongly density-dependent. In the middle and landward sites, mortality exceeds fecundity so that the populations should rapidly go extinct (as they do in simulation models—Watkinson, 1984). If we only take into account the *in situ* births and deaths of plants, we would predict that the population would not persist in the middle area, where in fact the species is most abundant. The actual abundance of this plant can only be explained when a landward migration of seeds, caused by both wave and wind action, is taken into account (Keddy, 1982; Watkinson, 1984). Plants are found only at the middle and landward ends of the gradient because of the high annual

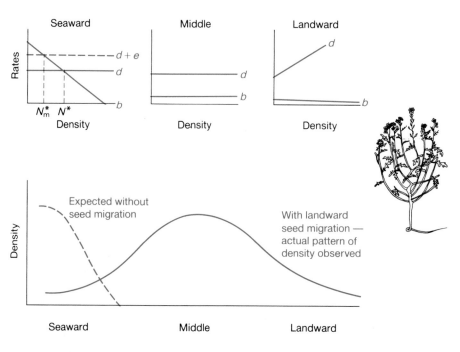

Figure 15.20. Idealized diagrammatic representation of variations in mortality (d) and seed production (births, b) of *Cakile edentula* in three areas along an environmental gradient from open sand beach (seaward) to densely vegetated dunes (landward). In contrast to other areas, seed production was prolific and density-dependent at the seaward site. Mortality was only density-dependent at the landward site. Births *in situ* exceeded deaths at the seaward site, in such a way that an equilibrium population density can be envisaged (N^*). Deaths always exceeded *in situ* births (new seeds) in the middle and landward sites so that populations should not in theory have persisted there (dashed line in lower graph). Persistence of populations in the middle and landward site occurs only because of the landward drift of the majority of seed produced by plants on the beach (seaward site). Thus the sum of *in situ* births (b) plus immigrating seeds (i) exceeds mortality (d) in the middle and landward sites. Note also that losses from the population at the seaward site actually include both deaths (d) and emigrating seeds (e) so that the equilibrium density there (N_m^*) is less than it would be in the absence of seed dispersal (solid line in lower graph).

dispersal of seeds landward from the beach (in effect, increasing the rate at which new individuals are added to the population above that produced by the resident plants) (Figure 15.20). Nevertheless, it is the density-dependent regulation of seed production at the seaward end of the gradient that regulates the overall population densities on the sand dunes.

15.6 The experimental perturbation of populations

The observation of nature can generate hypotheses about how it works. In ecology, field observation, theoretical models and the analysis of census data can all be used to *suggest* what the forces are that act in the field, ocean or forest. But just as we can only test what holds a pendulum at rest by setting it swinging, so we can only formally test ecological hypotheses by perturbing real ecological systems. If we suspect that predators or competitors determine the size of a population, we can ask what happens if we remove them. If they are absent, we may ask what would happen if we add them. If we suspect that a resource limits the size of a population we may add more of it. Besides indicating the adequacy of our hypotheses, the results of such experiments may show that we have ourselves the power to determine a population's size—to reduce the density of a pest or weed, or to increase the density of an endangered species. Ecology becomes a predictive science when it can forecast the future—it becomes a management science when it can determine the future.

15.6.1 Introduction of new species

There are many examples of species that have been accidentally introduced into alien habitats, but although consequences have often been ecologically dramatic, they have not been part of controlled experiments. On the other hand, a deliberate introduction of a species into a habitat can be a powerful means of beginning to understand the causes of commonness, rarity and absence. Such introductions are frowned on by conservation-minded ecologists who dislike any disturbance of the *status quo*, and by those who rightly fear that experimental introduction of a species may lead to its uncontrolled spread. Such experiments are, however, probably the most powerful tools available to the ecologist.

plantain seeds sown in
normal and 'abnormal'
habitats

Sagar and Harper (1960) sowed seeds of three species of plantain, *Plantago lanceolata*, *P. media* and *P. major*, into a variety of habitats in the neighbourhood of Oxford, England. The chosen habitats included some from which the species was naturally absent and others in which it was rare or abundant. In three of the habitats the seeds germinated and the radicle emerged but blackened and died almost immediately; no seedlings emerged at all. These were sites on acid heath or in a bog dominated by *Sphagnum* moss, in which the plantains are normally absent from the natural vegetation. Apparently the environments offered physical conditions outside the fundamental niches of all three species of plantain. In one habitat, a patch of limestone grassland within a woodland, all three species of plantain had been absent and all three formed established populations from the sown seed; the plants persisted through the year. Apparently the absence of the species from this site was due to the failure of seed to be dispersed naturally into it. The conditions in the site clearly lay within the fundamental niche of all three species and there was 'room' for plants to establish, but the fundamental niche had not been realized. On arable land, from which possible competitors were removed, all three species again formed established populations from seed.

On a trampled grassland only *P. major* was originally present, and adding extra seed did not change the size of its population. The other two species were absent initially, and although seedlings appeared from the sowings, the plants did not survive. This was the commonest situation in a number of habitats—the addition of seed of a species resulted in a flush of seedlings and, if the species was already a part of the community, the population settled down quite quickly to the same population size as in unseeded areas. In these habitats it was very difficult to change the *status quo*: both the species and their populations had stabilized at densities that could not easily be increased.

15.6.2 Augmenting resources

food supplementation in tree squirrels

If a population is limited in size by a resource that is in short supply, the addition of this resource should increase the abundance of the species. Tree squirrels of the genus *Tamiasciurus* are found in much of the boreal and temperate coniferous forest of North America. There is indirect evidence that the squirrel populations are limited by food—for example, it has been shown that the size of territories is inversely related to the availability of food. Population densities of the Douglas squirrel (*Tamiasciurus douglasii*) were estimated by live trapping in three areas of forest in British Columbia. Control areas received no additional food, but in experimental areas sunflower seeds and whole oats were distributed around each trap station every week from March to October 1977 and in March 1979. In the areas receiving food there was a five- to tenfold increase in the number of squirrels trapped. Control densities generally varied from three to ten squirrels per trapping area but the experimental population increased to 65 animals during the winter feeding (Figure 15.21). This irruption was produced by a combination of immigration, a greater rate of reproduction in females, and increased survival. After the food was withdrawn, the population declined to a level comparable with the controls (Sullivan & Sullivan, 1982).

To demonstrate that augmentation of a resource increased population size is not in itself a demonstration that competition for that resource had been occurring. In the food-rich environment the animals may simply have spent less energy in foraging and had more to spare for reproduction (Reynoldson & Bellamy, 1971). The experiment did, however, demonstrate that food supply was limiting population size.

15.6.3 Removal of possible competitors

competitor removal in coastal communities of plants

A population may be limited because potential resources are shared with individuals of other species. In such a situation we might expect that the removal of the competitors would allow the population to increase (several examples are given in Chapter 7). Silander and Antonovics (1982) asked whether populations of plants of different species would increase if their neighbours were removed. They made their study on a series of closely adjacent coastal communities along a gradient across Core Banks, a barrier island on the coast of North Carolina. Figure 15.22 shows the distribution of species along the environmental gradient. At each of the sites, the dominant and subdominant species were removed singly or in groups, within 1 m² plots, by weeding and by the action of selective herbicides. At each site, an additional treatment involved complete removal of all vegetation and finally there was an untreated control. Changes in vegetation were measured by using optical point cover—individual plants (tillers, shoots) were not counted, except for the grasses, but counts and point cover gave very similar

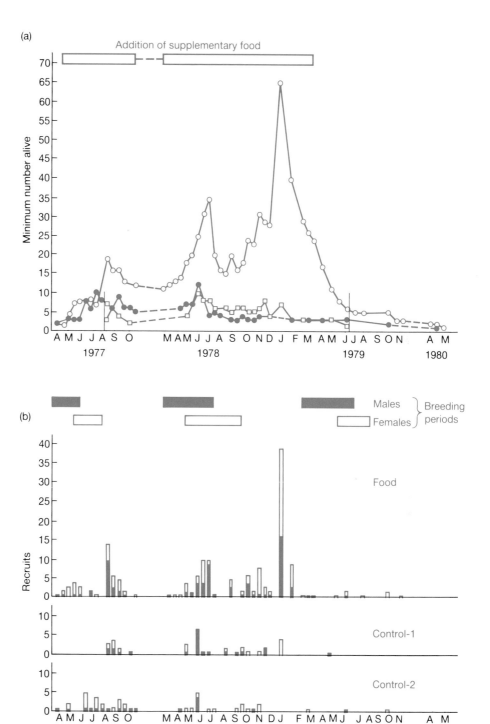

Figure 15.21. (a) Population density of the squirrel *Tamiasciurus douglasii* on two control (● and □) and one experimental area (supplementary food, ○) during 1977–1980. (b) Number of males (shaded) and females (unshaded) recruited into each population, and reproductive periods of males (scrotal testes) and females (lactation). (From Sullivan & Sullivan, 1982.)

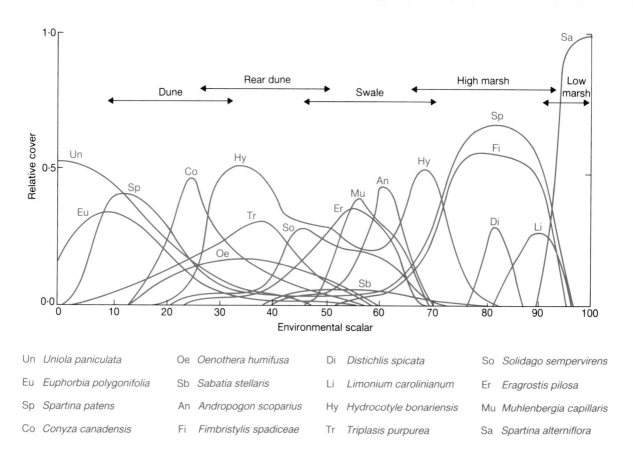

Un *Uniola paniculata*   Oe *Oenothera humifusa*   Di *Distichlis spicata*   So *Solidago sempervirens*

Eu *Euphorbia polygonifolia*   Sb *Sabatia stellaris*   Li *Limonium carolinianum*   Er *Eragrostis pilosa*

Sp *Spartina patens*   An *Andropogon scoparius*   Hy *Hydrocotyle bonariensis*   Mu *Muhlenbergia capillaris*

Co *Conyza canadensis*   Fi *Fimbristylis spadiceae*   Tr *Triplasis purpurea*   Sa *Spartina alterniflora*

Figure 15.22. Distribution of 16 herbaceous species along a gradient from dune to marsh at Core Banks, North Carolina. The 'environmental scalar' is directly related to the distance from the beach and inversely related to the depth of the water table. (From Silander & Antonovics, 1982.)

measures so the experiment can be viewed as a study of changes in relative abundance. These changes are shown in Figure 15.23.

In the high marsh, the removal of *Spartina patens* allowed a massive expansion of *Fimbristylis,* and the removal of *Fimbristylis* allowed an even more dramatic increase in *Spartina patens*. The expansion of each species after removal of the other suggests that each had been playing a major part in limiting the population size of the other—that they held habitable areas in common and realized the fundamental niche by sharing it between them. On the other hand, *Spartina patens* and *Uniola*, which have similar patterns of distribution on the dune, scarcely changed in abundance following the removal of either from the neighbourhood of the other. This suggests that their microhabitats are species-specific and that they do not share a fundamental niche. Silander and Antonovics suggest that this perturbation approach might be applied to a whole community in which all the species were reciprocally removed. Sets of species that react to each other reciprocally could then be identified and might be considered as belonging to the same guilds.

In the experimental treatments in which all the vegetation was removed, four species that had been absent or at low frequency in the community

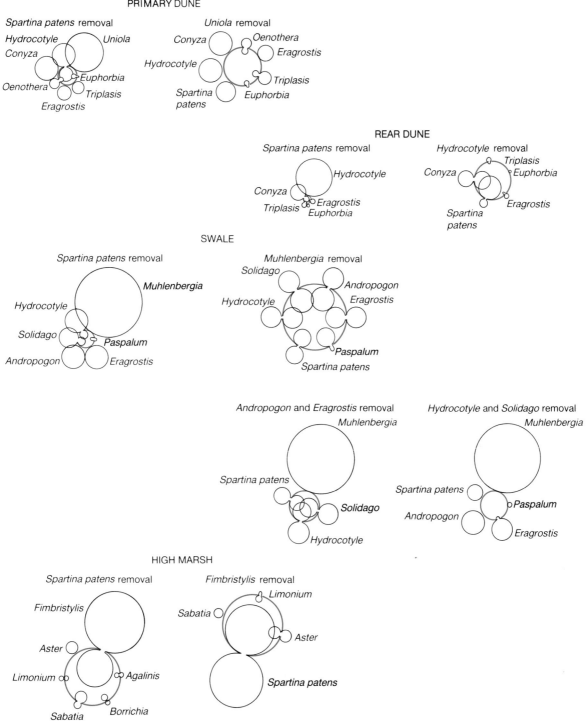

Figure 15.23. Perturbation response diagrams for species removal treatments applied to five sites from Core Banks, North Carolina. In each figure the area of the heavy central circle represents the relative abundance (percentage cover) of the species removed, *before* its removal. The areas of the outer circles represent the abundances of species in the control. The areas of the inner circles projecting into each heavy circle represents the increase in abundance of species following a removal.

irrupted as abundant populations. Thus *Sabetia, Salicornia, Triglochin* and *Setaria* can be regarded as fugitive species in the area. Their abundance appears to depend on the frequency with which disasters strike the rest of the vegetation, and their populations are then freed to expand. The habitable sites for these species are apparently normally occupied (realized) by other species.

15.6.4 Removal of predators

It is often relatively easy to set up experiments in which certain organisms, by virtue of their large size, can be excluded from a habitat by fencing. If the preferred prey species of such an organism is now free to grow, there may be repercussions extending through the community as it comes to occupy more of its habitable area. A striking example is provided by the changes in vegetation that followed the deliberate exclusion of rabbits from areas of the floristically very rich chalk grassland of southern Britain. The community rapidly degenerated to monotonous stands dominated by grasses which had previously been the preferred food of the rabbits (Tansley & Adamson, 1925). Subsequent to the experiment, the myxoma virus dramatically reduced the populations of rabbits and the same domination by grasses, with its associated loss of a diverse flora, followed over very large areas, to the consternation of conservationists.

the removal of rabbits from grassland

Other examples of the effects of removing predators are discussed in Chapter 19.

15.6.5 Introduction of a predator

There have been many occasions when aquatic plants have undergone massive population explosions after their introduction to new habitats, creating significant economic problems by blocking navigation channels and irrigation pumps and upsetting local fisheries. Notable among such species are Canadian pondweed (*Elodea canadensis*), water hyacinth (*Eichornia crassipes*) and the aquatic fern *Salvinia molesta*. The population explosions occur as a result of clonal growth accompanied by fragmentation and dispersal.

Salvinia molesta, which originated in south-eastern Brazil, has appeared since 1930 in various tropical and subtropical regions. It was first recorded in Australia in 1952 and spread very rapidly; under optimal conditions *Salvinia* has a doubling time of 2.5 days. Significant pests and parasites appear to have been absent. In 1978, Lake Moon Darra (Northern Queensland) carried an infestation of 50000 tonnes fresh weight of *Salvinia* covering an area of 400 hectares (Figure 15.24). Among possible control agents collected from *Salvinia*'s native range in Brazil, the black long-snouted weevil (*Cyrtobagous* sp.) was known to feed only on *Salvinia*. On 3 June 1980, 1500 adults were released in cages at an inlet to the lake and a further release was made on 20 January 1981. The weevil was free of any parasites or predators that might reduce its density and, by 18 April 1981, *Salvinia* throughout the lake was dark brown: samples taken 2 km apart contained 60 and 80 adult weevils m^{-2}

the introduction of insects to control pest aquatic plants

(suggesting a total population of 1 000 000 000 beetles on the lake). By August 1981, there was estimated to be less than 1 tonne of *Salvinia* left on the lake (Room *et al.*, 1981). This has been the most rapid success of any attempted biological control of one organism by the introduction of another. It was a controlled experiment to the extent that other lakes continued to

Figure 15.24. Lake Moon Darra (N. Queensland, Australia). Top: Covered by dense populations of the water fern (*Salvinia molesta*). Bottom: After introduction of weevil (*Cyrtobagous* sp.). (Photographs by P.M. Room.)

bear large populations of *Salvinia* and these will receive the weevil in due course.

The interaction of *Cyrtobagous* and *Salvinia* is just one example of a deliberate experiment in which an abundant plant has been brought to a state of rarity by the action of an introduced insect. Perhaps the greatest success story in the history of biological control was the virtual elimination of the prickly pear cactuses *Opuntia inermis* and *O. stricta* from many parts of Australia. Huge tracts of country, useless because of *Opuntia* infestation, were brought back into production after introduction of the moth *Cactoblastis cactorum* between 1928 and 1930. The moth multiplied rapidly, and by 1932 the original stands of the cactus had collapsed. By 1940 there was virtually complete control, and still today the cactus and moth exist only at a low, stable equilibrium (Dodd, 1940; Monro, 1967). Caughley and Lawton (1981) devised a model, based on the data of Dodd and Monro, which incorporates logistic plant growth, a type 2 functional response of the moth to cactus density, exponential herbivore population growth and intense competition between the larvae at high densities (see also sections 9.7.3 and 10.4.1). Their simple model simulated, in a gratifyingly precise way, the sequence of actually observed events (Figure 15.25).

the introduction of *Cactoblastis* to control prickly pear

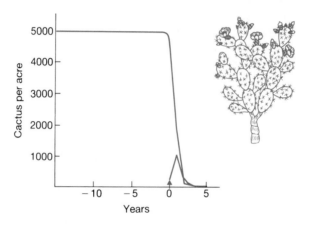

Figure 15.25. The control of *Opuntia* by *Cactoblastis*. The model derived by Caughley and Lawton (1981) accurately describes the crash of the cactus population from its pre-release level of 5000 plants per acre to a stable density of only 11 plants per acre within two years. Upper curve indicates cactus abundance; lower curve indicates moth numbers after introduction at year 0.

Almost without exception, the plant populations whose density has been dramatically reduced by an introduced predator had themselves been introduced, accidentally or deliberately, by man, usually from transoceanic origins. The animal intended to control the 'weed' has normally been brought deliberately from the natural home of the plant, having been passed through a strict quarantine procedure to ensure that it is free from its own predators, parasites and diseases, which might have been determining its abundance in the native habitat. Ideally, its population size is now determined by its interaction with its food supply.

In almost all cases, the plant species in its native habitat does not experience population explosions of the sort observed in its new habitat. And it is interesting to note that in successful cases of biological control, the plant in its exotic environment does not become extinct—its populations settle down to new but low densities at which it continues to support the introduced control organism. There are many cases in which a potential agent of biological control has been introduced without becoming established, but no cases in which a successful introduction has led to the extinction of the host plant. Finally, the successful biological control of plant infestations has almost always involved a perennial weed and an insect predator. Annual plants may too easily shrug off a specialist pest by disappearing as adult plants and reappearing from viable buried seed after the insect predator has become extinct.

The literature of biological control is a record of some remarkable ecological experiments (see, for examples, Huffaker, 1973). Nature may be unwilling to reveal its secrets unless we prod it.

15.7 Commonness and rarity

intensity and prevalence

Abundance is not just a question of density within an inhabited area—an aspect we may refer to as *intensity*. The concept must also take into account the number and size of inhabited areas within the region as a whole—an aspect we can call the *prevalence* (see also Chapter 4). Of the world's estimated three to ten million species of animal and perhaps 300 000 species of plant, most are absent from most places for most of the time; the majority are strongly clumped in their distributions. When we ask what determines the abundance of a species in a particular place we are really concerned with the tip of the tail of the problem. It is the zeros—the absences from most localities—that are of overwhelming importance in any species list that we prepare.

By distinguishing between the prevalence of a species and the intensity with which its populations are concentrated, it can be recognized that the terms 'common' or 'rare' are very unsatisfactory without qualification. A species may have a distribution that is (a) prevalent and of high intensity (widespread at high density), (b) prevalent and of low intensity, (c) localized at high intensity, and (d) localized at low intensity. Moreover, a statement about rarity and commonness of species is meaningless unless a specific area is referred to—a square metre, a hectare, a county, a national boundary, an island or a continent. A broad classification of types of rarity and commonness is given in Table 15.5.

15.7.1 The most common species

The French scientist Coquillat (1951) asked the question: what are the most abundant plants on the face of the earth? It is an interesting procedural problem in itself as to how an objective answer could be obtained to such a question. Coquillat used the extremely subjective procedure of writing to his many friends, widely spread over the globe, and asking each to produce his

Geographic range:	Large		Small	
Habitat specificity:	Wide	Narrow	Wide	Narrow
Local population size: Large, dominant somewhere	Locally abundant over a large range in several habitats	Locally abundant over a large range in a specific habitat	Locally abundant in several habitats but restricted geographically	Locally abundant in a specific habitat but restricted geographically
Plant examples	*Chenopodium album*	*Rhizophora mangle*	*Cupressus pygmaea*	*Lodoicea seychellarum* (the double coconut)
Animal examples	Rat, starling, house sparrow	Mud skipper (in mangrove swamps)		Aldabra tortoise, ostrich
Local population size: Small, non-dominant	Constantly sparse over a large range and in several habitats	Constantly sparse in a specific habitat but over a large range	Constantly sparse and geographically restricted in several habitats	Constantly sparse and geographically restricted in a specific habitat
Plant examples	*Setaria geniculata*	*Taxus canadensis*		*Torreya taxifolia*
Animal examples	Peregrine falcon	Osprey		Aldabra bush warbler Birds of paradise, Giant condor

Table 15.5. A classification of types of commonness and rarity (adapted from Rabinowitz, 1981).

a poll of plants

own list of species in order of abundance. He then generalized from their replies. The five most abundant species estimated in this way were in order: *Polygonum aviculare* (knot grass), *Capsella bursa-pastoris* (shepherd's purse), *Chenopodium album* (white goosefoot or fat hen), *Stellaria media* (chickweed) and *Poa annua* (annual meadow grass). If the list were extended to include ferns, the bracken fern, *Pteridium aquilinum*, would probably be added. All of these species (except the fern) are characteristic of disturbed ground, mainly near human dwellings. The list may have been biased because the plants most commonly seen by Coquillat's correspondents will have been those close to their homes, but all are also common in agricultural land and the list may not be far from the truth. All of these species are both world-wide in their distribution and often very common locally.

A corresponding list for the larger animals would probably include the brown rat and house mouse among mammals, the house sparrow among terrestrial birds, and perhaps an aphid among the insects. Again, the list is dominated by species that have been able to expand their populations in the environments of food and shelter created by the activities of man. The list excludes crop plants and domesticated animals, though wheat, sheep and cattle, must come high on any full list of the most abundant species. In marine environments, man's influence has been much less. The most abundant seabird is perhaps the fulmar petrel (Darwin thought it was the most abundant bird on the face of the globe). Among fish, probably the anchovy, and because of the enormous extent of the habitat, some species of the deep sea benthos, should probably also figure in the list of the most abundant organisms, but any list must be informed guesswork.

15.7.2 The most rare species

The rarest plants and animals on the face of the earth have been listed much more carefully, by conservationists interested in saving them from extinction. There are local nationalist interests in whether a particular species is rare within a national or county boundary but this has more to do with a stamp-collecting attitude to natural history than with the science of ecology. The International Union for the Conservation of Nature and Natural Resources (I.U.C.N.) is attempting to list those species that are so rare globally as to be at risk of extinction. They have not only published species lists, but also assembled the information about the particular causes of rarity and the steps that might be taken to prevent extinction or to build up populations of

the Red Data Books

endangered species. The information is published in a series of Red Data Books in which the emphasis—on flowering plants, ferns, mammals and birds—reflects the groups that most excite the conservationists' interest. Table 15.6, for example, summarizes information about the most extremely endangered species of American mammals, extracted from the I.U.C.N. Red Data Book (Thornback & Jenkins, 1982).

15.7.3 The causes of rarity

We can envisage the surface of the globe as providing a fine mosaic of conditions and resources that define different fundamental niches and represent places that are 'habitable' for different species. If we knew enough about the biology of individual species and the details of the earth's surface, we could imagine constructing a map of the world in which the *habitable area* of every species was drawn: the area within which each species could maintain a population provided that (a) it had the opportunity to colonize and (b) it was not excluded by competitors or by predators or parasites. The abundance of both plants and animals can be related to the frequency and distribution of such habitable areas.

(i) A species may be rare because its habitable areas are rare or small. Physicochemical conditions that are themselves uncommon in nature may bear a flora and fauna that is specialized for this rare condition. Serpentine rock, with its unusual mixture of generally toxic metals, provides such a rare environment on several continents, and specialist plant species occupying this are as rare as the serpentine outcrops to which they are limited. Similarly, the distribution of an insect that is a specialist on a single host plant species cannot be more than a flight's distance from the plant on which it depends. If the host plant is rare, so are the specialist parasites and predators that depend on it.

Table 15.6. Examples of endangered species of mammal in North and South America.

Species	Present status and habitat	Cause of present rarity
Insectivores, bats, lagomorphs and rodents		
Cuban solenodon (*Solenodon cubanus*)	Endemic to Cuba, where its numbers and range are contracting	Loss of forest habitat and predation from feral cats. Caught as a zoological rarity
Hispaniolan solenodon (*Solenodon paradoxus*)	Endemic to Hispaniola	Similar to Cuban solenodon
Volcano rabbit (*Romerolagus diazi*)	Slopes of volcanoes near Mexico City present as three subalpine populations	Habitat destruction, hunting for food and sport
Gray bat (*Myotis grisescens*)	Perhaps 1 500 000 individuals, of which half overwinter in a single cave in S.E. United States	Disturbance to roosting sites by cavers and vandals
Morrow Bay kangaroo rat (*Dipodomys heermanni morroensis*)	Morrow Bay, California. Census in 1957 suggested 8000 individuals; in 1971, 3000; and in 1977, 1200–1500	Dependent on early seral stages of chaparral succession, which are reduced by more efficient fire control. Loss of habitat to suburban housing. Predation by feral cats, and road-kills
Vancouver Island marmot (*Marmota vancouveriensis*)	Perhaps 50–100 individuals in isolated mountain colonies on Vancouver Island	Loss of habitat from ski developments and logging. Needs migration corridors which are destroyed. Hunting and collection for scientific purposes
Primates		
Cotton top tamarin (*Saguinus oedipus oedipus*)	N.W. Columbia. Unlikely that many forests remain large enough to support a breeding population	Loss of habitat from forest destruction. 14 000 imported to U.S.A. for pets and biomedical research. Estimated export from Columbia of 30–40 000 between 1960–75
Central American squirrel monkey (*Saimiri oerstedi*)	Panama and Costa Rica. Forests and shrubby areas. Great decline since 1950	Reduction of forest habitat. ? decline after pesticide spraying to control yellow fever and malaria. Capture and export as pets
Golden lion tamarin (*Leontopithecus rosalia*)	S.E. Brazil. By 1980 almost certainly less than 100 animals. Likely to be extinct by 1985–90	Habitat loss by forest destruction in favour of lumber, agriculture, pasture and housing. 2–300 were exported as pets between 1960 and '65
Carnivores		
Spectacled bear (*Tremarctos ornatus*)	From Venezuela to Bolivia. Numbers very uncertain—perhaps 100 in Venezuela, isolated individuals in Ecuador and Colombia, perhaps 850 in Peru	Habitat loss due to growth of human activity but also hunted for skins and meat. Shot as a marauder of crops.

Camelidae

Vicuna (*Vicugna vicugna*)	Rangelands of Central Andes. Reduced to 6000 animals in 1965. Conservation has allowed populations to rise to 80–85 000 by 1981	Extremely abundant before Columbus—but mass slaughter for wool. Could be increased and sustain a harvest (hides, wool, meat). May require culling to prevent habitat deterioration, but such culls are political dynamite among some conservationists

(ii) A species may be rare because habitable sites remain habitable for too short a time. Thus, the rarity of the Morrow Bay kangaroo rat (Table 15.6) is attributed to its dependence on early stages of chaparral succession in California; more efficient fire control means that such early successional phases are now less often renewed. A species may also be rare because some habitable sites are beyond its range of dispersal.

(iii) A species may be rare because other species make some sites uninhabitable, by driving the first species to extinction, by competitive exclusion or by heavy rates of predation or parasitism.

Cases (i) to (iii) above relate to the 'prevalence' aspect of abundance; they are all factors which determine the number and size of areas inhabited by a species. In contrast, cases (iv) and (v) below involve 'intensity'—the density of individuals within habitable areas.

(iv) A species may be rare because of the low availability of critical resources such as food, safe sites, etc., within habitable areas. For example, the food resources of vertebrate flesh-eaters are significantly less abundant than the food of their prey; and predatory birds and mammals are invariably rarer than the populations they attack (golden eagles are a good example).

(v) A species may be rare because the genetic variation amongst its members narrowly limits the range of areas that are habitable for it. Many of the 'rare and endangered' plant species listed in the Red Books of the I.U.C.N. do not reproduce sexually (they are apomictic or parthenogenetic)

(vi) A species may be rare because the phenotypic plasticity of its individuals limits the range of areas that are habitable.

(vii) A species may be rare because competitors, predators, parasites or human collectors maintain its populations below the level set by resources within habitable areas. It is interesting to note that in the case of humans, collected species become more valuable and more sought after as they become rarer. The final stage of decline of a rare species may be its extinction by collectors.

Changes in the abundance or rarity of species easily become matters of concern, especially when increasing abundance of a species threatens to make it a nuisance (classifying it as a pest) or when decline threatens its extinction and makes it a matter for the conservationist. Some of this concern is nostalgia; some is more rationally based. Ultimately, we might hope that ecology may develop to a stage at which we understand not only how abundance and rarity are determined, but how we may control them. Only then will we be in a position to make the faunas and floras that we want, by conserving them, re-creating them, or creating new reliable combinations for our own food or pleasure.

PART 4

Communities

Introduction

In nature, areas of land and volumes of water contain assemblages of different species, in different proportions and doing different things. These communities of organisms have properties that are the sum of the properties of the individual denizens plus their interactions. It is the interactions that make the community more than the sum of its parts. Just as it is a reasonable aim for a physiologist to study the behaviour of different sorts of cells and tissues and then attempt to use a knowledge of their interactions to explain the behaviour of the whole organism, so an ecologist may use his knowledge of interactions between organisms in an attempt to explain the behaviour and structure of a whole community. In fact, structural patterns can be expected to emerge in any entity composed of interacting components, whether living or non-living. For example, when crystals of potassium ferricyanide are dropped into a solution of copper sulphate, a forest of branch-like shapes grows, consumes resources (copper and ferricyanide ions) and develops a predictable structure. Community ecology is the study of emergent properties in the structure and behaviour of multi-species biological assemblages.

We consider first the nature of the community—what we mean by the term. Usually we mean some unit of the natural world that we (human investigators) can categorize according to features that mean something to us. We impose an anthropocentric process of selection in deciding what will be regarded as a community. It is vitally important to be aware that the categories erected may lack any relevance to the lives of the individual organisms within the communities. An oakwood or an estuary or the rumen of a cow may be recognized as communities by us, but they may be on a quite irrelevant scale in the life of a caterpillar, a shrimp or a rumen protozoan. To a caterpillar in an oakwood, the only community that matters in its life may consist of a few leaves on a branch, together with a handful of competitors and predators that visit its cluster of leaves. The organism's-eye view of the community in which it lives will differ from species to species. In Chapter 16 we impose our human perspective and describe patterns in the composition of communities in space and time.

In Chapter 17 we examine the ways in which arrays of feeders and their food may bind the inhabitants of a community into a web of interacting elements, through which energy and matter is moved. Later (Chapter 21) we return to this theme and consider the implications of food web structure for the dynamics of whole communities and, in particular, those aspects of structure that contribute to stability.

Chapters 18 and 19 return to the topics of earlier chapters in the book and examine the extent to which competition, predation and disturbance influence the patterns we recognize at the community level. Chapter 20 is concerned with an area of ecology that has been extraordinarily fruitful in producing new ecological insights—islands and island communities. Finally, we attempt a synthesis (Chapter 22). The science of ecology is young and a synthesis is only beginning to emerge.

To pursue an analogy we introduced earlier, the study of ecology at the community level is a little like making a study of watches and clocks. A collection can be made and the contents of each timepiece classified. We can recognize characteristics that they have in common in the way they are constructed and what they do. We can recognize patterns and hierarchy in the collection as a whole. But to understand how they work, they must be taken to pieces, studied and put back together again. We will have understood the nature of natural communities when we have taken them to pieces and *know* how to recreate them.

16 The Nature of the Community

16.1 Introduction

Physiological and behavioural ecologists are concerned primarily with *organisms*. They seek to understand the match between individuals and the environmental conditions in which they exist (Chapters 1–5 and 14). Coexisting individuals of a single species possess characteristics such as density, sex ratio, age-class structure, rates of natality and immigration, mortality and emigration, which are unique to *populations*. We explain the behaviour of a population in terms of the behaviour of the individuals that comprise it. Finally, activities at the population level have consequences for the next level up—that of the *community*. The community is an assemblage of species populations which occur together in space and time. The principal focus of the community ecologist is the manner in which groupings of species are distributed in nature and the ways in which these groupings can be influenced, or caused, by interactions between species and by the physical forces of their environment.

communities have emergent properties not possessed by the individual populations that comprise them

We have already seen that organisms of the same and different species interact with each other in processes of mutualism, parasitism, predation and competition. The nature of the community is obviously more than just the sum of its constituent species. It is their sum plus the interactions between them. Thus there are *emergent* properties that appear when the community is the focus of attention, as there are in many other cases in which we are concerned with the behaviour of complex mixtures. A cake has emergent properties of texture and flavour that are not apparent simply from a survey of the ingredients. A sandy beach has emergent properties in the arrangement of sand grains and pebbles of different sizes that give it pattern. In the case of ecological communities, the diversity of species, the limits to the similarity of competing species, the structure of the food web, community biomass and community productivity are examples of emergent properties. A primary aim of community ecology is to determine whether repeating patterns in such properties exist, even when there are great differences in the particular species that happen to be assembled together.

Traditionally, another category of ecological study has been set apart: the *ecosystem*. This comprises the biological community together with its physical environment. However, while the distinction between community and ecosystem may be helpful in some ways, the implication that communities and ecosystems can be studied as separate entities is wrong. No ecological system, whether individual, population or community, can be studied in isolation from the environment in which it exists. Thus we will not distinguish a separate ecosystem level of organization. Nor will discussions of ecological energetics and nutrient dynamics be segregated as ecosystem rather than community topics (as has usually been the practice in ecology textbooks). It is true that these phenomena depend explicitly on fluxes between living and non-living components of ecosystems, but the fundamental point is that they simply represent supplementary means of approaching an understanding of community structure.

the detection and description of patterns in communities

Doing science at the community level presents daunting problems because the database may be enormous and complex. A first step is usually to search for patterns in community structure and composition. The need to develop procedures for describing and comparing communities has dominated the development of community ecology. In essence this has been a search for simple ways to describe complex systems.

The recognition of patterns represents an important step in the development of all sciences (the Periodic Table in chemistry, movements of the heavenly bodies in astronomy, and so on). Patterns are repeated consistencies, such as the repeated grouping of the same species in different places (or the same growth forms, the same productivities, the same rates of nutrient turnover, etc.). Recognition of patterns leads in turn to the forming of hypotheses about the causes of these patterns. The hypotheses may then be tested by making further observations or by doing experiments.

communities can be recognized at a variety of levels—all equally legitimate

A community can be defined at any size, scale or level within a hierarchy of habitats. At one extreme, broad patterns in the distribution of community types can be recognized on a global scale. The temperate forest

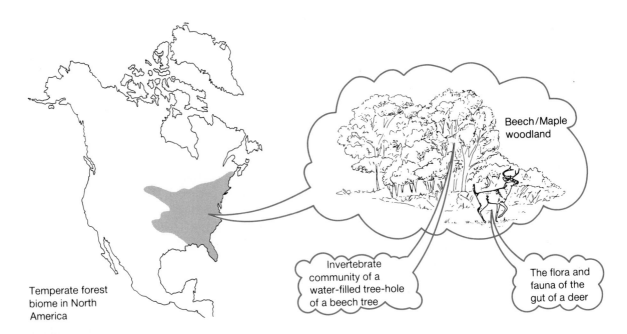

Figure 16.1. We can identify a hierarchy of habitats, nesting one into the other: temperate forest biome in North America; beech–maple woodland in New Jersey; water-filled tree-hole, or mammalian gut. The ecologist may choose to study the community which exists on any of these scales.

biome is one example; its range in North America is shown in Figure 16.1. At this scale, ecologists usually recognize climate as the overwhelming factor that determines the limits of vegetation types. At a finer scale, the temperate forest biome in parts of New Jersey is represented by communities of two species of tree in particular, beech and maple, together with a very large number of other, less conspicuous species of plants, animals and micro-organisms. Study of the community may be focused at this level. On an even finer habitat scale, the characteristic invertebrate community that inhabits water-filled holes in beech trees may be studied, or the flora and fauna in the gut of a deer in the forest.

Among these various levels of community study, no one is more legit-imate than another. The level appropriate for investigation depends on the sorts of questions that are being asked.

Community ecologists sometimes consider all of the organisms existing together in one area, though it is rarely possible to do this without a large team of taxonomists. Others restrict their attention within the community to a single taxonomic group (e.g. birds, insects or trees) or a group with a par-ticular activity (e.g. herbivores, detritivores). Thus we may refer to the bird community of a forest or the detritivore community of a stream.

16.2 The description of community composition

species richness: the number of species present in a community

One way to characterize a community is simply to count or list the species that are present. This sounds a straightforward procedure that enables us to describe and compare communities by their species richness. In practice, though, it is often surprisingly difficult, partly because of taxonomic prob-lems, but also because only a subsample of the organisms in an area can usually be counted. The number of species recorded then depends on the number of samples that have been taken, or on the volume of the habitat that has been explored. The commonest species are likely to be represented in the first few samples, and as more samples are taken, rarer species will be added to the list. At what point does one cease to take further samples? Ideally the investigator should continue to sample until the number of species reaches a

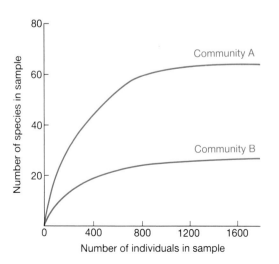

Figure 16.2. Relationship between species richness and number of individual organisms from two contrasting hypothetical communities. Community A has a total species richness considerably in excess of Community B.

plateau (Figure 16.2). In practice, this is not usually achieved. The species richness of different communities can then properly be compared only if they are based on the same sample sizes (in terms of area of habitat explored, time devoted to sampling or, best of all, number of individuals or modules included in the samples).

16.2.1 Diversity indices

diversity incorporates richness and common-ness and rarity

An important aspect of the numerical structure of communities is completely ignored when the composition of the community is described simply in terms of the number of species present. It misses the information that some species are rare and others common. Intuitively, a community of seven species with equal numbers in each seems more diverse than another, again consisting of seven species, but with 40% of individuals belonging to the commonest species and only 5% in each of the three rarest (Table 16.1). Yet each community has the same species richness.

Table 16.1. Examples of calculation of indices of diversity in four hypothetical communities. S, species richness; D, Simpson's diversity index; E, Simpson's equitability index; H Shannon diversity index; J, Shannon equitability index; P_i, proportion of total individuals in the ith species.

If the community we are interested in is clearly defined (e.g. the warbler community of a woodland), counts of the number of individuals in each species may suffice for many purposes. However, if we are interested in all the animals in the woodland, it makes very little sense to use the same sort of quantification for protozoa, woodlice, birds and deer. Their enormous disparity in size means that counts would be very misleading. We are also in great difficulty if we try to count plants (and other modular organisms). Do we count the number of shoots, leaves, stems or genets? One way round this problem is to describe the community in terms of the biomass (or the rate of production of biomass) per species per unit area.

The simplest measure of the character of a community that takes into

Community 1			Community 2			Community 3			Community 4		
P_i	P_i^2	$P_i \ln P_i$	P_i	P_i^2	$P_i \ln P_i$	P_i	P_i^2	$P_i \ln P_i$	P_i	P_i^2	$P_i \ln P_i$
0.143	0.0205	−0.278	0.40	0.16	−0.367	0.1	0.01	−0.23	0.40	0.16	−0.367
0.143	0.0205	−0.278	0.20	0.04	−0.322	0.1	0.01	−0.23	0.20	0.04	−0.322
0.143	0.0205	−0.278	0.15	0.0225	−0.285	0.1	0.01	−0.23	0.15	0.0225	−0.285
0.143	0.0205	−0.278	0.10	0.01	−0.230	0.1	0.01	−0.23	0.10	0.01	−0.230
0.143	0.0205	−0.278	0.05	0.0025	−0.150	0.1	0.01	−0.23	0.025	0.0006	−0.092
0.143	0.0205	−0.278	0.05	0.0025	−0.150	0.1	0.01	−0.23	0.025	0.0006	−0.092
0.143	0.0205	−0.278	0.05	0.0025	−0.150	0.1	0.01	−0.23	0.025	0.0006	−0.092
						0.1	0.01	−0.23	0.025	0.0006	−0.092
						0.1	0.01	−0.23	0.025	0.0006	−0.092
						0.1	0.01	−0.23	0.025	0.0006	−0.092

$S \quad = 7$

$D = \dfrac{1}{\Sigma P_i^2} \quad = 6.97$

$E = \dfrac{D}{S} \quad = 1.00$

$H = \Sigma P_i \ln P_i \quad = 1.95$

$J = \dfrac{H}{\ln S} \quad = 1.00$

$S = 7$

$D = 4.17$

$E = 0.60$

$H = 1.65$

$J = 0.85$

$S = 10$

$D = 10.00$

$E = 1.00$

$H = 2.30$

$J = 1.00$

$S = 10$

$D = 4.24$

$E = 0.42$

$H = 1.76$

$J = 0.76$

account both the abundance patterns and the species richness is Simpson's diversity index. This is calculated by determining for each species, the proportion of individuals or biomass that it contributes to the total in the sample, i.e. the proportion is P_i for the ith species:

Simpson's diversity index

$$\text{Simpson's index } D = \frac{1}{\displaystyle\sum_{i=1}^{S} P_i^2}$$

where S is the total number of species in the community (i.e. the richness). As required, the value of the index depends on both the species richness and the evenness (equitability) with which individuals are distributed among the species. Thus, for a given richness, D increases with equitability, and for a given equitability, D increases with richness. Note that it is possible for a species-rich but inequitable community to have a lower index than one that is less species-rich, but highly equitable.

Equitability can itself be quantified by expressing Simpson's index, D, as a proportion of the maximum possible value D would assume if individuals were completely evenly distributed among the species. In fact $D_{\max} = S$. Thus:

'equitability' and 'evenness'

$$\text{equitability } E = \frac{D}{D_{\max}} = \frac{1 \displaystyle\sum_{i=1}^{S} P_i^2}{S}$$

Equitability assumes a value between 0 and 1.

the Shannon diversity index

Another index that is frequently used is the Shannon diversity index, H. This again depends on an array of P_i values. Thus,

$$\text{diversity } H = -\sum_{i=1}^{S} P_i \ln P_i$$

$$\text{and equitability } J = \frac{H}{H_{\max}} = \frac{\displaystyle\sum_{i=1}^{S} P_i \ln P_i}{\ln S} .$$

Note that different authors use different logarithms (base 10, e or 2), which should obviously be specified when calculating H.

The behaviour of Simpson's index and the Shannon index are illustrated for a number of hypothetical communities in Table 16.1.

16.2.2 Rank–abundance diagrams

Of course, attempts to describe a complex community structure by one single attribute, such as richness, diversity or equitability, can be criticized because so much valuable information is lost. A more complete picture of the distribution of species abundances in a community makes use of the full array of P_i values by plotting P_i against rank. Thus the P_i for the most abundant species is

595   NATURE OF THE COMMUNITY

plotted first, then the next most common and so on until the array is completed by the rarest species of all. A rank–abundance diagram can be drawn for the number of individuals, or for the area of ground covered by different sessile species, or for the biomass contributed to a community by the various species.

Three of the many forms that a rank–abundance diagram can take are shown in Figure 16.3. The least equitable distribution of these is the *geometric series*. In its ideal form, this has the numbers of the most common species contributing a fraction d of the total of all species combined ($d = P/N = P_i$), the next most abundant species a fraction d of the remainder, and so on. Such a series produces a straight line when plotted as log abundance against rank. A more equitable distribution is represented by the 'broken stick' model, so called because each species contributes to the total a fraction that can be represented by the fragments of a stick that has been broken at random points. Intermediate curves generally approximate to a log-normal distribution (see May, 1975b, for a more detailed discussion and the mathematical details). Rank–abundance diagrams, like indices of richness, diversity and equitability, should be viewed simply as abstractions of the highly complex structure of communities which may be useful when making comparisons.

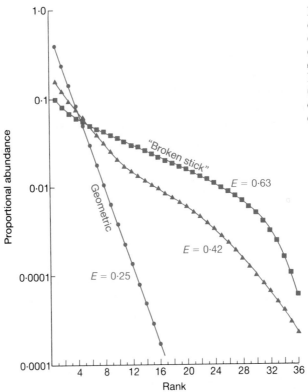

Figure 16.3. Examples of rank–abundance diagrams for three contrasting hypothetical communities. Equitability values (Simpson's *E*) corresponding to the curves are indicated.

Taxonomic composition and species diversity are just two of many possible ways of describing a community. Historically, most of the emphasis on diversity has been in terms of species, perhaps reflecting the dominance of taxonomists. It is important to note that there are other aspects of diversity that may be just as important, or even more important, when considering community structure. Thus many species have life cycles that make them diverse contributors to a community (tadpoles/frogs, caterpillars/butterflies) or have structure that offers a diversity of resources (a tree compared with a herb, a cow compared with a nematode worm). These kinds of diversity also deserve consideration.

the energetics approach: an alternative to taxonomic description

Another alternative (not necessarily better but quite different) is to describe communities in terms of their standing crop and the rate of production of biomass by plants, and its use and conversion by heterotrophic organisms. Studies that are oriented in this way may begin by describing the food web and then define the biomasses at each trophic level and the flow of energy and matter from the physical environment through the living organisms and back to the physical environment. Such an approach can (at least in theory) allow patterns to be detected among communities that may have no taxonomic features in common. This approach will be discussed in Chapter 17.

16.3 Community patterns in space

There may be communities that are separated by clear, sharp boundaries, where groups of species lie adjacent to, but do not intergrade into, each other. If they exist, they are exceedingly rare and exceptional. The meeting of terrestrial and aquatic environments might appear to be a sharp boundary but its ecological unreality is emphasized by the otters and frogs that regularly cross it and the many aquatic insects that spend their larval lives in the water but their adult lives as winged stages on land or in the air. Moreover, water levels change with the season, and because of this, together with the movements of water at its margins due to waves and splash, the sharp boundary between land and water is really a gradient of intermediate conditions.

are communities discrete entities with sharp boundaries?

On land, quite sharp boundaries occur between the vegetation types on acid and basic rocks where outcrops meet, or where serpentine (a term applied to a mineral rich in magnesium silicate) and non-serpentine rocks meet. However, even in such situations, minerals are leached across the boundaries, which become increasingly blurred. The safest statement we can make about community boundaries is probably that they do not exist, but that some communities are much more sharply defined than others. The ecologist is usually better employed looking at the ways in which communities grade into each other than in searching for sharp cartographic boundaries.

16.3.1 Gradient analysis

Figure 16.4 shows a variety of ways of describing the distribution of vegetation on the Great Smoky Mountains (Tennessee), where the species of tree present give the vegetation its main character. Figure 16.4a shows the characteristic associations of the dominant trees on the mountainside, drawn as if the communities had sharp boundaries. The mountainside itself provides a range of conditions for plant growth, and two of these, altitude and moisture, may be particularly important in determining the distribution of the various

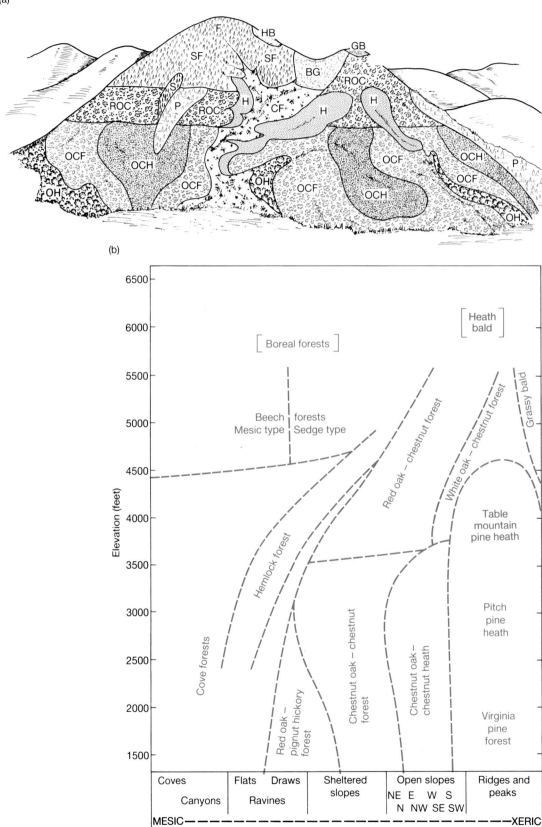

tree species. Figure 16.4b shows the dominant associations graphed in terms of these two environmental dimensions. Finally, Figure 16.4c shows the abundance of each individual tree species (expressed as a percentage of all tree stems present) plotted against the single gradient of moisture.

Figure 16.4a is a subjective analysis which acknowledges that the vegetation of particular areas differs in a characteristic way from that of other areas. It could be taken to imply that the various communities are sharply delimited. Figure 16.4b gives the same impression. Note that both Figures 16.4a and b are based on descriptions of the *vegetation*. However, Figure 16.4c sharpens the focus by concentrating on the pattern of distribution of the individual *species*. It is then immediately obvious that there is considerable overlap in

(c)

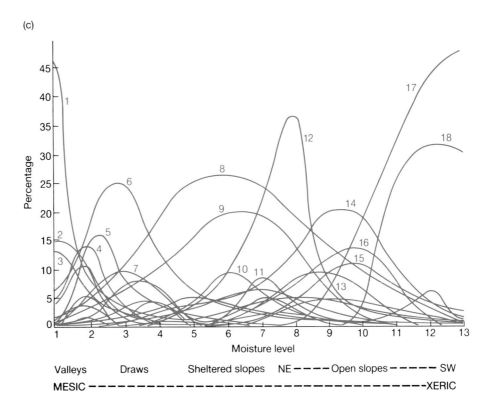

Figure 16.4. Three contrasting descriptions of distributions of the characteristic dominant tree species of the Great Smoky Mountains, Tennessee. Facing page. (a) Topographic distribution of vegetation types on an idealized west-facing mountain and valley. (b) Idealized graphical arrangement of vegetation types according to elevation and aspect. (c) Above. Distributions of individual tree populations (percentage of stems present) along the moisture gradient. (Modified from Whittaker, 1956.) Vegetation types: BG, beech gap; CF, cove forest; F, Fraser fir forest; GB, grassy bald; H, hemlock forest; HB, heath bald; OCF, chestnut oak–chestnut forest; OCH, chestnut oak–chestnut heath; OH, oak–hickory; P, pine forest and heath; ROC, red oak–chestnut forest; S, spruce forest; SF, spruce–fir forest; WOC, white oak–chestnut forest. Major species: 1, *Halesia monticola*; 2, *Aesculus octandra*; 3, *Tilia heterophylla*; 4, *Betula alleghaniensis*; 5, *Liriodendron tulipifera*; 6, *Tsuga canadensis*; 7, *Betula lenta*; 8, *Acer rubrum*; 9, *Cornus florida*; 10, *Carya alba*; 11, *Hamamelis virginiana*; 12, *Quercus montana*; 13, *Quercus alba*; 14, *Oxydendrum arboreum*; 15, *Pinus strobus*; 16, *Quercus coccinea*; 17, *Pinus virginiana*; 18, *Pinus rigida*.

the distributions of
species along gradients
end not with a bang but a
whimper

their abundance—there are no sharp boundaries. The various tree species are
now revealed as being strung out along the gradient with the tails of their
distributions overlapping. The limits of the distributions of each species end
'not with a bang but a whimper'.

Many other gradient studies have produced similar results. Figure 16.5a
shows ecological response curves for selected grass species along a pH
gradient in Britain, and Figure 16.5b shows the distribution of large invertebrates along an intertidal beach in Canada (here a gradient in particle size may
be the key condition that determines the pattern of distribution). Figure 15.22
provides a further example.

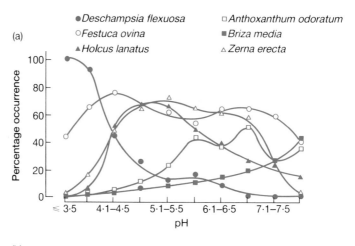

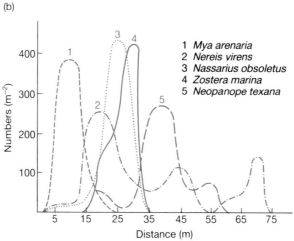

Figure 16.5. (a) Distribution gradients for selected grass species along a pH gradient in
England (after Grime & Lloyd, 1973). (b) Distribution gradients for macrofaunal
species along an oyster bed in Canada. 1. *Mya arenaria*; 2. *Nereis virens*; 3. *Nassarius
obsoletus*; 4. *Zostera marina*; 5. *Neopanope texana* (data from Hughes & Thomas, 1971).

The steeper the gradient of some key environmental condition, the more
sharply defined will be the edges of the distribution of contrasting groups of
species. In more extreme environments, arid or cold, gradients may be
particularly steep. Here, community dominants often appear to define sharp

edges to a community (Figure 16.6), and this is reflected in the relative ease with which ecologists working in such communities recognize, classify and even name distinct community types. In more temperate regions, this temptation is more easily resisted.

When a key environmental factor varies along a gradient, there are likely to be points where the relative competitive statuses of species switch, so that

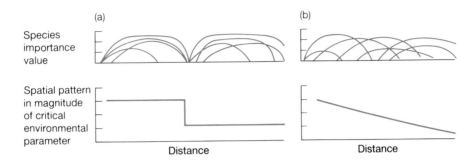

Figure 16.6. Whether or not predictable communities replace each other sharply depends on spatial pattern in important underlying environmental conditions. (a) If conditions vary in a stepwise manner groups of species may replace each other at the boundary; (b) if conditions vary continuously the relative importances of dominant species are also likely to vary continuously.

changes in competitive status along a gradient may generate sharp boundaries

one dominant species is replaced by another. We might then expect the interaction between species themselves to generate and intensify boundaries between them. One situation in which this could occur is in perennial grassland communities, where clonal growth may produce zones of tension between two dominant species; the composition of communities might then swing suddenly from one stable mixture to another at some point on the gradient. Similarly, a switch in competitive dominance of plants in response to decreasing grazing pressure at increasing distances from a water-hole or a rabbit warren could generate a sharp boundary. This problem would repay study. However, even if this type of sudden switch from one equilibrial vegetation type to another does occur, the difference between seasons and years will almost always mean that the position of the change-over point is continually shifting, so what might be sharp boundaries in theory become blurred gradients in practice. The idea that gradients may generate sudden switches in community composition has been explored by Pielou (1975) and by Noy-Meir (1975).

choice of gradient is almost always subjective

Perhaps the major criticism of gradient analysis as a way of detecting pattern in communities is that the choice of the gradient is almost always subjective. The investigator searches for some feature of the environment that appears to him to matter to the organisms. He then organizes the data that he has about the species concerned along a gradient of that factor. It is not necessarily the most appropriate factor to have chosen. The fact that the species from a community can be arranged in a sequence along a gradient of some environmental factor does not prove that this factor is the most important. It may only imply that the factor chosen is more or less loosely correlated with whatever really matters in the lives of the species involved. Gradient analysis is only a small step on the way to the objective description of communities.

16.3.2 The ordination and classification of communities

Formal statistical techniques have been defined to take the subjectivity out of community description. These techniques allow the data from community studies to sort themselves, without the investigator putting in any preconceived ideas about which species tend to be associated with each other or which environental variables correlate most strongly with the species distributions. One such technique is *ordination*.

in ordination, communities are displayed on a graph so that those most similar in composition are closest together . . .

Ordination is a mathematical treatment which allows communities to be organized on a graph so that those that are most similar in both species composition and relative abundance will appear closest together, while communities which differ greatly in the relative importance of a similar set of species, or which possess quite different species, appear far apart. The procedure is time consuming and is now invariably performed on a computer. Details of the methods are provided by Gauch (1982). Figure 16.7a shows the application of ordination to community data from 50 separate stands (communities) of vegetation on Welsh sand dunes. The data consisted of lists of the species present and their abundances.

The axes of the graphs are derived mathematically solely from the species compositions of the various stands; they represent dimensions that effectively summarize community patterns. The interpretation of these patterns in terms of environmental variables is a second step, in which the scatter of points in the ordination is examined to see if the axes correspond to ecologically meaningful gradients. Obviously the success of the method now depends

subsequently, it is necessary to ask what varies along the axes of an ordination graph

on our having sampled an appropriate variety of environmental variables. This is a major snag in the procedure—we may not have measured the qualities in the environment that are most relevant. In the plant ecologist's use of ordination it is usual to look at environmental factors such as moisture, nutrient levels, pH, rates of oxygen diffusion, etc. However, gradients in community composition may be produced by grazing pressure, disease, and a host of other interactions. There is no way in which the ecologist can determine in advance what factors will be most relevant, and there is no way in retrospect of discovering if a vital factor has been forgotten.

The sand dune study (Figure 16.7a) was quite successful in pin-pointing environmental conditions that were closely related to the species composition of the communities. On to Figures 16.7 b, c, d and e have been superimposed values corresponding to the soil moisture, oxygen diffusion rate, pH and sodium concentration at each site. The x-axis of the ordination is positively related to soil moisture; the lowest values for moisture tend to be to the left and the highest values to the right of the graph. In contrast, there is a tendency for the lowest values of oxygen diffusion rate to be towards the right (i.e. a negative relationship exists between the x-axis and diffusion rate). Neither pH nor sodium concentration exhibit any clear relationship with community structure in these sand dunes.

What do these results tell us? First, and most specifically, the correlations with environmental factors, revealed by the analysis, give us some specific hypotheses to test about the relationship between community composition and underlying environmental factors. (Remember that correlation does not necessarily imply causation. For example, soil moisture and community composition may vary together because of a common response to another environmental factor. A direct causal link can only be proved by controlled experimentation.)

A second, more general point is relevant to the discussion of the nature of the community. The results of this ordination emphasize that under a

ordination can generate hypotheses for subsequent testing

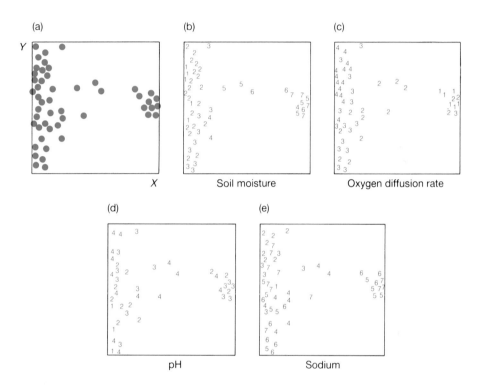

Figure 16.7. (a) Ordination of 50 stands (communities) of vegetation in Welsh sand dunes; (b) ordination with scores for soil moisture from each stand superimposed; (c) ordination with scores for oxygen diffusion rate from each stand superimposed; (d) ordination with scores for pH from each stand superimposed; (e) ordination with scores for sodium concentration from each stand superimposed. (From Pemadasa *et al.*, 1974.)

particular set of environmental conditions, a predictable association of species is likely to occur. It shows that community ecologists have more than just a totally arbitrary and ill-defined set of species to study.

The results of an ordination study on the invertebrate communities existing at 34 locations in streams in southern England are shown in Figure 16.8a. In this case, clear relationships were revealed between pH and the position of communities along the *x*-axis, and between average water temperature and their position along the *y*-axis. Once again, communities with

603   NATURE OF THE COMMUNITY

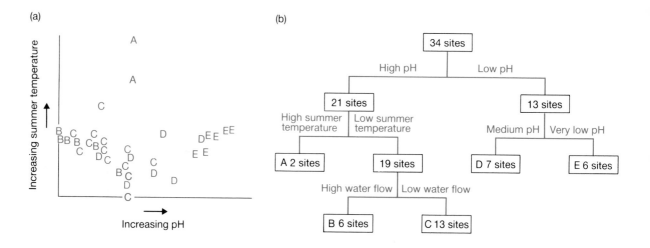

(a) [diagram: Increasing summer temperature vs Increasing pH, with letters A–E]

(b) [classification tree]
- 34 sites
  - High pH → 21 sites
    - High summer temperature → A 2 sites
    - Low summer temperature → 19 sites
      - High water flow → B 6 sites
      - Low water flow → C 13 sites
  - Low pH → 13 sites
    - Medium pH → D 7 sites
    - Very low pH → E 6 sites

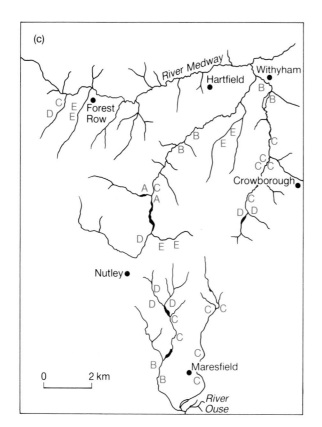

Figure 16.8. Analysis of 34 invertebrate communities in streams in southern England. (a) Ordination (letters A to E indicate community classes derived from classification). (b) Classification. (c) Distribution of community classes A to E in the stream catchment. (From Townsend *et al.*, 1983.)

predictable compositions occurred under specific sets of environmental conditions. If we knew the pH of a new stream in the area we could use the ordination data to predict the invertebrate fauna, and if we knew only the fauna we might predict the pH.

Classification, as opposed to ordination, begins with the assumption that communities consist of relatively discrete entities. It produces groups of related communities by a process conceptually similar to taxonomic classification. In taxonomy, similar individuals are grouped together in species, similar species in genera, and so on. In community classification, communities with similar species compositions are grouped together in sub-sets, and similar sub-sets may be further combined if desired (see Gauch, 1982 for details of the procedure).

The stream invertebrate communities ordinated in Figure 16.8a have also been subjected to classification (Figure 16.8b) to produce five groupings. At each division, which is based solely on species composition, a significant difference in environmental conditions exists and the important factors are indicated. The relationship between ordination and classification can be gauged by noting that communities falling into Classes A to E, derived from classification, are fairly distinctly separated on the ordination graph. In both cases, pH was implicated as a critical environmental condition, and communities in Class E (associated with very low stream pH) were characterized particularly by the presence of larvae of the stoneflies *Leuctra nigra* and *Nemurella picteti*, the caddis-fly *Plectrocnemia conspersa* and the alderfly *Sialis fuliginosa*, all of which are tolerant of these extreme conditions.

The spatial distribution of each class of community in the stream system itself is shown in Figure 16.8c. Note that there is little consistent spatial relationship; communities in each class are dotted about the stream catchment. This illustrates one of the strengths of classification and/or ordination. The methods show the order or structure in a series of communities without the necessity of picking out some supposedly relevant environmental variable in advance, a procedure that is necessary for gradient analysis.

A very detailed system of classification was applied to the benthic invertebrate communities of Oslofjord in Scandinavia (Figure 16.9). Seven reasonably distinct classes of communities were derived, and in this case there was consistent spatial arrangement in the fjord, with community structure varying according to position in relation to the city of Oslo and its output of pollutants. Thus, community Class A was characterized particularly by *Capitella*, *Polydora* and *Heteromastus* sp., invertebrates which can tolerate organically highly enriched sediments. This approach to community analysis shows its role in applied ecology, by revealing the extent of pollution. Further analyses of the communities through time could indicate whether anti-pollution measures were having the desired effects, or whether communities characteristic of polluted waters were becoming more common.

16.3.3 The problems of boundaries in community ecology

In the first quarter of this century there was considerable debate about the nature of the community. Clements (1916) conceived of the community as a sort of *superorganism* whose member species were tightly bound together both now and in their common evolutionary history. Thus individuals, populations and communities bore a relationship to each other which resembled that between cells, tissues and organisms.

In contrast, the *individualistic* concept devised by Gleason (1926) and others saw the relationship of coexisting species as simply the results of

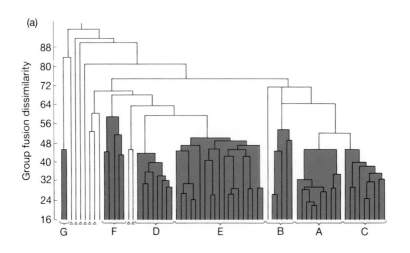

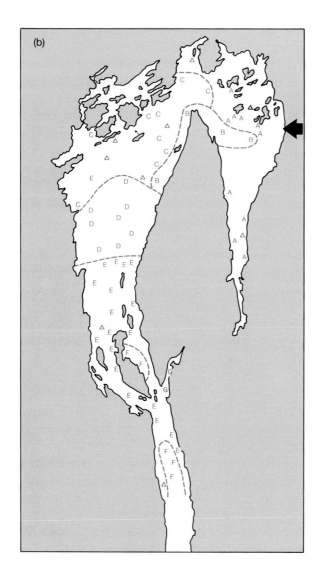

Figure 16.9. Analysis of 68 benthic invertebrate communities in Oslofjord, Scandinavia. (a) Classification in seven groups, A–G (small triangles indicate 'misclassifications' where groups are not clearly dissimilar from each other). (b) Distribution of community classes A–G in the fjord. Large arrow indicates the major output of pollution from the environs of Oslo. (Data of F.B. Mirza, presented in Gray, 1981.)

similarities in their requirements and tolerances (and partly the result of chance). Taking this view, community boundaries need not be sharp, and associations of species would be much less predictable than one would expect from the superorganism concept.

the community: not so much a super-organism . . .

The current view is close to the individualistic concept. Results of direct gradient analyis, ordination and classification all indicate that a given location, by virtue mainly of its physical characteristics, possesses a reasonably predictable association of species. However, a given species which occurs in one predictable association is also quite likely to occur with another group of species under different conditions elsewhere. This is illustrated in Figure 16.10. Such a state of affairs is certain to arise as long as (i) individuals have tolerance limits which encompass a range of conditions; (ii) different species have different tolerance limits; (iii) individuals within a species differ from each other in ecologically relevant respects; and (iv) conditions themselves vary as gradients in space. Thus, except where conditions vary in a dramatic, stepwise manner (e.g. land–water boundary, cave entrance), discrete community boundaries should not be expected to occur. Even when we seem to detect discrete boundaries to a community, we may simply be using too coarse a sampling procedure to detect what is really just a very steep gradient.

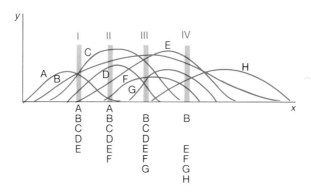

Figure 16.10. Community composition at four locations, each with five or six species, along a continuum. The *y*-axis corresponds to an index of importance such as percentage composition by numbers or biomass of species A to H. The *x*-axis describes continuous variation in a factor such as altitude, soil moisture, substrate grain size, etc.

Many ecologists have been preoccupied with the idea of community boundaries. Indeed, there has been much debate and concern over whether community ecology can legitimately be studied at all if communities do not exist as definable units. This problem has perhaps arisen because of a psychological need to be dealing with a readily defined entity. Whether or not communities have more or less clear boundaries is an important question, but it is not the fundamental consideration. Community ecology is the study of the *community level of organization* rather than of a spatially and temporally definable unit. It is concerned with the nature of the interactions between species and their environment, and with the structure and activities of the multi-species assemblage, usually at one point in space and time. It is not

. . . more a level of organisation

necessary to have discrete boundaries between communities to do community ecology.

16.3.4 Spatial patterns on a larger scale—biomes

There is much in common between our attempts to distinguish community types on a mountainside in Tennessee (Figure 16.4) and to discern patterns on a global scale. The communities characteristic of broad climatic regions are called *biomes*. The distinctions that biogeographers make are largely arbitrary (like communities, biomes grade into one another) and there is no general agreement on precisely how many biomes should be defined. We describe eight terrestrial biomes and illustrate their global distribution in Figure 16.11.

By performing the equivalent of gradient analysis, the distribution of these biomes can be plotted on a graph with axes that represent mean annual temperature and mean annual precipitation (Figure 16.12). This is undoubtedly an oversimplification, because many other factors certainly play a role. However, the figure illustrates the parallels between climate and community type.

Tundra occurs around the Arctic Circle, beyond the tree line. Small areas also occur on sub-antarctic islands in the Southern Hemisphere. Alpine tundra is found under similar climatic conditions but at high altitude (including some tropical mountains) rather than high latitude. The environment is characterized by the presence of permafrost, water permanently frozen in the soil. The typical flora includes lichens, mosses, sedges and dwarf trees. Insects are extremely seasonal in their appearance, and the native bird and mammal fauna is enriched by species which migrate from warmer latitudes in the summer.

Figure 16.11. Distribution of the major terrestrial biomes of the world (from Cox *et al.*, 1976).

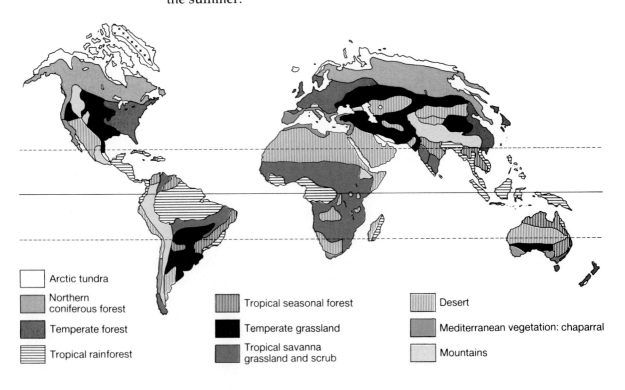

Arctic tundra

Northern coniferous forest

Temperate forest

Tropical rainforest

Tropical seasonal forest

Temperate grassland

Tropical savanna grassland and scrub

Desert

Mediterranean vegetation: chaparral

Mountains

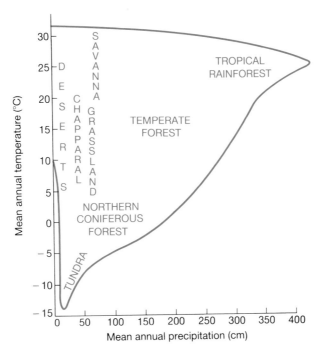

Figure 16.12. The distribution of eight major terrestrial biomes with respect to mean annual temperature and mean annual precipitation. It is not possible to draw sharp boundaries between biome types—local climatic effects, soil types and fire can each shift the balance. (Simplified from Whittaker, 1975.)

Northern coniferous forest (or *taiga*) occupies a broad belt across North America and Eurasia. The trees are mostly evergreen conifers and, characteristically, vast tracts of land are occupied by just one or two species of tree.

Temperate forests range in type from the mixed conifer and broad-leaved forest of much of North America and northern central Europe (where most habitat has been modified by man) to the moist, dripping forests of broad-leaved evergreen trees found, for example, in Florida and the southern tip of South Island, New Zealand (sometimes called temperate rainforest).

The extremely diverse *tropical rainforest* occurs between the tropics of Cancer and Capricorn in areas where annual rainfall is greater than about 200 cm and where at least 12 cm falls even in the driest month.

Grassland occupies somewhat drier parts of temperate and tropical regions. *Temperate grassland* has many local names: prairie, steppe, pampas, veld. Tropical grassland or *savanna* is the name applied to tropical vegetation ranging from pure grassland to some trees with much grass.

Chaparral occurs in Mediterranean-type climates (mild, wet winters and summer drought) in Europe, north-west Mexico and California, and in the southern hemisphere in a few small areas of southern Australia, southern Chile and South Africa. Chaparral receives less rainfall than grasslands. The vegetation is mainly drought-resistant, hard-leaved scrub composed of low-growing woody plants.

Finally, *deserts* are found in areas that experience extreme drought (rainfall is less than about 25 cm per year or, if higher, it is mainly lost quickly through evaporation). The desert biome spans a very wide range of temperatures from hot deserts such as the Sahara to very cold deserts like the Gobi in Mongolia.

These biomes are all terrestrial. One might add two aquatic biomes, *freshwater* and *marine,* which could of course be sub-divided further.

16.4 Community patterns in time—succession

Just as the relative importance of species varies in space, so their patterns of abundance may change with time. In either case, a species will occur only where and when (i) it is capable of reaching a location, (ii) appropriate conditions and resources exist there, and (iii) competitors and predators do not preclude it. A temporal sequence in the appearance and disappearance of species therefore seems to require that conditions, resources and/or the influence of enemies themselves vary with time.

For many organisms, and particularly short-lived ones, relative importance in the community changes with time of year as the individuals act out their life cycles against a background of seasonal change. The explanation for such temporal patterns is straightforward and will not concern us here. Nor will we dwell on the variations in abundance of species from year to year as individual populations respond to a multitude of factors that influence their reproduction and survival (dealt with in Chapters 6–13 and 15). Rather, our focus will be on the process of *succession,* defined as *the non-seasonal, directional and continuous pattern of colonization and extinction on a site by species populations.* This general definition encompasses a range of successional sequences which occur over widely varying time-scales and often as a result of quite different underlying mechanisms.

16.4.1 Degradative succession

some successions occur when a degradable resource is utilized successively by a number of species

One class of serial replacements may be termed *degradative successions,* and these occur over a relatively short time-scale of months or years. Any packet of dead organic matter, whether the body of an animal or plant, a shed skin of a snake or arthropod, or a faecal deposit, is exploited by microorganisms and detritivorous animals (see Chapter 11). Usually, different species invade and disappear in turn, as degradation of the organic matter uses up some resources and makes others available, while changes in the physical condition of the detritus favour first one species, then another. (Since heterotrophic organisms are involved in such sequences they have often been referred to as *heterotrophic successions.*) Ultimately, degradative successions terminate because the resource is completely metabolized and mineralized.

a degradative succession on pine needles

Figure 16.13 shows the sequence of fungal species colonizing pine needles which enter the litter beneath a forest of Scots pine (*Pinus sylvestris*). In this environment, litter is continually being deposited on the surface of the ground, and it accumulates in layers because there is no earthworm activity to incorporate it. The surface litter is therefore young, and the deeper layers are old—time and depth are directly related. The sequence begins even before the needles have been shed. *Coniosporum* is present on approximately 50% of

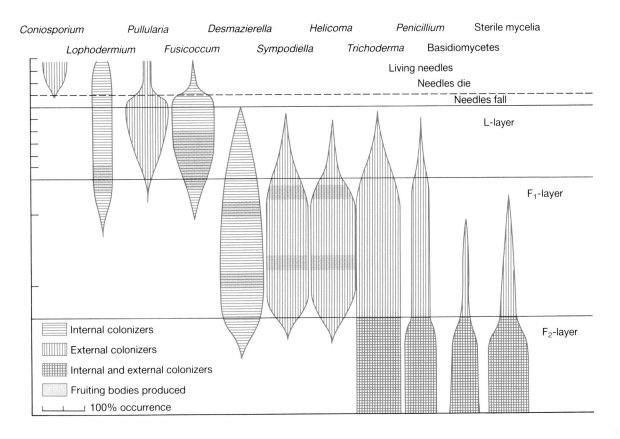

Figure 16.13. Temporal and spatial changes in fungal populations colonizing pine needles in litter layers beneath Scots pine (*Pinus sylvestris*) forest in England (from Richards, 1974, based on Kendrick & Burges, 1962).

living pine needles, but by the time the needles senesce and die this species can no longer be found. However, in August, when most needles fall, *Lophodermium* occurs on approximately 40% of needles, and *Fusicoccum* and *Pullularia* infect more than 80%. The needles fall to the ground to become part of the L layer ('litter' consisting of tough, light-brown needles which lay loose and uncompacted on the surface). During their six months in the L layer many of the needles are invaded by *Desmazierella*, and several other species also enter the sequence, gradually digesting and softening up the tissue.

The next stage of the succession is represented spatially by the F_1 layer (the upper sub-horizon of the 'fermentation' layer of the soil). The grey-black needles in this layer are more closely packed and have softened tissue with low tensile strength. The interior of the needle, particularly the phloem, is extensively attacked by *Desmazierella*, while its exterior, now surrounded by a much more humid atmosphere, is colonized by *Sympodiella* and *Helicoma*. Regions of internal attack are penetrated and further broken down by burrowing mites.

The character of the needles changes once more. After about 2 years in the F_1 layer they are tightly compressed, and fragments which have been attacked by *Lophodermium*, *Fusicoccum* and *Desmazierella* are invaded by various soil animals, mainly collembolans, mites and enchytraeid worms. These eventually bring about the complete physical breakdown of the needles. During this time the needle fragments are slowly attacked by basidiomycetes which can decompose cellulose and lignin. After about 7 years in the F_2 layer, the pine needles are no longer recognizable, and biological activity in the highly acid conditions of the H layer ('humus') falls to a very low level.

Most of the fungi encountered in this work do not show strong antagonisms, and the timing of different stages of the succession is probably controlled mainly by changing environmental conditions and modification of the nutritional status of the organic material in the needles. The fungi perform an initial softening and digestion of the needle tissues, facilitating attack and entry of the detritivorous animals. The majority of the food consumed by animals is probably fungal tissue rather than material of the needles themselves.

16.4.2 Allogenic succession

A new habitat is created when an elephant defaecates, a tree sheds a leaf or a small mammal dies (Chapter 11). Of more general interest to ecologists are those cases where the new habitat is an area of substrate opened up for invasion by green plants (referred to as *autotropic successions*) or other sessile organisms. In these cases, the new habitat does not become degraded and disappear but is merely occupied. A further disturbance may lead to another succession at the same location.

not all newly-created habitats degrade: some are simply occupied

It is important to distinguish between successions that occur as a result of biological processes modifying conditions and resources (known as *autogenic* successions) and other serial replacements of species, occurring as a result of changing external geophysiochemical forces (*allogenic* successions).

allogenic successions are driven by external influences which alter conditions: a salt-marsh–woodland transition

There have been a number of reports of an apparent allogenic transition between salt-marsh and woodland. One of the most thorough of these studies garnered evidence from historical maps, the present pattern of distribution of species in space, and a stratigraphic record of plant remains at different depths in the soil (Ranwell, 1974). The estuary of the River Fal in Cornwall, England, like many other estuaries, is subject to a quite rapid deposition of silt (increased somewhat due to china clay workings in its catchment). This accretion can occur at a rate of 1 cm per year on the mud flats which are found 15 km into the estuary. As a result, salt-marsh has extended 800 m seawards during the last century, while valley woodland has kept pace by invading the landward limits of the marshland (Figure 16.14a).

The vertical zonation (above sea level) of selected plant species is shown in Figure 16.14b. Species colonize at a particular height above sea level according to their tolerance of inundation by brackish water at high tides. Since the highest vertical levels correspond both to the landward edge of the study area and to the oldest stage in succession, Figure 16.14b illustrates both the horizontal distribution of species and the time sequence of their appearance in the succession. The lowest-lying pioneer species on these brackish mud-flats are *Scirpus* and *Agrostis*. These span the entire range of the marsh, right up to the seaward limit of oak trees (*Quercus*), a vertical range of only 2.26 m on the Fal estuary. The range of two obligate halophytes, *Puccinellia* and *Triglochin*, actually overlaps locally with the lower limits of the willow–alder–oak tidal woodland. Clearly, differences in level of as little as 0.2 m may separate salt-marsh from woodland.

That this transition from salt-marsh to woodland truly represents a succession in time, rather than simply a fixed zonation of species, is proved by analysis of the plant remains in a soil core taken within the woodland (Figure 16.14c). The species sequence described in space is faithfully recapitulated in time (= depth in the profile).

The external physical agency of silt deposition is no doubt primarily responsible for the succession observed. However, two biological processes are also capable of raising the level locally and may play some role: plants such as *Deschampsia* and *Carex* build tussocks, and ants construct mounds. Thus part of this succession may be autogenic, not simply allogenic; and, of course, the presence of vegetation may speed up silt deposition, further blurring the distinction.

16.4.3 Autogenic succession

primary and secondary successions

Successions which occur on newly exposed landforms, and in the absence of gradually changing abiotic influences, are known as *autogenic* successions. If the exposed landform has not previously been influenced by a community, the sequence of species is referred to as a primary succession. Freshly formed sand dunes, lava flows, and substrate exposed by the retreat of a glacier are examples of this. In cases where the vegetation of an area has been partially or completely removed, but where well-developed soil and seeds and spores remain, the subsequent sequence of species is termed a secondary succession. The loss of trees locally as a result of disease, high winds, fire or felling may lead to secondary successions.

Successions on newly exposed landforms typically take several hundreds of years to run their course. However, a precisely analogous process occurs among the seaweeds on recently denuded boulders in the rocky intertidal zone, and this succession takes less than a decade. The research life of an ecologist is sufficient to encompass the seaweed succession but not that following glacial retreat. Fortunately, however, information can sometimes be gained over the longer time-scale. Often successional stages in time are represented by community gradients in space—the transition from salt-marsh to woodland is a case in point. The use of historical maps, carbon dating or other techniques may enable the age of a community since exposure of the landform to be estimated. A series of communities currently in existence, but corresponding to different lengths of time since the onset of succession, can be inferred to reflect succession. However, whether or not different communities that are spread out in space really do represent various stages of succession must be judged with caution. We must remember, for example, that in northern temperate areas the vegetation we see may still be undergoing recolonization following the last ice age (Chapter 1).

facilitation: early
successional species
paving the way for later
ones—Glacier Bay

An early successional species may so alter conditions or the availability of resources in a habitat that the entry of new species is made possible. This process is known as *facilitation* (Connell & Slatyer, 1977). It may be particularly important in primary successions where conditions are initially severe, as for example on a substrate laid bare after the retreat of a glacier. Remarkably rapid glacial retreat has occurred at Glacier Bay in south-eastern Alaska. Since about 1750 the glaciers have retreated almost 100 km. As the glaciers retreat they leave moraines, the ages of which can be estimated by ring counts of the oldest trees present at the site. The oldest trees on the last morainic ridge are about 200 years old, and the maximum age of trees

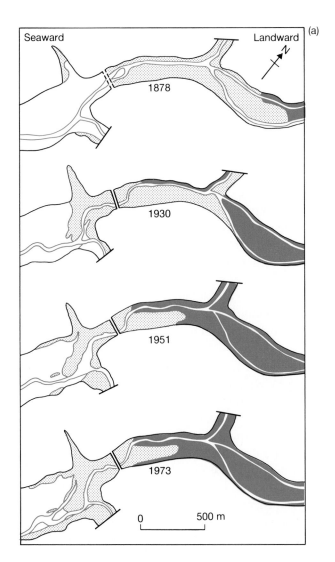

Figure 16.14. Above. (a) Seawards extension of salt-marsh (stippled) and tidal woodland (black) during the last century in the Fal estuary, Cornwall. Facing page. (b) Vertical zonation of selected species in the salt-marsh to tidal woodland succession. (c) Soil profile in tidal woodland with evidence from plant remains of direct succession from pioneer salt-marsh plants to tidal woodland trees. (From Ranwell, 1974.)

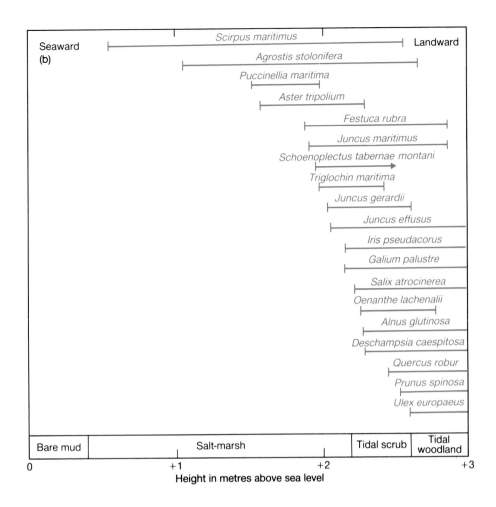

Seaward
(b)

| Scirpus maritimus |
| Agrostis stolonifera |
| Puccinellia maritima |
| Aster tripolium |
| Festuca rubra |
| Juncus maritimus |
| Schoenoplectus tabernae montani |
| Triglochin maritima |
| Juncus gerardii |
| Juncus effusus |
| Iris pseudacorus |
| Galium palustre |
| Salix atrocinerea |
| Oenanthe lachenalii |
| Alnus glutinosa |
| Deschampsia caespitosa |
| Quercus robur |
| Prunus spinosa |
| Ulex europaeus |

Landward

| Bare mud | Salt-marsh | Tidal scrub | Tidal woodland |

0   +1   +2   +3

Height in metres above sea level

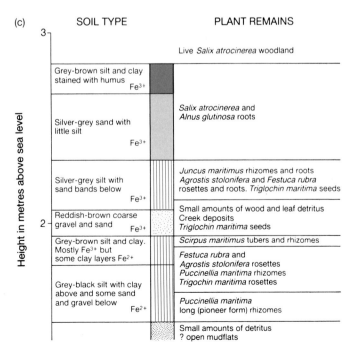

(c)

SOIL TYPE   PLANT REMAINS

Live *Salix atrocinerea* woodland

Grey-brown silt and clay stained with humus Fe^{3+}

Silver-grey sand with little silt Fe^{3+}

Salix atrocinerea and *Alnus glutinosa* roots

Silver-grey silt with sand bands below Fe^{3+}

Juncus maritimus rhizomes and roots
Agrostis stolonifera and *Festuca rubra* rosettes and roots. *Triglochin maritima* seeds

Reddish-brown coarse gravel and sand Fe^{3+}

Small amounts of wood and leaf detritus
Creek deposits
Triglochin maritima seeds

Grey-brown silt and clay. Mostly Fe^{3+} but some clay layers Fe^{2+}

Scirpus maritimus tubers and rhizomes

Festuca rubra and
Agrostis stolonifera rosettes
Puccinellia maritima rhizomes
Trigochin maritima rosettes

Grey-black silt with clay above and some sand and gravel below Fe^{2+}

Puccinellia maritima
long (pioneer form) rhizomes

Small amounts of detritus
? open mudflats

615   NATURE OF THE COMMUNITY

decreases as the glacier is approached. Information on the last 80 years has been gathered by direct observation.

The succession on the exposed boulder clay till, with its thin, nutrient-deficient clay soil, proceeds as follows. The first plants to colonize are mosses and a few shallow-rooted herbaceous species, notably *Dryas*. Next, several kinds of willow appear, prostrate species at first, but later shrubby types. Soon alder enters the succession and after about 50 years produces thickets up to 10 m tall with a scattering of cottonwood. The alder is invaded by sitka spruce, forming a dense mixed forest which continues to develop as western hemlock and mountain hemlock become established.

One of the principle forces that drives this succession is the change in soil conditions caused by the early colonists. Both *Dryas* and alder have symbionts which fix atmospheric nitrogen (see Chapter 13) and lead to an accumulation of a large nitrogen capital in the soil (Figure 16.15). Alder also has a strong acidifying effect, lowering the pH of the surface soil from approximately 8.0 to 5.0 within 50 years. Sitka spruce is then able to invade and displace the alder, using the accumulated nitrogen capital. The gradual accumulation of soil carbon leads to the development of crumb structure in the soil, and an increase in soil aeration and water-retaining capacity.

On less well drained slopes the succession does not always culminate in a spruce–hemlock forest. Where the ground is flat, or has only a shallow slope, drainage is impeded and the forest is invaded by *Sphagnum* mosses which accumulate water and greatly acidify the soil. The forest floor becomes water-logged and deficient in oxygen and most trees die. Occasional, scattered individuals of lodge-pole pine are the only trees that can tolerate the poor aeration of the resulting *muskeg* bog. The culminating vegetation of the succession clearly depends on local conditions.

successions where plants vary in their ability to colonize disturbed habitats, or to persist in the face of competition, predation or the prevailing physical conditions—algae on a rocky shore

Species of algae replace one another in a rather regular sequence on boulders in the low intertidal zone of the southern Californian coast (Sousa, 1979a). The major natural disturbance which clears space in this system is the overturning of boulders by wave action. Algae may recolonize cleared surfaces through vegetative regrowth of surviving individuals, or more generally by recruitment from spores. The typical successional pattern which occurs naturally on cleared boulders can be mimicked, either by artificially clearing boulders or by introducing concrete blocks to the intertidal zone. This made it possible for Sousa to obtain a precise description of succession and also, through various manipulations, to reveal the mechanisms underlying the successional pattern.

Bare substrate is colonized within the first month by a mat of the pioneer green alga *Ulva* (Figure 16.16). During the autumn and winter of the first year, several species of perennial red algae, including *Gelidium coulteri*, *Gigartina leptorhynchos*, *Rhodoglossum affine* and *Gigartina canaliculata* colonize the surface. Within a period of 2 or 3 years *Gigartina canaliculata* comes to dominate the community, covering 60–90% of the rock surface. In the absence of further disturbance this virtual monoculture persists through vegetative spread, resisting invasion by all other species.

We have already seen how some pioneer species in the succession that occurs after glacial retreat actually influence the environment to such an

Figure 16.15. Plant succession after glacial retreat in Glacier Bay, Alaska and changes in total nitrogen content of the soils (after Crocker & Major, 1955).

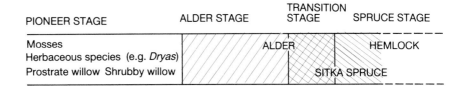

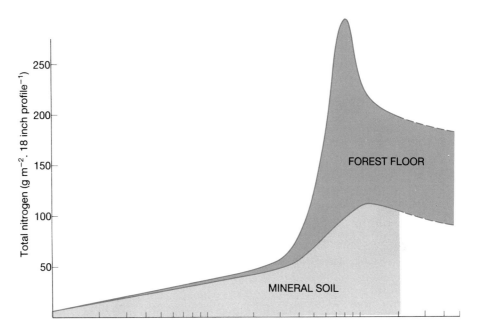

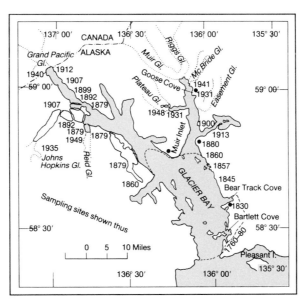

inhibition—the opposite of facilitation

extent that later species can enter the sequence and outcompete their predecessors. In contrast to this facilitation of succession, the pioneer rocky shore species *Ulva* actually inhibits further change. Inhibition was demonstrated conclusively by comparing the recruitment of *Gigartina* individuals (both species) to concrete blocks from which *Ulva* had or had not been

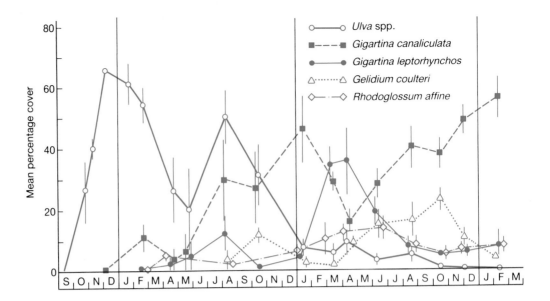

Figure 16.16. Mean percentage cover (±standard error) of five algal species which colonized concrete blocks introduced to the intertidal zone in September 1974 (from Sousa, 1979a).

removed. *Gigartina* recruitment was dramatically better in the absence of *Ulva* (Figure 16.17). Similarly, the middle successional red algal species *Gigartina leptorhynchos* and *Gelidium coulteri* inhibit, by their presence, the invasion and growth of the dominant late-successional *G. canaliculata*. Thus the speed with which open space is colonized is crucial in early succession. The relatively short-lived and rapidly growing *Ulva* is an efficient pioneer of open space because it reproduces throughout the year and can quickly establish itself on exposed substrate. The perennial red algae, with their highly seasonal recruitment and slow growth, are inhibited as long as the early colonists remain healthy and undamaged.

This succession is characterized by species, each of which inhibits its

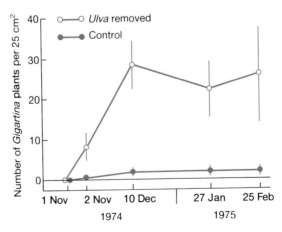

Figure 16.17. Effect of manually removing the early-successional species *Ulva* on the recruitment of *Gigartina* spp. over a 4-month period beginning on 15 October 1974. Data are mean numbers (±standard error) of sporelings of *Gigartina leptorhynchos* and *G. canaliculata* which were recruited to four 25 cm² plots on concrete blocks. (From Sousa, 1979a.)

replacement by another. The succession occurs only because the species which dominate early are more susceptible to the rigours of the physical environment. This was demonstrated by following the mortality of tagged plants during a 2-month period when low tides occurred in the afternoon, creating particularly harsh conditions of intense sunlight and exposure to the air and drying winds (Figure 16.18). In addition, selective grazing on *Ulva* by the crab *Pachygrapsus crassipes* accelerates succession to the tougher, longer-lived red algae. Late species colonize small openings and grow to maturity when individuals of the early species are killed. They gradually take over and dominate the space until a further large-scale opening of space occurs by wave action.

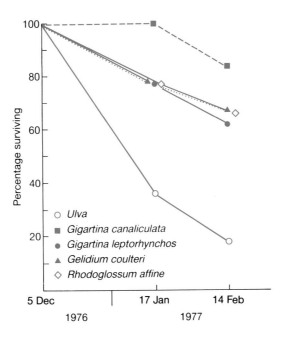

Figure 16.18. Survival curves for five species of algae over a 2-month period, beginning on 5 December 1976, when low tides occurred in the afternoon creating harsh physical conditions. Thirty plants of each species were tagged initially (From Sousa, 1979a.)

old field sequences— successions where facilitation is not obvious but inhibition is

Successions on old fields have been studied primarily along the eastern part of the United States, where many farms were abandoned by farmers who moved west after the frontier was opened up in the nineteenth century. Most of the pre-colonial mixed conifer–hardwood forest had been destroyed, but regeneration was swift. In many places, a series of sites that had been abandoned for different, recorded periods of time are available for study. The typical sequence of dominant vegetation is:

annual weeds → herbaceous perennials → shrubs → early successional trees → late successional trees

The early pioneers of the American West left behind exposed land which was colonized by pioneers of a very different kind! The pioneer species are those which can establish themselves quickly in the disturbed habitat of a recently cultivated field. Perhaps the most common annual of old field succession is *Ambrosia artemisiifolia*. Its germination is closely linked to disturbance, and this ensures the availability of resources by reducing the probability of competition with other species slower off the mark. The seeds of *Ambrosia* survive for many years in the soil and are most likely to germinate when movement of the soil brings them to the surface. Here they experience unfiltered light, reduced CO_2 concentration, and fluctuating temperatures— all conditions which induce germination in this and in many other annuals.

Summer annuals such as *Ambrosia* are often quickly eclipsed in importance by winter annuals (mostly other composites). These have small seeds with little or no dormancy, but disperse over long distances. Their seeds germinate soon after they land, usually in late summer or autumn, and develop into rosettes which overwinter. Next spring they have a head start over summer annuals, and thus they preempt the resources of light, water, space and nutrients. In contrast to the pioneer annuals, seeds of later-successional plants, especially those found in climax forest, do not require light for germination. Thus while seeds of pioneers could not be expected to germinate beneath a forest canopy, those of late-successional species are able to do so.

Early-successional plants have a fugitive life-style. Their continued survival depends on dispersal to other disturbed sites. They cannot persist in competition with later species, and thus they must grow and consume the available resources rapidly. A high relative growth rate is a crucial property of the fugitive. In connection with this, it has been observed that the rate of photosynthesis per unit of leaf area generally declines with position in the succession (Table 16.2), and relative growth rate, in turn, is slower in later-successional plants (perhaps because so much of the plant is old support tissue).

later species in old field successions tend to grow more slowly ...

Table 16.2. Some representative photosynthetic rates (mg CO_2 dm^{-2} h^{-1}) of plants in a successional sequence. Late-successional trees are arranged according to their relative successional position. (From Bazzaz, 1979.)

Plant	Rate	Plant	Rate
SUMMER ANNUALS		EARLY SUCCESSIONAL TREES	
Abutilon theophrasti	24	*Diospyros virginiana*	17
Amaranthus retroflexus	26	*Juniperus virginiana*	10
Ambrosia artemisiifolia	35	*Populus deltoides*	26
Ambrosia trifida	28	*Sassafras albidum*	11
Chenopodium album	18	*Ulmus alata*	15
Polygonum pensylvanicum	18		
Setaria faberii	38	LATE SUCCESSIONAL TREES	
		Liriodendron tulipifera	18
WINTER ANNUALS		*Quercus velutina*	12
Capsella bursa-pastoris	22	*Fraxinus americana*	9
Erigeron annuus	22	*Quercus alba*	4
Erigeron canadensis	20	*Quercus rubra*	7
Lactuca scariola	20	*Aesculus glabra*	8
		Fagus grandifolia	7
HERBACEOUS PERENNIALS		*Acer saccharum*	6
Aster pilosus	20		

Shade tolerance is one factor in the success of later species. Figure 16.19 illustrates idealized light saturation curves of early-, mid- and late-successional species. At low light intensities, later-successional species are able to grow, albeit quite slowly but still faster than the species they replace. Bazzaz (1979) has argued that individuals which arrive first may use up

... but tend also to be more shade-tolerant

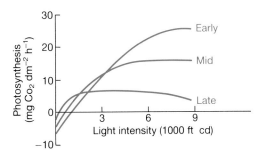

Figure 16.19. Idealized light saturation curves for early-, mid- and late-successional plants (from Bazzaz, 1979).

resources and starve later arrivals. However, competitive 'losers' become 'winners' if their physiology and life-history characteristics allow them to grow when their competitors become inactive. This is particularly significant when the individuals eventually acquire a stature sufficient to shade their predecessors and produce conditions on the forest floor where seeds of the earlier colonists cannot germinate.

The trees that take part in the later stages of an old field succession can themselves be grouped into early-successional and late-successional classes (see Table 16.1). A characteristic of many early-successional trees is that their foliage is multilayered. Leaves extend deep into the canopy, where they still receive enough light to be above the light compensation point. Species such as eastern red cedar (*Juniperus virginiana*) are thus able to use the abundant light that is available early in the tree stage of the succession. In contrast, sugar maple (*Acer saccharum*) and American beech (*Fagus grandifolia*) can be described as monolayered species. These possess a single layer of leaves in a shell round the tree and are more efficient in the crowded canopy of late succession (Horn, 1975). If a monolayered and a multilayered tree simultaneously colonize an open site, the multilayered tree will usually grow faster and dominate until it is crowded by its neighbours, when a slower growing monolayered tree will emerge to dominance.

> early and late trees—
> multilayered and
> monolayered species

The early colonists among the trees usually have efficient seed dispersal; this in itself makes them likely to be early on the scene. They are usually precocious reproducers and are soon ready to leave descendants in new sites elsewhere. The late colonists are those with larger seeds, poorer dispersal and long juvenile phases. The contrast is between the life styles of the 'quickly come, quickly gone' and 'what I have, I hold'.

16.4.4 Mechanisms underlying autogenic successions

> forest succession can be
> represented as a tree-
> by-tree replacement
> process—Horn's model

A model of succession developed by Horn (1975, 1981) sheds some light on the successional process. Horn recognized that in a hypothetical forest community it would be possible to predict changes in tree species composition given two things. First, one would need to know for each tree species the probability that, within a particular time interval, an individual would be replaced by another of the same species or of different species. Secondly, an initial species composition would have to be assumed.

Horn considered that the proportional representation of various species of saplings beneath an adult tree reflected the probability of the tree's replacement by each species. Using this information, he estimated the probability, after 50 years, that a site now occupied by a given species will be taken over by another species or will still be occupied by the same species (Table 16.3). Thus, for example, there is a 5% chance that a location now occupied by grey birch will still support grey birch in 50 years' time, whereas there is a 36% chance that blackgum will take over, a 50% chance for red maple and 9% for beech.

Table 16.3. A 50-year tree-by-tree transition matrix from Horn (1981). The table shows the probability of replacement of one individual by another of the same or different species 50 years hence.

Present occupant	Occupant 50 years hence			
	Grey birch	Blackgum	Red maple	Beech
Grey birch	0.05	0.36	0.50	0.09
Blackgum	0.01	0.57	0.25	0.17
Red maple	0.0	0.14	0.55	0.31
Beech	0.0	0.01	0.03	0.96

Beginning with an observed distribution of the canopy species in a stand in New Jersey known to be 25 years old, Horn modelled the changes in species composition over several centuries. The process is illustrated in simplified form in Table 16.4 (which deals with only four species out of those present). The progress of this hypothetical succession allows several predictions to be made. Red maple should dominate quickly, while grey birch disappears. Beech should slowly increase to predominate later, with blackgum and red maple persisting at low abundance. All these predictions are borne out by what happens in the real succession.

Table 16.4. The predicted percentage composition of a forest consisting initially of 100% grey birch (from Horn, 1981).

Age of forest: (years)	0	50	100	150	200	∞	Data from old forest
Grey birch	100	5	1	0	0	0	0
Blackgum	0	36	29	23	18	5	3
Red maple	0	50	39	30	24	9	4
Beech	0	9	31	47	58	86	93

The most interesting feature of Horn's matrix model is that, given enough time, it converges on a stationary, stable composition which is independent of the initial composition of the forest. The outcome is inevitable (it depends only on the matrix of replacement probabilities) and will be achieved whether the starting point is 100% grey birch or 100% beech (as long as adjacent areas provide a source of seeds of species not initially present), 50% black gum and 50% red maple, or any other combination.

Since the model seems to generate quite accurate predictions, it may prove to be a useful tool in formulating plans for forest management. Namkoong and Roberds (1974) have already used this modelling approach to support proposals for the management of Californian coastal redwood forests. However, an ideal theory of succession should not only predict, it should also explain. To do this, we need to consider the *biological* basis for the replacement values in the model, and here we have to turn to alternative models.

The examples of succession described in previous sections illustrate various underlying mechanisms, including facilitation and inhibition. An overview of mechanisms has been produced by Connell and Slatyer (1977). They propose three models, of which the first (facilitation) is the classical explanation most often invoked in the past, while the other two (tolerance and inhibition) may be equally important and have frequently been overlooked. The fundamental characteristics of these models are illustrated in Figure 16.20.

The essential feature of facilitation succession, in contrast with either the tolerance or inhibition models, is that changes in the abiotic environment are imposed by the developing community. Thus the entry and growth of the later species depends on earlier species preparing the ground. The most obvious examples are found in primary successions such as that occurring after glacial retreat.

Figure 16.20. Three models of the mechanisms which underlie successions (modified from Connell & Slatyer, 1977).

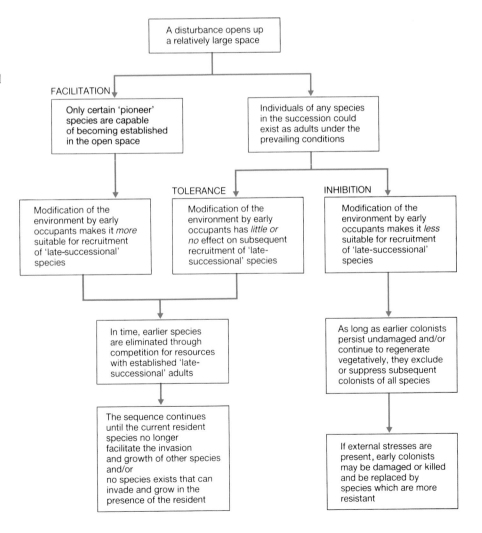

The tolerance model suggests that a predictable sequence is produced because different species have different strategies for exploiting resources. Later species are able to tolerate lower resource levels and can grow to maturity in the presence of early species, eventually outcompeting them. Old field successions provide our example.

The inhibition model applies when all species resist invasions of competitors, as described earlier for the seaweed–boulder succession. Later species gradually accumulate by replacing early individuals when they die. An important distinction between the models is in the cause of death of the early colonists. In the case of facilitation and tolerance, they are killed in competition for resources, notably light and nutrients. In the case of the inhibition model, however, the early species are killed by very local disturbances caused by extreme physical conditions or the action of predators.

the 'vital attributes' of species strongly influence their role in succession

Noble and Slatyer (1979, 1981) have attempted to formalize the role of different properties of species that determine their place in succession. They call these properties *vital attributes*. The two most important relate (i) to the method of recovery after disturbance (four classes are defined: vegetative spread, V; seedling pulse from a seed bank, S; seedling pulse from abundant dispersal from the surrounding area, D; no special mechanism with just moderate dispersal from only a small seed bank, N), and (ii) the ability of individuals to reproduce in the face of competition (defined in terms of tolerance T at one extreme and intolerance I at the other). Thus, for example, a species may be classed as SI if disturbance releases a seedling pulse from a seedbank, and if the plants are intolerant of competition (being unable to germinate or grow in competition with older or more advanced individuals of either their own or another species). Seedlings of such a species could establish themselves only immediately after a disturbance, when competitors are rare. Of course, a seedling pulse fits well with such a pioneer existence. An example is the annual *Ambrosia artemisiifolia*, already discussed in the context of old field succession.

In contrast, the American beech (*Fagus grandifolia*) could be classed as VT (being able to regenerate vegetatively from root stumps, and tolerant of competition since it is able to reproduce and establish itself in competition with older or more advanced individuals of either its own or other species) or NT (if no stumps remain, it would invade slowly via seed dispersal). In either case it would eventually displace other species and form part of the 'climax' vegetation. Noble and Slatyer argue that it should be possible to classify all the species in an area according to these two vital attributes (to which relative longevity might be added as a third). Given this information, quite precise predictions about successional sequences should be possible.

r and *K* species and succession

Consideration of vital attributes from an evolutionary point of view suggests that certain attributes are likely to occur together more often than by chance. We can envisage two alternatives that might increase the fitness of an organism in a succession (Harper, 1977): either (i) the species reacts to the competitive selection pressures and evolves characteristics which enable it to persist longer in the succession, that is, it responds to *K* selection; or (ii) it may develop more efficient mechanisms of escape from the succession, and discover and colonize suitable early stages of succession elsewhere, that is, it responds to *r* selection. Thus it may be that good colonizers will generally be poor competitors and vice versa. This is evident in Table 16.5, which lists some physiological characteristics that tend to go together in early- and late-successional plants.

Table 16.5. Physiological characteristics of early- and late-successional plants (Bazzaz, 1979).

Attribute	Early-successional plants	Late-successional plants
Seed dispersal in time	Well dispersed	Poorly dispersed
Seed germination enhanced by		
light	Yes	No
fluctuating temperatures	Yes	No
high NO_3^-	Yes	No
inhibited by		
far-red light	Yes	No
high CO_2 concentration	Yes	No?
Light saturation intensity	High	Low
Light compensation point	High	Low
Efficiency at low light	Low	High
Photosynthetic rates	High	Low
Respiration rates	High	Low
Transpiration rates	High	Low
Stomatal and mesophyll resistances	Low	High
Resistance to water transport	Low	High
Recovery from resource limitation	Fast	Slow
Resource acquisition rates	Fast	Slow?

necromass plays a crucial role in the late successional dominance of trees . . .

The structure of communities and the successions within them are usually treated as essentially botanical matters. There are obvious reasons for this. Plants commonly provide most of the biomass and the physical structure of communities; moreover, plants do not hide or run away and this makes it rather easy to assemble species lists, determine abundances and detect change. The massive contribution that plants make to determining the character of a community is not just a measure of their role as the primary producers, it is also a result of their slowness to decompose. The plant population not only contributes biomass to the community, but is also a major contributor of *necromass*. Thus, unless microbial and detritivore activity is fast, dead plant material accumulates as leaf litter or as peat. Moreover, the dominance of trees in so many communities comes about because they accumulate dead material; the greater part of a tree's trunk and branches is dead. The tendency in mesic habitats for shrubs and trees to succeed herbaceous vegetation comes largely from their ability to hold leaf canopies (and root systems) on an extending skeleton of predominantly dead support tissue (the heart wood).

Animal bodies decompose much more quickly, but there are situations in which animal remains, like those of plants, can determine the structure and succession of a community. This happens when the animal skeleton resists decomposition, as is the case in the accumulation of calcified skeletons during the growth of corals. A coral reef, like a forest or a peat bog, gains its structure,

. . . and of corals

and drives its successions, by accumulating its dead past. Reef-forming corals, like forest trees, gain their dominance in their respective communities by holding their assimilating parts progressively higher on predominantly dead support. In both cases, one result is that the organisms have an almost overwhelming effect on the abiotic environment, and they 'control' the lives of other organisms within it. The coral reef community (dominated by an animal, albeit one with a plant symbiont) is as structured, diverse and dynamic as a tropical rainforest.

The fact that plants dominate most of the structure and succession of communities does not mean that the animals always follow the communities that plants dictate. This will often be the case, of course, because the plants provide the starting point for all food webs and determine much of the character of the physical environment in which animals live. But it is also sometimes the animals that determine the nature of the plant community. We have already seen that part of the reason why the green seaweed *Ulva* is gradually replaced on boulders by tougher red species is because crabs feed preferentially on it. The terrestrial environment of northern Wisconsin provides another example of a similar phenomenon. A locally dense population of white-tailed deer (*Odocoileus virginianus*) browses on seedlings of both sugar maple and eastern hemlock (*Tsuga canadensis*), but the latter is more severely damaged. The rapid replacement of hemlock seedlings and saplings by sugar maple was reversed in trial plots from which deer had been excluded (reported in Packham & Harding, 1982). Finally, a dramatic example of the impact of grazing was revealed when rabbit populations in Britain were reduced by myxomatosis. Many grassland areas quickly changed their species composition; in particular there were dramatic increases in shrubs and trees which had previously been prevented by rabbit grazing from becoming established (Thomas, 1963).

More often though, animals are passive followers of successions amongst the plants. This is certainly the case for passerine bird species in an old field succession (Figure 16.21). But this does not mean that the birds, which eat seeds, do not influence the succession in its course. They probably do.

animals are often affected by, but may also affect, plant successions

16.4.5 The concept of the climax

Do successions come to an end? It is clear that a stable equilibrium will occur if individuals which die are replaced on a one-to-one basis by young of the same species. At a slightly more complex level, Horn's model (section 16.4.4) tells us that a stationary species composition should, in theory, occur whenever the replacement probabilities (of one species by itself or by any one of several others) remain constant through time.

The concept of the climax has a long history. One of the earliest students of succession, Frederic Clements (1916), argued that there was only one true climax in any given climatic region. This would be the end point of all successions, whether they happened to start from a sand dune, an abandoned old field, or even a pond filling in and progressing towards a terrestrial climax. Clements's rather extreme *monoclimax* theory was

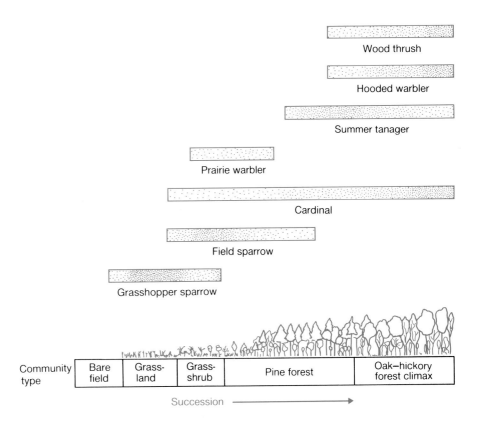

Figure 16.21. Bird species distributions along a plant succession gradient in the Piedmont region of Georgia. Differential stippling indicates relative abundance of the birds. (From Gathreaux, 1978, based on Johnston & Odum, 1956.)

eventually challenged by many ecologists, among whom Tansley (1939) was prominent. The *polyclimax* school of thought recognized that a local climax may be governed by one factor or a combination of factors: climate, soil conditions, topography, fire and so on. Thus a single climatic area could easily contain a number of specific climax types. Later still, Whittaker (1953) proposed his climax-pattern hypothesis. This conceives of a continuity of climax types, varying gradually along environmental gradients and not necessarily separable into discrete climaxes. (This is an extension of Whittaker's approach to gradient analysis of vegetation, discussed in section 16.3.1.)

In fact, it is very difficult to identify a stable climax community in the field. Usually we can do no more than point out that the rate of change of succession slows down to the point where any change is imperceptible to us. In this context, the seaweed–boulder succession discussed in section 16.4.3 is unusual in that convergence to a climax took only a few years. Old field successions might take 100–300 years to reach a 'climax', but in that time the probabilities of fire or hurricanes occurring (every 70 or so years in New England, for example) are so high that a process of succession may never go to completion. If we bear in mind that forest communities in northern temperate regions, and probably also in the tropics, are still recovering from the last

climaxes may be approached rapidly—or so slowly that they are rarely if ever reached

glaciation (Chapter 1), it is questionable whether climax vegetation is more than a wishful dream in the mind of the theorist.

Finally, we again return to issues of scale which feature in almost every chapter of this book. A forest, or a rangeland, that appears to have reached a stable community structure when studied on a scale of hectares, will always be a mosaic of miniature successions. Every time a tree falls or a grass tussock dies an opening is created in which a new succession starts. One of the most seminal papers published in the history of ecology was entitled 'Pattern and Process' (Watt, 1947). Part of the pattern of a community is caused by the dynamic processes of deaths, replacements and micro-successions that the broad view may conceal. We return to some of these themes in Chapter 19.

17

The Flux of Energy and Matter through Communities

17.1 Introduction

All biological entities require matter for their construction and energy for their activities. This is true for individual organisms, but also for the communities that they form in nature. The approach that we follow in this chapter takes the community as the unit of concern and, instead of considering the ways in which energy and matter are acquired and disposed of by the individual, we are concerned with the way they are processed by the community. In practice, this is tantamount to considering the 'activities' of unit areas of land or water.

the standing crop

The bodies of the living organisms within a unit area constitute a *standing crop* of biomass. By *biomass* we mean the mass of organisms per unit area of ground (or water) and this is usually expressed in units of energy (e.g. joules m^{-2}) or dry organic matter (e.g. tons ha^{-1}). The great bulk of the biomass in communities is almost always formed by the plants, which are the primary producers of biomass because of their almost unique ability to fix carbon in photosynthesis. (We have to say 'almost unique' because bacterial photosynthesis and chemosynthesis may also contribute to forming new biomass, but these are usually insignificant.) Biomass includes the whole bodies of the organisms even though parts of them may be dead. This needs to be borne in mind, particularly when considering woodland and forest communities in which the bulk of the biomass is dead heartwood and bark. We often need to

biomass and necromass

distinguish this mass of dead material, the *necromass*, from the living, active fraction of the biomass. The latter represents active capital capable of generating interest in the form of new growth, whereas the necromass is incapable of new growth. Resources may be tied up in necromass and not available for growth, but it is in a real sense capital that bears no interest. In practice we include in biomass all those parts, living or dead, that are attached to the living organism. They cease to be biomass when they fall off and become litter, humus or peat.

primary and secondary productivity

The *primary productivity* of a community is the rate at which biomass is produced *per unit area* by plants, the primary producers. It can be expressed either in units of energy (e.g. joules m^{-2} day^{-1}) or of dry organic matter (e.g. kg ha^{-1} yr^{-1}). The total fixation of energy by photosynthesis is referred to as *gross primary productivity* (GPP). A proportion of this is respired away by the plant itself and is lost from the community as respiratory heat (R). The difference between GPP and R is known as *net primary productivity* (NPP) and represents the actual rate of production of new biomass that is available for consumption by heterotrophic organisms (bacteria, fungi and animals). The rate of production of biomass by heterotrophs is called *secondary productivity*.

During the period 1964–74 there was a concerted worldwide attempt to gather information on the productivity of ecological communities and the factors that influence it. This was carried out under the aegis of the International Biological Programme (IBP, for short), set up after five years of discussion in which leading roles were played by Sir Rudolph Peters (President of the International Council for Scientific Unions, 1958–61) and G. Montalenti and C.H. Waddington (successive presidents of the International Union of Biological Sciences). The subject of the IBP was defined as 'The Biological Basis of Productivity and Human Welfare'. Recognizing the problem of a rapidly increasing human population, it was considered that sound scientific knowledge would be required as a basis for rational resource management. Work was directed towards the study of productivity on land, in freshwater, and in the seas, together with the processes of photosynthesis and nitrogen fixation (Worthington, 1975).

The initial emphasis of the IBP was on productivity, both primary and secondary, but the interdependence of the two soon came to be an important focus for the programme. The idea of the ecosystem dominated much ecological thinking at that time. This was seen as a functional unit within which the productivity of any particular component is influenced by, and influences, the productivity of others. A great deal of effort was put into obtaining quantitative descriptions of the pattern of flow of energy and matter through the communities. As the programme developed, a concentration on field measurements (the growth rates of plants and the population dynamics of animals) changed to an approach that tried to link the behaviour of organisms in the field to their behaviour under controlled (usually laboratory) conditions.

The IBP provided the first occasion on which biologists throughout the world were challenged to work together towards a common end. It was a mammoth effort. At the peak of its operational phase, an estimated $40 000 000 were being spent per year (with finance provided through the agencies of national academies, universities and research councils) and several thousand biologists were involved. Eventually, more than 30 books will be produced to synthesize the findings. Fortunately, much has already been published and a large proportion of this chapter is based on results produced through the IBP. In addition, earlier work and the results of studies outside the IBP are discussed.

17.2 Patterns in primary productivity

The best current estimate of global terrestrial net primary productivity is $110–120\times10^9$ tonnes dry weight per year, and in the sea $50–60\times10^9$ tonnes per year (Leith, 1975; Whittaker, 1975; Rodin *et al.*, 1975). Thus although oceans cover about two-thirds of the world's surface, they account for only one-third of its production. The fact that productivity is not evenly spread across the earth is further emphasized by Table 17.1, which provides estimates of annual net primary productivity and plant standing crop biomass for the principle biomes on land, for freshwater lakes and streams, and for marine systems.

Ecosystem type	Area (10^6 km^2)	Net primary productivity, per unit area $(\text{g m}^{-2}\ or\ \text{t km}^{-2})$ Normal range	Mean	World net primary production $(10^9\ \text{t})$	Biomass per unit area (kg m^{-2}) Normal range	Mean	World biomass $(10^9\ \text{t})$
Tropical rainforest	17.0	1000–3500	2200	37.4	6–80	45	765
Tropical seasonal forest	7.5	1000–2500	1600	12.0	6–60	35	260
Temperate evergreen forest	5.0	600–2500	1300	6.5	6–200	35	175
Temperate deciduous forest	7.0	600–2500	1200	8.4	6–60	30	210
Boreal forest	12.0	400–2000	800	9.6	6–40	20	240
Woodland and shrubland	8.5	250–1200	700	6.0	2–20	6	50
Savanna	15.0	200–2000	900	13.5	0.2–15	4	60
Temperate grassland	9.0	200–1500	600	5.4	0.2–5	1.6	14
Tundra and alpine	8.0	10–400	140	1.1	0.1–3	0.6	5
Desert and semidesert shrub	18.0	10–250	90	1.6	0.1–4	0.7	13
Extreme desert, rock, sand and ice	24.0	0–10	3	0.07	0–0.2	0.02	0.5
Cultivated land	14.0	100–3500	650	9.1	0.4–12	1	14
Swamp and marsh	2.0	800–3500	2000	4.0	3–50	15	30
Lake and stream	2.0	100–1500	250	0.5	0–0.1	0.02	0.05
Total continental	149		773	115		12.3	1837
Open ocean	332.0	2–400	125	41.5	0–0.005	0.003	1.0
Upwelling zones	0.4	400–1000	500	0.2	0.005–0.1	0.02	0.008
Continental shelf	26.6	200–600	360	9.6	0.001–0.04	0.01	0.27
Algal beds and reefs	0.6	500–4000	2500	1.6	0.04–4	2	1.2
Estuaries	1.4	200–3500	1500	2.1	0.01–6	1	1.4
Total marine	361		152	55.0		0.01	3.9
Full total	510		333	170		3.6	1841

Table 17.1. Net annual primary productivity and standing crop biomass estimates for contrasting communities of the world (after Whittaker, 1975).

below-ground productivity is almost certainly underestimated

The comparisons between terrestrial and aquatic systems may be biased against terrestrial communities because of the enormous technical problems in measuring the standing crop of underground parts—measuring productivity below ground is even more of a problem. It is said that one attempt to extract tree roots for measurement from a woodland site in California, by using dynamite, resulted in most of the root material landing in Nevada. Though this story may be a myth, it illustrates the problem. Most measurements of primary productivity in terrestrial systems represent measures of above-ground parts. Yet primary production below ground may equal or be greater than that above ground.

A large proportion of the globe produces less than 400 g m⁻² yr⁻¹. This includes over 30% of the land surface (Figure 17.1a) and 90% of the ocean (Figure 17.1b). The open ocean is, in effect, a marine desert. At the other extreme, the most productive systems are found amongst swamp and marshland, estuaries, algal beds and reefs, and cultivated land.

the productivity of forests, grasslands, crops and lakes follows a latitudinal pattern . . .

In the forest biomes of the world, there is a general latitudinal trend of increasing productivity from boreal, through temperate, to tropical conditions. The same trend is apparent in the productivity of tundra and grass-

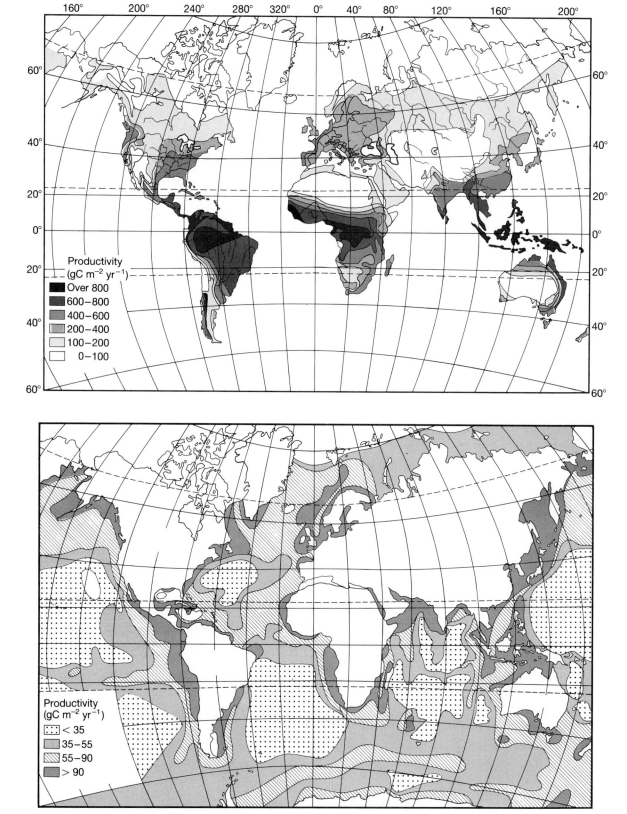

Figure 17.1. Top: Worldwide pattern of net primary productivity on land (from Reichle, 1970).

Bottom: Worldwide pattern of net primary productivity in the oceans (from Koblentz-Mishke *et al.*, 1970).

land communities (Figure 17.2a) and in various cultivated crops (Figure 17.2b). In aquatic communities this trend is clear in lakes (Figure 17.2c) but not in the oceans (Figure 17.1b). The trend with latitude suggests that radiation and temperature may be the factors usually limiting the productivity of communities. Other factors can, however, constrain productivity within even narrower limits. In the sea, the limit to productivity may more often be nutrient limitation—very high productivity occurs in marine communities where there are upwellings of nutrient-rich waters, even at high latitudes and low temperatures. In terrestrial systems, the overall trend of increasing productivity with decreasing latitude is upset in areas where water is in short supply, for instance the continental interior of Australia.

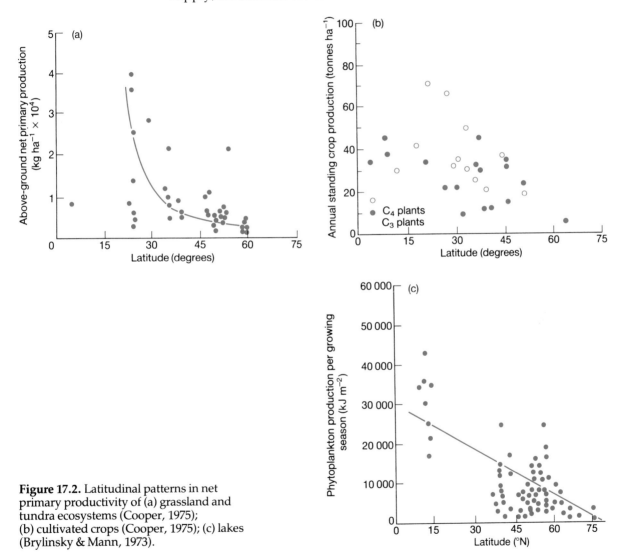

Figure 17.2. Latitudinal patterns in net primary productivity of (a) grassland and tundra ecosystems (Cooper, 1975); (b) cultivated crops (Cooper, 1975); (c) lakes (Brylinsky & Mann, 1973).

... which is frequently modified locally

Small differences in topography can result in large differences in community productivity. In tundra, for example, a distance of a few metres, from beach ridge to meadow with impeded drainage, can change primary production from <10 to 100 g m^{-2} yr^{-1} (Devon Island, Canada). In favourable sites in

Greenland and in South Georgia (Antarctica) productivities of tundra communities may reach 2000 g m^{-2} yr^{-1}, which is higher than that recorded in many temperate communities. Thus, although there is a general latitudinal trend, there is also a broad spectrum of variation at a given latitude resulting from contrasting microclimates.

17.2.1 Aquatic communities: autochthonous and allochthonous material

All biotic communities depend on a supply of energy for their activities. In most terrestrial systems this is contributed *in situ* by the photosynthesis of green plants. Organic matter (and fixed energy) generated within the community is called *autochthonous*. In aquatic communities, the autochthonous input is through the photosynthesis of large plants and attached algae in shallow waters, and by microscopic plankton in the open water. However, a substantial proportion of the organic matter (energetic resource) in such communities often comes into the community as dead organic material that has been formed outside it. Such *allochthonous* material arrives in rivers or is blown in by the wind. The relative importance of the two autochthonous sources and the allochthonous source of organic material in an aquatic system depends on the dimensions of the body of water and the types of terrestrial community that deposit organic material into it.

A small stream running through a wooded catchment derives almost all its energy input from litter shed by surrounding vegetation (Figure 17.3). Shading from the trees prevents any significant growth of planktonic or attached algae or aquatic higher plants. As the stream widens further downstream, shading by trees is restricted to the margins and autochthonous primary production increases. Still further downstream, in deeper waters,

variations along 'the river continuum' . . .

Figure 17.3. Longitudinal variation in the nature of the energy base in stream communities.

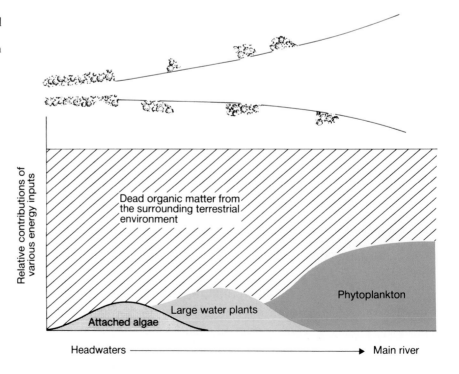

the contribution to production by the rooted higher plants becomes much less, and the role of the microscopic phytoplankton becomes more important.

The sequence from small, shallow lakes to large, deep ones shares some of the characteristics of the river continuum just discussed (Figure 17.4). A small lake is likely to derive quite a large proportion of its energy from terrestrial sources because its periphery, across which terrestrial litter passes, is large in relation to lake area. Small lakes are also usually shallow, so internal littoral production is more important than that by phytoplankton. In contrast, a large, deep lake will derive only limited organic matter from outside (small periphery relative to lake surface area), and littoral production, limited to the shallow margins, may also be low. The organic inputs to the community may then be due almost entirely to photosynthesis by the phytoplankton.

paralleled by lakes

Figure 17.4. Variation in the importance of terrestrial input of organic matter and littoral and planktonic primary production in contrasting aquatic communities.

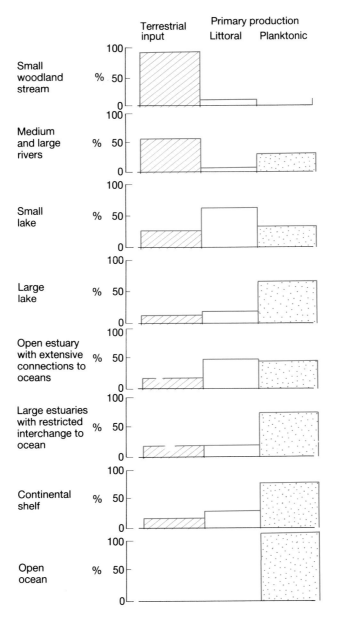

In a sense, the open ocean can be described as the largest, deepest 'lake' of all. Input of organic material from terrestrial communities is negligible, and the great depth precludes photosynthesis in the darkness of the sea bed. The phytoplankton are then all-important.

Estuaries are often highly productive systems, but the most important contribution to their energy-base varies. Phytoplankton tend to dominate in large estuarine basins with restricted interchange with the open ocean and with small marsh peripheries relative to basin area. In contrast, seaweeds dominate in some open basins with extensive connections to the sea.

the oceans, estuaries, and continental shelves

Finally, continental shelf communities derive a proportion of their energy from terrestrial sources (particularly via estuaries) and their shallowness often provides for significant production by littoral seaweed communities. Indeed, some of the most productive sytems of all are to be found among seaweed beds and reefs (Table 17.1).

17.2.2 Variations in the relationship of productivity to biomass

Figure 17.5. The relationship between average net primary productivity and average standing crop biomass for the community types listed in Table 17.1.

We can relate the productivity of a community to the standing crop that produces it (the interest rate on the capital). Alternatively, we can think of the standing crop as the biomass that is sustained by the productivity (the capital resource that is sustained by earnings). The average values of productivity (P)

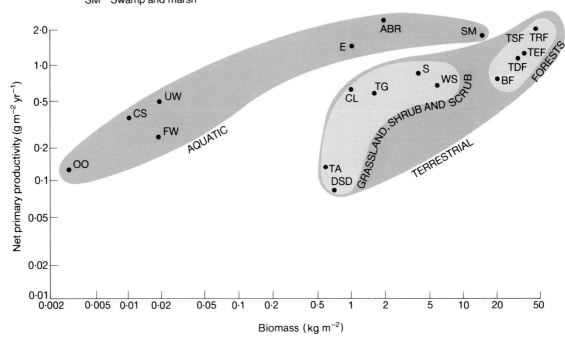

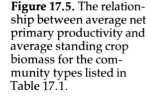

and standing crop biomass (B) for the community types listed in Table 17.1 are plotted against each other in Figure 17.5. It is evident that a given value of NPP is produced by a smaller biomass when non-forest terrestrial systems are compared with forests, and the biomass involved is smaller still when aquatic systems are considered. Thus P:B ratios (i.e. kg produced per year per kg of standing crop) average 0.042 for forests, 0.29 for other terrestrial systems, and 17 for aquatic communities. The major reason for this is almost certainly that a large proportion of forest biomass is dead (and has been so for a long time) and also that much of the living support tissue is not photosynthetic. In grassland and scrub, a greater proportion of the biomass is alive and involved in photosynthesis. In aquatic communities, particularly where productivity is due mainly to phytoplankton, there is no support tissue, dead cells do not accumulate (they are usually eaten before they die), and photosynthetic output per kg of biomass is very high indeed. Another factor that helps to account for high P:B ratios in phytoplankton communities is the rapid turn-over of biomass. The annual NPP shown in the figure is actually produced by a succession of overlapping phytoplankton generations, while the standing crop biomass is only the average present at an instant.

A general feature of autogenic successions (section 16.4.3) is that the pioneers are rapidly growing herbaceous species with relatively little support tissue. Thus, early in succession the production:biomass ratio is high. However, the species that come to dominate later are generally slow-growing, but they eventually achieve a large size and come to monopolize the supply of space and light. Their structure involves considerable investment in non-photosynthesizing and dead support tissue, and as a consequence their P:B ratio is low. This principle is illustrated in Figure 17.6.

There is another way of looking at the figures of productivity in relation

P:B ratios are very low in forests and very high in aquatic communities

P:B ratios tend to decrease during succession

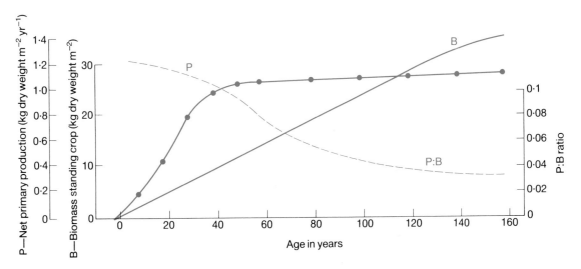

Figure 17.6. Annual above-ground net primary production (P), standing crop biomass (B), and production to biomass ratio (P:B) in forest succession following fire on Long Island, New York. Production increased rapidly through herb and shrub stages to a forest with a stable production of about 1.05 kg m⁻² yr⁻¹ after 40–50 years. Biomass was still increasing in this period and would be expected to achieve a value of about 40 kg m⁻² in a mature oak forest after about 200 years. The P:B ratio fell from more than 0.1 ten to twenty years after the fire to a value of only 0.03 after 160 years. (Data of Whittaker & Woodwell, 1968, 1969.)

to biomass. This is to say that biomass as we define it is a quite unrealistic way of measuring the standing crop of a community. Is it, for example, justifiable to treat the dead wood and bark of a tree as if they were in any direct way causal in determining the rate at which that tree makes new dry matter (or fixes energy)? Nearly all the differences in community productivity that we see *between* community types arise because we choose to define biomass in this way. It might be much more sensible to define biomass in terms of weight of living tissues (if we could find a way to measure it). It is certain that a large part of the difference between communities in their P:B ratio would then disappear. The differences that remain could be much more informative. Unfortunately, such measurements have not been made.

17.3 Factors limiting primary productivity

17.3.1 Terrestrial communities

Sunlight, carbon dioxide, water and soil nutrients are the resources required for primary production on land, while temperature, a condition, has a strong influence on the rate of photosynthesis. CO_2 is normally available at around 0.03% of atmospheric gases, and although variations occur in the vicinity of plants, CO_2 concentration usually plays no significant role in limiting productivity on land. On the other hand, the quality and quantity of light, the availability of water and nutrients, and temperature all vary dramatically from place to place. They are all candidates for the role of limiting factor. Which of them actually sets the limit to primary productivity?

Depending on location, something between 0 and 5 joules of solar energy strikes each square metre of the earth's surface every minute. If all this were converted by photosynthesis to plant biomass (that is, if photosynthetic efficiency were 100%) there would be a prodigious generation of plant material, one or two orders of magnitude greater than recorded values. Much of this solar energy is unavailable for use by plants. In particular, only about 44% of incident short-wave radiation occurs at wavelengths suitable for photosynthesis. However, even when this is taken into account, productivity still falls well below the maximum possible (see also Chapter 3, p. 76). Figure 17.7 plots on a log scale net photosynthetic efficiencies (percentage of incoming photosynthetically active radiation [PAR] incorporated into above-ground net primary productivity). The data were gathered during the IBP, in the United States, from seven coniferous forests, seven deciduous forests and eight desert communities. The conifer communities had the highest efficiencies, but these were only between 1% and 3%. For a similar level of incoming radiation, deciduous forests achieved 0.5% to 1%, and, despite their greater energy income, deserts were able to convert only 0.01% to 0.2% of PAR to biomass. These values can be compared with the peak efficiencies

Table 17.2. Maximum short-term growth rates and photosynthetic efficiencies of various crops throughout the world, expressed per unit area of land (from Cooper, 1975, where original references may be found).

		Crop growth rate $(g\,m^{-2}\,d^{-1})$	Total radiation $(J\,cm^{-2}\,d^{-1})$	Conversion of light energy (PAR) (%)
I. TEMPERATE				
C_3-SPECIES:				
Tall fescue (*Festuca arundinacea*)	UK	43	2201	7.8
Cocksfoot (*Dactylis glomerata*)	UK	40	2201	7.3
Ryegrass (*Lolium perenne*)	UK	28	1983	5.6
	Netherlands	20	1880	4.2
	New Zealand	19	2130	3.5
Red clover (*Trifolium pratense*)	New Zealand	23	2010	4.3
Potato (*Solanum tuberosa*)	Netherlands	23	1670	5.4
Sugar beet (*Beta vulgaris*)	UK	31	1230	9.5
	Netherlands	21	1460	5.6
Kale (*Brassica oleracea*)	UK	21	1598	4.9
	New Zealand	16	2130	2.9
Barley (*Hordeum vulgare*)	UK	23	2025	4.0
	Netherlands	18	1880	3.7
Wheat (*Triticum vulgare*)	Netherlands	18	1880	3.7
Rice (*Oryza sativa*)	Japan	36	—	7.1
Soybean (*Glycine max*)	Japan	27	—	9.8
Peas (*Pisum sativum*)	Netherlands	20	1880	4.2
C_4-SPECIES:				
Maize (*Zea mays*)	UK	24	1250	7.6
	Netherlands	17	1460	4.6
	New Zealand	29	1880	6.1
	Japan	52	—	10.2
	New York, USA	52	2090	9.8
	Kentucky, USA	40	2090	7.6
II. SUB-TROPICAL				
C_3-SPECIES:				
Alfalfa (*Medicago sativa*)	California, USA	23	2850	3.2
Potato	California, USA	37	2850	5.1
Cotton (*Gossypium hirsutum*)	Georgia, USA	27	2300	4.6
Rice	S. Australia	23	2720	3.0
C_4-SPECIES:				
Sudan grass (*Sorghum sp.*)	California, USA	51	2887	6.7
Maize	California, USA	52	3079	6.4
	California, USA	38	2694	5.6
III. TROPICAL				
C_3-SPECIES				
Cassava (*Manihot esculenta*)	Sierra Leone	15	1590	3.7
	Tanzania	17	1800	3.7
	Malaysia	18	1670	4.5
Oil palm (*Elaeis guineensis*)	Malaysia	11	1590	3.3
Rice	Philippines	27	1670	6.4
C_4-SPECIES:				
Pennisetum typhoides	NT, Australia	54	2134	9.5
Pennisetum purpureum	El Salvador	39	1674	9.3
Sugar cane (*Saccharinum sp.*)	Hawaii	37	1678	8.4
Maize	Thailand	31	2090	5.9

639   FLUX OF ENERGY AND MATTER THROUGH COMMUNITIES

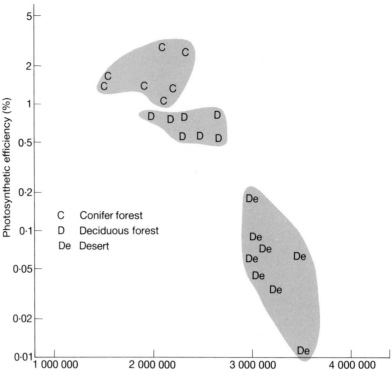

Figure 17.7. Photosynthetic efficiency (percentage of incoming photosynthetically active radiation converted to above-ground net primary production) for three sets of terrestrial communities in the United States (data from Webb *et al.*, 1983).

achieved by crop plants under ideal conditions, when values from 3% to 10% can be achieved (Table 17.2).

However, the fact that light is not used efficiently does not in itself imply that light does not limit community productivity. We would need to know whether at increased intensities of radiation the productivity increased or remained unchanged. Some of the evidence given in Chapter 3, p. 78, shows that the intensity of light during part of the day is below the optimum for canopy photosynthesis. Moreover, at peak light intensities, most canopies still have their lower leaves in relative gloom, and would almost certainly photosynthesize faster if the light intensity were higher. For C_4 plants (see Figure 3.3, p. 78) a saturating intensity of radiation never seems to be reached, and the implication is that productivity may in fact be limited by a shortage of photosynthetically active radiation even under the brightest natural radiation.

but productivity may still be limited by a shortage of PAR

There is no doubt, however, that what light is available would be used more efficiently if other resources were in abundant supply. The much higher values of community productivity from agricultural systems bear witness to this.

The general relationship between above-ground net primary productivity and precipitation for forests of the world is illustrated in Figure

17.8a. Water is an essential resource both as a constituent of cells and for photosynthesis. Large quantities of water are lost in transpiration—particularly because the stomata need to be open for much of the time for CO_2 to be taken in. It is not surprising that the rainfall of a region is quite closely correlated with its productivity. In arid regions, there is an approximately linear increase in NPP with increase in precipitation, but in the more humid forest climates there is a plateau beyond which productivity does not continue to rise.

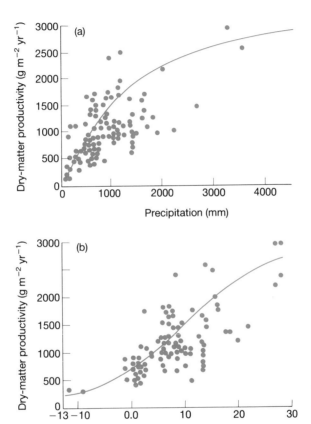

Figure 17.8. Relationships between forest net primary productivity and (a) annual precipitation, and (b) temperature (from Reichle, 1981).

There is a clear relationship between the above-ground NPP and mean annual temperature (Figure 17.8b), but the explanation is complex. Increasing temperature leads to increasing rates of gross photosynthesis, but there is a law of diminishing returns. Respiration, by contrast, tends to increase almost exponentially with temperature. The result is that net photosynthesis is maximal at temperatures well below those for gross photosynthesis (Figure 17.9). Higher temperatures are also associated with rapid transpiration, and thus they increase the rate at which water shortage may become important.

At several of the study sites illustrated in Figure 17.7, and also for a grassland site, estimates of annual precipitation (mm yr^{-1}) are available. In addition, *potential evapotranspiration* was calculated. This is an index of the theoretical maximum rate at which water might evaporate into the atmosphere (mm yr^{-1}) given the prevailing radiation, average vapour pressure

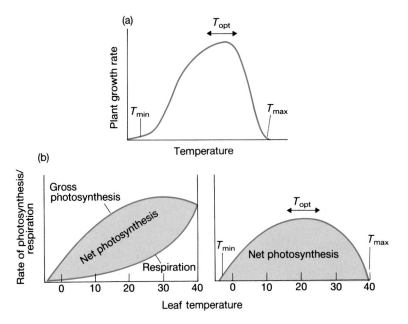

Figure 17.9. Schematic representations of plant responses to temperature. (a) A generalized diagram of the responses of plant growth rate to temperature, illustrating three critical temperatures, i.e. the minimum (T_{min}) and maximum (T_{max}) temperatures, and the optimum temperature range (T_{opt}) for growth. (b) The influence of temperature on gross photosynthesis, respiration and net photosynthesis in a typical plant. (Adapted from Pisek *et al.*, 1973, by Fitter & Hay, 1981.)

the interaction of temperature and precipitation

deficit in the air, windspeed and temperature. Potential evapotranspiration minus precipitation provides, for each site, a crude index of how far the water available for plant growth falls below what might be transpired by actively growing vegetation. Its relationship with above-ground NPP is shown in Figure 17.10. Clearly, drought is one of the characteristic features of some of the communities with low productivity.

Water shortage has direct effects on the rate of plant growth but also leads to the development of less dense vegetation. Vegetation that is sparse

productivity and the structure of the canopy

intercepts less light (much of which falls on bare ground). This wastage of solar radiation is the main cause of the low productivity in many arid areas, rather than the reduced photosynthetic rate of the droughted plants. This point is made by comparing the productivity per unit weight of leaf biomass instead of per unit area of ground. Coniferous forest produced 1.64 g g^{-1} yr^{-1}; deciduous forest 2.22 g g^{-1} yr^{-1}; grassland 1.21 g g^{-1} yr^{-1}; and desert 2.33 g g^{-1} yr^{-1}.

Leaf area index (LAI) is defined as the surface area of leaves per unit surface area of ground. Desert vegetation possesses a lower LAI than forest, and in the example above this accounts for much of the difference in productivity. In general, as leaves are added to the canopy, the increase in LAI may be expected to lead to greater productivity; but eventually, because of shading, as we have seen in Chapter 3, a point is reached where leaves low in the canopy do not receive enough light to photosynthesize at a rate equal to their rate of respiration. Beyond this level, increasing LAI leads to decreasing productivity.

LAI is not the only structural feature that influences canopy productivity.

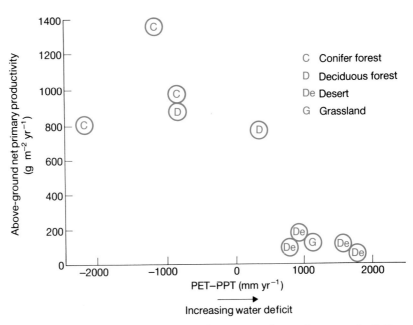

Figure 17.10. Relationship between above-ground net primary productivity and an index of water deficiency (potential evapotranspiration [PET] minus precipitation [PPT]) for several types of community in North America (data from Webb *et al.*, 1983).

Two other important attributes are the angle at which leaves are inclined and the pattern of leaf density with depth in the canopy (see Chapter 3). Their importance is illustrated in a series of nine contrasting hypothetical canopy arrangements in Table 17.3. At high light intensities, productivity is greater in a canopy in which leaves at the top are more steeply inclined. In the rich light environment of the surface layer, the consequent reduction in light absorption does not reduce their rate of photosynthesis, while more light is made available to lower levels in the canopy. High productivity is also associated with concentration of leaves in the upper canopy (except when leaves at the top are inclined horizontally).

The productivity of a community can be sustained only for that period of

Table 17.3. Simulated effect of canopy architecture on primary productivity (g m^{-2} d^{-1}) (from Duncan *et al.*, 1967).

Leaf inclination distribution	Leaf area distribution		
	A	B	C
a	36.2	37.3	38.8
b	33.8	34.0	34.2
c	32.2	31.8	31.6

The canopy used had a total leaf area index of 4.0, changing evenly in ten strata. In leaf area distribution A, the leaf area index was 0.2 in the top stratum and 0.6 in the bottom stratum. In B the leaf area was evenly distributed among strata. In C the leaf area was 0.6 at the top and 0.2 at the bottom. In leaf inclination distribution a, leaves were vertically inclined in the top stratum and horizontal in the bottom. In b, all leaves were at 45°. In c, leaves were horizontal at the top and vertical at the bottom.

the year when the plants bear photosynthetically active foliage. Deciduous trees have a self-imposed limit on the period of the year during which they bear foliage. Evergreen trees hold a canopy throughout the year, but during some seasons it may barely photosynthesize at all or even respire faster than it photosynthesizes. The latitudinal patterns in forest productivity seen earlier (Figure 17.2) are largely the result of differences in the number of days when there is active photosynthesis. Figure 17.11 shows the results of part of an IBP study, in which the productivity of North American deciduous forest is related to the number of days of photosynthesis along a transect from Wisconsin and New York in the north to Tennessee and Carolina in the south.

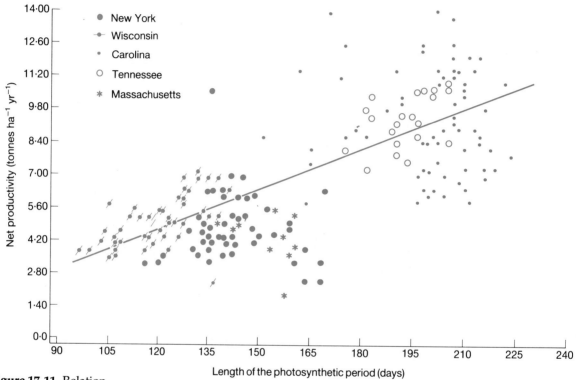

Figure 17.11. Relationship between net primary productivity of deciduous forests in North America and the length of the growing season (from Leith, 1975).

A contrast between the productivity of a deciduous beech forest (*Fagus sylvatica*) and the evergreen Norway spruce (*Picea abies*) is shown in Table 17.4. The forests were within 1 km of each other under very similar abiotic conditions. Beech leaves photosynthesize at a greater rate (per gram dry weight) than those of spruce; even the shade leaves of beech, deep in its canopy, have a higher photosynthetic capacity than one-year-old spruce needles. In addition, beech 'invests' a considerably greater amount of biomass in its leaves each year. However, these factors are easily outweighed by the greater length of the growing season of spruce, which shows positive CO_2 uptake during 260 days of the year compared with 176 days for beech.

No matter how brightly the sun shines and how often the rain falls, no matter how equable is the temperature, productivity must be low if there is no soil in a terrestrial community, or if the soil is deficient in essential mineral

Table 17.4.
Characteristics of
representative trees of
two contrasting species
growing within 1 km of
each other on the Solling
Plateau, Germany
(Schulze, 1970; Schulze *et
al.*, 1977a, 1977b).

	Beech	Norway spruce
Age	100 years	89 years
Height	27 m	25.6 m
Leaf shape	Broad	Needle
Annual production of leaves	Higher	Lower
Photosynthetic capacity per unit dry weight of leaf	Higher	Lower
Length of growing season	176 days	260 days
Primary productivity (tonnes carbon ha^{-1} yr^{-1})	8.6	14.9

NPP may be low because appropriate mineral resources are deficient

nutrients. The geological conditions that determine slope and aspect also determine whether a soil forms, and they have a large, though not wholly dominant, influence on the mineral content of the soil. For this reason, a mosaic of different levels of community productivity develops within a particular climatic regime. However, of all the mineral nutrients, the one that has the most overriding influence on community productivity is fixed nitrogen (and this is invariably partly or mainly biological, not geological, in origin, as a result of nitrogen fixation by microorganisms). There is probably no agricultural system that does not respond to applied nitrogen by increased primary productivity, and this may well be true of natural vegetation as well. Nitrogen fertilizers added to forest soils almost always stimulate forest growth (Spurr & Barnes, 1973).

The deficiency of other elements can hold the productivity of a community far below that of which it is theoretically capable. A classic example is deficiency of phosphate and zinc in South Australia, where the growth of commercial forest (Monterey pine, *Pinus radiata*) is made possible only when these nutrients are supplied artificially.

17.3.2 Resumé of factors limiting terrestrial productivity

The ultimate limit on the productivity of a community is determined by the amount of incident radiation that it receives—without this, no photosynthesis can occur.

Incident radiation is used inefficiently by all communities. The causes of this inefficiency can be traced to: (a) shortage of water restricting the rate of photosynthesis; (b) shortage of essential mineral nutrients which slows down the rate of production of photosynthetic tissue and its effectiveness in photosynthesis; (c) temperatures that are lethal or too low for growth; (d) an insufficient depth of soil; (e) incomplete canopy cover, so that much incident radiation lands on the ground instead of on foliage (this may be because of seasonality in leaf production and leaf shedding *or* because of defoliation by grazing animals, pests and diseases); and (f) the low efficiency with which leaves photosynthesize—even under ideal conditions, efficiencies of more than 10% (of photosynthetically active radiation) are hard to achieve even in the most productive agricultural systems. However, most of the variation in primary productivity of world vegetation is due to factors (a)–(e), and relatively little is accounted for by intrinsic differences between the photosynthetic efficiencies of the leaves of the different species.

In the course of a year, the productivity of a community may (and probably usually will) be limited by a succession of the factors (a)–(e). In a grassland community, for instance, the primary productivity may be far below the theoretical maximum because the winters are too cold and light intensity is low, the summers are too dry, the rate of nitrogen mobilization is too slow, and for periods grazing animals may reduce the standing crop to a level at which much incident light falls on bare ground.

17.3.3 The primary productivity of aquatic communities

The factors that most frequently limit the primary productivity of aquatic environments are the availability of nutrients, light, and the intensity of grazing. The most commonly limiting nutrients are nitrogen (usually as nitrate) and phosphorus (phosphate). Productive aquatic communities occur where, for one reason or another, nutrient concentrations are unusually high.

Lakes receive nutrients by weathering of rocks and soils in their catchment areas, in rainfall, and as a result of human activity (fertilizers and sewage input). They vary considerably in nutrient availability. An analysis of IBP data derived from lakes all over the world points to a significant role in limiting productivity for nutrients, particularly the estimated steady-state concentration of phosphorus (Figure 17.12). We noted earlier a positive relationship between lake primary productivity and latitude (Figure 17.2c). This must be due in part to the higher light intensities and longer growing seasons at lower latitudes. The supply of nutrients may also be greater in low latitudes, because of a more rapid rate of mineralization, and the concentrations of phosphorus compounds may be higher in tropical rainfall (Schindler, 1978).

Figure 17.12. Relationship between phytoplankton net primary productivity in lakes throughout the world and estimated steady state concentration of phosphorus (Schindler, 1978).

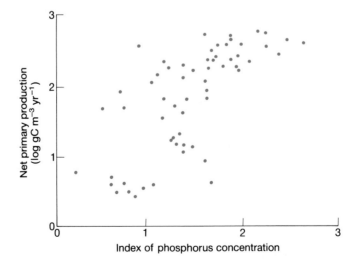

estuaries and upwellings provide rich supplies of nutrients in marine environments

In the oceans, locally high levels of primary productivity are associated with high nutrient inputs from two sources. First, nutrients may flow continuously into coastal shelf regions from estuaries. An example is provided in Figure 17.13. Productivity in the inner shelf region is particularly high both because of high nutrient concentrations and because the relatively clear water

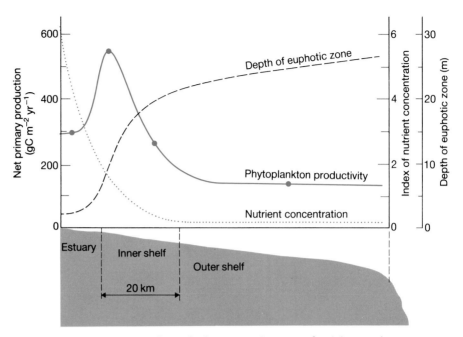

Figure 17.13. Variation in phytoplankton net primary productivity, nutrient concentration and euphotic depth on a transect from the coast of Georgia, U.S.A., to the edge of the continental shelf (after Haines, 1979).

provides a reasonable depth within which net photosynthesis is positive (the *euphotic zone*). Closer to land, the water is richer in nutrients but highly turbid and its productivity is less. The least productive zones are on the outer shelf (and open ocean) where, despite clear water and a very deep euphotic zone, nutrient concentrations are extremely low.

Ocean upwellings are a second source of locally high nutrient concentrations. These occur on continental shelves where the wind is consistently parallel to, or at a slight angle to, the coast. As a result, water moves offshore and is replaced by cooler, nutrient-rich water originating from the bottom, where nutrients have been accumulating by sedimentation. Strong upwellings can also occur adjacent to submarine ridges, as well as in areas of very strong currents. Where it reaches the surface, the nutrient-rich water sets off a bloom of phytoplankton production. A chain of heterotrophic organisms takes advantage of the abundant food, and the great fisheries of the world are located in these regions of high productivity. Figure 17.14 illustrates for the Pacific Ocean the general correspondence between phosphate concentration and phytoplankton productivity.

Although the concentration of a limiting nutrient usually determines the productivity of aquatic communities on an areal basis, in any given water body there is also considerable variation with depth as a result of attenuation of light intensity. Figure 17.15a shows how gross primary productivity declines with depth. The depth at which GPP is just balanced by phytoplankton respiration, R, is known as the compensation point. Above this, net primary productivity is positive. Light is absorbed by water molecules as well as by dissolved and particulate matter, and it declines exponentially with depth. Near the surface, light is superabundant, but at greater depths its

phytoplankton productivity varies with depth

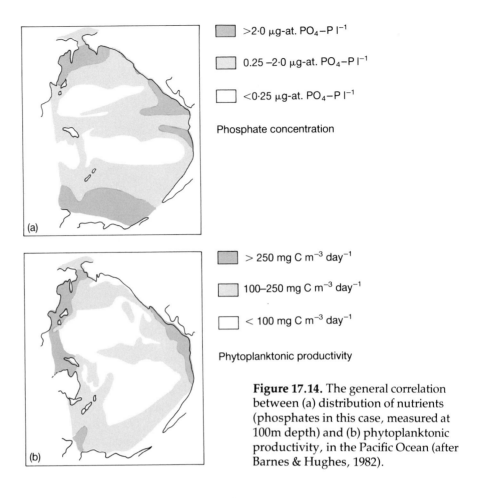

>2·0 μg-at. PO_4–P l^{-1}

0.25 –2·0 μg-at. PO_4–P l^{-1}

<0·25 μg-at. PO_4–P l^{-1}

Phosphate concentration

> 250 mg C m^{-3} day^{-1}

100–250 mg C m^{-3} day^{-1}

< 100 mg C m^{-3} day^{-1}

Phytoplanktonic productivity

Figure 17.14. The general correlation between (a) distribution of nutrients (phosphates in this case, measured at 100m depth) and (b) phytoplanktonic productivity, in the Pacific Ocean (after Barnes & Hughes, 1982).

supply is limited and light intensity ultimately determines the extent of the euphotic zone. Very close to the surface, particularly on sunny days, there may be photo-inhibition of photosynthesis. This seems to be due largely to radiation being absorbed by the photosynthetic pigments at such a rate that it cannot be used via the normal photosynthetic channels, and it overflows into destructive photo-oxidation reactions.

The more nutrient-rich a water body is, the shallower its euphotic zone is likely to be (Figure 17.15b). This is not really a paradox. Water bodies with higher nutrient concentrations usually possess greater biomasses of phytoplankton which absorb light and reduce its availability at greater depth. (This is exactly analogous to the shading influence of the tree canopy in a forest, which may remove up to 98% of the radiant energy before it can reach the ground layer vegetation.) Even quite shallow lakes, if sufficiently fertile, may be devoid of water weeds on the bottom because of shading by phytoplankton. The relationships shown in Figure 17.15a and b are derived from lakes but the pattern is qualitatively similar in ocean environments.

the productivity of phytoplankton varies with the seasons

Typical seasonal variations in phytoplankton primary productivity for arctic, north temperate and tropical lakes and seas are illustrated in Figure 17.16. For part of the year in arctic and temperate latitudes, conditions are

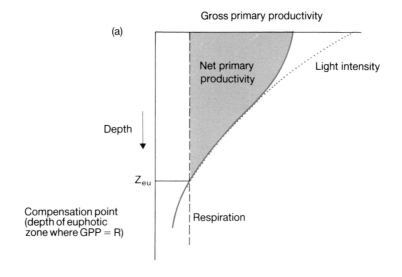

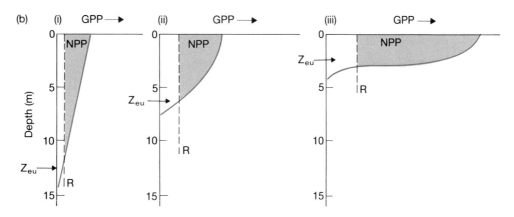

Figure 17.15. (a) The general relationship with depth, in a water body, of gross primary productivity (GPP), respiratory heat loss (R) and net primary productivity (NPP). The compensation point (or depth of the euphotic zone, Eu) occurs at the depth (Z_{eu}) where GPP just balances R and NPP is zero. (b) Total NPP increases with nutrient concentration in the water (lake iii > ii> i). Increasing fertility itself is responsible for greater biomasses of phytoplankton and a consequent decrease in the depth of the euphotic zone.

harsh and water turbulence, particularly in marine environments, constantly forces phytoplankton cells into deep water where GPP does not compensate for respiratory losses: NPP is then zero or negative. As the year progresses, increases in solar radiation, day-length and temperature, coupled with a decrease in wind-generated turbulence, permit phytoplankton production to commence in earnest. This happens progressively later in the year at higher latitudes. Production ceases again, first in arctic waters, later at temperate latitudes, either because nutrients have become exhausted or as a result of deteriorating conditions later in the year. At lower latitudes, production is maintained throughout the year.

Two peaks of biomass have sometimes been observed to develop within

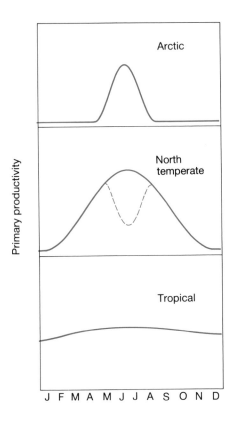

Figure 17.16. Generalized descriptions of seasonal patterns in phytoplankton net primary productivity in aquatic environments at different latitudes.

a season in certain temperate lakes and marine environments. There are two possible explanations. Primary productivity may itself pass through two peaks in the year (Figure 17.16, dashed line). Nutrients become limiting in the summer, but are recycled from deep water when the thermally stratified water becomes thoroughly mixed again in autumn. A second period of phytoplankton production then occurs while physical conditions are still favourable. Alternatively, there may be a single peak in the activity of the phytoplankton, but zooplankton grazing may be particularly intense during the summer, causing a dramatic decline in phytoplankton standing crop, followed by an autumnal increase when zooplankton grazing ceases—essentially a predator–prey cycle of the type discussed in Chapter 10. In freshwaters, nutrient depletion is probably the most important factor, while zooplankton grazing is generally thought to be more significant in the seas.

17.4 The fate of energy in communities

Secondary productivity is defined as the rate of production of new biomass by heterotrophic organisms. Unlike plants, the bacteria, fungi and animals cannot manufacture from simple molecules the complex, energy-rich compounds they need. They derive their matter and energy either directly by consuming plant material or indirectly from plants by eating other heterotrophs. Plants, the primary producers, comprise the first trophic level in a community; primary consumers occur at the second trophic level; secondary consumers (carnivores) at the third, and so on.

there is a general positive relationship between primary and secondary productivity

Since secondary productivity depends on primary productivity, we should expect a positive relationship between the two variables in communities. Figures 17.17a and b illustrate this general relationship in an aquatic and a terrestrial example. Secondary productivity by zooplankton, which principally consume phytoplankton cells, is positively related to phytoplankton productivity in a range of lakes situated in different parts of the world. Similarly, secondary productivity by large mammalian herbivores in a series of African game parks is positively related to estimates of the primary productivity of plants in these savanna systems.

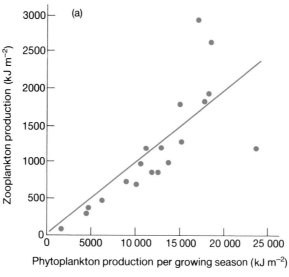

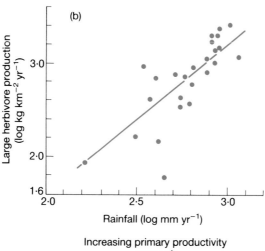

Figure 17.17. The relationship between primary and secondary productivity (a) in lake plankton communities (Brylinsky & Mann, 1973), (b) for large mammalian herbivores in African game parks—primary productivity in these communities is directly related to annual rainfall (Coe *et al.*, 1976).

In both examples, secondary productivity by the herbivores is approximately an order of magnitude less than the primary productivity upon which it is based. This is a consistent feature of all 'grazer' systems (that part of the trophic structure of a community which depends, at its base, on the consumption of *living* plant biomass). It results in a pyramidal structure in which the productivity of plants provides a broad base upon which a smaller productivity of primary consumers depends, with a still smaller productivity

651 FLUX OF ENERGY AND MATTER THROUGH COMMUNITIES

of secondary consumers above that. Trophic levels may also have a pyramidal structure when expressed in terms of density or biomass. (Elton (1927) was the first to recognize this fundamental feature of community architecture. The ideas were later elaborated by Lindemann (1942).) But there are many exceptions. Food chains based on trees will certainly have larger numbers (but *not* biomass) of herbivores per unit area than of plants, while chains dependent on phytoplankton production may give inverted pyramids of biomass, with a highly productive but small biomass of short-lived algal cells maintaining a larger biomass of longer-lived zooplankton.

The productivity of herbivores is invariably less than that of the plants on which they feed. Where has the missing energy gone? First, not all of the plant biomass produced is consumed alive by herbivores. Much dies without being grazed and supports the decomposer community (bacteria, fungi and detritivorous animals). Secondly, not all the plant biomass that is eaten by herbivores (nor herbivore biomass eaten by carnivores) is assimilated and available for incorporation into consumer biomass. Some is lost in faeces, and this also passes to the decomposers. Thirdly, not all the energy which has been assimilated is actually converted to biomass. A proportion is lost as respiratory heat. This occurs both because no energy conversion process is ever 100% efficient (some is lost as unusable random heat, consistent with the second law of thermodynamics) and also because animals do work which requires energy, again released as heat.

These three energy pathways occur at all trophic levels and are illustrated in Figure 17.18.

most of the primary productivity does not pass through the grazer system

17.4.1 A comprehensive model of the trophic structure of a community

Figure 17.19 provides a complete description of the trophic structure of a community. It consists of the grazer system pyramid of productivity, but with two additional elements of realism. Most importantly, it adds a *decomposer system* which is invariably coupled to the grazer system in communities. Secondly, it recognizes that there are subcomponents of each trophic level in each subsystem which operate in different ways. Thus a distinction is made between invertebrate and vertebrate categories, between microbes and detritivores which occupy the same trophic level and utilize dead organic matter, and between consumers of microbes (microbivores) and of detritivores (the carnivores in the decomposer system). Displayed in Figure 17.19 are the possible routes that a joule of energy, fixed in net primary production, can take through the community.

A joule of energy may be consumed and assimilated by an invertebrate herbivore which uses it to do work and loses it as respiratory heat. Or it might be consumed by a vertebrate herbivore and later be assimilated by a carnivore which dies and enters the dead organic matter compartment. Here the joule may be assimilated by a fungal hypha and consumed by a soil mite, which uses it to do work and loses it as heat. At each consumption step, the joule may fail to be assimilated and pass in the faeces to the dead organic matter, or it may be assimilated and respired, or assimilated and incorporated into growth of body tissue (or the production of offspring). The body may die and

there are many alternative pathways that energy can trace through a community

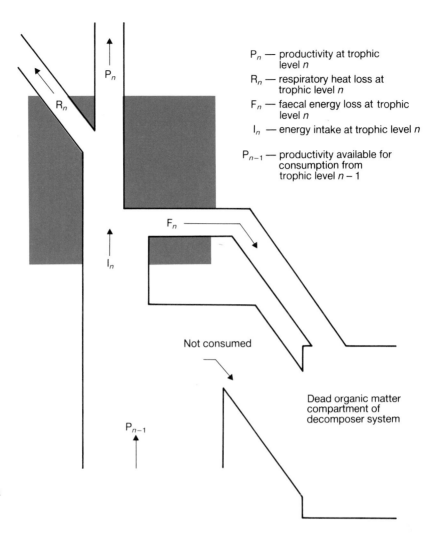

Figure 17.18. The pattern of energy flow through a trophic compartment.

P_n — productivity at trophic level n

R_n — respiratory heat loss at trophic level n

F_n — faecal energy loss at trophic level n

I_n — energy intake at trophic level n

P_{n-1} — productivity available for consumption from trophic level $n-1$

Not consumed

Dead organic matter compartment of decomposer system

the joule enter the dead organic matter compartment, or it may be captured alive by a consumer in the next trophic level where it meets a further set of possible branching pathways. Ultimately, each joule will find its way out of the community as respiratory heat from one or other of the trophic levels.

The possible pathways in the grazer and decomposer systems are the same, with one critical exception—faeces and dead bodies are lost to the grazer system (and enter the decomposer system), but faeces and dead bodies from the decomposer system are simply sent back to the dead organic matter compartment at its base. This has a fundamental significance. The energy available as dead organic matter may finally be completely metabolized—and all the energy lost as respiratory heat—even if this requires several circuits through the decomposer system. The exceptions to this are situations (i) where matter is exported out of the local environment to be metabolized elsewhere, for example detritus being washed out of a stream, and (ii) where local abiotic conditions are very unfavourable to decomposition processes, leaving pockets of incompletely metabolized high-energy matter, otherwise known as oil, coal and peat.

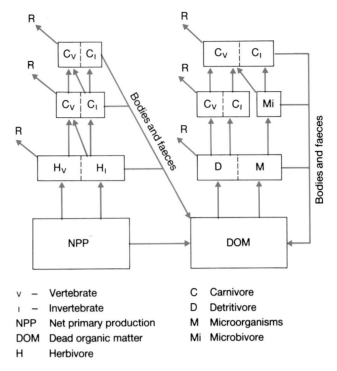

Figure 17.19. A generalized model of trophic structure and energy flow for a terrestrial community (after Heal & MacLean, 1975).

v — Vertebrate   C   Carnivore
ı — Invertebrate   D   Detritivore
NPP   Net primary production   M   Microorganisms
DOM   Dead organic matter   Mi   Microbivore
H   Herbivore

The proportions of net primary production that flow along each of the possible energy pathways depend on *transfer efficiencies* in the way energy is used and passed from one step to the next. A knowledge of the values of just three categories of transfer efficiency is all that is required to predict the pattern of energy flow. These are *consumption efficiency* (CE), *assimilation efficiency* (AE) and *production efficiency* (PE).

$$\text{Consumption efficiency, CE} = \frac{I_n}{P_{n-1}} \times 100.$$

In words, CE is the percentage of total productivity available at one trophic level (P_{n-1}) that is actually consumed ('ingested') by a trophic compartment one level up' (I_n). For primary consumers in the grazer system, CE is the percentage of joules produced per unit time as net primary productivity which finds its way into the guts of herbivores. In the case of secondary consumers, it is the percentage of herbivore productivity eaten by carnivores. The remainder dies without being eaten and enters the decomposer chain.

Various reported values for consumption efficiencies of herbivores are given in Table 17.5. All the estimates are remarkably low, perhaps reflecting the unattractiveness of much plant material because of its high proportion of structural support tissue, but more likely a consequence of generally low herbivore densities (because of the action of their natural enemies). The consumers of microscopic plants achieve greater densities and account for a greater percentage of primary production. When viewing Table 17.5, note that the estimates of consumption efficiency are for particular herbivore groups and do not necessarily represent the total consumed alive. Reasonable average figures for consumption efficiency are approximately 5% in forests, 25% in grasslands and 50% in phytoplankton-dominated communities.

consumption, assimilation and production efficiencies determine the relative importance of energy pathways

the variations in consumption efficiency

Table 17.5. Consumption efficiencies of various herbivore groups on their plant food (from Pimentel *et al.*, 1975, and Brylinski & Mann, 1973).

Host plant	Taxon of feeding animal	Percentage of productivity consumed
Beech trees	Invertebrates	8.0
Oak trees	Invertebrates	10.6
Maple–beech trees	Invertebrates	5.9–6.6
Tulip poplar trees	Invertebrates	5.6
Grass+forbs	Invertebrates	< 0.5–20
Alfalfa	Invertebrates	2.5
Sericea lespedeza	Invertebrates	1.0
Grass	Invertebrates	9.6
Aquatic plants	Bivalves	11.0
Aquatic plants	Herbivorous animals	18.9
Algae	Zooplankton	25.0
Phytoplankton	Zooplankton	40.0
Phytoplankton	Herbivorous animals	21.2
Marsh grass	Invertebrates	4.6–7.0
Meadow plants	Invertebrates	14.0
Sedge grass	Invertebrates	8.0

We know much less about the consumption efficiencies of carnivores feeding on their prey, and any estimates are speculative. Vertebrate predators may consume 50–100% of production from vertebrate prey but perhaps only 5% from invertebrate prey. Invertebrate predators consume perhaps 25% of available invertebrate prey production, but this is a very crude figure of uncertain reliability.

$$\text{Assimilation efficiency, AE} = \frac{A_n}{I_n} \times 100.$$

the variations in assimilation efficiency

Assimilation efficiency is the percentage of food energy taken into the guts of consumers in a trophic compartment (I_n) which is assimilated across the gut wall and becomes available for incorporation into growth or is used to do work (A_n). The remainder is lost as faeces and enters the base of the decomposer system. Assimilation efficiency is a term that cannot properly be ascribed to microorganisms. Much of the digestion of dead organic matter by bacteria and fungi is extracellular and the 'assimilation efficiency' of absorbed, digested matter is effectively 100%. Faeces are not produced.

Assimilation efficiencies are typically low for herbivores, detritivores and microbivores (20–50%) and high for carnivores (around 80%). In general, animals are poorly equipped to deal with dead organic matter (mainly plant material) and living vegetation, no doubt partly because of the very widespread occurrence of physical and chemical plant defences, but mainly as a result of the high proportion of complex structural chemicals such as cellulose and lignin in their make-up. As Chapter 11 describes, however, many contain a symbiotic gut microflora which produces cellulase and aids in the assimilation of plant organic matter. In one sense, these animals have harnessed their own personal decomposer system. The way that plants allocate production to roots, wood, leaves, seeds and fruits influences their usefulness to herbivores. Seeds and fruits may be assimilated with an efficiency as high as

60–70%, and leaves with about 50% efficiency, while the assimilation efficiency for wood may be as low as 15%.

The animal food of carnivores (and detritivores such as vultures which consume animal carcasses) poses less problem for digestion and assimilation.

$$\text{Production efficiency, PE} = \frac{P_n}{A_n} \times 100.$$

Production efficiency is the percentage of assimilated energy (A_n) which is incorporated into new biomass (P_n). The remainder is entirely lost to the community as respiratory heat. (Energy-rich secretory and excretory products, which have taken part in metabolic processes, may be viewed as production, P_n, and become available, like dead bodies, to the decomposers.)

the variations in production efficiency

PE varies mainly according to the taxonomic class of the organisms concerned. Invertebrates in general have high efficiencies (30–40%), losing relatively little energy in respiratory heat and converting more assimilate to production. Amongst the vertebrates, ectotherms (whose body temperature varies according to environmental temperature) have intermediate values for PE (around 10%), whilst endotherms, with their high energy expenditure associated with maintaining a constant temperature, convert only 1–2% of assimilated energy into production. The relationships between productivity and respiration rate for seven classes of consumer are given in Figure 17.20,

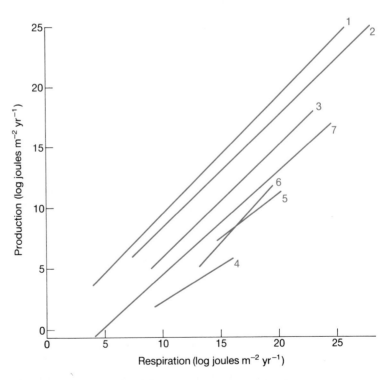

Figure 17.20. Graphs of animal production against animal respiration in seven groups of animals populations. 1—non-social insects; 2—invertebrates other than insects; 3—social insects and fish (not significantly different from each other); 4—insectivores; 5—birds; 6—small mammals; 7—mammals other than insectivores and other small mammals (after Humphreys, 1979)

and average values for PE in each class are shown in Table 17.6. The small-bodied endotherms have the lowest efficiencies, with the tiny insectivores (e.g. shrews) having the lowest production efficiencies of all.

Microorganisms, including protozoa, tend to have very high production efficiencies. They have short lives, small size and rapid population turnover. Unfortunately, available methods are not sensitive enough to detect population changes on scales of time and space relevant to microorganisms, especially in the soil. In general, efficiency of production increases with size in endotherms and decreases very markedly in ectotherms.

Table 17.6. Production efficiency (P/A×100) of various animal groups ranked in order of increasing efficiency (from Humphreys, 1979).

Group	P/A (%)
1. Insectivores	0.86
2. Birds	1.29
3. Small mammal communities	1.51
4. Other mammals	3.14
5. Fish and social insects	9.77
6. Non-insect invertebrates	25.0
7. Non-social insects	40.7
NON-INSECT INVERTEBRATES	
8. Herbivores	20.8
9. Carnivores	27.6
10. Detritivores	36.2
NON-SOCIAL INSECTS	
11. Herbivores	38.8
12. Detritivores	47.0
13. Carnivores	55.6

17.4.2 Energy flow through a model community

Given specified values for net primary productivity at a site, and CE, AE and PE for the various trophic groupings shown in the model in Figure 17.19, it is possible to map out the relative importance of different pathways. Heal and MacLean (1975) did this for a hypothetical grassland community. The values they used for assimilation and production efficiencies are shown in diagrammatic form in Figure 17.21, and the resulting fate of each 100 joules of NPP is given in Table 17.7. A consumption efficiency of 25% was assumed for the vertebrate herbivores and 4% for invertebrate herbivores. Thus for every 100J of NPP, 29 are ingested by herbivores. A greater proportion of NPP goes to the decomposer system, but, in addition, decomposers consume well in excess of 100J for every 100J of NPP! This comes about simply because energy that is not assimilated on its first trip through the decomposer chain is available for consumption again. The decomposers are responsible for 84.8% of the consumption of matter. However, they carry out 90.8% of the assimilation (mainly because of the importance of microbial activity and their assumed 100% assimilation efficiency). Once again the apparent discrepancy in total joules assimilated by the decomposers (157J per 100J of NPP) results from their ability to 'work over' organic matter on a number of occasions.

The most significant result of this study is the overwhelming importance

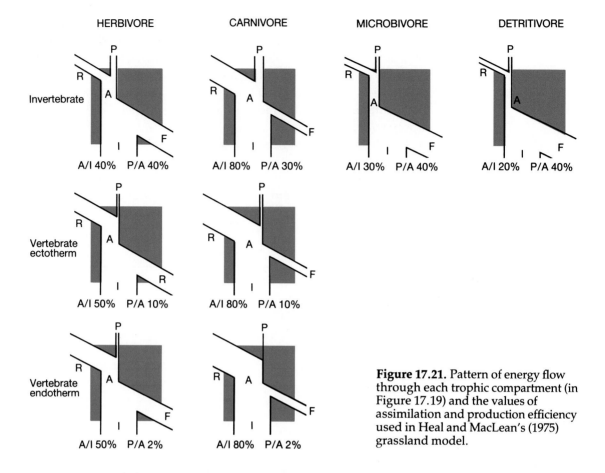

Invertebrate

A/I 40%   P/A 40%   A/I 80%   P/A 30%   A/I 30%   P/A 40%   A/I 20%   P/A 40%

Vertebrate
ectotherm

A/I 50%   P/A 10%   A/I 80%   P/A 10%

Vertebrate
endotherm

A/I 50%   P/A 2%   A/I 80%   P/A 2%

Figure 17.21. Pattern of energy flow through each trophic compartment (in Figure 17.19) and the values of assimilation and production efficiency used in Heal and MacLean's (1975) grassland model.

the decomposer system is responsible for 98% of secondary productivity in this grassland community

of the decomposer system. Even though the grazer system consumes 29% of NPP, it only accounts for 2% of secondary productivity. Of every 100J of NPP, more than 55J find their way into decomposer production per year but less than 1J into grazer production. Overall, in this steady state community, losses through animal respiration balance NPP so that standing crop biomass (not illustrated) stays the same.

the energy flow model receives support from IBP results

Recall that the data presented in Table 17.7 are for a hypothetical community, with the structure shown in Figure 17.19 and assumed (but realistic) values for transfer efficiencies. Is there any way to tell whether the model and assumptions are reasonable—in other words, do real systems operate as predicted? Heal and Maclean (1975) attempted a validation using real data on primary and secondary productivity from ten tundra, grassland and forest ecosystems. By feeding NPP data into the model (with appropriate herbivore consumption efficiencies: forest 5%, non-forest 25%), it was possible to predict secondary productivity of various trophic compartments. These *predicted* values are plotted against the *actual*, measured values in Figure 17.22, and there is good agreement (perfect agreement would have resulted in all the data points falling on the line at 45° to the origin).

In general then, the model receives some support from this exercise. Also in its favour, it incorporates most of the recognized fundamental features of community structure. It is faulty in at least one respect, however. As we saw

	Consumption	Assimilation	Egestion	Production	Respiration
GRAZER SYSTEM					
Herbivores					
vertebrate	25.00	12.50	12.50	0.25	12.25
invertebrate	4.00	1.60	2.40	0.64	0.96
Carnivores					
vertebrate	0.16	0.13	0.03	0.003	0.127
invertebrate	0.17	0.135	0.035	0.040	0.095
DECOMPOSER SYSTEM					
Decomposers+detritivores					
microbial decomposers	136.38	136.38	0	54.55	81.83
invertebrate detritivores	15.15	3.03	12.12	1.21	1.82
Microbivores					
invertebrates	10.91	3.27	7.64	1.31	1.96
Carnivores					
vertebrates	0.04	0.03	0.01	0.001	0.029
invertebrates	0.65	0.52	0.13	0.16	0.36
TOTAL	*192*	*157*	*35*	*58*	*99*
Percentage passing through:					
grazer system	15.2	9.2	42.9	1.6	13.5
decomposer system	84.8	90.8	57.1	98.4	86.5

Table 17.7. Calculated consumption, assimilation, egestion, production and respiration by heterotrophs per 100 J m^{-2} net annual primary production in a hypothetical grassland community (after Heal & MacLean, 1975).

in Chapter 11, many detritivorous animals consume both microbial biomass and dead organic matter, if inadvertently. Unfortunately, we have relatively few detailed data on these relationships, so the community structure illustrated in Figure 17.19 must stand for now. We need not be too concerned, however, because the overwhelming importance of microbial consumption and production means that fine adjustment in the model of relationships amongst detritivores and microbivores (often the same animals), microbes and dead organic matter will not affect the overall picture of productivity relationships to any great extent.

Like all generalizations about trophic structure, our model also suffers the shortcoming that not all consumers can be slotted neatly into a single compartment. Some herbivores eat dead matter on occasion, while some carnivores eat both herbivores and detritivores (and the occasional plant). In general, the productivity of omnivores can be split into the different compartments according to the magnitude of their involvement in each. Again, the general picture we have described is not dramatically affected by this technical problem.

17.4.3 Patterns of energy flow in contrasting communities

There have been only a very few studies in which all the community compartments have been studied together, and most of those which appeared in earlier ecological texts are unreliable because they over-stressed

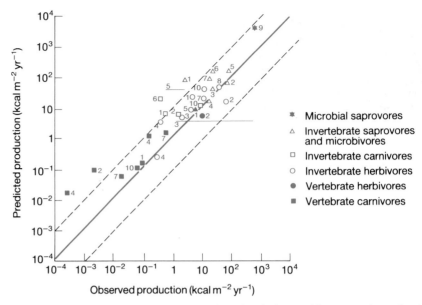

Figure 17.22. Relationship between predicted and observed heterotroph productivity values in a range of sites: tundra (3,4), cold temperate moorland (2,7), temperate grassland (1), temperate deciduous forest (5,6,8,9,10). (After Heal & MacLean, 1975.)

the role of the grazer system and generally failed to measure, or appreciate, the overwhelming importance of microbial production. However, some generalizations are possible if we compare the gross features of contrasting systems. Figure 17.23 illustrates the patterns of energy flow in a forest, a grassland, a plankton community (of the ocean or a large lake) and the community of a small stream or pond. The decomposer system is probably responsible for the majority of secondary production, and therefore respiratory heat loss, in every community in the world. However, the grazer system has its greatest role in plankton communities, where a large proportion of NPP may be consumed alive and assimilated at quite a high efficiency. The grazer system holds less sway in terrestrial communities because of low herbivore consumption and assimilation efficiencies, and it is almost non-existent in small streams and ponds simply because primary productivity is so low. The latter depend for their energy base on dead organic matter produced in the terrestrial environment which falls or is washed or blown into the water. The deep-ocean benthic community has a trophic structure very similar to that of streams and ponds (all can be described as heterotrophic communities). In this case, the community lives in water too deep for photosynthesis to be appreciable or even to take place at all, but it derives its energy base from dead phytoplankton, bacteria and animals which sink from the autotrophic community in the euphotic zone above. From a different perspective, the ocean bed is equivalent to a forest floor beneath an impenetrable forest canopy.

17.5 The fate of matter in communities

For much of the time, matter and energy follow the same pathways in communities, and in both cases the decomposer system plays a central role.

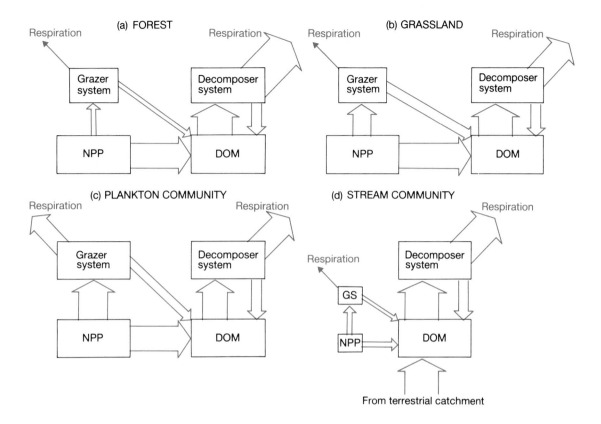

Figure 17.23. General patterns of energy flow for (a) a forest, (b) a grassland, (c) a plankton community in a large lake or the sea, (d) the community of a stream or small pond. Relative sizes of boxes and arrows are proportional to relative magnitudes of compartments and flows. NPP = net primary production, DOM = dead organic matter.

Carbon, for example, enters the trophic structure of a community when a simple molecule, CO_2 is recruited in photosynthesis. If it becomes incorporated in net primary productivity, it is available for consumption as part of a sugar, a fat, a protein or, very often, a cellulose molecule. It follows exactly the same route as energy, being successively consumed, defaecated, assimilated and perhaps incorporated into secondary productivity somewhere within one of the trophic compartments. When the high-energy molecule in which the carbon is resident is finally used to provide energy for work, the energy is dissipated as heat (as we have discussed above) while the carbon is released again to the atmosphere as CO_2, a product of tissue respiration. Here, the similarity between energy and carbon (or other nutrients) ends.

energy cannot be cycled and re-used—matter can   Once energy is transformed into the random form of heat, it can no longer be used by living organisms to do work or to fuel the synthesis of biomass. (Its only possible use is momentary, in helping to maintain a high body temperature.) The heat is eventually lost to the atmosphere, balancing the radiant energy income. Carbon in CO_2, on the other hand, can be used again in photosynthesis. Carbon, and all other nutrients (e.g. nitrogen, phosphorus, etc.) are available to plants as simple inorganic molecules either in the atmosphere (CO_2) or dissolved in water (nitrate, phosphate, etc.). Each can be incorporated into complex organic chemicals in biomass during photosynthesis. Ultimately, however, they become available again when the chemicals are metabolized, either within the living organisms (carbon released as CO_2) or as a result of the activity of the decomposer system (nitrogen and phosphorus released again in the form of simple inorganic molecules). The relationship between energy flow and nutrient cycling is illustrated in Figure 17.24.

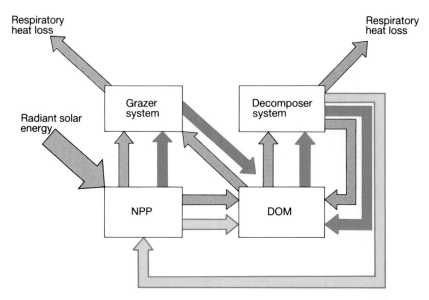

Figure 17.24. Diagram to show the relationship between energy flow (▨) and nutrient cycling. Nutients locked in organic matter (▨) are distinguished from the free inorganic state (▢).

By its very nature, energy cannot be recycled. It is made available to living organisms as solar radiation which can be fixed during photosynthesis. But once the resulting chemical energy is used, it is dissipated as useless heat. Although energy can pass back and forth between the dead organic matter compartment and the decomposer system this must not be described as cycling. It simply reflects the ability of the decomposer system to 'work over' organic matter several times; but each joule of energy can only be *used* once. Life on earth is possible because a fresh supply of solar energy is made available every day.

On the other hand, chemical nutrients, the building blocks of biomass, simply change the form of molecule of which they are part (e.g. nitrate-N→ protein-N→nitrate-N). They can be used again, and recycling is a critical feature. Unlike the energy of solar radiation, nutrients are not in unalterable supply, and the process of locking some up into living biomass reduces the supply remaining to the community. If plants, and their consumers, were not eventually decomposed, the supply of nutrients would become exhausted and life on earth would cease. The activity of heterotrophic organisms is crucial in bringing about nutrient cycling and maintaining productivity. Figure 17.24 shows the release of nutrients in their simple inorganic form as occurring only from the decomposer system. In fact, a proportion is also released from the grazer system, particularly in the case of carbon, but only a

matter (and nutrients) can therefore be 'locked-up' in the community

very small proportion is recycled through this pathway. The decomposer system plays a role of overwhelming importance.

The picture described in Figure 17.24 is an oversimplification in one important respect. Not all nutrients released during decomposition are necessarily taken up again by plants. Nutrient cycling is never perfect. Moreover, the community receives additional supplies of nutrient which do not depend directly on inputs from recently decomposed matter. The various categories of supply and loss of nutrients in terrestrial communities are listed in Table 17.8.

nutrient cycling is never perfect

Table 17.8. Major routes of import and export of nutrients in terrestrial communities.

Import	Export
Precipitation	Run-off and stream outflow
Particulate fallout from the atmosphere	Particulate loss by wind
Biotic immigration	Biotic emigration
Fixation from the atmosphere	Release to the atmosphere
Weathering of substrate	Loss by leaching
Fertilizer application and pollution	Human harvest

Just how important is nutrient cycling in relation to the through-put of nutrients? Is the amount of nutrients cycled per year small or large in comparison with external supplies and losses? The most thorough study of this question was carried out by Likens and his associates in the Hubbard Brook Experimental Forest, an area of temperate deciduous forest drained by small streams in the White Mountains of New Hampshire, U.S.A. The catchment area—the extent of terrestrial environment drained by a particular stream—was taken as the unit of study because of the role that streams play in nutrient export. Six small catchments were defined and their outflows were monitored. A network of precipitation gauges recorded the incoming amounts of rain, sleet and snow. Chemical analyses of precipitation and stream water made it possible to calculate the amounts of various nutrients entering and leaving the system, and these are shown in Table 17.9. A similar pattern is found each year. In most cases, the output of chemical nutrients in streamflow is greater than their input from rain, sleet and snow. The source of the excess chemicals is parent rock and soil, which are weathered and leached at a rate of about 70 $gm^{-2} yr^{-1}$.

In almost every case, the inputs and outputs of nutrients are small in comparison with the amounts held in biomass and recycled within the system. This is illustrated in Figure 17.25 for one of the most important nutrients—nitrogen. Nitrogen was added to the system not only in precipitation (6.5 kg ha^{-1} yr^{-1}) but also through atmospheric nitrogen fixation by

the Hubbard Brook project . . .

Table 17.9. Annual nutrient budgets for forested catchments at Hubbard Brook (kg ha^{-1} yr^{-1}). Inputs are for dissolved materials in precipitation or as dryfall. Outputs are losses in stream water as dissolved material plus particulate organic matter (From Likens *et al.*, 1971.)

	NH_4^+	NO_3^-	SO_4^{2-}	K^+	Ca^{2+}	Mg^{2+}	Na^+
Input	2.7	16.3	38.3	1.1	2.6	0.7	1.5
Output	0.4	8.7	48.6	1.7	11.8	2.9	6.9
Net change*	+2.3	+7.6	−10.3	−0.6	−9.2	−2.2	−5.4

*Net change is positive when the ecosystem gains matter and negative when it loses it.

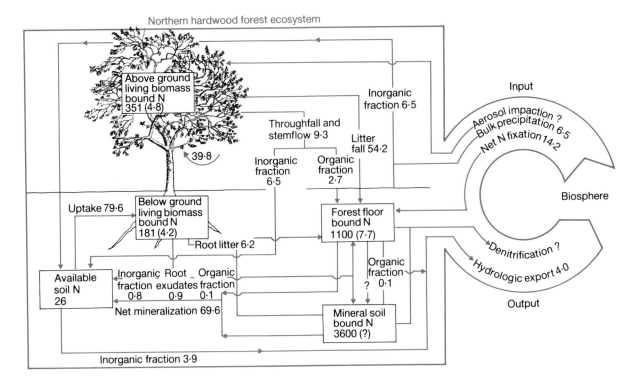

Figure 17.25. Annual nitrogen budget for the undisturbed Hubbard Brook Experimental Forest. Values in boxes are the sizes of the various nitrogen pools in kilograms of nitrogen per hectare. The rate of accretion of each pool (in parentheses) and transfer rates are expressed in kilograms of nitrogen per hectare per year. (From Bormann *et al.*, 1977.)

inputs and outputs of nutrients are typically low compared to the amounts cycling . . .

microorganisms (14 kg ha^{-1} yr^{-1}). (Note that denitrification by other microorganisms, releasing nitrogen to the atmosphere, will also have been occurring but was not measured.) The export in streams of only 4 kg ha^{-1} yr^{-1} emphasizes the tightness with which nitrogen is held and cycled within the forest biomass. Stream output represents only 0.1% of the total (organic) nitrogen standing crop held in living and dead forest organic matter.

Nitrogen was unusual in that its net loss in stream run-off was less than its input in precipitation, reflecting the complexity of inputs and outputs and the tightness of its cycling. However, despite the net loss to the forest of other nutrients, their export was also low in relation to the amounts bound in biomass. In other words, relatively tight recycling is the norm.

. . . though sulphur is an important exception (largely because of 'acid rain')

The major exception is sulphur. The amount of sulphur leaving the system annually (about 24 kg ha^{-1}) was far in excess of the amount in annual litter fall (5.5 kg ha^{-1}). It has been calculated that half the annual input of sulphur can be ascribed to pollution resulting from the burning of fossil fuels. Such pollution has led to *acid rain* (at Hubbard Brook, rain is essentially dilute sulphuric acid with a pH often below 4.0), and this is now recognized as one of the most widespread pollution problems in much of the Northern Hemisphere, although nitrogen oxides and ozone may also share the blame. So far, forestry resources (in Germany) and fish stocks (e.g. in Scandinavia and

Scotland) have apparently been adversely affected. The large proportions of sulphur leaving the Hubbard Brook system each year, in contrast to all the other macronutrients, must be seen as a consequence of the input of sulphur from pollution.

deforestation uncoupled cycling and led to a loss of nutrients

In a large-scale experiment, all the trees were felled in one of the Hubbard Brook catchments. The overall rate of export of dissolved inorganic sub-

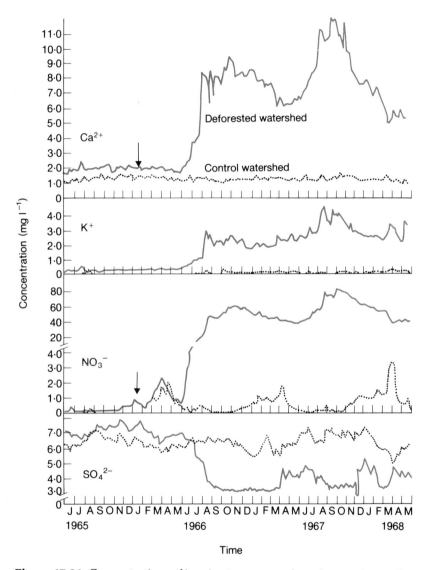

Figure 17.26. Concentrations of ions in stream water from the experimentally deforested catchment and a control catchment at Hubbard Brook. The timing of deforestation is indicated by arrows. Note that the 'nitrate' axis has a break in it. (From Likens & Borman, 1975.)

stances from the disturbed catchment was 13 times normal (Figure 17.26). Two phenomena were responsible. First, the enormous reduction in transpiring surfaces (leaves) led to 40% more precipitation passing through the groundwater to be discharged in the streams, and this increased outflow caused greater rates of leaching of chemicals and weathering of rock and soil. Secondly, and more significantly, deforestation has effectively broken the within-system nutrient cycling by uncoupling the decomposition process from the plant-uptake process. In the absence of nutrient uptake in spring, when the deciduous trees would have started production, the inorganic nutrients released by decomposer activity were available to be leached in the drainage water. Small wonder that large-scale deforestation, for example to create new agricultural land, can lead to loss of top-soil, nutrient impoverishment, and increased severity of flooding.

The main effect of deforestation was on nitrate-N, emphasizing the normally tight cycling to which inorganic nitrogen is subject. Stream output of nitrate increased 60-fold after the disturbance. Other biologically important ions were also leached faster as a result of the uncoupling of nutrient cycling mechanisms (potassium—14-fold increase; calcium—7-fold increase, magnesium—5-fold increase). However, the loss of sodium, an element of lower biological significance, showed a much less dramatic change following deforestation (2.5-fold increase). Presumably it is cycled less tightly in the forest, so that uncoupling had less effect. Once again, sulphur was exceptional. Its rate of loss actually decreased after deforestation. No satisfactory explanation for this is available.

terrestrial biomes differ in the distribution of nutrients between living and dead organic matter

Few studies have been made of the economy of a community that are as detailed as that at Hubbard Brook. It is hard to say how far we can generalize

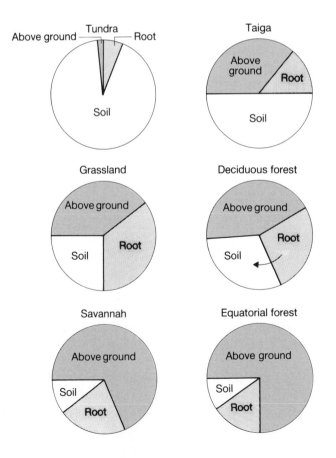

Figure 17.27. Disposition of nitrogen in the three organic matter compartments—above-ground, root, soil—for each of six biome types (after Swift et al., 1979).

from it to other communities. It may be unusual in the very high nitrogen outputs that followed felling. The subject is reviewed in detail by Gorham *et al.* (1979) and by Vitousek (1981). It seems likely that, in the majority of terrestrial environments, nutrient cycling will also be shown to be important relative to inputs and outputs to and from the systems. The major difference when biomes of the world are compared lies in the disposition of nutrients in the three organic matter compartments—above-ground, root and soil. The relative proportions in the case of nitrogen (and the pattern will be similar for other macronutrients) are shown in Figure 17.27.

The proportion of nutrients in living biomass increases from the poles to the equator. Decomposition processes are very slow in the cold boreal regions and nutrients build up in dead organic matter. In contrast, the climatic regime in a tropical rainforest favours rapid decomposition. Nutrients are rapidly released from dead organic matter and freed to be lost by leaching from the soil. In such a community, most nutrients are present in the living biomass. After land clearance in the tropics, it may be many centuries before the vegetational succession accumulates new nutrient resources that return it to its previous productivity.

The patterns outlined in Figure 17.27 emphasize the climatic control of decomposition. Temperature and moisture content of the litter are of prime importance. This is graphically illustrated in Figure 17.28, which shows the combined importance of the two factors on rates of decomposition, assessed as the respiration rate of the microbes and detritivores associated with the litter, for three plant categories.

When attention is switched from terrestrial to aquatic communities, there

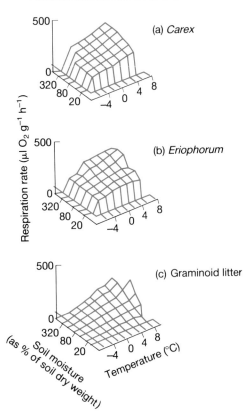

Figure 17.28. Regression surfaces showing the relationship between moisture content and temperature of three types of litter and their rate of decomposition (measured in terms of the respiration of the entire decomposer community) (from Bunnell *et al.*, 1977).

Table 17.10. Major routes of import and export of nutrients in aquatic communities. Precipitation and fallout are unimportant in marine ecosystems.

Import	Export
Stream inflow (from the catchment area)	Stream outflow
Precipitation	Biotic emigration
Particulate fallout from the atmosphere	Release to atmosphere
Biotic immigration	Loss to permanent sediments
Fixation from atmosphere	Human harvesting
Release from sediments	
Pollution	

stream flow and sedimentation are important phenomena in the flux of nutrients in aquatic systems

are several important distinctions to be made. First, the inputs and outputs of nutrients are not identical. In particular, aquatic systems receive the bulk of their supply of nutrients from stream inflow (Table 17.10). In stream and river communities, and also in lakes with a stream outflow, export in outgoing stream water is usually a major factor. By contrast, in lakes without an outflow (or where this is small relative to lake volume), and also in oceans, nutrients may accumulate in permanent sediments.

We noted in the case of Hubbard Brook that nutrient cycling within the forest was great in comparison to nutrient exchange through import and export. By contrast, only a tiny fraction of available nutrients takes part in biological interactions in stream and river communities. The vast majority

Figure 17.29. Main pathways of nitrogen transformation in a freshwater lake (from Moss, 1980).

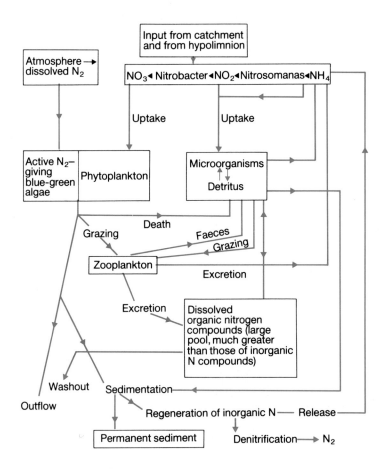

flow on, dissolved in the water, to be discharged into a lake or the sea. Thus nutrient exchange is great compared to internal cycling. Cycling can assume greater significance in lakes. Figure 17.29 provides a detailed picture of nitrogen flux, constructed on the basis of information gained in freshwater lakes. The pattern is similar in its essentials in marine communities.

17.6 A final word of caution

In Chapter 16 we cast aside the view that the community could be regarded as a sort of super-organism (the view of Clements). Yet in this chapter there is a sense in which we have taken just that view. A study of the trophic structure of a community treats it as a whole and asks what it does and how it compares with others. This can be done for the globe, a hectare, or a litre within a lake.

However, when we start asking questions about how it comes about that these community processes occur the way they do, we are forced back onto the behaviour of individual organisms for our explanations. It is they that are the material on which evolution works, and it is they, through their main effects and interactions together, that account for the activities of the community as a whole. We must not forget this.

18 The Influence of Competition on Community Structure

18.1 Introduction

In this chapter and the three that follow it, we will look at factors that shape and structure communities. We begin by looking at the role of interspecific competition. This was examined for pairs of species in Chapter 7; here the focus is not on pairs, but on whole assemblages of species.

Interspecific competition may play a central and powerful role in the shaping of communities. In the past, this view was fostered by the competitive exclusion principle (p. 259), which says that if two or more species compete for the same limiting resources, then all but one of the species will be driven to extinction. More recently, the same view has been underpinned by more sophisticated variants of the principle, namely the concepts of limiting similarity and niche packing (p. 262). These propose a limit to the similarity of competing species, and thus a limit to the number of species that can be fitted into a particular community before niche space is fully saturated. Within this theoretical framework, interspecific competition is obviously important, because it excludes particular species from some communities, and determines precisely which species coexist in others. The crucial question, however, is: how important are such theoretical effects in the real world? There is no argument about whether competition *sometimes* affects community structure; nobody doubts that it does. Equally, nobody claims that competition is of overriding importance in each and every case. (We will discuss some of the other important influences on community structure in the following few chapters.) The matters that we need to understand are the extent to which competition plays a powerful role in organizing communities and under what circumstances, and the precise effects that competition has.

18.2 The prevalence of current competition in natural communities

In a community where the species are competing with one another on a day-to-day or minute-by-minute basis, and where the environment is homogeneous, it is indisputable that competitive forces will have a powerful effect on community structure. Suppose instead, though, that other factors (predation, low-quality food, weather) keep densities at a low level where competition itself is negligible. By definition, competition *cannot* then be a potent force. One of the fundamental questions that community ecologists

670   CHAPTER 18

must try to answer, therefore, is: to what extent and in which cases *in practice* is interspecific competition an active force, shaping communities?

Perhaps the most direct way of answering this question is from the results of experimental field manipulations, in which one species is removed from or added to the community, and the responses of the other species are monitored (pp. 247–254). A number of workers have reviewed such experiments. Schoener (1983) examined the results of all the field experiments he could find on interspecific competition—164 studies in all. He found that approximately equal numbers of studies had dealt with terrestrial plants, terrestrial animals and marine organisms, but that studies of freshwater organisms amounted to only about half the number in the other groups. Amongst the terrestrial studies, however, he found that most were concerned with temperate regions and mainland populations, and that there were relatively few dealing with phytophagous insects (see below). Any conclusions were therefore bound to be subject to the limitations imposed by what ecologists had chosen to look at. Nevertheless, Schoener found that approximately 90% of the studies had demonstrated the existence of interspecific competition, and that the figures were 89, 91 and 94% for terrestrial, freshwater and marine organisms respectively. Moreover, if he looked at species or groups of species (of which there were 390) rather than at studies (which may have dealt with several groups of species), he found that 76% showed effects of competition at least sometimes, and 57% showed effects in all the conditions under which they were examined. Once again, terrestrial, freshwater and marine organisms gave very similar figures.

Schoener's review of field experiments on competition

Connell (1983) too carried out a review of field experiments on interspecific competition. His review was less extensive than Schoener's (being limited to six major journals, yielding 72 studies), but it was more intensive (dealing particularly with the relative importance of inter- and intraspecific competition). Connell's 72 studies dealt with a total of 215 species and 527 different experiments; interspecific competition was demonstrated in most of the studies, more than half of the species, and approximately 40% of the experiments. He did find, though, that in those cases where the intensities of inter- and intraspecific competition could be compared, interspecific competition was the more intense in only about one-sixth of the studies. In contrast with Schoener, he found that interspecific competition was more prevalent in marine than in terrestrial organisms, and also that it was more prevalent in large than in small organisms.

Connell's review of field experiments on competition

Taken together, Schoener's and Connell's reviews certainly seem to indicate that active, current interspecific competition is widespread. Its percentage occurrence amongst species is admittedly lower than its percentage occurrence amongst whole studies, but this is to be expected, since, for example, if four species were arranged along a single niche dimension and all adjacent species competed with each other, this would still be only three out of six (or 50%) of all possible pair-wise interactions. Moreover, the fact that intraspecific competition is typically more intense than interspecific competition does not mean that interspecific competition is unimportant; this pattern is to be expected whenever there is niche differentiation (p. 262).

competition appears to be widespread . . .

Connell also found, however, that in studies of just one pair of species,

. . . but are the data biased?

interspecific competition was almost always apparent, whereas with more species the prevalence dropped markedly (from more than 90% to less than 50%). This can be explained to some extent by the argument outlined above, but it may also indicate biases in the particular pairs of species studied, and in the studies that are actually reported (or accepted by journal editors). It is highly likely that many pairs of species are chosen for study because they are 'interesting', i.e. because competition between them is suspected, and if none is found this is simply not reported. Judging the prevalence of competition from such studies is rather like judging the prevalence of debauched clergymen from the 'gutter press'. This is a real problem, only partially alleviated in studies on larger groups of species when a number of 'negatives' can be conscientiously reported alongside one or a few 'positives'. Thus the results of surveys such as those by Schoener and Connell exaggerate, *to an unknown extent,* the frequency and importance of competition.

In addition, there is another kind of bias in such results, namely the relative contributions to the data made by different groups and different types of organism. This is discussed next.

18.2.1 Phytophagous insects and other possible exceptions

As Schoener himself noted, phytophagous insects were poorly represented in his data. This is particularly regrettable in view of the fact that they account for roughly one-quarter of all living species (Southwood, 1978b). In order to counter this neglect, Strong, Lawton and Southwood (1984) have reviewed studies reporting either the presence or absence of interspecific competition amongst phytophagous insects. Only 17 out of 41 studies (or 41%) showed its presence, and in many of these it was demonstrated for only a very small proportion of the possible pair-wise interactions.

In fact, there are other reasons too for thinking that interspecific competition between phytophagous insects may be relatively rare (Strong *et al.*, 1984). The proportion of populations exhibiting intraspecific competition, as indicated by k-value analysis, is only around 20%; and if organisms do not compete with members of their own species, they are unlikely to compete with the less similar members of other species. There are also many examples of 'vacant niches' for phytophagous insects: feeding-sites or feeding-modes on a widespread plant which are utilized by insects in one part of the world but not in another part of the world where the native insect fauna is different (Figure 18.1). This failure to saturate the niche space argues against a powerful role for interspecific competition.

On a more general level, it has been suggested that herbivores *as a whole* are seldom food-limited, and are therefore not likely to compete for common resources (Hairston *et al.*, 1960; Slobodkin *et al.*, 1967). The bases for this suggestion are the observations that green plants are normally abundant and largely intact, they are rarely devastated, and most herbivores are scarce most of the time. In a similar vein, it has been considered that phytophagous insects lie 'between the devil and the deep-blue sea' (Lawton & McNeill, 1979): between the devil of abundant predators and parasitoids, and the deep-blue sea of plant food which is generally low in quality as well as being

arguments against competition amongst phytophagous insects

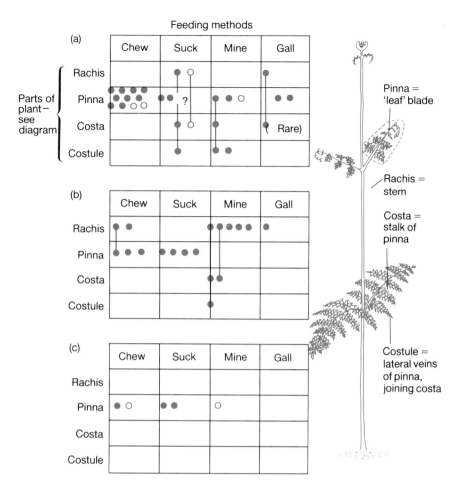

Figure 18.1. Feeding sites and feeding methods of herbivorous insects attacking bracken (*Pteridium aquilinum*) on three continents. (a) Skipwith Common in northern England; the data are derived from both a woodland and a more open site. (b) Hombrom Bluff, a savanna woodland in Papua New Guinea. (c) Sierra Blanca in the Sacramento Mountains of New Mexico and Arizona, U.S.A.; as at Skipwith, the data here are derived from both an open and a wooded site. Each bracken insect exploits the frond in a characteristic way. Chewers live externally and bite large pieces out of the plant; suckers puncture individual cells or the vascular system; miners live inside tissues; and gall-formers do likewise but induce galls. Feeding sites are indicated on the diagram of the bracken frond. Feeding sites of species exploiting more than one part of the frond are joined by lines. Key: ● open and woodland sites; ○ open sites only. (After Lawton, 1984.)

protected physically and chemically. Phytophagous insects may therefore only rarely be able to reach the sorts of densities at which competition becomes important. In this context, it is of considerable interest that Schoener found the proportion of herbivores exhibiting interspecific competition to be significantly lower than the proportions of plants, carnivores or detritivores.

the strength of competition is likely to vary from community to community

Taken overall, therefore, current interspecific competition has been reported in studies on a wide range of organisms and in some groups its incidence may be particularly obvious, for example amongst sessile organisms in crowded situations. However, in other groups of organisms, interspecific competition may have little or no influence. It appears to be relatively rare among herbivores generally, and particularly rare amongst the large and important group of phytophagous insects. In the next chapter, when the roles of predation and abiotic disturbance are discussed, it will be clear that there are many communities where the densities that induce competition are rarely achieved.

18.2.2 The intensity and the organizing power of competition are not always connected

Atkinson and Shorrocks's simulations

It is important to bear in mind that there may be situations where competition is potentially intense but where the species concerned nevertheless coexist. This has been highlighted by Atkinson and Shorrocks (1981) in their computer studies of model communities in which species compete for patchy and ephemeral resources, and in which the species themselves have aggregated distributions, with each species distributed independently of the others. The species exhibited 'current competition' (of the type that Schoener's and Connell's surveys documented), in that the removal of one species led to an increase in the abundance of others (Figure 18.2). But despite the fact that competition coefficients were high enough to lead to competitive exclusion in a uniform environment, the patchy nature of the environment and the aggregative behaviours of individuals of the two species made coexistence possible without any niche differentiation. Shorrocks *et al.* (1984) conclude that even if interspecific competition is actually affecting the abundance of populations, it need not necessarily influence the species composition of the community.

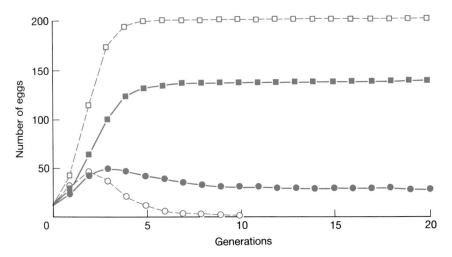

Figure 18.2. Coexistence of competitors without niche differentiation (computer simulations). The niche of species 2 is included entirely within the niche of species 1 and, in competition, species 2 has no realized niche. In a homogeneous environment species 1 (□) excludes species 2 (○). But in a patchy environment, over which both species are distributed in an aggregated manner, there is coexistence. In addition, both species 1 (■) and species 2 (●) have densities lower than they would achieve in the absence of interspecific competition (200 in both cases). The aggregated distributions, not niche differentiation, have led to coexistence. (After Atkinson & Shorrocks, 1981.)

On the other hand, even when interspecific competition is absent or difficult to detect, this does not necessarily mean that it is unimportant as an organizing force. Species may not compete at present because selection in the past favoured an avoidance of competition and thus a differentiation of niches (Connell's 'ghost of competition past'—p. 272). Alternatively, unsuccessful competitors may already have been driven to extinction; the present, observed species may then simply be those that are able to exist because they compete very little or not at all with other species. Furthermore, species may compete only rarely (perhaps during population outbreaks), but the results of such competition may be crucial to their continued existence at a particular location. In all of these cases, interspecific competition must be seen as a powerful influence on community structure, affecting which species coexist and the precise nature of those species. And yet this influence will not be reflected in the level of *current* competition. It could be claimed that in those types of organism where competition is typically rare at present (like phytophagous insects), competition is also likely to have been rare in the past and therefore unimportant as an organizing force. Nevertheless, the possibilities of past competition, added to the results of Atkinson and Shorrocks's simulations, stress that the intensity of *current* competition may sometimes be linked only weakly to the organizing power of competition within the community.

A further complication may arise if species 'appear' to compete for 'enemy-free space' (Holt, 1984; Jeffries & Lawton, 1984). As we saw in Chapter 7, there may be limits to the similarity of coexisting prey that share a predator, in just the same way as there may be limits to the similarity of coexisting consumers that share a resource. Any judgement as to the significance of these effects in the real world must await further study; but in the present context it is important to note that such 'competition' can occur (or could have occurred in the past) without any resource (other than enemy-free space) having been in short supply. Once again we see that competition may influence the structure of a community without its current power being at all obvious.

18.3 Evidence from community patterns

This weak link between current competition and the organizing power of competition has led a number of community ecologists to carry out studies on competition that do not rely on the existence of current competition. (In all honesty, many more ecologists have been led to such studies because this allows them to avoid the awkward question of whether there is current competition or not.) The approach has been firstly to predict what a community *should* look like if interspecific competition was shaping it or had shaped it in the past, and then to examine real communities to see whether they conform to these predictions.

The predictions themselves emerge readily from conventional competition theory (Chapter 7).

(i) Potential competitors that coexist in a community should, at the very least, exhibit niche differentiation.

(ii) This niche differentiation will often manifest itself as morphological differentiation.

(iii) Within any one community, potential competitors with little or no niche differentiation should be unlikely to coexist. Their distributions in space should therefore be negatively associated: each should tend to occur only where the other is absent.

There are very real problems in interpreting the data supposed to test these predictions. But the best way to appreciate the problems is to examine the data themselves.

18.3.1 Niche differentiation

The various types of niche differentiation in animals and plants were outlined in Chapter 7. On the one hand, resources may be utilized differentially. This may express itself directly within a single habitat, or as a difference in microhabitat, geographical distribution or temporal appearance if the resources are themselves separated spatially or temporally. Alternatively, species and their competitive abilities may differ in their responses to environmental conditions. This too can express itself as either microhabitat, geographical or temporal differentiation, depending on the manner in which the conditions themselves vary.

Pyke's bumblebees . . .

In one study of niche differentiation and coexistence in a community, Pyke (1982) examined a number of species of bumblebee in Colorado. During the summer (22 June to 8 September) he made visits every eight days to 17 sites along an altitudinal gradient (2860 m to 3697 m), and on each visit the numbers of each of seven common bumblebee species visiting the flowers of various plant species were recorded. The bumblebee species fell into four groups in terms of both their proboscis length and the corolla length of the plants they visited preferentially (Figure 18.3). The long-proboscis bumblebees (*Bombus appositus* and *B. kirbyellus*) clearly favoured plants with long corollas, particularly *Delphinium barbeyi* (61% and 72% of all observations of visits by the two species respectively). The short-proboscis species (*B. sylvicolla*, *B. bifarius* and *B. frigidus*) fed most frequently on various species of composite and on *Epilobium angustifolium*, all of which possess quite short corollas. The medium-proboscis species, *B. flavifrons*, fed over the entire range of corolla lengths. Finally, another short-proboscis species, *B. occidentalis*, fed as expected on plants with short corollas, but was also able to obtain nectar from long corollas. It did this by using its large, powerful mandibles to bite through the bases of the corollas and 'rob' them of their nectar. For this reason it was placed in a group of its own.

. . . which conform to the expectations of competition theory

The pattern in the proportional distribution of the two long-proboscis species is illustrated in Figure 18.4a. *B. appositus* predominated in the lower sites, to be replaced by *B. kirbyellus* at higher altitudes. In the case of the short-proboscis species (Figure 18.4b) a similar replacement was evident, with *B. bifarius* predominating in low altitude sites and being replaced by *B. sylvicola* at high altitudes, while *B. frigidus* was relatively most important in the middle of the altitudinal range. The single medium-proboscis species, *B. flavifrons*, was distributed throughout the transect and was usually the most

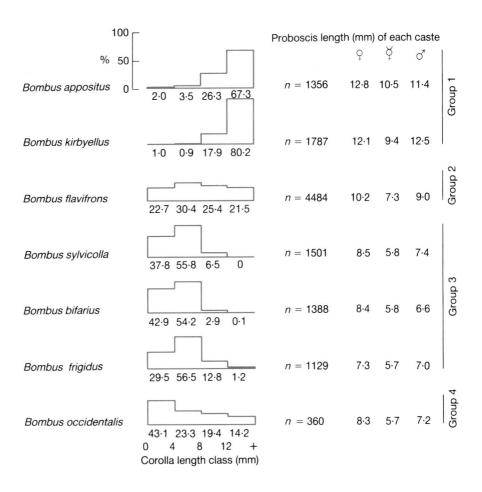

Figure 18.3. Percentage of all observations of each bumblebee species on plants in four classes of flower size. The total number of observations (*n*) is indicated in each case. Also shown for each species are the mean proboscis lengths of each caste (♀ queen, ⚥ sterile female worker, ♂ male). The seven species can be divided into three groups according to proboscis length. *Bombus occidentalis* is placed into a fourth group because of its unique mandible structure. (Modified from Pyke, 1982.)

abundant species. Finally, the nectar-robbing *B. occidentalis* was relatively most successful in sites where a particular plant, *Ipomopsis aggregata*, was common. Other bumblebee species were unable to gain access to the nectar of this plant, although humming-birds and moths both visited it.

There was a clear tendency, in any single locality, for the bumblebee community to be dominated by one long-proboscis, one medium-proboscis and one short-proboscis species, with *B. occidentalis* also being present if its exclusive nectar source was available. This pattern is consistent with what would be expected of communities moulded by competition (predictions (i), (ii) *and* (iii) on pp. 675–6). Indeed, the competition hypothesis is reinforced by the work of Inouye (1978) on two of the species that commonly coexisted— *B. appositus* and *B. flavifrons*. When Inouye temporarily removed one or other of the bee species, the remaining species quickly increased its own utilization of less-preferred flowers formerly exploited mainly by the other species. The

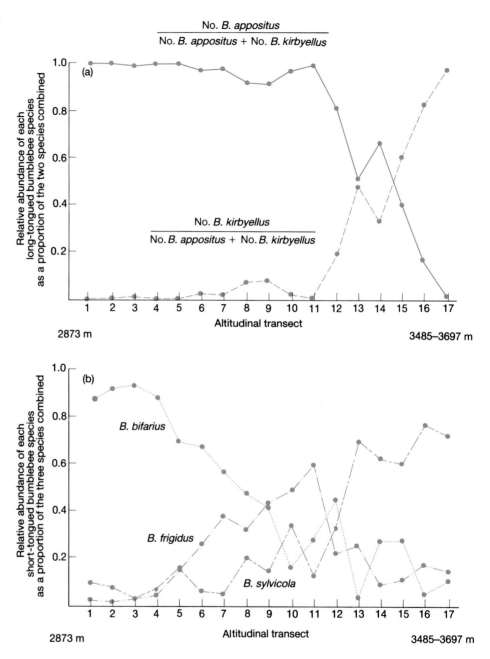

Figure 18.4. Proportional representation of bumblebee species along an altitudinal transect (a) for the two long-proboscis species and (b) for the three short-proboscis species (from Pyke, 1982).

competition and guilds

differences in resource utilization, between these species at least, appear to be actively maintained by current competition.

In addition, two further points are highlighted by Pyke's work which will be relevant throughout this chapter. First, the bumblebees can be considered to be a guild (Chapter 1), in that they are a group of species that exploit the same class of environmental resource in a similar way (Root, 1967). If interspecific competition is to occur at all, or if it has occurred in the past, then it will be most likely to occur, or to have occurred, within guilds. But this does *not* mean that guild members do necessarily compete or have necessarily competed: the onus is on ecologists to demonstrate that this is the case.

Moreover, although in the bumblebees guild membership is associated with taxonomic affiliation, there are other cases where closely related species do not belong to the same guild, and cases where guild members are not closely related (Chapter 1). It is not always easy to decide whether species belong to the same guild. Brown and Davidson (1977) have demonstrated competition between a number of seed-eating desert rodents and seed-eating desert ants. When either rodents or ants were removed, there was a statistically significant increase in the numbers of the other. The rodents and ants exploit the same resources but they differ, in several ways, in their mode of exploitation. For example, the ants take seeds of various sizes in proportion to their occurrence on the surface, whereas rodents prefer larger seeds and are particularly efficient at exploiting dense aggregations of buried seeds. Whether they belong to the same guild (exploiting the same resources in a similar way) is a moot point.

The second point about Pyke's results is that the bumblebees demonstrate *niche complementarity*, i.e. within the guild as a whole, niche differentiation involves several niche dimensions, and species that occupy a similar position along one dimension tend to differ along another dimension. Thus *B. occidentalis* differs from the other six species in terms of food-plant species (*Ipomopsis aggregata*) and feeding method (nectar-robbing). Amongst the others, species differ from each other either in the corolla length (and thus proboscis length) or in the altitude (i.e. habitat) favoured or in both. Where species compete with several others in several dimensions, the interaction have been termed *diffuse competition* (MacArthur, 1972).

Table 18.1. Summary of the main dimensions of resource partitioning by nine African species of tree squirrel in a lowland evergreen rainforest. Horizontal lines represent virtually complete ecological separation based on the character where the line is initiated. (From Emmons, 1980.)

In a study of nine species of African tree squirrels that coexist in a region of lowland evergreen rainforest, Emmons (1980) found evidence of differentiation along four dimensions. A summary of her findings is presented in Table 18.1. Two species in the genus *Funisciurus* occupied special habitats of their own: *F. isabella* lived in dense undergrowth, often associated with areas formerly cut by man, while *F. anerythrus* lived in seasonally or permanently

Species	Habitat type	Vegetation height	Food types distinguishing diet	Body size
Myosciurus pumilio	Mature and disturbed forest	Arboreal	Bark scrapings	Tiny
Aethosciurus poensis			Diverse arthropods	Small
Heliosciurus rufobrachium				Medium
Protoxerus stangeri			Hard nuts	Large
Funisciurus lemniscatus		Ground foraging	Many termites	Small
Funisciurus pyrrhopus				Medium
Epixerus ebii			Hard nuts	Large
Funisciurus isabella	Dense growth		Leaves Arthropods	Small
Funisciurus anerythrus	Flooded forests		Many ants	Medium

flooded forest. The remaining seven species, which occupied the main forest formation, could be divided into two groups according to whether they were active mainly in the forest canopy or at ground level. Most of the diet of all species consisted of fruits, except for *Myosciurus pumilio* which consumed an appreciable quantity of bark scrapings. However, the sizes of fruits taken varied to some extent. The two largest species, one arboreal and the other ground-foraging, were able to cope with the hardest nuts. Finally, the amount of animal matter consumed (arthropods of various kinds) varied inversely with squirrel body size, with two species in each of the arboreal and ground-foraging categories consuming appreciable quantities.

Complementary differentiation along the three dimensions of habitat, vertical space and food has also been extensively reported for lizards (Schoener, 1974) and birds (Cody, 1968).

A case of apparent seasonal and habitat partitioning has been reported for five species of amphipod crustaceans (genus *Gammarus*) which inhabit the brackish waters of Denmark's Limfjord (Fenchel & Kolding, 1979; Kolding & Fenchel, 1979). The geographical distribution of these species is related to salinity, as shown in Figure 18.5. *Gammarus duebeni*, for example, was found in the least saline areas (0–5‰), while at the opposite extreme *G. oceanicus* occurred only at the highest salinities (20–33‰). At any given location there were usually only two abundant species, and often one species predominated. These distribution patterns suggest a habitat selection maintained by interspecific competition, since in the laboratory all five species have been maintained through several generations at a salinity fluctuating between 23‰ and 27‰. It seems that the species realized niches are narrower than their fundamental niches with respect to salinity.

Figure 18.5. The number of individuals of each species of *Gammarus* as a percentage of the total number of all species combined from sampling stations with different salinities. due, *Gammarus duebeni*; zad, *G. zaddachi*; sal, *G. salinus*; loc, *G. locusta*; oce, *G. oceanicus*. (From Fenchel & Kolding, 1979.)

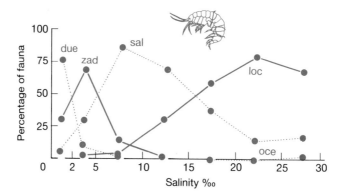

In considering the coexistence of the various pairs of species at the different locations, Kolding and Fenchel could find no evidence for differences in diet. All the species are omnivorous, and in the laboratory they feed and grow on a wide variety of food items, including fresh and decaying plant and animal matter. However, there was a marked pattern in the periods of peak reproductive activity of each species, as shown in Figure 18.6. The *Gammarus* species in the Limfjord go through one or two generations per year. Copulation occurs during the moulting of the female, after the male has been linked to her for some time in a precopula pairing. It is clear from Figure

18.6 that there are marked differences in the periods of peak reproductive activity between the pairs of species that normally coexist. These differences in the timing of reproduction mean that the development of juveniles of coexisting species will be staggered to some extent. This, in turn, is likely to relax interspecific competition for refuges in different-sized crevices, and it may allow different-sized food particles to be exploited by coexisting species at any given time. However, the timing of reproduction may well have a further significance. In the laboratory, interspecific precopulation followed by sterile mating can occur. Strong selective forces can be expected to act to prevent such matings, and the displaced breeding periods between neighbouring species along a salinity gradient may be interpreted as the result of such forces. In cases where partitioning of the time dimension occurs, this alternative hypothesis should always be considered.

...which may, however, lead to the avoidance of 'interbreeding' rather than competition

Figure 18.6. Periods of high precopulation activity in the five species of *Gammarus*. Within each species, a bar represents one generation. (From Kolding & Fenchel, 1979.)

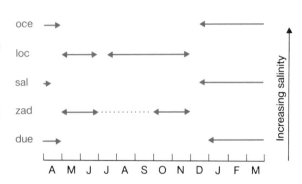

Many cases of apparent resource partitioning have been reported. It is likely, however, that studies failing to detect such differentiation have tended to go unpublished. It is always possible, of course, that these 'unsuccessful' studies are flawed and incomplete, and that they have failed to deal with the relevant niche dimensions; but a number, such as those by Rathke (1976) and Strong (1982) on leaf-eating insects, have been sufficiently beyond reproach to raise the possibility that in certain animal groups resource partitioning is not an important feature.

phytophagous insects that appear not to partition resources

Strong (1982) studied a group of hispine beetles (Chrysomelidae) that commonly coexist as adults in the rolled leaves of *Heliconia* plants. These long-lived tropical beetles are closely related; they eat the same food, and they occupy the same habitat. They would appear to be strong candidates for demonstrating resource partitioning. Yet Strong could find no evidence of segregation, except in the case of only one of the 13 species studied, which was segregated weakly from a number of others. The beetles lack any aggressive behaviour, either within or between species; their host specificity does not change as a function of co-occupancy of leaves with other species that might be competitors; and the levels of food and habitat are commonly not limiting for these beetles, which suffer heavily from parasitism and predation. In these species, resource partitioning associated with interspecific competition does not appear to structure the community. As we have seen, this may well be true of many phytophagous insect communities (Lawton & Strong, 1981; Strong *et al.*, 1984).

18.3.2 Niche differentiation in plant communities

the coexistence of competing plants is intrinsically difficult to understand

In examining the problems of coexistence and community structure in animals, there is no difficulty in providing a *possible* explanation in terms of interspecific competition. There is generally a variety of resources and habitats that could serve as the basis for niche differentiation. The difficulty comes in providing proof. With plants, on the other hand, it is often much more difficult to envisage even a possible explanation. Grubb (1977), for example, has asserted that the existence of 'a million or so animals can easily be explained in terms of the 300 000 species of plants (so many of which have markedly different parts such as leaves, bark, wood, roots, etc.), and the existence of three to four tiers of carnivores (Hutchinson, 1959)', but 'there is no comparable explanation for autotrophic plants; they all need light, carbon dioxide, water and the same mineral nutrients.'. Hutchinson (1959), in the sub-title of an influential early paper on niche differentiation and community structure, asked 'Why are there so many kinds of animals?'. A more challenging question, however, is 'Why are there so many kinds of plants?'.

resource competition *à la* Tilman can provide an explanation

A possible explanation framed in terms of interspecific competition has, in fact, been provided by Tilman (1982), whose approach to competition for resources was discussed in Chapters 3 and 7. Figure 7.25 from Chapter 7 is reproduced here as Figure 18.7a. It describes a range of outcomes (including coexistence) when two species compete for two resources; and it shows that the outcome depends on the zero net growth isoclines (ZNGIs) of the two species, and on the position of the supply-point for the two resources. Figure 18.7b describes a comparable situation, but in this case there are five species competing for the two resources. Not surprisingly, there is a wider range of outcomes, though for any one supply-point there can be only one or two coexisting species, or none. However, Figure 18.7b also includes an indication of the *range* of supply-points that might be expected, hypothetically but quite reasonably, within a small area of the environment. This range allows all five species to coexist on just two resources.

Thus a possible competition-based explanation for the coexistence of many plants with similar resource requirements can be provided by the combination of, on the one hand, Tilman's approach to resource competition, and on the other hand a recognition of a degree of environmental heterogeneity. A possible explanation therefore exists: proof of its validity and widespread importance in nature is still awaited (see Tilman, 1982).

niche differences and local coexistence in a Canadian pasture

The apparent niche differences between a number of plant species, and the patterns of co-occurrence of the plants on a very local scale in pastures in Ontario, Canada, have been studied by Turkington, Cavers and Aarssen (1977). These researchers measured the frequencies of physical contact between various species pairs, and compared these with expected frequencies derived simply from the abundances of the different species. They paid particular attention to six species—three grasses and three legumes— and their results for these six are presented in the first data column of Table 18.2. A '+' means that two species occurred in close contact more frequently than expected by chance; a '−' means that contacts were less frequent than expected.

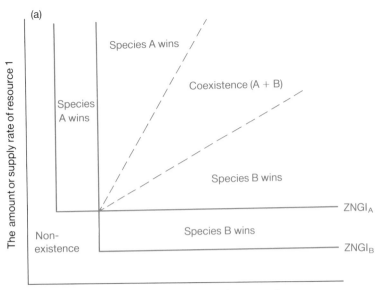

(a)

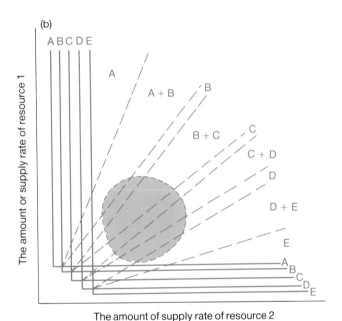

(b)

Figure 18.7. The outcome of competition between two species for two resources depends on the zero net growth isoclines (ZNGIs) of the two species, and the location of the supply-point of the two resources (see Chapter 7, pp. 285). (b) The outcome of competition between five species for two resources depends in a similar way on ZNGIs and the location of supply points. Any one supply point will support no species, one species or the coexistence of two species; but a natural, heterogeneous environment containing micro-environments with a range of supply points (shaded area) may support the coexistence of all five (or more) species. (Adapted from Tilman, 1982.)

A negative association could mean either that two species exclude one another through competition, or that they have quite different habitat requirements. A positive association could mean either that two species have similar habitat requirements (possibly bringing them into competition) or that

Table 18.2. Coexisting plant species which live in similar types of soil in a Canadian pasture tend not to occur together (possible competitive exclusion), but species living in different types of soil do occur together (coexistence with possible niche differentiation). The species are: TP, *Trifolium pratense*; TR, *T. repens*; PP, *Phleum pratense*; ML, *Medicago lupulina*; BI, *Bromus inermis*; and DG, *Dactylis glomerata*. Note that the soil characteristics could either reflect inherent properties of the soil, or they could be the consequences of that species growing in the soil and extracting nutrients differentially. (After Turkington *et al.*, 1977.)

Species pair	Distributions more (+) or less (−) congruent than expected by chance	Number of significant correlations between the soils of the two species examined for five characteristics		Balance of correlations
		+	−	
TR/TP	−	3	0	+
TR/ML	−	2	1	+
TR/BI	+	1	1	0
TR/DG	−	0	0	
TR/PP	+	0	1	−
TP/ML	−	2	0	+
TP/BI	−	0	0	
TP/DG	+	1	0	+
TP/PP	+	1	0	+
ML/BI	+	1	1	0
ML/DG	+	0	0	
ML/PP	−	1	0	+
BI/DG	−	2	0	+
BI/PP	−	2	1	+
DG/PP	−	2	0	+

they co-occur because they are indifferent to one another (i.e. they do not compete). In order to distinguish between these alternatives, Turkington and his colleagues examined the local habitats of each of the species by analysing the soil in the immediate vicinity of the roots. These analyses indicate which resources are left behind by the growing plant rather than their resource requirements. They tested samples for their phosphorus, potassium, magnesium and calcium content and for their pH. They then established, for each factor separately, whether particular species pairs tended to be found in the same or in different types of soil (Table 18.2). The final column in the table summarizes the overall similarity or dissimilarity of soil types for the various species pairs.

Overall, there was a marked tendency for species with similar soil resource requirements (and which depleted soil nutrients in the same way) to be negatively associated in their distribution, and for species living in different soil types to be positively associated. In particular, all three grass–grass pairs and all three legume–legume pairs lived in similar types of soil but were found only rarely in close contact. This certainly suggests an organizing role for competition in these localized plant communities. However, there are certain problems with the data. (i) Many pairs did not conform to the general tendency, and two pairs (*Trifolium pratense* with *Dactylis glomerata* and with *Phleum pratense*) ran counter to it: they lived in similar soils but were positively associated. (ii) It is important to know whether the number of pairs of species where the results suggested competition was greater than would be expected simply by chance. (iii) The soils were analysed in only a small number of ways; other factors (other niche dimensions) might have suggested relationships different to those observed. (iv) Indeed, it is not known whether the plants actually competed for any of the soil resources considered. Thus the results can be no more than suggestive of competition.

the results 'suggest' competition but can do no more

18.3.3 Negatively associated distributions

Several of the studies discussed so far have described negative associations in the distribution of potential competitors, consistent with apparent niche differentiation. But a number of studies have utilized distributions in their own right as evidence for the importance of interspecific competition. Foremost amongst these is Diamond's (1975) survey of the landbirds living on the islands of the Bismarck archipelago off the coast of New Guinea (Figure 18.8). The most striking evidence comes from distributions that constitute a 'checkerboard'. In these, two or more ecologically similar species (i.e. members of the same guild) have mutually exclusive but interdigitating distributions such that any one island supports only one of the species (or none at all). Figure 18.9 shows this for two small, ecologically similar cuckoo-dove species: *Macropygia mackinlayi* and *M. nigrirostris*.

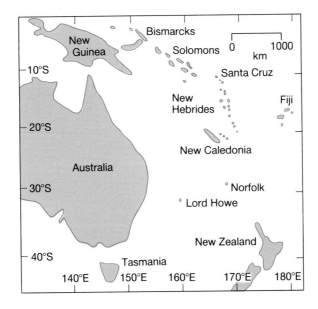

Figure 18.8. The Bismarck Archipelago, the Solomon Islands and the New Hebrides Archipelago off the east coasts of New Guinea and Australia.

Checkerboards, however, are relatively rare. Much more of Diamond's evidence suggests diffuse competition from a number of species (often unspecified). For a variety of species, Diamond plotted an *incidence function*, i.e. the relationship between island 'size', S (measured as the total number of bird species supported by the island), and the proportion of islands of that size, J, occupied by the species in question (Figure 18.10). Certain species, such as the flycatcher *Monarcha cinerascens* and the honeyeater *Myzomela pammelaena* (Figure 18.10a), appear to be excellent colonizers but poor at persisting in the diverse communities of 'large' islands. Diamond called these 'supertramps'. Other species, such as the pigeon *Chalcophaps stephani* (Figure 18.10b), seem to be competent colonizers but also persist in species-rich communities; while others (Figure 18.10c), especially those with specialized diets or restricted habitat requirements, appear to be confined to large islands. The absence of supertramps from large islands providing suitable resources and conditions implies strongly that competition is important.

(margin notes) checkerboard distributions

incidence functions and the possible role of diffuse competition in supertramps

685   COMPETITION AND COMMUNITY STRUCTURE

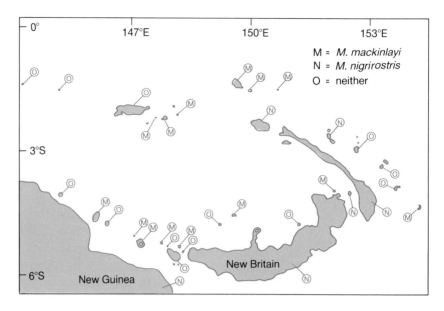

Figure 18.9. Checkerboard distribution of two small *Macropygia* cuckoo-dove species in the Bismarck region. Islands whose pigeon faunas are known are designated as M (*Macropygia mackinlay* resident), N (*M. nigrirostris* present) or O (neither species resident). Note that most islands have one of these two species, no island has both, and some islands have neither. (From Diamond, 1975.)

18.3.4 Conclusions

We can now draw a number of conclusions about the evidence for competition discussed in this section.

(i) Interspecific competition is a possible and indeed a plausible explanation for many aspects of the organization of many communities—but it is not often a proven explanation.

(ii) One of the main reasons for this is that active, current competition has been demonstrated in only a small number of communities. Its actual prevalence overall can be judged only imperfectly from the results and considerations discussed in section 18.2.

(iii) As an alternative to current competition, the ghost of competition past can always be invoked to account for present-day patterns. But it can be invoked so easily because it is impossible to observe and therefore difficult to disprove.

(iv) The communities chosen for study may not be typical. The ecologists observing them have usually been specifically interested in competition, and they may have selected appropriate, 'interesting' systems. Studies that fail to show niche differentiation are often considered 'unsuccessful' and are likely to have gone largely unreported.

(v) The community patterns uncovered, even where they appear to support the competition hypothesis, often have alternative explanations. Species may differ because of the advantages of avoiding interspecific breeding and the production of low-fitness hybrids (e.g. the *Gammarus* species). Alternatively, species that have negatively associated distributions may

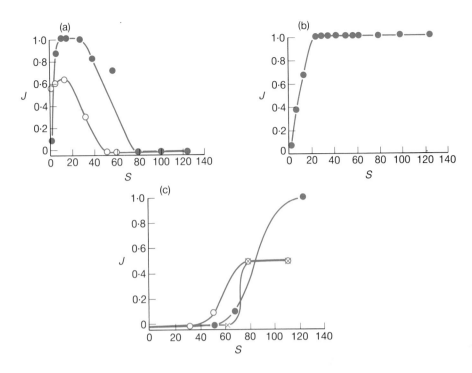

Figure 18.10. Incidence functions for various species in the Bismarcks in which *J*, the proportion of islands occupied by a given species, is plotted against *S*, a measure of island 'size' (actually the total number of bird species present).
(a) Incidence functions for two 'supertramps':
● — the flycatcher *Monarcha cinerascens*;
○ — the honeyeater *Myzomela pammelaena*
(b) Incidence function for the pigeon *Chalcophaps stephani*, a competent colonizer and, apparently, an effective competitor.
(c) Incidence functions for three species that are restricted to larger islands:
● — the hawk *Henicopernis longicauda*
○ — the rail *Rallina tricolor*
⊗ — the heron *Butorides striatus*.
(From Diamond, 1975.)

recently have speciated allopatrically (i.e. in different places), and their distributions may still be expanding into one another's ranges. Whenever there are alternative explanations, however, support for the competition hypothesis is strengthened greatly by demonstrations that there is contemporary competition between the species concerned (for example the bumblebees, although not the hispine beetles), or that the species occupy realized niches demonstrably smaller than their fundamental niches (the *Gammarus* species).

(vi) The recurring alternative explanation to competition as the cause of community patterns is that these have arisen simply by chance. Niche differentiation may occur because the various species have evolved independently into specialists, and their specialized niches happen to be different. Even niches arranged along a resource dimension at random are bound to differ to some extent. Similarly, species may differ in their distribution because each has been able, independently, to colonize and establish itself in only a small proportion of the habitats that are suitable for it. Ten blue and ten red balls thrown at random into 100 boxes are almost certain to end up with different distributions. Hence competition cannot be inferred from mere 'differences' alone. But what sorts of differences *do* allow the action of competition to be inferred? This problem is dealt with in the following section.

18.4 Neutral models and null hypotheses

A number of workers, notably Simberloff and Strong and their colleagues at Florida State University, have criticized what they see as a tendency to interpret 'mere differences' as confirming the importance of interspecific competition. On the other hand, competition theory actually does more than this. It predicts that there should be a limit to the similarity of competing species, so that their niches are arranged *regularly rather than randomly* in niche space (i.e. the niches should be overdispersed); and it predicts that species with very similar niches should exclude each other so that their distributions differ *more than would be expected from chance alone*. A more rigorous investigation of the role of interspecific competition, therefore, should address itself to the question: does the observed pattern, even if it appears to implicate competition, differ significantly from the sort of pattern that could arise in the community even in the absence of any interactions between species?

the aim of demonstrating that patterns are not generated merely by chance

Questions of this type have been the driving-force behind a number of analyses which have sought to compare real communities with so-called *neutral models*. These are models of actual communities that retain certain of the characteristics of their real counterparts, but reassemble the components at random (as described below), specifically excluding the consequences of biological interactions. In fact, the neutral model analyses are attempts to follow a much more general approach to scientific investigation, namely the construction and testing of *null hypotheses*. The idea (probably familiar to most readers in a statistical context) is that the data are rearranged into a form (the neutral model) representing what the data *would* look like in the absence of the phenomenon under investigation (in this case species interactions, particularly interspecific competition). Then, if the actual data show a significant statistical difference from the null hypothesis, the null hypothesis is rejected and the action of the phenomenon under investigation is strongly inferred. Rejecting (or falsifying) the absence of an effect is reckoned to be better than confirming its presence, because there are well-established statistical methods for testing whether things are significantly different (allowing falsification) but none for testing whether things are 'significantly similar'.

null hypotheses are intended to ensure statistical rigour

It will be clear by the end of this section that the application of null hypotheses to community structure is a difficult task. But, once again, the best way to examine these difficulties is by considering some examples of the neutral-model approach.

18.4.1 Neutral models and resource partitioning

Some of the less controversial applications of the approach have been in the field of differential resource utilization. Lawlor (1980), for example, has looked at ten North American lizard communities, consisting of four to nine species, for which he had estimates of the amounts of each of 20 food categories consumed by each species in each community (Pianka, 1973). A number of neutral models of these communities were created (see below), which were then compared with their real counterparts in terms of their patterns of overlap in resource use. If competition is or has been a significant force in determining community structure, the niches should be spaced out evenly, and overlap in resource use in the real communities should be less than predicted by the neutral models.

a neutral model of resource-use in lizard communities . . .

Lawlor's analysis was based on the 'electivities' of the consumer species, where, for instance, the electivity of species i for resource k was the proportion of the diet of species i which consisted of resource k. Electivities therefore ranged from 0 to 1. These electivities were in turn used to calculate, for each pair of species in a community, an index of resource use overlap, which itself varied between 0 (no overlap) and 1 (complete overlap). Finally, each community was characterized by a single value: the mean resource overlap for all pairs of species present.

The neutral models were of four types, generated by four 'reorganization algorithms' (RA1–RA4). Each retained a different aspect of the structure of the original community while randomizing the remaining aspects of resource use.

. . . based on four 'reorganization algorithms'

RA1 retained the minimum amount of original community structure. Only the original number of species and the original number of resource categories were retained. Observed electivities (including zeros) were replaced in every case by random values between 0 and 1. This meant that there were far fewer zeros than in the original community. The niche breadth of each species was therefore increased.

RA2 replaced all electivities, *except zeros*, with random values. Thus the qualitative degree of specialization of each consumer was retained, i.e. the number of resources consumed to any extent by each species was correct.

RA3 retained not only the original qualitative degree of specialization but also the original consumer niche breadths. No randomly generated electivities were used. Instead, the original sets of values were rearranged. In other words, for each consumer, all electivities, both zeros and non-zeros, were randomly reassigned to the different resource types.

RA4 reassigned only the non-zero electivities. Of all the algorithms, this one retained most of the original community structure.

Each of the four algorithms was applied to each of the ten communities. In every one of these 40 cases, 100 'neutral-model' communities were generated and the corresponding 100 mean values of resource overlap were calculated. If competition were important in the real community, these mean overlaps should have exceeded the real community value. The neutral model was therefore considered to have a *significantly* greater mean overlap than the real community ($P < 0.05$) if five or fewer of the 100 simulations gave mean overlaps less than the real value.

The results are shown in Figure 18.11. Increasing the niche breadths of all consumers (RA1) resulted in the highest mean overlaps (significantly higher than the real communities). Rearranging the observed non-zero electivities

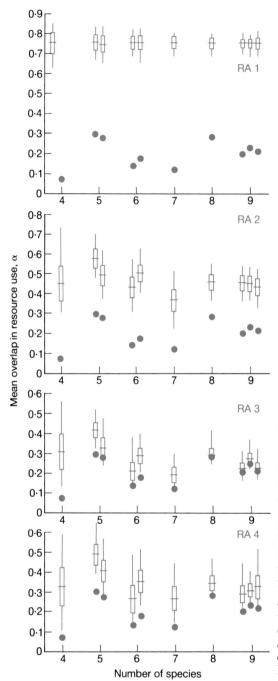

Figure 18.11. The mean indices of resource use overlap for each of Pianka's (1973) ten North American lizard communities are shown as solid circles. These can be compared, in each case, with the mean (horizontal line), standard deviation (vertical rectangle) and range (vertical line) of mean overlap values for the corresponding set of 100 randomly constructed communities. The analysis was performed using four different reorganization algorithms (RAs), as described in text. (From Lawlor, 1980.)

the lizards appear to pass the test

(RA2 and RA4) also always resulted in mean overlaps which were significantly higher than those actually observed. With RA3, on the other hand, where all electivities were reassigned, the differences were not always significant. But in all communities the algorithm mean was higher than the observed mean. In the case of these lizard communities, therefore, the observed low overlaps in resource use suggest that niches are segregated, and that interspecific competition plays an important role in community structure.

Joern and Lawlor (1980) carried out a similar analysis in a series of five arid grassland grasshopper communities, considering overlap in both food resources and microhabitat. In both cases (food and microhabitat) the results were similar: only complete randomization of utilization rates among all resource states (RA1) resulted in mean overlap values larger than those observed. While this may indicate some resource partitioning in real communities, the evidence is not convincing. Perhaps this is a further example of a community of phytophagous insects in which competition plays no significant role.

a similar model for grasshopper communities—with a different result

18.4.2 Neutral models and morphological differences

Niche differentiation in general is frequently manifested as morphological differentiation. In a similar way, the spacing out of niches can be expected to have its counterpart in a regularity in degree of morphological difference between species belonging to a guild. Specifically, a common feature claimed for animal guilds which appear to segregate strongly along a single resource dimension is that adjacent species tend to exhibit regular differences in body size or in the size of feeding structures. Hutchinson (1959) catalogued many examples, drawn from vertebrates and invertebrates, of sequences of potential competitors in which average individuals from adjacent species had weight ratios of approximately 2.0 or length ratios of approximately 1.3 (the cube root of 2.0). This also holds approximately for some of the animal guilds discussed earlier. Coexisting cuckoo-doves in the Bismarcks have a mean body-weight ratio of 1.9 for adjacent species (Diamond, 1975); locally coexisting bumblebees have a mean proboscis-length ratio of 1.32 (Pyke, 1982, calculated for worker bees); and the two groups of African tree squirrels which foraged at ground level or in the canopy have mean skull-length ratios of 1.45 and 1.27, respectively (Emmons, 1980). It has also been pointed out that this 'rule' appears to hold for body-length in the conventional musical ensembles of crumhorns and recorders (Figure 18.12), and for wheel size in larval instars of children's tricycles and bicycles (Horn & May, 1977).

The neutral-model approach to this apparent regularity of morphological differences has tended to concentrate on one particular aspect: the fact that

Hutchinson's ratio rules

Figure 18.12. The conventional musical ensemble of crumhorns and recorders, which appears to conform to Hutchinson's size-ratio rule (Horn & May, 1977). (Instruments kindly lent by Russell Acott, Oxford. Photograph: B. Roberts.)

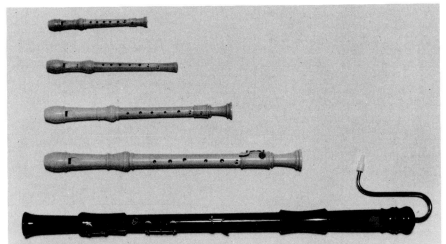

691 COMPETITION AND COMMUNITY STRUCTURE

coexisting potential competitors may be more different morphologically than might be expected by chance alone. In one example, Bowers and Brown (1982) investigated the relationship between body size and coexistence in granivorous (seed-eating) desert rodents. They did so on two scales—within individual local sites and within much larger geographic areas. They utilized data on species from 95 sites (< 5 ha in each case) in three major North American deserts: the Great Basin (33 sites), the Mojave (24 sites) and the Sonoran (38 sites). Analysis was restricted to common species (> 5% of total individuals at a site) except for very large rodents (> 80 g) which were always included. These criteria were applied in order to exclude from the analysis rare individuals of small species which were transients from other habitats, and to correct for under-representation of large rodents in the samples because of trapping bias. Bowers and Brown's analysis was straightforward. They determined the proportion of sites occupied by each species in each of the three deserts separately, and also in all three combined. They then calculated the expected frequency of co-occurrence within a site of any *pair* of species, assuming that all distributions were independent. These values were compared with the actual frequency of co-occurrence of the species pairs. Each pair was also scored as differing in body mass by either more or less than a ratio of 1.5, invoked as a conservative estimate of Hutchinson's body mass ratio. The competition hypothesis would predict that similar-sized species would co-occur less often than expected by chance, whereas this would not be so for different-sized species.

The results are shown in Table 18.3. In both the Great Basin and Sonoran Deserts, pairs of species which differed in body size by a ratio of less than 1.5 coexisted less frequently than expected on the basis of chance. The null

Table 18.3. 2×2 contingency tables testing the null hypothesis that local coexistence and geographic overlap of granivorous rodents in North American deserts are independent of body size. Association was scored as either positive (+) or negative (−) depending on whether observed frequencies of coexistence were greater or less than expected on the basis of chance. In each case the p value indicates the probability that the pattern would arise by chance. All values are statistically significant except for the Mojave Desert. (After Bowers & Brown, 1982.)

Location	Body mass ratio	Association −	Association +
Great Basin	< 1.5	6	0
	> 1.5	15	15
		$p < 0.005$	
Mojave	< 1.5	3	1
	> 1.5	11	5
		$p > 0.5$	
Sonoran	< 1.5	7	0
	> 1.5	23	15
		$p < 0.05$	
Local coexistence (all deserts combined)	< 1.5	27	0
	> 1.5	65	28
		$p < 0.01$	
Geographic overlap	< 1.5	44	13
	> 1.5	72	60
		$p = 0.01$	

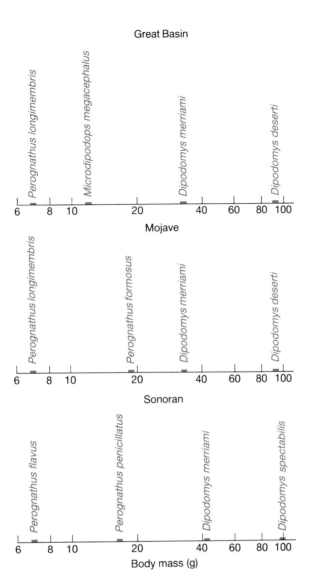

Figure 18.13. Distribution of body sizes of rodents (plotted on a logarithmic scale so that equal spacing represents equal ratios) in arbitrarily chosen diverse communities in three deserts. Note that the size distributions of species are very similar even though the identity of species in many cases is different. (From Bowers & Brown, 1982.)

hypothesis was not rejected for the Mojave Desert, perhaps because of the smaller sample size. But in the analysis of the largest, most complete set, that for all deserts combined, the null hypothesis was convincingly rejected. Figure 18.13 presents typical size distributions in diverse communities from each desert.

The overlap in overall geographic distributions also appeared to be related to body size. Pairs of species with size ratios less than 1.5 tended to be negatively associated, resulting in a strong rejection of the null hypothesis. Particularly striking cases of displaced geographic ranges involved rodents of extreme size. Figure 18.14 illustrates this for the tiny pocket mice (body mass < 11 g) and the larger kangaroo rats (> 100 g).

competition is apparently demonstrated for granivores. . . .

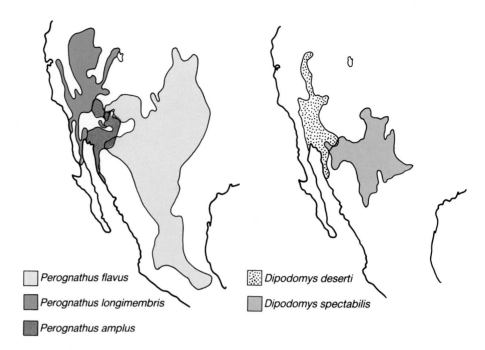

Figure 18.14. Geographic ranges of small pocket mice (*Perognathus* spp., body mass < 11g) and of large kangaroo rats (*Dipodomys* spp., body mass > 100g). Note the extremely small overlap in geographic ranges of species of similar size. (From Bowers & Brown, 1982.)

Perognathus flavus

Perognathus longimembris

Perognathus amplus

Dipodomys deserti

Dipodomys spectabilis

... but not for a mixture of guilds

Perhaps of equal significance, a further analysis carried out by Bowers and Brown considered a combination of guilds of desert rodents (granivores, folivores, insectivores and omnivores). In this case the null hypothesis could not be rejected, suggesting that important patterns will be missed if analyses are conducted on assemblages of consumer types which are too diverse, and within which competition is expected to be relatively rare.

A study with a similar aim was carried out by Strong, Szyska and Simberloff (1979), who examined three sets of data: the birds of the Tres Marías islands off the coast of Mexico, the birds of California Channel Islands and the finches of the Galapagos. In each case, any one island actually contained only a small proportion of the total number of species in the 'source pool'. (The source pools were all the species living on the nearby mainlands in the cases of the Tres Marías and California Channel Islands, and those living on the archipelago as a whole in the case of the Galapagos.) These actual communities were compared with 100 randomly assembled 'null' communities. The null communities had the same number of species per family on each island as did the actual communities, but the species themselves were drawn at random from the source pool. Using a number of different linear measurements, the length-ratios between pairs of contiguous species within families were compared. ('Contiguous species' were consecutive species in rank: longest compared with second longest, second longest with third longest and so on.) Interspecific competition was then inferred only if the length-ratios in the actual communities were, in general, significantly greater than those in the null communities.

neutral models of body size and coexistence of island vertebrates: ...

Strong, Szyska and Simberloff found that the ratios in the actual communities were not generally greater. They therefore found no reason to reject their null hypotheses, and no evidence for interspecific competition. However, their conclusions gave rise to considerable controversy (e.g. Grant & Abbott, 1980; Hendrickson, 1981; Strong & Simberloff, 1981; Simberloff & Boeklen, 1981; Simberloff, 1984; Schoener, 1984); and four aspects of the criticisms levelled at them are particularly worth considering.

. . . the generation of a controversy

(i) There were statistical flaws in the methods used: the general point being that the construction and testing of null hypotheses for things as complicated as biological communities is a technically difficult task.

(ii) Comparisons were made of species within families. But competition is expected to occur within *guilds*, which do not necessarily correspond with taxonomic groupings.

(iii) Conclusions were drawn on the basis of whole faunas (all bird species). This is likely to dilute the effects of any patterns that *are* generated by interspecific competition by lumping them together with other patterns that are not (compare this with Bowers and Brown's results).

(iv) The source pools are themselves communities which may have been influenced by interspecific competition. The 'null' communities may therefore have the effects of interspecific competition built into them. Lack of a difference between a real community and a null community may thus tell us nothing about interspecific competition.

Taking these points into account, the critics of Strong, Szyska and Simberloff found more evidence of interspecific competition than the original analysis had uncovered, though they could not demonstrate competition in every case. They also stressed that failure to reject a neutral model is not positive proof of the absence of interactions. Even at their best, these neutral models can never do more than either indicate that competition may be operating *or* infer that there is no good case for competition. Perhaps most important of all in the long term, however, is the fact that the critics disagreed with the details and the construction of the neutral models, but not with their aim. We will return to this point shortly.

18.4.3 Neutral models and distributional differences

The absence of coexistence of morphologically similar species is really only a special case of the failure of presumed competitors in general to coexist—as used by Diamond (1975), for example, to infer the action of competition amongst landbirds in the Bismarcks. Using neutral-model analyses, however, Connor and Simberloff (1979) disputed the conclusions drawn by Diamond and others from island distributions. Put simply, their neutral models retained three aspects of the actual community structure: the number of species on each island (retaining differentials in island size), the number of islands occupied by each species (retaining differentials in species' abilities to disperse and colonize), and the general form of each species' incidence function (p. 685). Within these constraints, they then reassembled the communities at random from the species pool for the archipelago as a whole. They used three data sets: the bird species of the New Hebrides (see Figure 18.8), and the birds

a neutral-model approach to Diamond's distributional evidence

and bats of the West Indies. In each case they concluded that the actual communities did not deviate significantly from random communities.

These conclusions, like those drawn by Strong, Szyska and Simberloff, gave rise to criticism and controversy (Alatalo, 1982; Diamond & Gilpin, 1982; Gilpin & Diamond, 1982, 1984). In the first place, the extent to which the neutral models were really free of interspecific competition was once again questioned. The species pools themselves, the abilities of species to colonize islands, and the incidence functions of the species were all parts of the neutral models—but they may all themselves have been affected by interspecific competition. It has also been claimed that there are serious statistical flaws in Connor and Simberloff's analysis (Gilpin & Diamond, 1984).

criticisms of this approach

In addition, Connor and Simberloff were impressed, for example, that for the West Indian birds the randomly assembled communities gave an expected number of exclusive pairs (i.e. pairs that never co-occur on an island) of 12 448, while the observed number is 12 757. However, the similarity is misleading. If each of the 211 species in question was competitively excluded by just one other species, this would still lead to no more than 106 exclusively distributed pairs. And if the 309 pairs (12 757 *minus* 12 448) were all attributable to competition, each species would be excluded, on average, by competition from three other species (Alatalo, 1982). Thus differences of this type for the West Indian birds and bats, despite being small numerically, actually suggest the action of something other than chance. On the other hand, the figures for the New Hebridean birds (63 observed exclusive pairs compared to 63 expected) seem to indicate that chance is a sufficient explanation. It is therefore interesting that the original workers on the distribution of this group made no strong claims for the action of interspecific competition (Diamond & Marshall, 1977).

Diamond and Gilpin (1982; Gilpin & Diamond, 1982) were particularly critical of one aspect of Connor and Simberloff's approach: 'By analysing a whole species pool rather than an ecologically defined guild, it buries distributions influenced by competition in an overwhelming mass of irrelevant data (the "dilution effect")'. This criticism clearly parallels points (ii) and (iii) on p. 695. Diamond and Gilpin's response, however, was to construct what they considered to be an improved neutral model of their own. Following Connor and Simberloff, they calculated for each pair of species a standard similarity value:

a modified neutral-model approach from the critics

$$\frac{\begin{array}{c}\text{observed number} \\ \text{of islands shared}\end{array} - \begin{array}{c}\text{expected number} \\ \text{of islands shared}\end{array}}{\text{standard deviation of expected number}}$$

If species are distributed as the random model predicts, then similarity values should be normally distributed with a mean of zero and a standard deviation of unity. For pairs of species influenced by competition, however, the similarity values should be markedly negative (the observed number of islands shared should be less than the expected number).

Figure 18.15 shows the observed and neutral-model distributions of similarity values for the birds of the Bismarck archipelago. There was a highly significant difference between the two ($P < 10^{-8}$), but overall there was a

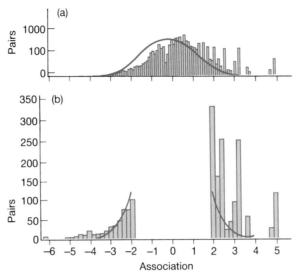

Figure 18.15. A comparison between the observed similarity values for the distribution of pairs of bird species in the Bismarck Archipelago (histograms) and the distributions expected from Gilpin and Diamond's (1982) neutral model (the curves). (a) All 11 325 species pairs on a logarithmic ordinate scale. (b) The tails of the distributions on an arithmetic ordinate scale.

considerable excess of *positively* associated pairs (with a mean similarity value of +0.69 standard deviations). Yet there was also a long tail of excess negatively associated pairs (best seen in Figure 18.15b). This shows that an overall, one-figure comparison of the actual and neutral-model communities conveys rather little information about the important differences between them. On the one hand, there are an enormous number of pairs in which owls are compared with humming-birds, ducks with warblers, and so on. These constitute the vast majority of pairs towards the centre of the distributions, but they have no relevance to the point at issue—the importance or otherwise of competition. On the other hand, there are also a large number of markedly deviant pairs (which the neutral-model analysis successfully pin-points); but competition is only one of several possible explanations for these deviants.

Gilpin and Diamond sought explanations for all species pairs with associations exceeding +3 or −3 standard deviations (Figure 18.15b). Of those with highly negative associations, the majority were ecologically dissimilar (little likelihood of competition), but they had very different incidence functions. For example, the fruit-eating starling *Aplonis metallica* is found on 28 large and medium-sized islands, while the small, supertramp honeyeater *Myzomela pammelaena* is found on 23 small and medium-sized islands; the two species share only five islands. A second factor underlying exclusive distributions was a difference in geographical origins. Some species are known to have colonized the Bismarcks from New Guinea in the west; others have come from the Solomon Islands in the east. A number of these have reached only the fringes of the Bismarcks, and they are therefore bound to generate exclusive distributions.

The third factor leading to these negative associations was interspecific competition. Gilpin and Diamond listed four pairs with similarity values less

competition is an
important determinant
of island distributions
... amongst many other
determinants

than -3 standard deviations in which the species were undoubtedly members of the same guild. Each of these had previously generated checkerboard distributions (Diamond, 1975).

The factors underlying the positive associations were shared incidence function, shared geographical origin, single-island endemics (several species were limited to just one large island in the archipelago), and shared habitat.

Overall, therefore, Gilpin and Diamond's application of a revised neutral-model approach uncovered a role for interspecific competition in the organization of the Bismarck bird communities, even though the revised version incorporated many of the shortcomings of the original. The role of competition was established clearly in only a few cases, but it was suggested in many more (where incidence functions were important and diffuse competition may have been too). Perhaps equally significant, however, was Gilpin and Diamond's demonstration that even when competition is important, it is only one important factor amongst several.

18.4.4 Verdict on the neutral-model approach

What then should be our verdict on the neutral model approach? Perhaps most fundamentally, its aim is undoubtedly worthy. We do need to guard against the temptation to see competition in a community simply because we are looking for it; even the critics of the earlier neutral models have tended to replace them with improved versions of their own. On the other hand, the approach is bound to be of limited use unless it is applied to groups (usually guilds) within which competition may be expected.

Secondly, the approach can only ever be valid when the effects of competition are truly eliminated from the neutral model itself. This has been highlighted in a computer study by Colwell and Winkler (1984). Using a program called GOD, they first generated a biota whose phylogenetic lineages obey specified rules. Subsets of the 'mainland biota' thus created then colonize archipelagos according to a program called WALLACE, and competition may or may not be introduced as a structuring influence. Unlike field workers, Colwell and Winkler *know* whether or not their communities are structured by competitive effects. The interesting and somewhat disconcerting conclusion is that some of the neutral-model approaches described above often failed to distinguish any effect of competition even though competition had been included in the program. There were several reasons for the difficulty in detecting 'the ghost of competition past'. The most important was that those species most vulnerable to competitive exclusion were already likely to have been lost from the biota of the entire archipelago. A search for more appropriate neutral models is an important task for the future.

In its favour, the neutral-model approach does concentrate the mind of the investigator, and it can stop him from jumping to conclusions too readily. Ultimately, though, it can never take the place of a detailed understanding of the field ecology of the species in question, or of manipulative experiments designed to reveal competition by increasing or reducing species abundances. It can only be part of the community ecologist's armoury.

a neutral model for
neutral models?

18.5 The role of competition: some conclusions

(i) The importance of interspecific competition in the organization of communities is easy to imagine but is usually difficult to establish. There are alternative explanations for many of the patterns that it might appear to generate, and much of its power is presumed to have been exercised in the past rather than at present. These difficulties make caution and scientific rigour particularly necessary.

(ii) Interspecific competition is certain to vary in importance from community to community: it has no single, all-purpose role. It appears, for example, frequently to be important in vertebrate communities, particularly those of stable, species-rich environments, and in communities dominated by sessile organisms such as plants and corals; while in phytophagous insect communities, for instance, it is less often important.

(iii) Even when interspecific competition is important, it may affect only a small proportion of the species interactions within a community. Mostly, it will affect only interactions between members of the same guild; and even within a guild, only the species closest together in niche space are likely to compete to a significant extent. Hence the effects of competition can easily and misleadingly be lost in sweeping overviews of large and heterogeneous assemblages of species.

The importance of competition will be discussed again in Chapter 22, after the other organizing (and dis-organizing) forces within communities have been examined.

19 The Influence of Predation and Disturbance on Community Structure

19.1 Introduction

Our attempts to understand the workings of the complex natural world involve simplification. We may do this under controlled conditions in experimental investigations of simple, ideal microcosms containing a single species or a pair of competing species, or by mimicking some small part of the real world in a mathematical model. We also do it when we examine and report on the natural world in an idealized way, for example by regarding a forest as inhabiting a homogeneous environment, or a succession as a smooth progression through time, or a zonation as a continuous change in conditions over space. The real world is not like this, but any attempt to record it in all its complexity may fail to uncover underlying principles and rules.

Biologists are sometimes accused of 'physics envy' because the systems that they work with seem not to reveal fundamental rules like the Laws of Thermodynamics, or neat and tidy order like the Periodic Table of the elements. One reason why simple laws are difficult to perceive in ecology is that the patterns keep changing; organisms are heritably variable and evolve, and thus at least some of the rules of behaviour and interaction themselves change. The life of the chemist or physicist would be very different if the Periodic Table kept changing or if gravitational force varied erratically in intensity.

Attempts to see in the living world the simplicity of the physicist's world may deny us a vision of the essence of ecological systems. They constantly undergo change—no environment is uniform, no year is exactly like any other. Moreover, the ultimate units that ecologists work with (individual organisms or their parts) exist in numbers so small compared with the number of molecules in a test-tube that we cannot disregard, as the chemist usually can, the chancy nature of events. One forest fire may make a rare species extinct.

If ecological communities are non-uniform, continually altering, and subject to the statistical events of random change, the fact cannot be ignored. A science has to be robust enough to take account of its realities. This is why 'disturbance' is allotted a whole chapter in this book. Among the forces of disturbance are included predators, earthquakes, fires, burrowing gophers, and even the fall of a raindrop. It is important not to impose a human sense of scale on the communities studied—a raindrop may be lethal to a seedling, and there can be no more serious disturbance to life than sudden death.

disturbance—a crucial aspect of ecological reality

A further reason for emphasizing disturbance in a text on ecology is that we ourselves are the source of particularly dramatic disturbances of nature, through agriculture and forestry, fertilizers and pollution, recreation and hunting (Figure 19.1). Most (perhaps now all) the communities of the world have been disturbed from their 'ideal' condition by the action of mankind. The ecologist who tries to study undisturbed communities (such as virgin forest) is likely to spend his whole life trying to find one!

19.1.1 Disturbance and the diversity of communities

The great diversity of species to be found in a community is one of the puzzles of ecology. In an ideal world the most competitive species (the one that is most efficient at converting limited resources into descendants) would be expected to drive less competitive species to extinction. The diversity of communities may then be explained by a partitioning of resources among species whose requirements do not overlap completely (resource partitioning was discussed in Chapter 18). However, this argument rests on two assumptions that are not necessarily always valid.

The first assumption is that the organisms are actually competing, which in turn implies that resources are limiting. But there are many situations where disturbance, such as grazing, storms on a rocky shore, or frequent fire, may hold down the densities of populations, so that resources are not limiting and individuals do not compete for them.

The second assumption is that when competition is operating and resources are in limited supply, one species will inevitably exclude another. But in the real world, when no year is exactly like another, and no square inch of ground exactly the same as another, the process of competitive exclusion may never proceed to its monotonous end (section 7.6). Any force that continually changes direction at least delays, and may prevent, an equilibrium or a stable conclusion being reached. Any force that simply interrupts the process of competitive exclusion may prevent extinction and enhance diversity.

19.1.2 What is disturbance?

According to the *Oxford English Dictionary*, disturbance is 'the interruption of tranquillity, peace, rest or a settled condition; interference with the due course of any action or process'. The implication is that disturbance is an unusual event in what is normality, and that it upsets normality. Some events and changes in a community are, however, a normal part of the environment; the interruption of day by night, the repeated interference with a rocky shore by the movement of the tides, and the falling of a tree in a forest, are regular, repeated and expected occurrences. Populations of organisms in a community will have been regularly and repeatedly exposed to these events. Nevertheless, for the individual shrimp stranded in a rock pool, or herb on the forest floor, these events that are everyday occurrences in the community may be disastrous. There are clearly issues of scale that determine what is a disturbance in the expected pattern of life.

One distinction that may be useful is between disasters and catastrophes.

the true meaning of disturbance is a matter of scale

Figure 19.1. Disturbed communities. (Facing page). Top: Aerial view of the Biology Gamma Forest. Death of the trees in the centre was caused by chronic exposure, 20 hours per day, to gamma radiation from a 9500-curie source of cesium-137, for about six months. This experiment, which began in 1961, is part of a programme in environmental biology established at Brookhaven National Laboratory specifically to investigate the effects of long-term chronic exposure of ecological systems to ionizing radiation. The cesium source is housed in a metal pipe in the centre of the forest. The plants of this forest vary greatly in sensitivity to damage from ionizing radiation. A sedge (*Carex pensylvanica*) is among the most resistant of the higher plants found here, having survived at exposure levels up to 350 r day^{-1}. On the other hand, the pitch pine (*Pinus rigida*) is among the least resistant of the higher plants found here, having been killed by exposures of 20 r day^{-1}.

Bottom: Copper mine on the island of Bougainville in Papua, New Guinea. The tailings from the mine, 80 000 tonnes per day, fill the valley of the Jaba river and flow out to sea. The effects of the toxicity of the tailings are not yet clear, but it is unlikely that the original rainforest vegetation will become re-established. (Photograph by courtesy of A.D. Bradshaw.)

Above: Disturbance on a different scale—molehills in an English churchyard. (Photograph: Heather Angel.)

'disasters' and 'catastrophes'

Disasters are events that happen so frequently in the lives of populations that they exert selection pressure and leave their record in evolutionary change. After disasters the population may have evolved, and the next time the disaster occurs, the population may react to it differently—it may even come to tolerate it. In contrast, we might recognize catastrophes as disturbances so infrequent that the populations have lost their 'genetic memory' of the event by the next time it recurs. When the next volcanic eruption happens in Mount St Helens (Washington) it is unlikely that the plant and animal populations that suffered from the eruption on 18 May 1980 (Baross *et al.*, 1982) will suffer any less. By contrast, hurricanes that fell forests in New England are a sufficiently frequent occurrence to be called disasters rather than catastrophes. An ecological (and perhaps evolutionary) consequence is that the characteristic forest species, white pine (*Pinus strobus*), has all the properties associated with an early-successful pioneer—precocious reproduction, effective seed dispersal, and so on.

We have grouped together predation and disturbance in this chapter because the activity of predators is often a disturbance in the 'normal' course of a succession, and because population explosions of pests and pathogens (which are included here with predators) often have dramatic effects in truly interfering with an otherwise settled condition. Moreover, the result of a predator opening up a gap for colonization is sometimes indistinguishable from that of battering by waves on a rocky shore or a hurricane in a forest. (In a related context, note that much of the information we have about predators in communities comes from disturbing them—by excluding, controlling or introducing them. The ecologist *uses* disturbance (or perturbation) as an experimental tool to discover how a community works, just as a physicist sets a pendulum swinging in order to discover how it is brought to rest.)

We discuss first the ways in which disturbances caused by predators, parasites and disease can affect community structure. We then consider the effects of temporal heterogeneity and physical disturbance. Finally, and more theoretically, a non-equilibrial theory of community structure is developed in which disturbance plays a key role. This contrasts with theories involving competitive equilibrium discussed in Chapters 7 and 18.

19.2 The effects of predation on community structure

In earlier chapters, it was emphasized that few if any predators or parasites are true generalists. All have some degree of preference in the prey they choose from those on offer, and some are extreme specialists. The effects of specialists and generalists on community structure can be quite different.

19.2.1 Generalist predators

Lawn-mowers and rabbits are relatively unselective predators. Both are capable of maintaining a close-cropped sward of vegetation over large areas of grassland. Darwin (1859) was the first to notice that the mowing of a lawn could maintain a higher diversity of species than occurred in its absence. He wrote that:

'If turf which has long been mown, and the case would be the same with turf closely browsed by quadrupeds, be let to grow, the most vigorous plants gradually kill the less vigorous, though fully grown plants; thus out of 20 species growing on a little plot of mown turf (3 feet by 4 feet) nine species perished from the other species being allowed to grow up freely.'.

Rabbits are more choosy than lawn-mowers, and this is clearly demonstrated by the occurrence in the neighbourhood of rabbit burrows of plants that are unacceptable to rabbits (including *Atropa belladonna, Urtica dioica,*

rabbit grazing: 'natural'
and artificial experi-
ments

Solanum dulcamara and *Sambuca nigra*). Nevertheless, they seem in many areas to have a similar general effect to lawn-mowers. The rabbit is not native to Britain, and its introduction (probably in the twelfth century) and subsequent spread must have been a major disturbance to the vegetation. Its presence and its effects subsequently became part of the normal state of affairs. Deliberate exclusion of rabbits from areas of vegetation was then a disturbance of the new *status quo*. Tansley and Adamson (1925) excluded rabbits from areas of species-rich chalk grassland on the English South Downs, and these areas quickly became dominated by just a few species of grass. In 1954, the viral disease myxomatosis was introduced to Britain and drastically reduced rabbit populations. The immediate response of the vegetation was an increase in the number of flowering perennial plants observed (e.g. orchids). These had been present but not recognized because they had been repeatedly nibbled. Later, a few species of grass became dominant (as they had in Tansley and Adamson's experiment), and later still the number of grass species also fell as the community became dominated by the taller, more severely self-shading types.

Grazing by rabbits had apparently kept the aggressive, dominant grasses in check and allowed a greater diversity of flora to persist. However, at very high intensities of grazing, diversity may actually be reduced as the rabbit is forced to turn from heavily grazed, preferred plant species to less preferred species. Plant species may then be driven to extinction. Figure 19.2 shows the relationship between the intensity of rabbit grazing (on a scale from 0 to 5) and the richness of the flora in a series of plots on sand dunes in the Frisian Islands of the Netherlands.

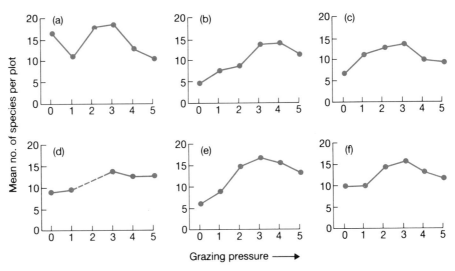

Figure 19.2. Relationship between plant species richness and intensity of rabbit grazing (on a scale of 0–5) in square metre plots on five sand dunes (a–e) and all dunes combined (f). (After Zeevalking & Fresco, 1977.)

A completely unselective grazer (like a mowing machine) will still have differential effects on the growth of different species; tall species will usually suffer more than those that are low growing, while those with large underground storage organs may recover more quickly. The effects on the community of unselective grazing depend on which groups of species suffer most. If it is subordinate species that suffer most, they may be driven to extinction and diversity will decline. If it is the competitive dominants that suffer most, the results of heavy grazing or mowing will usually be to free space and resources for other species, and diversity may then increase. When predation promotes the coexistence of species amongst which there would otherwise be competitive exclusion (because the densities of some or all of the species are reduced to levels at which competition is relatively unimportant) this is known generally as 'exploiter-mediated coexistence'.

exploiter-mediated coexistence

19.2.2 The influence of relatively selective predators

Along the rocky shores of New England in the United States, the most abundant and important herbivore in mid and low intertidal zones is the periwinkle snail *Littorina littorea*. *L. littorea* will feed on a wide range of algal species but shows a strong preference for small, tender species and in particular for the green *Enteromorpha* species. These algae apparently lack both physical and chemical deterrents to consumption by the snails. The least-preferred foods are much tougher (e.g. the perennial red alga *Chondrus crispus* and brown algae such as *Fucus* spp.). They are either never eaten by *L. littorea* or are eaten only if no other food has been available for some time.

herbivores: periwinkles and algae

Lubchenco (1978) noticed that the algal composition of tide pools in the rocky intertidal region varied from almost pure stands of *Enteromorpha intestinalis* to the opposite extreme, preponderance of *Chondrus crispus*. To test whether grazing by *L. littorea* was responsible for these differences, she removed all periwinkles from a *Chondrus* pool and added them to an *Enteromorpha* pool, monitoring subsequent changes for 17 months. During that time, periwinkles in a control *Chondrus* pool were observed to feed on microscopic plants and the young stages of many ephemeral algae that settled on the *Chondrus* (including *Enteromorpha*). No changes in algal composition occurred in the control pool. However, in the *Chondrus* pool from which periwinkles had been removed, *Enteromorpha* and several other seasonal, ephemeral algae immediately settled or grew from microscopic sporelings or germlings, and became abundant. *Enteromorpha* achieved dominance and outcompeted the *Chondrus*. Following settlement of *Enteromorpha* on *Chondrus*, the upright thalli of the latter became bleached and then disappeared. It is clear that the presence of *L. littorea* was responsible for the dominance of *Chondrus* in *Chondrus* pools. Addition of *L. littorea* to *Enteromorpha* pools led, in a year, to a decline in percentage cover of the green alga from almost 100% to less than 5%. *Chondrus* colonizes slowly, but eventually comes to dominate pools where *L. littorea* has eaten out its competitor.

Why then do some pools contain *L. littorea* while others do not? The periwinkle colonizes pools while it is in an immature, planktonic stage. Although planktonic periwinkles are just as likely to settle in *Enteromorpha* pools as in *Chondrus* pools, the crab *Carcinus maenas*, which can shelter in the *Enteromorpha* canopy, feeds on the young periwinkles and prevents them

. . . and crabs . . . and gulls

from establishing a new population. The final thread in this tangled web of interactions is the effect of gulls which prey on crabs where the dense green algal canopy is absent. Thus there is no bar to continuing periwinkle recruitment in *Chondrus* pools.

It is not hard to see that a disturbance that affected the numbers of *L. littorea*, or of crabs, or of gulls, would have repercussions on the communities of *Enteromorpha* and of *Chondrus*.

If we were to predict the effect that periwinkle grazing would have on algal species richness or diversity in tide pools, we might postulate that when *L. littorea* is absent or rare, *Enteromorpha* would competitively exclude several other species and algal diversity would be low. A survey of a number of pools with different densities of *L. littorea* demonstrated just this (Figure 19.3a). When *L. littorea* was present in intermediate densities, the abundance of *Enteromorpha* and other ephemeral algal species was reduced, competitive exclusion was prevented, and many species, both ephemeral and perennial, coexisted. But at very high densities of *L. littorea*, all palatable algal species were consumed to extinction and prevented from reappearing, leaving an almost pure stand of the tough *Chondrus*. Note the similarity between the humped curves in Figure 19.2 and Figure 19.3a.

predation and humped curves of diversity

Figure 19.3. Effect of *Littorina littorea* density on species richness (*S*) and species diversity (Shannon index *H* calculated on the basis of percentage cover, p. 601): (a) in tide pools and (b) on emergent substrata. (From Lubchenco, 1978.)

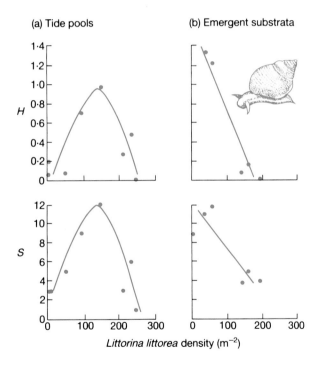

As a generalization, selective predation may be expected to favour a higher community diversity if the preferred prey are competitively dominant, at least below a threshold of predation. In a study of grazing on grassland, principally by sheep, Jones (1933) came to the same conclusion. Species-rich communities were maintained by continuous grazing because it was the competitive dominants, *Lolium perenne* and *Trifolium repens*, that were most severely grazed.

The picture is quite different when the preferred prey species is competitively inferior to other prey, and this is again well illustrated by Lubchenco's work on periwinkles. In the rocky intertidal zone Lubchenco studied in New England, the competitive dominance of the most abundant tide pool plants is actually reversed when the plant species interact on emergent substrata rather than in the tide pools. Here it is perennial brown and red algae that predominate, while a number of ephemeral species manage to maintain a precarious foothold, at least in sites where *L. littorea* is rare or absent. Any increase in the grazing pressure, however, decreased the algal diversity, as the preferred species were consumed totally and prevented from re-establishing themselves (Figure 19.3b).

A similar situation is suggested by the results of Milton's (1940, 1947) work on sheep grazing in upland habitats in Wales. The dominant plant species in these sites, *Festuca rubra*, *Agrostis tenuis* and *Molinia caerulea*, were relatively unpalatable and overgrazing consolidated the dominance of these few.

Note that the explanation offered above begs an important question. If the unpalatable species are truly competitively dominant, why do they not exclude the other, more palatable species in the absence of predation? The answer is that other mechanisms for coexistence must be operating. Resource partitioning among the plants is one possibility. Others will be discussed in section 19.3.

The interactions between grazers and plants on rocky shores and in sheep pastures are essentially two-dimensional, taking place across the substrate surface. The work of McCauley and Briand (1979) provides a contrast with these since it deals with zooplankton feeding on microscopic phytoplankton in a three-dimensional freshwater environment. McCauley and Briand postulated that in the absence of grazers, competition among phytoplankton species would be severe, especially in summer when nutrients are depleted in the surface waters of lakes. Experiments were carried out in Heney Lake, Quebec, by enclosing planktonic communities in 8000-litre polyethylene tubes (1 m in diameter, 10 m deep and sealed at the bottom). The density of grazers was manipulated so that comparisons could be made between lightly and heavily grazed communities. The results of one experiment are given in Figure 19.4. In the enclosure containing fewer herbivores, total species richness of phytoplankton was slightly reduced. This was the result of the reduction in number of inedible species, which more than counteracted the intermittent increases in edible species. Zooplankton feed selectively according to size, shape and taste of their prey. Thus, it appears that herbivory plays a role in maintaining high phytoplankton diversity. A reduced intensity of grazing amplifies competition among the phytoplankton, mainly favouring the edible at the expense of the many inedible species.

The rocky intertidal zone also provided the location for pioneering work by Paine (1966) on the influence of a top carnivore on community structure. The starfish *Pisaster ochraceus* preys on sessile filter-feeding barnacles and mussels, and also on browsing limpets and chitons and a small carnivorous whelk. These species, together with a sponge and four macroscopic algae,

diversity is decreased when competitive inferiors are preferred by predators

an experiment with plankton

carnivores: Paine's starfish on a rocky shore

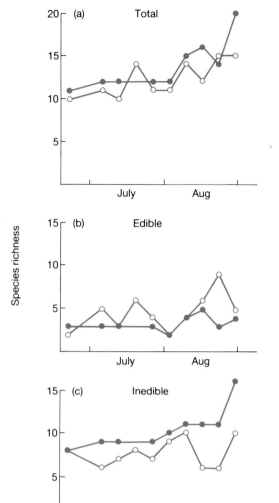

Figure 19.4. Species richness of: (a): the entire phytoplankton community, (b) edible phytoplankton and (c) inedible phytoplankton under low (○) and significantly higher grazing pressure (●) (from McCauley & Briand, 1979.)

form predictable associations on rocky shores of the Pacific coast of North America (Figure 19.5). Paine removed all starfish from a typical piece of shoreline about 8 m long and 2 m deep, and continued to exclude them for several years. At irregular intervals, the density of invertebrates and cover of benthic algae were assessed in the experimental area and in an adjacent control site. The latter remained unchanged during the study. Removal of *P. ochraceus*, however, had dramatic consequences. Within a few months, the barnacle *Balanus glandula* settled successfully. Later it was crowded out by mussels (*Mytilus californicus*), and eventually the site became dominated by these. All but one of the species of alga disappeared, apparently through lack of space, and the browsers tended to move away, partly because space was limited and partly due to lack of suitable food. Overall, the removal of starfish led to a reduction in number of species from fifteen to eight.

The main influence of the starfish *P. ochraceus* appears to be to make space available for competitively subordinate species. It cuts a swathe free of barnacles and, most importantly, free of the dominant mussels which would

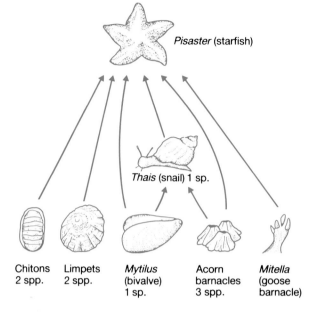

Figure 19.5. Paine's rocky shore community. (modified from Paine, 1966).

Chitons 2 spp. Limpets 2 spp. *Mytilus* (bivalve) 1 sp. Acorn barnacles 3 spp. *Mitella* (goose barnacle)

otherwise outcompete other invertebrates and algae for space. Once again, there is exploiter-mediated coexistence. Physical disturbance by storm waves can have a directly analogous effect by generating gaps which are invaded by rapidly colonizing but competitively inferior species (Paine & Levin, 1981; section 19.3.2).

19.2.3 Diet switching and frequency-dependent selection

Even relatively selective predators seldom simply take each potential prey species from a community in turn, bringing each to extinction before turning to the next. Selection is moderated by the time or energy spent in search for the preferred prey, and many species take a mixed diet. However, others switch sharply from one type of prey to another (a type 3 functional response—see section 9.5.3), taking disproportionately more of the commoner acceptable types of prey. In theory, such behaviour could lead to the coexistence of a large number of relatively rare species. In practice, it has been demonstrated only rarely. However, there is evidence that predation on seeds of tropical trees is often more intense where seeds are more dense (beneath and near the adult that produced them) (Connell, 1979); and wood-pigeons (*Columba palumbus*) in Britain change their diet from month to month, probably in relation to their rate of successful feeding, moving in turn from clover in winter and spring, to *Brassica* crops, weed leaves, sown cereals, weed seeds, and then to ripened cereals (Murton *et al.*, 1966). Another herbivore species, the butterfly *Battus philenor*, forms search images for leaf shape when foraging for its two larval host plants, and concentrates on whichever happens to be the more common (Rausher, 1978). Finally, the stream-dwelling invertebrate carnivore *Plectrocnemia conspersa* is most strongly associated with midge larvae when these are particularly abundant in summer, but it switches its attention to stonefly nymphs for the rest of the year (Townsend & Hildrew, 1978; Hildrew & Townsend, 1982). Despite the occurrence of such frequency-dependent selection in some consumers, it is not a general rule and may not be common.

19.2.4 The influence of specialist predators

Many species of predator are highly specific in the foods that they will consume. Giant pandas are specialists on bamboo shoots, the flea *Cediopsylla tepolita* feeds only on the blood of the rare and endangered Mexican volcano rabbit (*Romerolagus diazi*), and specialization in diet is equally extreme among many phytophagous insects (section 9.2). When the food resource of such predators becomes exhausted, they cannot switch to something else. These specialists, which may in turn support their own specialist predators and parasites, produce linear food chains in a community that is otherwise largely composed of branching food webs. This means that they are to some extent isolated and insulated from the rest of the community and its influences. If their food is overexploited, it will be they that suffer excessively. If they become superabundant, or suddenly become rare, it is their specialist prey alone that suffers or gains. It is this degree of insulation from the rest of the community that can make such specialists very suitable material for biological control. Specialist predators, introduced to kill a weed or an insect pest, do not spread on to other organisms when they have brought the enemy under control. Moreover, they rarely, if ever, bring their prey to extinction. Rather they enter into predator–prey interactions in which both persist, but often at low densities. When the prey form a dominant part of the community, the effects of introducing a biological control agent may be very dramatic (see, for example, the account of the moth *Cactoblastis* used to control prickly pear in Australia—section 10.4.1).

specialist predators, insulated from the rest of the community, are ideal for biological control . . .

In most of the cases in which biological control has been used, the pest to be controlled has itself been an introduced species. Thus there have usually been two disturbances. In the first an introduced organisms (such as prickly pear) expanded explosively to become a community dominant, and in the second a further introduced species brought the first under control. Almost certainly (though the data are hard to come by) the first disturbance resulted in a major loss of species from the communities that were invaded. The second disturbance largely eradicated the dominant invader and probably allowed diversity to increase again.

. . . where they may restore diversity

19.2.5 Outbreaks of parasites and disease

The incidence of a parasite, like that of other types of exploiter, may determine whether or not a host species occurs in an area. Once again, almost all the hard information about the role of parasites and disease in determining community structure comes from situations in which there has been a deliberate or natural disturbance of the community. For example, the extinction of nearly half the endemic bird fauna of the Hawaiian Islands has been attributed to the introduction of bird pathogens such as malaria and bird pox (Warner, 1968); and recent changes in the distribution of the North American moose (*Alces alces*) have been associated with the parasitic nematode *Pneumostrongylus tenuis* (Anderson, 1981). A similar explanation has been advanced for the restriction of natural stands of European larch trees to the Alps, where the fungal pathogen *Trichoscyphella willkommii* is of only minor importance. When larch is grown in extensive stands in lowland Europe, however, the pathogen is extremely damaging. The almost complete elimination of the English elm (*Ulmus campestris*) from England in the past 20 years has been caused by the spread of the introduced Dutch elm disease. Finally, probably the largest single change wrought in the structure of communities by a parasite has been the destruction of the chestnut (*Castanea dentata*) in North American forests, where it had been a dominant tree over large areas until the introduction of the fungal pathogen *Endothia parasitica*, probably from China.

the role of parasites is usually revealed only by deliberate or natural disturbance

The role of parasites in community structure can be seen in its full drama when an outbreak of disease occurs and sweeps through a community. Sooner or later, however, the community settles down to some new state in which the interactions between parasite and host become less obvious. The reduction in host density will itself reduce the rate of spread and intensity of damage. It is then very much more difficult to determine just how large a role the disease plays in the community structure. We need experiments in which fungicides or insecticides are applied to natural vegetation before we can really begin to see how important pests and diseases are in determining the structure of communities (Figure 19.6). Just as we can only measure the effect of predators on vegetation by excluding them, so we might gain a proper measure of the role of pests and diseases from exclusion experiments.

Another way to determine the effects that a parasite or disease might have on community structure is to assemble artificially simple communities, of one or two species, and to compare their behaviour with and without a

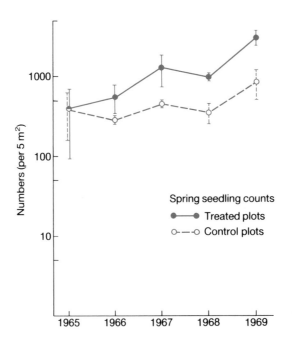

Figure 19.6. Insecticides were applied repeatedly to the ground beneath a Michigan woodland. Over four years the population of the herb *Melampyrum lineare* increased dramatically. The cause was traced to the beetle *Atlanticus testaceus,* which preys upon seedlings of *Melampyrum* and was destroyed by the insecticide. (Cantlon, 1969.)

parasite or disease. This can show us what a parasite or disease *might* do in nature. The presence of the barley pathogen *Erysiphe graminis* in glasshouse mixtures of barley and wheat significantly reduced the competitiveness of barley in relation to that of wheat (Figure 19.7; Burdon & Chilvers, 1977).

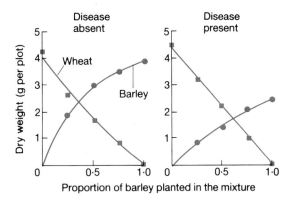

Figure 19.7. Replacement series for competition between barley and wheat showing the effect, after 67 days of growth, of powdery mildew. In the presence of disease, the performance of barley is adversely affected. (From Burdon & Chilvers, 1977.)

Similarly, the effect of crown rust on a mixed sward of ryegrass and white clover was to depress the yield of ryegrass by 84%, while the white clover component, which was not attacked, responded by an increase in yield of 87% (Latch & Lancashire, 1970). Finally, an elegant study of the effect of a parasite on competitive ability of animals was made by Park (1948). Park showed that in the laboratory, the protozoan parasite *Adelina triboli* was able to reverse the outcome of competition between two species of flour beetle, *Tribolium castaneum* and *T. confusum*.

It is worth noting that the rate of increase of parasites and disease is usually greater at high than at low host density. Thus their impact on a host species is likely to be *proportionately* greater when the host is abundant in a community than when it is rare. This is the phenomenon of frequency-dependence already referred to (p. 710), and increases the chance that host species may coexist.

19.2.6 Resumé of the effects of predators, parasites, and disease

(a) Selective predators are likely to act to enhance diversity in a community if their preferred prey are competitively dominant. It seems likely that there is some general correspondence between palatability to predators and high growth rates. If the production of chemical and physical defences by prey requires a sacrifice of resources used in growth and reproduction, we might expect species that are competitive dominants in the absence of predators (and hence which devote resources to competition rather than defence) to suffer excessively from their presence. Thus selective predators may frequently enhance diversity. If the predators act in a frequency-dependent manner their action should be even stronger.

(b) Even generalist predators may be expected to have the effect of increasing community diversity through exploiter-mediated coexistence, because even if the prey are attacked simply in proportion to their abundance, it will be those species that are assimilating resources and producing biomass and offspring most rapidly (the competitive dominants) that will be most abundant, and will therefore be most severely set back by predation.

(c) An intermediate intensity of predation is most likely to be associated with high prey diversity, since too low an intensity may not prevent competitive exclusion of inferior prey species, while too high an intensity may itself drive preferred prey to extinction (note, however, that 'intermediate' is difficult to define *a priori*).

(d) The role of predators, parasites and disease in shaping community structure is probably least significant in communities where physical conditions are more severe, variable or unpredictable (Connell, 1975). In sheltered coastal sites, predation appears to be a dominant force shaping community structure (Paine, 1966), but in exposed rocky tidal communities where there is direct wave action, predators seem to be scarce and to have negligible influence on community structure (Menge & Sutherland, 1976).

(e) The effects of animals on a community often extend far beyond just those due to the cropping of their prey. Burrowing animals (such as moles, gophers, earthworms, rabbits and badgers) and mound-builders (ants and termites) all create disturbances. Their activities provide local heterogeneities, including sites for new colonists to become established and for micro-successions to take place. Larger grazing animals introduce a mosaic of nutrient-rich patches, as a result of dunging and urinating, in which the local balance of other species is profoundly changed. Even the footprint of a cow in a wet pasture may so change the microenvironment that it is now colonized by species that would not be present were it not for the disturbance (Harper, 1977). The predator is just one of the many agents disturbing community equilibrium. Indeed disturbance of one sort or another is so much the norm in natural communities that it is an open question whether truly equilibrial situations ever occur in nature.

19.3 Temporal heterogeneity and physical disturbance

19.3.1 Disturbances and gaps

Disturbances which open up gaps are common in all kinds of community. In forests, they may be caused by high winds, lightning, earthquakes, elephants, lumberjacks, or simply by the death of a tree through disease or old age. Agents of disturbance in grassland include frost, burrowing animals, and the teeth, feet or dung of grazers. On rocky shores or coral reefs, gaps in algal or sessile animal communities may be formed as a result of severe wave action during hurricanes, tidal waves, battering by logs or moored boats, or by the action of predators (section 19.2). The formation of gaps is of considerable significance to sessile or sedentary species which have a requirement for open space. It matters much less in the lives of mobile animal species for whom space is not the limiting factor.

In some communities, gaps will be colonized by one or several species which proceed to pass through a more or less predictable mini-succession. In the absence of further disturbance, such gaps will revert to the climax state typical of the region. Alternatively, in cases where individual gaps can be occupied completely by an individual organism, the outcome of colonization may be much less predictable and a member of any of a number of species may come to dominate the space, at least for its lifetime. Such situations may be described in terms of a competitive lottery.

19.3.2 Competitive lottery for gaps

If a large number of species are approximately equivalent in their ability to invade gaps, are equally tolerant of the abiotic conditions, and can hold the gaps against all-comers during their lifetime, then the probability of competitive exclusion is much reduced in an environment where gaps are appearing continually and randomly. A further condition for coexistence is that the number of young which invade and occupy gaps should not be consistently greater for parent populations which produce more offspring, otherwise the most productive species would come to monopolize space even in a continuously disturbed environment.

If these conditions are met, it is possible to envisage how the occupancy of a series of gaps will change through time (Figure 19.8). On each occasion that an organism dies or is killed, the gap is re-opened for invasion. All conceivable replacements are possible and species richness will be maintained at a high level. Some tropical reef communities of fish may conform to this model (Sale, 1977, 1979; Sale & Douglas, 1984). They are extremely

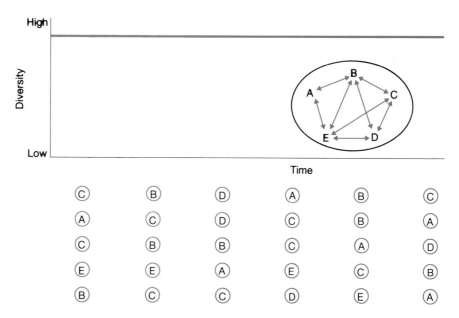

Figure 19.8. Hypothetical competitive lottery. Occupancy of gaps which periodically becomes available. Each of species A to E is equally likely to fill a gap, regardless of the identity of its previous occupant (this is illustrated in the inset). Species richness remains high and relatively constant.

diverse. For example, the number of species of fish on the Great Barrier Reef ranges from 900 in the south to 1500 in the north, and more than 50 resident species may be recorded on a single patch of reef 3 m in diameter. Only a proportion of this diversity can be attributed to resource partitioning of food and space—indeed the diets of many of the coexisting species are very similar. In this community, vacant living space seems to be a crucial limiting factor, and it is generated unpredictably in space and time when a resident dies or is killed. The life-styles of the species match this state of affairs. They breed often, sometimes year-round, and produce numerous clutches of dispersive eggs or larvae. It can be argued that the species compete in a lottery for living space in which larvae are the tickets, and the first arrival at the vacant space wins the site, matures quickly, and holds the space for its lifetime.

Three species of herbivorous pomacentrid fish co-occur on the upper slope of Heron Reef, part of the Great Barrier Reef off eastern Australia. Within rubble patches, the available space is occupied by a series of contiguous and usually non-overlapping territories, each up to 2 m² in area, held by individuals of *Eupomacentrus apicalis*, *Plectroglyphidodon lacrymatus* and *Pomacentrus wardi*. Individuals hold territories throughout their juvenile and adult life and defend them against a broad range of chiefly herbivorous species, including conspecifics. There seems to be no particular tendency for space initially held by one species to be taken up, following mortality, by the same species. Nor is any successional sequence of ownership evident (Table 19.1). *P. wardi* both recruited and lost individuals at a higher rate than the other two species, but all three species appear to have recruited at a sufficient level to balance their rates of loss and maintain a resident population of

Table 19.1. Numbers of individuals of each species observed occupying sites, or parts of sites, that had been vacated during the immediately prior inter-period between censuses through the loss of residents of each species. Sites vacated through loss of 120 residents have been reoccupied by 131 fish. Species of new occupant is not dependent on species of previous resident ($\chi^2 = 5.88$, 4 d.f., $P > 0.10$).

Resident lost	Reoccupied by		
	E. apicalis	*Pl. lacrymatus*	*Po. wardi*
E. apicalis	9	3	19
Pl. lacrymatus	12	5	9
Po. wardi	27	18	29

breeding individuals. Thus the maintenance of high diversity depends, at least in part, on the unpredictability of the supply of living space; and as long as all species win some of the time and in some places, they will continue to put larvae into the plankton and hence into the lottery for new sites.

An analogous situation has been postulated for the highly diverse chalk grasslands of Britain (Grubb, 1977). Any small gap which appears is rapidly exploited, very often by a seedling. In this case the tickets in the lottery are seeds, either in the act of dispersal or as components of a persistent seed bank in the soil. Which seeds develop to established plants, and therefore which species comes to occupy the gap, depends on a strong random element since the seeds of many species overlap in their requirements for germination. The successful seedling rapidly establishes itself amidst the short turf and retains the patch for its lifetime, in a similar way to the reef fish described above. However, the analogy with tropical reefs should not be taken too far. Three extra factors are important in the grassland.

(a) Gaps are not necessarily identical. They may vary in the nature of the soil surface, the amount of litter that is present, and whether the gap is also occupied by other organisms, including bacteria, fungi, viruses and animals. The orientation of the gap and even its size (Figure 19.9) can influence physical conditions such as temperature. Subtle differences in the germination requirements of particular species mean that not all those seeds that are present at the time the gap appears will stand an equal chance of establishing themselves.

. . . but not all gaps are equal

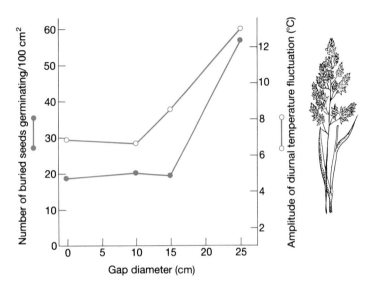

Figure 19.9. Comparison of soil temperature fluctuations (○) and germination (●) of the grass *Holcus lanatus* in gaps of different diameters in the canopy, in a sown pasture of low productivity in northern England (from Grime, 1979).

(b) Gaps which appear only a short time apart may be subjected to quite different microclimatic regimes in terms of temperature, light intensity and water potential of the atmosphere. Daily, even hourly, fluctuations are likely to result in a constant shifting of the identity of species for which the prevailing conditions are most nearly optimal for breaking dormancy, for photosynthesis and for growth. Identical gaps with identical arrays of seeds may nevertheless be taken over by different species as a result of very small fluctuations in weather (e.g. Figure 19.10).

(c) It is not surprising that in seasonal environments the species which which will colonize a gap varies with the season. For example, the colonists of bare ground which is created during the summer include small winter annuals and many grasses, such as *Festuca ovina, Helictotrichon pratense* and *Koeleria cristata,* which produce many seedlings in the early autumn; while colonization of gaps created during winter is often delayed until the appearance the following spring of quite different species such as *Linum catharticum, Pimpinella saxifraga* and *Viola riviniana* (Grime, 1979).

The formation of gaps is an inherent feature of all communities where space is monopolized by individual organisms—each is certain to die at some time. If interactions between species take the form of a race for unchallenged

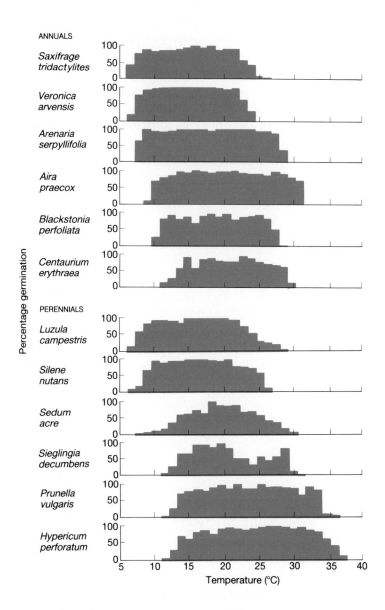

Figure 19.10. The relationship between temperature and probability of germination for several annual and perennial herbaceous plants of common occurrence in calcareous grassland in northern England (from Grime, 1979).

dominance in recently created openings, rather then direct competitive interactions between established adults, then species richness will be higher because the likelihood of competitive exclusion is reduced. An additional factor is that the identity of neighbouring individuals is unpredictable in space and time. This will also act to reduce the liklihood of competitive exclusion of one species by another.

19.3.3 Disturbances that interrupt succession

In contrast to the examples discussed above, in many communities the stages in the colonization of gaps are reasonably predictable. The effect of the

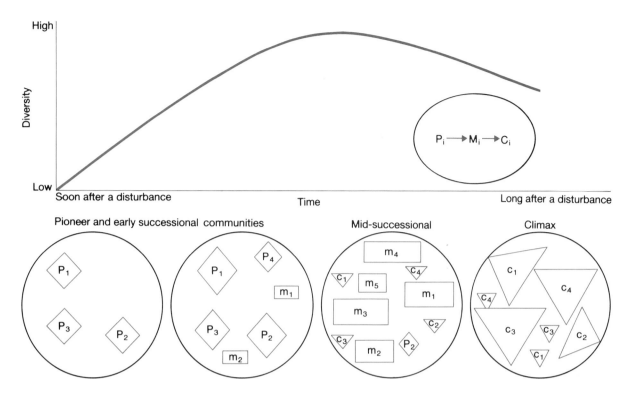

Figure 19.11. Hypothetical mini-succession in a gap. The occupancy of gaps is reasonably predictable. Diversity begins at a low level as a few pioneer p_i species arrive; reaches a maximum in mid-succession when a mixture of pioneer, mid-successional m_i and climax c_i species occur together; and drops again as competitive exclusion by climax species takes place (cf. Figure 19.8).

disturbance is to knock the community back to an earlier stage of succession (Figure 19.11). The open space is colonized by one or more of a group of opportunistic, early-successional species (p_1, p_2, etc., in Figure 19.11). As time passes, more species invade, often those with poorer powers of dispersal. These eventually reach maturity, dominating mid-succession (m_1, m_2, etc.), and many or all of the pioneer species are driven to extinction. Later still, the community regains the climax stage when the most efficient competitors (c_1, c_2, etc.) oust their neighbours. In this sequence, diversity starts at a low level, increases at the mid-successional stage and usually declines again at the climax. The gap essentially undergoes a mini-succession. The process is fundamentally different from a competitive lottery because species are replaced by other species in a reasonably predictable sequence, culminating in the climax. The other important difference is that many species are likely to be involved. In contrast, gaps in a competitive lottery are filled, and thus no longer available, when occupied by a single established individual.

Some disturbances are synchronized, or phased, over extensive areas. A forest fire may destroy a huge tract of a climax community. The whole area then proceeds through a more or less synchronous succession, with diversity increasing through the early colonization phase and falling again through competitive exclusion as the climax is approached. Other disturbances are much smaller and produce a patchwork of habitats. If these disturbances are

the importance of phasing, gap frequency and gap size

unphased, the resulting community comprises a mosaic of patches at different stages of succession. A climax mosaic, produced by unphased disturbances, is much more diverse in species than an extensive area undisturbed for a very long period and occupied by just one or a few dominant climax species.

The influence that disturbances have on a community depends on whether the timing of gap formation is phased or unphased, the frequency with which gaps are opened up, and the size of the gaps produced.

19.3.4 Frequency of gap formation

The intermediate disturbance hypothesis (Connell, 1978; see also the earlier account by Horn, 1975) proposes that the highest diversity is maintained at intermediate levels of disturbance. Soon after a severe disturbance, propagules of a few pioneer species arrive in the open space. If further disturbances occur frequently, gaps will not progress beyond the pioneer stage in Figure 19.11, and the diversity of the community as a whole will be low. As the interval between disturbances increases, the diversity will also increase because time is available for the invasion of more species. This is the situation at an intermediate frequency of disturbance. At very low frequencies of disturbance, most of the community for most of the time will reach and remain at the climax, with competitive exclusion having reduced diversity. This is shown diagrammatically in Figure 19.12, which plots the pattern of species richness to be expected as a result of unphased high, intermediate and low frequencies of gap formation, in separate patches and for the community as a whole.

The influence of the frequency of gap formation has been studied in southern California by Sousa (1979a,b), in an intertidal algal community associated with boulders of various sizes. Wave action disturbs small boulders more often than large. Using a sequence of photographs, Sousa estimated the probability that a given boulder would be moved during the course of a month. A class of mainly small boulders (which required a force < 49 newtons to move them) had a monthly probability of movement of 42%. An intermediate class (which required a force of 50–294 newtons) had a much smaller monthly probability of movement, 9%. Finally, the class of mainly large boulders (> 294 newtons) moved with a probability of only 0.1% per month. The 'disturbability' of the boulders had to be assessed in terms of the force required to move them, rather than simply in terms of top surface area, because some rocks which appeared to be small were actually stable portions of larger, buried boulders and a few large boulders with irregular shapes

moved when a relatively small force was applied. The three classes of boulder (< 49N, 50–294N, > 294N) can be viewed as patches exposed to a decreasing frequency of disturbance when waves caused by winter storms overturn them.

The successional sequence which occurs in the absence of disturbance was established by studying both large boulders which had been experimentally cleared and implanted concrete blocks (Figure 19.13). The basic succession has already been described in section 16.4.3. To recap, within the first month the surfaces were colonized by a mat of the ephemeral green alga,

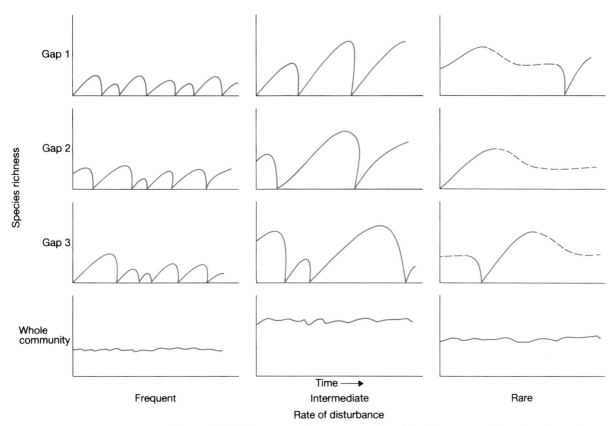

Figure 19.12. Diagrammatic representation of the time course of species richness in three gaps, and in the community as a whole, at three frequencies of disturbance. Dashed lines indicate the phase of competitive exclusion as the climax is approached.

Ulva spp.. In the autumn and winter of the first year, several species of perennial red alga became established, including *Gelidium coulteri, Gigartina leptorhynchos, Rhodoglossum affiine* and *Gigartina canaliculata*. The last-named species gradually came to dominate the community, holding 60–90% of the surface area after two to three years. What is essentially an algal monoculture persists through vegetative propagation, and resists invasion by all other species. Species richness increased during early stages of succession through a process of colonization, but declined again at the climax because of competitive exclusion by *G. canaliculata*. It is important to note that the same succession occurred in small boulders which had been artificially made stable. Thus variations in the communities associated with the surfaces of boulders of different size were not simply an effect of size, but rather of differences in the frequency with which they were disturbed.

Populations on unmanipulated boulders in each of the three size/ disturbability classes were assessed on four occasions. Table 19.2 shows that the percentage of bare space decreased from small to large boulders, indicating the effects of the greater frequency of disturbance of small boulders. Mean species richness was lowest on the regularly disturbed small boulders. These were dominated most commonly by *Ulva* spp. (and barnacles, *Chthamalus fissus*). The highest levels of species richness were consistently recorded on the intermediate boulder class. Most held mixtures

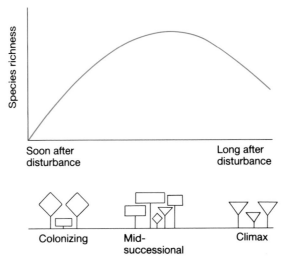

Figure 19.13. Simplified description of algal succession on experimentally cleared, stable rock surfaces. Diamond shapes represent pioneer species (U = *Ulva* spp.); rectangles represent later colonizers and mid-successional species (Ge = *Gelidium coulteri*, Gl = *Gigartina leptorhynchos*, R = *Rhodoglossum affine*); triangles represent climax species (Gc = *Gigartina canaliculata*). (After Sousa, 1979a.)

of three to five abundant species from all successional stages. The largest boulders had a lower mean species richness than the intermediate class, though a monoculture was achieved on only a few boulders. *Gigartina canaliculata* covered most of the rock surfaces.

... offering strong support for the hypothesis

These results offer strong support for the intermediate disturbance hypothesis as far as frequency of appearance of gaps is concerned. However, we must be careful not to lose sight of the fact that this is a highly stochastic process. By chance, some small boulders were not overturned during the period of study. These few were dominated by the climax species *G. canaliculata*. Conversely, two large boulders in the May census had been overturned, and these became dominated by the pioneer *Ulva*. However, on average, species richness and species composition followed the predicted pattern.

Table 19.2. Seasonal patterns in bare space and species richness on boulders in each of three classes, categorized according to the force (in newtons) required to move them (Sousa, 1979b).

Census date	Boulder class	Percentage bare space	Species richness		
			Mean	Standard error	Range
November 1975	< 49N	78.0	1.7	0.18	1–4
	50–294N	26.5	3.7	0.28	2–7
	> 294N	11.4	2.5	0.25	1–6
May 1976	< 49N	66.5	1.9	0.19	1–5
	50–294N	35.9	4.3	0.34	2–6
	> 294N	4.7	3.5	0.26	1–6
October 1976	< 49N	67.7	1.9	0.14	1–4
	50–294N	32.2	3.4	0.40	2–7
	> 294N	14.5	2.3	0.18	1–6
May 1977	< 49N	49.9	1.4	0.16	1–4
	50–294N	34.2	3.6	0.20	2–5
	> 294N	6.1	3.2	0.21	1–5

This study deals with a single community conveniently composed of identifiable patches (boulders) which become gaps (when overturned by waves) at short, intermediate or long intervals. Recolonization occurs mainly from propagules derived from other patches in the community. Because of the pattern of disturbance, this mixed boulder community is more diverse than one with only large boulders would be.

19.3.5 Size of gaps

Gaps of different sizes may influence community structure in different ways because of contrasting mechanisms of recolonization. The centres of very large gaps are most likely to be colonized by species producing propagules which travel relatively great distances. Such mobility is less important in small gaps, since most recolonizing propagules will be produced by adjacent established individuals. The smallest gaps of all may be filled simply by lateral movements of individuals around the periphery.

Paine and Levin (1981) have monitored the processes of formation and filling-in of gaps in intertidal beds of mussels, *Mytilus californianus*, on wave-swept rocky shores on the north-west coast of the United States. In the absence of disturbance, mussel beds may persist as extensive monocultures. More often, they are an ever-changing mosaic of many species which inhabit gaps formed by the action of waves. Gaps can appear virtually anywhere, and may exist for years as islands in a sea of mussels. The size of these gaps at the time of formation ranged from the dimensions of a single mussel to 38 m² (Figure 19.14). In general, a mussel or group of mussels becomes infirm or

gap formation and filling-in in a mussel bed

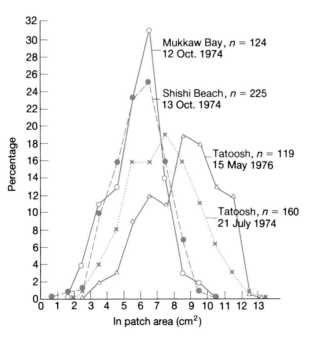

Figure 19.14. Frequency distributions of patch size at the time of their formation in mussel beds at four locations on the north-west coast of North America (from Paine & Levin, 1981).

damaged through disease, predation, old age or, most often, the effects of storm waves or battering by logs. Sometimes mussels are dislodged immediately. In other cases, the shells persist during the summer to be dislodged as wave action becomes more severe in winter. Once disruption has started, erosion of the bed is swift and the gap is formed. Gaps generated in summer tend to be smaller and less abundant than those formed in winter.

Gaps begin to fill as soon as they are formed. The recovery process is initiated by the mussels on the perimeter leaning towards the centre of the gap. Although it lasts only briefly, this process initially reduces the size of the gap very rapidly (0.2 cm d^{-1}), and it can completely fill very small gaps within days. The lifetime of small gaps (< 100 cm^2) is too short for any appreciable invasion by other species.

Intermediate-sized gaps in a mussel bed (< 3500 cm^2) may persist for several years before they are recolonized by mussels. In the meantime, the opportunity to invade the bare rock is taken by many species. In the first season after formation, a major portion of the gaps was held by the alga *Porphyra pseudolanceolata*. By the second year, the herbivore-resistant *Coralina vancouveriensis* had become established, together with barnacles (*Pollicipes polymerus*, *Balanus glandula* and *B. cariosus*) and mussels (*Mytilus edulis* and *M. californianus*). Within four years, bare space had been reduced to only 1%, and *Mytilus californianus* was again the dominant species.

In the largest gaps of all (> 3500 cm^2), leaning and lateral movement by peripheral mussels are of relatively little importance. Mussels recruit mainly as larvae from the plankton, although some dislodged adult mussels may be washed on to the patch and become entrapped. The filling process is mainly due to growth of individual mussels and takes many years. Associated with the opening of space, there is an initial burst of diversity, which persists for about four years, followed by a winnowing of species as they are eliminated through competition, eventually yielding a bed of mussels (Table 19.3).

Table 19.3. Frequency (fraction of patches occupied) of named species occupying $> 10\%$ of the primary space in quadrat samples from ten large, low intertidal patches from September 1970 to May 1979 (from Paine & Levin, 1981).

Age in months	3–6	15–18	27–30	39–42	51–54	63–66	≥ 67
Number of quadrat samples	43	50	83	66	60	36	161
Species							
Petalonia sp. (plant)	0.07						
Porphyra pseudolanceolata (plant)	0.37						
Halosaccion glandiforme (plant)	0.02	0.02					
Balanus cariosus (barnacle)	0.67	0.72	0.47	0.48	0.20	0.17	0.03
Iridaea lineare (plant)	0.09	0.06	0.01	0.03			
Alaria nana (alga)	0.02	0.02	0.12	0.09	0.02		
Mytilus edulis (mussel)		0.38	0.46	0.30	0.02		
Bossiella plumosa		0.04	0.06	0.11	0.07		
Pollicipes polymerus (barnacle)		0.62	0.84	0.94	0.93	0.72	0.22
Corallina vancouveriensis (plant)				0.08	0.02		
Mytilus californianus (mussel)		0.06	0.18	0.45	0.93	1.00	0.99
Lessoniopsis littoralis			0.01		0.02		0.01

The species which take advantage of the open space are essentially the same for both intermediate-sized and large gaps. However, large gaps provide islands of diversity which persist for longer, and this is their special contribution. Overall, it is the young stages of large gaps which contribute most to the richness of the community as a whole.

a similar picture from grasslands and forests

The pattern of colonization of gaps in Paine and Levin's mussel beds is repeated in almost every detail in the colonization of gaps in grassland caused by burrowing animals or patches killed by urine. Initially, leaves lean into the gap from plants outside it. Then colonization begins by vegetative spread from the edges, and a very small gap may close up quickly. In larger gaps, new colonists may enter as dispersed seed, or germinate from the seed bank in the soil. Over two or three years the vegetation begins to acquire the character that it had before the gap was formed. When a gap is formed in a forest by the death of a tree, the first response is the more rapid growth of the surrounding trees as they start to extend their branches into the sunlit space that has been created. Beneath the canopy in this space, colonization may largely take place by the rapid growth of saplings that were already present but suppressed. These are released from suppression and grow up rapidly into the gap. There is something of a race to fill the gap, and if the gap is small the race will usually be won by the lateral growth of the trees at its edge. The larger the gap the greater is the opportunity for new saplings to enter it. A fire creating a very large gap may have killed saplings as well as the main canopy trees. The course of succession is then determined by new colonization from seed dispersed into the gap or seedlings recruited from the seed bank.

19.3.6 Disturbances may be transient or a persistent change of conditions

Before we attempt to generalize about the effects of disturbance on the composition and structure of communities, it is important to distinguish between two main categories of disturbance. Most of the examples referred to in this chapter may be called *transient* disturbance. They are repeated interruptions in space and/or time that can be seen against a background of some kind of continuing 'normality'. But there is another class of disturbance that is represented by a more or less sudden change of conditions in which the change is then maintained. The overturning of rocks by storms and the death of trees in a forest fall into the first category—transient disturbance. They produce transient effects. The disturbances may happen so often that the community is held 'steady' in the 'transient' state, or so rarely that the effects of the last disturbance have disappeared by the time the next occurs.

Changes to a new, persistent condition occur when a new species is introduced to a community. This may produce a dramatic revolution in community structure as the species becomes established and later settles down to some kind of equilibrial condition, becoming part of a new 'normality'. Chestnut blight (*Endothia parasitica*) has done this in North America; and both the introduction of prickly pear to Australia and the subsequent introduction of the moth *Cactoblastis* to control it produced violent changes in communities which then settled down and absorbed the new arrivals into a new normality (Figure 19.15).

19.4 Non-equilibrium models of community diversity

The Lotka–Volterra model of interspecific competition predicts that in a uniform, constant environment, all but one of several competing species will be driven to extinction. Only when each species has less competitive effect on the other species than it has on itself is coexistence possible (Chapter 7). This will be the case whenever sufficient niche differentiation exists between the species. Our consideration of resource partitioning (Chapter 18) looked at the ways guilds of species sometimes differ in their niche requirements, permitting coexistence and enhancing diversity.

The view that in the absence of resource partitioning, communities will tend towards monocultures can, however, be challenged. No doubt, the general conclusion of the competitive exclusion principle will be valid if all the necessary assumptions are met: (i) competition occurs for precisely the same resource at the same time, (ii) in a stable, uniform environment, (iii) until equilibrium is reached. In fact though, since the physical environment is never constant in nature, the sizes of populations of competitors and the nature of the competitive interactions will also be subject to variation, such that in many natural communities competitive equilibrium may be the exception rather than the rule. This is illustrated for several hypothetical interactions in Figure 19.16. If the environment changes only slowly relative to the time competitive exclusion would take (several generations), then species 1 will exclude species 2 if their interaction occurs while the environment favours the former (Figure 19.16b). Conversely, if the species interact when the environment has changed to favour species 2, the outcome will be reversed (Figure 19.16b, case 2). In contrast, if the time-scale for environmental change is approximately equal to the period required for competitive exclusion, it is possible to envisage two or more species alternately gaining the ascendancy, with none being driven to extinction (Figure 19.16c).

The idea was first invoked by Hutchinson (1941, 1961) to account for the high species richness of phytoplankton communities in lakes and oceans (section 7.6.3). Such communities are often extremely diverse—more so than can reasonably be attributed to their limited opportunities for resource partitioning or to the influence of predators. Physical conditions such as temperature and light intensity, and chemical factors such as nutrient concentrations, are known to vary from hour to hour and from day to day. It seems reasonable to suppose that the high species richness of phytoplankton is partly due to repeated interruptions of the process of competitive exclusion. A similar phenomenon may occur among higher plants in communities whose conditions vary from year to year. For example, grasslands in the Rhine valley are subject to various intensities of flooding and silt deposition. This is reflected in changes in the relative importance of herbaceous species (Figure 19.17).

the speed of competitive exclusion relative to the speed of environmental change

Figure 19.15. Control of prickly pear in Australia by introduction of the moth *Cactoblastis cactorum*. Above: An abandoned selection at Chinchilla, Queensland, in May 1928. Below: The same view in October 1929, showing destruction of the cactus by *Cactoblastis*. (From Dodd 1940.)

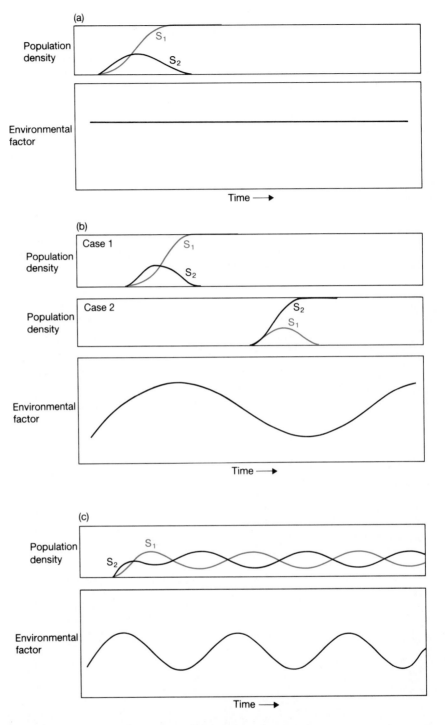

Figure 19.16. Idealized outcome of competition between two species. (a) When the environment is constant and favours species 1 over species 2. (b) When the environment remains favourable for species 1 (case 1) or for species 2 (case 2) for much longer than a single generation, competitive exclusion of one or other species occurs, depending on when the interaction commences. (c) When the environment remains favourable for no more than a few generations of either species 1 or 2 (specifically where the time that competitive exclusion would take is approximately equal to the time taken for a significant change in the environment), coexistence results.

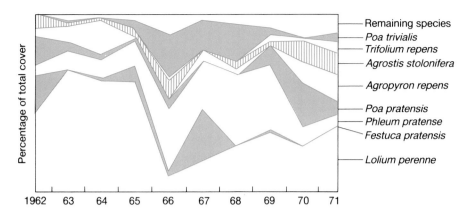

Figure 19.17. Changes in the species composition of a sown pasture in West Germany subject to varying intensities of flooding and silt deposition (after Muller & Foerster, 1974).

equilibrium and non-equilibrium theories

A basic distinction can thus be made between *equilibrium* and *non-equilibrium* theories. An equilibrium theory helps us to focus attention on the properties of a system at an equilibrium point—time and variation are not the central concern. A non-equilibrium theory, on the other hand, is concerned with the transient behaviour of a system away from an equilibrium point, and it specifically focuses our attention on time and variation. Of course, it would be naïve to think that any real community has a precisely definable equilibrium point, and it is wrong to ascribe this view to those who are associated with equilibrium theories. The truth is that investigators who focus attention on equilibrium points have in mind that these are merely states towards which systems tend to be attracted, but about which there may be greater or lesser fluctuation. In one sense, therefore, the contrast between equilibrium and non-equilibrium theories is a matter of degree. However, this difference of focus is instructive in unravelling the important role of disturbance in communities.

closed and open systems

A second distinction which must be made is between *closed* and *open* communities. A closed community is defined as a single unit. If it contains, for example, two species competing for the same resource, the outcome will be as predicted by the Lotka–Volterra model and only one species will persist. Extinction of the other occurs once and for all. Very often, however, real communities are better described as open systems, consisting of a mosaic of patches within which interactions proceed but with a vital extra feature— migration occurs from patch to patch. A single patch without migration is, by definition, a closed system, and any extinction would be final. However, extinction within a patch in an open system is not necessarily the end of the story because of the possibility of re-invasion from other patches. These distinctions provide a context within which to review the natural communities and processes highlighted in this chapter.

19.4.1 Closed non-equilibrium systems

The important variable in a non-equilibrium theory of competition is the rate

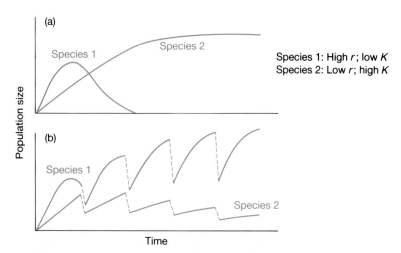

Figure 19.18. Effect of non-equilibrium conditions on the outcome of competition: (a) simulation in which competitive exclusion is reached; (b) competitive exclusion is prevented by periodic density-independent reductions of the population (after Huston, 1979).

the role of competitive exclusion must be compared with the death rate . . .

at which competitive exclusion occurs. Caswell (1978) has shown that the incorporation of predation into a simple competition model can lower the rate of competitive exclusion and increase the time to extinction to a point where competitors appear to coexist indefinitely. Periodic unselective reductions in population size (which could be caused either by predators or by physical disturbance) also appear to affect the outcome of competition. Figure 19.18a shows the results of a Lotka–Volterra simulation in which competitive equilibrium is quickly reached and one species goes extinct. Figure 19.18b is the result of a simulation in which competitive exclusion is prevented by a periodic density-independent reduction (by half) in both populations. Now

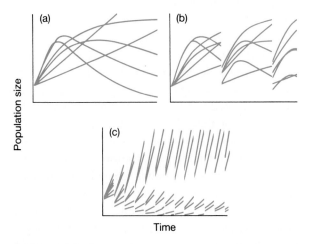

Figure 19.19. Effect of frequency of population reduction on the maintenance of diversity in a hypothetical community. (a) No reductions; diversity is reduced as the system approaches competitive equilibrium. (b) Periodic reductions; high diversity is maintained for longer than in (a). (c) High frequency of reductions; diversity is reduced as populations with low *r* are unable to recover between reductions. (After Huston, 1979.)

the outcome of competition is quite different. The species coexist for much longer, though eventually species 2 goes extinct because its low rate of increase (*r*) does not permit its population to recover sufficiently between disturbances. Interestingly, species 2 is the one to triumph under equilibrium

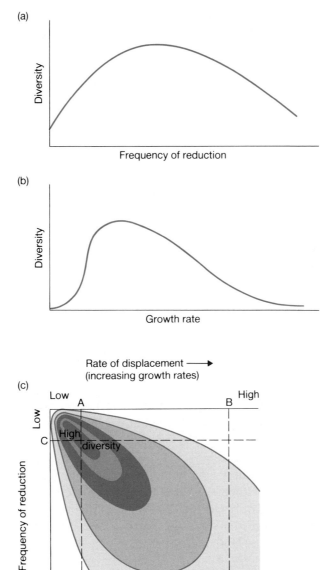

Figure 19.20. Relationships derived from Huston's model communities. (a) Predicted relationship between diversity and the frequency of population reduction.
(b) Predicted relationship between diversity and population growth rates in non-equilibrium systems with a low to intermediate frequency of population reduction.
(c) Generalized contour map of the relationship between the rate of competitive displacement (equivalent to population growth rates), the frequency of population reduction, and species diversity. Diversity is shown on the axis perpendicular to the page and is represented by contour lines. (After Huston, 1979.)

conditions because of its higher carrying capacity (K) (Huston, 1979; see also Shorrocks & Begon, 1975).

Huston (1979) has also modelled communities, consisting of six species, in which disturbances either never occur, occur with intermediate frequency, or are very common (Figure 19.19). Competitive exclusions takes place relatively rapidly in the absence of disturbances (one species has already gone extinct by the end of the run). At intermediate frequencies of disturbance, diversity is higher for much longer, because the rate of competitive exclusion is slowed dramatically. However, at a high frequency of disturbance, diversity is reduced by extinctions of the species unable to recover sufficiently between disturbances (Figure 19.19c). This provides a gratifying link with Connell's intermediate disturbance hypothesis (section 19.2.2). Note, however, that the process of colonization (presumed to be important in rocky shore communities) is not included, by definition, in a closed system such as that represented by Huston's model.

... and depends on the rate of population growth

Not unexpectedly, the rate of competitive exclusion is significantly higher when rates of population increase are themselves high (Figure 19.20b). In other words, coexistence may in theory be extended indefinitely at appropriate rates of disturbance, but only when population growth rates are not too high. An increase in growth rates will tend to negate the effect of fluctuating conditions. Does this prediction gain any support from real communities? In fact, it offers an explanation for the reduction in species richness noted in certain productive environments (Chapter 22). On this view, highly productive environments contain populations which approach and reach the stage of competitive exclusion much more reliably than in unproductive environments. Thus unproductive environments may be expected to be richer in species.

The relationships shown in Figures 19.20a and 19.20b are combined in Figure 19.20c. The dashed line transects on Figure 19.20c demonstrate the predicted changes in diversity when one parameter is held constant and the other is varied.

Huston's simple model provides some reassuring links with the real world. However, it fails to correspond closely to many communities in one important respect: it does not describe the community as a patchwork with migration between the patches. For this, we need to resort to an open, non-equilibrium model.

19.4.2 Open, non-equilibrium systems

A model of an open community should consist of a set of cells within which species interact, possibly going extinct, and between which migration can occur. This is a special case of a more general family of models, in which the system is composed of a large number of similar components, each of which interacts with (has high 'connectedness' with) some of its fellows. In general, even for a small number of very simple components, high levels of connectedness may lead to astronomically long delays in reaching equilibrium.

an analogy with light bulbs

Consider first a simple physical analogue for a community. Ashby (1960) describes a system of 100 light bulbs, each existing in one of two states (on or off). For each bulb, in each second of time, the transition on → off occurs with a probability of 0.5. For each bulb, in each second, the reverse transition (off → on) also occurs with a probability of 0.5, but only if at least one bulb in the 'on' state is connected to the bulb in question. The stable equilibrium for this system is the state where all bulbs are off. If the system is started with all bulbs on, how long will it take to reach equilibrium? If there are no connections amongst the bulbs (i.e. each is a closed system) the expected time to equilibrium is 2^1 s = 2 seconds. At the opposite extreme, when every bulb is connected to every other, the time to equilibrium is of the order of 2^{100} s = 10^{22} years. This is a long time. The age of the universe is only 10^{110} years. To all intents and purposes, this entirely open system would never reach equilibrium at all.

Caswell (1978) has developed an equivalent model, which explores the influence of one species of predator on competition between two prey species in an open community consisting of 50 cells (patches) with colonization of cells possible by any of the species. Exploiter-mediated coexistence was readily demonstrated in this system. All three species persisted for 1000 generations, until the run was terminated, and the same result was obtained in ten replicates. In the absence of predation, however, the inferior of the two competitor species went extinct after an average of only 64 generations (in ten runs this ranged from 53 to 80).

exploiter-mediated coexistence in patches

In Caswell's open, non-equilibrium model, the rate of competitive exclusion can be slowed down to such an extent that indefinite coexistence of the species results. The structure of the model is biologically quite realistic; for example, it seems to correspond closely to Paine's starfish-dominated community (section 19.2.2). In the model, and in reality, the predator opens up otherwise closed cells for colonization by the inferior competitor. In real communities it is not only predators which can have this effect. As we have seen, gaps are generated in many kinds of community by physical disturbances. As Caswell has pointed out, when a physical disturbance acts as if it were a 'predator', the coexistence effect should be even more dramatic because there is no chance of the 'predator' becoming extinct.

19.5 The relevance of disturbance theory and experiment to ecological management

Good theory not only helps to explain, it also predicts and can thus be used to control events. Disturbance (non-equilibrium) theory suggests ways in which communities might be manipulated to desired ends—such as nature conservation, agriculture, forestry and wildlife management. In particular, it suggests that if we are keen to preserve natural diversity, we should not prevent disturbances. Indeed, disturbance may be the most powerful way in which we can generate diversity. Just as the recurrent disturbances of the ice ages, and the break-up of continents and formation of islands, appear to have been powerful forces in the origin of species diversity, so the creation of gaps, new successions, and patchwork mosaics within communities may be the most powerful way in which we might generate and maintain ecological diversity.

The disturbances that are most effective in allowing a diversity of species to invade an area are those that hurt community dominants. This suggests that patchwork deforestation should increase species diversity—though there will be an optimal size for the patches.

Agriculture involves repeated disturbance. It is not surprising that it brings with it such species diversity that it keeps chemical manufacturers in lucrative business, developing the variety of herbicides needed to control the diversity of weeds. Practices that reduce the frequency of disturbance, such as cropping without cultivation, result in much reduced problems in the long term. Disturbances are less prominent in forestry practice, and forestry usually involves establishing a potential community dominant. Commercial forestry short-circuits the normal progress of succession, taking the community direct to the dominant. The communities that are supported beneath such forests tend to be poor in species, especially when the monotony of a single-species canopy is exaggerated by having a uniform age-structure. A mixed-age population generates some of its own diversity through the formation of gaps and through regeneration cycles.

We can begin to see dimly that we might find ways of measuring the force of different types of disturbance—is a forest fire a stronger perturbation than the introduction of deer? We might then begin to compare the response of different types of community, and measure their resistance to change or compare their rates of recovery. In the meantime, the theoretical models suggest persuasively that we may be on the edge of an understanding of the role of the accidents, uncertainties and hazards that dominate the life of most communities.

Islands, Areas and Colonization

20.1 Introduction: species–area relationships

species–area
relationships on
oceanic islands . . .

It has long been recognized that islands contain fewer species than apparently comparable pieces of mainland. Not only that, it is also well established that the number of species on islands decreases as island area decreases. Such *species–area* relationships can be seen in Figure 20.1 for a variety of groups of organisms on a variety of islands. They are usually displayed with both species number and area plotted on log scales, though plotting species number itself against log area sometimes gives the best straight line (fully discussed by Williamson, 1981). In all cases, however, a plot of species number against area is curved, with the number of species increasing more slowly at larger areas (e.g. Figure 20.1c).

habitat islands and areas
of mainland

'Islands' need not mean islands of land in a sea of water. Lakes are islands in a 'sea' of land; mountain tops are high-altitude islands in a low-altitude ocean; gaps in a forest canopy where a tree has fallen (section 19.3) are islands in a sea of trees; and there can be islands of particular geological types, soil types or vegetation types surrounded by dissimilar types of rock, soil or vegetation. Species–area relationships can be equally apparent for these types of islands (Figure 20.2).

In fact, species–area relationships are not restricted to islands at all. They can also be seen when the numbers of species occupying different-sized arbitrary areas of the same geographical region are compared (Figure 20.3). The question therefore arises: is the impoverishment of species on islands more than would be expected in comparably small areas of mainland? In other words, does the characteristic isolation of islands contribute to their impoverishment of species? And if the answer to these questions is 'yes', another question follows: can we understand how the degree of impoverishment is related to the degree of isolation or any other quality of the island?

'island effects' and
community structure

These are important questions for an understanding of community structure. There are many oceanic islands, many lakes, many mountain tops, many woodlands surrounded by fields, and many isolated trees. Even an individual animal or a leaf is an island from the point of view of its parasites (Chapter 12). In short, there can be few natural communities lacking at least some element of 'islandness'. Hence, we cannot hope to understand community structure without an understanding of *island biogeography*. In addition, as we shall see, island biogeography is of considerable potential importance for an enlightened approach to nature conservation.

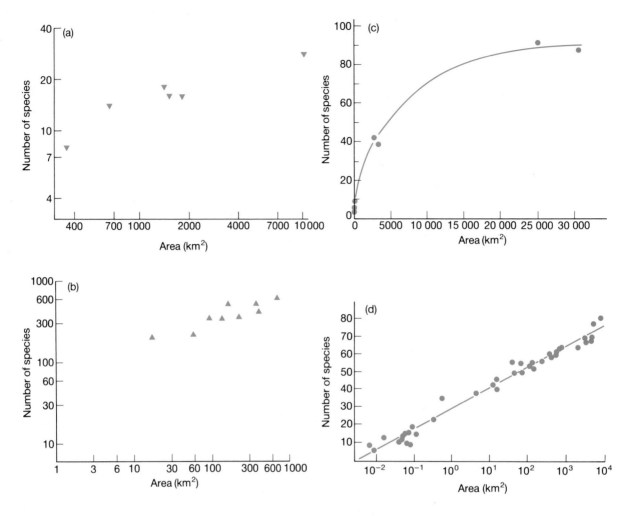

Figure 20.1. Species–area relationships for organisms on land islands. (a) Native land and freshwater birds on Hawaiian islands (Hawaiian Audobon Society, 1975). (b) Vascular plants on the Azores (Eriksson *et al.*, 1974). (c) Amphibians and reptiles on West Indian islands—unlike the others, plotted simply as species number against area (MacArthur & Wilson, 1976). (d) Birds on the Solomon Islands (Diamond & Mayr, 1976).

In this chapter we examine three approaches to the understanding of island biogeography. These have sometimes been treated as alternative explanations for the species richness of island communities, but they are really complementary. The first concentrates on the suitability of islands as habitats for various species ('habitat diversity'—sections 20.2.1 and 20.2.2). The second concentrates on the balance between the rate at which islands are colonized by species new to the island and the rate at which resident species go extinct on the island ('the equilibrium theory'—sections 20.2.3 and 20.2.4). The third takes a more evolutionary approach, and considers the balance between colonization from outside the island and evolution within it (section 20.4).

These three approaches will be followed for islands of various types and, where appropriate, for arbitrarily defined areas of mainland that are not islands. The approaches will also be applied, in a slightly different way, to one

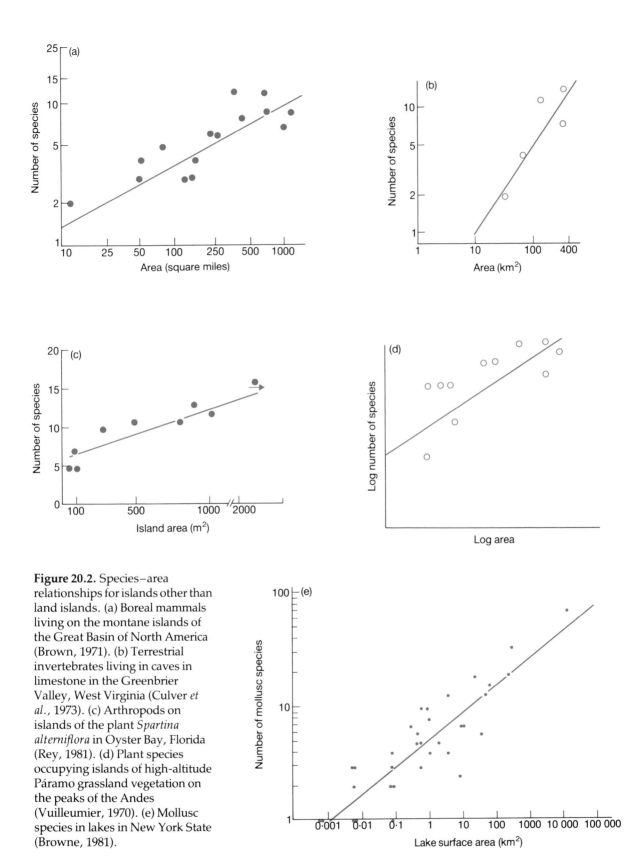

Figure 20.2. Species–area relationships for islands other than land islands. (a) Boreal mammals living on the montane islands of the Great Basin of North America (Brown, 1971). (b) Terrestrial invertebrates living in caves in limestone in the Greenbrier Valley, West Virginia (Culver *et al.*, 1973). (c) Arthropods on islands of the plant *Spartina alterniflora* in Oyster Bay, Florida (Rey, 1981). (d) Plant species occupying islands of high-altitude Páramo grassland vegetation on the peaks of the Andes (Vuilleumier, 1970). (e) Mollusc species in lakes in New York State (Browne, 1981).

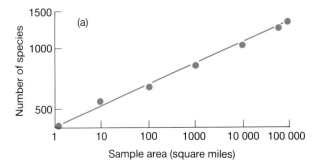

Figure 20.3. The species–area curve for the number of flowering plants found in different-sized sample areas of England (Williams, 1964, after Gorman, 1979).

particular group of organisms, namely communities of insects feeding on the living tissues of higher plants. These comprise an estimated 35% of all animal species, and they have received careful and considerable attention from ecologists (fully reviewed by Strong *et al.*, 1984). The straightforward 'island' approach to phytophagous insect communities views patches of one plant species, or even individual plants, as islands in a sea of other species. As an alternative, however, Figure 20.4 shows a number of examples that are superficially like those in Figures 20.1–20.2, but in this case the 'area' of a plant refers to how widespread it is, i.e. in how many grid squares on a map it is found. 'Islands of differing size' are therefore represented by 'plant species of differing range'. Yet, as we shall see, the searches for an understanding of the structures of the communities of animals on such plants, and on real islands, run remarkably parallel courses.

analogous relationships amongst phytophagous insects

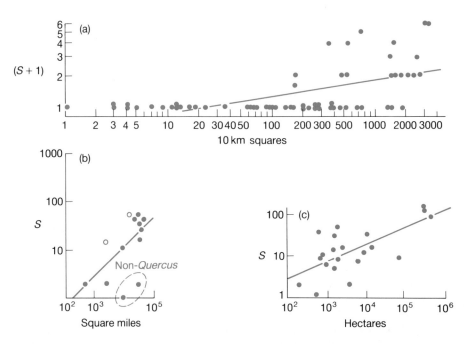

Figure 20.4. Species–area relationships for phytophagous insects on their host plants—the areas are actually the ranges of different plant species. S = number of species. (a) Agromyzid flies on British Umbelliferae (Lawton & Price, 1979). (b) Cynipid gall wasps on Californian oaks—all oaks are species of *Quercus* except those indicated (Cornell & Washburn, 1979). (c) Insect pests on cacao—each point is for a different country (Strong, 1974).

20.2 Ecological theories of island communities

20.2.1 Habitat diversity

Probably the most obvious reason why larger areas should contain more species is that larger areas typically encompass more different habitat types. In the particular context of island biogeography, the major proponent of this view was Lack (1969b, 1976, discussed by Williamson, 1981). Lack argued that the number of species found on an island simply reflects the 'type' of island, within which he included its climate and the habitats it provides. He believed essentially that large islands contain more species because they contain more habitats. This must obviously be true to some degree, although the conjecture has received few critical tests (these are discussed in section 20.3.1).

larger islands contain more habitats and support more species

Lack was himself concerned only with birds, and this is important since he was insistent that the failure of species to establish populations on a particular island comes not from a failure to disperse to the island but from a failure to find the right habitat. The major problem with Lack's view, therefore, is that it takes no account of the rather low probability for many organisms of actually dispersing to all suitable islands. Even amongst birds, it takes no account of the limitations that undoubtedly exist in the dispersal abilities of many species. In addition, this theory is a purely ecological rather than an evolutionary one. It does not concern itself with the extent to which the community on an island may reflect evolution that has occurred *within* the island itself (cf. section 20.4)

20.2.2 Habitat diversity and phytophagous insects

There are two ways in which the habitat diversity argument can be applied to species–area relationships for insects on plants. The first, related directly to data sets like those in Figure 20.4, proposes simply that widely distributed plants themselves live in a wide variety of habitats. Thus they in turn offer a wide variety of habitats to insects, since the habitat of the insect includes not only the presence of the plant itself, but also climatic conditions, the presence of other plants and so on.

two ways in which plants can vary in their habitat diversity

The second way of applying the argument concentrates on the size, the structure, the variety of parts, or the 'architecture' of different plant species. It treats plant species as areas differing in habitat diversity, rather than treating them as entities differing in the area that they occupy: 'complex' plant species might be expected to support more insect species than 'simple' ones.

20.2.3 MacArthur and Wilson's 'equilibrium' theory

The essence of MacArthur and Wilson's (1967) 'equilibrium theory of island biogeography' is very simple. It is that the number of species on an island is determined by a balance between immigration and extinction, and that this balance is dynamic, with species continually going extinct and being replaced (through immigration) by the same or by different species.

The theory is depicted in Figure 20.5. Taking immigration first, imagine

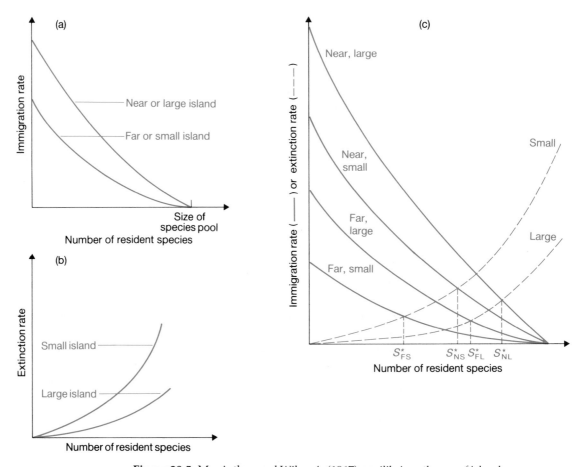

Figure 20.5. MacArthur and Wilson's (1967) equilibrium theory of island biogeography. (a) The rate of species immigration on to an island, plotted against the number of resident species on the island, for large and small islands and for close and distant islands. (b) The rate of species extinction on an island, plotted against the number of resident species on the island, for large and small islands. (c) The balance between immigration and extinction on small and large and on close and distant islands. In each case, S^* is the equilibrium species richness. For further details see text.

an island that as yet contains no species whatsoever. The rate of immigration of *species* will be high, because any colonizing individual represents a species new to that island. However, as the number of resident species rises, the rate of immigration of new, unrepresented species diminishes. The immigration rate reaches zero when all species from the 'source pool' (i.e. from the mainland or from other nearby islands) are present on the island in question (Figure 20.5a).

MacArthur and Wilson's immigration curves
The immigration graph is drawn as a curve, because immigration rate is likely to be particularly high when there are low numbers of residents and many of the species with the greatest powers of dispersal are absent. In fact, the curve should really be a blur rather than a single line, since the precise curve will depend on the exact sequence in which species arrive, and this will vary by chance. In this sense, the immigration curve can be thought of as the 'most probable' curve.

The exact immigration curve will depend on the degree of remoteness of the island from its pool of potential colonizers (Fig. 20.5a). The curve will always reach zero at the same point (when all members of the pool are resident); but it will generally have higher values on islands close to the source of immigration than on more remote islands, since colonizers have a greater chance of reaching an island, the closer it is to the source. It is also likely (though not specified in the original formulation of MacArthur and Wilson's theory) that immigration rates will generally be higher on a large island than on a small island, since the larger island represents a larger 'target' for the colonizers (Figure 20.5a).

MacArthur and Wilson's extinction curves

The rate of species extinction on an island (Figure 20.5b) is bound to be zero when there are no species there, and it will generally be low when there are few species. However, as the number of resident species rises, the extinction rate is assumed by the theory to increase, probably at a more than proportionate rate. This is thought to occur because with more species, competitive exclusion becomes more likely, and the population size of each species is on average smaller, making it more vulnerable to chance extinction. Similar reasoning suggests that extinction rates should be higher on small than on large islands—population sizes will typically be smaller on small islands (Figure 20.5b). As with immigration, the extinction curves are best seen as 'most probable' curves.

the balance between immigration and extinction, and the predictions of the equilibrium theory....

In order to see the net effect of immigration and extinction, their two curves can be superimposed (Figure 20.5c). The number of species where the curves cross (S^*) is a dynamic equilibrium, and should be the characteristic species richness for the island in question. Below S^*, richness increases (immigration rate exceeds extinction rate); above S^*, richness decreases (extinction exceeds immigration). The theory, then, makes a number of predictions.

(i) The number of species on an island should eventually become roughly constant through time.

(ii) This should be a result not of stasis but of a continual *turnover* of species, with some becoming extinct and others immigrating.

(iii) Large islands should support more species than small islands.

(iv) Species number should decline with the increasing remoteness of an island.

which are not all characteristic of this theory alone

It is important to realize that several of these predictions could also be made without any reference to the equilibrium theory. An approximate constancy of species number would be expected if richness were determined simply by island type (section 20.2.1). Similarly, a higher richness on larger islands would be expected as a consequence of larger islands having more habitat types. One test of the equilibrium theory, therefore, would be whether richness increases with area at a rate greater than could be accounted for by increases in habitat diversity alone.

The effect of island remoteness can be considered quite separately from the equilibrium theory. Merely recognizing that many species are limited in their dispersal ability, and have yet to colonize all islands, leads to the prediction that more remote islands are less likely to be saturated with potential colonizers. However, the final prediction arising from the equilibrium theory—constancy as a result of turnover—is truly characteristic of the equilibrium theory.

Comparing the equilibrium theory and the habitat diversity theory, the equilibrium theory deals mostly with *numbers* of species and lays relatively little stress on the identity of species' requirements. It does, though, recognize that dispersal abilities are generally limited and vary from species to species. Like the habitat diversity theory, it pays no attention to evolution.

20.2.4 The equilibrium theory and phytophagous insects

Applying the equilibrium theory to insects on plants was suggested first by Janzen (1968). Relatively widespread plants can be seen as relatively 'large' islands in a sea of other vegetation. Indeed, Southwood (1961) had previously proposed that more widespread plants represented a larger target for potential colonizing insects. In addition, a plant species may be seen as 'remote' from other species if it is morphologically, biochemically or otherwise biologically unusual. Thus, the equilibrium theory predicts that insect species richness will be higher for plants with large ranges, lower for plants that are geographically isolated or rare, and lower for plants that are morphologically or biochemically 'isolated'. However, the problems of disentangling the predictions of the equilibrium theory from those of other theories are the same here as they are for islands generally. In particular, widespread species are not only 'large islands', but also occupy a wide diversity of habitats.

20.3 Evidence for the ecological theories

20.3.1 Habitat diversity alone—or a separate effect of area?

The most fundamental question in island biogeography is whether there is an 'island effect' as such, or whether islands simply support few species because they are small areas containing few habitats. Some studies have attempted to partition species–area variation on islands into that which can be entirely accounted for in terms of environmental heterogeneity, and that which remains and must be accounted for by island area in its own right. At one extreme, the diversity of fish species in northern Wisconsin lakes is significantly correlated both with lake area and with vegetation diversity (Figure 20.6a). However, lake area and vegetation diversity are themselves closely correlated, and in this case it seems that area itself does not exert a strong direct effect on species richness (Tonn & Magnuson, 1982). At the opposite extreme, Abbott (1978) found a significant relationship between bird species richness and the area of islands off the coast of Western Australia, but he found no relationship between species richness and habitat diversity (Figure 20.6b). In fact, there was no relationship between island area and habitat diversity. In this case, area *per se* appears to play a very major role. Watson (1964, described in Williamson, 1981) examined the relationship between species number, island area and habitat diversity for birds of the Aegean islands. He concluded that habitat diversity was more important than area. Finally, Johnson (1975) related the number of birds on coniferous islands in

partitioning variation between habitat diversity and area itself

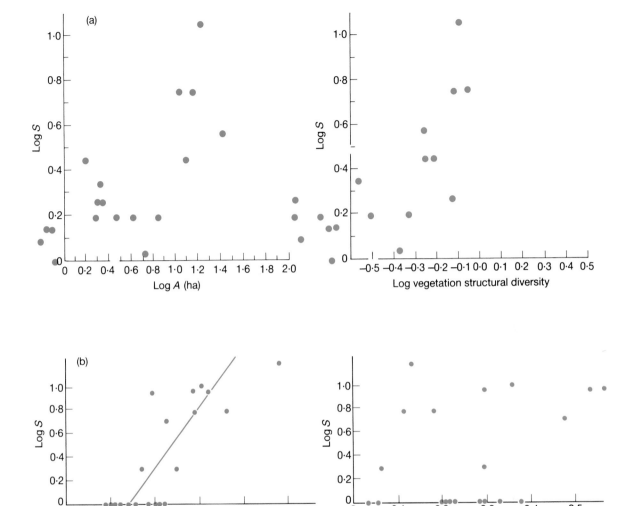

Figure 20.6. (a) The relationship between species richness (*S*) of fish in lakes in northern Wisconsin, U.S.A., and both lake area and vegetation structural diversity (all on log scales) (after Tonn & Magnusson, 1982). (b) The relationship between species richness (*S*) of birds on islands off the coast of western Australia and both island area and structural habitat diversity (all on log scales) (after Abbott, 1978).

the Great Basin (western U.S.A.) to a variety of habitat variables, and was able to show a relationship between bird species diversity and environmental heterogeneity.

An attempt to separate experimentally the effects of habitat diversity and area was made by Simberloff (1976) on some small mangrove islands in the Bay of Florida. These consist of pure stands of the mangrove species *Rhizophora mangle*, which support arboreal arthropod communities of insects, spiders, scorpions and isopods. After a preliminary faunal survey, some islands were reduced in size by means of a power saw and brute force! Habitat diversity was not affected, but arthropod species richness on three

experimental reductions in the size of mangrove islands

islands nonetheless diminished over a period of two years (Figure 20.7). A control island, the size of which was unchanged, showed a slight *increase* in richness over the same period, presumably as a result of random events.

Islands that used to be part of a land-bridge to the mainland, sections of which have since become submerged in the ocean, provide data consistent with the equilibrium theory. If an equilibrium number of species is determined partly by a relationship between extinction rate and island area, these

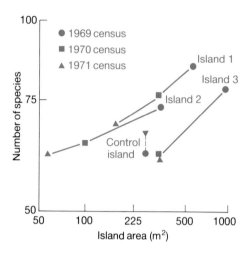

Figure 20.7. The effect on the number of arthropod species of artificially reducing the size of mangrove islands. Islands 1 and 2 were reduced in size after both the 1969 and 1970 censuses. Island 3 was reduced only after the 1969 census. The control island was not reduced, and the change in its species richness was attributable to random fluctuations. (After Simberloff, 1976.)

the 'relaxation' of richness on land-bridge islands

new islands would be expected to lose species until they reached a new equilibrium appropriate to their size. The process is usually referred to as 'relaxation' and several instances have been reported. For example, Wilcox (1978) described a relationship between the length of time for which islands in the Gulf of California have been separated from the mainland and the number of lizard species they support (Figure 20.8). Having made corrections for island area and latitude, Wilcox was able to show a reduction in species richness from 50–75 species down to 25 or so during a period of 4000 years. Of course, such data must be interpreted with caution. Since such long time periods are involved, we need to be confident that major climatic changes have not played a role. Another concern is the influence man has exerted, by destroying habitat or by his introduction of competitors and predators. Nevertheless, these studies lend weight to the 'equilibrium' interpretation of island diversity (i.e. they support a separate effect of area).

species–area relationships for islands are, on average, steeper than those for mainland areas

Another way of trying to distinguish a separate effect of island area is to compare species–area graphs for islands with those for arbitrarily defined areas of mainland. The species–area relationships in the latter should be due almost entirely to habitat diversity alone. All species will be well able to 'disperse' between such areas, and the continual flow of individuals across the arbitrary boundaries will therefore mask local extinctions (i.e. what would

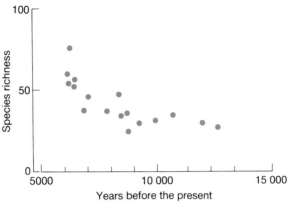

Figure 20.8. Species richness of lizards on former land-bridge islands in the Gulf of California, plotted against the length of time for which they have been isolated (after Wilcox, 1978).

be an extinction on an island is soon reversed by the exchange of individuals between local areas). An arbitrarily defined area of mainland should thus contain more species than an otherwise equivalent island, and this is usually interpreted as meaning that the slopes of species–area graphs for islands should be steeper than those for mainland areas (since the effect of island isolation should be most marked on small islands, where extinctions are most likely). The difference between the two types of graph would then be attributable to the island effect in its own right. Table 20.1 shows that despite considerable variation, the island graphs do typically have steeper slopes.

Note that a reduced number of species per unit area on islands should also lead to a lower value for the intercept on the *S* axis of the species–area graph. Figure 20.9 illustrates both an increased slope and a reduced value for the intercept for the species–area graph for pomerine ant species on isolated Pacific islands, compared to the graph for progressively smaller areas of the very large island of New Guinea.

Overall, it is clear that smaller areas typically contain a lower diversity of habitats and thus support fewer species. But it is also apparent that there is often a recognizable island-effect, reducing still further the species richness of communities that are isolated from other, similar communities.

Figure 20.9. The species–area graph for pomerine ants on various Moluccan and Melanesian islands compared to a graph for different-sized sample areas on the very large island of New Guinea (from Wilson, 1961).

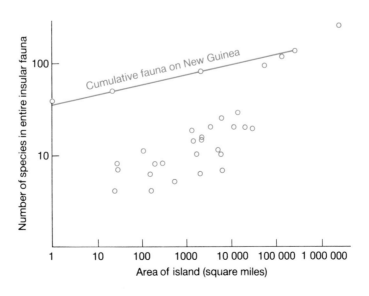

Table 20.1. Values of the slope z of species–area curves ($\log S = \log C + z\log A$ where S is species richness A is area and C is a constant giving the number of species when A has a value of 1) for (a) arbitrary areas of mainland, (b) oceanic islands, and (c) habitat islands (data from Preston, 1962; May, 1975b; Gorman, 1979; and Browne, 1981).

Taxonomic group	Location	z
(a) ARBITRARY AREAS OF MAINLAND		
Flowering plants	England	0.10
Land plants	Britain	0.16
Birds	Mediterranean	0.13
Birds	Neotropics	0.16
Birds	Nearctic	0.12
Savanna vegetation	Brazil	0.14
(b) OCEANIC ISLANDS		
Birds	New Britain Islands	0.18
Birds	New Guinea Islands	0.22
Birds	West Indies	0.24
Birds	East Indies	0.28
Birds	East Central Pacific	0.30
Ants	Melanesia	0.30
Beetles	West Indies	0.34
Land plants	Galapagos	0.31
Land plants	Californian islands	0.37
(c) HABITAT ISLANDS		
Zooplankton (lakes)	New York State	0.17
Snails (lakes)	,, ,, ,,	0.23
Fish (lakes)	,, ,, ,,	0.24
Birds (Paramo vegetation)	Andes	0.29
Mammals (mountains)	Great Basin, U.S.A.	0.43
Terrestrial invertebrates (caves)	West Virginia	0.72

20.3.2 Remoteness

It follows from the above argument that the island effect and the species impoverishment of an island should be greater for more remote islands. (Indeed, the comparison of islands with mainland areas is only an extreme example of a comparison of islands varying in remoteness, since local mainland areas can be thought of as having minimal remoteness.) Remoteness, however, can mean two things. First, it can simply refer to the degree of physical isolation. Alternatively, a single island can also itself vary in remoteness, depending on the type of organism being considered: the same island may be remote from the point of view of land mammals but not from the point of view of birds.

The effects of remoteness can be demonstrated either by plotting species richness against remoteness itself, or by comparing the species–area graphs of groups of islands or groups of organisms that differ in their remoteness. In either case, there can be considerable difficulty in extricating the effects of remoteness from all the other characteristics by which two islands may differ (see discussion in Williamson, 1981). Nevertheless, the direct effect of remoteness can be seen in Figure 20.10 for non-marine, lowland birds on

bird-species richness on Pacific islands decreases with remoteness

tropical islands in the south-west Pacific. With increasing distance from the large source island of New Guinea, there is a decline in the number of species, expressed as a percentage of the number present on an island of similar area but close to New Guinea. Species richness decreases exponentially with distance, approximately halving every 2600 km. The species–area graphs in Figure 20.11a also show that remote islands of a given size possess fewer species than their counterparts close to a land mass. In addition, Figure 20.11b contrasts the species–area graphs of two classes of organisms in two regions: the relatively remote Azores (in the Atlantic, far to the west of Portugal) and the Channel Islands (close to the north coast of France). Whereas the Azores are indeed far more remote than the Channel Islands from the point of view of the birds, the two island groups are apparently equally remote for ferns, which are particularly good dispersers because of their light, wind-blown spores. Finally, it is noteworthy that the two steepest slopes in Table 20.1 relate to (a) mammals from mountains surrounded by desert in the Great Basin of North America, and (b) terrestrial invertebrates inhabiting limestone caves in West Virginia. In both cases, rates of migration are dramatically low, and the habitat islands are therefore extremely isolated from one another, irrespective of their degree of physical separation. Thus, on the basis of all these examples, the species impoverishment caused by the island effect does indeed appear to increase as the degree of isolation of the island increases.

the effects of remoteness on species–area relationships

A more direct effect on the species impoverishment of islands, especially remote islands, arises from the fact that many islands lack species that they could potentially support, simply because there has been insufficient time for the species to colonize the island. This phenomenon is illustrated in Figure

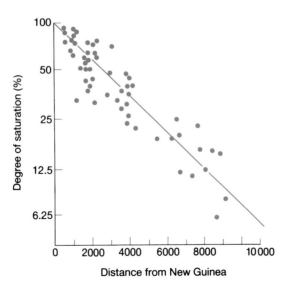

Figure 20.10. The number of resident, non-marine, lowland bird species on islands more than 500 km from the larger source island of New Guinea expressed as a proportion of the number of species on an island of equivalent area but close to New Guinea, and plotted as a function of island distance from New Guinea (from Diamond, 1972).

747 ISLANDS, AREAS AND COLONIZATION

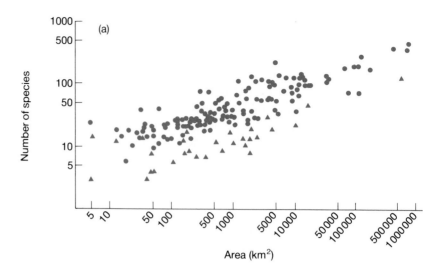

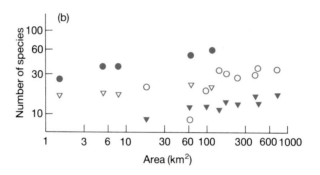

Figure 20.11. Remoteness increases the species impoverishment of islands. (a) A species–area plot for the land birds of individual islands in tropical and sub-tropical seas. ▲ = islands more than 300 km from the next largest land mass or the very remote Hawaiian and Galapagos archipelagos; ● islands less than 300 km from source . (From Williamson, 1981, after Slud, 1976.) (b) Species–area plots in the Azores and the Channel Islands for land and freshwater breeding birds (Azores= ▼, Channel Islands = ●) and for native ferns (Azores= ○, Channel Islands = ▽). The Azores are more remote for birds but not for ferns. (After Williamson, 1981.)

islands may lack species simply because there has been insufficient time for colonization

20.12 over a time-scale of only 2–3 years. In the summer of 1971, a bloom of toxic dinoflagellate algae killed 97% of all macroscopic benthic invertebrates in Old Tampa Bay, Florida. Recolonization commenced immediately. There was an initial rapid influx of opportunistic polychaetes, while molluscs and amphipods, with their poorer powers of dispersal, colonized somewhat more slowly. After one year, the faunal richness was similar to levels prior to defaunation, but additional species continued to arrive.

An example with a longer time-scale is provided by the island of Surtsey, which emerged in 1963 as a result of a volcanic eruption. The new island, 40 km south-west of Iceland, was reached by bacteria and fungi, some seabirds, a fly, and seeds of several beach plants within six months of the start of the

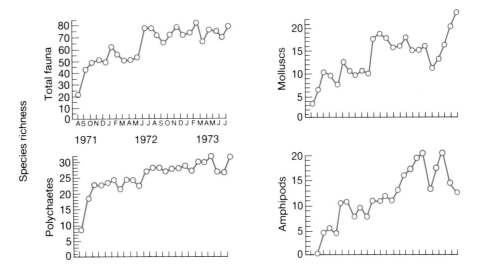

Figure 20.12. The pattern of increase in species richness of different elements of the benthic fauna in Old Tampa Bay after natural defaunation (after Simon & Dauer, 1977).

eruption. Its first established vascular plant was recorded in 1965, and the first moss colony in 1967. By 1973, 13 species of vascular plant and more than 66 mosses had become established (Figure 20.13). Colonization is continuing still.

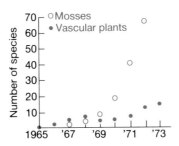

Figure 20.13. The number of species of mosses and vascular plants recorded on the new island of Surtsey from 1965 to 1973 (after Fridriksson, 1975).

The general importance of these examples is that the communities of many islands can be understood *neither* in terms of simple habitat suitability *nor* as a characteristic equilibrium richness. Rather, they stress that many island communities have not reached equilibrium and are certainly not fully 'saturated' with species.

20.3.3 Diversity, area and remoteness for phytophagous insects

We can now retrace our steps, and look at species–area graphs, habitat diversity and remoteness in phytophagous insect communities. In the first place, widespread plant species undoubtedly owe much of their rich insect fauna to the high diversity of habitats within which they grow. For example, the principal mechanism generating the species–area relationship in Figure 20.4a appears to be the fact that umbellifers growing in several types of habitat

widespread plants support many insects because they occupy a variety of habitats. . .

support more species of leaf-mining agromyzid flies than those confined to only one or two habitats (Fowler & Lawton, 1982). In fact, there is an important parallel between the habitat diversity of plant species and the habitat diversity of islands generally. Many of the insect species supported by widespread plants are confined to only a small part of their host plant's range. Thus a number of studies have shown that widespread and common plant species do not necessarily support more insect species at a particular locality even though they support more species overall (Claridge & Wilson, 1976, 1978; Futuyma & Gould, 1979; Karban & Ricklefs, 1983). This argues against a separate effect of area in addition to the simple effect of habitat diversity. But it also means that a *particular* community of phytophagous insects will not necessarily be richer just because it exists on a widespread plant. In a similar way, a particular community within an island is likely to be comparatively less impoverished than the community of the whole island. In short, the effect of island biogeography on community structure depends crucially on the scale of the community under study, relative to the size of the island of which it is part.

. . . but there is also evidence for a separate 'area effect'

Some evidence for the existence of a separate 'area effect' amongst phytophagous insects comes from the fact that a number of plants that have increased their abundance (i.e. their 'area') recently, apparently within a single habitat, have increased their species richness; while others that have decreased in abundance have lost species (Strong *et al.*, 1984). However, it is possible that *from the point of view of the insects* the change in abundance has also led to a change in habitat diversity.

Better evidence for an area effect comes from Lawton's (1984) study of phytophagous insect communities associated with bracken (*Pteridium aquilinum*), already touched on in section 18.2.1. Communities were compared at both open and wooded sites in northern England and New Mexico, at a wooded site in Papua New Guinea, and an open site in Hawaii. At all these locations, bracken occurs naturally. A total of 27 species fed on bracken in Britain, with another eight possibly or occasionally doing so. Approximately 30 species have been recorded from Papua New Guinea. However, only five species fed on bracken in New Mexico, and only one or two in Hawaii. Lawton attempted to define the niches occupied by each species both in terms of the method of feeding (chewers, suckers, miners and gall-formers) and the part of the plant upon which the insects feed (main stem, pinnae (leaves), main stalks arising from the stem, main leaf-veins of the pinnae). His results are presented in Figure 18.1 (p. 673). They indicate a large number of apparently unexploited niches in New Mexico. (Equivalent data for the two Hawaiian species have not been presented, but this community must have the greatest number of unexploited niches of all.)

Figure 20.14 gives the species–area graph for herbivorous insects on bracken from the four continents. 'Area' refers to Lawton's assessment of the area occupied by bracken in each case. Although based on only four points, the species–area relationship is statistically highly significant. In short, there seem to be few phytophagous insects on bracken in New Mexico and Hawaii because the plant itself is comparatively rare. And since the effect is apparent

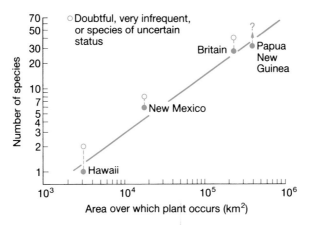

Figure 20.14. Species–area relationship for herbivorous insects on bracken. Those parts of the world where bracken grows more widely appear to support more species. The regression line is fitted to the solid dots (open circles include species which are infrequent or of uncertain host plant status) (from Lawton, 1984).

in comparisons of individual sites, it seems to be rarity itself that is important and not just habitat diversity.

where bracken is comparatively rare, the community of insects on it is comparatively unsaturated

Lawton's study also indicates that the New Mexico and Hawaiian communities, at least, are limited in richness not by competitive exclusion but by exhaustion of the species-pool (itself small because the plant is not widespread). This then is another example in which the island nature of a community plays an important part in ensuring that it is far from fully saturated with species.

The effect of plant *structural* diversity on phytophagous insect communities can be seen in Figure 20.15a and b. From Figure 20.15a it is clear that, comparing plants with similar geographic ranges, trees support more insect species than woody shrubs, which support more species than herbs (Strong & Levin, 1979). This is a sequence of plant types showing a decrease in size, a decrease in the variety of microclimates offered, a decrease in the number of modules (for instance leaves) per individual, a decrease in the variety of plant parts, and a decrease in the range of resources available to insects. Figure

the diversity of plant architecture influences the richness of phytophagous insects

20.15b plots the number of phytophagous insect species known to attack different species of American *Opuntia* cactus against the 'architectural rating' of each cactus species (Moran, 1980). Cacti with a high rating (larger, more modules, and a greater variety of plant parts) clearly supported more insect species.

Plant architecture obviously has a major influence on the richness of the phytophagous insect community, but, as with species on islands, it is difficult to decide how much is due simply to size (including the *number* of modules) and how much is due to environmental and resource heterogeneity. Moran estimated for the cacti that size alone accounted for 35% of the variance in richness, whereas all components of plant architecture together accounted for 69%. However, the cacti cover a size range much smaller than the tree–herb contrast in Figure 20.15a, and a partitioning between size and heterogeneity in these cases awaits further work.

751   ISLANDS, AREAS AND COLONIZATION

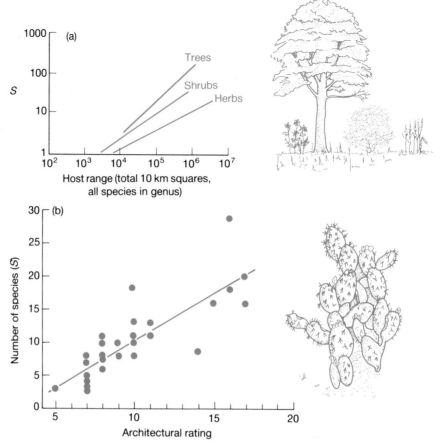

Figure 20.15. The species richness of phytophagous insects on plants increases with the complexity of 'plant architecture'. (a) The 'species–area' relationships for genera of British trees, shrubs and herbs; host range is expressed in terms of the number of 10 km map grid squares in which the plant is found (Strong & Levin, 1979). (b) The number of species associated with *Opuntia* cactus species in North and South America as a function of their 'architectural rating'. Architectural rating is the sum of five variables each measured on a scale from 1 to 4: height of mature plant, mean number of cladodes on a large plant, cladode size, degree of development of woody stems, and cladode complexity from smooth to strongly tuberculate (Moran, 1980).

biologically 'remote' plants support fewer species

The influence of taxonomic or biochemical 'remoteness' on the species impoverishment of plants has not always been apparent in comparisons of different species, but some studies suggest such an effect. For instance, Cornell and Washburn (1979) found that two 'non-*Quercus*' oaks in California supported fewer cypinid gall-formers than might be expected from the data on the more numerous *Quercus* oak species (Figure 20.4b); while Lawton and Schröder (1977) found that British monocotyledonous plant species with fewer other species in the same genus supported fewer species of insect than otherwise expected (Figure 20.16).

In summary then, a greater number of phytophagous insect species are to be found on larger plant species with a more complex architecture, on common and widespread plants, and perhaps on plants that live in the same area as related species.

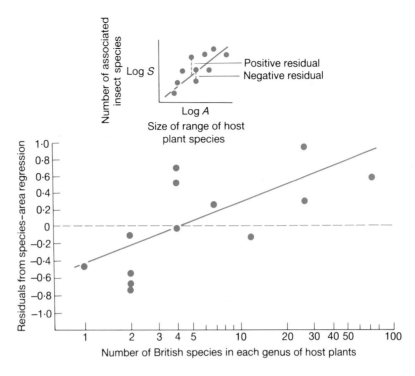

Figure 20.16. Taxonomic isolation of the host plant can reduce the species richness of associated phytophagous insect fauna. Plotted here are the residuals from the species–area relationship (inset) against the number of congeneric host plants, for insects on monocotyledons in Britain. As taxonomic isolation increases (i.e. number of congeneric species *decreases*), so the residuals decrease (i.e. there are fewer species than expected on the basis of the average species–area relationship). (From Strong *et al.*, 1984, after Lawton & Schröder, 1977.)

20.3.4 Which species?—Turnover

So far, we have seen that smaller islands support fewer species, partly but not entirely because of their lower habitat diversity; and we have seen that this species impoverishment tends to increase with island remoteness. But this emphasis on *species number* deflects attention from *which* species live on different islands, *which* species immigrate, and *which* go extinct. We turn to these important questions next.

species turnover

MacArthur and Wilson's equilibrium theory predicts not only a characteristic species richness for an island, but also a *turnover* of species in which new species continually colonize while others become extinct. This implies a significant degree of chance regarding precisely which species are present at any one time. However, studies of turnover itself are rare, because communities have to be followed over a period of time (usually difficult and costly). Good studies of turnover are rarer still, because it is necessary to count every species on every occasion so as to avoid 'pseudo-immigrations' and 'pseudo-extinctions'. Note that any results are bound to be underestimates of actual turnover, because an observer cannot be everywhere all the time.

One study that can be examined confidently has used yearly censuses

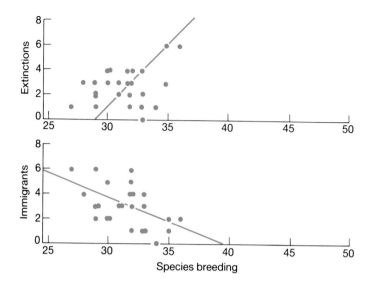

Figure 20.17. Immigration and extinction of breeding birds at Eastern Wood. The line in the extinction diagram is at 45°. The line in the immigration diagram is the calculated regression line with a slope of −0.38 ($P <$ 0.01). (From Williamson, 1981, after Beven, 1976.)

from 1949 to 1975 of the breeding birds in a small oak wood in southern England: Eastern Wood, Bookham Common (Beven, 1976, discussed by Williamson, 1981). In all, 44 species bred in the wood over this period, of which 16 species bred every year. The number breeding in any one year varied between 27 and 36, with an average of 32 species. The immigration and extinction 'curves' are shown in Figure 20.17. Their most obvious feature is the scattering of points compared to the assumed simplicity of the MacArthur–Wilson model (Figure 20.5). Nevertheless, while the positive correlation in the extinction graph is statistically insignificant, the negative correlation in the immigration graph is significant at the 1% level; and the two lines do seem to cross at roughly 32 species, with three new immigrants and three extinctions each year. There is clearly a considerable turnover of species, and consequently considerable year-to-year variation in the bird community of Eastern Wood despite its approximately constant species richness.

Experimental evidence of turnover and indeterminacy is provided by the work of Simberloff and Wilson (1969), who exterminated the invertebrate fauna on a series of small mangrove islands in the Florida Keys and monitored recolonization. Within about 200 days, species richness had stabilized around the level prior to defaunation but with many differences in species composition. Since then, the rate of turnover of species on the islands has been estimated as 1.5 extinctions or colonizations per year (Simberloff, 1976).

Thus the idea that there is a turnover of species leading to a characteristic equilibrium richness on islands—but an indeterminacy regarding particular species—appears to be correct, at least approximately.

20.3.5 Which species?—Disharmony

It has long been recognized (Hooker, 1866, discussed by Turrill, 1964) that one of the main characteristics of island biotas is 'disharmony', by which it is meant that the relative proportions of different taxa are not the same on islands as they are on the mainland. We have already seen from the species–area relationships on p. 747 that groups of organisms with good powers of dispersal (like ferns and, to a lesser extent, birds) are more likely to colonize remote islands than are groups with relatively poor powers of dispersal (most mammals); even within groups, powers of dispersal vary. Figure 20.18, for example, shows the eastern, seaward limits in the Pacific of various groups of

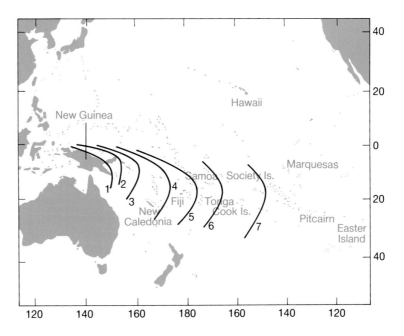

Figure 20.18. Eastern limits of families and subfamilies of land and freshwater breeding birds found in New Guinea. The decline in taxa is fairly smooth, and shows both differences in dispersal ability and a general decline in island size to the east. 1. Not beyond New Guinea: pelicans, storks, larks, pipits, birds of paradise and nine others. 2. Not beyond New Britain and the Bismark Islands: cassowaries, quails and pheasants. 3. Not beyond the Solomon Islands: owls, rollers, hornbills, drongos and six others. 4. Not beyond Vanuatu and New Caledonia: grebes, cormorants, ospreys, crows and three others. 5. Not beyond Fiji and Niufo'ou: hawks, falcons, turkeys and wood swallows. 6. Not beyond Tonga and Samoa: ducks, thrushes, waxbills and four others. 7. Not beyond the Cook and Society Islands: barn owls, swallows and starlings. Beyond 7, the Marquesas and Pitcairn group: herons, rails, pigeons, parrots, cuckoos, swifts, kingfishers, warblers and flycatchers. (From Williamson, 1981, after Firth & Davidson, 1945.)

some groups, and some species, are better suited than others to reaching islands and persisting on them

land and freshwater birds found in New Guinea. On a more anecdotal level, most of the species of land snail on Pacific islands are very small, and so easily transported (Vagvolgyi, 1975), and most of the beetles on the island of St Helena are wood-borers or bark-clingers that are most likely to have been carried across the sea on floating trees (MacArthur & Wilson, 1967).

However, variation in dispersal ability is not the only factor leading to disharmony. Species may vary in their liability to extinction. In particular, species that naturally have low densities per unit area are bound to have only small populations on islands, and a chance fluctuation in a small population is quite likely to eliminate it altogether. Predators, which generally have relatively small populations, are notable for their absence on many islands. For example, the birds on the Atlantic island of Tristan da Cunha have no bird, mammal or reptile predators apart from those released by man.

Predators are also liable to be absent from islands because their immigration can only lead to colonization if their prey have arrived first (whereas for the prey there is no reciprocal dependence on their predators). Similar arguments apply to parasites, mutualists and so on. In other words, for many species an island is only suitable if some other species is present, and disharmony arises because some types of organism are more 'dependent' than others.

The construction by Diamond (1975) of *incidence functions* and *assembly rules* for the birds of the Bismarck archipelago (section 18.3.3) is probably the fullest attempt to understand island communities by combining ideas on dispersal and extinction differentials with those on sequences of arrival and habitat suitability. Constructing such incidence functions (Figure 18.10) allowed Diamond to contrast 'super-tramp' species (high rates of dispersal but a poorly developed ability to persist in communities with many other species), with 'high-*S*' species (only able to persist on large islands with many other species), and to contrast these in turn with intermediate categories. Faaborg (1976) made similar detailed studies of the birds on West Indian islands. Such work illustrates particularly clearly that it takes far more than a single species-richness value to characterize the community of an island.

In the context of phytophagous insect communities, it is found, not surprisingly, that an unusually large proportion of the species that colonize a newly introduced plant are already polyphagous, and that chewing and

incidence functions and assembly rules emphasize that island communities are characterized by more than just species richness

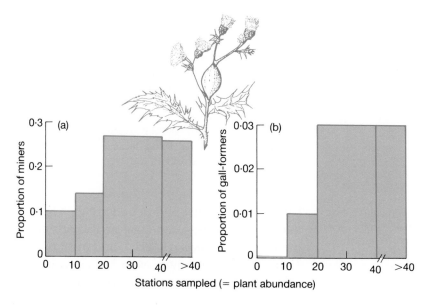

Figure 20.19. As the abundance of host plants within the European Cynareae (thistles) increases, so the proportion of (a) leaf- and stem-miners, and (b) gall-formers also increases (Lawton & Schröder, 1978).

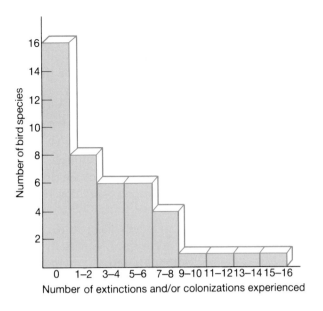

Figure 20.20. In Eastern Wood many species never become extinct, but others become extinct and re-colonize repeatedly. The frequency histogram shows the number of extinctions *plus* re-colonizations over 26 years. The average number per species of bird is 3.4. (After Beven, 1976.)

sucking insects, feeding externally on the plant, are more likely to colonize new hosts than are leaf-miners and gall-formers (Strong *et al.*, 1984). It is particularly pertinent that at least within the European Cynareae (the thistles and their relatives) the proportion of miners and gall-formers is higher on more widespread species (Figure 20.19; Lawton & Schröder, 1978). Here too, therefore, 'small islands' (plants with small ranges) support a community which is disharmonic compared to those of larger 'islands'.

Island communities then are not merely impoverished—the impoverishment affects particular types of organism disproportionately. Those with poor dispersal ability, a high probability of extinction and/or a requirement for the prior arrival of some other species all tend to be relatively rare. Predators are therefore especially affected, and this is bound to have repercussions throughout the whole community (Chapter 19). Moreover, although turnover introduces an element of chance into the community structure of an island, this tends to affect only a proportion of the species, while a core of other species remain unaffected. This can be seen in Figure 20.20 for the birds of Eastern Wood (p. 754).

20.4 Evolution and island communities

No aspect of ecology can be fully understood without reference to evolutionary processes taking place over evolutionary time-scales, but this is particularly true for an understanding of island communities. On isolated islands, the rate at which new species evolve may be comparable with or even faster than the rate at which they arrive as new colonists. Indeed, Williamson (1981) has suggested that 'oceanic' islands should be defined as those on which evolution of species is faster than immigration, while on 'continental' islands immigration is faster than evolution (and thus the same island may be oceanic for one taxon and continental for another). Clearly, the communities of many islands will be incompletely understood by reference only to ecological processes.

To begin with an extreme example, the remarkable number of *Drosophila* species found on the remote Hawaiian islands and described in Chapter 1 (Figure 1.7) have evolved, almost entirely, on the islands themselves. The communities of which they are a part are clearly much more strongly affected by evolution than by the processes described in previous sections of this chapter.

A more modest but more widespread aspect of the influence of evolution is the very common occurrence, especially on 'oceanic' islands, of *endemic* species, i.e. species that are found nowhere else. The Hawaiian *Drosophila* themselves are endemics, as are all the species of land-birds on the island of Tristan da Cunha (p. 756). A more complete illustration of the balance between colonization and the evolution of endemics is shown in Figure 20.21. Norfolk Island is a small island (about 70 km²) approximately 700 km from New Caledonia and New Zealand but about 1200 km from Australia. The ratio of Australian species to New Zealand and New Caledonian species within a group can therefore be used as a measure of the group's dispersal ability, and as Figure 20.21 shows, the proportion of endemics on Norfolk Island is highest in groups with poor dispersal ability and lowest in groups with good dispersal ability.

In a similar vein, Lake Tanganyika, one of the ancient and deep Great Rift lakes of Africa, contains 214 species of cichlid fish, many of which show

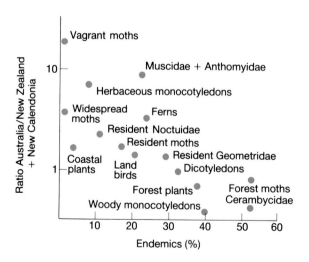

Figure 20.21. Poorly dispersing groups on Norfolk Island have a higher proportion of endemic species, and are more likely to contain species that have reached Norfolk Island from either New Caledonia or New Zealand than species from Australia, which is further away. The converse holds for good dispersers. (After Holloway, 1977.)

exquisite specializations in the manner and location of their feeding. Of these 214 species, 80% are endemic. By contrast, Lake Rudolph, which has only been an isolated water body for 5000 years, since its connection to the River Nile system was broken, contains only 37 species of cichlid of which just 16% are endemic (Fryer & Iles, 1972).

Communities of phytophagous insects show particularly clearly that low rates of colonization combined with insufficient time for evolution can lead to especially marked impoverishment on 'isolated' plant species. One study examined three kinds of tree that are found in both Britain and South Africa: *Betula pendula*, *Quercus robur* and *Buddleia* spp. (Southwood *et al.*, 1982). *Betula* and *Quercus* are native to Britain but were introduced to South Africa, while for *Buddleia* the converse is true. It is apparent that the species richness of phytophagous arthropods in each case is greater in the location where the tree is native (Figure 20.22). A community may be less than fully saturated not only because there has been insufficient time for colonization, but also because there has been insufficient time for evolution. Islands, therefore, because of their isolation, are particularly good at reminding us that the natural world is in a state of ecological and evolutionary flux.

communities may be unsaturated because of insufficient time for evolution

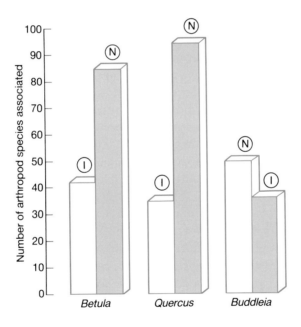

Figure 20.22. Relative species richness of phytophagous arthropods associated with three species of tree in Britain (tinted) and South Africa. In each case fewer species are associated with the tree where it has been introduced (I) compared to where it is native (N). (After Southwood *et al.*, 1982.)

20.5 Islands and conservation

Perhaps the most surprising application of island biogeography theory is in nature conservation. This is because many conserved areas and nature reserves are surrounded by an 'ocean' of habitat made unsuitable, and therefore hostile, by man. Can the study of islands in general provide us with 'design principles' that can be used in the planning of nature reserves? The answer is a qualified 'yes' (see discussions in Simberloff & Abele, 1976a; Soulé & Wilcox, 1980; Diamond & May, 1981). Certain general recommendations can be made, but these generalizations must be tempered by the realization that there is no substitute for a painstaking study of the ecology of the species or community whose conservation is intended.

The most obvious point to emerge from this chapter in the present context is that an area that is part of a continuum will, if it is turned into an isolated reserve, support fewer species than it had previously. We can also expect a large reserve to support more species (and allow fewer extinctions) than a small reserve. However, even these apparently obvious points require some qualification. The creation of an isolated reserve means the creation of a boundary and therefore of 'boundary' habitats. Species suited to these habitats will then be supported by the reserve, whereas they may previously have been absent in the continuous habitat from which the reserve was created. In addition, certain species fare best on small islands, and they may therefore be supported by small reserves but not by large ones. These qualifications, however, are unlikely to out-weigh the general trends.

A less straightforward question is whether to construct one large reserve or several small ones adding up to the same total area. If each of the small reserves supported the same species, then it would clearly be preferable to construct the larger reserve, and hence conserve more species. However, this conclusion ignores epidemiological aspects of management; many scattered reserves are less susceptible to the ravages of an epidemic disease. Moreover, if the region as a whole is heterogeneous, then each of the small reserves may support a different group of species and the total number conserved might exceed that in a large reserve. Even in a homogeneous area, the smaller reserves could support more species in total provided that (a) the species' characteristics ensure considerable (perhaps random) variation between the small reserves and (b) the species–area curve is sufficiently shallow to ensure that the equilibrium richness of the large reserve is not too much larger than that of the small reserves. Of course, if either or both of these conditions were not met, the larger reserve would support more species.

Perhaps more significantly, conservation effort is often directed at low-density species (large predators, for example) that can only be supported by large areas. In this common circumstance, the single large reserve clearly has more to recommend it. Another consideration, adding to the complexity of the conservationist's problems, involves genetic questions concerning the minimum critical population sizes necessary to avoid inbreeding.

A few further, fairly uncontroversial recommendations can also be made. First, since a certain number of extinctions in island reserves are inevitable, the conservationist can help by increasing the amount of counterbalancing immigration. This can be done by the careful juxtaposition of scattered reserves, or by providing corridors or stepping-stones of natural habitat between them. In fact, long, thin reserves may be optimal for catching immigrants, but other reasoning suggests that reserves should be approximately circular, since this minimizes the chances of creating 'islands within islands' on semi-isolated peninsulars, and reduces 'edge effects' to a minimum.

Apart from these recommendations, it seems likely that little of a general nature can be gleaned from island biogeography for the benefit of conservation. What is required is a detailed knowledge of individual species' ecologies—and a will to conserve.

we require a will rather than a way

Community Stability and Community Structure

21.1 Introduction

Ecologists have shown an interest in community stability for two reasons. The first is practical, and indeed pressing. Since modern man is perturbing natural and agricultural communities at an ever-increasing rate, it is essential to know how the communities respond to such perturbations and how they are likely to respond in future. The stability of a community measures its sensitivity to disturbance.

The second reason for being interested in stability is less practical but more fundamental. Stable communities are, by definition, those that persist. The communities we actually see are thus likely to possess properties conferring stability. The most fundamental question in community ecology is: why are communities the way they are? At least part of the answer is therefore likely to be: because they possess certain stabilizing properties.

So, for two quite different reasons it is important to understand what determines the stability or instability of a community. Before this question can be addressed, however, it is necessary to define 'stability', or rather to identify the various different types of stability.

resilience and resistance

The first distinction that can be made is between the resilience of a community (or any other system) and its resistance. *Resilience* describes the speed with which a community returns to its former state after it has been perturbed and displaced from that state. *Resistance* describes the ability of the community to avoid displacement in the first place. (Figure 21.1 provides a figurative illustration of these and other aspects of stability.)

local and global stability

The second distinction is between local stability and global stability. *Local stability* describes the tendency of a community to return to its original state (or something close to it) when subjected to a small perturbation. *Global stability* describes this tendency when the community is subjected to a large perturbation.

fragility and robustness

A third aspect of stability is related to the local/global distinction, but concentrates more on the environment of the community. The stability of any community depends on the environment in which it exists, as well as on the densities and characteristics of the component species. A community which is stable only within a narrow range of environmental conditions, or for only a very limited range of species characteristics, is said to be *dynamically fragile*. Conversely, one which is stable within a wide range of conditions and characteristics is said to be *dynamically robust*.

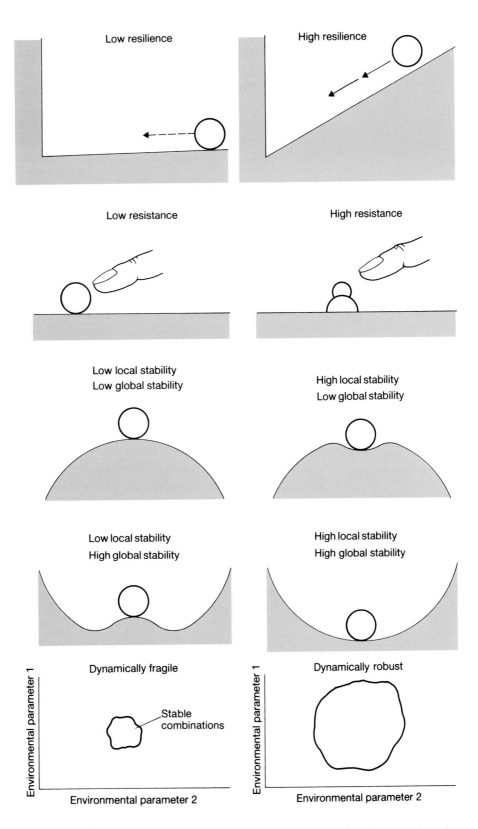

Figure 21.1. Stability. Various aspects of stability, used in this chapter to describe communities, illustrated here in a figurative way.

Lastly, it remains for us to specify the aspect of the community which is having its stability assessed. Usually, ecologists have taken a demographic approach. They have concentrated on the composition of a community: the number and identity of the component species, and their densities. However, as we shall see below (section 21.5), it is also possible to deal with other properties of a community: for example the rate of biomass production, or the amount of calcium it contains. We begin though, in sections 21.2–21.4, with the demographic approach.

21.2 Complexity and stability

21.2.1 The 'conventional wisdom'

The aspect of community structure that has received most attention from the point of view of stability has been community complexity. During the 1950s and '60s, the 'conventional wisdom' in ecology was that increased complexity within a community leads to increased stability. Increased complexity, then as now, was variously taken to mean more species, more interactions between species, greater average strength of interaction, or some combination of all of these things. Elton (1958), amongst others, brought *a long-standing belief that complexity leads to stability* together a variety of empirical and theoretical observations in support of the view that more complex communities are more stable (Table 21.1). Nowadays though, the points Elton made can be seen as either untrue or else liable to some other plausible interpretation. (Indeed, Elton himself pointed out that more extensive analysis was necessary.) At about the same time, MacArthur (1955) proposed another argument in favour of the conventional wisdom. He suggested that the more possible pathways there were by which energy passed through a community, the less likely it was that the densities of constituent species would change in response to an abnormally raised or lowered density of one of those species. In other words, the greater the complexity (more pathways), the greater is the stability (less numerical change) in the face of a perturbation. However, the conventional wisdom has by no means always received support from more recent work, and has been undermined particularly by the analysis of mathematical models.

21.2.2 Complexity and stability in model communities

There have been a number of attempts to explore mathematically the relationship between community complexity and community stability (reviewed by May, 1981), but most of these have come to similar conclusions. The approach taken by May (1972) can be used to illustrate the general nature of this mathematical work, the conclusions that can be drawn from such work, and its shortcomings. May constructed model food webs comprising a number of species, and concerned himself with the way in which the population size of each species changed in the neighbourhood of its equilibrium abundance (i.e. he was concerned with *local* stability of a type similar to the 'stability' of one- and two-species models examined in Chapter 6, 7 and 10). Each species was influenced by its interaction with all other species, and the

	STATEMENT	ASSESSMENT
Table 21.1. Summary of Elton's arguments (1958) in support of the 'conventional wisdom' prior to 1970 that complexity begets stability in communities. Each observation is consistent with the complexity/stability hypothesis. However, each can be explained in terms of reasonable alternative hypotheses or questioned because of lack of a control.	1. Mathematical models of interactions between two or a few species are inherently unstable	No longer held to be true after recent development of two-species models (see Chapter 10). In any case, there was no evidence at the time that multispecies models would be any more stable (none had been developed)
	2. Simple laboratory communities of two or a few species are difficult to maintain without extinctions	True, but no evidence that multispecies laboratory communities would be more stable. The probable reason that laboratory cultures are difficult to set up and maintain is because it is virtually impossible to reproduce the natural environmental conditions
	3. Islands, which usually possess few species, are more vulnerable to invading species than are species-rich continents	There have also been well-documented and remarkable examples of introduced species assuming pest proportions on continents (see Chapter 15). Nevertheless, statement 3 may be true. However, vulnerability to invasion has only a tenuous link with more conventional definitions of stability and resilience
	4. Crop monocultures are peculiarly vulnerable to invasions and destruction by pests	Differences between natural and agricultural communities could derive from the long periods of coevolution to which natural associations may have been subject (Maynard Smith, 1974). Note also that arable crops are typically early-successional species, naturally subject to rapid change. Natural monocultures such as salt-marsh and bracken seem to be stable (May, 1972)
	5. Species-rich tropical communities are not noted for insect outbreaks when compared to their temperate and boreal counterparts	Such a pattern, if real, could be due to the destabilizing effects of climatic fluctuations in temperate and boreal communities (Maynard Smith, 1974). Wolda (1978) has produced evidence suggesting that insect abundances fluctuate just as markedly in the tropics

term β_{ij} was used to measure the effect of species j's density on species i's rate of increase. Thus β_{ij} would be zero when there was no effect, both β_{ij} and β_{ji} would be negative for two competing species, and β_{ij} would be positive and β_{ji} negative for a predator (i) and its prey (j).

May 'randomly assembled' food webs. He set all self-regulatory terms (β_{ii}, β_{jj}, etc.) at -1, but distributed all other β-values at random, including a certain number of zeros. The model webs so constructed could then be described by three parameters: S, the number of species; C, the 'connectance' of the web (the fraction of all possible pairs of species which interacted directly, i.e. with $\beta_{ij} \neq 0$); and $\overline{\overline{\beta}}$, the average 'interaction strength' (i.e. the average of the non-zero β-values, disregarding sign). What May found was that these food webs were only likely to be stable (i.e. the populations would return to equilibrium after a small disturbance) if:

in randomly assembled food webs, local stability declines with complexity

$$\overline{\overline{\beta}}\,(SC)^{1/2} < 1.$$

Otherwise they tended to be unstable (i.e. disturbed populations failed to return to equilibrium).

In other words, increase in number of species, increase in connectance, and increase in interaction strength all tend to increase instability (because they increase the left-hand side of the inequality above). Yet each of these represents an increase in complexity. Thus this model (along with others) suggests that complexity leads to *instability*. This clearly runs counter to the conventional wisdom of Elton and MacArthur, and it certainly indicates that there is no necessary, unavoidable connection linking stability to complexity. It is possible, though, that the connection between complexity and instability is an artefact arising out of the particular characteristics of the model communities or the way they have been analysed.

... but if the models are altered, the conclusions become less clear-cut

In the first place, randomly assembled food webs often contain biologically unreasonable elements (for instance, loops of the type: A eats B eats C eats A). Analyses of food webs that are constrained to be reasonable (Lawlor, 1978; Pimm, 1979a) show (a) that they are more stable than their unreasonable counterparts, and (b) that while stability still declines with complexity, there is no sharp transition from stability to instability (compared with the inequality in the equation above.)

Secondly, the results of model analyses are altered if it is assumed that consumer populations are affected by their food supply, but that the food supply is not affected by the consumers ($\beta_{ij} > 0$, $\beta_{ij} = 0$: so-called 'donor-controlled' systems). In this type of food web, stability is unaffected by or actually increases with complexity (DeAngelis, 1975). In practice, the only group of organisms to which this condition commonly applies are the detriti-

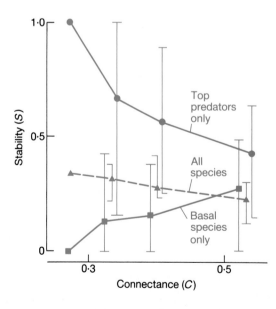

Figure 21.2. Species-deletion stability plotted against connectance (C) for six-species models. Results are for systems where (i) only top predators are deleted, (ii) only basal species are deleted, and (iii) all species are deleted in turn. The bars show the range in species-deletion stability for all models of a given connectance. (After Pimm, 1979b.)

vores. Nevertheless, as Chapter 11 made clear, this is not an unimportant group of 'exceptions'.

The picture also alters if, instead of considering local stability, larger perturbations are examined. Pimm (1979b) has attempted to do this by using 'species-deletion stability'. His model communities were subjected to a single, large and persistent perturbation: the deletion of one species. The system was then said to be species-deletion stable if all of the remaining species were retained at locally stable equilibria. Figure 21.2 shows the results of simulated deletions of species from a simple six-species community containing two top predators, two intermediate predators and two basal species (plants or categories of dead organic matter). Interaction strengths were ascribed at random, while connectance was varied systematically; complexity could therefore be equated with connectance. In general, stability decreased with increasing complexity; but this trend was reversed when basal species were removed. These simulations therefore agree with most other models when perturbing 'from above' (removal of a top predator), but they conform with the conventional wisdom when perturbing 'from below' (removal of a species from the base of the food web). Interestingly, MacArthur's original argument, since it dealt with energy flowing through a community, also envisaged perturbations coming from below.

Finally, the relationship between complexity and stability in models becomes more complicated if attention is focused on the resilience of those communities which *are* stable. Pimm (1979a), like other modellers, found that as complexity (actually connectance) increased, the proportion of stable com-

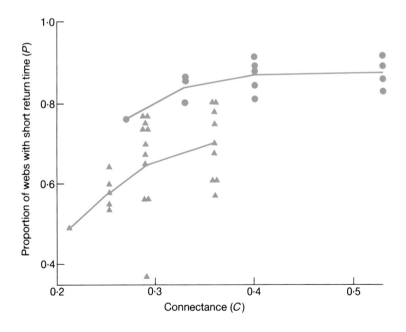

Figure 21.3. Complexity and resilience in model food webs. As complexity (actually connectance) increases, the proportion of stable food web models which have short return times (P) after a perturbation also increases (a short return time indicates a high resilience). ● = models with three trophic levels; ▲ = models with four trophic levels. (After Pimm, 1982.)

munities *decreased*. However, within the small sub-set of stable communities, resilience (a crucial aspect of stability) actually increased with complexity (Figure 21.3).

on balance, though, the conventional wisdom is undermined

Overall, most models indicate that stability tends to decrease as complexity increases. This is sufficient to undermine the conventional wisdom prior to 1970. However, the conflicting results amongst the models at least suggest that no single relationship will be appropriate in all communities. It would be wrong to replace one sweeping generalization with another.

Even if complexity and instability are connected, this does not mean that we should expect to see an association between complexity and instability in real communities. For one thing, the models often refer to randomly constructed communities. Real communities are far from randomly constructed; their complexity may be of a particular type and *could* enhance stability. One task facing theoretical ecologists is to suggest ways in which special structures in food webs may enhance stability. Furthermore, unstable communities will fail to persist when they experience environmental conditions which reveal their instability. But the range and predictability of environmental conditions will vary from place to place. In a stable and predictable environment, a community will experience only a limited range of conditions, and a community which is dynamically fragile may still persist. However, in a variable and unpredictable environment, only a community which is dynamically robust will be able to persist. These theoretical arguments require that predictability can be defined in a sensible way, and measured. This is far from being a straightforward task. However, what we might expect to see are (a) complex and fragile communities in stable and predictable environments, with simple and robust communities in variable and unpredictable environments; and (b) approximately the same recorded stability (in terms of population fluctuations, etc.) in all communities, since this will depend on the inherent stability of the community combined with the variability of the environment.

stability may vary between environments . . .

This line of argument, however, carries a further, very important implication for the likely effects of unnatural, man-made perturbations on communities. We might expect these to have their most profound effects on the dynamically fragile, complex communities of stable environments, which are relatively un-used to perturbations. Conversely, they should have least effect on the simple, robust communities of variable environments, which have previously been subjected to repeated (albeit natural) perturbations. In this sense too, complexity and instability may be associated.

. . . and the fragile, complex communities of stable environments may be particularly prone to disturbance by man

What is the evidence from real communities?

21.2.3. Complexity and stability in practice

A number of studies have sought to investigate the general importance of dynamic/stability constraints on community structure by examining the relationship between S, C and $\bar{\bar{\beta}}$ in real communities. The argument they use runs as follows. If communities are only stable for

$$\bar{\bar{\beta}} \, (SC)^{1/2} < 1,$$

then increases in S will lead to decreased stability unless there are compensatory decreases in C and/or $\bar{\bar{\beta}}$. Data on interaction strengths for whole communities are unavailable. It is therefore usually assumed, for simplicity, that $\bar{\bar{\beta}}$ is constant (i.e. it does not vary with S). In such a case, communities with more species will only retain stability if there is an associated decline in average connectance, which means that the product SC should remain approximately constant.

Rejmanek and Stary (1979) studied a number of geographically distinct, plant–aphid–parasitoid communities in central Europe. They painstakingly compiled the food webs from field observations, and confirmed many of the supposed interactions in laboratory studies. Connectance was calculated to provide both minimum and maximum estimates in each case. The minimum estimate simply expressed the number of predator–prey links as a proportion of the total possible number of pairwise interactions. It ignored competitive interactions. The maximum estimate assumed that all pairs of predators which share a prey species are in competition. Both C_{min} and C_{max} are plotted against species number in Figure 21.4a. In these communities C does decrease with an increase in S, and the product SC lies between 2 and 6 (i.e. approximately constant).

A large group of 40 food webs has been gleaned from the literature by Briand (1983), including terrestrial, freshwater and marine examples (Table 21.2). For each community, a single value for connectance has been calculated on the basis of both predator–prey and presumed competitive interactions (i.e. C_{max}). It is plotted against S in Figure 21.4b. Once again, connectance decreases with species number; SC is approximately equal to 7.

McNaughton (1978) collected data on plant species in 17 grassland communities in the Serengeti National Park in Tanzania, where the interactions were purely competitive. He made the sweeping assumption that pairs of species were competing if there were significant negative correlations in their distributions (see p. 682), and calculated connectance as the proportion of possible pairwise interactions which were non-zero. Lawton and Rallison (1979) have pointed out that the results of this work are clouded somewhat by a statistical artefact. Nevertheless, there was a remarkably constant relationship between species richness and connectance with $SC = 4.7 \pm 0.7$ (95% confidence limits) (Figure 21.4c).

There is evidence, then, that the product SC remains constant when different communities are compared. However, as Pimm (1980) has pointed out, this need have nothing to do with stability constraints on community structure. SC will also remain approximately constant as long as the average number of species with which any one species interacts remains constant, irrespective of the total number of species. This is not an unreasonable supposition (although it is untested), and the studies of S and C are therefore merely consistent with the stability hypothesis rather than being a confirmation of it.

On the other hand, stability considerations suggest that for a given number of species, connectance should be lower (and stability therefore higher) in fluctuating environments compared with constant environments. Briand (1983) was able to classify his communities on just this basis: any

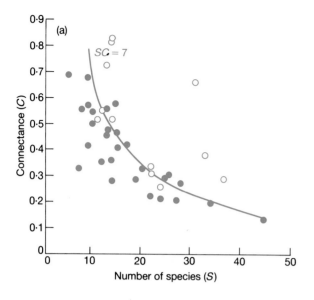

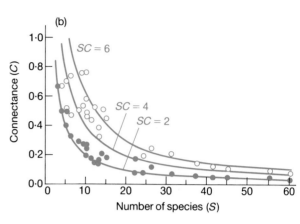

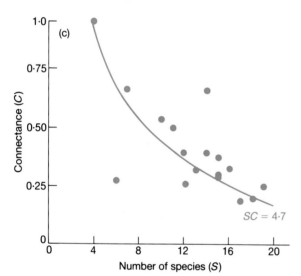

Figure 21.4 (a) The relationship between connectance and species richness in 40 food webs from terrestrial, freshwater and marine environments. The hyperbola corresponding to a constant product of SC of 7 is also shown. Closed circles are for communities judged to be from fluctuating environments and open circles for those from constant environments. (After Briand, 1983.) (b) The relationship between C and S in plant–aphid–parasitoid communities. $\bullet = C_{min}$, minimum estimates of connectance; $\circ = C_{max}$, maximum estimates of connectance. Also shown are the hyperbolae corresponding to constant products of SC of 2, 4 and 6. (After Rejmanek & Stary, 1979. (c) The relationship between C and S for several African grassland communities. The hyperbola corresponding to a constant product of SC of 4.7 is shown. (After McNaughton, 1977.)

community described in the original published report as subject to substantial variations in temperature, salinity, pH, water availability, or any other major factor, was labelled 'fluctuating'. Others were considered constant. Significantly, Briand did find that values of C were generally lower in fluctuating environments (Figure 21.4b).

a complex community that is less resilient when perturbed

The prediction that complex communities are less stable when given an experimental perturbation was tested by McNaughton (1977) on two sets of plant communities. In the first experiment, the perturbation consisted of an addition of plant nutrients to the soil, while in the second it involved the action of grazing animals. In each case, the effect of the perturbation was monitored in both species-rich and species-poor plant communities. The results are presented in Table 21.3. Each perturbation significantly reduced the diversity of the species-rich but not the species-poor community. This is

Table 21.2. Main structural parameters of 40 food webs from terrestrial, freshwater and marine environments. The column headed 'Trophic structure' indicates how many kinds of organisms occupy each trophic level from producers to top predators (e.g. 2-4-1-1-1 means two taxa in the first trophic level, four in the second level, and one each in the third, fourth and fifth. Species feeding on more than one trophic level are recorded at their highest position in the web. S and C denote species richness and connectance, respectively, and L is the mean number of links in all food chains in the community (mean maximal food chain length). (After Briand 1983.)

Case no.	Community	Trophic structure (producers→ top predators)	S	C	L
A. FLUCTUATING ENVIRONMENTS					
1	Cochin estuary	2-4-1-1-1	9	0.69	4.13
2	Knysna estuary	3-7-4-1	15	0.48	3.78
3	Long Island estuary	4-10-6-4	24	0.21	3.30
4	California salt-marsh	2-2-6-2-1	13	0.56	3.57
5	Georgia salt-marsh	3-2-2	7	0.33	3.00
6	California tidal flat	2-7-5-3-2-5-1	25	0.30	4.82
7	Narragansett Bay	3-5-7-4-1	20	0.33	3.97
8	Bissel Cove marsh	4-4-2-4-1	15	0.42	3.44
9	Lough Ine rapids	2-3-4-1	10	0.51	3.89
10	Exposed intertidal (New England)	2-2-1	5	0.70	3.00
11	Protected intertidal (New England)	3-4-1	8	0.43	3.00
12	Exposed intertidal (Washington)	3-7-1-2	13	0.46	3.32
13	Protected intertidal (Washington)	3-6-2-2	13	0.49	3.50
14	Mangrove swamp (station 1)	1-3-3-1	8	0.57	3.40
15	Mangrove swamp (station 3)	1-5-2-1	9	0.58	3.20
16	Pamlico River	4-4-5-1	14	0.36	3.14
17	Marshallese reefs	3-3-3-3-1-1	14	0.29	4.57
18	Kapingamarangi atoll	8-9-2-3-5	27	0.20	3.07
19	Moosehead Lake	2-3-8-3-1	17	0.43	4.00
20	Antarctic pack ice zone	3-3-5-3-4-1	19	0.30	4.26
21	Ross Sea	3-2-1-1-1-1-1	10	0.56	4.26
22	Bear Island	6-10-2-3-2-2-2-1	28	0.28	4.69
23	Canadian prairie	1-5-4-4-1	15	0.59	3.40
24	Canadian willow forest	4-3-1-3-1	12	0.36	3.71
25	Canadian aspen communities	3-11-8-2-1	25	0.31	3.15
26	Aspen parkland	9-10-6-4-3-1-1	34	0.20	4.00
27	Wytham Wood	4-6-4-5-3	22	0.27	3.89
28	New Zealand salt-meadow	7-19-10-9	45	0.14	2.90
B. 'CONSTANT' ENVIRONMENTS					
29	Arctic seas	2-3-6-6-3-2	22	0.31	4.34
30	Antarctic seas	1-2-3-2-3-2-1	14	0.53	4.34
31	Black Sea epiplankton	2-3-5-1-1-1-1	14	0.84	4.90
32	Black Sea bathyplankton	2-3-5-1-1-1-1	14	0.85	4.86
33	Crocodile Creek	5-16-6-4-2	33	0.39	2.85
34	River Clydach	4-4-1-2-1	12	0.56	3.56
35	Morgan's Creek	2-4-2-2-3	13	0.74	3.71
36	Mangrove swamp (station 6)	8-7-4-2-1	22	0.35	3.00
37	California sublittoral	6-10-3-5	24	0.26	2.76
38	Lake Nyasa rocky shore	3-10-9-9	31	0.67	3.13
39	Lake Nyasa sandy shore	5-15-12-5	37	0.30	2.87
40	Malaysian rainforest	3-3-4-1	11	0.53	2.88

consistent with the hypothesis that more complex communities are less likely to return to their state prior to perturbation.

The idea that the effect of complexity on stability depends on which trophic level is perturbed gains some support from a study of simple, laboratory protozoan communities set up by Hairston *et al.* (1968). The first trophic level comprised one, two or three species of bacterium. These were

the importance of trophic level

Table 21.3. The influence of (a) nutrient addition on species richness equitability ($H/\ln S$) and diversity (H) in two fields, and (b) grazing by African buffalo on species diversity in two areas of vegetation (from McNaughton, 1977).

	Control plots	Experimental plots	Statistical significance
(a) NUTRIENT ADDITION			
Species richness per 0.5m² plot			
Species-poor plot	20.8	22.5	not significant
Species-rich plot	31.0	30.8	n.s.
Equitability			
Species-poor plot	0.660	0.615	n.s.
Species-rich plot	0.793	0.740	$p < 0.05$
Diversity			
Species-poor plot	2.001	1.915	n.s.
Species-rich plot	2.722	2.532	$p < 0.05$
(b) GRAZING			
Species diversity			
Species-poor plot	1.069	1.357	n.s.
Species-rich plot	1.783	1.302	$p < 0.005$

consumed by one, two or three *Paramecium* species. The percentage of cultures showing no extinctions of *Paramecium* increased with diversity of bacteria. However, the extinction of the rarest *Paramecium* species was most likely when three species of *Paramecium* were cultured together as opposed to only two.

Finally, there have been a number of studies directed at the question of whether the level of 'perceived stability' (i.e. observed fluctuations in abundance) is roughly the same in all communities, or whether there are any noticeable trends. Wolda (1978), for instance, drew together an extensive array of data on annual fluctuations in the abundance of tropical, temperate and sub-arctic insect populations. The old conventional wisdom assumed that the more diverse insect communities of the tropics were more stable (i.e. fluctuated less) than their depauperate temperate and sub-arctic counterparts. However, Wolda concluded that, on average, populations of tropical insects have the same annual variability as those of insects from temperate zones. Even the few sub-arctic studies show a similar spread of values. Similar conclusions have been reached by Leigh (1975) for herbivorous vertebrates and Bigger (1976) for crop pests.

'perceived stability' is apparently constant

21.2.4 Appraisal

The relationship between the complexity of a community and its inherent stability is not clear-cut. It appears to vary with the precise nature of the community, with the way in which the community is perturbed, and with the way in which stability is assessed. Nonetheless, there appears to be an *overall* tendency for inherent stability to increase as complexity decreases.

There is an associated tendency for relatively stable environments to support complex but fragile communities, while relatively variable environments only allow the persistence of simpler, more robust communities.

As a consequence, there is no clear trend by which the level of population fluctuation varies from simple to complex communities.

It also seems likely (and we have presented some evidence to support this) that the complex, fragile communities of relatively constant environments (e.g. the tropics) are more susceptible to outside, unnatural disturbance (and are more in need of protection) than the simpler, more robust communities that are more 'used' to disturbance (e.g. in more temperate regions).

Lastly, it is worth noting that there is likely to be an important parallel between the properties of a community and the properties of its component populations. In stable environments, populations will be subject to a relatively high degree of K selection (Chapter 14); in variable environments they will be subject to a relatively high degree of r selection. The K-selected populations (high competitive ability, high inherent survivorship but low reproductive output) will be *resistant* to perturbations, but once perturbed will have little capacity to recover (low resilience). The r-selected populations, by contrast, will have little resistance but a higher resilience. (Note this important inverse relationship between resistance and resilience.) The forces acting on the component populations will therefore reinforce the properties of their communities, namely fragility (low resilience) in stable environments and robustness in variable ones.

the properties of component populations may reinforce the properties of communities

21.3 Compartments in communities

Related to the complexity–stability issue, a number of theoretical studies have suggested that communities will have increased stability if they are 'compartmentalized' (e.g. May, 1972; Goh, 1979). In other words, for given values of S, C and $\bar{\bar{\beta}}$, a community will be more stable if it is organized into sub-units within which interactions are strong but between which interactions are weak. On the other hand, when Pimm (1979a) excluded biologically unreasonable elements from his model food webs he found no clear connection between stability and compartmentalization.

*some* models suggest that compartmentalization promotes stability . . .

Pimm (1982) has also pointed out that compartmentalization is to be expected in all large communities, simply because these communities include several different habitats within which different sets of species operate, and interactions are bound to be stronger within habitats than between habitats. It is, of course, impossible to draw absolute lines of distinction within communities between one habitat and another: what constitutes a habitat will vary from species to species. Nevertheless, there is still likely to be a tendency for compartments to reflect habitats. Pimm and Lawton (1980) and Pimm (1982) have reviewed the evidence for compartments both between and within habitats.

In studies where the habitat divisions are major and unequivocal, there is a clear tendency for compartments to parallel habitats. Figure 21.5, for instance, shows the major interactions within and between three interconnected habitats on Bear Island in the Arctic Ocean (Summerhayes & Elton, 1923). Pimm and Lawton (1980) demonstrated that there is a significantly smaller number of interactions between habitats than would be expected by

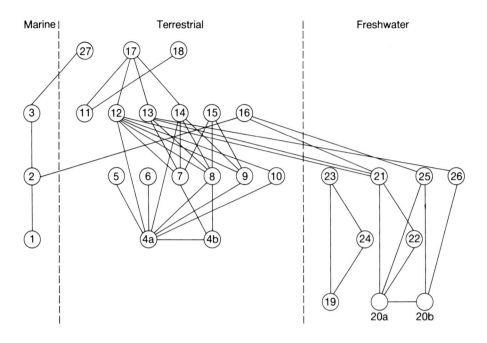

Figure 21.5. The major interactions within and between three interconnected habitats on Bear Island in the Arctic Ocean. (1) Plankton, (2) marine animals, (3) seals, (4a) plants, (4b) dead plants, (5) worms, (6) geese, (7) Collembola, (8) Diptera, (9) mites, (10) Hymenoptera, (11) seabirds, (12) snow bunting, (13) purple sandpiper, (14) ptarmigan, (15) spiders, (16) ducks and divers, (17) arctic fox, (18) skua and glaucous gull, (19) planktonic algae, (20a) benthic algae, (20b) decaying matter, (21) protozoa a, (22) protozoa b, (23) invertebrates a, (24) Diptera, (25) invertebrates b, (26) microcrustacean, (27) polar bear. (From Pimm & Lawton, 1980.)

. . . but there is evidence only for compartments *between* habitats

chance. On the other hand, when the habitat divisions are more subtle, the evidence for compartments is typically poor. For example, Shure (1973) studied two plants that were dominant in the initial stages of an old-field succession: *Ambrosia artemisiifolia* and *Raphanus raphanistrum*. He labelled the plants with radioactive ^{32}P, and measured the amount of this label found in herbivorous and predatory insects over subsequent weeks. The results shown in Figure 21.6 are difficult to interpret with certainty, because there is no sharp distinction that can be drawn between a system with compartments and one without. But since an average of 50% or more of the label goes to species shared by the two plants at each consumer level, the data do not seem to support the hypothesis of two obvious compartments.

There are even greater difficulties in providing a clear demonstration of compartments (or the lack of them) *within* habitats. Such analyses as have been carried out, however, suggest that food webs within habitats are approximately as compartmentalized as would be expected by chance alone (Pimm & Lawton, 1980).

It appears that food webs show some tendency to form compartments between habitats, but there is no clear tendency within habitats. The predictions from model food webs are equivocal; but there is certainly no weight of evidence to support the idea that compartmentalization has promoted the persistence of particular food webs because of the stability it confers.

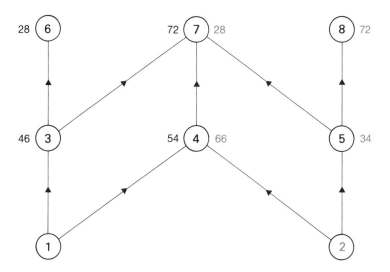

Figure 21.6. Compartmentalization of a food web? The pattern of flow of a ^{32}P tracer from two plant species through an old-field community. The two plants are (1) *Ambrosia* and (2) *Raphanus*; (3) = herbivores restricted to (1); (4) = herbivores common to both species; (5) = herbivores restricted to (2); (6) = predators restricted to (3); (7) = predators feeding on (4) and/or predators feeding on both (3) and (5); (8) = predators restricted to (5). The small numbers are the percentages of the tracer following the routes indicated: black figures indicate tracer originally in *Ambrosia*, red figures tracer originally in *Raphanus*. (After Pimm & Lawton, 1980, following Shure, 1973.)

21.4 The number of trophic levels

A further important feature of food webs is the number of trophic links in pathways from basal species to top predators. Table 21.2 contains, for each community, the average number of trophic levels present (Briand, 1983). These values were derived as follows. A *maximal food chain* is defined as a sequence of species running from a basal species to another species which feeds on it, to yet another species which feeds on the second, and so on up to a top predator (fed on by no other species). Starting with basal species 1 in Figure 21.7, for instance, we can trace four possible trophic pathways via species 4 to a top predator. These are 1–4–11–12, 1–4–11–13, 1–4–12 and 1–4–13. This provides four values for the length of a maximal food chain: 4,4,3 and 3. Figure 21.7 also lists a total of 21 further estimates for chains, starting from basal species 1, 2 and 3. The average of all the possible maximal food chains lengths is 3.32. This parameter defines the number of trophic levels which can be assigned to the food web in question. Inspection of the values in Table 21.2 shows that communities typically consist of between two and five trophic levels, and they most often have three or four. This is in line with other estimates of food chain length. Ecologists interested in community structure have asked why food chains are as short as this.

most communities have 3 or 4 trophic levels

Once again, the stability properties of various model communities may tell us something about the pattern to be expected in real food webs.

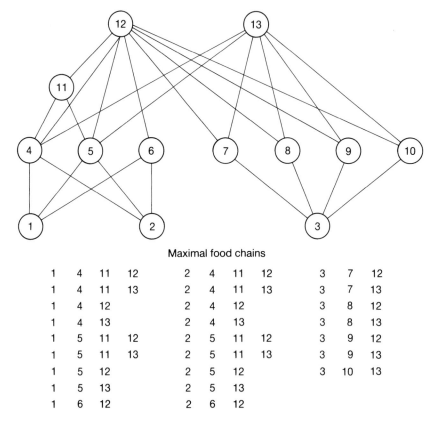

Maximal food chains

1	4	11	12		2	4	11	12		3	7	12
1	4	11	13		2	4	11	13		3	7	13
1	4	12			2	4	12			3	8	12
1	4	13			2	4	13			3	8	13
1	5	11	12		2	5	11	12		3	9	12
1	5	11	13		2	5	11	13		3	9	13
1	5	12			2	5	12			3	10	13
1	5	13			2	5	13					
1	6	12			2	6	12					

Figure 21.7. Community matrix for an exposed intertidal rocky shore in Washington State, U.S.A. (from Menge & Sutherland, 1976). (This is matrix number 12 from Table 21.2.) The pathways of all possible maximal food chains are listed. (1) Detritus, (2) plankton, (3) benthic algae, (4) acorn barnacles, (5) *Mytilus edulis*, (6) *Pollicipes*, (7) chitons, (8) limpets, (9) *Tegula*, (10) *Littorina*, (11) *Thais*, (12) *Pisaster*, (13) *Leptasterias*. (After Briand, 1983.)

However, the 'stability' hypothesis is not the only one to consider. Other plausible hypotheses based on fundamental ecological principles must also be examined.

21.4.1 The energy flow hypothesis

It has long been argued that energetic considerations set a limit to the number of trophic levels which an environment can support. Of the radiant energy which reaches the earth, only a small fraction is fixed by photosynthesis and is available either as live food for herbivores or as dead food for detritivores. However, the amount of energy available for consumption is considerably less than that fixed by the plants, because of work done by the plants (in growth and maintenance) and because of losses due to inefficiencies in all energy-conversion processes (Chapter 17). Each feeding link among the heterotrophs is characterized by the same phenomena: at most 30%, and sometimes as little as 1%, of energy consumed at one trophic level is available as food to the next. In theory therefore, the observed pattern of just three or four trophic levels could be due to energetic limitations—a further trophic level just could not be supported by the available energy.

an apparently testable prediction . . .

This hypothesis appears to provide a testable prediction. Systems which possess a greater primary productivity should be able to support a larger number of trophic levels. In fact, though, despite the fact that primary productivity worldwide varies over three or four orders of magnitude in terrestrial, freshwater and marine communities (Pimm, 1982), there is no firm evidence for higher numbers of trophic levels in more productive environments. Food chains are not noticeably shorter in unproductive arctic lakes and tundra than in their temperate and tropical lake and grassland counterparts. Only at very low levels of primary productivity does energy seem to set an upper limit to the number of trophic levels. For example, Lake Vangalin in the Antarctic has a remarkably low productivity and apparently lacks a third trophic level (Goldman et al., 1967).

. . . which is not supported

It appears as if the classical energy explanation for the lengths of food chains should be rejected, and it has been rejected by a number of ecologists. However, it should be borne in mind that species diversity is usually significantly higher in productive regions (Chapter 22), and that each consumer probably feeds on only a limited range of species at a lower trophic level. The amount of energy flowing up a single food chain in a productive region (a large amount of energy, but divided amongst many sub-systems) may not be very different from that flowing up a single food chain in an unproductive region (having been divided amongst fewer sub-systems). Thus the energy explanation cannot be accepted uncritically—but neither should it be rejected altogether.

but the energy explanation should not be rejected altogether

21.4.2 The dynamic fragility of model food webs

In their study of the stability properties of variously structured Lotka–Volterra models, Pimm and Lawton (1977) showed that webs with long food chains (more trophic levels) typically underwent population fluctuations so severe that the extinction of top predators was more likely than with shorter food chains. Some of their results are shown in Figure 21.8. Return times after a perturbation were very much shorter in four-species models with only two trophic levels than in those arranged into three or four levels. Because less resilient systems are unlikely to persist in an inconstant environment, Pimm and Lawton suggested that only systems with few trophic levels will commonly be found in nature. Note that the three model food webs constructed by Pimm and Lawton differ not only in the number of trophic levels but also in the number of links between levels 1 and 2. Short return times may conceivably result from the larger number of links rather than the smaller number of trophic levels.

21.4.3 Constraints on predator design and behaviour

To feed on prey at a given trophic level, a predator has to be large enough, manoeuvrable enough and fierce enough to effect a capture. In general, predators are larger than their prey (not true, though, of grazing insects and

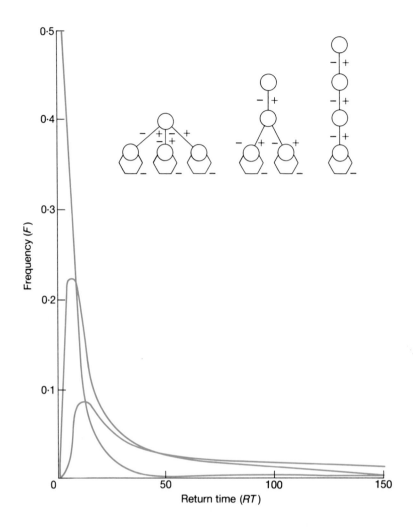

Figure 21.8. Frequency (*F*) of return times (*RT*, arbitrary scale) for the two, three and four trophic level models illustrated. A small number of trophic levels is associated with shorter return times—i.e. the systems are more resilient. (After Pimm & Lawton, 1977.)

parasites), and body size increases at successive trophic levels. There may well be a limit above which design constraints rule out another link in the food chain. In other words, it may be impossible to design a predator that is both fast enough to catch an eagle and big and fierce enough to kill it. Of course, it is not possible to carry out a critical test of this hypothesis since we can only judge what is possible in evolution from what has actually evolved.

A further question relates to the density of top predators. Since they are generally large and less than 100% efficient in finding and converting prey into predator biomass, they are necessarily much rarer than their prey and need to forage over a larger home range. There must be some limit to the necessary home range size and density which would still permit members of the opposite sex to meet occasionally. If the length of food chains were limited by home range considerations, we might expect communities which are restricted to small areas, such as islands, to have fewer trophic levels. In fact, there have been a number of reports of an absence of large predatory birds on small islands—and other birds often seem remarkably tame as a result.

We need also to consider the question of the optimal choice of diet. Consider the arrival in a community of a new carnivore species. Would it do best to feed on the herbivores or the carnivores already there? The herbivore trophic level offers the richer energy environment because there have been only a minimum number of energy-losing steps. Thus an advantage to feeding low down in the food chain can readily be seen. On the other hand, if all species did this, competition would intensify, and feeding higher in the food chain could reduce competition.

In more general terms, it is difficult to imagine a top predator sticking religiously to a rule that it should prey only on the trophic level immediately below it, especially as the prey there are likely to be larger, fiercer and rarer than species at lower levels. In fact, it is unlikely that animals conceive of their world as the same rigid trophic framework that ecologists see. More probably carnivores will attack a potential prey item that happens to be in the right size range and in the right place, regardless of its trophic level. If this is so, food chains will frequently be shorter than they would otherwise.

21.4.4 Appraisal

(another) pattern with no agreed explanation

Several plausible hypotheses to account for the shortness of food chains are on offer. Is it possible to determine which is correct? The simple answer is 'no'. As May (1981) has remarked, the situation for food chain length is typical of the situation for community structure generally: 'the empirical patterns are widespread and abundantly documented, but instead of an agreed explanation there is only a list of possibilities to be explored.'

One potentially testable prediction of the stability hypothesis is that food chains should be shorter in less predictable environments, since only the most resilient food webs would be expected to persist there, and short food chains are more resilient. When Briand (1983) assembled his 40 food webs, he divided them into those associated with fluctuating environments (cases 1–28 in Table 21.2) and those associated with constant environments (cases 29–40). There is no significant difference in the mean length of maximal food chains in the two sets: 3.66 and 3.60 trophic levels, respectively (Figure 21.9). Further critical tests of these ideas are required.

21.5 Non-demographic stability

Lastly, as has been suggested already, community stability can be viewed from perspectives other than a purely demographic one.

First, recall McNaughton's (1978) studies from p. 770. Table 21.3 showed how species-rich grassland communities responded to perturbations with a significant drop in species diversity, not recorded in the case of species-poor communities. The effect of the perturbations was quite different, however, when viewed in terms of an aspect of functioning (primary productivity) and a composite aspect of structure (standing-crop biomass). The addition of

the stability of productivity and biomass

fertilizer significantly increased primary productivity in the species-poor field in New York State (+53%), but only slightly and insignificantly changed productivity in the species-rich field (+16%). Similarly, grazing in the Serengeti significantly reduced the standing-crop biomass in the species-poor grassland (−69%), but only slightly reduced that of the species-rich field

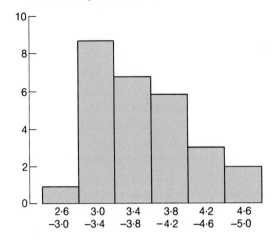

(a) Fluctuating environments

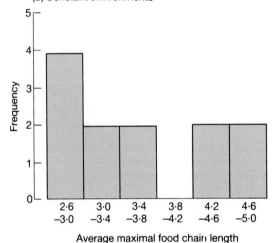

(b) Constant environments

Average maximal food chain length

Figure 21.9. Frequency histograms of average maximal food chain lengths for communities from (a) fluctuating environments and (b) constant environments. The overall means are not significantly different (3.66 links in food chains from fluctuating environments and 3.60 from constant environments). (Data from Briand, 1983.)

(−11%). In other words, a more complex structure seems to have enhanced the stability of these communities when criteria other than demographic ones are used.

resilience in the face of energy and nutrient perturbations

Several studies have revealed that the structure of the food web can influence its resilience (speed of return to equilibrium) in response to perturbations of energy and nutrient supplies. O'Neill (1976) considered the community as a three-compartment system consisting of active plant tissue (P), heterotrophic organisms (H) and inactive dead organic matter (D) (illustrated in Figure 21.10). The rate of change in standing crop in these compartments depends on transfers of energy between them. Thus the rate of change in P depends on one input (net primary productivity) and two outputs (fraction consumed by heterotrophs and fraction lost to D as litter). The rate of

Figure 21.10. A simple model of a community. The three boxes represent components of the system and arrows represent transfers of energy between the system components. (From O'Neill, 1976.)

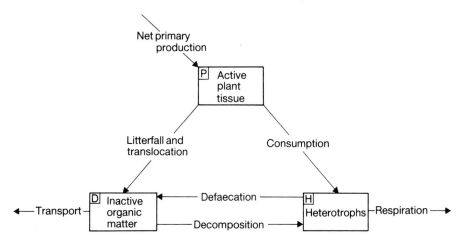

change of H depends on two inputs (consumption of living plant biomass and consumption of dead organic matter) and two outputs (defaecation and respiratory heat loss). Finally, the rate of change of D depends on two inputs (litterfall and defaecation) and two outputs (consumption by heterotrophs and physical transport out of the system). Inserting real data from six communities representing tundra, tropical forest, temperate deciduous forest, a salt-marsh, a freshwater spring and a pond, O'Neill subjected the *models* of these communities to a standard perturbation, consisting of a 10% decrease in the initial standing crop of active plant tissue. He then monitored the rates of recovery towards equilibrium, and plotted these as a function of the energy input per unit standing crop of living tissue (Figure 21.11).

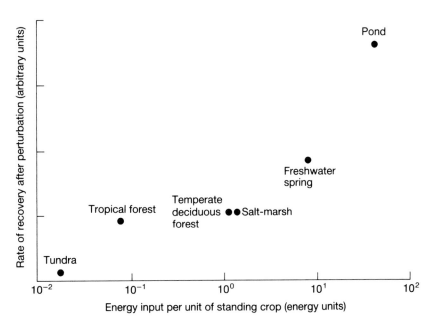

Figure 21.11. The rate of recovery (index of resilience) after perturbation (as a function of energy input per unit standing crop) for models of six contrasting communities. The pond community was most resilient to perturbation, tundra least so. (Based on data from O'Neill, 1976.)

The pond system, with a relatively low standing crop and a high rate of biomass turnover, was the most resilient to the standard perturbation. Most of its populations doubtless have short lives and rapid rates of population increase. The salt-marsh and forests had intermediate values, while tundra had the slowest rate of recovery and hence the lowest resilience. There is a clear relationship between resilience and energy input per unit standard crop. This seems to depend in part on the relative importance of heterotrophs in the systems. The most resilient system, the pond, has a biomass of heterotrophs 5.4 times that of autotrophs (reflecting the short life and rapid turnover of phytoplankton, the dominant plants in this system), while the least resilient tundra has a heterotroph to autotroph ratio of only 0.004.

resilience seems to depend on the rate of flux through the community . . .

Thus the flux of energy through the system has an important influence on resilience. The higher this flux, the more quickly will the effects of a perturbation be 'flushed' from the system. An exactly analogous conclusion has been reached by DeAngelis (1980), but for nutrient cycling rather than energy flow.

. . . and on the nutrient concerned

Finally, a striking pattern has emerged from studies based on residence times of various nutrients in various components of woodland communities. Table 21.4 shows the residence times of nitrogen and calcium in four components: soil, forest biomass, litter and detritivorous heterotrophic organisms. For example, a unit of nitrogen will typically exist in the soil for an average of 109 years, be taken up into forest biomass for 88 years, drop to the litter and remain there for up to 5 years, pass through a detritivore in a matter of days, only to be taken up into the forest biomass for a second time. Nitrogen, because of its tight recycling, tends to reside in the system for a very long time (1815 years) before being leached out and lost. Calcium is also a long-term resident (445 years). Both of these contrast with less tightly cycled elements, such as caesium.

Table 21.4. Estimated residence times, in years, of a unit of nitrogen and of calcium in four compartments, and in total, in temperate deciduous forest (from Reichle *et al.*, 1975).

Component	Nitrogen	Calcium
Soil	109	32
Forest Biomass	88	8
Litter	< 5	< 5
Decomposers	0.02	0.02
Total	1815	445

Models involving both calcium and caesium, and based on the structure and measured flux rates in a Puerto Rican tropical rainforest, were subjected to a perturbation consisting of a doubling of the steady-state nutrient input for 10 years. Their return to equilibrium was then simulated for a further 30 years (Jordan *et al.*, 1972). The extra caesium in the system was quickly flushed out (Figure 21.12). This conclusion has received validation from a study of the half-life of radioactive [137]caesium from fallout, which was essentially lost from the recycling system of a tropical rainforest as soon as the leaves fell (Kline, 1970). By contrast with caesium, the calcium model showed much less resilience (Figure 21.12).

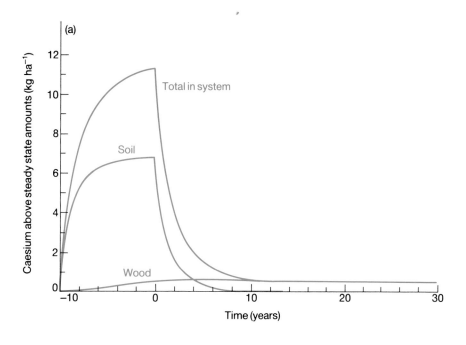

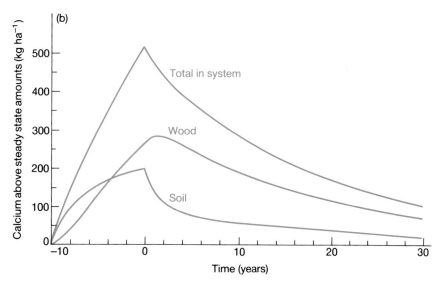

Figure 21.12. (a) Response of the caesium content of components of a model of the Puerto Rican tropical rainforest during 30 years following a perturbation (for 10 years) involving a doubling of the steady-state input rate of this element. (b) Response of the calcium content when subjected to the equivalent perturbation. (From Jordan *et al.*, 1972.)

It is clear, in short, that there is no such thing as the stability of a community. Stability varies with the aspect of the community under study and the nature of the perturbation.

Patterns of Species Diversity

22.1 Introduction

In this final chapter, we turn to some central questions which present themselves not only to community ecologists but to anybody who observes and ponders the natural world. Why do some communities contain more species than others? Are there patterns or gradients of species diversity? If so, what are the reasons for these patterns? There are plausible and sensible answers to these questions, but conclusive answers are virtually lacking. Yet this is not so much a disappointment as a challenge to ecologists of the future. Much of the fascination of ecology lies in the fact that many of the problems are blatant and obvious for everybody to see, while the solutions have as yet eluded us.

geographic factors

There are a number of factors to which the species richness of a community can be related, and these are of several different types. First, there are factors that can be referred to broadly as 'geographic', notably latitude, altitude, and, in aquatic environments, depth. These have often been correlated with species richness, as we shall discuss below, but they cannot presumably be causal agents in their own right. If species richness changes with latitude, then there must be some other factor changing with latitude exerting a direct effect on the communities.

other primary factors

A second group of factors does indeed show a tendency to be correlated with latitude and so on—but they are not perfectly correlated. To the extent that they are correlated at all, they may play a part in explaining latitudinal and other gradients. But because they are not perfectly correlated, they serve also to blur the relationships along these gradients. Such factors include the productivity of the environment, climatic variability, possibly the 'age' of the environment, and the 'harshness' of the environment (though this latter can be hard to define—see below).

A further group of factors vary geographically but not in a consistent manner, i.e. they vary quite independently of latitude, etc. They therefore tend to blur or counteract relationships between species richness and other factors. This is true of the amount of physical disturbance a habitat experiences (Chapter 19), the isolation or 'islandness' of a habitat (Chapter 20), and the extent to which it is physically and chemically heterogeneous.

secondary factors

Finally, there is a group of factors that are biological attributes of a community, but are also important influences on the structure of the community of which they are part. Notable amongst these are the amount of predation in a community (Chapter 19), the amount of competition (Chapter

18), the spatial or architectural heterogeneity generated by the organisms themselves, and the successional status of a community (Chapter 16). These should be thought of as 'secondary' factors in that they are themselves the consequences of influences outside the community. Nevertheless, they can all play powerful roles in the final shaping of a community.

A number of these factors of various types have been examined in detail in previous chapters: competition, predation, disturbance and 'islandness'. In this chapter we continue (section 22.3) by examining the relationships between species richness and a number of other factors which can be thought of as exerting an influence in their own right, namely productivity, spatial heterogeneity, climatic variation, environmental harshness, and the age of the environment (i.e. the time over which evolution/colonization has been possible). We will then be in a position (section 22.4) to consider trends with latitude, altitude, depth, succession and position in the fossil record. We begin, though, by constructing a simple theoretical framework (following MacArthur, 1972) to help us think about variations in species richness.

22.2 A simple model

We will assume, for simplicity, that the resources available to a community can be depicted as a one-dimensional continuum, R units long (Figure 22.1). Each species in the community can utilize only a portion of this resource continuum, and these portions define the niche breadths (n's) of the various species: the average niche breadth within the community is $\bar{n}$. Some, at least, of these niches are likely to overlap, and each overlap between adjacent species can be measured by a value o. The average niche overlap within the community is then $\bar{o}$.

With this simple background, it is possible to consider why some communities should contain more species than others. First, for given values of $\bar{n}$ and $\bar{o}$, a community will contain more species the larger the value of R, i.e. the greater the range of resources (Figure 22.1a). This is true when the community is dominated by competition and the species 'partition' the resources (Chapter 18). But it will also presumably be true when competition is relatively unimportant. Wider resource spectra provide the means for existence of a wider range of species, whether or not those species interact with one another.

Secondly, for a given range of resources, more species will be accommodated if $\bar{n}$ is smaller, i.e. if the species are more specialized in their use of resources (Figure 22.1b). Alternatively, if species overlap to a greater extent in their use of resources (greater $\bar{o}$) then more can coexist along the same resource continuum (Figure 22.1c). Finally, a community will contain more species the more fully saturated it is; conversely it will contain fewer species when more of the resource continuum is unexploited (Figure 22.1d).

the role of competition

We can now reconsider the various factors and processes described in previous chapters. If a community is dominated by interspecific competition (Chapter 18), the resources are likely to be fully exploited. Species richness will then depend on the range of available resources, the extent to which species are specialists, and the permitted extent of niche overlap (Figure

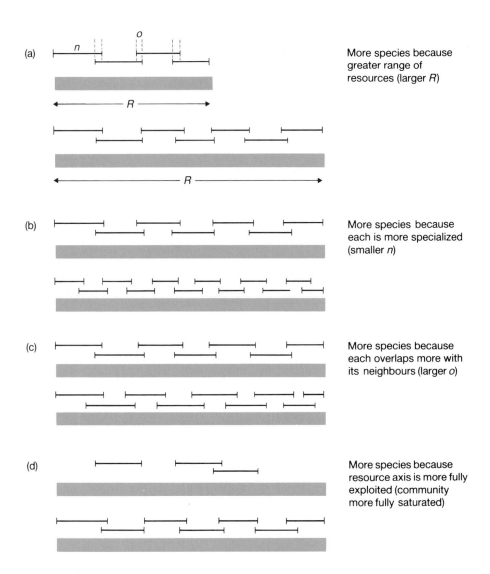

Figure 22.1. A simple model of species richness. Each species utilizes a portion *n* of the available resource dimension *R*, overlapping with adjacent species by an amount *o*. More species may occur in one community than in another (a) because a greater range of resources is present (larger *R*), (b) because each species is more specialized (smaller *n*), (c) because each species overlaps more with its neighbours (larger *o*), (d) because the resource dimension is more fully exploited.

22.1a, b and c). We will examine presently certain influences on these three things.

the role of predation

Predation is capable of exerting a variety of effects in communities. First, many field studies from a wide range of habitats have demonstrated that predators can exclude certain prey species (Holt, 1984; Jeffries & Lawton, 1984, 1985). A community in the absence of these species may then be less than fully saturated, in the sense that some available resources may go unexploited (Figure 22.1d). Secondly, predation may tend to keep species below their carrying capacities for much of the time, reducing the intensity and importance of direct interspecific competition for resources, and permitting much more niche overlap and a greater richness of species than in a community dominated by competition (Figure 22.1c). (Note that physical disturbances can exert a similar effect—Chapter 19.) Thirdly, predation may generate patterns in the structure of communities similar to those produced by competition. Holt (1977, 1984) and Jeffries and Lawton (1984, 1985) have shown how, in theory, prey species may compete for 'enemy-free space'. Such 'apparent competition' means that invasion and stable coexistence of prey in a habitat are favoured by prey being sufficiently different from other prey species already present. In other words, there may be a limit to the similarity of prey that can coexist (equivalent to the presumed limits to similarity of coexisting competitors; p. 678).

the role of islands

Finally, the depauperate communities of islands (Chapter 20) can be seen in the present context as partly reflecting the reduced range of resources offered by smaller areas (Figure 22.1a), and partly reflecting a reduced level of saturation (Figure 22.1d) as a result of a higher tendency for species to go extinct, combined with a higher probability that not all supportable species will have colonized the island.

22.3 Richness relationships

In the following sections, we examine the relationships between species richness and a number of factors that may, in theory, influence the composition of communities. It will become clear that it is often extremely difficult to come up with unambiguous predictions and 'clean' tests of hypotheses when dealing with something as complex as a community.

22.3.1 Productivity

variations in productivity

For plants, the productivity of the environment can depend on whichever resource or condition is most limiting to growth. There is a general increase in primary productivity from the poles to the tropics (Chapter 17) as light levels, average temperatures and the length of the growing season all increase. With increasing altitude in terrestrial environments, the declines in temperature and the length of the growing season lead to a general drop in productivity; while in aquatic environments, productivity typically declines with depth as temperature and light levels fall. In addition, there is often a striking decrease in productivity with aridity, especially in relatively dry environments where water-supply may limit growth; and there is always likely to be an increase in productivity associated with an increase in the supply rates of essential nutrients like nitrogen, phosphorus and potassium. Broadly speaking, the productivity of the environment for animals follows these same trends, both as a result of the changes in resource levels at the base of the food chain, and as a result of the changes in temperature and other conditions.

If an increase in productivity leads to an increased range of available resources, then this is likely to lead to an increase in species richness (Figure 22.1a). However, a more productive environment may contain larger amounts or supply rates of resources without this affecting the variety of resources. This might lead to more individuals per species rather than more

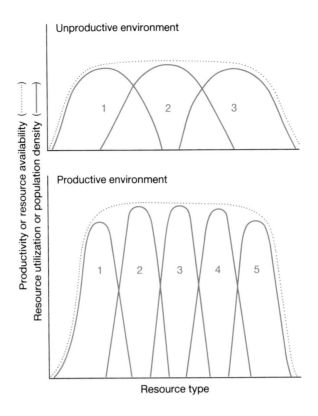

Figure 22.2. A more productive environment can support a greater number of more specialized species (smaller *n* at equilibrium).

increased productivity
might be expected to lead
to increased richness . . .

species. On the other hand, it is possible, even if the overall variety of resources is unaffected, that rare resources or low-productivity segments of the resource spectrum, which are insufficient to support species in an unproductive environment, may become abundant enough in a productive environment for extra species to be added. A similar line of argument would suggest that if the community were dominated by competition, then an increase in the quantity of resources would allow greater specialization (i.e. a decrease in $\bar{n}$) without the individual specialist species being driven to very low densities (Figures 22.2 and 22.1b).

In general then, we might expect species richness to increase with

Figure 22.3. Patterns of species richness of seed-eating rodents (▲) and ants (●) inhabiting sandy soils in a geographic gradient of precipitation and productivity (after Brown & Davidson, 1977).

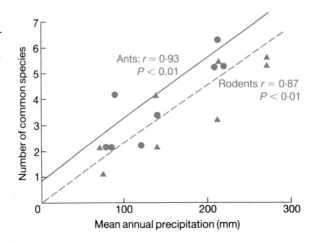

productivity. This has certainly been demonstrated by Brown and Davidson (1977), who found very strong correlations between species richness and precipitation for both seed-eating ants and seed-eating rodents in the south-western deserts of the United States (Figure 22.3). In such arid regions, it is well established that mean annual precipitation is closely related to primary productivity and thus to the amount of seed resource available. It is particularly noteworthy (Davidson, 1977) that in the species-rich sites, the ant communities contain more very large species (which consume large seeds) and more very small species (which take small seeds). There are also more very small rodents at these sites. It seems that either the range of sizes of seeds is greater in the more productive environments, or that the abundance of seeds becomes sufficient to support extra consumer species.

It is difficult to provide other unambiguous associations between species richness and productivity, because although the two often vary in parallel (for instance, with latitude or altitude) there are usually further factors that vary

. . . and there is evidence to support this

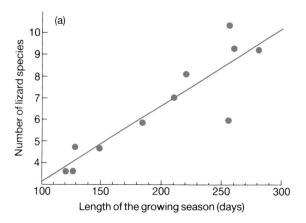

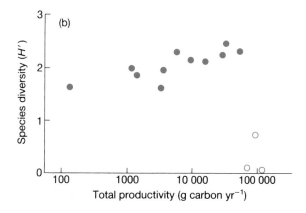

Figure 22.4. (a) Species richness of lizards at 11 sites in the south-western United States plotted against length of the growing season (after Pianka, 1967). (b) Relationship between species diversity (H) of chydorid cladocerans and primary productivity (expressed as g carbon yr^{-1} per entire lake) in 14 lakes in Indiana, U.S.A. Open circles show three heavily polluted lakes (data of Whiteside & Harmsworth, 1967, re-analysed in Brown & Gibson, 1983).

alongside them (see below). The correlation need not therefore imply a causal connection. Nevertheless, Pianka (1967) has described a positive relationship between the species richness of lizards in the deserts of the south-western United States and the length of the growing season—an important aspect of productivity in desert environments (Figure 22.4a); while Brown and Gibson (1983, using data from Whiteside & Harmsworth, 1967) have demonstrated that the diversity of chydorid cladocerans (a kind of zooplankton) in 11 unpolluted Indiana lakes is positively related to the total productivity of those lakes, measured as grams of carbon per entire lake per year (Figure 22.4b).

but other evidence shows richness declining with productivity . . .

On the other hand, an increase in diversity with productivity is by no means universal, as shown for example by the unique 'Parkgrass' experiment which has been running from 1856 to the present day at Rothamsted in England. An eight-acre pasture was divided into 20 plots, two serving as controls and the others receiving a fertilizer treatment once a year. Figure 22.5 shows how species diversity (*H*—the Shannon index, p. 595) and the equitability (*J*) of grass species changed between 1856 and 1949 for both control plots and those receiving complete fertilization. While the unfertilized areas remained essentially unchanged, the fertilized areas show a progressive decline in diversity and equitability. This decline in diversity (referred to by Rosenzweig (1971) as 'the paradox of enrichment') has also been found in several other studies of plant communities. It can be seen when the cultural eutrophication of lakes, rivers, estuaries and coastal marine regions consistently leads to a decrease in diversity of phytoplankton (as well as an increase in primary productivity); and it is paralleled by the fact that two of the most species-rich plant communities in the world occur on very nutrient-poor soils (the Fynbos of South Africa and the heath scrublands of Australia), whereas nearby communities on more nutrient-rich soils have much lower plant richness (all reviewed by Tilman, 1982).

and further evidence suggests a 'humped' relationship

Moreover, there are an increasing number of studies suggesting that diversity or richness may be highest at intermediate levels of productivity. For instance, there are humped curves when the number of Malaysian rainforest woody species is plotted against an index of phosphorus and potassium concentration, expressing the resource richness of the soil (Figure 22.6a), and

Figure 22.5. Species diversity (*H*) and equitability (*J*) of a control plot and a fertilized plot in the Rothamsted Parkgrass experiment (after Tilman, 1982).

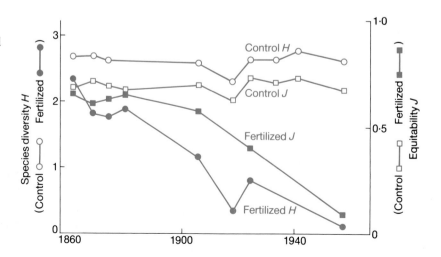

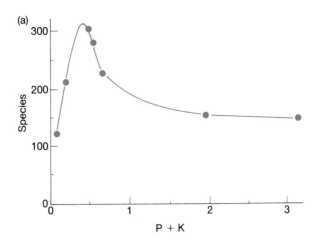

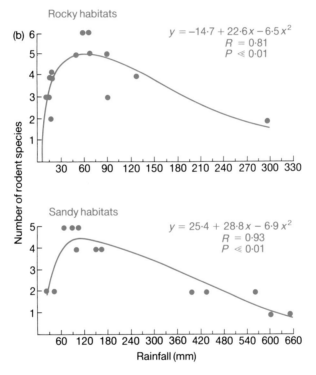

Figure 22.6. (a) Species richness of woody species in several Malaysian rainforests plotted against an index of phosphorus and potassium concentration (from Tilman, 1982). (b) Relationship between species richness of desert rodents in Israel plotted against rainfall for both rocky and sandy habitats (from Abramsky & Rosenzweig, 1983).

also when the species richness of desert rodents is plotted against precipitation (and thus productivity) along a gradient in Israel (Figure 22.6b). Similarly, Whittaker and Niering (1975) have reported that the highest diversities of vascular plants along an elevation gradient in Arizona coincided with intermediate positions along the moisture gradient and intermediate levels of primary productivity.

explanations for declines
in richness with
increasing productivity

As we have already explained, it is easy to see why diversity might increase with productivity (either generally or in the left-hand limb of a humped curve). But a decrease in diversity with productivity is less easy to explain. One possible solution is that high productivity leads to high rates of population growth, bringing a speedy conclusion to any potential competitive exclusion (Huston, 1979). At lower productivities, the environment is more likely to have changed before competitive exclusion is achieved.

A second explanation is that some other factor varies in parallel with productivity, and it is this, not productivity itself, which has a causal influence on species richness. Indeed, in discussing the data in Figure 22.6b, Abramsky and Rosenzweig (1983) noted that the gradient of precipitation and productivity was also a gradient of disturbance. The data may therefore actually reflect a peak in richness at intermediate disturbance (p. 720).

A third explanation for a decline in richness with productivity (and thus for humped curves too) is provided by the work of Tilman (1982). Figure 22.7 describes, in Tilman's terms (p. 682), a situation in which five species compete for two resources. A relatively unproductive environment is depicted by the circle to the lower-left, encompassing a range of concentrations of the two resources, all of which are fairly low. This range of supply points allows all five species to coexist. By contrast, the circle to the upper-right depicts a relatively productive environment: this allows the existence of only two of the five species. It does so because the circle is the same size as for the unproductive environment. In other words, although the average concentrations of resources in the productive environment are higher, the ranges of concentrations are the same, and the range of resource *ratios* is therefore smaller.

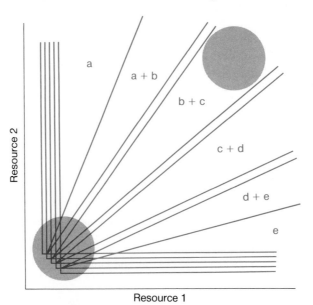

Figure 22.7. Richness can decline with increasing productivity. The supply points of two resources (1 and 2) that permit the existence or coexistence of five species (a–e), based on the work of Tilman (1982) as described in Chapter 18 (Figure 18.7b). The shaded zone to the lower-left represents the supply points to be found in a relatively unproductive environment, permitting all five species to coexist. The zone to the upper-right is from a relatively productive environment (with, however, the same *range* of supply points)—only two species can coexist.

Since it is the range of resource ratios which promotes the coexistence of several species, richness is lower in more productive environments.

Tilman's explanation, then, relies on the increase in the range of resource concentrations in productive environments being somewhat less than the increase in the average concentrations. This seems to be a reasonable supposition. His explanation is also easier to apply when the resources are both essential (and not, say, substitutable—p. 118). It is easier to apply too when the organisms are immobile; high mobility would make a single, all-encompassing supply point most appropriate. Tilman's explanation is therefore more appropriate for plants than for animals.

more resources, or a wider range of resources?

Overall, it seems that when increased productivity means an increased range of resources, an increase in species richness is to be expected (and has, at least sometimes, been found). In particular, a more productive and diverse community of plants is likely to support a greater richness of herbivores, and so on up the food chain. On the other hand, if increased productivity means an increased supply but not a greater range of resources, then there are theoretical grounds for expecting both an increase and a decrease in species richness. And the evidence, especially from plants, suggests that a decrease in species richness with resource enrichment is most common, or at least that a humped curve of species richness will be found if the whole productivity range is examined.

more light may lead to more light regimes

In this context, it is worthwhile reconsidering the nature of light as a resource for plants. In productive environments (like tropical forests), with light supplied at high rates, the light is reflected and diffused down a long column of vegetation. Hence there is not only a high rate of supply, but also a long and gradual gradient of light intensities (extending to very high intensities) and probably also a wide range of frequency spectra (p. 77). A high rate of supply, therefore, seems to lead necessarily to a wide range of light regimes, to an increased opportunity for specialization, and thus to an increased species diversity. A further consequence is that the tallest species must be able to operate over the whole range of light intensities, as they grow up from ground level to the upper canopy.

22.3.2 Spatial heterogeneity

We have already seen how the patchy nature of an environment, coupled with aggregative behaviour, can lead to coexistence of competing species (Atkinson & Shorrocks, 1981). In addition, environments which are more spatially heterogeneous can be expected to accommodate extra species, precisely because they provide a greater variety of microhabitats, a greater range of microclimates, more types of places to hide from predators, and so on. In effect, the extent of the resource spectrum is increased (Figure 21.1a). (Environments which experience patchy disturbance are likely to be spatially heterogeneous as a consequence—but it is convenient to think of this as 'disturbance' [Chapter 19]. Here we deal separately with spatial heterogeneity in its own right.)

In some cases, it has been possible to relate species richness to the spatial heterogeneity of the abiotic environment. Harman (1972), for instance, studied freshwater molluscs in a total of 348 sites, including road-side

richness and the
heterogeneity of the
abiotic environment

ditches, swamps, rivers and lakes, and related species richness to an estimate of the number of types of mineral and organic substrates present. As Figure 22.8 shows, he uncovered a convincing positive relationship. Similarly, a plant community covering a range of soil types and a variety of topographies is almost certain (other things being equal) to contain more species than one covering a flat area of homogeneous soil.

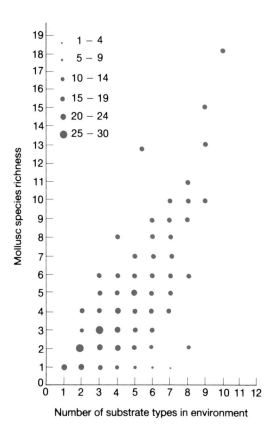

Figure 22.8. Relationship between species richness of freshwater molluscs and spatial heterogeneity, assessed in terms of the number of substrate types present. The size of the plotted points reflects the number of study sites which fall into that category. (From Harman, 1972.)

Most studies of spatial heterogeneity, however, have related the species richness of animals to the structural diversity of the plants in their environment. For example, Figure 22.9a shows the positive correlation between the species richness of freshwater fish and an index of plant spatial heterogeneity in lakes in northern Wisconsin; while Figure 22.9b shows a similar correlation for birds in Mediterranean-type habitats in California, central Chile and south-west Africa. The difficulty with such studies is that the diversity of the animals may actually have been caused by the same factor that led to the diversity of the vegetation. For instance, for the data in Figure 22.9a, vegetation diversity was itself correlated with both lake size and plant nutrient concentration.

Thus evidence that spatial heterogeneity is important in its own right in animal communities is much more convincing when the correlation of animal richness with plant *structural* diversity is far stronger than the correlation with plant *species* diversity. A number of studies of birds have shown this to be the case, and the same pattern has been described for lizards in the deserts of the

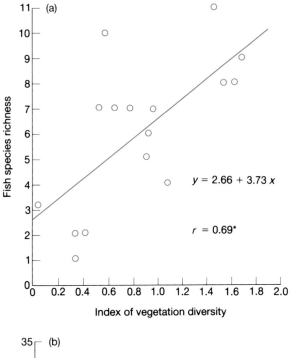

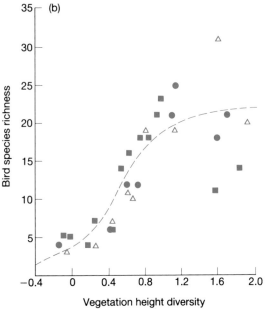

Figure 22.9. Relationships between animal species richness and an index of structural diversity of the vegetation for: (a) freshwater fish in 18 Wisconsin lakes (after Tonn & Magnuson, 1982); and (b) birds in Mediterranean-type habitats in California (▲) central Chile (●) and south-west Africa (■) (after Cody, 1975).

south-west United States (Figure 22.10). Lizards of different species forage at different heights in the vegetation, and use perches at characteristic locations to watch for food, competitors, mates and predators. Whether spatial heterogeneity arises intrinsically from the abiotic environment, or from the other biological components of the community, it is capable of promoting an increase in species richness.

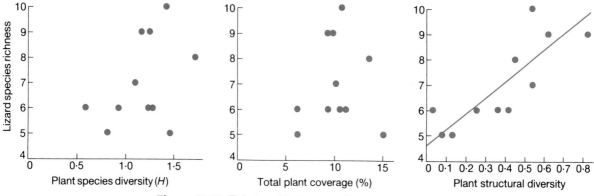

Figure 22.10. Relationship between lizard species richness and three measures of vegetation in desert sites in the south-west United States (after Pianka, 1967).

22.3.3 Climatic variation

The effects of climatic variation on species richness depend on whether the variation is predictable or unpredictable (measured on time-scales that matter to the organisms involved). In a predictable, seasonally changing environment, different species may be suited to conditions at different times of the year. More species might therefore be expected to coexist in a seasonal environment than in a completely constant one. Different annual plants in temperate regions, for instance, germinate, grow, flower and produce seeds at different times during a seasonal cycle; while phytoplankton and zooplankton pass through a seasonal succession in large, temperate lakes, with a variety of species dominating in turn as changing conditions and resources become suitable for each.

On the other hand, there are opportunities for specialization in a non-seasonal environment that do not exist in a seasonal environment. For example, it would be difficult for a long-lived obligate fruit-eater to exist in a seasonal environment when fruit is available for only a very limited portion of the year. But such specialization is found repeatedly in non-seasonal, tropical environments where fruit of one type or another is available continuously.

Unpredictable climatic variation (climatic instability) could have a number of effects on species richness. On the one hand: (i) stable environments may be able to support specialized species that would be unlikely to persist where conditions or resources fluctuated dramatically (Figure 22.1b); (ii) stable environments are more likely to be saturated with species (Figure 22.1d); and (iii) theoretical considerations (p. 803) suggest that a higher degree of niche overlap will be found in more stable environments (Figure 22.1c). All these processes could increase species richness. On the other hand, populations in a stable environment are more likely to reach their carrying capacities, the community is more likely to be dominated by competition, and species are therefore more likely to be excluded through competition ($\bar{o}$ smaller—Figure 22.1c). It is therefore reasonable to argue that unpredictable climatic variation is a form of disturbance, and species richness may be highest at 'intermediate' levels, i.e. species richness may increase *or* decrease with climatic instability.

temporal niche differentiation in seasonal environments

specialization in non-seasonal environments

climatic instability may increase or decrease richness . . .

Some studies have seemed to support the notion that species richness increases as climatic variation decreases. For example, MacArthur (1975) examined the birds, mammals and gastropods that inhabit the west coast of North America (from Panama in the south to Alaska in the north) and found a significant negative relationship between species richness and the range of monthly mean temperatures (Figure 22.11). However, there are many other things that change between Panama and Alaska, and MacArthur's correlation therefore cannot be said to prove causation. Other studies of climatic variation have been similarly unable to provide unequivocal conclusions.

. . . but there is no good evidence either way

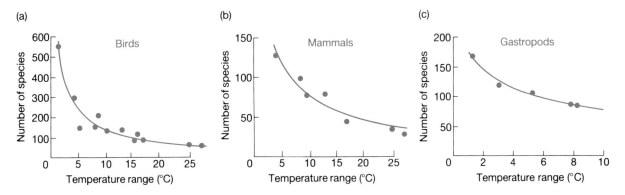

Figure 22.11. Relationships between species richness and temperature range at sites along the west coast of North America for (a) birds, (b) mammals, and (c) gastropods (after MacArthur, 1975).

22.3.4 Environmental harshness

Environments dominated by an extreme abiotic factor—often called harsh environments—are more difficult to recognize than might be immediately apparent. An anthropocentric view might describe as 'extreme' both very cold and very hot habitats, unusually alkaline lakes and grossly polluted rivers. However, species have evolved which live in all such environments; and what is very cold and extreme for us must seem benign and unremarkable to a penguin.

what is harsh?

A less arbitrary definition might state that for factors that can be classified along a continuum, values close to the minimum and maximum are extreme. But is a relative humidity close to 100% (saturated air) as extreme as one of 0%? And is the minimum concentration of a pollutant 'extreme'? Certainly not.

We might get round the problem by 'letting the organism decide'. An environment may be classified as 'extreme' if organisms, by their failure to live there, show it to be so. But if the claim is to be made—as it often is—that species richness is lower in extreme environments, then this definition is circular, and it is designed to prove the very claim we wish to test.

Perhaps the most reasonable definition of an extreme condition is one that requires, of any organism tolerating it, a morphological structure or biochemical mechanism which is not found in most related species, and is costly, either in energetic terms, or in terms of the compensatory changes in the biology of the organism that are needed to accommodate it. For example, plants living in highly acidic soils (low pH) may be affected through direct injury by hydrogen ions, or indirectly via deficiencies in the availability and uptake of important resources such as phosphorus, magnesium and calcium. In addition, aluminium, manganese and heavy metals may have their solubility increased to toxic levels, and mycorrhizal activity and nitrogen fixation may be impaired. Plants can only tolerate low pH if they have specific structures or mechanisms allowing them to avoid or counteract these effects.

In unmanaged grasslands in northern England, the mean number of plant species recorded per 1 m² quadrat was lowest in soils of low pH (Figure 22.12a). Similarly, the diversity of benthic stream invertebrates in the Ashdown Forest (southern England) was markedly lower in the more acidic streams (Figure 22.12b). Further examples of extreme environments which are associated with low species diversity include hot springs, caves, and highly saline water bodies such as the Dead Sea. The problem with these examples, however, is that they are also characterized by other features associated with low species richness. Many are unproductive and (perhaps as a consequence) most have low spatial heterogeneity. In addition, many occupy small areas (caves, hot springs) or are at least rare compared to other types of habitat (only a small proportion of the streams in southern England are acidic). Hence 'extreme' environments can often be seen as small and isolated islands. Although it appears reasonable that intrinsically extreme environments should as a consequence support few species, this has proved an extremely difficult proposition to establish.

it seems that harsh environments are species-poor—but the concept of 'harsh' is difficult to define and the evidence is difficult to interpret

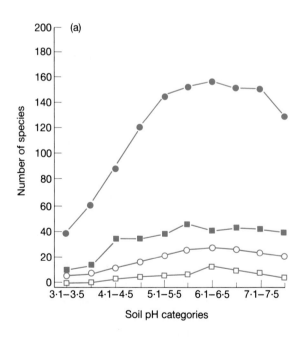

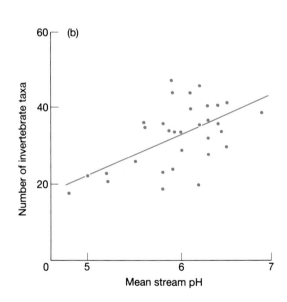

Figure 22.12. (a) Mean (○), maximum (■) and minimum (□) plant species richness per square metre, and total number (●) of species in categories of surface soil pH in unmanaged grasslands in northern England (from Grime, 1973). (b) Number of taxa of invertebrates in streams in Ashdown Forest, southern England, plotted against pH of the stream water (from Townsend *et al.*, 1983).

22.3.5 Environmental age: evolutionary time

The idea that a community may have a relatively low species richness because there has been insufficient time for colonization or evolution has been discussed already in the context of island communities (Chapter 20). In addition, we have seen that the non-equilibrium nature of many communities in disturbed habitats is the result of incomplete re-colonization following a disturbance (Chapter 19). It has often been suggested, however, that communities occupying large areas, and which are 'disturbed' only on very extended time-scales, may also lack species because they have yet to reach an ecological or an evolutionary equilibrium (e.g. Stanley, 1979). Thus communities may differ in species richness because some are closer to equilibrium and are therefore more saturated than others (Figure 22.1d).

variable recovery following the glaciations

The idea has been proposed most frequently with regard to the recovery of communities following the Pleistocene glaciations. For example, the low diversity of temperate forest trees in Europe compared to North America has been attributed to the fact that the major mountain ranges in Europe run east to west (the Alps and the Pyrenees), whereas in North America they run north to south (the Appalachians, the Rockies and the Sierra Nevada). The suggestion is that in Europe the trees were trapped by the glaciers against the mountains and became extinct, whereas in North America they simply retreated southwards and survived. There has subsequently been insufficient evolutionary time for the European trees to recover their equilibrium richness. In fact, even in North America it is unlikely that equilibrium is regained during interglacial periods, because of the slow rate of postglacial spread of the displaced species (section 1.2.2).

the unchanging tropics and the recovering temperate zones?

More generally, it has often been proposed that the tropics are richer in species than are more temperate regions (see below) at least in part because the tropics have existed over long and uninterrupted periods of evolutionary time, whereas the temperate regions are still recovering from the Pleistocene glaciations (or even more ancient events). It seems, however, that the long-term stability of the tropics has in the past been greatly exaggerated by ecologists. While the climatic and biotic zones of the temperate region moved towards the equator during the glaciations (Figure 22.13), the tropical forest appears to have contracted to a limited number of small refuges surrounded by grasslands (Figure 22.13b; see section 1.2.2 for a more complete discussion). A simplistic contrast between the unchanging tropics and the disturbed and recovering temperate regions is therefore untenable. If the lower species richness of communities nearer the poles is to be attributed partly to their being well below an evolutionary equilibrium, then some complex (and unproven) argument must be constructed. (Perhaps the movements of temperate zones to a very different latitude caused many more extinctions than the contractions of the tropics into smaller areas at the same latitude.) The matter could be settled if there were a detailed fossil record showing that the tropics have always had approximately the same species richness, and that the temperate regions either had a very much higher species richness in the past, or that their species richness is currently increasing markedly. Unfortunately, no such fossil record has been un-

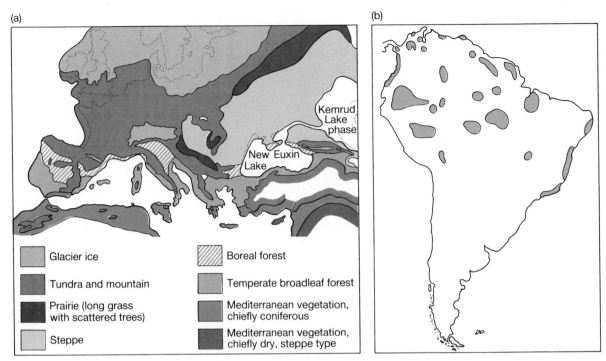

(a)

(b)

Figure 22.13. (a) Reconstructed zones of vegetation in Europe during the glacial maximum (18 000 years before present). Major vegetation types were shifted southward of their present locations by 10° to 20° of latitude, and the precursors of the Black Sea and the Caspian Sea were interconnected (after Flint, 1971).

(b) Pleistocene forest refugia in Amazonia. During glacial episodes most of the Amazon basin was grassland or savanna habitat, and tropical rainforest was restricted to isolated patches as shown here (after Prance, 1981).

covered. Thus, while it seems likely that some communities are further below equilibrium than others, it is impossible at present to pin-point those communities with certainty or even confidence.

22.4 Gradients of richness

As section 22.3 makes plain, *explanations* for variations in species richness are difficult to formulate and test. However, *patterns* in species richness are easy to document and these are discussed below.

22.4.1 Latitude

richness decreases with latitude

Perhaps the most widely recognized pattern in species diversity is the increase that occurs from the poles to the tropics. This can be seen in a wide variety of groups, including trees, marine bivalves, ants, lizards and birds (Figure 22.14). The pattern can be seen, moreover, in terrestrial, marine and freshwater habitats. For instance, Stout and Vandermeer (1975) found that these were usually 30–60 species of insects in streams in tropical America, compared with 10–30 species in the temperate United States. Such an increase in diversity is apparent not only over extensive geographic regions, but also in small communities. A single hectare of tropical rainforest may

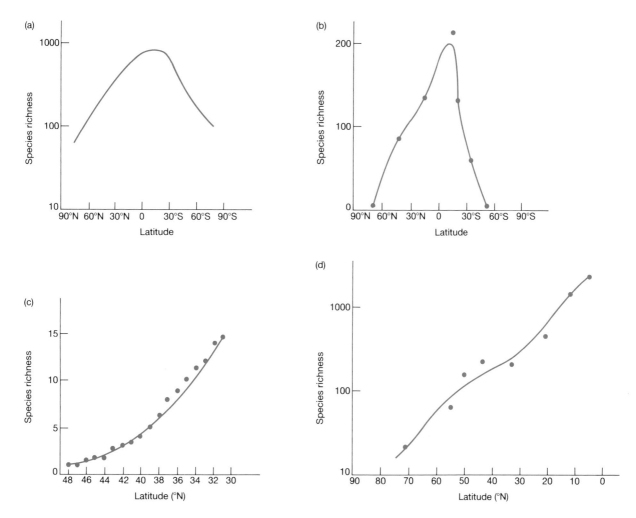

Figure 22.14. Latitudinal patterns in species richness of: (a) marine bivalve molluscs (after Stehli *et al.*, 1967); (b) ants (after Kusnezov, 1957); (c) lizards in the United States (Pianka, 1983); (d) breeding birds in North and Central America (after Dobzhansky, 1950).

contain 40–100 different tree species, while comparable areas of deciduous forest in eastern North America and coniferous forest in northern Canada usually have 10–30 species and 1–5 species, respectively (Brown & Gibson, 1983). There are, of course, exceptions. Particular groups like penguins and seals are most diverse in polar regions; while coniferous trees and ichneumonid parasitoids both reach peaks of diversity at temperate latitudes. For each of these exceptions, however, there are many groups found only in the tropics, for instance the New World fruit bats and the Indo-Pacific giant clams.

A number of explanations have been put forward for the general latitudinal trend in diversity, but not one of these is without problems. In the first place, the richness of tropical communities has been attributed to a greater intensity of predation. Both Janzen (1970) and Connell (1971) proposed that natural enemies could be a key factor in maintaining a high richness of tree species in tropical forests. Their prediction, that disproportionately high mortality of young trees will occur close to adults because these harbour host-specific consumers, has received support from a number of studies (Clark & Clark, 1984; see, for example, Figure 22.15). If there is a low probability of recruitment close to conspecific adults, the

predation as a 'secondary' explanation

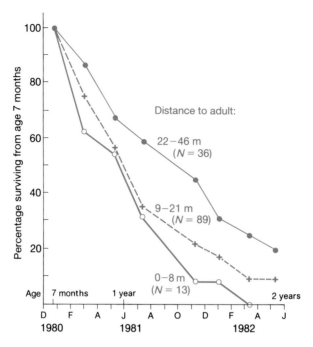

Figure 22.15.
Survivorship of seedlings of the tropical rainforest tree *Dipteryx panamensis*, from 7 months to 2 years after germination, when situated at different distances from the nearest adult (from Clark & Clark, 1984).

likelihood of establishment of non-conspecifics is enhanced, and a rich diversity of tree species can be expected to result. Note, however, that while host-specific predation may contribute to tropical diversity, it cannot be the root cause, since predation is itself an attribute of the community.

Secondly, and more persuasively, diversity has been related to the increase in productivity from the poles to the tropics. This seems reasonable for the heterotrophic elements of the community: decreasing latitude means a wider range of resources, i.e. more different types of resource existing in exploitable amounts. But if this is true for the animals, then it must depend on there also being an increase in the diversity of plants nearer the tropics. Can this be explained by increasing productivity?

If increasing productivity nearer the tropics means 'more of the same', as it does for light, then this would be expected to lead to a lower not a higher species richness (section 22.3.1). More light may, however, mean a wider range of light regimes, and thus greater richness—but this is only conjecture (p. 793). On the other hand, light is not the only determinant of plant productivity. Tropical soils tend, on average, to have lower concentrations of plant nutrients than temperate soils. The species-rich tropics might therefore be seen as reflecting their *low* productivity. The tropical soils are poor in nutrients because most of the nutrients are locked up in the large tropical biomass, and because decomposition and release of nutrients are relatively rapid in the tropics (Chapter 17). A 'productivity' argument would therefore have to run as follows. The light, temperature and water regimes of the tropics lead to high plant biomass (not necessarily diverse). This leads to nutrient-poor soils and perhaps a wide range of light regimes. These in turn lead to high plant species richness. There is certainly no simple 'productivity explanation' for the latitudinal trend in diversity.

productivity as a complex explanation

Some ecologists have invoked the climate of the tropics as a reason for their high species richness. Certainly equatorial regions are less seasonal than temperate regions (though rainfall may follow a marked seasonal cycle in the tropics in general), and for many organisms they are probably also more predictable (though this conjecture is extremely difficult to test because of the effects of body-size and generation time on what constitutes 'predictability' to a particular species). The argument that a less seasonally variable climate allows species to be more specialized (i.e. have narrower niches, Figure 21.1b) has now received a number of tests. Karr (1971), for example, has compared the bird communities in temperate Illinois and tropical Panama. Both tropical shrub and tropical forest habitats contained very many more breeding species than their temperate counterparts, and between 25 and 50% of the increase in richness consisted of specialist fruit-eaters in the tropical habitats. Further

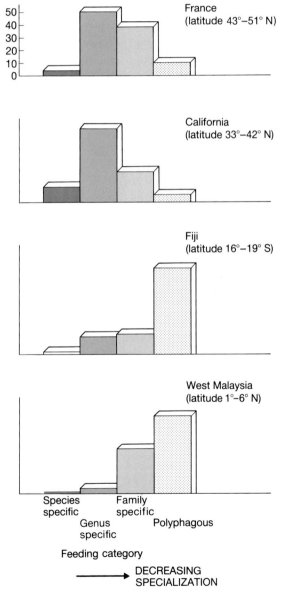

Figure 22.16. Combined number of species of ambrosia beetles and bark beetles, in various feeding categories. These species are generally more specialized in their diets at high latitudes. (From data in Beaver, 1979.)

species exploited large insects, which were available throughout the year only in the tropics. Thus the consistent availability of certain food resources creates extra opportunities for specialization in the tropical avian communities. In contrast, a group of insects, the bark and ambrosia beetles (Coleoptera: Scotytidae and Platypodidae), are less host-specific in the tropics than in temperate regions, even though there are considerably more species present in the tropics (Figure 22.16). Finally, contrasting patterns can be discerned among the parasites of marine fish. One group, the digenean trematodes, are more specialized closer to the equator, as indicated by the proportion of species which are restricted to a single host (Figure 22.17). However, monogenean parasites are just as specific at all latitudes, despite the greater species richness nearer the equator. There is, then, a plausible place for climatic variation in any explanation of the latitudinal trend in diversity, but its exact role is uncertain.

Finally, the greater evolutionary 'age' of the tropics has been proposed as a reason for their greater species richness. As already discussed, this theory too is plausible but very far from proven.

Overall, the latitudinal gradient lacks a clear and unequivocal explanation. This is hardly surprising. The components of a possible explanation—trends with productivity, climatic stability and so on—are themselves understood only in an incomplete and rudimentary way, and the latitudinal gradient intertwines these components with one another, and with other, often opposing forces. Nevertheless, the explanation may ultimately prove to be simple, for the following reason. If there was a single extrinsic factor promoting a latitudinal gradient in richness amongst the plants, for example, then the increases in resource richness and diversity, and in heterogeneity, would promote increased richness amongst herbivores. This would increase the predation pressure on the plants (promoting further increased richness) while providing a rich and diverse resource for carnivores—which would in turn increase the predation pressure on the herbivores, and so on. In short, a subtle extrinsic force could promote a cascade effect, leading eventually to a marked diversity gradient. As yet, however, we lack a convincing explanation that could 'set the ball rolling'.

Part of the problem lies with the many exceptions to the general latitudinal trend. It is, of course, as important to explain the exceptions as it is to explain the generalities. Islands are one large class of communities coming

the full explanation must be complex . . .

. . . or perhaps simple

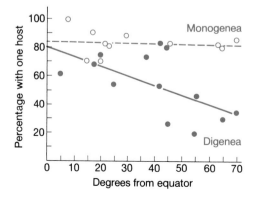

Figure 22.17. The percentage of monogenean and digenean parasite species that are restricted to a single species of fish host, at various locations on a latitudinal gradient (from data in Rohde, 1978).

into this category. In addition, deserts, even near the tropics, are species-poor, probably because they are highly unproductive (being arid) and because the climate is very extreme. Salt-marshes and hot springs are relatively poor in species, despite being highly productive, seemingly because they represent harsh or extreme abiotic environments (and in the case of the springs, because they are small islands). And as Chapter 19 made clear, neighbouring communities may differ in richness simply because of differences in the amount of physical disturbance they experience.

22.4.2 Altitude

In terrestrial environments, a decrease in species diversity with altitude is a phenomenon almost as widespread as a decrease with latitude. Someone climbing a mountain in an equatorial region will pass through a tropical habitat at the base of the mountain, and then through climatic and biotic zones that have much in common with Mediterranean, temperate and arctic

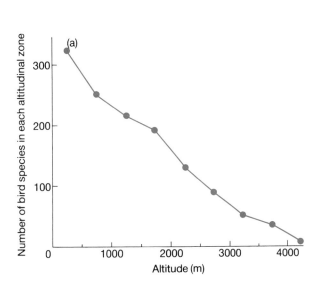

Figure 22.18.
Relationship between species richness and altitude for (a) bird species in New Guinea (after Kikkawa & Williams, 1971), and (b) vascular plants in the Nepalese Himalayas (data of K. Yoda, after Whittaker, 1977).

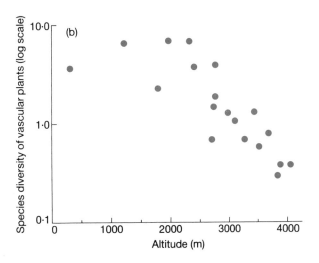

environments. If the mountaineer was also an ecologist, he would probably notice a reduction in species richness as he climbed higher. Examples are given in Figure 22.18 for birds in New Guinea and for vascular plants in the Himalayan mountains of Nepal.

This suggests that at least some of the factors instrumental in the latitudinal trend in diversity are also important as explanations for this altitudinal trend (though this is unlikely for evolutionary age, and less likely for climatic stability). Of course, the problems that exist in explaining the latitudinal trend are also equally pertinent to altitude. In addition, however, high-altitude communities almost invariably occupy smaller areas than low-lands at equivalent latitudes; and they will usually be more isolated from similar communities than lowland sites, which will often form part of a continuum. These effects of area and isolation are certain to contribute to the decrease in species richness with altitude.

On a much smaller altitudinal scale, Figure 22.19 shows how species richness can vary dramatically between the troughs and peaks of ridge-and-furrow grassland in Britain. It is worth emphasizing that great variations in community composition and diversity can occur over very small distances within what may be classified as a single community.

Figure 22.19. Histogram showing the relationship between number of species present and topography of ridge-and-furrow grassland at Marston, England.

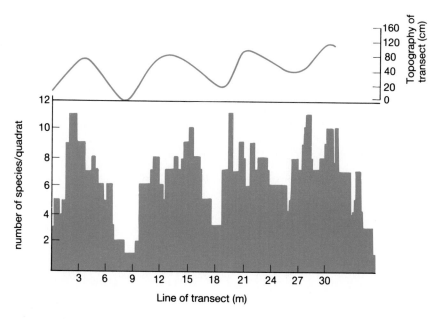

22.4.3 Depth

In aquatic environments, the change in species diversity with depth shows strong similarities to the terrestrial gradient with altitude. Certainly, in the larger lakes, the cold, dark, oxygen-poor abyssal depths contain fewer species than the shallow surface waters. Likewise, in marine habitats, plants are confined to the photic zone (where they can photosynthesize) which rarely extends below 30 m. In the open ocean, therefore, there is a rapid decrease in diversity with depth, reversed only by the variety of often bizarre

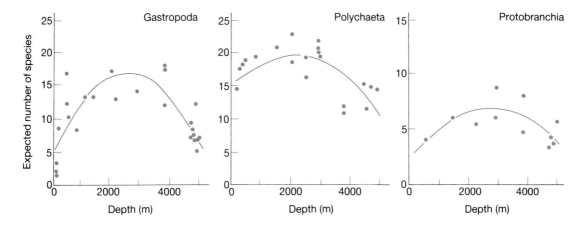

Figure 22.20. Variation in species richness (for samples of 50 individuals) of bottom-dwelling animals along the depth gradient of the ocean (after Rex, 1981).

animals living on the ocean floor. Interestingly, however, the effect of depth on the species richness of benthic invertebrates is to produce not a single gradient, but a peak of richness at about 2000 m—the approximate limit of the continental slope (Figure 22.20). This has been said to reflect an increase in the predictability of the environment between 0 and 2000 m (Sanders, 1968). At greater depths, beyond the continental slope, species richness declines again, probably because of the extreme paucity of food resources in abyssal regions.

22.4.4 Succession

Several studies of plant communities have indicated a gradual increase in species richness during succession. This increase either continues through to the climax or reverses to an extent as some late-successional species disappear. Figure 22.21 shows how richness during an old-field succession changes with time since abandonment. The increase in species richness is accompanied by a shift from a geometric type of rank–abundance curve, with a high level of dominance, to a more even distribution of ground cover among the species, and lowered dominance. The few studies which have been carried out on animals in successions indicate a parallel increase in species richness. Figure 22.22 illustrates this for birds and insects associated with different old-field successions.

To a certain extent, the successional gradient is a necessary consequence of the gradual colonization of an area by species from surrounding communities which are at later successional stages, i.e. later stages are more fully saturated with species (Figure 22.1d). However, this is only a small part of the story, since succession is essentially a process of replacement of species rather than mere addition of species.

a cascade effect
As with the other gradients, there is bound to be a cascade effect with succession. In fact, it is possible to think of succession as being this cascade effect in action. The earliest species will be those that are the best colonizers and the best competitors for open space. They immediately provide resources that were not previously present, and they introduce heterogeneity that was not previously present. For example, the earliest plants generate resource-

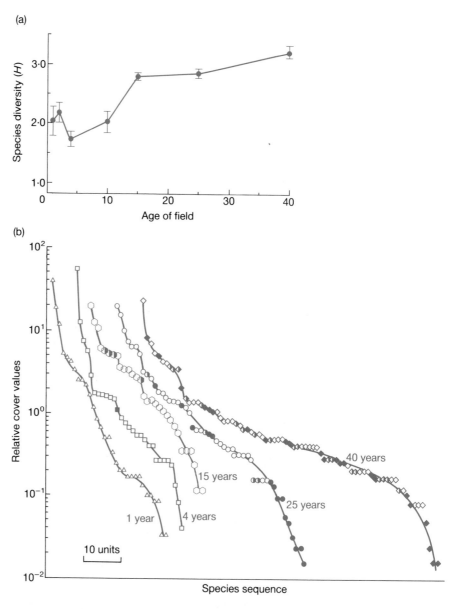

Figure 22.21.
(a) Relationship between plant species diversity (*H*) and time in succession in old fields in southern Illinois, U.S.A. (b) Rank–abundance curves for plants in old fields of five different ages. Symbols are open for herbs, half-open for shrubs and closed for trees. (From Bazzaz, 1975.)

depletion zones in the soil which inevitably increase the spatial heterogeneity of plant nutrients. The plants themselves provide a new variety of micro-habitats, and for the animals that might feed on them they provide a much greater range of food resources. The increase in herbivory and predation may then feed back to promote further increases in species richness, which provides further resources and more heterogeneity, and so on. In addition, temperature, humidity and wind speed are much less variable within a forest than in an exposed early-successional stage, and the enhanced constancy of the environment may provide a stability of conditions and resources which permits specialist species to build up populations and persist. Indeed, there is some evidence in support of this contention (e.g. Parrish & Bazzaz, 1979, 1982; Brown & Southwood, 1983).

As with the other gradients, the interaction of many factors makes it

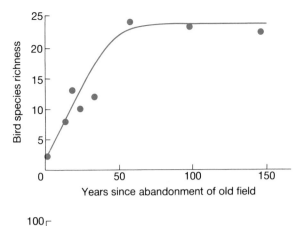

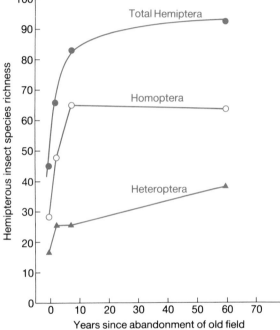

Figure 22.22. The increase in species richness during old-field successions of (a) birds (after Johnston & Odum, 1956), and (b) hemipterous insects (from data in Brown & Southwood, 1983).

difficult to disentangle cause from effect. But with the successional gradient of diversity, the tangled web of cause and effect appears to be of the essence.

22.4.5 Patterns in faunal and floral richness in the fossil record

The imperfection of the fossil record has always been the greatest impediment to the palaeontological study of evolution. Incomplete preservation, inadequate sampling, and taxonomic problems combine to make the accurate assessment of long-term evolutionary patterns in diversity a formidable task. Nevertheless, some general patterns have emerged, and our knowledge of three important groups of organisms is summarized in Figure 22.23.

About 600 million years ago, almost all the phyla of marine invertebrates entered the fossil record within the space of only a few million years (Figure 22.23a). Before that, for the previous 2500 million years, the world was populated virtually only by bacteria and algae. Of course, we can never be

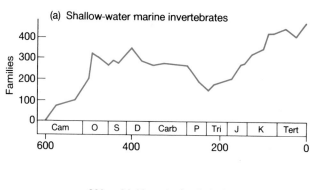

(a) Shallow-water marine invertebrates

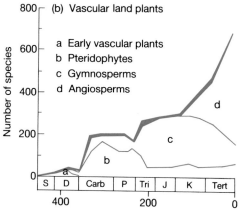

(b) Vascular land plants

a Early vascular plants
b Pteridophytes
c Gymnosperms
d Angiosperms

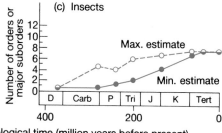

(c) Insects

Geological time (million years before present)

Figure 22.23. Curves showing patterns in diversity through the fossil record for: (a) families of shallow-water invertebrates (from Valentine, 1970); (b) species of vascular land plants in four groups—early vascular plants, pteridophytes, gymnosperms and angiosperms (from Niklas *et al,*. 1983); and (c) major orders and suborders of insects. The minimum values are derived from definite fossil records while the maximum values include 'possible' records (from Strong *et al.*, 1984). Key to geological periods: Cam, Cambrian; O, Ordovician; S, Silurian; D, Devonian; Carb, Carboniferous; P, Permian; Tri, Triassic; J, Jurassic; K, Cretaceous; Tert, Tertiary.

the Cambrian explosion—exploiter-mediated coexistence?

sure why this Cambrian 'explosion' took place. Stanley (1976), recognizing that the introduction of a higher trophic level can increase diversity at a lower level, sees the first single-celled herbivorous protist as the 'hero' of the story. The opening up of space by cropping of the algal monoculture, coupled with the availability of recently evolved eukaryotic cells, may have caused the biggest burst of evolutionary activity the world has known (Gould, 1981). Speculating on the reason for the equally dramatic decline in the number of

the Permian decline—
a species–area
relationship?

families of shallow-water invetebrates during the Permian (Figure 22.23a), Schopf (1974) noted that the late Permian was the time when the earth's continents coalesced to produce the single super-continent of Pangaea. This joining of continents produced a marked reduction in the area occupied by shallow seas (which occur around the periphery of continents) and thus a marked decline in the area of habitat available to shallow-water invertebrates. Thus the well known species–area relationship may be invoked to account for the reduction in richness of this fauna. The increase in faunal richness after the Permian may similarly reflect the increasing area of shallow seas as the continents spread apart again.

The analysis of fossil remains of vascular land plants (Figure 22.23b) reveals four distinct evolutionary phases: (i) a Silurian–mid-Devonian proliferation of early vascular plants, (ii) a subsequent late-Devonian–Carboniferous radiation of fern-like lineages, (iii) the appearance of seed plants in the late Devonian and the adaptive radiation to a gymnosperm-dominated flora, and (iv) the appearance and rise of flowering plants in the Cretaceous and Tertiary. It seems that following initial invasion of the land, the diversification of each group coincides with a decline in species numbers of the previously dominant group. Niklas *et al.* (1983) believe that in two of the transitions (early plants to gymnosperms, and gymnosperms to angiosperms), this pattern reflects the competitive displacement of older, less specialized taxa by newer and presumably more specialized taxa. On the other hand, the transition from pteridophytes to gymnosperms coincides with major environmental changes, and it may be a function of the partial extinction of the pteridophyte group and the radiation of the gymnosperms into the vacated ecological space.

competitive displace-
ment amongst the
major plant groups?

The first undoubtedly phytophagous insects are known from the Carboniferous. Thereafter, modern orders appeared steadily (Figure 22.23c) with the Lepidoptera arriving last on the scene, coincident with the rise of the angiosperms. Strong *et al.* (1984) point out that this continuous rise in the number of orders, implying a concomitant increase in the number of species, provides no reason to believe that insect species richness has reached a present day plateau or equilibrium level. Reciprocal evolution and counter-evolution between plants and phytophagous insects has almost certainly been, and still is, an important mechanism driving the increase in richness observed in both land plants and insects through their evolution.

What does the fossil record tell us about the extent to which present-day communities are 'saturated' with species? We can compare the observed patterns against two extreme perspectives. First, does evolution tend steadily to enrich communities with more and more species (arguing against any limits to species diversity or an important role for species interactions)? Alternatively does the number of niches stay roughly constant as different species fill them in turn (a pattern that is consistent with ecological interactions constraining species richness)? Taken overall, the answer appears to lie between these alternatives. Evolutionary 'breakthroughs' may have opened up major new niche dimensions, permitting dramatic jumps in species richness, followed by remarkably long periods of approximately constant species diversity.

22.5 Relative abundance of small and large species

A different kind of community pattern, one that has received little attention, is the tendency among animal species for there to be many more species of small animals than of large ones (Stearns, 1977; May, 1978; Huston, 1979). For example, May (1978) has compiled a rough estimate of the overall number of species of terrestrial animals as a function of their physical size (specifically length) (Figure 22.24). Given that a very large number of small arthropods and other invertebrates are yet to be discovered and named, the overall impression is one of a remarkable rarity of large animal species.

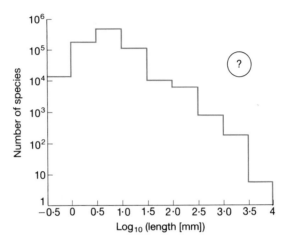

Figure 22.24. The number of species of terrestrial animals as a function of their characteristic body-length. The question mark serves to emphasize that this is a tentative figure, based on very crude approximations. (From May, 1978.)

In one sense, this is paradoxical. Theory relating to the evolution of life-history traits suggests that at the level of the individual, natural selection will tend to produce large-bodied species, with long life-spans; and, within given evolutionary lineages, there is indeed a general tendency for body size to increase through evolutionary time ("Cope's rule"). However, acting against this may be a tendency for larger species with longer generation times to suffer higher extinction rates. Fowler and MacMahon (1982) argue that extinction rates will be higher among species with low evolutionary plasticity, in this case those with long generation times. Since body size is positively correlated with generation time large-bodied species will tend to have higher extinction rates than smaller-bodied species. The notion that greater evolutionary plasticity is a property of smaller species with short generation times has a further consequence. We expect that in addition to having lower extinction rates, the rate of speciation of smaller species will be higher. Fowler and MacMahon conclude that the distribution of animal species richness according to body size may be explained as the dynamic balance between *evolution* and *selective extinction*.

22.6 Finalé

The search for patterns in species diversity, and for explanations of those patterns, has emphasized some of the general difficulties we have in testing ecological theories. To answer these problems, we will need to use evolutionary arguments (e.g. Chapters 1–3, 14), and to invoke our knowledge of population dynamics (Chapters 4–7, 10, 15) and of population interactions (Chapters 8, 9, 11–13, 16–21). The study of community ecology is one of the most difficult and challenging areas of modern ecology. Clear, unambiguous predictions and tests of ideas are often very difficult to devise and will require great ingenuity on the part of future generations of ecologists.

References

The section(s) in which each reference appears is given after the reference

Abbott, I. (1978) Factors determining the number of land bird species on islands around south-western Australia. *Oecologia,* **33,** 221–223.
20.3.1

Abrahamson, W.G. (1975) Reproductive strategies in dewberries. *Ecology,* **56,** 721–726.
8.2.3

Abrams, P. (1976) Limiting similarity and the form of the competition coefficient. *Theoretical Population Biology,* **8,** 356–375.
7.5

Abramsky, Z. & Rosenzweig, M.L. (1983) Tilman's predicted productivity–diversity relationship shown by desert rodents. *Nature* (London), **309,** 150–151.
22.3.1

Abramsky, Z. & Sellah, C. (1982) Competition and the role of habitat selection in *Gerbillus allenbyi* and *Meriones tristrami*: a removal experiment. *Ecology,* **63,** 1242–1247.
7.10.1

Abramsky, Z. & Tracy, C.R. (1979) Population biology of a 'noncycling' population of prairie voles and a hypothesis on the role of migration in regulating microtine cycles. *Ecology,* **60,** 349–361.
15.4.2

Agricultural Research Council (1965) *The Nutritional Requirements of Farm Livestock. 2. Ruminants.* Agricultural Research Council, London.
8.4

Alatalo, R.V. (1982) Bird species distributions in the Galápagos and other archipelagoes: competition or chance? *Ecology,* **63,** 881–887.
18.4.3

Albertson, F.W. (1937) Ecology of mixed prairie in west central Kansas. *Ecological Monographs,* **7,** 481–547.
3.3.2

Alicata, J.E. & Jindrak, K. (1970) *Angiostrongylosis in the Pacific and Southeast Asia.* C.C. Thomas, Springfield, Illinois.
12.4

Allee, W.C. (1931) *Animal Aggregations. A Study in General Sociology.* University of Chicago Press, Chicago.
10.5

Allen, K.R. (1972) Further notes on the assessment of Antarctic fin whale stocks. *Report of the International Whaling Commission,* **22,** 43–53.
6.4

Alphey, T.W. (1970) Studies on the distribution and site location of *Nippostrongylus brasiliensis* within the small intestine of laboratory rats. *Parasitology,* **61,** 449–460.
12.4

Anderson, J.M. (1975) Succession, diversity and trophic relationships of some soil animals in decomposing leaf litter. *Journal of Animal Ecology,* **44,** 475–495.
11.3.1

Anderson, J.M. (1978) Inter- and intra-habitat relationships between woodland Cryptostigmata species diversity and diversity of soil and litter microhabitats. *Oecologia,* **32,** 341–348.
11.2.2

Anderson, R.M. (1979) The influence of parasitic infection on the dynamics of host population growth. In: *Population Dynamics* (R.M. Anderson, B.D. Turner & L.R. Taylor eds), pp. 245–281. Blackwell Scientific Publications, Oxford.
12.5.5, 12.6.5

Anderson, R.M. (1981) Population ecology of infectious disease agents. In: *Theoretical Ecology: Principles and Applications,* 2nd edn. (R.M. May ed.), pp. 318–355. Blackwell Scientific Publications, Oxford.
12.6.5, 19.2.5

Anderson, R.M. (1982) Epidemiology. In: *Modern Parasitology* (F.E.G. Cox ed.), pp. 204–251. Blackwell Scientific Publications, Oxford.
12.3.1, 12.3.2, 12.4, 12.6.1, 12.6.3

Anderson, R.M. & May, R.M. (1978) Regulation and stability of host–parasite population interactions. I. Regulatory processes. *Journal of Animal Ecology,* **47,** 219–249.
12.6.1

Anderson, R.M. & May, R.M. (1980) Infectious diseases and population cycles of forest insects. *Science, N.Y.,* **210,** 658–661.
12.6.5

Anderson, R.M., Jackson, H.C., May, R.M. & Smith, A.M. (1981) Population dynamics of fox rabies in Europe. *Nature* (London), **289,** 765–771.
12.6.5

Andrewartha, H.G. & Birch, L.C. (1954) *The Distribution and Abundance of Animals.* University of Chicago Press, Chicago.
15.2.1

Andrewartha, H.G. & Birch, L.C. (1960) Some recent contributions to the study of the distribution and abundance of insects. *Annual Review of Entomology,* **5,** 219–242.
15.2.1

Andrews, R.V., Ryan, K., Strohben, R. & Ryan-Klein, M. (1975) Physiological and demographic profiles of brown lemmings during their cycle of abundance. *Physiological Zoology,* **48,** 64–83.
15.4.2

Antonovics, J. & Bradshaw, A.D. (1970) Evolution in closely adjacent plant populations. VIII. Clinal patterns at a mine boundary. *Heredity,* **25,** 349–362.
1.5.1

Arnold, G.W. (1964) Factors within plant associations affecting the behaviour and performance of grazing animals. In: *Grazing in Terrestrial and Marine Environments* (D.J. Crisp ed.), pp. 133–154. Blackwell Scientific Publications, Oxford.
9.2.2

Arthur, W. (1982) The evolutionary consequences of interspecific competition. *Advances in Ecological Research,* **12,** 127–187.
7.10.2

Ashby, W.R. (1960) *Design for a Brain,* 2nd edn. Chapman & Hall, London.
19.4.1

Askew, R.R. (1961) On the biology of the inhabitants of oak galls of the Cynipidae (Hymenoptera) in Britain. *Transactions of the British Society for Entomology,* **14,** 237–268.
12.4, 12.5.4

Askew, R.R. (1971) *Parasitic Insects.* American Elsevier, New York.
12.5.4

Ashmole, N.P. (1971) Sea bird ecology and the marine environment. In: *Avian Biology, Volume 1* (D.S. Farner & J.R. King eds). Academic Press, New York.
14.14

Atkinson, W.D. & Shorrocks, B. (1981) Competition on a divided and ephemeral resource: a simulation model. *Journal of Animal Ecology,* **50,** 461–471.
7.6.5, 11.4, 18.2.2, 22.3.2

Ausmus, B.S., Edwards, N.T. & Witkamp, M. (1976) Microbial immobilisation of carbon, nitrogen, phosphorus and potassium: implications for forest ecosystem processes. In: *The Role of Terrestrial and Aquatic Organisms in Decomposition Processes* (J.M. Anderson & A. MacFadyen eds), pp. 397–416. Blackwell Scientific Publications, Oxford.
11.2.3

Bailey, J.A. & Mansfield, J.W. (1982) *Phytoalexins.* Blackie, Glasgow.
12.5.3

Bailey, W.S. (1972) *Spirocerca lupi*: a continuing enquiry. *Journal of Parasitology,* **58,** 3–22.
12.4

Baker, H.G. (1972) Seed weight in relation to environmental conditions in California. *Ecology,* **53,** 997–1010.
14.7.4

Baker, R.R. (1978) *The Evolutionary Ecology of Animal Migration.* Hodder & Stoughton, London.
5.4.1

Baker, R.R. (1982) *Migration: Paths Through Time and Space.* Hodder & Stoughton, London.
5.3.3, 5.4.1

Bakker, K. (1961) An analysis of factors which determine success in competition for food among larvae of *Drosophila melanogaster. Archives néerlandaises de Zoologie,* **14,** 200–281.
6.6

Baltensweiler, W., Benz, G., Bovey, P. & Delucchi, V. (1977) Dynamics of larch budmoth populations. *Annual Review of Ecology and Systematics,* **22,** 79–100.
8.2.3

Barkalow, F.S., Hamilton, R.B. & Soots, R.F. (1970) The vital statistics of an unexploited gray squirrel population. *Journal of Wildlife Management,* **34,** 489–500.
14.3

Bärlocher, F. & Kendrick, B. (1975) Assimilation efficiency of *Gammarus pseudolimnaeus* (Amphipoda) feeding on fungal mycelium or autumn-shed leaves. *Oikos,* **26,** 55–59.
11.3.1

Barnard, C. & Thompson, D.B.A. (1985) *Gulls and Plovers: The Ecology and Behaviour of Mixed Species Feeding Groups.* Croom Helm, London.
9.6

Barnes, H. (1957) The northern limits of *Balanus balanoides* (L.). *Oikos,* **8,** 1–15.
2.4

Barnes, R.S.K. & Hughes, R.N. (1982) *An Introduction to Marine Ecology.* Blackwell Scientific Publications, Oxford.
17.3.3

Baross, J.A., Dahm, C.N., Ward, A.K., Lilley, M.D. &

Sedell, J.R. (1982) Initial microbial response in lakes to the Mt. St. Helens eruption. *Nature* (London), **296**, 49–52.
19.1.2

Bartholomew, G.A. (1982) Body temperature and energy metabolism. In: *Animal Physiology* (M.S. Gordon ed.). Macmillan, New York.
2.2.1, 2.2.2, 2.2.11

Baskin, J.M. & Baskin, C.C. (1979) Studies on the autecology and population biology of the monocarpic perennial *Grindelia lanceolata*. *American Midland Naturalist*, **41**, 290–299.
4.8

Bateson, P.P.G. (1978) Sexual imprinting and optimal outbreeding. *Nature* (London), **273**, 659–660.
5.6

Bateson, P.P.G. (1980) Optimal outbreeding and the development of sexual preferences in the Japanese quail. *Zeitschrift für Tierpsychologie*, **53**, 231–244.
5.6

Battarbee, P.W. (1984) Diatom analysis and the acidification of lakes. *Philosophical Transactions of the Royal Society of London*, B, **305**, 451–477.
2.6

Batzli, G.O. (1983) Responses of arctic rodent populations to nutritional factors. *Oikos*, **40**, 396–406.
10.3, 15.4.2

Batzli, G.O., White, R.G., MacLean, S.F., Pitelka, F.A. & Collier, B.D. (1980) The herbivore based trophic system. In: *An Arctic Ecosystem: The Coastal Tundra at Barrow, Alaska* (J. Brown, P.C. Miller, L.L. Tieszen & F.L. Bunnell eds). Dowden, Hutchinson and Ross, Stroudsburg, Pennsylvania.
15.4.2

Batzli, G.O., Jung, H.G. & Guntenspergen, G. (1981) Nutritional ecology of microtine rodents: linear forage-rate curves for brown lemmings. *Oikos*, **37**, 112–116.
9.5.2

Bazzaz, F.A. (1975) Plant species diversity in old-field successional ecosystems in southern Illinois. *Ecology*, **56**, 485–488.
22.4.4

Bazzaz, F.A. (1979) The physiological ecology of plant succession. *Annual Review of Ecology and Systematics*, **10**, 351–371.
16.4.3, 16.4.4

Beaver, R.A. (1979) Host specificity of temperate and tropical animals. *Nature* (London), **281**, 139–141.
22.4.1

Beddington, J.R., Free, C.A. & Lawton, J.H. (1978) Modelling biological control: on the characteristics of successful natural enemies. *Nature* (London), **273**, 513–519.
10.4.1

Begon, M. (1976) Temporal variations in the reproductive condition of *Drosophila obscura* Fallén and *D. subobscura* Collin. *Oecologia*, **23**, 31–47.
5.7.1

Begon, M. (1979) *Investigating Animal Abundance: Capture–Recapture for Biologists*. Edward Arnold, London.
4.3

Begon, M. (1985) A general theory of life-history variation. In: *Behavioural Ecology* (R.M. Sibly & R.H. Smith eds), pp. 91–97. Blackwell Scientific Publications, Oxford.
14.6.2

Begon, M. & Mortimer, M. (1981) *Population Ecology: A Unified Study of Animals and Plants*. Blackwell Scientific Publications, Oxford.
10.2

Begon, M., Firbank, L. & Wall, R. (1986) Is there a self-thinning rule for animal populations? *Oikos* **46**, 122–124.
6.12

Bekoff, M. (1977) Mammalian dispersal and the ontogeny of individual behavioural phenotypes. *American Naturalist*, **111**, 715–732.
5.5.4

Bellows, T.S. Jr. (1981) The descriptive properties of some models for density dependence. *Journal of Animal Ecology*, **50**, 139–156.
6.3, 6.8.2, 6.8.4

Benjamin, R.K. & Shorer, L. (1952) Sex of host specificity and position specificity of certain species of *Laboulbenia* on *Bembidion picipes*. *American Journal of Botany*, **39**, 125–131.
12.4

Benson, J.F. (1973a) The biology of Lepidoptera infesting stored products, with special reference to population dynamics. *Biological Reviews*, **48**, 1–26.
6.6

Benson, J.F. (1973b) Population dynamics of cabbage root fly in Canada and England. *Journal of Applied Ecology*, **10**, 437–446.
6.6

Bentley, B.L. (1977) Extrafloral nectaries and protection by pugnacious bodyguards. *Annual Review of Ecology and Systematics*, **8**, 407–427.
13.2.5

Beringer, J.E., Brewin, N., Johnston, A.W.B., Schulman, H.M. & Hopwood, D.A. (1979) The *Rhizobium*–legume symbiosis. *Proceedings of the Royal Society of London*, B, **204**, 219–233.
13.10.1

Bevan, G. (1976) Changes in breeding bird populations of an oak-wood on Bookham Common, Surrey, over twenty-seven years. *London Naturalist*, **55**, 23–42.
20.3.4

Beverley, S.M. & Wilson, A.C. (1985). Ancient origin for Hawaiian Drosophilinae inferred from protein comparisons. *Proceedings of the National Academy of Sciences of the U.S.A.*, **82**, 4753–4757.
1.2.3

Beverton, R.J.H. & Holt, S.J. (1957) On the dynamics of exploited fish populations. *Fishery Investigations, London* (Series II), **19**, 1–533.
10.8.6

Bigger, M. (1976) Oscillations of tropical insect populations. *Nature* (London), **259**, 207–209.
21.2.3

Biggers, J.D. (1981) *In vitro* fertilisation and embryo transfer in human beings. *New England Journal of Medicine*, **304**, 336–342.
4.5.1

Black, J.N. (1963) The interrelationship of solar radiation and leaf area index in determining the rate of dry matter production of swards of subterranean clover (*Trifolium subterraneum*). *Australian Journal of Agricultural Research*, **14**, 20–38.
3.2.3

Black, J.N. (1964) An analysis of the potential production of swards of subterranean clover (*Trifolium subterraneum* L.) at Adelaide, South Australia. *Journal of Applied Ecology*, **1**, 3–18.
8.3

Blueweiss, L., Fox, H., Kudzma, V., Nakashima, D., Peters, R. & Sams, S. (1978) Relationships between body size and some life history parameters. *Oecologia*, **37**, 257–272.
14.14.1

Boray, J.C. (1969) Experimental fascioliasis in Australia. *Advances in Parasitology*, **7**, 85–210.
2.5.5

Bormann, F.H., Likens, G.E. & Melillo, J.M. (1977) Nitrogen budget for an aggrading northern hardwood forest ecosystem. *Science, N.Y.*, **196**, 981–983.
17.5

Botkin, D.B., Jordan, P.A., Dominski, A.S., Lowendorf, H.S. & Hutchinson, G.E. (1973) Sodium dynamics in a northern ecosystem (moose, wolves, plants). *Proceedings of the National Academy of Science of the U.S.A.*, **70**, 2745–2748.
14.4.5

Boucher, D.H., James, S. & Kesler, K. (1984) The ecology of mutualism. *Annual Review of Ecology and Systematics*, **13**, 315–347.
13.1

Bowers, M.A. & Brown, J.H. (1982) Body size and coexistence in desert rodents: chance or community structure? *Ecology*, **63**, 391–400.
18.4.2

Box, E.O. (1981) *Macroclimate and Plant Forms: An Introduction to Predictive Modelling in Phytogeography*. Junk, The Hague.
1.4.1

Bradley, D.J. (1977) Human pest and disease problems: contrasts between developing and developed countries. In: *Origins of Pest, Parasite, Disease and Weed Problems* (J.M. Cherrett & G.R. Sagar eds), pp. 329–346. Blackwell Scientific Publications, Oxford.
12.5.5

Bradshaw, A.D. (1959) Population differentiation in *Agrostis tenuis* Sibth. II. The incidence and significance of infection by *Epichloe typhina*. *New Phytologist*, **58**, 310–315.
12.5.4

Bradshaw, A.D. (1965) Evolutionary significance of phenotypic plasticity in plants. *Advances in Genetics*, **13**, 115–155.
1.6

Bradshaw, A.D. (1972) Some of the evolutionary consequences of being a plant. *Evolutionary Biology*, **5**, 25–47.
1.5

Bradshaw, A.D. & McNeilly, T. (1981) *Evolution and Pollution*. Edward Arnold, London.
2.11

Branch, G.M. (1975) Intraspecific competition in *Patella cochlear* Born. *Journal of Animal Ecology*, **44**, 263–281.
6.5, 6.6, 6.10

Breznak, J.A. (1975) Symbiotic relationships between termites and their intestinal biota. In: *Symbiosis* (D.H. Jennings & D.L. Lee eds), pp. 559–580. Symposium 29, Society for Experimental Biology. Cambridge University Press, Cambridge.
13.5.2

Briand, F. (1983) Environmental control of food web structure. *Ecology*, **64**, 253–263.
21.2.2, 21.2.3, 21.4, 21.4.4

Brower, L.P. & Corvino, J.M. (1967) Plant poisons in a terrestrial food chain. *Proceedings of the National Academy of Science of the U.S.A.*, **57**, 893–898.
3.4.3

Brown, J., Everett, K.R., Webber, P.J., MacLean, S.F. & Murray, D.F. (1980) The coastal tundra at Barrow. In: *An Arctic Ecosystem: The Coastal Tundra at Barrow, Alaska* (J. Brown, P.C. Miller, L.L. Tieszen & F.L. Bunnell eds). Dowden, Hutchinson and Ross, Stroudsburg, Pennsylvania.
15.4.2

Brown, J.H. & Davidson, D.W. (1977) Competition between seed-eating rodents and ants in desert ecosystems. *Science, N.Y.*, **196**, 880–882.
7.10.2, 18.3.1, 22.3.1

Brown, J.H. & Gibson, A.C. (1983) *Biogeography*. C.V. Mosby, St. Louis.
22.3.1, 22.4.1

Brown, J.H., Reichman, O.J. & Davidson, D.W. (1979) Granivory in desert ecosystems *Annual Review of Ecology and Systematics* **10**, 201–227.

Brown, K.M. (1982) Resource overlap and competition

in pond snails: an experimental analysis. *Ecology*, **63**, 412–422.
7.6.4

Brown, V.K. & Southwood, T.R.E. (1983) Trophic diversity, niche breadth and generation times of exopterygote insects in a secondary succession. *Oecologia*, **56**, 220–225.
22.4.4

Browne, R.A. (1981) Lakes as islands: biogeographic distribution, turnover rates, and species composition in the lakes of central New York. *Journal of Biogeography*, **8**, 75–83.
19.1, 20.3.2

Bryant, J.P. & Kuropat, P.J. (1980) Selection of winter forage by subarctic browsing vertebrates: the role of plant chemistry. *Annual Review of Ecology and Systematics*, **11**, 261–285.
8.2.3, 9.2.2

Brylinski, M. & Mann, K.H. (1973) An analysis of factors governing productivity in lakes and reservoirs. *Limnology and Oceanography*, **18**, 1–14.
17.2, 17.4, 17.4.1

Buchanan, G.A., Crowley, R.H., Street, J.E. & McGuire, J.A. (1980) Competition of sicklepod (*Cassia obtusifolia*) and redroot pigweed (*Amaranthus retroflexus*) with cotton (*Gossypium hirsutum*). *Weed Science*, **28**, 258–262.
7.9.2

Bunnell, F.L., Tait, D.E.N., Flanagan, P.W. & Van Cleve, K. (1977) Microbial respiration and substrate weight loss. I. A general model of the influences of abiotic variables. *Soil Biology and Biochemistry*, **9**, 33–40.
17.5

Burdon, J.J. (1980) Intraspecific diversity in a natural population of *Trifolium repens*. *Journal of Ecology*, **68**, 717–735.
1.5.2

Burdon, J.J. & Chilvers, G.A. (1975) Epidemiology of damping off disease (*Pythium irregulare*) in relation to density of *Lepidium sativum* seedlings. *Annals of Applied Biology*, **81**, 135–143.
12.3.1

Burdon, J.J. & Chilvers, G.A. (1977) The effect of barley mildew on barley and wheat competition in mixtures. *Australian Journal of Botany*, **25**, 59–65.
19.2.5

Burnett, T. (1958) Dispersal of an insect parasite over a small plot. *Canadian Entomologist*, **90**, 279–283.
9.6

Buss, L.W. (1979) Bryozoan overgrowth interactions— the interdependence of competition for food and space. *Nature* (London), **281**, 475–477.
7.3

Cain, A.J. & Sheppard, P.M. (1954). Natural selection in *Cepaea*. *Genetics*, **39**, 89–116.
1.5.2

Callaghan, T.V. (1976) Strategies of growth and population dynamics of plants: 3. Growth and population dynamics of *Carex bigelowii* in an alpine environment. *Oikos*, **27**, 402–413.
4.6.4

Calow, P. (1981) Resource utilization and reproduction. In: *Physiological Ecology: An Evolutionary Approach to Resource Use*. (C.R. Townsend & P. Calow eds), pp. 245–270. Blackwell Scientific Publications, Oxford.
14.7.3

Cantlon, J.E. (1969) The stability of natural populations and their sensitivity to technology. In: *Diversity and Stability in Ecological Systems* (G.M. Woodwell ed.), *Brookhaven Symposium in Biology*, **22**, 197–203.
19.2.5

Carne, P.B. (1969) On the population dynamics of the eucalypt-defoliating chrysomelid *Paropsis atomaria* Ol. *Australian Journal of Zoology*, **14**, 647–672.
8.3

Carson, H.L. & Kaneshiro, K.Y. (1976) *Drosophila* of Hawaii: systematics and ecological genetics. *Annual Review of Ecology and Systematics*, **7**, 311–345.
1.2.3

Caswell, H. (1978) Predator-mediated coexistence: a non-equilibrium model. *American Naturalist*, **112**, 127–154.
19.4.1

Caughley, G. & Lawton, J.H. (1981) Plant–herbivore systems. In: *Theoretical Ecology: Principles and Applications*, 2nd edn. (R.M. May ed.), pp. 132–166. Blackwell Scientific Publications, Oxford.
10.3, 10.4.1, 15.6.5

Charlesworth, B. (1980) *Evolution in Age-Structured Populations*. Cambridge University Press, London.
14.3

Charnov, E.L. (1976a) Optimal foraging: attack strategy of a mantid. *American Naturalist*, **110**, 141–151.
9.3.1

Charnov, E.L. (1976b) Optimal foraging: the marginal value theorem. *Theoretical Population Biology*, **9**, 129–136.
9.11.1

Charnov, E.L. & Finnerty, J. (1980) Vole population cycles; a case for kin-selection? *Oecologia*, **45**, 1–2.
5.5.4, 15.4.2

Chater, E.H. (1931) A contribution to the study of the natural control of gorse. *Bulletin of Entomological Research*, **22**, 225–235.
8.2.2

Chew, R.M. & Chew, A.E. (1965) The primary productivity of a desert shrub (*Larrea tridentata*) community. *Ecological Monographs*, **35**, 355–375.
4.6.4

Chitty, D. (1960) Population processes in the vole and their relevance to general theory. *Canadian Journal of Zoology*, **38**, 99–113.
15.4.2

Christensen, B.M. (1978) *Dirofilaria immitis*: Effects on the longevity of *Aedes trivittatus*. *Experimental Parasitology*, **44**, 116–123.
12.5.5

Christian, J.J. (1950) The adreno-pituitary system and population cycles in mammals. *Journal of Mammalogy*, **31**, 247–259.
15.4.2

Christian, J.J. (1970) Social subordination, population density, and mammalian evolution. *Science, N.Y.*, **168**, 84–90.
5.5.4

Clapham, W.B. (1973) *Natural Ecosystems*. Collier-Macmillan, New York.
7.2.4

Claridge, M.F. & Wilson M.R. (1976) Diversity and distribution patterns of some mesophyll-feeding leafhoppers of temperate woodland canopy. *Ecological Entomology*, **1**, 231–250.
20.3.3

Claridge, M.F. & Wilson, M.R. (1978) British insects and trees: a study in island biogeography or insect/plant coevolution? *American Naturalist*, **112**, 451–456.
20.3.3

Clark, C.W. (1976) *Mathematical Bioeconomics: The Optimal Management of Renewable Resources*. Wiley-Interscience, New York.
10.8.1

Clark, C.W. (1981) Bioeconomics. In: *Theoretical Ecology: Principles and Applications*, 2nd edn. (R.M. May, ed.), pp. 387–418. Blackwell Scientific Publications, Oxford.
10.8.1, 10.8.2, 10.8.5

Clark, D.A. & Clark, D.B. (1984) Spacing dynamics of a tropical rain forest tree: evaluation of the Janzen–Connell model. *American Naturalist*, **124**, 769–788.
22.4.1

Clark, L.R. (1962) The general biology of *Cardiaspina albitextura* (Psyllidae) and its abundance in relation to weather and parasitism. *Australian Journal of Zoology*, **10**, 537–586.
10.6.1

Clark, L.R. (1964) The population dynamics of *Cardiaspina albitextura* (Psyllidae). *Australian Journal of Zoology*, **12**, 362–380.
10.6.1

Clausen, J., Keck, D.D. & Hiesey, W.M. (1941) Experimental studies on the nature of species. IV Genetic structure of ecological races. *Carnegie Institute of Washington Publication, No. 242*. Washington D.C.
1.5.1

Clements, F.E. (1916) Plant succession: Analysis of the development of vegetation. *Carnegie Institute of Washington Publication, No . 242*. Washington D.C.
16.3.3, 16.4.5

Clutton-Brock, T.H. & Harvey, P.H. (1979) Comparison and adaptation. *Proceedings of the Royal Society of London*, B, **205**, 547–565.
14.14.4

Clutton-Brock, T., Guinness, F.E. & Albon, S.D. (1983) The costs of reproduction to red deer hinds. *Journal of Animal Ecology*, **52**, 367–383.
14.4.3

Cody, M.L. (1968) On the methods of resource division in grassland bird communities. *American Naturalist*, **102**, 107–148.
18.3.1

Cody, M.L. (1975) Towards a theory of continental species diversities. In: *Ecology and Evolution of Communities* (M.L. Cody & J.M. Diamond eds), pp. 214–257. Belknap, Cambridge, Massachusetts.
22.3.2

Cody, M.L. & Mooney, H.A. (1978) Convergence versus non-convergence in Mediterranean climate ecosystems. *Annual Review of Ecology and Systematics*, **9**, 265–321.
1.4.1

Coe, M.J. (1978) The decomposition of elephant carcases in the Tsavo (East) National Park, Kenya. *Journal of Arid Environments*, **1**, 71–86.
11.3.2

Coe, M.J., Cumming, D.H. & Phillipson, J. (1976) Biomass and production of large African herbivores in relation to rainfall and primary production. *Oecologia*, **22**, 341–354.
17.4

Colwell, R.K. & Winkler, D.W. (1984) A null model for null models in biogeography. In: *Ecological Communities: Conceptual Issues and the Evidence* (D.R. Strong, D. Simberloff, L.G. Abele & A.B. Thistle eds), pp. 344–359. Princeton University Press, Princeton, New Jersey.
18.4.3

Connell, J.H. (1961) The influence of interspecific competition and other factors on the distribution of the barnacle *Chthamalus stellatus*. *Ecology*, **42**, 710–723.
7.2.3

Connell, J.H. (1963) Territorial behaviour and dispersion in some marine invertebrates. *Researches on Population Ecology*, **5**, 87–101.
6.10

Connell, J.H. (1970) A predator-prey system in the marine intertidal region. I. *Balanus glandula* and several predatory species of *Thais*. *Ecological Monographs*, **40**, 49–78.
4.6.1, 4.7

Connell, J.H. (1971) On the role of natural enemies in preventing competitive exclusion in some marine animals and in rain forest trees. In: *Dynamics of Populations* (P.J. den Boer & G.R. Gradwell eds), pp. 298–310. *Proceedings of the Advanced Study Institute*

in Dynamics of Numbers in Populations, Oosterbeck. Centre for Agricultural Publishing and Documentation, Wageningen.
22.4.1

Connell, J.H. (1975) Some mechanisms producing structure in natural communities: a model and evidence from field experiments. In: *Ecology and Evolution of Communities* (M.L. Cody & J.M. Diamond eds). Belknap, Cambridge, Massachusetts.
19.2.6

Connell, J.H. (1978) Diversity in tropical rainforests and coral reefs. *Science, N.Y., 199,* 1302–1310.
19.3.4

Connell, J.H. (1979) Tropical rainforests and coral reefs as open non-equilibrium systems. In: *Population Dynamics* (R.M. Anderson, B.D. Turner & L.R. Taylor eds). Blackwell Scientific Publications, Oxford.
19.2.3

Connell, J.H. (1980) Diversity and the coevolution of competitors, or the ghost of competition past. *Oikos, 35,* 131–138.
7.10.2

Connell, J.H. (1983) On the prevalence and relative importance of interspecific competition: evidence from field experiments. *American Naturalist, 122,* 661–696.
7.9, 18.2

Connell, J.H. & Slatyer, R.O. (1977) Mechanisms of succession in natural communities and their role in community stability and organisation. *American Naturalist, 111,* 1119–1144.
16.4.3, 16.4.4

Connor, E.F. & Simberloff, D.S. (1979) The assembly of species communities: chance or competition? *Ecology, 60,* 1132–1140.
18.4.3

Cooper, J.P. (ed.) (1975) *Photosynthesis and Productivity in Different Environments.* Cambridge University Press, Cambridge.
17.2, 17.3.1

Coquillat, M. (1951) Sur les plantes les plus communes à la surface du globe. *Bull. Soc. Mens. Linnéene Lyon, 20,* 165–170.
15.7.1

Cornell, H.V. & Washburn, J.O. (1979) Evolution of the richness–area correlation for cynipid gall wasps on oak trees: a comparison of two geographic areas. *Evolution, 33,* 257–274.
19.1, 20.3.3

Cowie, R.J. (1977) Optimal foraging in great tits *Parus major. Nature* (London), *268,* 137–139.
9.11.2

Cox, C.B., Healey, I.N. & Moore, P.D. (1976) *Biogeography,* 2nd edn. Blackwell Scientific Publications, Oxford.
2.4, 2.7, 16.3.4

Cox, F.E.G. (1982) Immunology. In: *Modern Parasitology* (F.E.G. Cox ed.), pp. 173–203. Blackwell Scientific Publications, Oxford.
12.2.2, 12.5.2

Crawley, M.J. (1983) *Herbivory: the dynamics of animal–plant interactions.* Blackwell Scientific Publications, Oxford.
8.2.1, 8.2.6, 8.3, 8.4, 9.2.2, 9.5.2, 9.7, 9.8, 10.1

Crisp, D.J. (1961) Territorial behaviour in barnacle settlement. *Journal of Experimental Biology, 38,* 429–446.
6.11

Crisp, D.J. (1976) The role of the pelagic larva. In: *Perspectives in Experimental Biology* (P. Spencer Davies ed.), Vol. I (Zoology), pp. 145–155. Pergamon Press, Oxford.
5.4.7

Crisp, M.D. & Lange, R.T. (1976) Age structure distribution and survival under grazing of the arid zone shrub *Acacia burkittii.* Oikos, *27,* 86–92.
4.6.1, 4.6.2

Crocker, R.L. & Major, J. (1955) Soil development in relation to vegetation and surface age at Glacier Bay, Alaska. *Journal of Ecology, 43,* 427–448.
16.4.3

Croll, N.A., Anderson, R.M., Gyorkos, T.W. & Ghadirian, E. (1982) The population biology and control of *Ascaris lumbricoides* in a rural community in Iran. *Transactions of the Royal Society of Tropical Medicine and Hygiene, 76,* 187–197.
12.4.1

Cromartie, W.J. (1975) The effect of stand size and vegetational background on the colonization of cruciferous plants by herbivorous insects. *Journal of Applied Ecology, 12,* 517–533.
9.7.3

Crombie, A.C. (1945) On competition between different species of graminivorous insects. *Proceedings of the Royal Society of London,* B, *132,* 362–395.
6.4

Culver, D., Holsinger, J.R. & Baroody, R. (1973) Toward a predictive cave biogeography: the Greenbriar Valley as a case study. *Evolution, 27,* 689–695.
19.1

Cummins, K.W. (1974) Structure and function of stream ecosystems. *BioScience, 24,* 631–641.
11.2.2, 11.3.1

Dahl, E. (1979) Deep-sea carrion feeding amphipods: evolutionary patterns in niche adaptation. *Oikos, 33,* 167–175.
11.3.2

Daly, H.V., Doyer, J.T. & Ehrlich, P.R. (1978) *Introduction to Insect Biology and Diversity.* McGraw Hill, New York.
3.4.1

Darlington, C.D. (1960) Cousin marriage and the evol-

ution of the breeding system in man. *Heredity*, **14**, 297–332.
5.6

Darwin, C. (1859) *The Origin of Species by Means of Natural Selection*, 1st edn. John Murray, London.
1.1.1, 8.2.4, 19.2.1

Darwin, C. (1888) *The Formation of Vegetable Mould through the Action of Worms*. John Murray, London.
11.2.2

Dash, M.C. & Hota, A.K. (1980) Density effects on the survival, growth rate, and metamorphosis of *Rana tigrina* tadpoles. *Ecology*, **61**, 1025–1028.
6.5

Davidson, D.W. (1977) Species diversity and community organization in desert seed-eating ants. *Ecology*, **58**, 711–724.
7.10.2, 22.3.1

Davidson, D.W. (1978) Size variability in the worker caste of a social insect (*Veromessor pergandei* Mayr) as a function of the competitive environment. *American Naturalist*, **112**, 523–532.
7.10.2

Davidson, J. (1938) On the growth of the sheep population in Tasmania. *Transactions of the Royal Society of South Australia*, **62**, 342–346.
4.3

Davidson, J. (1944) On the relationship between temperature and rate of development of insects at constant temperatures. *Journal of Animal Ecology*, **13**, 26–38.
2.2.4

Davidson, J. & Andrewartha, H.G. (1948a) Annual trends in a natural population of *Thrips imaginis* (Thysanoptera). *Journal of Animal Ecology*, **17**, 193–199.
15.2, 15.2.1

Davidson, J. & Andrewartha, H.G. (1948b) The influence of rainfall, evaporation and atmospheric temperature on fluctuations in the size of a natural population of *Thrips imaginis* (Thysanoptera). *Journal of Animal Ecology*, **17**, 200–222.
15.2, 15.2.1

Davies, N.B. (1977) Prey selection and social behaviour in wagtails (Aves: Motacillidae). *Journal of Animal Ecology*, **46**, 37–57.
9.2.2

Davies, N.B. (1978) Ecological questions about territorial behaviour. In: *Behavioural Ecology: An Evolutionary Approach* (J.R. Krebs & N.B. Davies eds), pp. 317–350. Blackwell Scientific Publications, Oxford.
6.11

Davies, N.B. & Houston, A.I. (1984) Territory economics. In: *Behavioural Ecology: An Evolutionary Approach*, 2nd edn. (J.R. Krebs & N.B. Davies eds), pp. 148–169. Blackwell Scientific Publications, Oxford.
6.11

Davis, M.B., Brubaker, L.B. & Webb, T. III (1973) Calibration of absolute pollen influx. In: *Quaternary Plant Ecology* (H.J.B. Birks & R.G. West eds), pp. 9–25. Blackwell Scientific Publications, Oxford.
1.2.2

Davis, M.G. (1976) Pleistocene biogeography of temperate deciduous forests. *Geoscience and Man*, **13**, 13–26.
1.2.2

Dawkins, R. (1976) *The Selfish Gene*. Oxford University Press, Oxford.
5.3.6

DeAngelis, D.L. (1975) Stability and connectance in food web models. *Ecology*, **56**, 238–243.
21.5

DeAngelis, D.L. (1980) Energy flow, nutrient cycling and ecosystem resilience. *Ecology*, **61**, 764–771.
21.2.2

DeBach, P. (ed.) (1964) *Biological Control of Insect Pests and Weeds*. Chapman and Hall, London.
9.1

Deevey, E.S. (1947) Life tables for natural populations of animals. *Quarterly Review of Biology*, **22**, 283–314.
4.5.1

Dempster, J.P. (1983) The natural control of populations of butterflies and moths. *Biological Reviews*, **58**, 461–481.
10.2.3

Dempster, J.P. & Lakhani, K.H. (1979) A population model for cinnabar moth and its food plant, ragwort. *Journal of Animal Ecology*, **48**, 143–164.
10.1

Denholm-Young, P.A. (1978) Studies on decomposing cattle dung and its associated fauna. D. Phil. thesis, Oxford University.
11.3.2

Diamond, J.M. (1972) Biogeographic kinetics: estimation of relaxation times for avifaunas of Southwest Pacific islands. *Proceedings of the National Academy of Science of the U.S.A.* **69**, 3199–3203.
20.3.2

Diamond, J.M. (1973) Distributional ecology of New Guinea birds. *Science, N.Y.*, **179**, 759–769.
5.4.1

Diamond, J.M. (1975) Assembly of species communities. In: *Ecology and Evolution of Communities* (M.L. Cody & J.M. Diamond eds). Belknap, Cambridge, Massachusetts.
7.10.1, 18.3.3, 18.4.2, 18.4.3, 20.3.5

Diamond, J.M. (1983) Taxonomy by nucleotides. *Nature* (London), **305**, 17–18.
1.2.1

Diamond, J.M. & Gilpin, M.E. (1982) Examination of the 'null' model of Connor and Simberloff for species co-occurrences on islands. *Oecologia*, **52**, 64–74.
18.4.3

Diamond, J.M. & Marshall, A.G. (1977) Distributional ecology of new Hebridean birds: a species kaleidoscope. *Journal of Animal Ecology, 46,* 703–727.
18.4.3

Diamond, J.M. & May, R.M. (1976) Island biogeography and the design of natural reserves. In: *Theoretical Ecology: Principles and Applications* (R.M. May ed.), pp. 228–252. Blackwell Scientific Publications, Oxford.
5.4.1, 20.4

Diamond, J.M. & Mayr, E. (1976) The species–area relation for birds of the Solomon Archipelago. *Proceedings of the National Academy of Science of the U.S.A., 73,* 262–266.
19.1

Dirzo, R. & Harper, J.L. (1982) Experimental studies of slug–plant interactions. IV. The performance of cyanogenic and acyanogenic morphs of *Trifolium repens* in the field. *Journal of Ecology, 70,* 119–138.
1.5.2, 3.4.3

Dixon, A.F.G. (1971) The role of aphids in wood formation. II. The effect of the lime aphid, *Eucallipterus tiliae* L. (Aphididae), on the growth of lime *Tilia×vulgaris* Hayne. *Journal of Applied Ecology, 8,* 383–399.
8.2.2

Dobben, W.H. van (1952) The food of the cormorants in the Netherlands. *Ardea, 40,* 1–63.
12.5.5

Dobzhansky, T. (1950) Evolution in the tropics. *American Scientist, 38,* 209–221.
22.4.1

Dodd, A.P. (1940) *The Biological Campaign Against Prickly Pear.* Commonwealth Prickly Pear Board. Government Printer, Brisbane.
8.2.2, 15.6.5

Donald, C.M. (1951) Competition among pasture plants. I. Intra-specific competition among annual pasture plants. *Australian Journal of Agricultural Research, 2,* 335–376.
6.5

Donaldson, A.I. (1983) Quantitative data on airborne foot-and-mouth disease virus: its production, carriage and deposition. *Philosophical Transactions of the Royal Society of London,* B, **302,** 529–534.
12.3.1

Douglas, A. & Smith, D.C. (1984) The green hydra symbiosis. VIII. Mechanisms in symbiont regulation. *Proceedings of the Royal Society of London,* B, **221,** 291–319.
13.8

Downs, C. & McQuilkin, W.E. (1944) Seed production of Southern Appalachian oaks. *Journal of Forestry,* **42,** 913–920.
4.6.3

Drift, J. van der & Witkamp, M. (1959) The significance of the breakdown of oak litter by *Enoicyla pusilla* Burm. *Archives néerlandaises de Zoologie,* **13,** 486–492.
11.2.3

Duncan, W.G., Loomis, R.S., Williams, W.A. & Honau, R. (1967) A model for simulating photosynthesis in plant communities. *Hilgardia,* **38,** 181–205.
17.3.1

Dunn, E. (1977) Predation by weasels (*Mustela nivalis*) on breeding tits (*Parus* spp.) in relation to the density of tits and rodents. *Journal of Animal Ecology,* **46,** 634–652.
6.11

Dyne, G.M. van, Brockington, N.R., Szocs, Z., Duek, J. & Ribic, C.A. (1980) Large herbivore subsystem. In: *Grasslands, Systems Analysis and Man* (A.I. Breymeyer & G.M. van Dyne eds), pp. 269–537. Cambridge University Press, Cambridge.
10.8.5

Edmunds, M. (1974) *Defence in Animals: A Survey of Anti-predator Defences.* Longman, Harlow, Essex.
3.4.3

Einarsen, A.S. (1945) Some factors affecting ringnecked pheasant population density. *Murrelet,* **26,** 39–44.
6.4

Eis, S., Garman, E.H. & Ebel, L.F. (1965) Relation between cone production and diameter increment of douglas fir (*Pseudotsuga menziesii* (Mirb.) Franco), grand fir (*Abies grandis* Dougl.), and western white pine (*Pinus monticola* Dougl.). *Canadian Journal of Botany,* **43,** 1553–1559.
14.4.1

Eisner, T. & Meinwald, J. (1966) Defensive secretions of arthropods. *Science, N.Y.,* **153,** 1341–1350.
3.4.3

Elliott, J.M. (1976) Body composition of brown trout (*Salmo trutta* L.) in relation to temperature and ration size. *Journal of Animal Ecology,* **45,** 273–289.
4.2.1

Elliott, J.M. (1984) Numerical changes and population regulation in young migratory trout *Salmo trutta* in a Lake District stream 1966–83. *Journal of Animal Ecology,* **53,** 327–350.
6.4

Elner, R.W. & Hughes, R.N. (1978) Energy maximisation in the diet of the shore crab, *Carcinus maenas* (L.). *Journal of Animal Ecology,* **47,** 103–116.
9.2.2

Elner, R.W. & Raffaelli, D.G. (1980) Interactions between two marine snails, *Littorina radis* Maton and *Littorina nigrolineata* Gray, a predator, *Carcinus maenas* (L.), and a parasite, *Microphallus similis*

Jagerskiold. *Journal of Experimental Marine Biology and Ecology,* **43,** 151–160.
14.11

Elton, C. (1927) *Animal Ecology.* Sidgwick and Jackson, London.
17.4

Elton, C.S. (1958) *The Ecology of Invasion by Animals and Plants.* Methuen, London.
21.2.1

Elton, C.S. & Nicholson, A.J. (1942) The ten year cycle in numbers of the lynx in Canada. *Journal of Animal Ecology,* **11,** 215–244.
15.4.1

Emiliani, C. (1966) Isotopic palaeotemperatures. *Science, N.Y.,* **154,** 851–857.
1.2.2

Emmons, L.H. (1980) Ecology and resource partitioning among nine species of African rain forest squirrels. *Ecological Monographs,* **50,** 31–54.
18.3.1, 18.4.2

Emson, R.H. & Faller-Fritsch, R.J. (1976) An experimental investigation into the effect of crevice availability on abundance and size structure in a population of *Littorina radis* (Maton): Gastropoda; Prosobranchia. *Journal of Experimental Marine Biology and Ecology,* **23,** 285–297.
14.11

Eriksson, O., Hansen, A. & Sunding, P. (1974) *Flora of Macronesia: Checklist of vascular plants.* University of Umeå, Sweden.
19.1

Erlinge, S., Göransson, G., Högstedt, G., Jansson, G., Liberg, O., Loman, J., Nilsson, I.N., von Schantz, T. & Sylvén, M. (1984) Can vertebrate predators regulate their prey? *American Naturalist,* **123,** 125–133.
10.5

Errington, P.L. (1945) Some contributions of a fifteen-year local study of the northern bobwhite to a knowledge of population phenomena. *Ecological Monographs,* **15,** 1–34.
10.6.1

Errington, P.L. (1946) Predation and vertebrate populations. *Quarterly Review of Biology,* **21,** 144–177.
8.3

Evans, L.T. (1971) Evolutionary, adaptive and environmental aspects of the photosynthetic pathway: assessment. In: *Photosynthesis and Photorespiration* (M.D. Hatch, C.B. Osmond & R.O. Slatyer eds). John Wiley & Sons, New York.
3.3.2

Faaborg, J. (1976) Patterns in the structure of West Indian bird communities. *American Naturalist,* **111,** 903–916.
20.3.5

Feeny, P. (1976) Plant apparency and chemical defence.

Recent Advances in Phytochemistry, **10,** 1–40.
3.4.3

Fenchel, T. (1975) Character displacement and coexistence in mud snails (Hydrobiidae). *Oecologia,* **20,** 19–32.
7.10.2

Fenchel, T. & Kofoed, L. (1976) Evidence for exploitative interspecific competition in mud snails (Hydrobiidae). *Oikos,* **27,** 367–376.
7.10.2

Fenchel, T. & Kolding, S. (1979) Habitat selection and distribution patterns of five species of the amphipod genus *Gammarus. Oikos,* **33,** 316–322.
18.3.1

Fenner, F. (1983) Biological control, as exemplified by smallpox eradication and myxomatosis. *Proceedings of the Royal Society of London,* B, **218,** 259–285.
12.7

Ferguson, R.G. (1958) The preferred temperature of fish and their midsummer distribution in temperate lakes and streams. *Journal of the Fisheries Research Board of Canada,* **15,** 607–624.
2.2.1

Fernando, M.H.J.P. (1977) Predation of the glasshouse red spider mite by *Phytoseiulus persimilis* A.-H. Ph.D. thesis, University of London.
9.6

Finlayson, L.H. (1949) Mortality of *Laemophloeus* (Coleoptera, Cucujidae) infected with *Mattesia dispora* Naville (Protozoa, Schizogregarinaria). *Parasitology,* **40,** 261–264.
12.6.3

Firth, R. & Davidson, J.W. (1945) *Pacific Islands. Vol. 1. General Survey.* Naval Intelligence Division, London.
20.3.5

Fisher, R.A. (1930) *The Genetical Theory of Natural Selection.* Clarendon Press, Oxford.
14.3

Fitter, A.H. & Hay, R.K.M. (1981) *Environmental Physiology of Plants.* Academic Press, London.
3.2.4, 17.3.1

Flint, R.F. (1971) *Glacial and Quarternary Geology.* John Wiley & Sons, New York.
22.3.5

Flor, H.H. (1960) The inheritance of X-ray induced mutations to virulence in a urediospore culture of race 1 of *Melamspora lini. Phytopathology,* **50,** 628–634.
12.7

Flor, H.H. (1971) Current status of the gene-for-gene concept. *Annual Review of Phytopathology,* **60,** 628–634.
12.7

Flower, R.J. & Battarbee, R.W. (1983) Diatom evidence for recent acidification of two Scottish lochs. *Nature*

(London), **305,** 130–133.
2.6

Ford, E.B. (1940) Polymorphism and taxonomy. In: *The New Systematics* (J. Huxley ed.) pp. 493–513. Clarendon Press, Oxford.
1.5.2

Ford, E.B. (1945) *Butterflies.* William Collins, London.
13.3.2

Ford, R.G. & Pitelka, F.A. (1984) Resource limitation in the California vole. *Ecology,* **65,** 122–136.
15.4.2

Forman, R.T.T. (1964) Growth under controlled conditions to explain the hierarchical distributions of a moss, *Tetraphis pellucida. Ecological Monographs,* **34,** 1–25.
2.5

Fowler, C.W. (1981) Density dependence as related to life history strategy. *Ecology,* **62,** 602–610.
6.3

Fowler, C.W. & MacMahon, J.A. (1982) Selective extinction and speciation: their influence on the structure and functioning of communities and ecosystems. *American Naturalist,* **119,** 480–498.
22.5

Fowler, S.V. & Lawton, J.H. (1982) The effects of host-plant distribution and local abundance on the species richness of agromyzid flies attacking British umbellifers. *Ecological Entomology,* **7,** 257–265.
20.3.3

Franklin, R.T. (1970) Insect influences in the forest canopy. In: *Analysis of Temperate Forest Ecosystems* (D.E. Reichle ed.), pp. 86–99. Springer-Verlag, New York.
8.2.5

Free, C.A., Beddington, J.R. & Lawton, J.H. (1977) On the inadequacy of simple models of mutual interference for parasitism and predation. *Journal of Animal Ecology,* **46,** 543–554.
9.7.2

Fretwell, S.D. & Lucas, H.L. (1970) On territorial behaviour and other factors influencing habitat distribution in birds. *Acta Biotheoretica,* **19,** 16–36.
9.8

Fricke, H.W. (1975) The role of behaviour in marine symbiotic animals. In: *Symbiosis* (D.H. Jennings & D.L. Lee eds), pp. 581–594. Symposium 29, Society for Experimental Biology. Cambridge University Press, Cambridge.
13.2.2, 13.2.3

Fridriksson, S. (1975) *Surtsey: Evolution of Life on a Volcanic Island.* Butterworths, London.
20.13

Fryer, G. & Iles, T.D. (1972) *The Cichlid Fishes of the Great Lakes of Africa.* Oliver & Boyd, Edinburgh.
20.4

Futuyma, D.J. (1973) Evolutionary interactions among herbivorous insects and plants. In: *Coevolution* (D.J. Futuyma & M. Slatkin eds), pp. 207–31. Sinauer Associates, Sunderland, Massachusetts.
3.4.3

Futuyma, D.J. & Gould, F. (1979) Associations of plants and insects in a deciduous forest. *Ecological Monographs,* **49,** 33–50.
20.3.3

Futuyma, D.J. & Slatkin, M. (eds) (1983) *Coevolution.* Sinauer Associates, Sunderland, Massachusetts.
9.2.4

Gadgil, M. (1971) Dispersal: population consequences and evolution. *Ecology,* **52,** 253–61.
5.4.2

Gadgil, P.M. & Solbrig, O.T. (1972) The concept of *r* and *K* selection. Evidence from some wild flowers and theoretical considerations. *American Naturalist,* **106,** 14–31.
14.7.2

Gaines, M.S. & McClenaghan, L.R. (1980) Dispersal in small mammals. *Annual Review of Ecology and Systematics,* **11,** 163–96.
5.4.3, 5.5.1, 5.5.4

Gaines, M.S., Vivas, A.M. & Baker, C.L. (1979) An experimental analysis of dispersal in fluctuating vole populations: demographic parameters. *Ecology,* **60,** 814–828.
15.4.2

Garrod, D.J. & Jones, B.W. (1974) Stock and recruitment relationship in the N.E. Atlantic cod stock and the implications for management of the stock. *Journal Conseil International pour l'Exploration de la Mer,* **173,** 128–144.
10.8.6

Gartside, D.W. & McNeilly, T. (1974) The potential for evolution of heavy metal tolerance in plants. II. Copper tolerance in normal populations of different plant species. *Heredity,* **32,** 335–348.
2.11

Gass, C.L., Angehr, G. & Centa, J. (1976) Regulation of food supply by feeding territoriality in the rufous hummingbird. *Canadian Journal of Zoology,* **54,** 2046–2054.
6.11

Gates, D.M. & Porter, W.P. (1970) The energy budget of animals. In: *Physiological and Behavioural Temperature Regulation.* (J.D. Hardy, A.P. Gagge & J.A.J. Stolwijk eds). C.C. Thomas, Springfield, Illinois.

Gauch, H.G. (1982) *Multivariate Analysis in Community Ecology.* Cambridge University Press, Cambridge.
16.3.2

Gause, G.F. (1934) *The Struggle for Existence.* Williams & Wilkins, Baltimore (Reprinted 1964 by Hafner, New York).
7.2.4

Gause, G.F. (1935) Experimental demonstration of

Volterra's periodic oscillation in the numbers of animals. *Journal of Experimental Biology,* **12,** 44–48.
7.2.4

Gauthreaux, S.A. (1978) The structure and organization of avian communities in forests. In: *Proceedings of the Workshop on Management of Southern Forests for Nongame Birds* (R.M. DeGraaf ed.), pp. 17–37. Southern Forest Station, Asheville, North Carolina.
16.4.5

Geiger, R. (1955) *The Climate Near the Ground.* Harvard University Press, Cambridge, Massachusetts.
2.3

Gibson, A.H. & Jordan, D.C. (1983) Ecophysiology of nitrogen fixing systems. In: *Physiological Plant Ecology III. Encyclopedia of Plant Physiology. New Series* 12 c. (O.L. Lange, P.S. Nobel, C.B. Osmond & H. Ziegler eds), pp. 300–390. Springer-Verlag, Berlin.
13.10

Gilbert, L.E. (1984) Biology of butterfly communities. In: *The Biology of Butterflies* (R. Vane-Wright & P. Ackery eds). Academic Press. New York.
7.7

Gilbert, N. (1984) Control of fecundity in *Pieris rapae* II. Differential effects of temperature. *Journal of Animal Ecology,* **53,** 589–597.
2.2.4

Gill, L.S. (1957) Dwarf mistletoe on lodgepole pine. *United States Department of Agriculture and Forest Service Forest Pest Leaflet,* 18.
12.2.2

Gilpin, M.E. & Diamond, J.M. (1982) Factors contributing to non-randomness in species co-occurrences on islands. *Oecologia,* **52,** 75–84.
18.4.3

Gilpin, M.E. & Diamond, J.M. (1984) Are species co-occurrences on islands non-random, and are null hypotheses useful in community ecology? In: *Ecological Communities: Conceptual Issues and the Evidence* (D.R. Strong, D. Simberloff, L.G. Abele & A.B. Thistle eds), pp. 297–315. Princeton University Press, Princeton, New Jersey.
18.4.3

Gleason, H.A. (1926) The individualistic concept of the plant association. *Torrey Botanical Club Bulletin,* **53,** 7–26.
16.3.3

Goh, B.S. (1979) Robust stability concepts for ecosystem models. In: *Theoretical Systems in Ecology* (E. Halfon ed.), pp. 467–487. Academic Press, New York.
21.3

Goldman, C.R., Mason, D.T. & Hobbie, J.E. (1967) Two antarctic desert lakes. *Limnology and Oceanography,* **12,** 295–310.
21.4

Goodall, D.W. (1967) Computer simulation of changes in vegetation subject to grazing. *Journal of the Indian Botanical Society,* **46,** 356–362.
3.3.1

Gorham, E., Vitousek, P.M. & Reiners, W.A. (1979) The regulation of chemical budgets over the course of terrestrial ecosystem succession. *Annual Review of Ecology and Systematics,* **10,** 53–84.
17.5

Gorman, M.L. (1979) *Island Ecology.* Chapman and Hall, London.
19.1

Goss-Custard, J.D. (1970) Feeding dispersion in some overwintering wading birds. In: *Social Behaviour in Birds and Mammals* (J.H. Crook ed.), pp. 3–34. Academic Press, London.
9.7.1

Gould, S.J. (1966) Allometry and size in ontogeny and phylogeny. *Biological Reviews,* **41,** 587–640.
14.14.2, 14.14.3

Gould, S.J. (1981) Palaeontology plus ecology as palaeobiology. In: *Theoretical Ecology: Principles and Applications,* 2nd edn. (R.M. May ed.), pp. 295–317. Blackwell Scientific Publications, Oxford.
22.4.5

Gould, S.J. & Lewontin, R.C. (1979) Spandrels of San Marco and the Panglossian paradigm—a critique of the adaptationist program. *Proceedings of the Royal Society,* B, **205** (1161), 581–598.
1.1.2

Grace, J. (1983) *Plant–Atmosphere Relationships.* Chapman and Hall, London.
3.2.4

Grace, J.B. & Wetzel, R.G. (1981) Habitat partitioning and competitive displacement in cattails (*Typha*): experimental field studies. *American Naturalist,* **118,** 463–474.
7.3

Grant, V. (1963) *The Origin of Adaptations.* Columbia University Press, New York.
13.4

Gray, J.S. (1981) *The Ecology of Marine Sediments: An Introduction to the Structure and Function of Benthic Communities.* Cambridge University Press, Cambridge.
16.3.3

Grant, P.R. & Abbott I. (1980) Interspecific competition, island biogeography and null hypotheses. *Evolution,* **34,** 332–341.
18.4.2

Green, T.R. & Ryan, C.A. (1972) Wound-induced proteinase inhibitor in plant leaves: a possible defense mechanism against insects. *Science, N.Y.,* **175,** 776–777.
8.2.3

Greenbank, D.O. (1957) The role of climate and dispersal in the initiation of outbreaks of the spruce

budworm in New Brunswick. II. The role of dispersal. *Canadian Journal of Zoology,* **35,** 385–403.
5.4.3

Greenwood, D.J. (1969) Effect of oxygen distribution in the soil on plant growth. In: *Root Growth* (W.J. Whittingham ed.), pp. 202–223. *Proceedings of the 15th Easter School of Agricultural Science, Nottingham, 1968.*
3.3.4

Greenwood, P.J. (1980) Mating systems, philopatry and dispersal in birds and mammals. *Animal Behaviour,* **28,** 1140–1162.
5.5.2

Greenwood, P.J., Harvey, P.H. & Perrins, C.M. (1978) Inbreeding and dispersal in the great tit. *Nature* (London), **271,** 52–4.
5.4.3, 5.6

Gregor, J.W. (1930) Experiments on the genetics of wild populations. I. *Plantago maritima. Journal of Genetics,* **22,** 15–25.
1.5.1

Greig-Smith, P. (1983) *Quantitative Plant Ecology,* 3rd edn. Blackwell Scientific Publications, Oxford.
5.2

Griffiths, K.J. (1969) Development and diapause in *Pleolophus basizonus* (Hymenoptera: Ichneumonidae). *Canadian Entomologist,* **101,** 907–914.
9.5.1

Griffiths, M. & Barker, R. (1966) The plants eaten by sheep and by kangaroos grazing together in a paddock in south-western Queensland. *C.S.I.R.O. Wildlife Research,* **11,** 145–167.
1.3

Grime, J.P. (1973) Control of species density in herbaceous vegetation. *Journal of Environmental Management,* **1,** 151–167.
22.3.4

Grime, J.P. (1974) Vegetation classification by reference to strategies. *Nature* (London), **250,** 26–31.
14.10.2

Grime, J.P. (1979) *Plant Strategies and Vegetation Processes.* John Wiley & Sons, Chichester.
14.10.2, 19.3.2

Grime, J.P. & Lloyd, P.S. (1973) *An Ecological Atlas of Grassland Plants.* Edward Arnold, London.
16.3.1

Groves, R.H. & Williams, J.D. (1975) Growth of skeleton weed (*Chondrilla juncea* L.) as affected by growth of subterranean clover (*Trifolium subterraneum* L.) and infection by *Puccinia chondrilla* Bubak and Syd. *Australian Journal of Agricultural Research,* **26,** 975–983.
7.3

Grubb, P. (1977) The maintenance of species richness in plant communities: the importance of the regener-

ation niche. *Biological Reviews,* **52,** 107–145.
18.3.2, 19.3.2

Gulland, J.A. (1971) The effect of exploitation on the numbers of marine animals. In: *Dynamics of Populations* (P.J. den Boer & G.R. Gradwell eds), pp. 450–468. Centre for Agricultural Publishing and Documentation, Wageningen.
4.3, 10.8.2

Haefner, P.A. (1970) The effect of low dissolved oxygen concentrations on temperature-salinity tolerance of the sand shrimp *Crangon septemspinosa. Physiological Zoology,* **43,** 30–37.
2.12

Haines, E. (1979) Interaction between Georgia salt marshes and coastal waters: a changing paradigm. In: *Ecological Processes in Coastal and Marine Systems* (R.J. Livingston ed.). Plenum Press, New York.
17.3.3

Hainsworth, F.R. (1981) *Animal Physiology.* Addison-Wesley, Reading, Massachusetts.
2.2.2, 2.2.11

Hairston, N.G. (1980) The experimental test of an analysis of field distributions: competition in terrestrial salamanders. *Ecology,* **61,** 817–826.
7.2.1

Hairston, N.G., Smith, F.E. & Slobodkin, L.B. (1960) Community structure, population control, and competition. *American Naturalist,* **44,** 421–425.
18.2.1

Hairston, N.G., Allan, J.D., Colwell, R.K., Futuyma, D.J., Howell, J., Lubin, M.D., Mathias, J. & Vandermeer, J.H. (1968) The relationship between species diversity and stability: an experimental approach with protozoa and bacteria. *Ecology,* **49,** 1091–1101.
21.2.3

Hagner, S. (1965) Cone crop fluctuations in Scots Pine and Norway Spruce. *Stud. For. suec. Skogshogski.,* **33,** 1–21.
4.6.3

Haldane, J.B.S. (1949) Disease and evolution. Symposium sui fattori ecologici e genetici della speciazione negli animali. *Rio. Sci.,* **19** (suppl.), 3–11.
4.5.1

Haldane, J.B.S. (1953) Animal populations and their regulation. *New Biology* (published by Penguin Books, London), **15,** 9–24.
15.2.2

Hamilton, W.D. (1971) Geometry for the selfish herd. *Journal of Theoretical Biology,* **31,** 295–311.
5.3.6

Hamilton, W.D. & May, R.M. (1977) Dispersal in stable habitats. *Nature* (London), **269,** 578–81.
5.4.2

Hanek, G. (1972) Microecology and spatial distribution

of the gill parasites infesting *Lepomis gibbosus* (L.) and *Ambloplitis rupestris* (Raf.) in the Bay of Quinto, Ontario. Ph.D. thesis, University of Waterloo.
12.4

Hanlon, R.D.G. & Anderson, J.M. (1980) The influence of macroarthropod feeding activities on fungi and bacteria in decomposing oak leaves. *Soil Biology and Biochemistry*, **12**, 255–261.
11.2.3

Hanski, I. & Kuusela, S. (1977) An experiment on competition and diversity in the carrion fly community. *Annales Zoologi Fennici*, **43**, 108–115.
7.6.5

Harcourt, D.G. (1971) Population dynamics of *Leptinotarsa decemlineata* (Say) in eastern Ontario. III. Major population processes. *Canadian Entomologist*, **103**, 1049–1061.
5.4.3, 15.3.1

Harley, J.L. & Smith, S.E. (1983) *Mycorrhizal Symbiosis*. Academic Press, London.
13.7

Harman, W.N. (1972) Benthic substrates: their effect on freshwater molluscs. *Ecology*, **53**, 271–272.
22.3.2

Harper, J.L. (1961) Approaches to the study of plant competition. In: *Mechanisms in Biological Competition* (F.L. Milthorpe ed.) pp. 1–39. Symposium No. 15, Society for Experimental Biology. Cambridge University Press, Cambridge.
6.5, 7.6.1, 7.6.2

Harper, J.L. (1977) *The Population Biology of Plants*. Academic Press, London.
5.4.4, 5.7, 6.4.3, 7.6.1, 7.6.2, 8.2.2, 12.4, 14.4.5, 14.9.2, 16.4.4, 19.2.6

Harper, J.L. (1981) The meanings of rarity. In: *The Biological Aspects of Rare Plant Conservation* (H. Synge ed.), pp. 189–203. John Wiley & Sons, Chichester.
15.2.2

Harper, J.L. (1982) After description. In: *The Plant Community as a Working Mechanism* (E.I. Newman ed.). *British Ecological Society Special Publications Series*, **1**, 11–25. Blackwell Scientific Publications, Oxford.
1.1.2

Harper, J.L. & Ogden, J. (1970) The reproductive strategy of higher plants: I. The concept of strategy with special reference to *Senecio vulgaris* L. *Journal of Ecology*, **58**, 681–698.
14.5.2

Harper, J.L. & White, J. (1974) The demography of plants. *Annual Review of Ecology and Systematics*, **5**, 419–463.
14.5.2, 14.9.2

Harper, J.L., Williams, J.T. & Sagar, G.R. (1965) The behaviour of seeds in soil. Part 1. The heterogeneity of soil surfaces and its role in determining the establishment of plants from seed. *Journal of Ecology*, **53**, 273–286
2.9

Harper, J.L., Lovell, P.H. & Moore, K.G. (1970) The shapes and sizes of seeds. *Annual Review of Ecology and Systematics*, **1**, 327–356.
5.4.5

Harrison, R.G. (1980) Dispersal polymorphisms in insects. *Annual Review of Ecology and Systematics*, **11**, 95–118.
5.5.3

Hart, A. & Begon, M. (1982) The status of general reproductive-strategy theories, illustrated in winkles. *Oecologia*, **52**, 37–42.
14.5.2, 14.11

Hart, J.S. (1956) Seasonal changes in insulation of the fur. *Canadian Journal of Zoology*, **34**, 53–57.
1.6

Hassall, M. & Jennings, J.B. (1975) Adaptive features of gut structure and digestive physiology in the terrestrial isopod *Philoscia muscorum* (Scop.). *Biological Bulletin*, **149**, 348–364.
11.3.1

Hassell, M.P. (1971) Mutual interference between searching insect parasites. *Journal of Animal Ecology*, **40**, 473–486.
9.7.1

Hassell, M.P. (1976) *The Dynamics of Competition and Predation*. Edward Arnold, London.
6.3

Hassell, M.P. (1978) *The Dynamics of Arthropod Predator–Prey Systems*. Princeton University Press, Princeton.
8.4, 9.5.1, 9.6, 10.2

Hassell, M.P. (1981) Arthropod predator–prey systems. In: *Theoretical Ecology: Principles and Applications*, 2nd edn. (R.M. May ed.), pp. 105–131. Blackwell Scientific Publications, Oxford.
10.5

Hassell, M.P. (1982) Patterns of parasitism by insect parasitoids in patchy environments. *Ecological Entomology*, **7**, 365–377.
9.7.1

Hassell, M.P. (1985) Insect natural enemies as regulating factors. *Journal of Animal Ecology*, **54**, 323–334.
10.2.3, 10.4.1

Hassell, M.P. & May, R.M. (1974) Aggregation of predators and insect parasites and its effect on stability. *Journal of Animal Ecology*, **43**, 567–594.
9.7.1

Hassell, M.P. & Varley, G.C. (1969) New inductive population model for insect parasites and its bearing on biological control. *Nature* (London), **223**, 1133–1136.
9.6

Hassell, M.P., Lawton, J.H. & Beddington, J.R. (1977)

Sigmoid functional responses by invertebrate predators and parasitoids. *Journal of Animal Ecology*, **46**, 249–262.
9.5.3

Hastings, J.R. & Turner, R.M. (1965) *The Changing Mile*. University of Arizona Press, Tucson.
2.4

Hatto, J. & Harper, J.L. (1969) The control of slugs and snails in British cropping systems, specially grassland. International Copper Research Association Project, **115**(A), 1–25.
9.5.1

Hawkes, C. (1974) Dispersal of adult cabbage root fly (*Erioischia brassicae* (Bouché)) in relation to a brassica crop. *Journal of Applied Ecology*, **11**, 83–94.
3.4.3

Heads, P.A. & Lawton, J.H. (1983) Studies on the natural enemy complex of the holly leaf-miner: the effects of scale on the detection of aggregative responses and the implications for biological control. *Oikos*, **40**, 267–276.
10.4.1

Heal, O.W. & MacLean, S.F. (1975) Comparative productivity in ecosystems—secondary productivity. In: *Unifying Concepts in Ecology* (W.H. van Dobben & R.H. Lowe–McConnell eds), pp. 89–108. Junk, The Hague.
17.4.1, 17.4.2, 17.4.3

Heed, W.B. (1968) *Ecology of Hawaiian Drosophilidae*. University of Texas Publications, **6861**, 387–419.
1.2.3

Heinrich, B. (1977) Why have some animals evolved to regulate a high body temperature? *American Naturalist*, **111**, 623–640.
2.2.7

Hendrickson, J.A. Jr. (1981) Community-wide character displacement reexamined. *Evolution*, **35**, 794–810.
18.4.2

Hendrix, S.D. (1979) Compensatory reproduction in a biennial herb following insect defloration. *Oecologia*, **42**, 107–118.
8.2.1

Henttonen, H., Vaheri, A., Lähdevirta, J. & Brummer–Korvenkontio, M. (1981) The epidemiology of *Nephropathia epidemica* in wild rodents. *Abstracts of the 5th International Congress of Virology, Strasbourg*, p. 201.
15.4.2

Henkelem, W.F. van (1973) Growth and lifespan of *Octopus cyanea* (Mollusca: Cephalopoda). *Journal of Zoology*, **169**, 299–315.
4.9

Hibbs, D.E. & Fischer, B.C. (1979) Sexual and vegetative reproduction of striped maple (*Acer pensylvanicum* L.). *Bulletin of the Torrey Botanical Club*, **106**, 222–227.
4.6.4

Hildebrand, M. (1974) *Analysis of Vertebrate Structure*.
J. Wiley & Sons, New York.
1.3

Hildrew, A.G. (1985) A quantitative study of the life history of a fairy shrimp (Branchiopoda: Anostraca) in relation to the temporary nature of its habitat, a Kenyan rainpool. *Journal of Animal Ecology*, **54**, 99–110.
4.5.2

Hildrew, A.G. & Townsend, C.R. (1980) Aggregation, interference and foraging by larvae of *Plectrocnemia conspersa* (Trichoptera: Polycentropodidae). *Animal Behaviour*, **28**, 553–560.
9.10

Hildrew, A.G. & Townsend, C.R. (1982) Predators and prey in a patchy environment: a freshwater study. *Journal of Animal Ecology*, **51**, 797–816.
19.2.3

Hildrew, A.G., Townsend, C.R., Francis, J. & Finch, K. (1984) Cellulolytic decomposition in streams of contrasting pH and its relationship with invertebrate community structure. *Freshwater Biology*, **14**, 323–328.
2.6

Hill, R.B. (1926) The estimation of the number of hookworms harboured by the use of the dilution egg count method. *American Journal of Hygiene*, **6**, (Suppl.), 19–41.
12.4.1

Holling, C.S. (1959) Some characteristics of simple types of predation and parasitism. *Canadian Entomologist*, **91**, 385–398.
9.5

Holloway, J.D. (1977) *The Lepidoptera of Norfolk Island, their Biogeography and Ecology*. Junk, The Hague.
20.4

Holmes, M.G. (1983) Perception of shade. In: *Photoperception by Plants* (P.F. Wareing & H. Smith eds). The Royal Society, London.
1.6, 3.2.2

Holt, R.D. (1977) Predation, apparent competition and the structure of prey communities. *Theoretical Population Biology*, **12**, 197–229.
7.7, 22.2

Holt, R.D. (1984) Spatial heterogeneity, indirect interactions, and the coexistence of prey species. *American Naturalist*, **124**, 377–406.
7.7, 22.2

Horn, H.S. (1975) Markovian processes of forest succession. In: *Ecology and Evolution of Communities* (M.L. Cody & J.M. Diamond eds), pp. 196–213. Belknap, Cambridge, Massachusetts.
16.4.3, 16.4.4, 19.3.4

Horn, H.S. (1978) Optimal tactics of reproduction and life-history. In: *Behavioural Ecology: An Evolutionary Approach* (J.R. Krebs & N.B. Davies eds), pp. 411–429. Blackwell Scientific Publications, Oxford.
14.8, 14.10.1

Horn, H.S. (1981) Succession. In: *Theoretical Ecology:*

Principles and Applications (R.M. May ed.), pp. 253–271. Blackwell Scientific Publications, Oxford. 16.4.4

Horn, H.S. & May, R.M. (1977) Limits to similarity among coexisting competitors. *Nature* (London), **270**, 660–661. 18.4.2

Horton, K. (1964) Deer prefer jack pine. *Journal of Forestry*, **62**, 497–499. 9.2.1

Howarth, S.E. & Williams, J.T. (1972) *Chrysanthemum segetum*. *Journal of Ecology*, **60**, 473–584. 14.5.2

Hubbard, S.F. & Cook, R.M. (1978) Optimal foraging by parasitoid wasps. *Journal of Animal Ecology*, **47**, 593–604. 9.10, 9.11.2

Hudson, P.J. (1984) The effect of a parasitic nematode on the breeding production of red grouse. *Journal of Animal Ecology*, **55**, 85–92. 12.6.5

Huffaker, C.B. (1958) Experimental studies on predation: dispersion factors and predator–prey oscillations. *Hilgardia*, **27**, 343–383. 9.9

Huffaker, C.B. (1973) Biological control in the management of pests. *Agroecosystems*, **2**, 15–31. 15.6.5

Huffaker, C.B., Shea, K.P. & Herman, S.G. (1963) Experimental studies on predation. *Hilgardia* **34**, 305–330. 9.9

Hughes, R.G. & Thomas, M.L. (1971) The classification and ordination of shallow-water benthic samples from Prince Edward Island, Canada. *Journal of Experimental Marine Biology and Ecology*, **7**, 1–39. 16.3.1

Humphreys, W.F. (1979) Production and respiration in animal populations. *Journal of Animal Ecology*, **48**, 427–454. 17.4.1

Hungate, R.E. (1975) The rumen microbial ecosystem. *Annual Review of Ecology and Systematics*, **6**, 39–66. 13.5.1

Hungate, R.E., Reichl, J. & Prins, R.A. (1971) Parameters of rumen fermentation in a continuously fed sheep: evidence of a microbial rumination pool. *Applied Microbiology*, **22**, 1104–1113. 13.5.1

Huston, M. (1979) A general hypothesis of species diversity. *American Naturalist*, **113**, 81–101. 19.4.1, 22.3.1, 22.5

Hutchinson, G.E. (1957) Concluding remarks. *Cold Spring Harbor Symposium on Quantitative Biology*, **22**, 415–427. 2.12

Hutchinson, G.E. (1959) Homage to Santa Rosalia, *or* why are there so many kinds of animals? *American Naturalist*, **93**, 145–159. 18.3.2, 18.4.2

Hutchinson, G.E. (1961) The paradox of the plankton. *American Naturalist*, **95**, 137–145. 7.6.3

Iles, T.D. (1973) Interaction of environment and parent stock size in determining recruitment in the Pacific sardine as revealed by analysis of density-dependent O-group growth. *Rapports et Procès-verbaux, Conseil international pour l'Exploration de la Mer*, **164**, 228–240. 10.8.4

Iles, T.D. (1981) A comparison of two stock/recruitment hypotheses as applied to North Sea herring and Gulf of St. Lawrence cod. *International Council for the Exploration of the Sea*, C.M. 1981/H: 48. 10.8.4

Inglesfield, C. & Begon, M. (1981) Open ground individuals and population structure in *Drosophila subobscura* Collin. *Biological Journal of the Linnean Society*, **15**, 259–278. 2.5

Inglesfield, C. & Begon, M. (1983) The ontogeny and cost of migration in *Drosophila subobscura* Collin. *Biological Journal of the Linnean Society*, **19**, 9–15. 14.4.1

Inouye, D.W. (1978) Resource partitioning in bumblebees: experimental studies of foraging behaviour. *Ecology*, **59**, 672–678. 18.3.1

Inouye, D.W. & Taylor, O.R. (1979) A temperate region plant–ant–seed predator system: Consequences of extra-floral nectary secretion by *Helianthella quinquinervis*. *Ecology*, **60**, 1–7. 13.2.5

Ives, J.D. (1974) Biological refugia and the nunatak hypothesis. In: *Arctic and Alpine Environments* (J.D. Ives & R.D. Berry eds). Methuen, London. 1.2.2

Jackson, J.B.C. (1979) Overgrowth competition between encrusting cheilostome ectoprocts in a Jamaican cryptic reef environment. *Journal of Animal Ecology*, **48**, 805–823. 7.3

Jacob, F. (1977) Evolution and tinkering. *Science, N.Y.*, **196**, 1161–1166. 1.1.2

Janzen, D.H. (1967) Interaction of the bull's-horn acacia (*Acacia cornigera* L.) with an ant inhabitant (*Pseudomyrmex ferruginea* F. Smith) in Eastern Mexico. *University of Kansas Science Bulletin*, **47**, 315–558. 13.2.5

Janzen, D.H. (1968) Host plants in evolutionary

and contemporary time. *American Naturalist,* **102,** 592–595.
12.3.1, 20.2.4

Janzen, D.H. (1970) Herbivores and the number of tree species in tropical forests. *American Naturalist,* **104,** 501–528.
22.4.1

Janzen, D.H. (1973) Host plants as islands. II. Competition in evolutionary and contemporary time. *American Naturalist,* **107,** 786–790.
12.3.1

Janzen, D.H. (1976) Why bamboos wait so long to flower. *Annual Review of Ecology and Systematics,* **7,** 347–391.
5.3.6

Janzen, D.H. (1979) How to be a fig. *Annual Review of Ecology and Systematics,* **10,** 13–51.
13.4

Janzen, D.H. (1980) Specificity of seed-eating beetles in a Costa Rican deciduous forest. *Journal of Ecology,* **68,** 929–952.
9.2

Janzen, D.H. (1981) Evolutionary physiology of personal defence. In: *Physiological Ecology: An Evolutionary Approach to Resource Use* (C.R. Townsend & P. Calow eds). Blackwell Scientific Publications, Oxford.
11.3.1

Janzen, D.H., Juster, H.B. & Bell, E.A. (1977) Toxicity of secondary compounds to the seed-eating larvae of the bruchid beetle *Callosobruchus maculatus. Phytochemistry,* **16,** 223–227.
3.6.2

Jeffries, M.J. & Lawton, J.H. (1984) Enemy-free space and the structure of ecological communities. *Biological Journal of the Linnean Society,* **23,** 269–286.
7.7

Jeffries, M.J. & Lawton, J.H. (1985) Predator–prey ratios in communities of freshwater invertebrates: the role of enemy free space. *Freshwater Biology,* **15,** 105–112.
22.2

Jenkins, D., Watson, A. & Miller, C.R. (1967) Population fluctuations in the red grouse *Lagopus lagopus scoticus. Journal of Animal Ecology,* **36,** 97–122.
10.3

Joern, A. & Lawlor, L.R. (1980) Food and microhabitat utilization by grasshoppers from arid grasslands: comparisons with neutral models. *Ecology,* **61,** 591–599.
18.4.1

Johnson, C.G. (1967) International dispersal of insects and insect-borne viruses. *Netherlands Journal of Plant Pathology,* **73** (Suppl. 1), 21–43.
5.4.3

Johnson, C.G. (1969) *Migration and Dispersal of Insects by Flight.* Methuen, London.
5.4.1, 5.4.4

Johnson, N.K. (1975) Controls of number of bird species on montane islands in the Great Basin. *Evolution,* **29,** 545–567.
20.3.1

Johnston, D.W. & Odum, E.P. (1956) Breeding bird populations in relation to plant succession on the piedmont of Georgia. *Ecology,* **37,** 50–62.
16.4.5, 22.4.4

Jones, M.G. (1933) Grassland management and its influence on the sward. *Empire Journal of Experimental Agriculture,* **1,** 43–367.
19.2.2

Jordan, C.F., Kline, J.R. & Sasscer, D.S. (1972) Effective stability of mineral cycles in forest ecosystems. *American Naturalist,* **106,** 237–253.
21.5

Juhren, M., Went, F.W. & Phillips, E. (1956) Ecology of desert plants. IV. Combined field and laboratory work on germination of annuals in the Joshua Tree National Monument, California. *Ecology,* **37,** 318–330.
4.5.3

Kalela, O. (1962) On the fluctuations in the numbers of arctic and boreal small rodents as a problem of production biology. *Annales Academiae Scientiarum Fennicae (A IV),* **66,** 1–38.
15.4.2

Kaplan, R.H. & Salthe, S.N. (1979) The allometry of reproduction: an empirical view in salamanders. *American Naturalist,* **113,** 671–689.
14.14.2, 14.14.4

Karban, R. & Ricklefs, R.E. (1983) Host characteristics, sampling intensity, and species richness of Lepidopteran larvae on broad-leaved trees in southern Ontario. *Ecology,* **64,** 636–641.
20.3.3

Karplus, I., Tsurnamal, M. & Szlep, M. (1972) Associative behaviour of the fish *Cryptocentrus cryptocentrus* (Gobiidae) and the pistol shrimp *Alpheus djiboutensis* (Alpheidae) in artificial burrows. *Marine Biology,* **15,** 95–104.
13.2.2

Kays, S. & Harper, J.L. (1974) The regulation of plant and tiller density in a grass sward. *Journal of Ecology,* **62,** 97–105.
4.2.1, 6.5

Keddy, P.A. (1981) Experimental demography of the sand-dune annual, *Cakile edentula,* growing along an environmental gradient in Nova Scotia. *Journal of Ecology,* **69,** 615–630.
15.5

Keddy, P.A. (1982) Population ecology on an environmental gradient: *Cakile edentula* on a sand dune.

Oecologia, **52,** 348–355.
15.5

Keith, L.B. (1983) Role of food in hare population cycles. *Oikos,* **40,** 385–395.
10.2.3, 10.4.1

Kendrick, W.B. & Burges, A. (1962) Biological aspects of the decay of *Pinus sylvestris* leaf litter. *Nova Hedwiga,* **4,** 313–342.
16.4.1

Kershaw, K.A. (1973) *Quantitative and Dynamic Plant Ecology,* 2nd edn. Edward Arnold, London.
4.3, 5.2

Kigel, J. (1980) Analysis of regrowth patterns and carbohydrate levels in *Lolium multiflorum* Lam. *Annals of Botany,* **45,** 91–101.
8.2.1

Kikkawa, J. & Williams, W.T. (1971) Altitudinal distribution of land birds in New Guinea. *Search,* **2,** 64–69.
22.4.2

Kingston, T.J. (1977) Natural manuring by elephants in the Tsavo National Park, Kenya. D.Phil. thesis, University of Oxford.
11.3.2

Kingston, T.J. & Coe, M.J. (1977) The biology of a giant dung-beetle (*Heliocopris dilloni*) (Coleoptera: Scarabaeidae). *Journal of Zoology* (London), **181,** 243–263.
11.3.2

Kira, T., Ogawa, H. & Shinozaki, K. (1953) Intraspecific competition among higher plants. I. Competition-density-yield inter-relationships in regularly dispersed populations. *Journal of the Polytechnic Institute, Osaka City University* 4(4), 1–16.
6.5

Kitting, C.L. (1980) Herbivore–plant interactions of individual limpets maintaining a mixed diet of intertidal marine algae. *Ecological Monographs,* **50,** 527–550.
9.2.2

Klemow, K.M. & Raynal, D.J. (1981) Population ecology of *Melilotus alba* in a limestone quarry. *Journal of Ecology,* **69,** 33–44.
4.8

Klepac, D. (1955) Effect of *Viscum album* on the increment of silver fir stands. *Sumarski List,* **79,** 231–244.
12.2.2

Kline, J.R. (1970) Retention of fallout radionuclides by tropical forest vegetation. In: *A Tropical Rain Forest* (H.T. Odum ed.). Division of Technical Information, U.S. Atomic Energy Commission, Washington, D.C.
21.5

Koblentz-Mishke, I.J., Volkovinsky, V.V. & Kabanova, J.B. (1970) Plankton primary production of the world ocean. In: *Scientific Exploration of the South Pacific*

(W.S. Wooster ed.). National Academy of Sciences, Washington.
17.2

Kok, L.T. & Surles, W.W. (1975) Successful biocontrol of musk thistle by an introduced weevil, *Rhinocyllus conicus. Environmental Entomology,* **4,** 1025–1027.
8.2.6

Kolding, S. & Fenchel, T.M. (1979) Coexistence and life-cycle characteristics of five species of the amphipod genus *Gammarus. Oikos,* **33,** 323–327.
18.3.1

Koller, D. & Roth, N. (1964) Studies on the ecological and physiological significance of amphicarpy in *Gymnarhena micrantha* (Compositae). *American Journal of Botany,* **51,** 26–35.
5.5.3

Krebs, C.J. (1972) *Ecology.* Harper & Row, New York.
Introduction

Krebs, C.J. (1978) A review of the Chitty Hypothesis of population regulation. *Canadian Journal of Zoology,* **56,** 2463–2480.
5.5.4

Krebs, C.J. (1985) Do changes in spacing behaviour drive population cycles in small mammals? In: *Behavioural Ecology* (R.M. Sibly & R.H. Smith eds). Blackwell Scientific Publications, Oxford.
15.4.2

Krebs, C.J. & Myers, J.H. (1974) Population cycles in small mammals. *Advances in Ecological Research,* **8,** 267–399.
15.4.2

Krebs, C.J., Keller, B.L. & Tamarin, R.H. (1969) *Microtus* population biology: demographic changes in fluctuating populations of *M. ochrogaster* and *M. pennsylvanicus* in southern Indiana. *Ecology,* **50,** 587–607.
5.4.3, 15.4.2

Krebs, C.J., Gaines, M.S., Keller, B.L., Myers, J.M. & Tamarin, R.H. (1973) Population cycles in small rodents. *Science, N.Y.,* **179,** 35–41.
15.4.2

Krebs, J.R. (1971) Territory and breeding density in the great tit, *Parus major* L. *Ecology,* **52,** 2–22.
6.11

Krebs, J.R. (1978) Optimal foraging: decision rules for predators. In: *Behavioural Ecology: An Evolutionary Approach* (J.R. Krebs & N.B. Davies eds), pp. 23–63. Blackwell Scientific Publications, Oxford.
9.2.2, 9.3, 9.3.1, 9.11.2

Krebs, J.R. & McCleery, R.H. (1984) Optimization in behavioural ecology. In: *Behavioural Ecology: An Evolutionary Approach,* 2nd edn. (J.R. Krebs & N.B. Davies eds), pp. 91–121. Blackwell Scientific Publications, Oxford.
9.3.1, 9.11.3

Krebs, J.R., Stephens, D.W. & Sutherland, W.J. (1983) Perspectives in optimal foraging. In: *Perspectives in*

Ornithology (G.A. Clark & A.H. Brush eds). Cambridge University Press, New York.
9.3.1, 9.5.1

Krebs, J.R., Erichsen, J.T., Webber, M.I. & Charnov, E.L. (1977) Optimal prey selection in the great tit (*Parus major*). *Animal Behaviour*, **25**, 30–38.
9.3.1

Krueger, D.A. & Dodson, S.I. (1981) Embryological induction and predation ecology in *Daphnia pulex*. *Limnology and Oceanography*, **26**, 219–223.
3.4.3

Kullenberg, B. (1946) Über verbreitung und Wanderungen von vier *Sterna*—Arten. *Arkiv för Zoologie: utgifvet af K. Svenska vetenskapsakademien*, bd 1, **38**, 1–80.
5.3.3

Kusnezov, M. (1957) Numbers of species of ants in faunas of different latitudes. *Evolution* **11**, 298–299.
22.4.1

Lack, D. (1947) *Darwin's Finches*. Cambridge University Press, Cambridge.
12.3

Lack, D. (1954) *The Natural Regulation of Animal Numbers*. Clarendon Press, Oxford.
5.3.3

Lack, D. (1969a) Subspecies and sympatry in Darwin's finches. *Evolution*, **23**, 252–263.
1.2.3

Lack, D. (1969b) The numbers of bird species on islands. *Bird Study*, **16**, 193–209.
20.2.1

Lack, D. (1971) *Ecological Isolation in Birds*. Blackwell Scientific Publications, Oxford.
7.8

Lack, D. (1976) *Island Birds*. Blackwell Scientific Publications, Oxford.
20.2.1

Laine, K. & Henttonen, H. (1983) The role of plant production in microtine cycles in northern Fennoscandia. *Oikos*, **40**, 407–418.
15.4.2

Lanciani, C.A. (1975) Parasite-induced alterations in host reproduction and survival. *Ecology*, **56**, 689–695.
12.5.5

Lappe, F. (1971) *Diet for a Small Planet*. Ballantine, New York.
3.6.2

Larcher, W. (1980) *Physiological Plant Ecology*, 2nd edn. Springer Verlag, Berlin.
2.6, 3.2, 3.2.2, 3.2.4

Latch, G.C.M. & Lancashire, J.A. (1970) The importance of some effects of fungal disease on pasture yield and composition. *Proceedings of the XIth International Grassland Congress*.
19.2.5

Law, R. & Lewis, D.H. (1983) Biotic environments and the maintenance of sex—some evidence from mutualistic symbiosis. *Biological Journal of the Linnean Society*, **20**, 249–276.
13.13

Law, R., Bradshaw, A.D. & Putwain, P.D. (1977) Life history variation in *Poa annua*. *Evolution*, **31**, 233–246.
14.9.2

Lawlor, L.R. (1976) Molting, growth and reproductive strategies in the terrestrial isopod, *Armadillidium vulgare*. *Ecology*, **57**, 1179–1194.
14.4.3

Lawlor, L.R. (1978) A comment on randomly constructed ecosystem models. *American Naturalist*, **112**, 445–447.
21.2.2

Lawlor, L.R. (1980) Structure and stability in natural and randomly constructed competitive communities. *American Naturalist*, **116**, 394–408.
18.4.1

Lawrence, W.H. & Rediske, J.H. (1962) Fate of sown douglas-fir seed. *Forest Science*, **8**, 211–218.
8.3

Laws, R.M. (1984) Seals. In: *Antarctic Ecology* (R.M. Laws ed.). Academic Press, London.
1.4.2

Lawton, J.H. (1984) Non-competitive populations, non-convergent communities, and vacant niches: The herbivores of bracken. In: *Ecological Communities: Conceptual Issues and the Evidence* (D.R. Strong, D. Simberloff, L.G. Abele & A.B. Thistle eds). Princeton University Press, Princeton, New Jersey.
18.2.1, 20.3.3

Lawton, J.H. & Hassell, M.P. (1981) Asymmetrical competition in insects. *Nature* (London), **289**, 793–795.
7.3

Lawton, J.H. & S. McNeill (1979) Between the devil and the deep blue sea: on the problems of being a herbivore. In: *Population Dynamics* (R.M. Anderson, B.D. Turner & L.R. Taylor eds), pp. 223–244. Blackwell Scientific Publications, Oxford.
18.2.1

Lawton, J.H. & May, R.M. (1984) The birds of Selborne. *Nature* (London), **306**, 732–733.
15.2

Lawton, J.H. & Price, P.W. (1979) Species richness of parasites on hosts: agromyzid flies on the British Umbelliferae. *Journal of Animal Ecology*, **48**, 619–637.
19.1

Lawton, J.H. & Rallison, S.P. (1979) Stability and diversity in grassland communities. *Nature* (London), **279**, 351.
21.2.3

Lawton, J.H. & Schröder, D. (1977) Effects of plant type,

size of geographical range and taxonomic isolation on number of insect species associated with British plants. *Nature* (London), **265,** 137–140.
20.3.3, 20.3.5

Lawton, J.H. & Strong, D.R. Jr. (1981) Community patterns and competition in folivorous insects. *American Naturalist,* **118,** 317–338.
18.3.1

Lawton, J.H., Beddington, J.R. & Bonser, R. (1974) Switching in invertebrate predators. In: *Ecological Stability* (M.B. Usher & M.H. Williamson eds), pp. 141–158. Chapman and Hall, London.
9.2.3

Le Cren, E.D. (1973) Some examples of the mechanisms that control the population dynamics of salmonid fish. In: *The Mathematical Theory of the Dynamics of Biological Populations* (M.S. Bartlett & R.W. Hiorns eds), pp. 125–135. Academic Press, London.
6.3

Lee, K.E. & Wood, T.G. (1971) *Termites and Soils.* Academic Press, London.
11.3.1

Leigh, E. (1975) Population fluctuations and community structure. In: *Unifying Concepts in Ecology* (W.H. van Dobben & R.H. Lowe–McConnell eds), pp. 67–88. Junk, The Hague.
21.2.3

Leith, H. (1975) Primary productivity in ecosystems: comparative analysis of global patterns. In: *Unifying Concepts in Ecology* (W.H. van Dobben & R.H. Lowe–McConnell eds). Junk, The Hague.
17.2, 17.3.1

Leverich, W.J. & Levin, D.A. (1979) Age-specific survivorship and reproduction in *Phlox drummondii. American Naturalist,* **113,** 881–903.
4.5.1, 14.3

Levin, D.A. & Kerster, H.W. (1974) Gene-flow in seed plants. *Evolutionary Biology,* **7,** 179–220.
5.6

Levin, S.A. & Pimentel, D. (1981) Selection of intermediate rates of increase in parasite–host systems. *American Naturalist,* **117,** 308–315.
12.7

Lewis, D.H. (1974) Microorganisms and plants. The evolution of parasitism and mutualism. *Symposium of the Society for General Microbiology,* **24,** 367–392.
13.7.3

Lewis, J.R. (1976) *The Ecology of Rocky Shores.* Hodder & Stoughton, London.
2.4, 2.10

Lidicker, W.Z. Jr. (1975) The role of dispersal in the demography of small mammal populations. In: *Small Mammals: Their Productivity and Population Dynamics* (K. Petruscwicz, F.B. Golley & L. Ryszkowski eds), pp. 103–128. Cambridge University Press, New York.
5.5.4

Likens, G.E. & Bormann, F.G. (1975) An experimental approach to New England landscapes. In: *Coupling of Land and Water Systems* (A.D. Hasler ed.), pp. 7–30. Chapman and Hall, London.
17.5

Likens, G.E., Bormann, F.G., Pierce, R.S. & Fisher, D.W. (1971) Nutrient-hydrologic cycle interaction in small forested watershed-ecosystems. In: *Productivity of Forest Ecosystems* (P. Duvigneaud ed.). UNESCO, Paris.
17.5

Limbaugh, C. (1961) Cleaning symbiosis. *Scientific American,* **205,** 42–49.
13.2.4

Lindemann, R.L. (1942) The trophic-dynamic aspect of ecology. *Ecology,* **23,** 399–418.
14.4

Lonsdale, W.M. & Watkinson, A.R. (1982) Light and self-thinning. *New Phytologist,* **90,** 431–435.
6.12

Lonsdale, W.M. & Watkinson, A.R. (1983) Plant geometry and self-thinning. *Journal of Ecology,* **71,** 285–297.
6.12

Lotka, A.J. (1907) Studies on the mode of growth of material aggregates. *American Journal of Science,* **24,** 199–216.
4.7

Lotka, A.J. (1925) *Elements of Physical Biology.* Williams & Wilkins, Baltimore.
7.4.1, 10.2

Lovett Doust, J. (1980) Experimental manipulation of patterns of resource allocation in the growth cycle and reproduction of *Smyrnium olusatrum* L. *Biological Journal of the Linnean Society,* **13,** 155–166.
14.12

Lovett Doust, L. & Lovett Doust, J. (1982) The battle strategies of plants. *New Scientist,* **95,** 81–84.
5.8

Lowe, V.P.W. (1969) Population dynamics of the red deer (*Cervus elaphus* L.) on Rhum. *Journal of Animal Ecology,* **38,** 425–457.
4.6.1, 4.6.2

Lubchenko, J. (1978) Plant species diversity in a marine intertidal community: importance of herbivore food preference and algal competitive abilities. *American Naturalist,* **112,** 23–39.
19.2.2

Luxmoore, R.J., Stolzy, L.H. & Letey, J. (1970) Oxygen diffusion in the soil plant system. IV. Oxygen concentration profiles, respiration rates, and radial oxygen losses predicted for rice roots. *Agronomy Journal,* **62,** 329–332.
3.3.4

MacArthur, J.W. (1975) Environmental fluctuations and species diversity. In: *Ecology and Evolution of Communities* (M.L. Cody & J.M. Diamond eds), pp.

74–80. Belknap, Cambridge, Massachusetts.
22.3.3

MacArthur, R.H. (1955) Fluctuations of animal populations and a measure of community stability. *Ecology*, **36**, 533–536.
21.2.1

MacArthur, R.H. (1972) *Geographical Ecology*. Harper & Row, New York.
2.3, 2.4, 18.3.1, 22.1

MacArthur, R.H. & Levins, R. (1967) The limiting similarity, convergence and divergence of coexisting species. *American Naturalist*, **101**, 377–385.
7.5

MacArthur, R.H. & Pianka, E.R. (1966) On optimal use of a patchy environment. *American Naturalist*, **100**, 603–609.
9.3.1

MacArthur, R.H. & Wilson, E.O. (1967) *The Theory of Island Biogeography*. Princeton University Press, Princeton, New Jersey.
14.8, 15.2.2, 19.1, 20.2.3, 20.3.5

McBrayer, J.F. (1973) Exploitation of deciduous litter by *Apheloria moutana* (Diplopoda: Eurydesmidae). *Pedobiologia*, **13**, 90–98.
11.3.2

McCauley, E. & Briand, F. (1979) Zooplankton grazing and phytoplankton species richness: Field tests of the predation hypothesis. *Limnology and Oceanography*, **24**, 243–252.
19.2.2

McClure, P.A. & Randolph, J.C. (1980) Relative allocation of energy to growth and development of homeothermy in the eastern wood rat (*Neotoma floridana*) and hispid cotton rat (*Sigmodon hispidus*). *Ecological Monographs*, **50**, 199–219.
14.13

Mackerras, M.I. & Sandars, D.F. (1955) The life history of the rat lung-worm, *Angiostrongylus cantonensis* (Chen) (Nematoda: Metastrongylidae). *Australian Journal of Zoology*, **3**, 1–21.
12.4

Mackie, G.L., Qadri, S.U. & Reed, R.M. (1978) Significance of litter size in *Musculium securis* (Bivalvia: Sphaeridae). *Ecology*, **59**, 1069–1074.
6.3

McLachlan, A.J., Pearce, L.J. & Smith, J.A. (1979) Feeding interactions and cycling of peat in a bog lake. *Journal of Animal Ecology*, **48**, 851–861.
11.3.2

MacLulick, D.A. (1937) Fluctuations in numbers of the varying hare (*Lepus americanus*). *University of Toronto Studies, Biology Series*, **43**, 1–136.
10.1

McMahon, T. (1973) Size and shape in Biology. *Science, N.Y.*, **179**, 1201–1204.
14.14.2

McMillan, C. (1957) Nature of the plant community. III. Flowering behaviour within two grassland communities under reciprocal transplanting. *American Journal of Botany*, **44**, 143–153.
1.5.1

McNaughton, S.J. (1975) r- and k-selection in *Typha*. *American Naturalist*, **109**, 251–261.
14.9.2

McNaughton, S.J. (1977) Diversity and stability of ecological communities: a comment on the role of empiricism in ecology. *American Naturalist*, **111**, 515–525.
21.2.2, 21.2.3, 21.4.4

McNeill, S. (1973) The dynamics of a population of *Leptoterna dolobrata* (Heteroptera: Miridae) in relation to its food resources. *Journal of Animal Ecology*, **42**, 495–507.
6.6

Madge, D.S. (1969) Field and laboratory studies on the activities of two species of tropical earthworms. *Pedobiologia*, **9**, 188–214.
11.2.2

Margulis, L. (1975) Symbiotic theory of the origin of eukaryotic organelles. In: *Symbiosis* (D.H. Jennings & D.L. Lee, eds), pp. 21–38. Symposium 29, Society for Experimental Biology. Cambridge University Press, Cambridge.
13.11

Marshall, C. & Sagar, G.R. (1968) The interdependence of tillers in *Lolium multiflorum* Lam.: a quantitative assessment. *Journal of Experimental Botany*, **19**, 785–794.
8.2.1

Marshall, D.R. & Jain, S.K. (1969) Interference in pure and mixed populations of *Avena fatua* and *A. barbata*. *Journal of Ecology*, **57**, 251–270.
7.9.1

Marzusch, K. (1952) Untersuchungen über die Temperaturabhängigkeit von Lebensprozessen bei Insekten unter besonderer Berücksichtigung winterschlatender Kartoffelkäfer. *Zeitschrift für vergleicherde Physiologie*, **34**, 75–92.
2.2.3

May, R.M. (1972) Will a large complex system be stable? *Nature* (London), **238**, 413–414.
21.2.2, 21.3

May, R.M. (1973) *Stability and Complexity in Model Ecosystems*. Princeton University Press, Princeton.
7.5

May, R.M. (1975a) Biological populations obeying difference equations: stable points, stable cycles and chaos. *Journal of Theoretical Biology*, **49**, 511–524.
6.8.4

May, R.M. (1975b) Patterns of species abundance and diversity. In: *Ecology and Evolution of Communities* (M.L. Cody & J.M. Diamond eds), pp. 81–120.

Belknap, Cambridge, Massachusetts.
16.2.2, 20.3.2

May, R.M. (1976) Estimating *r*: a pedagogical note. *American Naturalist*, **110**, 496–499.
4.7

May, R.M. (1977) Thresholds and breakpoints in ecosystems with a multiplicity of stable states. *Nature* (London), **269**, 471–477.
10.6

May, R.M. (1978) The dynamics and diversity of insect faunas. In: *Diversity of Insect Faunas* (L.A. Mound & N. Waloff eds). *Symposium of the Royal Entomological Society of London*, **9**, 188–204. Blackwell Scientific Publications, Oxford.
22.5

May, R.M. (1980) Nations and numbers, 1980. *Nature* (London), **287**, 482–483.
4.7, 4.10

May, R.M. (1981a) Models for single populations. In: *Theoretical Ecology: Principles and Applications*, 2nd edn. (R.M. May ed.), pp. 5–29. Blackwell Scientific Publications, Oxford.
10.2.2

May, R.M. (1981b) Models for two interacting populations. In: *Theoretical Ecology: Principles and Applications*, 2nd edn. (R.M. May, ed.), pp. 78–104. Blackwell Scientific Publications, Oxford.
7.5

May, R.M. (1981c) Patterns in multi-species communities. In: *Theoretical Ecology: Principles and Applications*, 2nd edn. (R.M. May ed.). Blackwell Scientific Publications, Oxford.
21.2.2

May, R.M. (1981d) Population biology of parasitic infections. In: *The Current Status and Future of Parasitology* (K.S. Warren & E.F. Purcell eds), pp. 208–235. Josiah Macy, Jr., Foundation, New York.
12.6.1

May, R.M. (1982) Mutualistic interactions among species. *Nature* (London), **296**, 803–804.
13.1, 13.12, 13.13

May, R.M. (1983) Parasitic infections as regulators of animal populations. *American Scientist*, **71**, 36–45.
12.6.5

May, R.M. & Anderson, R.M. (1979) Population biology of infectious diseases. *Nature* (London), **280**, 455–461.
12.2

May, R.M. & Anderson, R.M. (1983) Epidemiology and genetics in the coevolution of parasites and hosts. *Proceedings of the Royal Society of London*, B, **219**, 281–313.
12.7

Maynard Smith, J. (1972) *On Evolution*. Edinburgh University Press, Edinburgh.
5.4.2

Maynard Smith, J. (1974) *Models in Ecology*. Cambridge University Press, Cambridge.
21.2.2

Maynard Smith, J. (1978) *The Evolution of Sex*. Cambridge University Press, Cambridge.
13.4

Maynard Smith, J. & Slatkin, M. (1973) The stability of predator–prey systems. *Ecology*, **54**, 384–391.
6.8.2

Mead, C.J. & Harrison, J.D. (1979) Sand martin movements within Britain and Ireland. *Bird Study*, **26**, 73–86.
5.4.1

Menge, B.A. & Sutherland, J.P. (1976) Species diversity gradients: synthesis of the roles of predation, competition, and temporal heterogeneity. *American Naturalist*, **110**, 351–369.
19.2.6

Milinski, M. (1979) An evolutionarily stable feeding strategy in sticklebacks. *Zeitschrift für Tierpsychologie*, **51**, 36–40.
9.8

Miller, D. (1970) *Biological Control of Weeds in New Zealand 1927–48*. New Zealand Department of Scientific and Industrial Research Information Series, vol. 74, pp. 1–104.
10.1

Miller, M.R. (1975) Gut morphology of mallards in relation to diet quality. *Journal of Wildlife Management*, **39**, 168–173.
9.2.3

Milne, L.J. & Milne, M. (1976) The social behaviour of burying beetles. *Scientific American*, August, 84–89.
11.3.2

Milton, W.E.J. (1940) The effect of manuring, grazing and liming on the yield, botanical and chemical composition of natural hill pastures. *Journal of Ecology*, **28**, 326–356.
19.2.2

Milton, W.E.J. (1947) The composition of natural hill pasture, under controlled and free grazing, cutting and manuring. *Welsh Journal of Agriculture*, **14**, 182–195.
19.2.2

Mithen, R., Harper, J.L. & Weiner, J. (1984) Growth and mortality of individual plants as a function of 'available area'. *Oecologia*, **62**, 57–60.
6.10

Monro, J. (1967) The exploitation and conservation of resources by populations of insects. *Journal of Animal Ecology*, **36**, 531–547.
9.7.3, 9.8, 10.4.1, 15.6.5

Moran, P.A.P. (1952) The statistical analysis of gamebird records. *Journal of Animal Ecology*, **21**, 154–158.
15.4.1

Moran, P.A.P. (1953) The statistical analysis of the

Canadian lynx cycle. *Australian Journal of Zoology*, **1**, 163–173.
15.4.1

Moran, V.C. (1980) Interactions between phytophagous insects and their *Opuntia* hosts. *Ecological Monographs*, **5**, 153–164.
20.3.3

Moreau, R.E. (1952) The place of Africa in the palaearctic migration system. *Journal of Animal Ecology*, **21**, 250–271.
5.3.3

Morris, R.F. (1959) Single-factor analysis in population dynamics. *Ecology*, **40**, 580–588.
15.3.1

Morris, R.F. (ed.) (1963) The dynamics of epidemic spruce budworm populations. *Memoirs of the Entomological Society of Canada*, **31**, 1–332.
10.6.1

Morrison, G. & Strong, D.R. Jr. (1981) Spatial variations in egg density and the intensity of parasitism in a neotropical chrysomelid (*Cephaloleia consanguinea*). *Ecological Entomology*, **6**, 55–61.
9.7.1

Moss, B. (1980) *Ecology of Fresh Waters*. Blackwell Scientific Publications, Oxford.
17.5

Moss, G.D. (1971) The nature of the immune response of the mouse to the bile duct cestode, *Hymenolepis microstoma*. *Parasitology*, **62**, 285–294.
12.4.1

Moss, R., Welch, D. & Rothery, P. (1981) Effects of grazing by mountain hares and red deer on the production and chemical composition of heather. *Journal of Applied Ecology*, **18**, 487–496.
9.7

Müller, G. & Foerster, E. (1974) Entwicklung von Weideansaaten im Überflutungsbereich des Rheines bei Kleve. *Acker und Pflanzenban*, **140**, 161–174.
19.4

Müller, H.J. (1970) Formen der Dormanz bei Insekten. *Nova Acta Leopoldina*, **191**, 1–27.
5.7

Murdoch, W.W. (1969) Switching in general predators: experiments on predator specificity and stability of prey populations. *Ecological Monographs*, **39**, 335–354.
9.2.3

Murdoch, W.W. & Oaten, A. (1975) Predation and population stability. *Advances in Ecological Research*, **9**, 1–131.
9.2.3, 9.9

Murdoch, W.W., Reeve, J.D., Huffaker, C.B. & Kennett, C.E. (1984) Biological control of olive scale and its relevance to ecological theory. *American Naturalist*, **123**, 371–392.
10.4.1

Murphy, G.J. (1977) Clupeoids. In: *Fish Population Dynamics* (J.A. Gulland ed.), pp. 283–308. Wiley-Interscience, New York.
10.8.2, 10.8.4

Murton, R.K. (1971) The significance of a specific search image in the feeding behaviour of the wood pigeon. *Behaviour*, **40**, 10–42.
9.2.3

Murton, R.K., Isaacson, A.J. & Westwood, N.J. (1966) The relationships between wood pigeons and their clover food supply and the mechanism of population control. *Journal of Applied Ecology*, **3**, 55–93.
5.3.6, 9.7.1, 19.2.3

Murton, R.K., Westwood, N.J. & Isaacson, A.J. (1974) A study of wood-pigeon shooting: the exploitation of a natural animal population. *Journal of Applied Ecology*, **11**, 61–81.
8.3

Muscatine, L. & Pool, R.R. (1979) Regulation of numbers of intracellular algae. *Proceedings of the Royal Society of London*, B, **204**, 115–139.
13.8

Myers, J.M. & Krebs, C.J. (1971) Genetic, behavioural and reproductive attributes of dispersing field voles *Microtus pennsylvanicus* and *Microtus ochrogaster*. *Ecological Monographs*, **41**, 53–78.
5.5.1

Namkoong, G. & Roberds, J.H. (1974) Extinction probabilities and the changing age structure of redwood forests. *American Naturalist*, **108**, 355–368.
16.4.4

Naylor, R. & Begon, M. (1982) Variations within and between populations of *Littorina nigrolineata* Gray on Holy Island, Anglesey. *Journal of Conchology*, **31**, 17–30.
14.11

Nicholson, A.J. (1933) The balance of animal populations. *Journal of Animal Ecology*, **2**, 131–178.
15.2.1

Nicholson, A.J. (1954a) Compensatory reactions of populations to stress, and their evolutionary significance. *Australian Journal of Zoology*, **2**, 1–8.
15.2.1

Nicholson, A.J. (1954b) An outline of the dynamics of animal populations. *Australian Journal of Zoology*, **2**, 9–65.
6.6, 8.3, 10.8, 15.2.1

Nicholson, A.J. (1957) The self adjustment of populations to change. *Cold Spring Harbor Symposium of Quantitative Biology*, **22**, 153–172.
15.2.1

Nicholson, A.J. (1958) Dynamics of insect populations. *Annual Review of Entomology*, **3**, 107–136.
15.2.1

Nicholson, A.J. & Bailey, V.A. (1935) The balance of animal populations. *Proceedings of the Zoological*

Society of London, **3,** 551–598.
10.2

Nielsen, B.O. & Ejlerson, A. (1977) The distribution pattern of herbivory in a beech canopy. *Ecological Entomology,* **2,** 293–299.
8.2.1

Niklas, K.J., Tiffney, B.H. & Knoll, A.H. (1983) Patterns in vascular land plant diversification. *Nature* (London), **303,** 614–616.
22.4.5

Nixon, M. (1969) The lifespan of *Octopus vulgaris* Lamarck. *Proceedings of the Malacological Society of London,* **38,** 529–540.
4.8

Noble, I.R. & Slatyer, R.O. (1979) The effect of disturbance on plant succession. *Proceedings of the Ecological Society of Australia,* **10,** 135–145.
16.4.4

Noble, I.R. & Slatyer, R.O. (1981) Concepts and models of succession in vascular plant communities subject to recurrent fire. In: *Fire and the Australian Biota* (A.M. Gill, R.H. Graves & I.R. Noble eds). Australian Academy of Science, Canberra.
16.4.4

Noble, J.C., Bell, A.D. & Harper, J.L. (1979) The population biology of plants with clonal growth. I. The morphology and structural demography of *Carex arenaria. Journal of Ecology,* **67,** 983–1008.
4.2.1, 5.7.3

Norton, I.O. & Sclater, J.G. (1979) A model for the evolution of the Indian Ocean and the breakup of Gondwanaland. *Journal of Geophysics Research,* **84,** 6803–6830.
1.2.1

Noy-Meir, I. (1975) Stability of grazing systems: an application of predator–prey graphs. *Journal of Ecology,* **63,** 459–483.
10.6, 16.3.1

Noyes, J.S. (1974) The biology of the leek moth, *Acrolepia assectella* (Zeller). Ph.D. thesis, University of London.
9.6, 9.7.1

Nye, P.H. & Tinker, P.B. (1977) *Solute Movement in the Soil–Root System.* Blackwell Scientific Publications, Oxford.
3.3.3

Oakeshott, J.G., May, T.W., Gibson, J.B. & Willcocks, D.A. (1982) Resource partitioning in five domestic *Drosophila* species and its relationships to ethanol metabolism. *Australian Journal of Zoology,* **30,** 547–556.
11.3.1

Obeid, M., Machin, D. & Harper, J.L. (1967) Influence of density on plant to plant variations in fiber flax, *Linum usitatissimum. Crop Science,* **7,** 471–473.
6.10

O'Dor, R.K. & Wells, M.J. (1978) Reproduction versus somatic growth: hormonal control in *Octopus vulgaris. Journal of Experimental Biology,* **77,** 15–31.
4.8

Ødum, S. (1965) Germination of ancient seeds; floristical observations and experiments with archaeologically dated soil samples. *Dansk Botanisk Arkiv,* **24,** 1–70.
5.7.2

Ødum, S. (1978) *Dormant Seeds in Danish Ruderal Soils.* Horsholm Arboretum, Denmark.
4.5.2

Ogden, J. (1968) Studies on reproductive strategy with particular reference to selected composites. Ph.D. thesis, University of Wales.
14.9.2

Ollason, J.G. (1980) Learning to forage—optimally? *Theoretical Population Biology,* **18,** 44–56.
9.11.3

O'Neill, R.V. (1976) Ecosystem persistence and heterotrophic regulation. *Ecology,* **57,** 1244–1253.
21.5

Ong, C.K., Marshall, C. & Sagar, G.R. (1978) The physiology of tiller death in grasses. 2. Causes of tiller death in a grass sward. *Journal of the British Grassland Society,* **33,** 205–211.
8.2.1

Orshan, G. (1963) Seasonal dimorphism of desert and mediterranean chamaephytes and its significance as a factor in their water economy. In: *The Water Relations of Plants* (A.J. Rutter & F.W. Whitehead eds), pp. 207–222. Blackwell Scientific Publications, Oxford.
1.6

Osborn, R.G. (1975) Models of lemming demography and avian predation near Barrow, Alaska. M.S. thesis, San Diego State University.
15.4.2

Packham, J.R. & Harding, D.T.L. (1982) *Ecology of Woodland Processes.* Edward Arnold, London.
16.4.4

Paine, R.T. (1966) Food web complexity and species diversity. *American Naturalist,* **100,** 65–75.
19.2.2, 19.2.6

Paine, R.T. (1979) Disaster, catastrophe and local persistence of the sea palm *Postelsia palmaeformis. Science, N.Y.,* **205,** 685–687.
7.6.1

Paine, R.T. & Levin, S.A. (1981) Intertidal landscapes: Disturbance and the dynamics of pattern. *Ecological Monographs,* **51,** 145–178.
19.2.2, 19.3.5

Painter, E.L. & Detling, J.K. (1981) Effects of defoliation on net photosynthesis and regrowth of western wheatgrass. *Journal of Range Management,* **34,** 68–71.
8.2.1

Palmblad, I.G. (1968) Competition studies on experimental populations of weeds with emphasis on the regulation of population size. *Ecology*, **49**, 26–34.
6.6

Paris, O.H. & Pitelka, F.A. (1962) Population characteristics of the terrestrial isopod *Armadillidium vulgare* in California grassland. *Ecology*, **43**, 229–248.
14.2.1

Park, T. (1948) Experimental studies of interspecific competition. I. Competition between populations of the flour beetles *Tribolium confusum* Duval and *Tribolium castaneum* Herbst. *Ecological Monographs*, **18**, 267–307.
12.6.3, 19.2.5

Park, T. (1954) Experimental studies of interspecific competition. II. Temperature, humidity and competition in two species of *Tribolium*. *Physiological Zoology*, **27**, 177–238.
7.4.3

Park, T. (1962) Beetles, competition and populations. *Science, N.Y.*, **138**, 1369–1375.
7.4.3

Park, T., Mertz, D.B., Grodzinski, W. & Prus, T. (1965) Cannibalistic predation in populations of flour beetles. *Physiological Zoology*, **38**, 289–321.
7.4.3

Parker, G.A. (1970) The reproductive behaviour and the nature of sexual selection in *Scatophaga stercoraria* L. (Diptera: Scatophagidae) II. The fertilization rate and the spatial and temporal relationships of each sex around the site of mating and oviposition. *Journal of Animal Ecology*, **39**, 205–228.
9.8

Parker, G.A. (1984) Evolutionarily stable strategies. In: *Behavioural Ecology: An Evolutionary Approach*, 2nd edn. (J.R. Krebs & N.B. Davies eds), pp. 30–61. Blackwell Scientific Publications, Oxford.
5.4.2

Parker, G.A. & Stuart, R.A. (1976) Animal behaviour as a strategy optimizer: evolution of resource assessment strategies and optimal emigration thresholds. *American Naturalist*, **110**, 1055–1076.
9.11.1

Parrish, J.A.D. & Bazzaz, F.A. (1979) Differences in pollination niche relationships in early and late successional plant communities. *Ecology*, **60**, 597–610.
22.4.4

Parrish, J.A.D. & Bazzaz, F.A. (1982) Competitive interactions in plant communities of different successional ages. *Ecology*, **63**, 314–320.
22.4.4

Parsons, P.A. & Spence, G.E. (1981) Ethanol utilization: threshold differences among three *Drosophila* species. *American Naturalist*, **117**, 568–571.
11.3.1

Pearl, R. (1927) The growth of populations. *Quarterly Review of Biology*, **2**, 532–548.
6.4

Pearl, R. (1928) *The Rate of Living*. Knopf, New York.
4.5.1

Pearsall, W.H. & Bengry, R.P. (1940) The growth of *Chlorella* in darkness and in nutrient solution. *Annals of Botany*, **4**, 365–377.
6.4

Pemadasa, M.A., Greig-Smith, P. & Lovell, P.H. (1974) A quantitative description of the distribution of annuals in the dune system at Aberffraw, Anglesey. *Journal of Ecology*, **62**, 379–402.
16.3.2

Perrins, C. (1964) Survival of young swifts in relation to brood-size. *Nature* (London), **201**, 1147–1149.
14.4.4

Perrins, C.M. (1965) Population fluctuations and clutch size in the great tit, *Parus major* L. *Journal of Animal Ecology*, **34**, 601–647.
4.6.3

Perrins, C.M. (1979) *British Tits*. Wm. Collins, London.
6.11

Person, C. (1966) Genetic polymorphism in parasitic systems. *Nature* (London), **212**, 266–267.
12.7

Persson, L. (1983) Food consumption and competition between age classes in a perch *Perca fluviatilis* population in a shallow eutrophic lake. *Oikos*, **40**, 197–207.
6.10

Peterman, R.M., Clark, W.C. & Holling, C.S. (1979) The dynamics of resilience: shifting stability domains in fish and insect systems. In: *Population Dynamics* (R.M. Anderson, B.D. Turner & L.R. Taylor eds), pp. 321–341. Blackwell Scientific Publications, Oxford.
10.6.1, 10.8.4

Peters, B. (1980) The demography of leaves in a permanent pasture. Ph.D. thesis, University of Wales.
3.4.3

Peters, R.H. (1983) *The Ecological Implications of Body Size*. Cambridge University Press, Cambridge.
14.14.3

Pianka, E.R. (1967) On lizard species diversity: North American flatland deserts. *Ecology*, **48**, 333–351.
22.3.1, 22.3.2

Pianka, E.R. (1970) On *r*- and *k*-selection. *American Naturalist*, **104**, 592–597.
14.8

Pianka, E.R. (1973) The structure of lizard communities. *Annual Review of Ecology and Systematics*, **4**, 53–74.
18.4.1

Pianka, E.R. (1983) *Evolutionary Ecology*, 3rd edn. Harper & Row, New York.
22.4.1

Pielou, E.C. (1975) *Ecological Diversity*. John Wiley & Sons, New York.
16.3.1

Pijl, L. van der (1969) *Principles of Dispersal in Higher Plants*. Springer Verlag, Berlin.
5.4.6

Pillemer, E.A. & Tingey, W.M. (1976) Hooked trichomes: a physical barrier to a major agricultural pest. *Science, N.Y.*, **193**, 482–4.
3.4.3

Pimentel, D., Levins, S.A. & Soans, A.B. (1975) On the evolution of energy balance in some exploiter–victim systems. *Ecology*, **56**, 381–390.
17.4.1

Pimm, S.L. (1979a) Complexity and stability: another look at MacArthur's original hypothesis. *Oikos*, **33**, 351–357.
21.2.2, 21.3

Pimm, S.L. (1979b) The structure of food webs. *Theoretical Population Biology*, **16**, 144–158.
21.2.2

Pimm, S.L. (1980) Bounds on food web connectance. *Nature* (London), **284**, 591.
21.2.2

Pimm, S.L. (1982) *Food Webs*. Chapman and Hall, London.
11.1, 21.2.2, 21.3

Pimm, S.L. & Lawton, J.H. (1977) The number of trophic levels in ecological communities. *Nature* (London), **268**, 329–331.
21.4.2

Pimm, S.L. & Lawton, J.H. (1980) Are food webs divided into compartments? *Journal of Animal Ecology*, **49**, 879–898.
21.3, 21.4.2

Pisek, A., Larcher, W., Vegis, A. & Napp-Zin, K. (1973) The normal temperature range. In: *Temperature and Life* (H. Precht, J. Christopherson, H. Hensel & W. Larcher eds), pp. 102–194. Springer-Verlag, Berlin.
2.12, 17.3.1

Pitcher, T.J. & Hart, P.J.B. (1982) *Fisheries Ecology*. Croom Helm, London.
10.8.3, 10.8.6

Podoler, H. & Rogers, D.J. (1975) A new method for the identification of key factors from life-table data. *Journal of Animal Ecology*, **44**, 85–114.
15.3.1

Pojar, T.M. (1981) A management perspective of population modelling. In: *Dynamics of Large Mammal Populations* (D.W. Fowler & T.D. Smith eds), pp. 241–261. Wiley-Interscience, New York.
10.8.3

Pond, C.M. (1981) Storage. In: *Physiological Ecology: An Evolutionary Approach to Resource Use* (C.R. Townsend & P. Calow eds), pp. 190–219. Blackwell Scientific Publications, Oxford.
14.4.4

Poole, R.W. (1978) *An Introduction to Quantitative Ecology*. McGraw Hill, New York.
15.4.1

Potts, G.R., Tapper, S.C. and Hudson, P.J. (1984) Population fluctuations in red grouse: Analysis of bag records and a simulation model. *Journal of Animal Ecology*, **53**, 21–36.
12.5.5, 12.6.5, 15.4.1

Prance, G.T. (1981) Discussion. In: *Vicariance Biogeography: A Critique* (G. Nelson & D.E. Rosen eds), pp. 395–405. Columbia University Press, New York.
22.3.4

Pratt, D.M. (1943) Analysis of population development in *Daphnia* at different temperatures. *Biological Bulletin*, **85**, 116–140.
4.3

Preston, F.W. (1962) The canonical distribution of commonness and rarity. *Ecology*, **43**, 185–215, 410–432.
20.3.2

Price, M.V. & Waser, N.M. (1979) Pollen dispersal and optimal outcrossing in *Delphinium nelsoni*. *Nature* (London), **277**, 294–7.
5.6

Price, P.W. (1980) *Evolutionary Biology of Parasites*. Princeton University Press, Princeton, New Jersey.
8.1

Pugh, G.J.F. (1980) Strategies in fungal ecology. *Transactions of the British Mycological Society*, **75**, 1–14.
11.2.1

Pulliam, H.R. & Curaco, T. (1984) Living in groups: is there an optimal group size? In: *Behavioural Ecology: An Evolutionary Approach*, 2nd edn. (J.R. Krebs & N.B. Davies eds), pp. 122–147. Blackwell Scientific Publications, Oxford.
5.3.6

Putman, R.J. (1978a) Patterns of carbon dioxide evolution from decaying carrion. Decomposition of small mammal carrion in temperate systems 1. *Oikos*, **31**, 47–57.
11.2.3

Putman, R.J. (1978b) Flow of energy and organic matter from a carcass during decomposition. Decomposition of small mammal carrion in temperate systems 2. *Oikos*, **31**, 58–68.
11.3.2

Putman, R.J. (1983) *Carrion and Dung: The Decomposition of Animal Wastes*. Edward Arnold, London.
11.3.2

Pyke, G.H. (1982) Local geographic distributions of bumblebees near Crested Butte, Colorado: Competition and community structure. *Ecology*, **63**, 555–573.
18.3.1, 18.4.2

Rabinowitz, D. (1981) Seven forms of rarity. In: *The Biological Aspects of Rare Plant Conservation* (H.

Synge ed.), pp. 205–217. John Wiley & Sons, Chichester.
15.7

Rabotnov, T.A. (1964) On the biology of monocarpic perennial plants. *Bulletin of the Moscow Society of Naturalists*, **71**, 47–55. (In Russian.)
5.7.3

Rafes, P.M. (1970) Estimation of the effects of phytophagous insects on forest production. In: *Analysis of Temperate Forest Ecosystems* (D.E. Reichle ed.), pp. 100–106. Springer-Verlag, New York.
8.2.5

Raffaelli, D.G. & Hughes, R.N. (1978) The effects of crevice size and availability on populations of *Littorina rudis* and *Littorina neritoides*. *Journal of Animal Ecology*, **47**, 71–83.
14.11

Randall, M.G.M. (1982) The dynamics of an insect population throughout its altitudinal distribution: *Coleophora alticolella* (Lepidoptera) in northern England. *Journal of Animal Ecology*, **51**, 993–1016.
2.4

Ranwell, D.S. (1972) *Ecology of Salt Marshes and Sand Dunes*. Chapman and Hall, London.
2.7

Ranwell, D.S. (1974) The salt marsh to tidal woodland transition. *Hydrobiological Bulletin (Amsterdam)*, **8**, 139–151.
16.4.2, 16.4.3

Rathcke, B.J. (1976) Competition and coexistence within a guild of herbivorous insects. *Ecology*, **57**, 76–87.
18.3.1

Rathcke, B.J. & Poole, D.O. (1975) Coevolutionary race continues: butterfly larval adaptations to plant trichomes. *Science, N.Y.*, **187**, 175.
3.4.3

Raunkiaer, C. (1934) *The Life Forms of Plants*. Oxford University Press, Oxford. (Translated from the original published in Danish, 1907.)
1.4.1

Rausher, M.D. (1978) Search image for leaf shape in a butterfly. *Science, N.Y.*, **200**, 1071–1073.
19.2.3

Raushke, E., Haar, T.H. von der, Bardeer, W.R. & Pasternak, M. (1973) The annual radiation of the earth–atmosphere system during 1969–70 from Nimbus measurements. *Journal of the Atmospheric Sciences*, **30**, 341–346.
3.2

Reichle, D.E. (1970) *Analysis of Temperate Forest Ecosystems*. Springer-Verlag, New York.
17.2, 17.3.1

Rejmanek, M. & Stary, P. (1979) Connectance in real biotic communities and critical values for stability in model ecosystems. *Nature* (London), **280**, 311–313.
21.2.2

Rex, M.A. (1981) Community structure in the deep sea benthos. *Annual Review of Ecology and Systematics*, **12**, 331–353.
22.4.3

Rey, J.R. (1981) Ecological biogeography of arthropods on *Spartina* islands in northwest Florida. *Ecological Monographs*, **51**, 237–265.
19.1

Reynoldson, T.B. & Bellamy, L.S. (1971) The establishment of interspecific competition in field populations with an example of competition in action between *Polycelis nigra* and *P. tenuis* (Turbellaria, Tricladidae) In: *Dynamics of Populations* (P.J. den Boer & G.R. Gradwell eds), pp. 282–297. Proceedings of the Advanced Study Institute in Dynamics of Numbers in Populations, Oosterbeck. Centre for Agricultural Publishing and Documentation, Wageningen.
15.6.2

Rhoades, D.F. & Cates, R.G. (1976) Towards a general theory of plant antiherbivore chemistry. *Recent Advances in Phytochemistry*, **10**, 168–213.
3.4.3

Richards, B.N. (1974) *Introduction to the Soil Ecosystem*. Longman, Harlow, Essex.
16.4.1

Richards, O.W. & Davies, R.G. (1977) *Imm's General Textbook of Entomology*, Vol. 1. Structure, Physiology and Development. Vol. 2. Classification. Biology. John Wiley & Sons, New York.
3.4.1

Richards, O.W. & Waloff, N. (1954) Studies on the biology and population dynamics of British grasshoppers. *Anti-Locust Bulletin*, **17**, 1–182.
4.5.1, 5.7.1, 6.4

Richman, S. (1958) The transformation of energy by *Daphnia pulex*. *Ecological Monographs*, **28**, 273–291.
8.4

Ridley, H.N. (1930) *The Dispersal of Plants Throughout the World*. L. Reeve and Company, Ashford, Kent.
5.4.6

Rieck, A.F., Belli, J.A. & Blaskovics, M.E. (1960) Oxygen consumption of whole animal tissues in temperature acclimated amphibians. *Proceedings of the Society of Experimental Biology and Medicine*, **103**, 436–439.
2.2.6

Rigler, F.H. (1961) The relation between concentration of food and feeding rate of *Daphnia magna* Straus. *Canadian Journal of Zoology*, **39**, 857–868.
9.5.2

Roberts, H.A. (1964) Emergence and longevity in cultivated soil of seeds of some annual weeds. *Weed Research*, **4**, 296–307.
4.5.2

Rodin, L.E. *et al.* (1975) Primary productivity of the main world ecosystems. In: *Proceedings of the First International Congress of Ecology*, pp. 176–181. Centre for

Agricultural Publications, Wageningen.
17.2

Rohde, K. (1978) Latitudinal differences in host-specificity of marine Monogenea and Digenea. *Marine Biology,* **47,** 125–134.
22.4.1

Rohmeder, E. (1967) Beziehungen zwischen Frucht-bzw. Samenerzeugung und Holzerzeugung der Waldbäume. *Augemeine Forstzeitung,* **22,** 33–39.
8.4

Room, P.M., Harley, K.L.S., Forno, I.W. and Sands, D.P.A. (1981) Successful biological control of the floating weed *Salvinia. Nature* (London), **294,** 78–80.
15.6.4

Root, R. (1967) The niche exploitation pattern of the blue-grey gnatcatcher. *Ecological Monographs,* **37,** 317–350.
1.3, 18.3.1

Rosenzweig, M.L. (1971) Paradox of enrichment: destabilization of exploitation ecosystems in ecological time. *Science, N.Y.,* **171,** 385–387.
22.3.1

Rosenzweig, M.L. & Abramsky, Z. (1980) Microtine cycles: the role of habitat heterogeneity. *Oikos,* **34,** 141–146.
15.4.2

Rosenzweig, M.L. & MacArthur, R.H. (1963) Graphical representation and stability conditions of predator–prey interactions. *American Naturalist,* **97,** 209–223.
10.2

Ross, M.A. & Harper, J.L. (1972) Occupation of biological space during seedling establishment. *Journal of Ecology,* **60,** 77–88.
6.10

Ross, R. & Thomson, D. (1910) A case of sleeping sickness studied by precise enumerative methods: Regular periodical increase of the parasite described. *Proceedings of the Royal Society of London,* B, **82,** 411–415.
12.7

Roth, G.D. (1981) *Collins Guide to the Weather.* Wm. Collins, London.
2.3

Rotheray, G.E. (1979) The biology and host searching behaviour of a cynipid parasite of aphidophagous syrphid larvae. *Ecological Entomology,* **4,** 75–82.
9.10

Rovira, A.D. & Campbell, R.E. (1974) Scanning electron microscopy of microorganisms on the roots of wheat. *Microbial Ecology,* **1,** 15–23.
11.2.1

Rubenstein, D.I. (1981) Individual variation and competition in the Everglades pygmy sunfish. *Journal of Animal Ecology,* **50,** 337–350.
6.10

Russell, E.W. (1973) *Soil Conditions and Plant Growth,* 10th edn. Longman, London.
3.2.2

Ryle, G.J.A. (1970) Partition of assimilates in an annual and a perennial grass. *Journal of Applied Ecology,* **7,** 217–227.
8.2.1

Sagar, G.R. & Harper, J.L. (1960) Factors affecting the germination and early establishment of plantains (*Plantago lanceolata, P. media* and *P. major*). In: *The Biology of Weeds* (J.L. Harper ed.), pp. 236–244. Blackwell Scientific Publications, Oxford.
15.6.1

Sakai, A. & Otsuka, K. (1970) Freezing resistance of alpine plants. *Ecology,* **51,** 665–671.
2.2.8

Sakai, K.I. (1958) Studies on competition in plants and animals. IX. Experimental studies on migration in *Drosophila melanogaster. Evolution,* **12,** 93–101.
5.5.1

Sale, P.F. (1977) Maintenance of high diversity in coral reef fish communities. *American Naturalist,* **111,** 337–359.
19.3.2

Sale, P.F. (1979) Recruitment, loss and coexistence in a guild of territorial coral reef fishes. *Oecologia,* **42,** 159–177.
19.3.2

Sale, P.F. & Douglas, W.A. (1984) Temporal variability in the community structure of fish on coral patch reefs and the relation of community structure to reef structure. *Ecology,* **65,** 409–422.
19.3.2

Salisbury, E.J. (1942) *The Reproductive Capacity of Plants.* Bell, London.
14.9.2

Salonen, K., Jones, R.I. and Arvola, L. (1984) Hypolimnetic retrieval by diel vertical migrations of lake phytoplankton. *Freshwater Biology,* **14,** 431–438.
5.3.1

Sanders, H.L. (1968) Marine benthic diversity: a comparative study. *American Naturalist,* **102,** 243–282.
22.4.3

Sarukhán, J. (1974) Studies on plant demography: *Ranunculus repens* L., *R. bulbosus* L. and *R. acris.* II. Reproductive strategies and seed population dynamics. *Journal of Ecology,* **62,** 151–177.
8.2.6, 15.2

Sarukhán, J. & Harper, J.L. (1973) Studies on plant demography: *Ranunculus repens* L., *R. bulbosus* L. and *R. acris* L. I. Population flux and survivorship. *Journal of Ecology,* **61,** 675–716.
15.2

Schaffer, W.M. (1974) Optimal reproductive effort in

fluctuating environments. *American Naturalist*, **108**, 783–790.
14.10.1

Schildknecht, H. (1971) Evolutionary peaks in the defensive chemistry of insects. *Endeavour*, **30**, 136–141.
8.4.3

Schindler, D.W. (1978) Factors regulating phytoplankton production and standing crop in the world's freshwaters. *Limnology and Oceanography*, **23**, 478–486.
17.3.1, 17.3.3

Schmidt-Nielsen, K. (1983) *Animal Physiology*. Cambridge University Press, Cambridge.
2.2.3

Schmidt-Nielsen, K. (1984) *Scaling: Why is Animal Size so Important?* Cambridge University Press, Cambridge.
14.14.3

Schoener, T.W. (1974) Resource partitioning in ecological communities. *Science, N.Y.*, **185**, 27–39.
18.3.1

Schoener, T.W. (1983) Field experiments on interspecific competition. *American Naturalist*, **122**, 240–285.
7.9, 18.2

Schoener, T.W. (1984) Size differences among sympatric, bird-eating hawks: a worldwide survey. In: *Ecological Communities: Conceptual Issues and the Evidence* (D.R. Strong, D. Simberloff, L.G. Abele & A.B. Thistle eds), pp. 254–281. Princeton University Press, Princeton, New Jersey.
18.2, 18.4.2

Schopf, Y.J.M. (1974) Permo-Triassic extinctions: relation to sea-floor spreading. *Journal of Geology*, **82**, 129–143.
22.4.5

Schultz, A.M. (1964) The nutrient recovery hypothesis for arctic microtine cycles. II. Ecosystem variables in relation to arctic microtine cycles. In: *Grazing in Terrestrial and Marine Environments* (D.J. Crisp ed.). Blackwell Scientific Publications, Oxford.
15.4.2

Schultz, A.M. (1969) A study of an ecosystem: The arctic tundra. In: *The Ecosystem Concept in Natural Resource Management* (G.M. van Dyne ed.). Academic Press, New York.
15.4.2

Schulze, E.D. (1970) Der CO_2-Gaswechsel de Buche (*Fagus sylvatica* L.) in Abhängigkeit von den Klimafaktoren in Freiland. *Flora, Jena*, **159**, 177–232.
17.3.1

Schulze, E.D., Fuchs, M.I. & Fuchs, M. (1977a) Spatial distribution of photosynthetic capacity and performance in a mountain spruce forest in northern Germany. I. Biomass distribution and daily CO_2 uptake in different crown layers. *Oecologia*, **29**, 43–61.
17.3.1

Schulze, E.D., Fuchs, M. & Fuchs, M.I. (1977b) Spatial distribution of photosynthetic capacity and performance in a mountain spruce forest of northern Germany. III. The significance of the evergreen habit. *Oecologia*, **30**, 239–248.
17.3.1

Schwerdtfeger, F. (1941) Über die Ursachen des Massenwechsels der Insekten. *Zeitschrift für angewardte Entomologie*, **28**, 254–303.
10.6.1

Sherman, P.W. (1981) Reproductive competition and infanticide in Belding's ground squirrels and other animals. In: *Natural Selection and Social Behaviour: Recent Research and New Theory* (R.D. Alexander & D.W. Tinkle eds), pp. 311–331. Chiron Press, New York.
6.16

Shorrocks, B. & Begon, M. (1975) A model of competition. *Oecologia*, **20**, 363–367.
19.4.1

Shorrocks, B., Rosewell, J., Edwards, K. & Atkinson, W. (1984) Competition may not be a major organising force in many communities of insects. *Nature* (London), **310**, 310–312.
18.2.2

Shryock, H.S., Siegel, J.S. & Stockwell, E.G. (1976) *The Methods and Materials of Demography*. Academic Press, New York.
4.10

Shure, D.J. (1973) Radionuclide tracer analysis of trophic relationships in an old-field ecosystem. *Ecological Monographs*, **43**, 1–19.
21.3

Sibma, L., Kort, J. & de Wit, C.T. (1964) Experiments on competition as a means of detecting possible damage by nematodes. *Jaarboek, Instituut voor biologischen scheikundig onderzoek van Landbouwgewassen*, 1964, 119–124.
8.2.2

Sih, A. (1982) Foraging strategies and the avoidance of predation by an aquatic insect, *Notonecta hoffmanni*. *Ecology*, **63**, 786–796.
9.4

Silander, J.A. & Antonovics, J. (1982) A perturbation approach to the analysis of interspecific interactions in a coastal plant community. *Nature* (London), **298**, 557–560.
15.6.3

Silvertown, J.W. (1980) The evolutionary ecology of mast seeding in trees. *Biological Journal of the Linnean Society*, **14**, 235–250.
8.4

Silvertown, J.W. (1982) *Introduction to Plant Population Ecology*. Longman, London.
15.3.2

Simberloff, D.S. (1976) Experimental zoogeography of islands: effects of island size. *Ecology*, **57**, 629–648.
20.2.4

Simberloff, D. (1984) Properties of coexisting bird species in two archipelagoes. In: *Ecological Communities: Conceptual Issues and the Evidence* (D.R. Strong, D. Simberloff, L.G. Abele & A.B. Thistle eds), pp. 234–253. Princeton University Press, Princeton, New Jersey.
18.4.2

Simberloff, D.S. & Abele, L.G. (1976) Island biogeography theory and conservation practice. *Science, N.Y.*, **191**, 285–286.
20.4

Simberloff, D. & Boecklen, W. (1981) Santa Rosalia reconsidered: size ratios and competition. *Evolution*, **35**, 1206–1228.
18.4.2

Simberloff, D.S. & Wilson, E.O. (1969) Experimental zoogeography of islands: the colonization of empty islands. *Ecology*, **50**, 278–296.
20.3.4

Simon, J.L. & Dauer, D.M. (1977) Re-establishment of a benthic community following natural defaunation. In: *Ecology of Marine Benthos* (B.C. Coull ed.), pp. 139–154. University of South Carolina Press, Columbia.
20.3.2

Sinclair, A.R.E. (1973) Regulation, and population models for a tropical ruminant. *East African Wildlife Journal*, **11**, 307–16.
15.3.2

Sinclair, A.R.E. (1975) The resource limitation of trophic levels in tropical grassland ecosystems. *Journal of Animal Ecology*, **44**, 497–520.
8.4

Skidmore, D.I. (1983) Population dynamics of *Phytophthora infestans* (Mont.) de Bary. Ph.D. thesis, University of Wales.
12.3.2

Skogland, T. (1983) The effects of density dependent resource limitation on size of wild reindeer. *Oecologia*, **60**, 156–168.
6.5

Slobodkin, L.B. & Richman, S. (1956) The effect of removal of fixed percentages of the newborn on size and variability in populations of *Daphnia pulicaria* (Forbes). *Limnology and Oceanography*, **1**, 209–237.
10.8.6

Slobodkin, L.B., Smith, F.E. & Hairston, N.G. (1967) Regulation in terrestrial ecosystems, and the implied balance of nature. *American Naturalist*, **101**, 109–124.
18.2.1

Smith, D.C. (1979) From extracellular to intracellular: the establishment of a symbiosis. *Proceedings of the Royal Society of London*, B, **204**, 131–139.
13.1

Smith, F.E. (1961) Density dependence in the Australian thrips. *Ecology*, **42**, 403–407.
15.2.1

Smith, J.N.M. (1974) The food searching behaviour of two European thrushes. II. The adaptiveness of the search patterns. *Behaviour*, **49**, 1–61.
9.10

Smith, J.N.M. & Dawkins, R. (1971) The hunting behaviour of individual great tits in relation to spatial variations in their food density. *Animal Behaviour*, **19**, 695–706.
9.7.1

Smith, R.H. & Bass, M.H. (1972) Relation of artificial pod removal to soybean yields. *Journal of Economic Entomology*, **65**, 606–608.
8.2.1

Snaydon, R.W. & Bradshaw, A.D. (1969) Differences between natural populations of *Trifolium repens* L. in response to mineral nutrients. II. Calcium, magnesium and potassium. *Journal of Applied Ecology*, **6**, 185–202.
1.5.1

Snell, T.W. & King, C.E. (1977) Lifespan and fecundity patterns in rotifers: the cost of reproduction. *Evolution*, **31**, 882–890.
14.4.1

Snodgrass, R.E. (1944) The feeding apparatus of biting and sucking insects affecting man. *Smithsonian Miscellaneous Collections*, **104** (7), 113.
3.4.1

Snyman, A. (1949) The influence of population densities on the development and oviposition of *Plodia interpunctella* Hubn. (Lepidoptera). *Journal of the Entomological Society of South Africa*, **12**, 137–171.
6.6

Solbrig, O.T. & Simpson, B.B. (1974) Components of regulation of a population of dandelions in Michigan. *Journal of Ecology*, **62**, 473–486.
14.7.2

Solbrig, O.T. & Simpson, B.B. (1977) A garden experiment on competition between biotypes of the common dandelion (*Taraxacum officinale*). *Journal of Ecology*, **65**, 427–430.
14.7.2

Solomon, M.E. (1949) The natural control of animal populations. *Journal of Animal Ecology*, **18**, 1–35.
9.5

Sorensen, A.E. (1978) Somatic polymorphism and seed dispersal. *Nature* (London), **276**, 174–176.
5.4.5

Soulé, M.E. & Wilcox, B.A. (eds) (1980) *Conservation Biology: an Evolutionary–Ecological Perspective*.

Sinauer, Sunderland, Massachusetts.
20.4

Sousa, W.P. (1979a) Experimental investigation of disturbance and ecological succession in a rocky intertidal algal community. *Ecological Monographs,* **49,** 227–254.
15.4.3

Sousa, W.P. (1979b) Disturbance in marine intertidal boulder fields: the nonequilibrium maintenance of species diversity. *Ecology,* **60,** 1225–1239.
19.3.4

Southern, H.N. (1970) The natural control of a population of tawny owls (*Strix aluco*). *Journal of Zoology,* **162,** 197–285.
10.1, 16.3.2

Southwood, T.R.E. (1961) The number of species of insect associated with various trees. *Journal of Animal Ecology,* **30,** 1–8
20.2.4

Southwood, T.R.E. (1977) Habitat, the templet for ecological strategies? *Journal of Animal Ecology,* **46,** 337–365.
14.1, 14.6.1

Southwood, T.R.E. (1978a) *Ecological Methods,* 2nd edn. Chapman and Hall, London.
4.3, 5.2

Southwood, T.R.E. (1978b) The components of diversity. *Symposium of the Royal Entomological Society of London,* **9,** 19–40.
18.2.1

Southwood, T.R.E., Moran, V.C. & Kennedy, C.E.J. (1982) The richness, abundance and biomass of the arthropod communities on trees. *Journal of Animal Ecology,* **51,** 635–649.
20.4

Spooner, G.M. (1947) The distribution of *Gammarus* species in estuaries. *Journal of the Marine Biological Association,* **27,** 1–52.
2.7

Spradbery, J.P. (1970) Host finding of *Rhyssa persuasoria* (L.), an ichneumonid parasite of siricid woodwasps. *Animal Behaviour,* **18,** 103–114.
9.6

Sprague, M.A. (1954) The effect of grazing management on forage and grain production from rye, wheat and oats. *Agronomy Journal,* **46,** 29–33.
8.3

Sprent, J.I. (1979) *The Biology of Nitrogen Fixing Organisms.* McGraw Hill, London.
13.10.1

Staaland, H., White, R.G., Luick, J.R. & Holleman, D.F. (1980) Dietary influences on sodium and potassium metabolism of reindeer. *Canadian Journal of Zoology,* **58,** 1728–1734.
9.2.2

Stafford, J. (1971) Heron populations of England and Wales 1928–70. *Bird Study,* **18,** 218–221.
10.6.1

Stahler, A.N. (1960) *Physical Geography,* 2nd edn. John Wiley & Sons.
1.6

Stanley, S.M. (1976) Ideas on the timing of metazoan diversification. *Paleobiology,* **2,** 209–219.
22.4.5

Stanley, S.M. (1979) *Macroevolution.* W.H. Freeman, San Francisco.
22.3.5

Stapledon, R.G. (1928) Cocksfoot grass (*Dactylis glomerata* L.): ecotypes in relation to the biotic factor. *Journal of Ecology,* **16,** 72–104.
1.5.1

Stearns, S.C. (1976) Life history tactics: a review of the ideas. *Quarterly Review of Biology,* **51,** 3–47.
14.10.1

Stearns, S.C. (1977) The evolution of life history traits. *Annual Review of Ecology and Systematics,* **8,** 145–171.
14.9.3, 22.5

Stearns, S.C. (1980) A new view of life-history evolution. *Oikos,* **35,** 266–281.
14.13

Stehli, F.G., McAlester, A.L. & Helsley, C.E. (1967) Taxonomic diversity of recent bivalves and some implications for geology. *Geological Society of America Bulletin,* **78,** 455–466.
22.4.1

Stemberger, R.S. & Gilbert, J.J. (1984) Spine development in the rotifer *Keratella cochlearis*: induction by cyclopoid copepods and *Asplachna. Freshwater Biology,* **14,** 639–648.
3.4.3

Stenseth, N.C. (1983) Causes and consequences of dispersal in small mammals. In: *The Ecology of Animal Movement* (I.R. Swingland & P.J. Greenwood eds), pp. 63–101. Oxford University Press, Oxford.
15.4.2

Stephens, G.R. (1971) The relation of insect defoliation to mortality in Connecticut forests. *Connecticut Agricultural Experimental Station Bulletin,* **723,** 1–16.
8.2.4

Stern, W.R. & Donald, C.M. (1962) Light relationships in grass–clover swards. *Australian Journal of Agricultural Research,* **13,** 599–614.
3.2.3

Stout, J. & Vandermeer, J. (1975) Comparison of species richness for stream-inhabiting insects in tropical and mid-latitude streams. *American Naturalist,* **109,** 263–280.
22.4.1

Stribley, D.P., Tinker, P.B. & Snellgrove, R.C. (1980) Effect of vesicular–arbuscular mycorrhizal fungi on the relations of plant growth, internal phosphorus

concentration and soil phosphate analysis. *Journal of Soil Science,* **31,** 655–672.
13.7.2

Strobel, G.A. & Lanier, G.N. (1981) Dutch elm disease. *Scientific American,* **245,** 40–50.
8.2.2

Strong, D.R. Jr. (1974) Rapid asymptotic species accumulation in phytophagous insect communities: the pests of Cacao. *Science, N.Y.,* **185,** 1064–1066.
19.1

Strong, D.R. Jr. (1982) Harmonious coexistence of hispine beetles on *Heliconia* in experimental and natural communities. *Ecology,* **63,** 1039–1049.
18.3.1

Strong, D.R. Jr. & Levin, D.A. (1979) Species richness of plant parasites and growth form of their hosts. *American Naturalist,* **114,** 1–22.
20.3.3

Strong, D.R. Jr. & Simberloff, D.S. (1981) Straining at gnats and swallowing ratios: character displacement. *Evolution,* **35,** 810–812.
18.4.2

Strong, D.R. Jr., Lawton, J.H. & Southwood, T.R.E. (1984) *Insects on Plants: Community Patterns and Mechanisms.* BlackwellScientific Publications, Oxford.
18.2.1, 18.3.1, 19.1, 22.4.5

Strong, D.R. Jr., Szyska, L.A. & Simberloff, D.S. (1979) Tests of community-wide character displacement against null hypotheses. *Evolution,* **33,** 897–913.
18.4.2

Stubbs, M. (1977) Density dependence in the life-cycles of animals and its importance in k- and r-strategies. *Journal of Animal Ecology,* **46,** 677–688.
15.3.2

Suberkropp, K., Godshalk, G.L. & Klug, M.J. (1976) Changes in the chemical composition of leaves during processing in a woodland stream. *Ecology,* **57,** 720–727.
11.2.1

Sullivan, T.P. & Sullivan, D.S. (1982) Population dynamics and regulation of the Douglas squirrel (*Tamiasciurus douglasii*) with supplemental food. *Oecologia,* **53,** 264–270.
15.6.2

Summerhayes, V.S. & Elton, C.S. (1923) Contributions to the ecology of Spitsbergen and Bear Island. *Journal of Ecology,* **11,** 214–286.
21.3

Sunderland, K.D., Hassall, M. & Sutton, S.L. (1976) The population dynamics of *Philoscia muscorum* (Crustacea, Oniscoidea) in a dune grassland ecosystem. *Journal of Animal Ecology,* **45,** 487–506.
4.5.3

Sunderland, N. (1960) Germination of the seeds of angiospermous root parasites. In: *The Biology of Weeds* (J.L. Harper ed.). Blackwell Scientific Publications, Oxford.
5.7.4

Sutcliffe, J. (1977) *Plants and Temperature.* Edward Arnold, London.
2.2.7, 2.2.8

Sutherland, W.J. (1982) Do oystercatchers select the most profitable cockles? *Animal Behaviour,* **30,** 857–861.
9.5.1

Swift, M.J., Heal, O.W. & Anderson, J.M. (1979) *Decomposition in Terrestrial Ecosystems.* Blackwell Scientific Publications, Oxford.
11.2.2, 11.2.4, 11.3.1, 17.5

Symonides, E. (1977) Mortality of seedlings in the natural psammophyte populations. *Ekologia Polska,* **25,** 635–651.
15.3.2

Symonides, E. (1979a) The structure and population dynamics of psammophytes on inland dunes. II. Loose-sod populations. *Ekologia Polska,* **27,** 191–234.
6.6, 15.2, 15.3.2, 15.4.3

Symonides, E. (1979b) The structure and population dynamics of psammophytes on inland dunes. IV. Population phenomena as a phytocenose-forming factor. (A summing-up discussion.) *Ekologia Polska,* **27,** 259–281.
15.3.2, 15.4.3

Takahashi, F. (1968) Functional response to host density in a parasitic wasp, with reference to population regulation. *Researches in Population Ecology,* **10,** 54–68.
9.5.3

Tamarin, R.H. (1978) Dispersal, population regulation, and *K*-selection in field mice. *American Naturalist,* **112,** 545–555.
15.4.2

Tamm, C.O. (1956) Further observations on the survival and flowering of some perennial herbs. *Oikos,* **7,** 274–292.
6.10

Tansley, A.G. (1917) On competition between *Galium saxatile* L. (*G. hercynicum* Weig.) and *Galium sylvestre* Poll. (*G. asperum* Schreb.) on different types of soil. *Journal of Ecology,* **5,** 173–179.
7.2.2

Tansley, A.G. (1939) *The British Islands and their Vegetation.* Cambridge University Press, Cambridge.
16.4.5

Tansley, A.G. & Adamson, R.S. (1925) Studies of the vegetation of the English chalk. III. The chalk grasslands of the Hampshire–Sussex border. *Journal of Ecology,* **13,** 177–223.
15.6.4, 19.2.1

Taylor, D.L. (1975) Symbiotic dinoflagellates. In: *Symbiosis* (D.H. Jennings & D.L. Lee eds). Symposium

29, Society for Experimental Biology. Cambridge University Press, Cambridge.
13.8

Taylor, F.J.R. (1982) Symbioses in marine microplankton. *Annales de l'Institut Oceanographique, Paris,* **58**(S), 61–90.
13.8

Temple, S.A. (1977) Plant–animal mutualism: coevolution with dodo leads to near extinction of plant. *Science, N.Y.,* **197,** 885–886.
8.2.6

Thiegles, B.A. (1968) Altered polyphenol metabolism in the foliage of *Pinus sylvestris* associated with European pine sawfly attack. *Canadian Journal of Botany,* **46,** 724–725.
8.2.3

Thomas, A.S. (1963) Further changes in vegetation since the advent of myxomatosis. *Journal of Ecology,* **51,** 151–183.
16.4.4

Thompson, D.B.A. (1983) Prey assessment by plovers (Charadriidae): net rate of energy intake and vulnerability to kleptoparasites. *Animal Behaviour,* **31,** 1226–1236.
9.4

Thompson, D.B.A. & Barnard, C. (1984) Prey selection by plovers: optimal foraging in mixed-species groups. *Animal Behaviour,* **32,** 554–563.
9.4

Thompson, D.J. (1975) Towards a predator–prey model incorporating age-structure: the effects of predator and prey size on the predation of *Daphnia magna* by *Ischnura elegans. Journal of Animal Ecology,* **44,** 907–916.
9.5.1

Thompson, J.N. (1982) *Interaction and Coevolution.* Wiley-Interscience, New York.
8.1

Thompson, K. & Grime, J.P. (1979) Seasonal variation in seed banks of herbaceous species in ten contrasting habitats. *Journal of Ecology,* **67,** 893–921.
4.5.2

Thornback, J. & Jenkins, M. (1982) *The I.U.C.N. Mammal Red Data Book, Part I. Threatened mammalian taxa of the Americas and the Australasian zoogeographic region (excluding Cetacea).* I.U.C.N., Gland, Switzerland.
15.7.2

Tilman, D. (1977) Resource competition between planktonic algae: an experimental and theoretical approach. *Ecology,* **58,** 338–348.
7.11.1

Tilman, D. (1982) *Resource Competition and Community Structure.* Princeton University Press, Princeton, New Jersey.
3.1, 3.6, 7.11.1, 18.3.2, 22.3.1

Tilman, D., Mattson, M. & Langer, S. (1981) Competition and nutrient kinetics along a temperature gradient: an experimental test of a mechanistic approach to niche theory. *Limnology and Oceanography,* **26,** 1020–1033.
7.2.5

Tinbergen, L. (1960) The natural control of insects in pinewoods. 1: Factors influencing the intensity of predation by songbirds. *Archives néerlandaises de Zoologie,* **13,** 266–336.
9.2.3

Tinker, P.H.B. (1975) Effects of vesicular–arbuscular mycorrhizas on higher plants. In: *Symbiosis* (D.H. Jennings & D.L. Lee eds). Symposium 29, Society for Experimental Biology. Cambridge University Press, Cambridge.
13.7.1

Tisdale, W.H. (1919) Physoderma disease of corn *Journal of Agricultural Research,* **16,** 137–154.
12.2.1

Tonn, W.M. & Magnuson, J.J. (1982) Patterns in the species composition and richness of fish assemblages in northern Wisconsin lakes. *Ecology,* **63,** 1149–1166.
20.3.1, 22.3.1

Toth, J.A., Papp, L.B. & Lenkey, B. (1975) Litter decomposition in an oak forest ecosystem (*Quercetum petreae* Cerris) in northern Hungary studied in the framework of 'Sikfökut Project'. In: *Biodegradation et Humification* (G. Kilbertus, O. Reisinger, A. Mourey, J.A. Cancela da Foneseca eds), pp. 41–58. Pierrance Editeur, Sarregue Mines.
11.2.1

Townsend, C.R. (1980) *The Ecology of Streams and Rivers.* Edward Arnold, London.
2.8, 2.9, 5.4.7

Townsend, C.R. & Hildrew, A.G. (1978) Predation strategy and resource utilisation by *Plectrocnemia conspersa* (Curtis) (Trichoptera: Polycentropodidae). *Proceedings of the Second International Symposium on Trichoptera,* pp. 299–307. Junk, The Hague.
9.10, 19.2.3

Townsend, C.R. & Hildrew, A.G. (1980) Foraging in a patchy environment by a predatory net-spinning caddis larva: A test of optimal foraging theory. *Oecologia,* **47,** 219–221.
9.10

Townsend, C.R. & Hughes, R.N. (1981) Maximizing net energy returns from foraging. In: *Physiological Ecology: An Evolutionary Approach to Resource Use* (C.R. Townsend & P. Calow eds), pp. 86–108. Blackwell Scientific Publications, Oxford.
9.3.1, 9.11.3

Townsend, C.R. & Winfield, I.J. (1985) The application of optimal foraging theory to feeding behaviour in fish. In: *Fish Energetics—A New Look* (P. Calow &

P. Tytler eds). Croom Helm, Beckenham.
9.4

Townsend, C.R., Hildrew, A.G. & Francis, J.E. (1983) Community structure in some southern English streams: the influence of physicochemical factors. *Freshwater Biology*, **13**, 521–544.
16.3.2, 22.3.4

Trench, R.K. (1975) Of 'leaves that crawl': functional chloroplasts in animal cells. In: *Symbiosis* (D.H. Jennings & D.L. Lee eds), pp. 229–266. Symposium 29, Society for Experimental Biology. Cambridge University Press, Cambridge.
13.9

Tripp, M.R. (1974) A final comment on invertebrate immunity. In: *Contemporary Topics in Immuno-biology*, Vol. 4 (E.L. Cooper ed.), pp. 289–290. Plenum Press, New York.
12.5.2

Turesson, G. (1922) The genotypical response of the plant species to the habitat. *Hereditas*, **6**, 147–236.
1.5.1

Turkington, R. & Harper, J.L. (1979) The growth, distribution and neighbour relationships of *Trifolium repens* in a permanent pasture. IV. Fine scale biotic differentiation. *Journal of Ecology*, **67**, 245–254.
1.5.2

Turkington, R.A., Cavers, P.B. & Aarssen, L.W. (1977) Neighbour relationships in grass–legume communities: I. Interspecific contacts in four grassland communities near London, Ontario. *Canadian Journal of Botany*, **55**, 2701–2711.
18.3.2

Turnbull, A.L. (1962) Quantitative studies of the food of *Linyphia triangularis* Clerck (Aranaea: Linyphiidae). *Canadian Entomologist*, **96**, 568–579.
8.4

Turrill, W.B. (1964) *Joseph Dalton Hooker*. Nelson, London.
2.4

Urquhart, F.A. (1960) *The Monarch Butterfly*. University of Toronto Press, Toronto.
5.3.4

Utida, S. (1957) Cyclic fluctuations of population density intrinsic to the host–parasite system. *Ecology*, **38**, 442–449.
10.2.3

Vagvolgyi, J. (1975) Body size, aerial dispersal, and origin of the Pacific land snail fauna. *Systematic Zoology*, **24**, 465–488.
20.3.5

Vandermeer, J.H. (1972) Niche theory. *Annual Review of Ecology and Systematics*, **3**, 107–132.
2.12

Vandermeer, J.H. (1980) Indirect mutualism: variations on a theme by Stephen Levine. *American Naturalist*, **116**, 441–448.
13.1

Vandermeer, J.H. & Boucher, D.H. (1978) Varieties of mutualistic interaction in population models. *Journal of Theoretical Biology*, **74**, 549–558.
13.1

Varley, G.C. (1947) The natural control of population balance in the knapweed gall-fly (*Urophora jaceana*). *Journal of Animal Ecology*, **16**, 139–187.
10.2.3

Varley, G.C. & Gradwell, G.R. (1968) Population models for the winter moth. *Symposium of the Royal Entomological Society of London*, **9**, 132–142.
6.8.2, 15.3.1, 15.3.2

Varley, G.C. & Gradwell, G.R. (1970) Recent advances in insect population dynamics. *Annual Review of Entomology*, **15**, 1–24.
4.5.1

Varley, G.C., Gradwell, G.R. & Hassell, M.P. (1973) *Insect Population Ecology*. Blackwell Scientific Publications, Oxford.
10.2.3, 15.2.1

Varley, M.E. (1967) *British Freshwater Fishes*. Fishing News Books, London.
2.4

Verhulst, P.F. (1838) Notice sur la loi que la population suit dans son accroissement. *Correspondences Math. Phys.*, **10**, 113–121.
6.9

Vickerman, K. (1970) Morphological and physiological considerations of extracellular blood protozoa. In: *Ecology and Physiology of Parasites* (A.M. Fallis ed.), pp. 58–89. University of Toronto Press, Toronto.
12.2.1

Vickerman, K. & Cox, F.E.G. (1967) *The Protozoa*. John Murray, London.
12.2.1

Vitousek, P.M. (1981) Clear-cutting and the nitrogen cycle. In: *Terrestrial Nitrogen Cycles* (F.E. Clark & T. Rosswall eds), pp. 631–642. *Ecological Bulletin (Stockholm)*, **33**.
17.5

Vitt, L.J. & Congdon, J.D. (1978) Body shape, reproductive effort and relative clutch mass in lizards: resolution of a paradox. *American Naturalist*, **112**, 595–608.
14.7.3

Volterra, V. (1926) Variations and fluctuations of the numbers of individuals in animal species living together. (Reprinted in 1931. In: R.N. Chapman, *Animal Ecology*. McGraw-Hill, New York.)
7.4.1, 10.2

Vuilleumier, F. (1970) Insular biogeography in continental regions: the northern Andes of South America. *American Naturalist*, **104**, 373–388.
19.1

Waage, J.K. (1979) Foraging for patchily-distributed hosts by the parasitoid *Nemeritis canescens*. *Journal of*

Animal Ecology, **48**, 353–371.
9.11.3

Wall, R. (1985) Competition and the individual: intra-specific competition in the common field grasshopper, *Chorthippus brunneus* Thunberg (Orthoptera: Acrididae). Ph.D. thesis, University of Liverpool.
14.4.2

Wall, R. & Begon, M. (1985) Competition and fitness. *Oikos* **44**, 356–360.
6.2

Wallace, B. (1960) Influence of genetic systems on geographical distribution. *Cold Spring Harbor Symposium on Quantitative Biology*, **24**, 193–204.
2.2.9

Walley, K., Khan, M.S.I. & Bradshaw, A.D. (1974) The potential for evolution of heavy metal tolerance in plants. I. Copper and zinc tolerance in *Agrostis tenuis*. *Heredity*, **32**, 309–319.
2.2.9, 2.11

Walsby, A.E. (1980) A square bacterium. *Nature* (London), **283**, 69–71.
1.4

Wang, J.Y. (1960) A critique of the heat unit approach to plant response studies. *Ecology*, **41**, 785–790.
2.2.6

Warkowska-Dratnal, H. & Stenseth, N. C. (1985) Dispersal and the microtine cycle: comparison of two hypotheses. *Oecologia*, **65**, 468–477.
15.4.2

Warner, R.E. (1968) The role of introduced diseases in the extinction of endemic Hawaiian avifauna. *Condor*, **70**, 101–120.
19.2.5

Waterbury, J.B., Calloway, C.B. & Turner, R.D. (1983) A cellulolytic nitrogen-fixing bacterium cultured from the gland of Deshayes in shipworms (Bivalvia: Teredinidae). *Science, N.Y.*, **221**, 1401–1403.
11.3.1

Waterhouse, D.F. (1974) The biological control of dung. *Scientific American*, **230**, 100–108.
11.3.2

Watkins, C.V. & Harvey, L.A. (1942) On the parasites of silver foxes on some farms in the South West. *Parasitology*, **34**, 155–179.
12.3.3

Watkinson, A.R. (1981) Interference in pure and mixed populations of *Agrostemma githago*. *Journal of Applied Ecology*, **18**, 967–976.
7.9.2

Watkinson, A.R. (1984) Yield–density relationships: the influence of resource availability on growth and self-thinning in populations of *Vulpia fasciculata*. *Annals of Botany*, **53**, 469–482.
6.5

Watkinson, A.R. (1985) On the abundance of plants along an environmental gradient. *Journal of Ecology*, **73**, 569–578.
15.5

Watkinson, A.R. & Davy, A.J. (1985) Population biology of salt marsh and sand dune annuals. *Vegetatio* **62**, 487–497.
6.3

Watkinson, A.R. & Harper, J.L. (1978) The demography of a sand dune annual: *Vulpia fasciculata*. I. The natural regulation of populations. *Journal of Ecology*, **66**, 15–33.
6.3, 15.2.2

Watson, A. (1967) Territory and population regulation in the red grouse. *Nature* (London), **215**, 1274–1275.
6.11

Watson, A. & Moss, R. (1972) A current model of population dynamics in red grouse. In: *Proceedings of the XVth International Ornithological Congress* (K.H. Voous ed.), pp. 139–149.
10.3

Watson, A. & Moss, R. (1980) Advances in our understanding of the population dynamics of red grouse from a recent fluctuation in numbers. *Ardea*, **68**, 103–111.
15.4.1

Watson, D.J. (1958) The dependence of net assimilation rate on leaf area index. *Annals of Botany*, **22**, 37–54.
3.2.3

Watson, G.E. (1964) Ecology and evolution of passerine birds on the islands of the Aegean Sea. Ph.D. thesis, Yale University (Dissertation microfilm 65-1956).
20.3.1

Watt, A.S. (1947) Pattern and process in the plant community. *Journal of Ecology*, **35**, 1–22.
16.4.5

Way, M.J. & Cammell, M. (1970) Aggregation behaviour in relation to food utilization by aphids. In: *Animal Populations in Relation to their Food Resources* (A. Watson ed.), pp. 229–247. Blackwell Scientific Publications, Oxford.
9.7.3

Weaver, J.E. & Albertson, F.W. (1943) Resurvey of grasses, forbs and underground plant parts at the end of the great drought. *Ecological Monographs*, **13**, 63–117.
3.2.2

Webb, W.L., Lauenroth, W.K., Szarek, S.R. & Kinerson, R.S. (1983) Primary production and abiotic controls in forests, grasslands and desert ecosystems in the United States. *Ecology*, **64**, 134–151.
17.3.1

Webster, J. (1970) *Introduction to Fungi*. Cambridge University Press, Cambridge.
11.2.1, 12.3

Wegener, A. (1915) *Die Entstehung der Kontinente und Ozeane*. Braunschweig, Vieweg. (Other editions

1920, 1922, 1924, 1929, 1936.)
1.2.1

Weiser, C.J. (1970) Cold resistance and injury in woody plants. *Science, N.Y.*, **169**, 1269–1278.
2.2.9

Werner, E.E. & Hall, D.J. (1974) Optimal foraging and the size selection of prey by the bluegill sunfish *Lepomis macrochirus. Ecology*, **55**, 1042–1052.
9.3.1

Werner, E.E., Mittelbach, G.G., Hall, D.J. & Gilliam, J.F. (1983a) Experimental tests of optimal habitat use in fish: the role of relative habitat profitability. *Ecology*, **64**, 1525–1539.
9.4

Werner, E.E., Gilliam, J.F., Hall, D.J. & Mittelbach, G.G. (1983b) An experimental test of the effects of predation risk on habitat use in fish. *Ecology*, **64**, 1540–1550.
9.4

Werner, P.A. (1975) Predictions of fate from rosette size in teasel (*Dipsacus fullonum* L.). *Oecologia*, **20**, 197–201.
4.8

Wesson, G. & Wareing, P.F. (1969) The induction of light sensitivity in weed seeds by burial. *Journal of Experimental Botany*, **20**, 413–425.
5.7.2

Whatley, J.W. & Whatley, F.R. (1980) *Light and Plant Life.* Edward Arnold, London.
3.2.4

White, J. (1980) Demographic factors in populations of plants. In: *Demography and Evolution* (O.T. Solbrig ed.). Blackwell Scientific Publications, Oxford.
6.12

White, J. (1981) The allometric interpretation of the self-thinning rule. *Journal of Theoretical Biology*, **89**, 475–500.
6.12

White, J. & Harper, J.L. (1970) Correlated changes in plant size and number in plant populations. *Journal of Ecology*, **58**, 467–485.

White, R.G., Bunnell, F.L., Gare, E., Skogland, T. & Hubert, B. (1981) Ungulates on arctic ranges. In: *Tundra Ecosystems: A Comparative Analysis* (L.C. Bliss, O.W. Heal & J.J. Moore eds), pp. 397–483. Cambridge University Press, Cambridge.
9.5.2

White, T.C.R. (1978) The importance of relative shortage of food in animal ecology. *Oecologia*, **33**, 71–86.
8.4

White, T.C.R. (1984) The abundance of invertebrate herbivores in relation to the availability of nitrogen in stressed food plants. *Oecologia*, **63**, 90–105.
8.4

Whiteside, M.C. & Harmsworth, R.V. (1967) Species diversity in chydorid (Cladocera) communities. *Ecology*, **48**, 664–667.
22.3.1

Whitfield, P.J. (1982) *The Biology of Parasitism: An Introduction to the Study of Associating Organisms.* Edward Arnold, London.
12.2.2

Whittaker, J.B. (1979) Invertebrate grazing, competition and plant dynamics. In: *Population Dynamics* (R.M. Anderson, B.D. Turner & L.R. Taylor eds), pp. 207–222. Blackwell Scientific Publications, Oxford.
8.2.2

Whittaker, R.H. (1953) A consideration of climax theory: the climax as a population and pattern. *Ecological Monographs*, **23**, 41–78.
16.4.5

Whittaker, R.H. (1956) Vegetation of the Great Smoky Mountains. *Ecological Monographs*, **26**, 1–80.
16.3.1

Whittaker, R.H. (1975) *Communities and Ecosystems*, 2nd edn. Macmillan, London.
2.5, 16.3.4, 17.2

Whittaker, R.H. (1977) Evolution of species diversity in land communities. *Evolutionary Biology*, **10**, 1–67.
22.4.2

Whittaker, R.H. & Niering, W.A. (1975) Vegetation of the Santa Catalina Mountains, Arizona. V. Biomass, production and diversity along the elevation gradient. *Ecology*, **56**, 771–790.
22.3.1

Whittaker, R.H. & Woodwell, G.M. (1968) Dimension and production relations of trees and shrubs in the Brookhaven Forest, New York. *Journal of Ecology*, **56**, 1–25.
17.2.2

Whittaker, R.H. & Woodwell, G.M. (1969) Structure production and diversity of the oak–pine forest at Brookhaven, New York. *Journal of Ecology*, **57**, 157–176.
17.2.2

Wiebes, J.T. (1979) Coevolution of figs and their insect pollinators. *Annual Review of Ecology and Systematics*, **10**, 1–12.
13.4

Wiebes, J.T. (1982) Fig wasps (Hymenoptera). *Monographiae Biologicae*, **42**, 735–755.
13.4

Wiens, D. (1984) Ovule survivorship, life history, breeding systems and reproductive success in plants. *Oecologia*, **64**, 47–53.
4.5.1

Wiklund, C. & Ahrberg, C. (1978) Host plants, nectar source plants, and habitat selection of males and females of *Anthocharis cardamines* (Lepidoptera). *Oikos*, **31**, 169–183.
9.8

Wilbur, H.M. & Collins, J.P. (1973) Ecological aspects

of amphibian metamorphosis. *Science, N.Y.,* **182,** 1305–1314.
6.10

Wilcox, B.A. (1978) Supersaturated island faunas: A species–age relationship for lizards on post-pleistocene land-bridge islands. *Science, N.Y.,* **199,** 996–998.
20.3.1

Williams, C.B. (1944) Some applications of the logarithmic series and the index of diversity to ecological problems. *Journal of Ecology,* **32,** 1–44.
12.3.3

Williams, C.B. (1964) *Patterns in the Balance of Nature and Related Problems in Quantitative Ecology.* Academic Press, New York.
19.1

Williams, G.C. (1975) *Sex and Evolution.* Princeton University Press, Princeton, New Jersey.
13.4

Williamson, M.H. (1972) *The Analysis of Biological Populations.* Edward Arnold, London.
7.2.4

Williamson, M.H. (1981) *Island Populations.* Oxford University Press, Oxford.
1.2.3, 20.1, 20.2.1, 20.3.2, 20.3.4, 20.4

Wilson, E.O. (1961) The nature of the taxon cycle in the Melanesian ant fauna. *American Naturalist,* **95,** 169–193.
20.3.2

Winfield, I.J., Peirson, G., Cryer, M. & Townsend, C.R. (1983) The behavioural basis of prey selection by underyearling bream (*Abramis brama* (L.)) and roach (*Rutilus rutilus* (L.)). *Freshwater Biology,* **13,** 139–149.
3.4.3

Wit, C.T. de (1960) On competition. *Verslagen van landbouwkundige onderzoekingen,* **660,** 1–82.
7.9.1

Wit, C.T. de (1965) Photosynthesis of leaf canopies. *Verslagen van landbouwkundige onderzoekingen,* **663,** 1–57.
3.2.2

Wit, C.T. de, Tow, P.G. & Ennik, G.C. (1966) Competition between legumes and grasses. *Verslagen van landbouwkundige onderzoekingen,* **687,** 3–30.
7.11, 13.10.1

Wolda, H. (1978) Fluctuations in abundance of tropical insects. *American Naturalist,* **112,** 1017–1045.
21.2.2, 21.2.3

Woodwell, G.M., Whittaker, R.H. & Houghton, R.A. (1975) Nutrient concentrations in plants in the Brookhaven oak pine forest. *Ecology,* **56,** 318–322.
3.3.3

Woolhead, A.S. (1983) Energy partitioning in semelparous and iteroparous triclads. *Journal of Animal Ecology,* **52,** 603–620.
14.12

Worthington, E.B. (ed.) (1975) *Evolution of I.B.P.* Cambridge University Press, Cambridge.
17.1

Wright, J.L. & Lemon, E.R. (1966) Photosynthesis under field conditions. IX. Vertical distribution of photosynthesis within a corn crop. *Agronomy Journal,* **58,** 265–268.
3.3.1

Wynne-Edwards, V.C. (1962) *Animal Dispersion in Relation to Social Behaviour.* Oliver and Boyd, Edinburgh.
6.11

Wynne-Edwards, V.C. (1977) Intrinsic population control and introduction. In: *Population Control by Social Behaviour* (F.J. Ebling & D.M. Stoddart eds), pp. 1–22. Institute of Biology, London.
6.11

Yoda, K., Kira, T., Ogawa, H. & Hozumi, K. (1963) Self thinning in overcrowded pure stands under cultivated and natural conditions. *Journal of Biology, Osaka City University,* **14,** 107–129.
6.3, 6.12

Yorke, J.A. & London, W.P. (1973) Recurrent outbreaks of measles, chickenpox and mumps. II. Systematic differences in contact rates and stochastic effects. *American Journal of Epidemiology,* **98,** 469–482.
12.6.1

Zadoks, J.S. & Schein, R.D. (1979) *Epidemiology and Disease Management.* Oxford University Press, Oxford.
12.6.3

Zahirul Islam (1981) The influence of cinnabar moth on reproduction of ragwort. M.Sc. thesis, University of London.
8.4

Zeevalking, H.J. & Fresco, L.F.M. (1977) Rabbit grazing and diversity in a dune area. *Vegetatio,* **35,** 193–196.
19.2.1

Organism Index

Canada lynx (*Lynx canadensis*), 349, 356–7, 560–1, 562
Canadian pondweed (*Elodea canadensis*), 578
Cape sugarbird (*Promerops cafer*) 469
Capitella, 605
Capsella bursa-pastoris, 142, 222, 582, 620
Carabodes labynnthicus, 406
Carassius auratus, 42
Carcinus maenas, 706
Cardamine pratensis, 337
Cardiaspina albitextura, 370–2
Carduus nutans, 299
Carex, 613
    C. arenaria, 129, 194
    C. bigelowii, 150–1
    C. pensylvanica, 703
Caribou, 169, 311, 562
Carp, 59
Carrion flies, 268
Caruus nutans, 299
Cassava (*Manihot esculenta*), 640
Cassia obtusifolia, 276
Cassowaries, 7, 755
Castanea dentata, 712
Cattail, *see* Typha
Cattle, 472
    longhorn, 208
Cauliflower, 22
Cauliflower mosaic virus, 417, 420
Cedar, red (*Juniperus virginiana*), 621
Cediopsylla tepolita, 711
Centaurium erythraea, 718
Centipedes, 393
Centropus violaceus, 174–5
Cepaea nemoralis, 35
Cercariae, 452, 453
Certhidea oliracea, 17
Cervus elaphus, 145–8, 510
Chaerophyllum prescottii, 194
Chaetoceros lorenzianum, 483
    C. tetrastichon, 483
Chalcophaps indica, 278–9
    C. stephani, 278–9, 685
Chamaephytes, 26
Chamobates cuspidatus, 406
Chandrus crispus, 69
Chaoborus, 109
Chenopodiaceae, 87
Chenopodium album, 128, 193, 582, 620
Chervill, Prescott (*Chaerophyllum prescottii*), 194
Chestnut, 12
    Castanea dentata, 712
Chickweed (*Stellaria media*), 142, 582
Chironomus, 396
    C. lugubris, 408
Chlorella, 37, 214, 243, 481

Chondrilla juncea, 254
Chondrus crispus, 706–73
Choristoneura fumiferana, 370–3
Chorthippus brunneus, 11, 134–41, 191, 210–11, 510
Chrysanthemum segetum, 516
Chrysocharis gemma, 365
    C. pubicornis, 365
Chthamalus, 252, 260
    C. fissus, 721
    C. stellatus, 24–50
Chydorus sphaericus, 408
Cichlid fish (*Genychromis meuto*), 99
Ciliates, 473, 476
    Mesodinum rubrum, 484
    Vaginicola, 483
Cinnabar moth (*Tyria jacobaeae*), 304, 347–8, 349
Circulifer tenellus, 185
Cistus, 423
Citrus limonum, 74
Cladochytrium replicatum, 390
Clam, fingernail (*Musculium securis*), 208
Clavicularia, 493
Clostridium, 488
    C. locheadii, 473
Clouded yellow butterfly (*Colias croceus*), 171
Clover, 30, 83, 87, 710
    hare's foot (*Trifolium arvense*), 126
    red (*Trifolium pratense*), 640
    subterranean (*Trifolium subterraneum*), 84, 301–2, 215–17
    white (*Trifolium repens*), 31, 32, 33, 34–5, 87, 113, 707
Clown fish (*Amphiprion*), 464
Coal tit (*Parus ater*), 271
Coccinella septempunctata, 331
Cockroach, 404, 477
    Blatella germanica, 477
    wood-eating (*Cryptocercus*), 476
Cocksfoot (*Dactylis glomerata*), 81, 87, 234, 640, 684
Cockspur (*Echinochlos crus-galli*), 87
Cod, arctic, 384
Codium fragile, 486
Coelenterates, 494
Coffee, 466
Coleophora alticolella, 59
Colias croceus, 171
Colinus virginianus, 372
Collema, 487
Colorado potato beetle (*Leptinotarsa decemlineata*), 45, 177, 552–6, 558
Columba palumbus, 300, 312, 331, 710
Compositae, 22
Condor, giant, 582
Coniosporum, 610
Conyze canadensis, 576

Copepod (*Ergasilus caeruleus*), 436
Coral, 30, 124, 482, 559
Corallina officinalis, 69
    C. vancouveriensis, 724
Cordulegaster, 396
Cordyline, 22
Cormorant, 445, 755
Corn (*Zea mays*), 81, 87, 218, 419, 454, 640
Cornus stolonifera, 51
Corophium volutator, 331
Cotton (*Gossypium hirsutum*), 640
Cotton top tamarin (*Saguinus oedipus oedipus*), 584
Crab, 30, 310, 518
    Carcinus maenas, 706
    Pachygrapsus crassipes, 619
Crangon septemspinosa, 74
Creosote bush (*Larrea tridentata*), 152
Cress (*Lepidium sativum*), 430
Crow, 411, 755
Crowfoot, water (*Ranunculus batrachium*), 38
Cruciferae, 477
Cryptocentrus cryptocentrus, 463–4
Cryptocercus, 476
Cryptophytes, 26
Cryptosula, 126
Cuban solenodon (*Solenodon cubanus*), 583
Cuckoo, 755
    Centropus violaceus, 174, 175
Cuckoo-dove (*Macropygia*), 174, 175, 585, 691
Cupressus, 196
    C. pygmaea, 582
Cuscuta epithymum, 422
Cycad (*Encephalartos*), 22, 493
Cyclops vicinus, 117
Cyclotelia arenti, 66
    C. kutzingiana, 66
    C. meneghiniana, 287–8
Cymadothea, 33?
Cymbalai gracilis, 66
Cynosurus cristatus, 71
Cyperus papyrus, 87
    C. rotundus, 87
Cypress (*Cupressus*), 196
Cyrtobagous, 578–80
Cytinus, bladder (*Cytinus hypocistis*), 423
Cytinus hypocistis, 423

Dactylis glomerata, 71, 87, 233, 640, 684,
Dacus tryoni, 337
Damselfly, 323
Danaus plexippus, 170
Dandelion (*Taraxacum officinale*), 519, 520
Daphnia, 318, 323, 383

D. *magna*, 131, 325
D. *pulex*, 109, 303
Deer, 145–8, 472
  mule, 38
  red (*Cervus elaphus*), 510, 518
Delia brassicae, 114
Delphinium barbeyi, 676–9
  D. *nelsoni*, 190
Dendroceras, 34
Depressaria pastinacella, 294
Deschampsia, 613
  D. *caespitosa*, 615
  D. *flexuosa*, 143, 600
Desert bighorn sheep, 381
Desmazierella, 611
Desulfotomaculum, 488
Diadromus pulchellus, 331
Diatoms, 65–6
  Chaetoceros lorenzianum, 483
  Chaetoceros tetrastichon, 483
  Leptocylindricus mediterraneus, 483
Dicaeum livundinaceum, 181
Dichothrix, 487
Dicrostonyx torquatus, 564
Digitalis purpurea, 143, 510
Dimorphotheca pluvialis, 186
Dinophyceae, flagellate, 482
Diospyros virginiana, 620
Diphyllobothrium latum, 424
Dipodomys deserti, 693–4
  D. *heermanni morroensis*, 583
  D. *merriami*, 693
  D. *spectabillis*, 693–4
Dipsacus fullonum, 159–60
  D. *sylvestris*, 179
Dipteryx panamensis, 802
Dirofilaria immitis, 445
Distichlis spicata, 576
Ditylenchus, 392
Dodder, 419
  lesser (*Cuscuta epithymum*), 422
Dodo, 299
Dogwood (*Cornus stolonifera*), 51
Douglas fir (*Pseudotsuga menziesii*), 303, 508, 509
Douglas squirrel (*Tamiasciurus douglasii*), 574–5
Dove, ground
  Chalcophaps, 278–9, 686
  Gallicolumba, 278–9
Dracaena, 22
Drongo, 755
Drosophila, 14, 74, 313, 758
  D. *adiastola*, 16
  D. *attigua*, 15
  D. *busckii*, 406, 407
  D. *hydei*, 405
  D. *immigrans*, 405
  D. *melanogaster*, 183–4, 213, 220, 222, 406–7
  D. *obscura*, 192

D. *primaeva*, 15
D. *setosimentum*, 16
D. *simulans*, 406
D. *subobscura*, 62, 508, 509
  picture-winged, 14, 15
Dryas, 493, 616
  D. *octopetala*, 60
Duck, 755
  eider, 518
  mallard, 312
Duckweed (*Lemna*), 126, 196
Dugesia lugubris, 532
Duiker, 56
Dung beetle, 409
Dungfly (*Scatophaga stercoraria*), 410

Earthworm, 393–4
Echinocactus fendleri, 87
Echinochloa crus-galli, 87
Echinococcus granulosus, 424, 425
Echium, 22
Ecotypes, 30–32
Ectomycorrhizae, 477
Eel, 167, 170
Eelworm (*Heterodera avenae*), 296
Eichornia, 126
  E. *crassipes*, 578
Elaesis guineensis, 640
Elassonia evergladei, 235
Elephant, 56
Elk, American, 168
Elm (*Ulmus capestris*), 712
Elodea canadensis, 578
Elysia atro-viridis, 486
  E. *viridis*, 485
Emdymion non-scriptus, 143
Empoasca fabae, 109
Emu, 7
Encarsia formosa, 329
Encephalartos, 493
Endogone, 480
Endosymbionts, 495
Endothia parasitica, 712, 725
Engraulis ringens, 377
Eniochthonius minutissimus, 406
Enoicyla pusilla, 399–400
Entamoeba histiolytica, 418
Enterobacteriaceae, 488
Enteromorpha intestinalis, 706
Entodiniomorphs, 473, 474
Eptetranychus sexmaculatus, 337–8
Ephemera, 396
  E. *simulans*, 67
Ephestia cautella, 220, 331–2, 345–7
Epichloe typhina, 444
Epilobium angustifolium, 676–9
Epixerus ebii, 679
Eragrostis pilosa, 576
Ergasilus caeruleus, 436
Ericaceae, 481
Erichthonius braziliensis, 236, 238

Erigeron annuus, 620
  E. *canadensis*, 620
Erioischia brassicae, 221
Eriophyllum wallacei, 144
Erithacus rubecula, 169
Ermine (*Mustela erminea*), 568–9
Erysiphe graminis, 713
Escherichia coli, 493
Espeletia, 20, 22
Eucallipterus tiliae, 295–6
Eucalyptus, 39, 194, 300
  E. *regnans*, 179
Eucalyptus psyllid (*Cardiaspina albitextura*), 370–2
Eunotia alpina, 66
  E. *arcus*, 66
  E. *pectinalis*, 66
  E. *tenelia*, 66
  E. *veneris*, 65, 66
Euphorbia corollata, 87
  E. *maculata*, 87
  E. *polygonifolia*, 576
Euphrasia, 422
Eupomacentrus apicalis, 716
Eutermes, 404
Eutintinnus pinguis, 483

Fagus, 87
  F. *grandifolia*, 620–2
  F. *sylvatica*, 74, 305, 644
Fairy shrimp (*Streptocephalus vitreus*), 141, 142
Falcon, 755
Fasciola hepatica, 439, 445
Fat hen (*Chenopodium album*), 128, 193, 582, 620
Fern, 56, 477
  Azolla, 493
  bracken (*Pteridium aquilinum*), 113, 176, 195, 582, 673, 750
  Salvinia molesta, 578–80
Ferocactus acanthodes, 87
Festuca arundinacea, 640
  F. *octoflora*, 126
  F. *ovina*, 600, 717
  F. *rubra*, 615, 708
Ficus, 470, 471
Field bean (*Vicia faba*), 87, 440
Fig (*Ficus*), 470, 471
Filariae, 427
Fimbristylis, 574, 576
  F. *spadiceae*, 576
Finch, 16–18
Fingernail clam (*Musculium securis*), 208
Fin whale, 377
Fir
  balsam, 372
  Douglas (*Pseudotsuga menziesii*), 303
  silver, 424
Flagellates, 476

anaerobic, 476
Ruttnera pringsheimii, 483
Flatworm
Bdellacephala punctata, 532
Dugesia lugubris, 532
Flax (*Linum usitatissimum*), 232–3
Flea, 419–20, 435
Cediopsylla tepolita, 711
Flour beetle
Tribolium castaneum, 227, 260–2, 455, 713
T. confusum, 206–7, 227, 260–2, 713
Flour moth (*Ephestia cautella*), 220, 331–2, 345–7
Fluke, liver (*Fasciola hepatica*), 439, 445
Fly
agromyzid, 443
terphritid, 466
tsetse, 418
Flycatcher, 755
Monarcha cinerascens, 685–6
Foot and mouth disease parasite, 431
Forest buffalo, 56
Fox, 457
arctic (*Alopax lagopus*), 411, 568–9, 774
Foxglove (*Digitalis purpurea*), 142, 510
Fragilana construens, 66
F. virescens, 66
Frankia, 488, 492
Fraxinus americana, 620
French bean (*Phaseolus multiflorus*), 87
Frog, 436–8
Rana pipiens, 47
Rana tigrina, 215
Fruit-fly, *see Drosophila*
Fucus, 706–7
F. serratus, 69, 70
F. spiralis, 69
F. vesiculosis, 69
Fulmar, 538
Fungus
ambrosia, 467
Ascomycete, 477
Basidiomycete, 477
biotrophic, 442
Epichloe typhina, 444
honey (*Armillaria mellea*), 430
Laboulbenia, 436, 438
mycorrhizal, 462
rust, 39
sheath-forming mycorrhizal, 495
Funisciurus anerythrus, 679
F. isabella, 679
F. lemniscatus, 679
F. pyrrhopus, 679
Fusicoccum, 611

Gale, sweet (*Myrica*), 488, 493
Galinsoga parviflora, 186
Galium hercynicum, 248, 259
G. palustre, 615
G. pumilum, 248, 259
G. saxatile, 248
G. sylvestre, 248
Gallicolumba rufigula, 278–9
Gall wasp, 107
Gambusia affinis, 533
Gammarus, 395, 396, 687
G. duebeni, 680
G. locusta, 65, 680
G. oceanicus, 680
G. pseudolimnaeus, 407
G. pulex, 65, 67
G. salinus, 680
G. zaddachi, 65, 67, 680
Gelidium coulteri, 616–9, 721–2
Genychromis meuto, 99
Geocryptophytes, 26
Geophytes, 26
Geospiza conirostris, 17
G. difficilis, 17
G. fortis, 17
G. fuliginosa, 7
G. magnirostris, 17
G. scandens, 17
Geotrupes spiniger, 408
Gerbil
Gerbillus allenbyi, 279
Meriones, 279
Gerris, 187
Geum reptans, 74
Giant condor, 582
Giardia intestinalis, 418
Gibbula cineraria, 69
Gigartina canaliculata, 616–8, 721–2
G. leptorhynchos, 616–8, 721–2
G. stellata, 69
Gigaspora calospora, 479
Glaucous gull, 569, 774
Glomus, 480
G. mosseae, 479, 480
Glossina, 420
Glossiphomia, 396
Glossosoma, 396
Glycine, max, 293, 640
G. soja, 207, 492
Goby (*Cryptocentrus cryptocentrus*), 463–4
Golden lion tamarin (*Leontopithecus rosalia*), 584
Goldfish (*Carassius auratus*), 42
Gomphocerripus rufus, 60
Goose barnacle (*Mitella*), 710
Goosefoot, white (*Chenopodium album*), 128, 193, 582, 620
Gordius, 444
Gorgonia, 126
Gorse (*Ulex europaeus*), 295, 348, 615

Gossypium hirsutum, 640
Granulosis virus, 457
Grape phylloxera (*Viteus vitifoliae*), 186
Grass, 83, 298
'annual' meadow (*Poa annua*), 141, 142, 298, 530–1, 582
Bromus, 266–7
Festuca ovina,, 600, 717
Helictotrichon, 717
Holcus lanatus, 600, 717
Lolium multiflorum, 292–3
Paspalum, 491, 492
rye (*Lolium perenne*), 31, 71, 87, 142, 217–18, 241–2, 323, 474, 640, 707, 717
Grasshopper, 45, 60, 203
Austroicetes cruciata, 46
Chorthippus brunneus, 11, 134–41, 191, 210–211, 509, 510
Gray bat (*Myotis grisescens*), 584
Great tit, (*Parus major*), 149, 176, 189, 239, 240, 271, 318, 331, 344–7
Grebe, 755
Grey birch, 622
Grindelia lanceolata, 158–9
Ground dove
Chalcophaps, 278–9, 686
Gallicolumba, 278–9
Ground squirrel, 361
Belding's (*Spermophilus beldingi*), 240
Groundsel, (*Senecio vulgaris*), 141–2, 516
Grouse, 311
red (*Lagopus scoticus*), 361, 446, 458, 562–3
ruffed, 356, 562
spruce, 356, 562
Gull, glaucous, 569, 774
Gunnera, 493
Guppy, 312–13
Gurania, 120
Gymnarrhena micrantha, 185
Gymnosperm, 477
Gypsy moth (*Lymantria dispar*), 297

Haematopus ostralegus, 323
Halidrys siliquosa, 69
Halipegus eccentricus, 436–8
Halosaccion glandiforme, 724
Hare, 311
snowshoe (*Lepus americanus*), 297, 348, 349, 356–7, 562
Harvester ant (*Veromessor pergandei*), 280–1
Harvest spider, 393
Hawk, 755
Henicopernis longicauda, 686
Heather (*Calluna vulgaris*), 361
Heath rush (*Juncus squarrosus*), 59

Plankton, 484
Plantain (*Plantago*), 573
  P. lanceolata, 573
  P. major, 221, 573
  P. maritima, 32
  P. media, 573
Plasmodiophora, 447
P. brassicae, 417, 418, 419, 443
Plasmodium, 417, 419, 420, 429, 435, 449
Plate limpet (*Acumaea scutum*), 311
Platyhelminths, 419
Platynothrus peltifer, 406
Platypus, duck-billed, 19
Platyspiza crassirostris, 17
Plectrocnemia conspersa, 340–1, 605, 710
Plectroglyphidodon lacrymatus, 716
Pleolophus basizonus, 322, 324
Plethodon glutinosus, 247–8
  P. jordani, 247–8
Plodia interpunctella, 220, 327, 333
Pneumostrongylus tenuis, 711
Poa alpina, 32
  P. annua, 141, 143, 298, 526–7, 582
  P. trivialis, 71, 143
Poecilochirus necrophori, 414
Polar bear, 774
Pollicipes, 776
  P. polymerus, 724
Polydora, 605
Polygonum aviculare, 143, 582
  P. pensylvanicum, 620
Polypodium, 126
Pomacea, 309
Pomacentrus wardi, 716
Pomarine jaegars, 569
Poppy
  opium, 466
Papaver, 111
Populus deltoides, 620
  P. tremuloides, 195
Porphyra pseudolanceolata, 724
  P. umbilicalis, 69
Portulacaceae, 87
Postelsia palmaeformis, 266
Potato (*Solanum tuberosa*), 417, 432, 640
  Pentland Crown, 432
  Pentland Dell, 432
Potato beetle, 177
Potato leaf roll virus, 420
Praying mantis, 444
Prescot chervill (*Chaerophyllum prescottii*), 194
Prickly pear cactus, *see Opuntia*
Procellariformes, 533
Prochloron, 462
Prognathus penicillatus, 693
Promerops cafer, 469
Prostrate willow, 617
Protea eximia, 469

Protohemicrytophytes, 26
Protoxerus stangeri, 679
Protozoa, 474, 475, 476, 483
Prunus aurocerasus, 74
  P. spinosa, 615
Pseudomyrmex ferruginea, 464–5
Pseudopeziza medicaginis, 440
Pseudotsuga menziesii, 302–3, 508, 509
Psoralia tenuiflora, 91
Ptarmigan, 311, 744
  willow, 562
Pteridium, 95
  P. aquilinum, 113, 582, 673, 750
Puccinellia, 613
  P. maritima, 615
Puccinia graminis, 428
  P. striiformis, 452
Pullularia, 611
Pythium, 99, 430–1, 440
  P. irregulare, 430

Quail, 755
  bobwhite (*Colinus virginianus*), 372
Quercus, 87, 126, 613, 738
  Q. alba, 96, 150, 391, 620
  Q. carris, 107
  Q. cerris, 391
  Q. ilex, 74
  Q. nigra, 150
  Q. prinus, 150
  Q. pubescens, 74
  Q. rober, 443, 615, 759
  Q. rubra, 620
  Q. varifolia, 96
  Q. velutina, 620

Rabbit, 459, 460, 626, 704–5
  Mexican volcano (*Romerolagus diazi*), 583, 711
Radianthus, 464
Radish virus, 417
Rafflesia arnoldii, 422
Ragwort (*Senecio jacobaea*), 179, 298, 304, 347–8, 349
Rail (*Rallina tricolor*), 686, 755
Rallina tricolor, 686
Rana pipiens, 47
  R. sylvatica, 234
  R. tigrina, 215
Ranunculaceae, 469
Ranunculus, 298
  R. aquatilis, 39
  R. batrachium, 38
  R. circinatus, 38
  R. ficaria, 143, 469
  R. flabellaris, 39
  R. fluitans, 38
  R. glacialis, 74
  R. hederaceus, 38
  R. omiophylus, 38

  R. repens, 541–3, 560
  R. richophyllus, 38
Raphanus raphanistrum, 774–5
Rapistrum rugosum, 186
Raspberry beetle (*Byturus tomentosus*), 101
Raspberry moth (*Lampronia rubiella*), 101
Rat, 436, 438, 582
  eastern wood (*Neotoma floridana*), 533
  hispid cotton (*Sigmodon hispidus*), 533
  kangaroo, 693–4
  Morrow Bay kangaroo (*Dipodomys heermanni morroensis*), 583
Red admiral butterfly (*Vanessa atlanta*), 171
Red algae
  Chondrus crispus, 706–7
  Gelidium, 616–9, 721–2
  Gigartina, 616–9, 721–2
  Rhodoglosseum, 616–9
Red cedar (*Juniperus virginiana*), 620–1
Red clover (*Trifolium pratense*), 640
Red deer (*Cervus elaphus*), 145–8, 510
Red grouse (*Lagopus lagopus scoticus*), 361, 562–3
Red maple, 622
Red-osier dogwood (*Cornus stolonifera*), 51, 60
Red pine, 309
Redroot pigweed (*Amaranthus retroflexus*), 276, 620
Redshank (*Tringa totanus*), 331
Redwood (*Sequoia sempervirens*), 243
Reedmace (*Typha*), 252–3, 524–6
Reindeer, 172, 215, 326, 562
Reticulitermes flavipes, 475
Rhea, 7
Rhinanthus minor, 422
Rhinocyllus conicus, 298
Rhizobiaceae, 488
Rhizobium, 488–93
Rhizopertha dominica, 214
Rhizophora mangle, 582, 743
Rhizopus, 388
Rhodnius prolixus, 477
Rhodoglossum affine, 616–8, 721–2
Rhodymenia palmata, 69
Rhyssa persuasoria, 329
Rice (*Oryza sativa*), 466, 640
Riparia riparia, 173
Roach, 445, 446
Robin (*Erithacus rubecula*), 169
Rock hyrax, 56
Rodolia cardinalis, 348
Roller, 755

tree (*Tamiasciurus*), 574–50, 691
Squirrel monkey (*Saimiri oerstedi*), 584
Starfish, 338–9
 Pisaster, 708–10, 776
Starling, 582, 755
 Aplonis metallica, 697
Steer, 401
Steganacarus magnus, 406
Stegobium panaceum, 227
Stellaria media, 142, 582
Stenophylax, 395
Sterna paradisaea, 168–9
Stickleback, 336
 three-spined, 421
Stigonema, 487
Stonefly, 605, 710
 Leuctra nigra, 605
 Nemurella picteti, 605
Stork, 755
Strawberry, 18
Streptocephalus vitreus, 142
Strix aluco, 349, 556
Subterranean clover (*Trifolium subterraneum*), 84, 216–17, 254, 301–2
Sudan grass (*Sorghum*), 81, 87, 640
Sugar beet, 449
 Beta vulgaris, 640
Sugarbird, Cape (*Promerops cafer*), 469
Sugar cane (*Saccharum officinale*), 87, 640
Sugar maple (*Acer saccharum*), 620–1, 626
Sunfish
 bluegill (*Lepomis macrochirus*), 318, 320–1
 Everglades pygmy (*Elassonia evergladei*), 235
Sunflower, aspen (*Helianthella quinquenervis*), 466
Swallow, 168, 755
Sweet gale (*Myrica*), 488, 493
Sweet white clover (*Melilotus alba*), 157–8
Swift (*Apus apus*), 512, 513, 541, 755
Sympodiella, 611
Syncerus caffer, 556, 558
Synchytrium endobioticum, 417
Synedra, 259
 S. ulna, 251–2

Tabellaria binalis, 66
 T. flocculosa, 65, 66
 T. quadriseptata, 65, 66
Taenia saginata, 425
Tall fescue (*Festuca arundinacea*), 640
Tamarin
 cotton top (*Saguinus oedipus oedipus*), 583

golden lion (*Leontopithecus rosalia*), 583
Tamiasciurus douglasii, 574–53
Tapeworm, 419, 424, 520, 524
 Ancylostoma, 422, 439
 Hymenolepis microstoma, 438
 Ligula intestinalis, 445
Taraxacum officinale, 519, 520
Tawny owl (*Strix aluco*), 369, 555, 571
Taxus baccata, 74
 T. canadensis, 582
Tectocepheus velatus, 406
Tegula, 776
Tenebrio mollitor, 331
Termite, 404, 475, 476
 Reticulitermes flavipes, 475
Tern, arctic (*Sterna paradisaea*), 168–9
Tetranchyus urticae, 329
Tetraphis pellucida, 62–3
Teucrium polium, 86
Thais, 710, 776
 T. lapillus, 69
Theileria parasites, 435
Therophytes, 26
Thistle, 757
 nodding (*Carduus nutans*), 299
Thlaspi arvense, 142
Thrips imaginis, 543–8
Thrush, 36, 755
Thyme, wild, 466
Tick, 435
 hedgehog (*Ixodes hexagonus*), 421
Tilia vulgaris, 295
Tillandsia recurvata, 87
Tilletia tritici, 436
Tinamou, 7
Tintinnid (*Eutintinnus pinguis*), 483
Tipula, 396
Tobacco, 466
Tomato yellow net virus, 417
Tomocerus, 405
Torreya taxifolia, 582
Tortoise, Aldabra, 582
Toxocara canis, 434
Trebouxia, 486
Tree fern, 56
Tree-groundsel, 56
Tremarctos ornatus, 583
Trematodes, 436
Trialeurodes vaporariorum, 329
Tribolium castaneum, 227, 260–2, 455, 713
 T. confusum, 206–7, 227, 260–2, 713
Trichoderma, 611
Trichogramma pretiosum, 333
Trichomonas termopsidis, 476
Trichoscyphella willkommii, 711
Trichostrongylus tenuis, 444, 458
Tridachia crispata, 485

Trifolium pratense, 640, 684
 T. repens, 31, 33, 34–5, 83, 87, 112, 113, 707
 T. subterraneum, 84, 215–17, 254, 301–2
Triglochin, 578, 613
 T. maritima, 615
Tringa totanus, 331
Triplasis purpurea, 576
Tripleurospermum maritimum, 143
Triticum vulgare, 87, 640
Trout, 124, 207
 brown (*Salmo trutta*), 210–11
Trypanosoma brucei, 418
Trypanosomes, 417, 418–20, 459
Tsetse fly (*Glossina*), 418, 420
Tsuga canadensis, 626
Tubifex, 396
Tuncrium pollium, 39
Tunny, 169
Turdus merula, 339
 T. philomelos, 339
Turkey, 755
 domestic, 179
Tussilago farfara, 179
Typha, 524
 T. angustifolia, 252–3, 524–6
 T. domingensis, 524–6
 T. latifolia, 252–3
Typhlodromus occidentalis, 337–8
Typhoid bacterium, 417
Tyria jacobaeae, 304, 347–8, 349

Ulex europaeus, 295, 348, 615
Ulmus alata, 620
 U. campestris, 712
Ulva, 616–8, 721–2
Unio, 396
Uniola paniculata, 576
Urtica dioica, 143
Ustilago tritici, 436

Vaccinium vacillans, 96
Vaginicola, 483
Vampyrella, 03
Vancouver island marmot (*Marmota vancouveriensis*), 584
Vanessa atalanta, 171
 V. caddui, 171
Venturia canescens, 328, 331–2, 345–7
Verbascum thapsiforme, 142
Veromessor pergandei, 280–1
Veronica, 141
 V. arvensis, 718
Verrucaria maura, 69
Viceroy butterfly, 115
Vicia faba, 87, 440
Vicugna vicugna, 583
Viola riviniana, 717
Virus

beet, 417
cauliflower, 417, 420
granulosis, 457
lettuce necrotic yellow, 420
measles, 417
myxoma, 459, 460
pea mosaic, 417
potato leaf roll, 420
radish, 417
tomato yellow net, 417
Viteus vitifoliae, 186
Volcano rabbit (*Romerolagus diazi*),
   584, 711
Vole (*Microtus*), 361, 559–71
  M. agrestis, 367, 559–71
  M. ochrogaster, 176
  M. pennsylvanicus, 176, 184, 187
Vulpia fasciculata, 208, 216–7, 551

Warbler, 755
  Aldabra bush, 582
Wasp
  cecidomyid, 443
  cynipid, 443
  fig, 470–1
  gall, 37–8
  wood, 296
Water bug (*Hydrometra myrae*), 444
Water crowfoot, 67
Water flea (*Daphnia*), 382
  D. magna, 131, 325
  D. pulex, 109, 303
Water hyacinth, 196
  Eichornia, 126
Water lettuce (*Pistia*), 126

Waxbill, 755
Weasel (*Mustela*), 568–9
Webworm, parsnip (*Depressaria
   pastinacella*), 294
Weevil
  Apion ulicis, 348
  Callosobruchus chinensis, 355–6
  Cyrtobagous, 578–80
  Phyllobius argentatus, 292
  Rhinocyllus conicus, 298
Whale, 131, 132
Whale, baleen, 29, 169, 377
Wheat, 81, 277, 425, 438, 450, 713
  Triticum vulgare, 87, 640
White clover (*Trifolium repens*), 31,
   33, 34–5, 83, 87, 108, 112, 113,
   707
Whitefly (*Trialeurodes
   vaporariorum*), 329
White goosefoot (*Chenopodium
   album*), 128, 582, 620
White pine, 309
  Pinus strobus, 704
White spruce, 309, 371
White-tailed deer (*Odocoileus
   virginianus*), 626
Wildebeest, 306
Wild madder (*Rubia peregrina*), 58
Willow, 617
Willow ptarmigan, 562
Willow tit (*Parus montanus*), 271
Winkle
  Littorina negrolineata, 529–31
  L. rudis, 529
Winter moth (*Operophtera brumata*),

112, 185, 227, 354, 556
Woodlouse, 393, 404
Woodpecker, acorn (*Melanerpes
   formicovorus*), 181–2
Woodpigeon (*Columba palumbus*),
   172, 300, 312, 331, 710
Wood swallow, 755
Wood wasp, 296
Worm
  digenean, 426, 436–8
  nematode, 420, 427, 438
  parasitic helminth, 420, 436
  ectoparasitic platyhelminth, 420
  gordian, 444
  monogenean, 436
  trematode, 424
Wuchereria bancrofti, 425

Xanthorrhoeaceae, 22
Xyleborus xylographus, 467

Yeast, 389
  Saccharomyces cerevisiae, 325
Yellow perch (*Perca flavescens*), 42
Yew, 18
Yucca, 22

Zea mays, 87, 218, 419, 640
Zeiraphera diniana, 296, 457
Zerna erecta, 600
Zoothamnium pelagicum, 483
Zorotypus hubbardi, 186
Zostera marina, 600

Subject Index

Emigration, 176, 329, 548, 553–6
  affecting abundance, 571
Endangered species, 583
Endemic species, 758
Endogenous rhythms, 37
Endotherms, 43, 52–3, 656
Energy
  community, 629–69
    flow of, 597, 654, 657–60
    related to trophic levels, 776–7
    units of, 629
Enforced dormancy, 191–3
Environments(s)
  fluctuating, 36–9, 266–7, 769
  harshness of, 797–8
  heterogeneity of, 27, 797–8
Environmental age, species richness
    related to, 799, 802
Environmental factors, ordination
    of communities involving,
    602
Enzyme(s)
  activity affected by acidity, 64
  alcohol dehydrogenase, 407
  carbohydrase, 411
  cellulase, 390
  collagenase, 412
  effect of temperature on, 48
  in decomposition, 388
  lipase, 411
  protease, 411
Ephemeral annuals, 143–4
Epidemiology, 427, 429–33, *see also*
    Parasites
  and population dynamics, 454
  contact rates, 429
  distance between hosts, 430
Epidermal hairs, 109
Epilimnion, 166
Equilibrium
  in level of abundance, 550
  in population size, 229
  multiple, 369–72, 379
  stable, 209
  theories, 727–9
    of island biogeography, 739,
    744–5
Equitability, 595
Essential resources, 120
Estuaries, primary productivity and
    biomass of, 631
Ethanol, as a resource, 407
Ethylene, 98
Eukaryotes, evolution of,
    mutualism and, 494
Euphotic zone, 647–8
Evaporation, 61
Evapotranspiration, potential,
    641–2
Even distributions, 166, 337
Evenness, 595
Evergreen forest, primary

productivity and biomass of,
    631
Evolution, 108
  and natural selection, 5–6
  convergent and parallel, 18–23
  diet width related to, 314
  in island communities, 757–9
  in relation to pollution, 70
  of territorial behaviour, 239
Evolutionary phases, species
    diversity associated with, 811
Evolutionary time, species richness
    related to, 799, 802
Exclusion, competitive, 585, 670
Exploitation, 204–5, 252
Explosion, population, 578
Exponential growth, 223
Exposure, affecting zonation, 70
External rumen, 404–5
Extinction, 250–1, 369, 373, 376,
    379–80, 549, 581, 583–5, 610,
    670, 719, 729
  local, 281
  of predators and prey, 337–8
  related to food chain length, 777
  selective, 813
Extinction rate
  in relation to species turnover,
    753–4
  on islands, 741

Facilitation, 614
  social, 330
Facilitation succession, 623
Facultative annuals, 143–4
Facultative diapause, 192
Facultative mutualism, 404
Faeces, 653
  as a resource, 388
  detritivore, 401
  feeding on, 397, 407–10
Farmland, succession on, 619
Fats, 103
  water from metabolism of, 86
Feathers, digestion of, 412
Fecundity, 138, 204, 249, 558
  affected by herbivory, 298–9
  density-dependent, 206–9, 218,
    235
  schedules, 132, 136–7, 149–50,
    155, 157
Feeding
  efficiency, 320
  rate, 328, 341
Fencing, 578
Fermentation, 389, 406
  in decomposition, 389
Fermentation layer of soil, 611
Fertilizers
  affecting leming population,
    567–8
  affecting species diversity, 789

nitrogen, 217, 645
Fibre, 103, 297
Field capacity of soil, 88, 90
Field succession, 619, 627
Fire
  effect on dormancy, 194
  successions following, 613
Fish
  marking of, 540
  temperature effect on oxygen
    requirement, 59
Fishery-protection vessels, 379
Fishing, 373–85, 377
  intensities, affecting yield, 384–5
Fish scales, feeding on, 101
Fitness, 6–7, 191, 236, 239, 272
Fixed quotas, 374–8, 380
Flavonoids, 112
Fleet size, 370
Flightless birds, distributions of, 7, 9
Flocks, 369
Flower abortion, 294
Flowerings, delay in, 298
Fluctuations of populations,
    228–30, 540–2
Food, role in microtine cycles, 566
Food chains, 98, 652, 775–9
Food density, 322–8, 327
Food patches, 330–5
Food preferences, 309–14
Food supply, limiting population
    size, 574
Food web, 589, 765–9, 775–7
  as emergent property, 591
  complexity, 387
Foraging approach, 341–7
Foraging theory, 315–22
Forest, 609
  biomass, 637
  climax, 620
  communities, 629
  floor, nitrogen availability in, 402
  northern temperate, 12
  primary productivity and biomass
    of, 631
  tropical, 13
Fossil fuels, burning of, 65
Fossil record, richness in, 809–12
Fragility, community, 762
Free distribution, ideal, 335–7
Frequency-dependent selection, 710
Freshwater lakes, primary
    productivity of, 630–1
Frost, 56
  damage due to, 49, 298
  dependence of germination on, 47
  mortality induced by, 553
  resistance to, 49
Fruit
  assimilation of, 656
  rotting, 405
Fugitive species, 265–6

Hypersensitive reactions in plants, 442
Hypervolume, *n*-dimensional, 73, 121
Hyphae
  feeding on, 392
Hypolimnion, 166

Ice crystals in cells, 49
Ideal free distribution, 335–7
Immigration, 176, 548
  affecting abundance, 571
  island, 739–42
  rate, in relation to species turnover, 753–4
Immobilization, 387
Immune response, 441–2
Immunity, 433
Inbreeding, 189–90
Incidence functions, 685, 756
Index of abundance, 132
Induced dormancy, 191, 193
Infant deaths, in under-developed countries, 446
Inhibition model of succession, 624
Innate dormancy, 191–2
Inorganic molecules, as resources, 85–98
Insecticides, 712
Insects, phytophagous, 672–4
Instability of harvested populations, 379–80
Intensity of species, 581
Interaction, in communities, 589
Interference, 204–5
  associated with aggregation, 335–7
  coefficient of, 329
  competition, 118
  mutual, 328–30, 358
International Biological Programme, 630
International Whaling Commission, 377
Interspecific competition, 247–88, 725
  influencing community structure, 670–99
  related to species diversity, 785
Intertidal zones, 249, 601, 613, 616, 706–8
Intraspecific competiton, 203–46, 284, 300, 358–60, 370, 374, 383, 545, 556, 672
Invertebrate(s)
  drift, 182
  faeces, 407–8
  functional responses, 328
Iodine, 94
Iron, availability affected by pH, 64
Islands, 735–61
  biogeography of, 735

biotas, 13–18
  conservation, 759–61
  in relation to species diversity, 787
  remoteness, 745–9
Isoclines of growth, 119, 283–4
Isoenzymes, 184
Isopods, terrestial, 62
Isotherm, species distribution associated with, 56
Iteroparity, 134, 141, 504, 516–19
  continuous, 161–3
  overlapping, 145–53
  in perennials, 150–3

Juvenile mortality, 558

K-factor analysis, 559
K-selection, 521–31
K-value, 136, 554–9, 678
Keratin, 388
  digestion of, 412
Key-factor analysis, 552–9
Killing-power, 136
Killing response, 415
Kin selecton, 570
Kleptoparasitism, 322

Lactobacilli, 389
Lakes
  cycling in, 668
Land masses, movement of, 7–8
Larvae, starvation of, 555–6
Laterals, root, 91
Latitude, species diversity correlated with, 800–5
Leaching, 389, 666
Lead, as pollutant, 70
Leaf(ves)
  angle to light, 78–9, 82
  area index (LAI), 82–4, 212, 301–2, 642
  biomass, productivity related to, 642
  decomposition of, 400
  digestion of, 407
  litter, 389
    affecting nitrogen availability, 402
  polymorphism, 87
  pores of, *see* Stomata
  production of, 298, 302
  surface of, 86
Leghaemoglobin, 489
Leptokurtic transmission of parasites, 431
Lichens, 484–8
Life cycle, 132–4, 157–63
  and environmental change, 37–8
  and seasons, 38
  phases in, 559
  problems in census, 540
Life-history, 501–43

allometric constraints, 533–8
  alternatives of, 520–5
  as compromise, 508–13
  comparisons between, 542–3
  components of, 502–5
  Grime's classification of, 528–9
  interactions with physiological demands, 533
  optima, 512–13
  phylogenetic constraints, 533–8
  short-term responses to, 531–2
  size constraints, 538–9
  study of, 502
Life-table, 132
  cohort, 134, 137, 145–7
  from census data, 552
  static, 147–9
Light
  affecting dormancy, 193
  as resource, 76–85
  extinction curve, 82
  intensity, 78–82, 243–5
    in aquatic communities, 647
  levels, in succession, 620
  productivity related to, 638
  species diversity related to, 793
  stimulating tiller production, 298
  thinning associated with variations of, 245
Lignin, 101, 106, 388, 655
  decomposition of, 390–1, 612
  digestive problems associated with, 403
  in wood, 404
  resistance to decomposition, 389
Limit cycles, stable, 229
Limiting similarity, 262
Linamarin, 112
Lipase, 411
Lipid(s)
  content of detritivores, 402
  degradation of, 391
Litter, 629
  as energy source, 634
  surface, 610
Littoral seaweed communities, 611, 636
Local stability of community, 762
Logistic equation, 230–2, 255, 353, 374
Logistic growth, plant, 580
Longevity, 506
Lotka–Volterra model, 255, 350–3, 358–64, 725, 729
Low-temperature tolerance, 49
Lysine, 121

Macrofauna, associated with decomposition, 393
Macroparasites, 419–27
  directly transmitted, 420, 450–1
  indirectly transmitted, 424, 451–4

Northern coniferous forests, 609
Noxious plant chemicals, 112
Null community, 694
'Nunataks', 10
Nutrient(s)
    availability affected by pH, 64
    cycle, 662–8
    depletion, 650
    levels in ordination, 602
    limitation of productivity, 633
Nutrient-recovery hypothesis, 567
Nutrition, under-, 558

Obligate fungal parasites, 99
Obligate mutualism, 403–4
Obligatory diapause, 191
Oceans, 636
    biomass of, 630–1
    productivity of, 630–1, 646,
        646–50
Offspring size, 504, 521
Oil, 415
Oligophagous consumers, 309
Omnivores, 289
    food preference in, 311
Open communities, 729, 732–3
Opportunism, 388
Ordination of communities, 602–5
Organic acids, from decomposition,
    389
Organic matter
    dead, 387, 389
    types of, 634
Organisms as food resources,
    98–117
Organization, community level of,
    607
Oscillations in populations, 229
    divergent, 368
    predator–prey, coupled, 348, 355,
        360
Osmoregulation affected by acidity,
    64
Osmotic forces in soil, 89
Outbreeding, in relation to
    dispersal, 189–91
Overcrowding, 550, 565
Overdispersal, 166
Overexploitation, 379, 384–5
Overgrowth competition, 253
Oviposition, 552
    rate of, 322–3
Oxalic acid, 112
Oxygen
    as propellant, 113
    as resource, 98
    consumption
        of Colorado beetle, 45
        of frog, 47
    diffusion in ordination, 602
    dissolved, relationship between
        temperature and, 59

Ozone, 664

Pampas, 609
Parallel evolution, 18–23
Parasite burden, 562
Parasites(ism), 98, 200, 289–90, 386,
        417–60, 556, 558, 563
    see also Epidemiology; Hosts
    affecting community structure,
        711–14
    altering host behaviour, 444
    and host morphogenesis, 443–4
    biotrophic, 440
    causing microtine mortality, 569
    definition of term, 416
    density-dependence within
        hosts, 439–40
    distribution of, 433–4
    diversity of, 417–27
    gene-for-gene relation between
        virulence and resistance,
        458–9
    genetic change in, 458–60
    habitat specificity within hosts,
        436
    mean intensity, 433, 435
    necrotrophic, 388, 440–1
    polymorphism in, 458–60
    population dynamics of, 447–58
    prevalence of, 433, 435
    pupal, 555
    rarity due to, 585
    tolerance of, 442–3
    transmission of, 427–33
Parasitoid(s), 289–91, 308–9, 322,
        364
    aggregative response, 331
Parasitoid–host model, 354
Parthenogenesis, 187, 585
Partial refuges, 331–5, 361–6
Passive dispersal, 177
Patch, food, 330–5
Pathogens, plant, 290, 295
Peat, 415, 629
    sphagnum, 408
Pectins, decomposition of, 389
Pelagic larvae, 182
Pentosans, in wood, 404
Peptide chains, in keratin, 412
Perennials, modular iteroparous,
        150–3
Periodic fluctuation, 562
Permafrost, 608
Permanent wilting point, 88
Peroxidase, 113
Pests
    control of, 364
    forestry, 295
pH, 682
    effect on decomposition, 389
    in ordination, 602–5
    of soil, 64–5, 616

of water, 64–5
    plants affected by, 798
Phalanx growth, 196
Phenol metabolism, 296
Phenotypic change, 566
Phenotypic plasticity, 585
    gradient, 600
Phloem, 295
    sap, 103, 295
Phosphate
    availability affected by pH, 64
    consumption of, 288
    diffusion coefficient, 96
    in soil, 96–7
    shortage of, 567, 645
Phosphorus, 166, 311, 405, 646, 682
    in detritivores, 402
    species diversity in relation to, 789
Photoinhibition, 648
Photoperiod, 192
    interaction with temperature, 47
Photosynthate redistribution, 292–3
Photosynthesis, 87, 245, 292, 387,
        629–30, 634–46, 647–8
    efficiency of, 638
    radiant energy fixed in, 76
Photosynthetically active radiation
        (PAR), 77, 638
Photosynthetic period, length of,
        644
Photosynthetic rate, 79
Phyllosphere, 388
Physiological time, 44–6
Phytoalexins in plants, 442
Phytophagous insects, 672–4,
        738–9, 742, 749–52, 759, 811
Pioneer species, 619
Pipecolic acid, 120
Planktotrophic strategy, 182
Plant(s)
    architecture of, affecting diversity
        of phytophagous insects,
        751
    communities, Raunkiaer's
        classification of, 23–7
    compensation for effects of
        herbivory, 292–4
    competition, interacting with
        herbivory, 295
    detritus, 407
        consumption of, 403–7
    fecundity affected by herbivory,
        289–9
    growth, herbivory affecting, 298,
        350, 358
    hormones, 298
    macroparasites
        epidemic, 451
        latent period, 451
    pathogens, 290, 295
    polysaccharides, digestion of, 404
    toxins, 112–14

Reproduction
    affected by population density,
        235
    as component of life-history, 504
    cost of, 510–12
    delayed, 516–19
    effect of conditions on, 42
    precocious, 516–19
Reproductive allocation, 504–5
    defined, 504
    measurement of, 504–5
    relation to cost of reproduction,
        519–21
Reproductive effort
Reproductive isolation, 14
Reproductive output, 204
Reproductive rate, 153–7, 230
    affected by intraspecific
        competition, 224
    basic, 136, 447
Reproductive value, 506–7, 521
    formal definition of, 507
Resampling, 540
Reserves, 760
Resilience of community, 762
Resin, 297
Resistance of community, 762
Resource(s)
    as basis for ecological niche, 72
    augmenting, 574
    classification of, 118–21
    depletion zone, 79, 96–7, 204
    exploitation of, 252
    levels, changes in, 283–5
    partitioning of, 281–2, 708
    ratios of, species diversity related
        to, 792–3
    renewal of, 386
    utilization of, differential, 282
Respiration
    balance with photosynthesis, 292
    temperature effect, 641
    rate of
        effect of conditions on, 42
        productivity related to, 656
Respiratory heat, 629, 652, 656
Response, functional, 322–8
Rhizomes, 150–1
    dormant, 194
Rhizosphere, 388
Richness relationships, 787–99
Ring-barking, 294
Ring counts, in trees, 614
Rings, leg, 540
r/K selection, 521–31
    assessment of concept, 527
    demographic forces beyond,
        529–31
    evidence for, 524–7
    in relation to 'alternatives', 527–9
Robustness of community, 762

Roots, 88–93, 254
    caps, 388
    feeders, 101
    growth, affected by herbivory,
        292
    hairs, 96
    suckers, 195
    uptake of mineral resources in, 96
    uptake of oxygen in, 98
Rosettes, 541
Rumen, 103
    ecosystem of, 472–5
    external, 404
    microbial digestion within, 473
Ruminants, 105, 401

Saline soils, 89
Salinity, 65–7, 680
    in relation to ecological niche,
        72–3
Salivary glands, 103
Salt-marsh, 65–6
    allogenic transition involving,
        612–13
Salt pans, 88
Salt tolerance, 65
Sampling, 540
    of populations, 131
Sanctuary zones, 379
Sand dunes, 572, 602, 613
Sap, removal of, 295
Sap-sucking insects, 294
Saponins, 112
Satiation, of seed predators, 304
Savanna, 609
    primary productivity and biomass
        of, 631
Scavengers, 411
Schistosomes, 424, 426
Scramble competition, 220
Sea
    bed of, carrion feeders on, 414
    shore zonation, 68–70
    temperature of, 57–8
Search, area-restricted, 339–40
Searching efficiency, 324, 329, 350,
    363
Search time, 317
Seasonal closures, 379
Seasonal movement, 167
Seasonal partitioning, 680
Secondary productivity, 629, 650
Secondary succession, 613
Sedimentation and accumulation of
    nutrients, 647
Sediment core, 65–6
Seed(s)
    abortion of, 299
    assimilation of, 656
    bank, 141–3, 267, 624, 725
    defensive chemicals, 121

density of sowing, 217
    dispersal of, 112, 172–96, 621
    dormancy of, 192–4
    marking of, 540
    mortality of, 558
    polymorphism of, 186–7
    predation of, 304
    production of, 277, 302, 577
        following defoliation, 294
    size of, as specification for eating,
        280
Selection
    frequency-dependent, 710
    r/K, see r/K selection
Selective extinction, 813
Selective predation, 706–10
Selenium, 93
Self-limitation, 358
Self-thinning, 241–6
Selfish herd, 171–2
Semelparity, 134, 141, 504, 516–19
    continuous, 160
    overlapping, 157–60
Septate fungi, 390–1
Serpentine rock, 585, 597
Sessile organisms, 596
    contrasted with mobile
        organisms, 29–30
Sexual differences in dispersal, 184
Shade tolerance, 620
Shading, 292
Shadow leaves, 81
Shadow species, 80–1
Shannon diversity index, 595
Shells, 110–11
Shoot(s), 254
    growth of, 292, 373–85
    volume of, relationship to age,
        152
Shredders, in decomposition, 395
Shrubland, primary productivity
    and biomass of, 631
Silage making, 389
Silicate, consumption of, 288
Silicon, 93
Silt, 612
Silurian, 811
Simpson's diversity index, 595
Size
    affected by competition, 281
    as component of life-history,
        502–3
    distribution of, in relation to
        competition, 232
    relative species abundance
        according to, 812–13
Skin, as resource, 388
Sleeping sickness, 418–20
Social cohesion hypothesis for
    dispersal, 188
Social facilitation, 330